CONTROL OF PESTS AND WEEDS
BY NATURAL ENEMIES

CONTROL OF PESTS AND WEEDS BY NATURAL ENEMIES

AN INTRODUCTION TO BIOLOGICAL CONTROL

Roy Van Driesche, Mark Hoddle, and Ted Center

BLACKWELL PUBLISHING
350 Main Street, Malden, MA 02148-5020, USA
9600 Garsington Road, Oxford OX4 2DQ, UK
550 Swanston Street, Carlton, Victoria 3053, Australia

First published 2008 by Blackwell Publishing Ltd

1 2008

Library of Congress Cataloging-in-Publication Data

Van Driesche, Roy.
Control of pests and weeds by natural enemies : an introduction to biological control / Roy Van Driesche, Mark Hoddle, and Ted Center. – 1st ed.
p. cm.
Includes bibliographical references and index.
ISBN 978-1-4051-4571-8 (pbk. : alk. paper) 1. Pests–Biological control. 2. Weeds–Biological control.
I. Hoddle, Mark. II. Center, Ted D. III. Title.

SB975.V38 2008
632'.96–dc22 2007038529

ISBN: 978-1-4051-4571-8 (paperback)

A catalogue record for this title is available from the British Library.

Set in 9/11pt Photina MT
by Graphicraft Limited, Hong Kong
Printed and bound in Singapore
by C.O.S. Printers Pte Ltd

The publisher's policy is to use permanent paper from mills that operate a sustainable forestry policy, and which has been manufactured from pulp processed using acid-free and elementary chlorine-free practices. Furthermore, the publisher ensures that the text paper and cover board used have met acceptable environmental accreditation standards.

For further information on
Blackwell Publishing, visit our website at
www.blackwellpublishing.com

CONTENTS

PREFACE

This book replaces another on the same subject published in 1996 by the senior author and Thomas Bellows, Jr., of the University of California, whose earlier contributions we acknowledge. This new book builds on and updates the view of biological control that was presented in that earlier book. One important change has been an extensive effort to treat insect and weed biological control with equal depth in all of the book's topic areas. This was facilitated immeasurably by Ted Center of the USDA-ARS invasive plants laboratory. While superficially similar, weed and insect biological control differ profoundly in a long list of particulars, not least of which being that plants rarely respond to attack by sudden death (the universal currency for scoring arthropod biological control), but by a wide range of lesser impacts that accumulate and interact. We have covered topics such as natural enemy host-range estimation, agent colonization, and impact evaluation, to name a few, in ways that work for both pest insects and invasive weeds. We have also included a chapter (Chapter 12) that is distinctly focused on classical weed biological control.

Another major change is our effort to fully confront both the non-target impacts associated with biological control and the technical features of host-range measurement and prediction that are the tools for better future practice. Three chapters address these aspects. Chapter 16 provides a summary of important historical stages in the development of classical biological control relevant to non-target impacts, including discussions of many widely emphasized cases. Chapter 17 summarizes issues and techniques relevant to predicting host ranges of new agents and Chapter 18 considers indirect effects and whether, as a potential means to limit such effects, it might be feasible to predict the efficacy of an agent before its release.

Of the four general methodologies through which biological control might be implemented (natural enemy importation, augmentation, conservation, and the biopesticidal method), we have devoted most space to classical biological control, the approach most useful as a response to invasive species. Because species invasions are one of the most important crises in conservation biology and because classical biological control is the only biological control method with an expansive historical record of proven success against invasive pests, it has been emphasized in this book.

Conversely, we have de-emphasized biopesticides, which have largely failed to play major roles in pest control. In Chapter 23, we review the principles of biopesticides and the biology of insect pathogens. In Chapter 24, we discuss the current and potential uses of nematodes and each pathogen group. Separately, in Chapter 21, we discuss Bt crop plants, which have dramatically reduced pesticide use in cotton and corn, greatly supporting conservation biological control.

We view augmentation and conservation biological control as largely unproven approaches, mainly of research interest, with, however, some notable exceptions that we discuss. We cover augmentative control (releases of insectary-reared natural enemies) in two chapters: one on use in greenhouse crops and one in outdoor crops or other contexts. In Chapter 25, we explore the success of augmentative biological control in greenhouse crops, particularly vegetables, which we consider a proven technology. Outdoor releases of parasitoids and predators (Chapter 26), however, have largely been a failure, often for economic reasons. Enthusiasm for the method in some sectors has outstripped reality, and we attempt to delineate the likely extent of its future use, which we view as more limited than do its proponents.

Conservation biological control is covered in two chapters. Chapter 21 covers methods for the integration of natural enemies into pesticide-dominated crop pest-management systems. Chapter 22 treats aspects of conservation biological control that are more aligned with the organic farming movement, although not limited to it, such as cover crops, intercrops, refuges, and planting of natural enemy resource strips. This area is currently extremely popular but so far has had few practical successes. However, active research is underway and the method requires time for evaluation before a clearer view can be had of both its biological potential and the willingness of farmers to employ it, given the associated costs.

Finally, we end the book with two chapters that cover outliers and new directions. In Chapter 27, we consider vertebrate biological control, including new developments in immunocontraception. In Chapter 28, we consider the potential to apply classical biological control to pests of conservation importance and to taxa of organisms not previously targeted for biological control. We consider both applications to be critical future contributions of biological control to the solution of environmental and economic problems caused by invasive species.

Instructors using this textbook to teach courses on biological control will find the Powerpoint presentations of Dr Van Driesche's course on biological control at the University of Massachusetts at the following URL (click on Resources on the homepage): www.invasiveforestinsectandweedbiocontrol.info/index.htm. The Powerpoint files are downloadable and may be used in whole or in part for any educational, non-commercial purpose. They will be updated periodically. In addition, all photographs that appear in this textbook are posted on this website in downloadable form for classroom use.

We hope this book will help train a new generation of biological control practitioners, who will be problem-solvers and skilled ecologists. The faults of classical biological control have been widely discussed, and in our view exaggerated, in recent years. We hope this text will instill in students a sense of the power of this tool to combat invasive plants and arthropods, both for protection of agriculture and nature.

Reviews of one or more chapters were provided by the following colleagues, whom we thank: David Briese, Naomi Cappacino, Kent Daane, Brian Federici, Howard Frank, John Goolsby, Matthew Greenstone, George Heimpel, Kevin Heinz, John Hoffmann, Michael Hoffmann, Keith Hopper, Frank Howarth, David James, Marshall Johnson, Harry Kaya, David Kazmer, Armand Kuris, Edward Lewis, Lloyd Loope, Alec McClay, Jane Memmot, Russell Messing, Judy Myers, Cliff Moran, Joseph Morse, Steve Naranjo, Robert O'Neil, Timothy Paine, Robert Pfannenstiel, Robert Pemberton, Charles Pickett, Paul Pratt, Marcel Rejmanek, Les Shipp, Grant Singleton, Lincoln Smith, Peter Stiling, Phil Tipping, Serguei Triaptisyn, Talbot Trotter, Robert Wharton, Mark Wright, and Steve Yaninek. We are also grateful for the contributed chapters by Joe Elkinton (Chapter 10) and Richard Stouthamer (Chapter 15) and the final reading of the whole manuscript by Judy Myers and George Heimpel. Geoff Attardo of Keypoint Graphics assisted with assessing images selected for inclusion in the book and Ruth Vega of the Applied Biological Control Laboratory of the University of California helped in preparing materials for figures.

Roy Van Driesche
Mark Hoddle
Ted Center

Part 1

SCOPE OF BIOLOGICAL CONTROL

Chapter 1

INTRODUCTION

Biological control can be approached by several means for somewhat different purposes. When permanent suppression of a pest (usually a non-native invasive species) over a large area is the goal, the only feasible method is **classical biological control**. This approach seeks to cause permanent, ecological change to the natural enemy complex (i.e. parasitoids, predators, pathogens, herbivores) attacking the pest by introducing new species from the pest's homeland (or, in the case of native pests or exotic pests of unknown origin, from related species or ecologically similar species). This approach was historically the first method of manipulating natural enemies that was dramatically successful as a form of pest control. In the past century it has been used to suppress over 200 species of invasive insects and 40 species of weeds in many countries around the world, and is arguably the most productive and economically important form of biological control. This strategy can be applied against pests of natural areas (forests, grasslands, wetlands), urban areas, and outdoor agricultural production areas. Classical biological control must be a community-level, government-regulated activity conducted for regional benefit rather than for the benefit of a few individuals.

Additional forms of biological control (**conservation of natural enemies, release of commercially reared natural enemies, microbial pesticides**) exist that can temporarily suppress pests, either native or invasive, in crops. These approaches make sense when pest control is needed only at some specific location and time. The cost to implement these practices is borne by the farmer in order to reduce losses from pest damage. Such approaches must be cost-effective to be useful, paying for themselves in reduced pest losses and doing so more conveniently or economically than other available methods of control. They depend on the interest of the grower and his or her willingness to pay the associated costs.

On public lands, government funds can support natural enemy releases to protect forests or achieve other pest-management goals if a clear consensus exists on the need and the government is willing and able to pay. The microbial pesticide *Bacillus thuringiensis* Berliner subsp. *kurstaki*, for example, is used by Canadian forestry agencies as an alternative to spraying forests with chemical pesticides to suppress outbreaks of insects such as spruce budworm (*Choristoneura fumiferana* [Clemens]). However, these non-classical biological control methods are used mostly in private farms, orchards, or greenhouses to supplement natural control.

Biological control of vertebrate pests has been attempted, and recently the use of genetically engineered vertebrate pathogens has been investigated. There is an emerging need for biological control of non-traditional invasive pests such as crabs, starfish, jellyfish, marine algae, snakes, and freshwater mussels, for which experience with insects and plants provides little direct guidance. Finally, we examine the constraints on each of the four major approaches to biological control (importation, conservation, augmentation, and biopesticides) and speculate on the likely degree of their future use.

Chapter 2

TYPES OF BIOLOGICAL CONTROL, TARGETS, AND AGENTS

WHAT IS BIOLOGICAL CONTROL?

The definition of biological control hinges on the word **population**. All biological control involves the use, in some manner, of **populations of natural enemies** to suppress pest populations to lower densities, either permanently or temporarily. In some cases, populations of natural enemies are manipulated to cause permanent change in the food webs surrounding the pest. In other cases, the natural enemies that are released are not expected to reproduce, and only the individuals applied have any effect. Some approaches to biological control are designed to enhance natural enemy densities by improving their living conditions.

Methods that do not act through populations of live natural enemies are not biological control. Biologically based, non-pesticidal methods, which include the release of sterile males to suppress insect reproduction, use of pheromones to disrupt pest mating, pest-resistant crops, biorational chemicals, and transgenic pest-resistant plants, are not biological control. However, if these methods replace toxic pesticides, they can bolster biological control by conserving existing natural enemies.

PERMANENT CONTROL OVER LARGE AREAS

When pests are to be controlled over large areas, the only long-term effective approach is introduction of natural enemies. If the target pest is an invasive non-native species and its natural enemies are introduced, the approach is called **classical biological control**. If the target is a native pest (or an exotic species of unknown origin) and the natural enemies released against it come from a different species, the approach is called **new-association biological control**. Classical and new-association projects are similar in operation, but differ in whether or not the natural enemies employed have an evolutionary association with the target pest.

Classical biological control

Many of the important arthropod pests of agriculture and natural areas are non-native invasive species (Sailer 1978, Van Driesche & Carey 1987). In the USA, for example, 35% of the 700 most important insect pests are invasive species, even though invasive insects comprise only 2% of US arthropods (Knutson et al. 1990). Vigorous invaders (ones well adapted to the climate and competition in the invaded community) often remain high-density pests because local natural enemies are not specialized to feed on unfamiliar species. Consequently, the level of attack is too limited to adequately control the pest. In such cases, introductions of specialized natural enemies that have an evolutionary relationship with the pest are needed for control. Since 1888, natural enemy introductions have provided complete or partial control of more than 200 pest arthropods and about 40 weeds (DeBach 1964a, Laing & Hamai 1976, Clausen 1978, Goeden

1978, Greathead & Greathead 1992, Nechols et al. 1995, Hoffmann 1996, Julien & Griffiths 1998, McFadyen 1998, Waterhouse 1998, Olckers & Hill 1999, Waterhouse & Sands 2001, Mason & Huber 2002, Van Driesche et al. 2002a, Neuenschwander et al. 2003).

Effective natural enemies of invasive species are most likely to occur in the native range of the pest, where species specialized to exploit the target pest have evolved. In some cases, effective natural enemies may already be known from earlier projects. When pink hibiscus mealybug (*Maconellicoccus hirsutus* [Green]) invaded the Caribbean in the 1990s (Kairo et al. 2000), previous control of the same mealybug in Egypt provided considerable information on which natural enemies might be useful (Clausen 1978). As a group, mealybugs are well known to be controlled by parasitoids, especially Encyrtidae (Neuenschwander 2003). The only mealybugs that have been difficult to control have been those tended by ants, which protect them (e.g. the pineapple mealybug, *Dysmicoccus brevipes* [Cockerell], in Hawaii, USA; González-Hernandez et al. 1999) or those that feed underground on plant roots and thus are not reachable by parasitoids (e.g. the vine mealybug, *Planococcus ficus* [Signoret], on Californian grapes; Daane et al. 2003).

Classical biological control projects require the collection of natural enemies from the area of origin of the invader, their shipment to the invaded country, and (after appropriate quarantine testing to ensure correct identification and safety) their release and establishment. In the case of pink hibiscus mealybug (native to Asia), the encyrtid *Anagyrus kamali* Moursi, originally collected in Java for release in Egypt, was quickly identified as a candidate for release in the Caribbean. Before the mealybug was controlled, a wide range of woody plants in the Caribbean were heavily damaged, including citrus, cocoa, cotton, teak, soursop, and various ornamental plants (Cock 2003). Inter-island trade was restricted to check the pest's spread, causing further economic losses. Within a year of introduction, *A. kamali* reduced pink hibiscus mealybug to non-economic levels in the Caribbean, and later was introduced into Florida and California, USA.

Rapid suppression of an invasive plant by an introduced insect is illustrated by the case of the floating fern *Azolla filiculoides* Lamarck (McConnachie et al. 2004). *Azolla filiculoides*, a native of the Americas, appeared in South Africa in 1948 at a single location. By 1999 it had infested at least 152 sites, mostly water reservoirs and small impoundments. It formed thick floating mats that interfered with water management, increased siltation, reduced water quality, harmed local biodiversity, and even occasionally caused drowning of livestock (Hill 1997). Biological control provided the only option for suppression because no herbicides were registered for use against this plant (Hill 1997). Fortunately, potentially effective plant-feeding insects were known from the USA and one of these, the weevil *Stenopelmus rufinasus* Gyllenhal, was imported from Florida. Hill (1997) confirmed that it was a specialist and fed only on species of *Azolla*, so it was approved for release (Hill 1998). South African scientists released it at 112 sites beginning in 1997 (McConnachie et al. 2004) and it extirpated *A. filiculoides* from virtually all release sites (except those destroyed by flooding or drainage) within 7 months. The fern was controlled throughout the country within 3 years, with a cost/benefit ratio expected to reach 15:1 by 2010 (McConnachie et al. 2003).

Introduction as a method of biological control has a major advantage over other forms of biological control in that it is self-maintaining and less expensive over the long term. On farms or tree plantations, after new natural enemies are established, conservation measures (such as avoidance of damaging pesticides) may be required for the new species to be fully effective. Because classical biological control projects produce nothing to sell, and require considerable initial funding and many trained scientists, they are usually conducted by public institutions, using public resources to solve problems for the common good.

New-association biological control

This term applies if the target pest is a native species or an invasive species of unknown origin. In both cases, natural enemies are collected from different species that are related either taxonomically or ecologically to the pest. Use against a native species is illustrated by efforts against the sugarcane borer (*Diatraea saccharalis* [Fabricius]) in Barbados. This borer is a New World pest of sugarcane that is not readily controlled with pesticides. The braconid parasitoid *Cotesia flavipes* Cameron was found in India attacking stem borers of other large grass species and imported to Barbados, where it reduced the incidence of sugarcane borer from 16 to 6% (Alam et al. 1971).

A current example of a new-association project is the effort to reduce bud and fruit feeding by native *Lygus* bugs in North America with parasitoids of European *Lygus* (Day 1996). The braconid *Peristenus digoneutis* Loan was successfully established in the eastern USA and reduced densities of tarnished plant bug (*Lygus lineolaris* [Palisot de Beauvois]) in alfalfa, its major reservoir crop, by 75% (Day 1996). Reduction of *Lygus* populations in alfalfa should lead to fewer immigrants reaching high-value crops such as apples and strawberries (Day et al. 2003, Tilmon & Hoffmann 2003).

The same general approach can be used against invasive species whose areas of origin remain undiscovered. For example, the coconut moth (*Levuana iridescens* Bethune-Baker) in Fiji was believed to be an invasive species from somewhere west of Fiji, but the source population was never found. Tothill et al. (1930) introduced the tachinid *Bessa remota* (Aldrich) after encountering it as a parasitoid of other zygaenid moths, making this a likely case of new association against an invasive species (see Chapter 16 for outcomes).

New-association biological control of native species differs from classical biological control in several important ways. First, the ecological justification for classical biological control (restoring disturbed ecosystems to pre-invasion conditions) is missing when native species are targeted. For some pests, human society deems permanent lowering of the density of a native species as acceptable because of the economic damage caused. This is clearly true for pests such as the tarnished plant bug (*L. lineolaris*). New-association biological control is not advisable for native plants, even those that become weeds. A number of such projects were proposed in the past against such native plants as mesquite (*Prosopis glandulosa* Torrey and *Prosopis velutina* Wooten) and snake weeds (*Guiterrezia* spp.) in the southwestern USA (DeLoach 1978). If biological control of a native plant were attempted, success would also affect many species dependent in various ways on the plant.

Another way in which new-association biological is different from classical biological control, regardless of whether the target is a native species or an invasive species of unknown origin, is that, by definition, natural enemies are not located by finding the pest overseas and collecting its natural enemies. Rather, one has to select surrogates from another biogeographic region that are enough like the pest (based on shared taxonomy, ecology, morphology, etc.) to have natural enemies that would attack the pest. In some cases, congeneric species have similar life histories and (for insect targets) attack the same genera of plants as the pest. The geographic ranges of such species then indicate the available places from which to collect potential natural enemies, provided climates and day-length patterns of the donor and recipient regions are similar. In other cases, however, there may be no obvious related species from which to collect natural enemies.

TEMPORARY PEST SUPPRESSION IN PRODUCTION AREAS

Whereas classical biological control has been used extensively to suppress pest insects attacking crops, biological control in production systems does not have to be permanent or wide-ranging. The goal can be merely to suppress pest densities enough to protect the current year's harvest. Biological control in crops begins with practices to enhance **natural control** by conserving whatever natural enemies live in the crop fields. These may be generalist predators or specialized parasitoids (either of native pests or parasitoids previously introduced for control of invasive insects). These species may be enhanced by a variety of manipulations of the crop, the soil, or the non-crop vegetation in or around the crop field (**conservation biological control**). If pest suppression from these natural enemies is insufficient, additional natural enemies can be released (**augmentation biological control**), providing the right species are available and able to offer cost-effective pest control. Commercial products containing pathogens (**biopesticides**) may be sprayed on crops to kill additional pests.

Conservation biological control

Farming practices greatly influence the extent to which natural enemies actually suppress pest insects and mites. Conservation biological control is the study and manipulation of such influences. Its goal is to minimize factors that harm beneficial species and enhance features that make agricultural fields suitable habitat for natural enemies. This approach assumes that the natural enemies already present can potentially suppress the pest if given an opportunity to do so. This assumption is likely to be true for many native insect pests, but is not true for weeds. Nor is it usually true for invasive insects unless a program of classical biological control has imported effective specialized natural enemies.

In non-organic farm fields, pesticide use is the most damaging influence affecting natural enemies

(Croft 1990). Other negative forces can be dust on foliage (DeBach 1958, Flaherty and Huffaker 1970) and ants that defend honeydew-producing insects (DeBach & Huffaker 1971). Farming practices that may harm natural enemies include use of crop varieties with unfavorable features, date and manner of cultivation, destruction of crop residues, size and placement of crop patches, and removal of vegetation that provides natural enemy overwintering sites or food.

In principle, crop fields and their margins can be enhanced as natural enemy habitats by manipulating the crop, the farming practices, or the surrounding vegetation. Useful practices might include creation of physical refuges needed by natural enemies, provision of places for alternative hosts to live, planting flowering plants as nectar sources, or planting ground covers between crop rows to moderate temperature and relative humidity. Even the manner or timing of harvest or post-harvest treatment of crop residues can influence populations of natural enemies (van den Bosch et al. 1967, Hance and Gregoire-Wibo 1987, Heidger & Nentwig 1989). The conscious inclusion of such features in farming systems has been called ecological engineering (Gurr et al. 2004).

Conservation methods depend on knowing how effective a particular conservation practice will be under local conditions. This requires extensive local research in farmers' fields. The method often can be implemented on individual farms independently of the actions of the community as a whole after such information becomes available.

Releases of commercially reared natural enemies

When natural enemies are missing (as in greenhouses), or arrive too late for new plantings (some row crops), or simply are too scarce to provide control (in large monocultures), their numbers may be increased artificially by releasing insectary-reared individuals (King et al. 1985). Release of commercially produced natural enemies is called augmentation biological control. Augmentation covers several situations. **Inoculative releases** are those in which small numbers of a natural enemy are introduced early in the crop cycle with the expectation that they will reproduce in the crop and their offspring will continue to provide pest control for an extended period of time. For example, an early release of *Encarsia formosa* Gahan can assist whitefly control in greenhouse tomato crops throughout the growing season. **Inundation**, or **mass release**, is used when insufficient reproduction of the released natural enemies is likely to occur, and pest control will be achieved mostly by the released individuals themselves. For example, *Eretmocerus eremicus* Rose and Zolnerowich must be released weekly for continuous suppression of whiteflies in greenhouse-grown poinsettia.

Augmentation, suitable for use against both native and invasive pests, is limited principally by cost, agent availability and quality, and field effectiveness of the reared organisms. Costs limit the use of reared natural enemies to situations where: (1) the natural enemy is inexpensive to rear, (2) the crop has high cash value, and (3) cheaper alternatives such as insecticides are not available. Only in such circumstances can private companies recoup production costs and compete economically with alternative methods. Somewhat broader use is possible when public institutions rear the necessary natural enemies. In both cases, production of high-quality natural enemies is essential, as are research studies determining the best release strategies and assessing the degree of pest control provided by the reared agent under field conditions.

Application of biopesticides

Inundation with nematodes or pathogens differs from mass release of parasitoids and predators. **Biopesticides** resemble chemical pesticides in their packaging, handling, storage, and application methods, as well as their curative-use strategy and requirement (except for nematodes) for government registration. Use of the bacterium *Bacillus thuringiensis* Berliner is the best-known example of a biopesticide. Such pathogens, however, while present in the marketplace for over 65 years, have remained niche products and currently make up less than 1% of insecticide use. Transgenic plants that express the toxins of this bacterium (known as Bt plants), however, have exploded in use, with more than 40 million ha of Bt crops planted around the world by 2000, mainly of cotton, soybeans, and corn (Shelton et al. 2002), a figure that is increasing rapidly. These insect-resistant plants usually replace conventional pesticides and improve the crop as habitat for natural enemies, thus supporting conservation biological control (see Chapter 21).

KINDS OF TARGETS AND KINDS OF AGENTS

Biological control has been used primarily for the control of weeds, insects, and mites. In a few instances pest vertebrates or snails have been targeted. Need exists for biological control of new kinds of pests, such as marine algae, starfish, mussels, and jellyfish, but these are non-traditional targets about whose potential for suppression by natural enemies we know relatively little (see Chapter 28). For the principal targets of biological control, several groups of natural enemies have been widely used. For biological weed control, natural enemies have been mainly insects and plant pathogenic fungi. For insect targets, parasitoids and predaceous insects are the natural enemies used, together with some pathogens formulated for use as biopesticides. For pest mites, predatory mites have been widely manipulated by conservation methods. To develop a better appreciation of how these groups are manipulated for biological control, in the opening part of this book we consider the taxonomic diversity and ecology of the key natural enemy groups (Chapters 3–6) before discussing methods for their manipulation.

Part 2

KINDS OF NATURAL ENEMIES

Chapter 3

PARASITOID DIVERSITY AND ECOLOGY

Natural enemies are the fundamental resource of biological control. Agents come from many groups, differing widely in their biology and ecology. A detailed knowledge of natural enemy taxonomy, biology, and ecology is a great asset to practitioners of biological control. For pest insects, parasitoids are often the most effective natural enemies.

WHAT IS A PARASITOID?

Parasitoids have been the most common type of natural enemy introduced against pest insects (Hall & Ehler 1979, Greathead 1986a). Unlike true parasites, parasitoids kill their hosts and complete their development on a single host (Doutt 1959, Askew 1971, Vinson 1976, Vinson & Iwantsch 1980, Waage & Greathead 1986, Godfray 1994). Most parasitoids are Diptera or Hymenoptera, but a few are Coleoptera, Neuroptera, or Lepidoptera. Pennacchio and Strand (2006) discuss the evolution of parasitoid life histories in the Hymenoptera. Of some 26 families of parasitoids, the groups used most frequently in biological control are Braconidae, Ichneumonidae, Eulophidae, Pteromalidae, Encrytidae, and Aphelinidae (Hymenoptera), and Tachinidae (Diptera) (Greathead 1986a).

TERMS AND PROCESSES

All insect life stages can be parasitized. Trichogrammatid wasps that attack eggs are called **egg parasitoids**. Species that attack caterpillars are **larval parasitoids**, and so on. Parasitoids whose larvae develop inside the host are called **endoparasitoids** (Figure 3.1a) and those that develop externally are **ectoparasitoids**.

Ectoparasitoids often attack hosts in leafmines, leaf rolls, or galls, which prevent the host and parasitoid from becoming separated. If parasitoids permit hosts to grow after being attacked they are called **koinobionts.** The koinobiont group includes the internal parasitoids that attack young larvae or nymphs and a few ectoparasitoids, such as some pimpline ichneumonids on spiders and most ctenopelmatine ichneumonids (Gauld & Bolton 1988). In contrast, **idiobionts** allow no growth after attack. These are either internal parasitoids of egg, pupae, or adults (which do not grow), or external parasitoids that paralyze larvae (Godfray 1994). Internal parasitoids of stages other than eggs must suppress the host's immune system, whereas egg and external parasitoids do not. Parasitoids that must overcome host immune systems are often more specialized than groups that do not. Egg parasitoids such as species of *Trichogramma*, for example, have much broader host ranges than internal larval parasitoids such as braconid *Cotesia* species.

Terms to describe the number of parasitoid individuals or species that develop in a single host include **solitary parasitoid**, which denotes that only a single parasitoid can develop to maturity per host, and **gregarious parasitoid** (Figure 3.1b), for which several can do so.

Superparasitism occurs when more eggs, of one species, are laid than can survive, whereas the presence of two or more individuals of different species is called **multiparasitism**. When one parasitoid attacks another, **hyperparasitism** occurs, which is generally thought

(a)

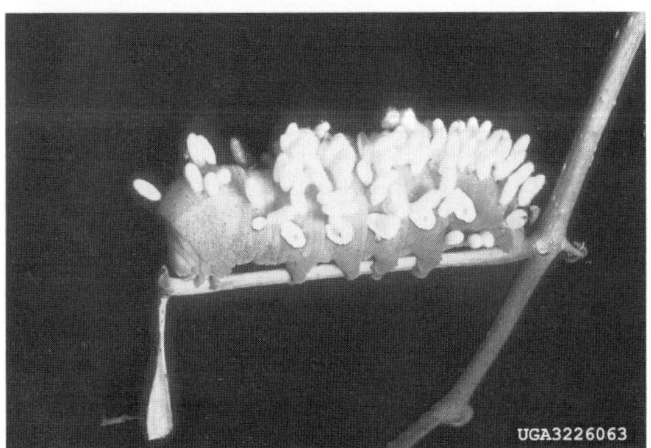

(b)

Figure 3.1 (a) Pupa (dark body) of the endoparasitoid *Encarsia luteola* Howard inside the integument of its whitefly host. Photograph courtesy of Jack Kelly Clark, University of California IPM Photo Library. (b) Cocoons of a gregarious parasitoid on a luna caterpillar (*Actias luna* [L.]). Photograph courtesy of Ron Billings, www.Forestryimages.org.

to be unfavorable for biological control, except in special cases such as adelphoparasitism of whiteflies.

The pattern of egg maturation over the lifetime of a parasitoid affects the potential ways in which a parasitoid can be used in biological control. **Pro-ovigenic** species emerge with their lifetime supply of eggs present, allowing rapid attack on many hosts. Conversely, eggs of **synovigenic** species develop gradually over the female's lifetime. An **ovigeny index (OI)** is the proportion of a parasitoid's lifetime egg supply that is present upon emergence (Jervis & Ferns 2004), with strictly pro-ovigenic species scored as 1.0. Synovigenic parasitoids need protein to mature eggs. Some synovigenic species feed on nectar or honeydew, but others consume host hemolymph. This is obtained by puncturing the host's integument with the ovipositor and consuming hemolymph as it bleeds from the wound (Figure 3.2). This process is called **host feeding**, a behavior found in many hymenopteran parasitoids (Bartlett 1964a, Jervis & Kidd 1986).

(a)

(b)

(c)

Figure 3.2 Host feeding by an aphelinid parasitoid (*Physcus* sp.) on the armored scale *Aonidiella aurantii* (Maskell), showing ovipositor insertion in scale (a), exuded hemolymph (b), and feeding by parasitoid (c). Photographs courtesy of Mike Rose, reprinted from Van Driesche and Bellows (1996) with permission from Kluwer.

SOME REFERENCES TO PARASITOID FAMILIES

For general information about parasitoid families see Clausen (1962; useful but dated), Askew (1971), Waage and Greathead (1986), Gauld and Bolton (1988), Grissell and Schauff (1990), Godfray (1994), Hanson and Gauld (1995), Quicke (1997), and Triplehorn and Johnson (2005). For some information on host records, see Fry (1989). Further information is available in regional catalogs such as Krombein et al. (1979). Townes (1988) lists sources of taxonomic literature for parasitic Hymenoptera. A key to families in the Hymenoptera of the world is provided by Goulet and Huber (1993); a key to the families of Neartic Chalcidoidea is given by Grissell and Schauff (1990), and to the genera by Gibson et al. (1997). An electronic database to the chalcidoids is maintained by Noyes at www.nhm.ac.uk/jdsml/research-curation/projects/chalcidoids/. The material is available on CD-ROM at www.nhm.ac.uk/publishing/pubrpch.html. Yu and van Achterberg have an electronic catalog to all Ichneumonoidea (www.taxapad.com/). Wharton et al. (1997) present a key to braconid genera of the western hemisphere. Shaw and Huddleston (1991) summarize information on biology of braconids. Current world catalogs exist for the Evaniidae (Deans 2005) and Proctotrupoidea (Johnson 2005). For a review of the Scelionidae, see Austin et al. (2005).

GROUPS OF PARASITOIDS

Parasitic flies

Thirteen fly families include species parasitic on arthropods or snails (Cecidomyiidae, Acroceridae, Nemestrinidae, Bombyliidae, Phoridae, Pipunculidae, Conopidae, Pyrgotidae, Sciomyzidae, Cryptochetidae, Calliphoridae, Sarcophagidae, and Tachinidae), but the most important are the Tachinidae, Phoridae, and Cryptochetidae. See Feener and Brown (1997) for a review of Diptera as parasitoids.

Phoridae

These flies have been reared from termites, bees, crickets, caterpillars, moth pupae, and fly larvae, but are currently of interest as parasitoids of invasive fire ants (Williams & Banks 1987, Feener & Brown 1992, Williams et al. 2003, Porter et al. 2004; Figure 3.3).

Figure 3.3 Adult fly of the phorid *Pseudacteon litoralis* Borgmeier attacking a worker of the imported fire ant, *Solenopsis invicta* (Burden). Photograph courtesy of S.D. Porter and L.A. Calcaterra, USDA-ARS.

Cryptochetidae

All species are in the genus *Cryptochetum* and all parasitize margarodid scales. *Cryptochetum iceryae* (Williston) was introduced into California, USA, from Australia and controls the cottony cushion scale (*Icerya purchasi* Maskell), a major citrus pest (Bartlett 1978).

Tachinidae

These (Plate 3.1a) are the most important Diptera for classical biological control. Most are solitary endoparasitoids and none are hyperparasitic (Askew 1971). *Lydella thompsoni* Herting was introduced to the USA to control the European corn borer, *Ostrinia nubilalis* (Hübner) (Burbutis et al. 1981). In Canada, introduction of *Cyzenis albicans* (Fallén) controlled the invasive winter moth *Operophtera brumata* L. (Embree 1971). *Trichopoda giacomellii* (Blanchard) was introduced to Australia, where it controlled an important vegetable pest, *Nezara viridula* (L.) (Coombs & Sands 2000). Tachinids such as *Lixophaga diatraeae* (Townsend) have been used for augmentative releases (Bennett 1971), and other species have been of interest as indigenous parasitoids of native pests; for example, *Bessa harveyi* (Townsend), which is a parasitoid of the larch sawfly, *Pristiphora erichsonii* (Hartig) (Thompson et al. 1979). Grenier (1988) reviews the role of the tachinids in applied biological control and Stireman et al. (2006) discuss tachinid evolution, behavior, and ecology. Tachinids vary in how they attack hosts (O'Hara

1985). Adults of some species deposit their eggs on or in their hosts, whereas others retain their eggs and deposit first-instar larvae on, near, or in their hosts. Still others place eggs or larvae on foliage or soil. Eggs laid on foliage are placed where they are likely to be consumed later by a host. In such cases, plant volatiles from herbivore-damaged plant tissue may attract ovipositing flies (Roland et al. 1989). Eggs laid on foliage are often very small (microtype) and deposited in greater numbers than the larger (macrotype) eggs of species which oviposit directly on their hosts (Askew 1971).

Tachinids vary from narrowly specific species, such as *T. giacomellii* (Sands & Combs 1999), to extremely polyphagous ones, such as *Compsilura concinnata* (Meigen), introduced to suppress gypsy moth [*Lymantria dispar* (L.)] and browntail moth [*Euproctis chrysorrhoea* (L.)] in North America. While providing highly effective control of browntail moth, this tachinid causes high rates of mortality to native silkworm moths (Saturniidae) (Boettner et al. 2000).

Parasitic wasps

Parasitoids occur in at least 36 families of Hymenoptera, but these vary greatly in the degree to which they have been used in biological control, due to family size and the types of insects they attack. The parasitoids of greatest importance to biological control are in two superfamilies, the Chalcidoidea and Ichneumonoidea.

The **Chalcidoidea** includes 16 families with parasitoids, of which Encyrtidae and Aphelinidae have been used most frequently in biological control.

Pteromalidae

These attack a wide range of hosts with some distinctions occurring by subfamily or tribe. For example, muscoid fly pupae, wood-boring beetles, or stem- or mud-nesting wasps are attacked by the Cleonyminae; flies in the Agromyzidae, Cecidomyiidae, Tephritidae, and Anthomyiidae (Miscogastrini); and various Lepidoptera, Coleoptera, Diptera, and Hymenoptera (Pteromalinae). Species of *Muscidifurax* and *Spalangia* are reared for augmentative releases against manure-breeding flies (Patterson et al. 1981).

Encyrtidae

These parasitize scales, mealybugs, and either eggs or larvae of various Blattaria, Coleoptera, Diptera, Lepidoptera, Hymenoptera, Neuroptera, Orthoptera, spiders, and ticks. This family, together with the Aphelinidae, accounts for half of the cases of successful classical biological control. Important genera in the family include *Anagyrus*, *Apoanagyrus*, *Comperiella*, *Hunterellus*, and *Ooencyrtus*. The South American encyrtid *Apoanagyrus* (formerly *Epidinocarsis*) *lopezi* (De Santis) controlled the invasive mealybug *Phenacoccus manihoti*, which devastated cassava crops throughout much of tropical Africa (Neuenschwander et al. 1989). *Anagyrus kamali* Moursi (Plate 3.1b) controlled the pink hibiscus mealybug [*Maconellicoccus hirsutus* (Green)] in the Caribbean.

Eulophidae

This family is of major importance to biological control, attacking a wide range of hosts, including scales, thrips, and species of Coleoptera, Lepidoptera, Diptera, and Hymenoptera. Some species attack leafminers or wood-boring insects.

Aphelinidae

Members of this family are important parasitoids of armored scales, mealybugs, whiteflies, aphids, psyllids, and eggs of various insects. Genera of major importance include *Aphelinus*, *Aphytis*, *Encarsia*, and *Eretmocerus* (Rosen & DeBach 1979). *Aphytis melinus* DeBach (Plate 3.1c) controlled the California red scale [*Aonidiella aurantii* (Maskell)] on citrus. Viggiani (1984) reviews the bionomics of the Aphelinidae. Some species such as *Encarsia formosa* Gahan and *Eretmocerus eremicus* Rose and Zolnerowich are mass-reared for use in greenhouse crops against whiteflies.

Trichogrammatidae

All trichogrammatids are egg parasitoids. Species names in older literature (<1970s) are often incorrect because of difficulty in accurately identifiying species without DNA-based molecular tools (Pinto & Stouthamer 1994). About 10 *Trichogramma* species are mass-reared extensively for augmentative releases against pest Lepidoptera in corn, cotton, and other crops (Plate 3.1d).

Mymaridae

All mymarids are egg parasitoids, attacking species of Hemiptera, Psocoptera, Coleoptera, Diptera, and Orthoptera. Release of *Anaphes flavipes* (Förster) in the USA helped suppress the cereal leaf beetle, *Oulema melanopus* (L.) (Maltby et al. 1971). *Gonatocerus ashmeadi* Girault (Plate 3.1e) controlled the glassy-winged sharpshooter, *Homalodisca coagulata* Say, in French Polynesia.

The superfamily **Platygastroidea** includes the Scelionidae and Platygasteridae, which are of interest in biological control.

Scelionidae

All species in this large family are egg parasitoids, and some, such as *Trissolcus basalis* (Wollaston), a parasitoid of the southern green stink bug, *N. viridula* (Jones 1988), have been used in biological control. Other important genera are *Telenomus* and *Scelio*.

The superfamily **Ichneumonoidea** is comprised of the Ichneumonidae and Braconidae. Aphidiinae are sometimes elevated to family level but here are kept within Braconidae.

Ichneumonidae

Members of this large family (Townes 1969, Yu & Horstmann 1997) parasitize many different kinds of hosts. Many species have long antennae and long ovipositors that are always visible, but in some groups ovipositors are short and not visible. The most important subfamilies can, in general, be grouped by type of host (after Askew 1971): ectoparasitoids of larvae or pupae of diverse orders in plant tissue (Pimplinae, e.g. *Pimpla*); ectoparasitoids of exposed larvae of Lepidoptera and sawflies (Typhoninae, e.g. *Phytodietus*); ectoparasitoids of insects in cocoons, hyperparasitoids (Cryptinae, e.g. *Gelis*); endoparasitoids of lepidopteran larvae (Banchinae, e.g. *Glypta*; Porizontinae, e.g. *Diadegma*; Ophioninae, e.g. *Ophion*); endoparasitoids of lepidopteran pupae (Ichneumoninae, e.g. *Ichneumon*); endoparasitoids of sawfly larvae (Ctenopelmatinae, e.g. *Perilissus*); and endoparasitoids of syrphid larvae (Diplazontinae, e.g. *Diplazon*).

Braconidae

These have been widely used in biological control, especially against aphids, Lepidoptera, Coleoptera, and Diptera. Braconids often pupate inside silk cocoons outside the body of their host, but Aphidiinae pupate inside mummified aphids. Wharton (1993) discusses the bionomics of the Braconidae. *Aphidius colemani* Viereck is sold commercially for control of aphids in greenhouses (Plate 3.1f). Most workers recognize 35–40 subfamilies. The main subfamilies and types of hosts they attack (after Askew 1971; Shaw & Huddleston 1991) include endoparasitoids of aphids (Aphidiinae, e.g. *Aphidius*, *Trioxys*; for biology of this group, see Starý 1970); endoparasitoids of larvae of Lepidoptera and Coleoptera (Meteorinae, e.g. *Meteorus*; Blacinae, e.g. *Blacus*; Microgasterinae, e.g. *Cotesia*, *Microplitis*; Rogadinae, e.g. *Aleiodes*); endoparasitoids of adult beetles or nymphal Hemiptera (Euphorinae, e.g. *Microctonus*); egg-larval endoparasitoids of Lepidoptera (Cheloninae, e.g. *Chelonus*); egg-larval and larval endoparasitoids of cyclorrhaphous Diptera (Alysiinae, e.g. *Dacnusa*; Opiinae, e.g. *Opius*); and ectoparasitoids of lepidopteran and coleopteran larvae in concealed places (Braconinae, e.g. *Bracon*; Doryctinae, e.g. *Heterospilus*).

The superfamily **Chrysidoidea** includes seven families. For biological control, the Bethylidae is the most important, though several species of Dryinidae have also been released against crop and ornamental pests.

Bethylidae attack larvae of beetles and Lepidoptera, often those in confined habitats such as leaf rolls and under bark. Species used as biological control agents include parasitoids of the coffee berry borer, *Hypothenemus hampei* (Ferrari) (Abraham et al. 1990), and *Goniozus legneri* Gordh, which controls the pyralid moth *Amyelois transitella* (Walker) in almond [*Prunus dulcis* (Miller) D.A. Webb var. *dulcis*] orchards in California (Legner & Gordh 1992).

The superfamily **Vespoidea** includes seven families with parasitic members: Tiphiidae, Mutillidae, Scoliidae, Bradynobaenidae, Pompilidae, Rhopalosomatidae, and Sapygidae, of which the Tiphiidae and Scoliidae are likely to be the most important for biological control projects.

Tiphiidae are parasitoids of beetle larvae. Species of the subfamily Tiphiinae burrow into soil to attack scarabaeid larvae in earthen cells. *Tiphia popilliavora* Rohwer and *Tiphia vernalis* Rohwer were introduced into the USA against the Japanese beetle, *Popillia japonica* Newman. Parasitism levels were high initially, but ultimately declined and both parasitoids are now rare while their host is still common (King 1931, Ladd & McCabe 1966).

FINDING HOSTS

Compared to other groups of natural enemies, parasitoids have a relatively coherent set of distinguishing features, being mostly Hymenoptera. Even so, the 100,000 or so known parasitoids are diverse in the details of their biology (see Askew 1971, Doutt et al. 1976, Waage & Greathead 1986, Godfray & Hassell 1988, Godfray 1994, Jervis & Kidd 1996, and Hochberg & Ives 2000). Aspects of parasitoid biology crucial to biological control include (1) finding hosts, (2) host recognition and assessment, (3) defeating host defenses, (4) regulating host physiology, and (5) patch-time allocation, and these will be dealt with in this and the following sections.

Overview

Host-finding by parasitoids has been investigated intensely and is now understood at both the behavioral and chemical levels (Vinson 1984, Tumlinson et al. 1993, Kidd 2005). Initially, a parasitoid must find the host's habitat (Vinson 1981). Sometimes, the parasitoid simply emerges in the right place and begins to seek hosts. In other cases, the parasitoid leaves the habitat to seek resources like nectar or emerges where hosts have died out. Host habitats are usually found by detecting signals perceptible at a distance, not by random search. Vision likely plays an important role in habitat location in the broadest sense (forest or grassland, etc.), but microhabitat location (plant species likely to support hosts) is frequently a response to volatile chemicals, such as: (1) odors from the uninfested host plants, (2) materials (pheromones, feces) produced by the host, or (3) plant volatiles induced and released in response to herbivore feeding. Parasitoids can use odors to locate hosts either by moving upwind when perceiving the odor plumes (Figure 3.4) or, on surfaces, by following gradients of increasing odor strength. In some cases, sights and sounds associated with hosts may be cues attracting parasitoids. Tachinids that attack crickets, for example, literally hear the cricket chirping and fly toward the sound (Cade 1975).

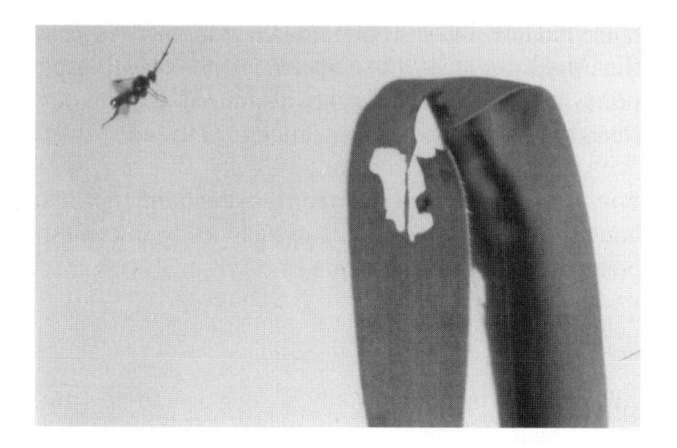

Figure 3.4 Parasitoid flying to odors emitted from caterpillar-damaged corn leaf. Photograph courtesy of Ted Turlings, reprinted from Van Driesche and Bellows (1996) with permission from Kluwer.

Figure 3.5 Parasitoid using antennae to detect chemical cues in frass to help localize a potential host. Photograph courtesy of Joe Lewis, reprinted from Van Driesche and Bellows (1996) with permission from Kluwer.

After parasitoids find infested plants, they find hosts by detecting non-volatile chemicals (Figure 3.5) and other cues (scales, other body parts) on the plant surface (Lewis et al. 1976, Vinson 1984, van Alphen & Vet 1986, Bell 1990, Lewis & Martin 1990, Vet & Dicke 1992). These materials are perceived by touching them with the antennae or tarsi of the legs. Parasitoids attacking hosts concealed inside wood, fruits, or leafmines detect vibrations. Chemicals associated with host presence are called **kairomones**. Discovery of kairomones or host vibrations causes parasitoids to engage in intensified local search, which consists of arrestment and circuitous walking, both of which cause the local area to be searched more thoroughly. For concealed hosts, detection of vibrations from hosts arrests the parasitoid where vibrations are strongest and induces increased probing with the ovipositor.

Long-distance orientation

Habitat and host-finding are parts of a continuum of responses that occur at various spatial scales. For convenience of discussion, we define long-distance orientation as movement that depends on signals, like volatile odors, that are perceived at a distance. Flight is often, but not always, the means of locomotion towards the signal. In contrast, short-distance orientation, for our purposes, will refer to motion, often walking, that takes place on surfaces on which non-volatile signals are perceived by touch, rather than olfaction. This framework accurately fits many, but not all, natural enemies. Better understanding of what host-location odors or signals a parasitoid responds to improves understanding of its ecology and makes its manipulation for biological control easier.

Finding uninfested host plants

Attraction to uninfested host plants is not widespread, but some parasitoids do respond to odors of uninfested plants in olfactometers (Elzen et al. 1986, Martin et al. 1990, Wickremasinghe and van Emden 1992). *Leptopilina heterotoma* (Thompson), a parasitoid of drosophilid larvae in rotting fruits, responds to odors from yeasts, common in rotting materials (Dicke et al. 1984).

Direct location of hosts

Some parasitoids are attracted to insect sex or aggregation pheromones. The aphelinid *Encarsia* (formerly *Prospaltella*) *perniciosi* (Tower), for example, was caught in larger numbers on sticky traps baited with the synthetic pheromone of its host [*Quadraspidiotus perniciosus* (Comstock)] than on unbaited traps (Rice & Jones 1982). *Trichogramma pretiosum* Riley in olfactometers responded to sex pheromone of *Helicoverpa zea* (Boddie) (Lewis et al. 1982, Noldus et al. 1990). The scelionids *Telenomus busseolae* (Gahan) and *Telenomus isis* (Polaszek) were attracted to calling females (emitting pheromones) of the African pink stemborer, *Sesamia calamistis* Hampson (Fiaboe et al. 2003). Tachinid parasitoids of adult southern green stink bugs (*N. viridula*) (Harris & Todd 1980) and a scelionid attacking eggs of the predaceous bug *Podisus maculiventris* (Say) (Aldrich et al. 1984) were attracted to their host's aggregation pheromone. Attraction to specific host odors rather than to host-damaged plants has an obvious advantage for egg parasitoids, which might arrive after egg hatch if only attracted to odors from larval-damaged plants.

Sights and sounds may also attract parasitoids. The tachinid *Ormia ochracea* (Bigot) flew to and attacked dead crickets placed on speakers emitting cricket songs (Cade 1975), but not to dead crickets associated with other noises. The sarcophagid *Colcondamyia auditrix* Shewell locates cicadas [*Okanagana rimosa* (Say)] by their characteristic buzzing (Soper et al. 1976).

Attraction to infested plants

Parasitoids of plant-feeding life stages might be attracted to volatile host products like pheromones, but these are associated with reproduction, not larvae, and might induce larval parasitoids to arrive too early. In theory, larvae or their feces might emit volatile compounds. However, many studies have shown they are either not attractive from a distance or only slightly so. In most cases, larval parasitoids are attracted by volatiles emitted by plants infested with actively feeding insects (Nadel and van Alphen 1987, McCall et al. 1993). Many plants respond to herbivore feeding by increasing emissions of volatiles. Emissions are a mix of pre-formed compounds (green-leaf volatiles) and other compounds synthesized in specific response to herbivore feeding (Paré & Tumlinson 1996; Figure 3.6). Plants are induced to synthesize new volatiles by caterpillar regurgitate (spit) landing on damaged tissue (Potting et al. 1995). This mechanism is widespread, found not only in hymenopteran parasitoids attacking chewing insects like caterpillars, but also parasitoids of sucking insects such as mealybugs (Nadel and van Alphen 1987) and pentatomids (Moraes et al. 2005). Tachinid flies have similar responses (Stireman 2002) and even egg parasitoids sometimes respond to cues from feeding damage (Moraes et al. 2005).

Attractive volatiles are emitted not just from infested plant parts, but also from non-infested ones via a systemic response (Potting et al. 1995), and even from those of non-infested plants adjacent to damaged ones (Choh et al. 2004). Jasmonic acid is a key compound influencing the signaling pathway between plants and natural enemies (Lou et al. 2005). Artificial application of either inductive compounds or directly attractive compounds has potential to draw natural enemies into crop fields (James 2005).

Parasitoids also respond to volatiles from organisms associated with hosts or their habitats (Dicke 1988). For example, a fungus associated with tephritid fly larvae in fruits produces acetaldehyde, which attracts *Biosteres longicaudatus* Ashmead [now *Diachasmimorpha longicaudata* (Ashmead)] (Hymen.: Braconidae) (Greany et al. 1977). Similarly, *Ibalia leucospoides* (Hockenwarth) (Hymen.: Ibaliidae) responds to odors of the wood-digesting fungus *Amylostereum* sp. that is a

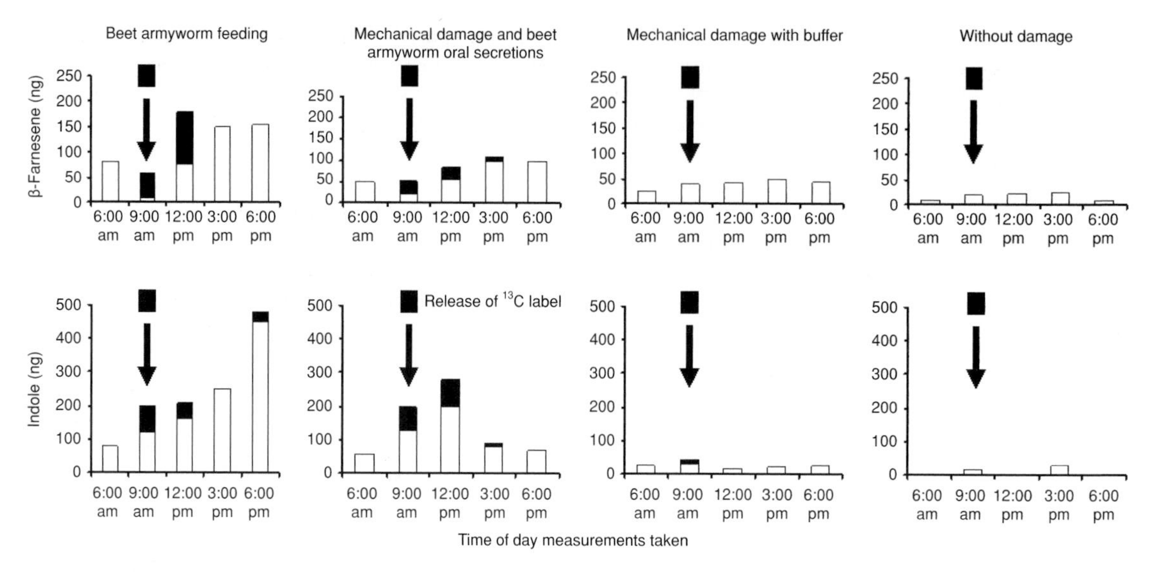

Figure 3.6 Herbivore feeding induces release of a wider range and increased amount of volatile compounds, some of which are the result of *de novo* synthesis stimulated by herbivore attack. Here *de novo* synthesis is demonstrated by release of compounds incorporating a ^{15}C label (black bars) introduced into the reaction vessel at the start of herbivore attack (arrows). Redrawn with permission from Paré and Tumlinson (1996).

symbiont of its woodwasp host, *Sirex noctilio* (Fabricius) (Hymen.: Siricidae) (Madden 1968).

Finding hosts over short distances

Once on a host-infested plant, parasitoids use various materials shed by hosts or emitted by infested plants (collectively called kairomones) to track hosts down. Such materials include chemicals found at feeding sites, waste products (frass, honeydew), body parts (scale, setae, cast skins), and secretions (silk, salivary gland or mandibular secretions, marking pheromones). Kairomones found on plant surfaces promote host discovery by altering parasitoid behavior, producing: (1) arrestment, (2) trail-following, and/or (3) intensified local search.

Arrestment

Parasitoids that hunt for concealed hosts such as those in wood or fruit may stop when they contact kairomones on the item's surface. Arrestment is also produced in some parasitoids by detection of host vibrations (Vet & Bakker 1985). Increased ovipositor probing follows arrestment and helps locate host (Vinson 1976, Vet & Bakker 1985). *Leptopilina* sp., a vinegar fly parasitoid, hunts for hosts inside rotting fruits or mushrooms by remaining stationary on infested structures to detect larval movement (Vet & Bakker 1985). The braconid *Dapsilarthra rufiventris* (Nees), after detecting a host's (*Phytomyza ranunculi* Schrank) leafmine uses sound to locate larvae within mines (Sugimoto et al. 1988).

Trail-following

Kairomones deposited as a line can evoke trail-following. The bethylid *Cephalonomia waterstoni* Gahan follows chemicals that escape from larvae of rusty grain beetles, *Cryptolestes ferrugineus* (Stephens), as they crawl to pupation sites (Howard & Flinn 1990).

Intensified local search

Kairomone-induced behaviors can cause moving parasitoids to search a local area more thoroughly, by staying longer or limiting the areas searched (Figure 3.7). These behaviors increase the number of parasitoids on a host patch and the average time spent there (Prokopy & Webster 1978, Vet 1985, Nealis 1986).

Host feeding damage causes the braconid *Cotesia rubecula* (Marshall) to remain longer on infested cabbages (Nealis 1986). The eucoilid *Leptopilina clavipes* (Hartig)

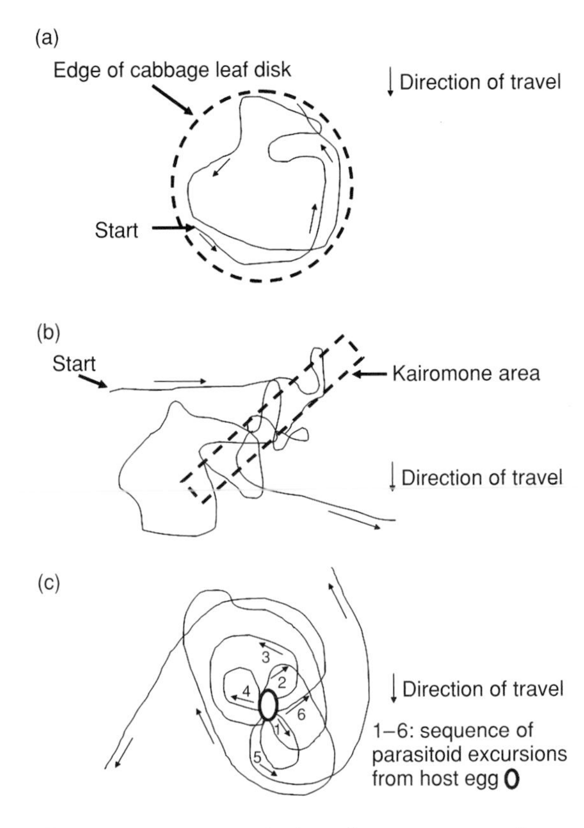

Figure 3.7 Foraging trails of a *Trichogramma* wasp under three different circumstances: (a) when no host kairomone is present the walking path is spread over whole leaf surface; (b) when kairmone is artificially applied to a rectangular area, the search path folds back on itself, concentrating on the kairomone-treated area; and (c) when a host egg is detected, search paths are focused tightly around the egg but departures from the egg occur in random directions (numbers 1–6 represent six departure events). Redrawn with permission from Gardener and van Lenteren (1986) *Oecologia* **68**, 265–70.

searches longer on areas treated with extracts of mushrooms infested with host larvae than on untreated patches (Vet 1985). The parasitoid *Utetes canaliculatus* (Gahan) (formerly *Opius lectus* Gahan) remains on apples longer and antennates more if host-marking pheromone is present (Prokopy & Webster 1978). Honeydew increases the time spent on plants by the aphid parasitoid *Ephedrus cerasicola* Starý (Hågvar & Hofsvang 1989). Parasitoids are held to a smaller area during search by several behaviors stimulated by kairomones, including reduced walking speed (Waage 1978), a change from straight-line walking to paths that loop back often (Waage 1979, Loke & Ashley 1984, Kainoh et al. 1990), and reversal of direction at kairomone boundaries (Waage 1978).

HOST RECOGNITION AND ASSESSMENT

The "quality" of discovered hosts must be judged before they are accepted for oviposition. Host quality is determined by host species and size (or life stage), physiological condition, and state of parasitism. Assessments are influenced by internal and external chemical cues. Some responses are genetically fixed but others can be modified by recent experience. Understanding determinants of host recognition helps scientists choose highly specific natural enemies for introduction and reduces non-target risk.

Assessment of host quality also increases the efficiency of a parasitoid's egg allocation, allowing for larger, fitter progeny. In response to host size, parasitoids may choose to lay female or male eggs. Placing female eggs in larger hosts increases progeny fitness. Superparasitization is generally less profitable than exploiting an unparasitized host because of lower offspring survival. But if better options are lacking, even the low return from attacking parasitized hosts may be valuable.

Host species recognition

How is a parasitoid to know if a potential host can be parasitized successfully? When parasitoids encounter a prospective host, some general features of host size, position, shape, and location in the habitat suggest that the encountered life stage might be an appropriate host. Egg size affects host acceptance for *Trichogramma minutum* Riley. Females assess egg size by sensing the scapal-to-head angle while walking on host eggs (Schmidt & Smith 1986, 1987). Other parasitoids, respond to a host's surface chemistry. *Telenomus heliothidis* Ashmead (Scelionidae) judges whether eggs might be *Heliothis virescens* (Fabricius) with its antennae and ovipositor (Strand & Vinson 1982, 1983a, 1983b, 1983c; Figure 3.8). Antennal drumming on the egg's surface allows the wasps to detect two proteins produced by the moth's accessory glands (Strand & Vinson 1983c). Glass beads coated with these proteins stimulate oviposition attempts (Strand

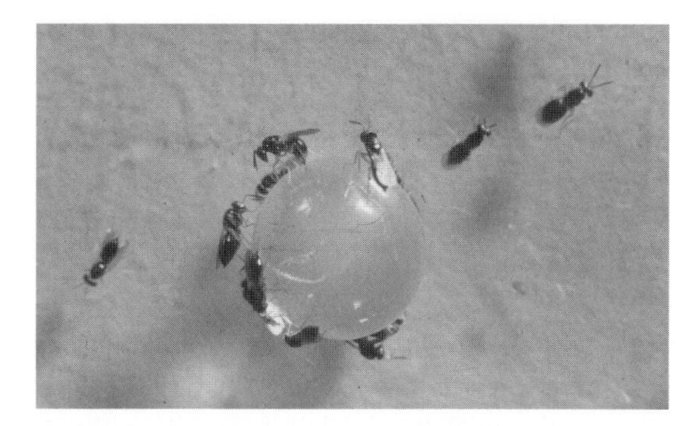

Figure 3.8 Females of *Aprostocetus hagenowii* (Ratzeburg) searching a glass bead treated with calcium oxalate and other materials from host glands that serve, along with a curved surface, to elicit host recognition. Photograph courtesy of Brad Vinson, reprinted from Van Driesche and Bellows (1996) with permission from Kluwer.

& Vinson 1983b). When these proteins are placed on eggs of non-hosts such as *Spodoptera frugiperda* (J.E. Smith) and *Phthorimaea operculella* Zeller, oviposition is induced (Strand & Vinson 1982).

Other such examples include: (1) use of the oöethecal glue of brown-banded cockroaches [*Supella longipalpa* (Fabricius)] by its host-specific egg parasitoid, *Comperia merceti* Compere (Van Driesche & Hulbert 1984), (2) response by aphelinid armored scale parasitoids to chemicals in the host's wax covering (Luck & Uygun 1986, Takahashi et al. 1990), (3) recognition by *Cotesia melanoscela* (Ratzeburg) (Braconidae) of gypsy moth caterpillars based on dense groups of long setae and chemicals in the larval integument (Weseloh 1974), (4) stimulation of *Lemophagus pulcher* (Szepligeti) (Ichneumonidae) by fecal shields of lily leaf beetle [*Lilioceris lilii* (Scopoli)], even when on unnatural hosts or dummies (Schaffner & Müller 2001).

Internal parasitoids gain more information from their ovipositors while probing before oviposition. These cues are less specific (Kainoh et al. 1989), consisting of amino acids, salts, and trehalose (Vinson 1991), which stimulate oviposition and can provide information about prior parasitism.

Assessment of host quality

After recognizing a host's species and life stage, parasitoids must assess quality to determine the number and sex of eggs to lay. Important attributes of quality are host size (and associated nutritional aspects) and previous parasitism.

Host size

Size means different things depending on whether or not hosts grow after parasitism. Some parasitoids attack small hosts and allow them to grow before killing them, increasing the resource for the parasitoid's progeny. *Cotesia glomerata* (L.) oviposits in first- or second-instar caterpillars, but kills fifth instars. Ovipositing in small *Pieris* larvae is advantageous because they are less able to encapsulate parasitoid eggs than later instars (Van Driesche 1988). When hosts do not grow after being parasitized, host size may be judged to decide the number and sex of eggs to lay. The mealybug parasitoid *Anagyrus indicus* Shafee et al., for example, lays up to three eggs in adults but only one in first-instar nymphs (Nechols & Kikuchi 1985). Scale parasitoids typically lay more male eggs in smaller scales (see below). Mechanisms for judging size vary with parasitoid species and may depend on the past experience of individual parasitoids.

Previous parasitism

When examining a host, a parasitoid must learn whether it is parasitized or not, and decide to attack or reject it. Pre-existing parasitism may be from members of the same or different species. When potential repeated parasitism involves conspecifics (a process called **superparasitism**), detection frequently leads to quick rejection. The braconid *Orgilus lepidus* Muesebeck quickly rejects already-parasitized potato tuberworms, *P. operculella* (Greany & Oatman 1972). Parasitoids may, however, obtain some advantage by superparasitism if unparasitized hosts are very scarce or the parasitoid has a high egg load. Rejection is less routine when repeated parasitism is among different species (called **multiparasitism**), but rather depends on the intrinsic competitiveness of the second parasitoid relative to the first. Rejection occurs in some species combinations (Bai & Mackauer 1991), but not in others. Highly competitive species may have little reason to reject previously parasitized hosts (Scholz & Höller 1992).

In either case, cues used to detect parasitism include external marks and internal changes in host hemolymph or tissues. External marks typically last only a few days. For example, the scelionid *Trissolcus*

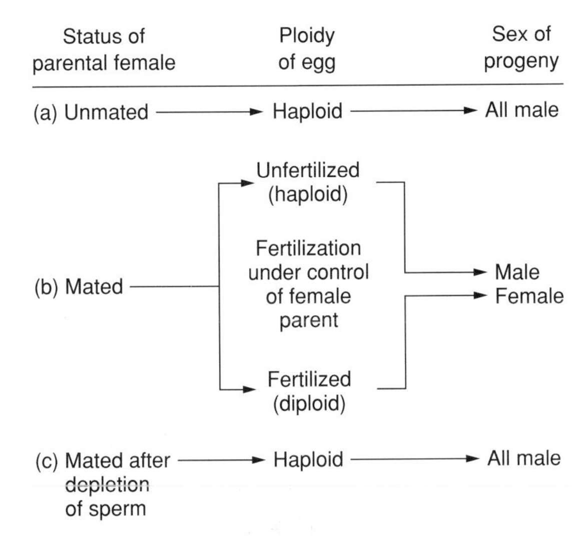

Figure 3.9 Parasitic Hymenoptera, if unmated (a) or depleted of sperm (c), produce only haploid male offspring; if sperm are available in the spermatheca (b), females can control fertilization to produce either female or male offspring based on evaluation of the host. Reprinted from Van Driesche and Bellows (1996) with permission from Kluwer.

euschisti (Ashmead) marks host eggs with a water-soluble chemical (Okuda & Yeargan 1988), and the braconid larval parasitoid *Microplitis croceipes* (Cresson) uses secretions from its alkaline gland (Vinson & Guillot 1972). If superparasitism does occur, larvae compete. In some cases, each merely tries to outgrow the other, using available resources faster. In other combinations, parasitoids seek to eliminate competitors by physical attack, using mandibles (Hymenoptera) or mouth hooks (Diptera), or by physiological means such as anoxia, poisons, or cytolytic enzymes (Vinson & Iwantsch 1980).

Choosing the sex ratio of offspring

Many hymenopteran parasitoids are **arrhenotokous**, having **haplodiploid reproduction**. Females of such species can selectively control egg fertilization. Fertilized diploid eggs yield females and unfertilized haploid eggs produce males (Figure 3.9). This allows parasitoids to put female eggs in the best hosts, reserving male eggs for less-than-optimal hosts.

Aphytis lingnanensis Compere (Aphelinidae) puts male eggs more often in small scales, whereas larger ones receive female eggs (Opp & Luck 1986; Figure 3.10). Previously parasitized hosts often receive more male eggs because they provide fewer resources (Waage & Lane 1984). Sex ratios in laboratory colonies can become male-biased due to encounters with too many parasitized or small hosts, lowering colony productivity. More frequent encounters with conspecific ovipositing females increase the percentage of male eggs laid. However, even under ideal conditions, females on small patches lay at least some male eggs in large hosts to ensure fertilization of their daughters.

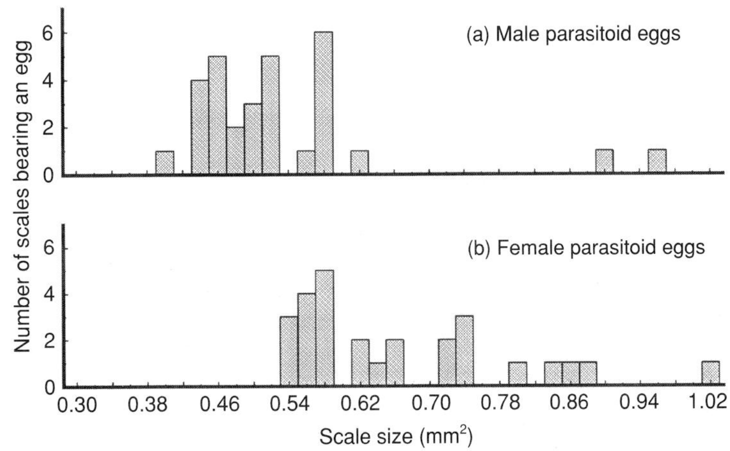

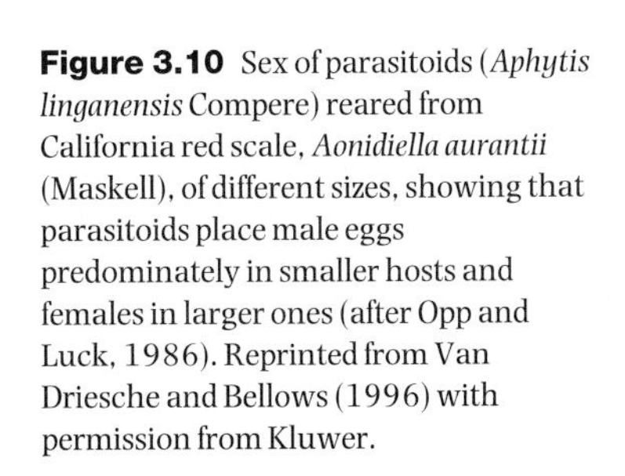
Figure 3.10 Sex of parasitoids (*Aphytis linganensis* Compere) reared from California red scale, *Aonidiella aurantii* (Maskell), of different sizes, showing that parasitoids place male eggs predominately in smaller hosts and females in larger ones (after Opp and Luck, 1986). Reprinted from Van Driesche and Bellows (1996) with permission from Kluwer.

Conditioning and associative learning

Parasitoids learn and use what they learn to help find hosts. Both **conditioning** and **associative learning** have been demonstrated amply for parasitoids. Conditioning occurs when prior experience with a host strengthens the response to that species. Strengthening of an innate response is illustrated by *Brachymeria intermedia* (Nees), which in olfactometer tests walked upwind more often, moved more rapidly, and probed more often in air streams containing kairomones of a host experienced previously (Cardé & Lee 1989). Prior experience can also influence preference for one host over another. Many adult parasitoids contact host kairomones during emergence. If a parasitoid's preferences are weakly fixed genetically, contact with the natal host or its products can strengthen preference for that species. Consequently, parasitoids reared on alternative hosts may perform less well against the pest (van Bergeijk et al. 1989). For specialist parasitoids, whose host preferences are strongly fixed genetically, conditioning may have little effect.

Associative learning occurs when experience links two stimuli that are experienced together (Lewis et al. 1991; Figure 3.11). Secondary stimuli that are often learned as associated with hosts include: (1) form, color, or odor of the host's habitat (Wardle & Borden 1989, 1990), (2) plant species inhabited by the host (Kester & Barbosa 1992), (3) odors from infested host plants (Lewis et al. 1991), and (4) odors associated with nectar or other food sources (Lewis & Takasu 1990).

Parasitoids can also simultaneously associate two or more cues, such as odor and color, with hosts (Wäckers & Lewis 1994). Learned responses cease to affect parasitoid behavior after a few days (Papaj & Vet 1990, Poolman Simons et al. 1992), allowing parasitoids to continually adjust their search image towards recently useful cues.

Learning has several practical implications for biological control. Establishing new species may be easier if parasitoids are exposed first to the pest on the host plant. Similarly, exposure of mass-reared natural enemies to the target pest before release may correct any loss of efficacy (Hérard et al. 1988) from rearing on an alternative host (Matadha et al. 2005). In conservation biological control, non-crop reservoirs are used to produce parasitoids on alternative hosts on border vegetation, but these efforts may be less effective than assumed if natural enemies are conditioned to prefer the non-crop plant or alternative host.

DEFEATING HOST DEFENSES

For a parasitoid larva to successfully mature in a host, it must defeat the host defenses. Hosts defend themselves from parasitism by reducing the chance of being found, physically resisting attack if discovered, and killing parasitoid eggs or larvae if attacked (Gross 1993). Below we present a generalized discussion of these processes, with special reference to Lepidoptera and their parasitoids.

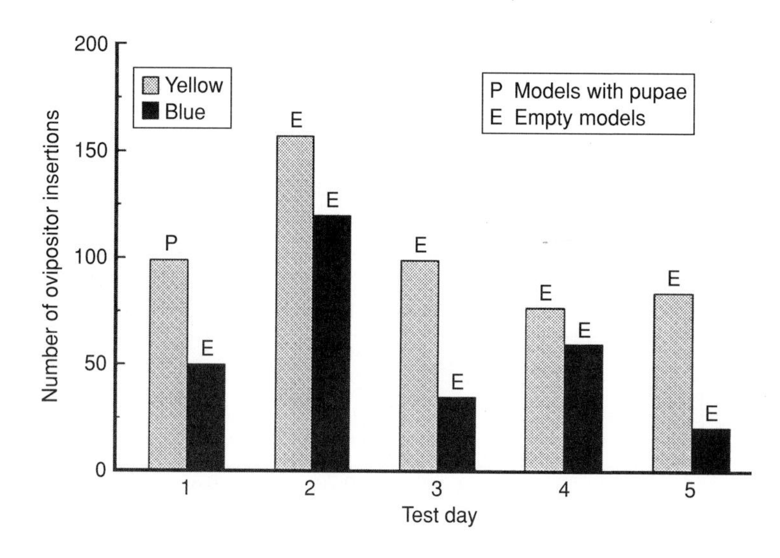

Figure 3.11 *Pimpla instigator* Fabricius wasps, conditioned to the presence of hosts inside yellow cocoon models on day 1, probed yellow models more than blue models for up to four additional days, demonstrating the persistence of associative learning (after Schmidt et al. 1993). Reprinted from Van Driesche and Bellows (1996) with permission from Kluwer.

Reducing the chance of being found

One way for insects to reduce their rate of discovery by parasitoids is to disassociate themselves from kairomones. Some caterpillars frequently change positions during feeding or flick frass away from feeding sites. For concealed feeders (leafminers, borers, etc.) vibrations can be a critical cue revealing host location and periodic cessation of feeding or movement can reduce their apparency to parasitoids.

Over evolutionary time, herbivores may escape parasitoids by exploiting new host plants, a process called occupying **enemy-free space**. This process must meet three criteria (Berdegue et al. 1996), which are illustrated by the shift of the potato tuberworm (*P. operculella*) moth from potato to tomato in Ethiopia (Mulatu et al. 2004). First, the herbivore must be natural enemy limited on the initial plant (here shown as a decrease in mortality on potato when protected by cages). Second, natural enemy impact must be reduced on the new host plant (here, shown as lower mortality on uncaged tomato than on uncaged potato). Third, the new host must not convey a nutritional advantage (here, tomato is an inferior host nutritionally compared to potato, as shown by lower survival on caged tomato than on caged potato).

Preventing attack if found

Some herbivores, if found by a parasitoid, mount a chemical defense (Pasteels et al. 1983). Some species forcefully eject noxious chemicals at attackers. Others concentrate defensive compounds in their outer tissues and become distasteful. *Trogus pennat*or (Fabricius) (Ichneumonidae) does not parasitize larvae of the butterfly *Battus philenor* (L.), even though it has attractive frass, because the caterpillar's integument contains distasteful artistolochic acids sequestered from the host plant (Sime 2002).

Insects may also escape parasitism by: (1) possessing defensive structures, (2) engaging in evasive or aggressive behaviors, or (3) employing ants or parents as bodyguards (Gross 1993).

Defensive structures can be as simple as grouping eggs into a pile. For example, parasitism of gypsy moth (*L. dispar*) eggs by *Ooencyrtus kuvanae* (Howard) is greater in small egg masses, presumably because a higher fraction is physically accessible (Weseloh 1972). Thicker cuticles can be a defensive structure, which likely contributes to the general absence of parasitism in adult insects. Euphorine braconids are one of the few groups that efficiently attack adult insects, and do so by oviposting specifically in lightly sclerotized regions (Shaw 1988).

Behaviors also help hosts evade parasitism. Older aphid nymphs partially deter parasitism by kicking (Gerling et al. 1988). Caterpillars of *Euphydryas phaeton* (Drury) (Nymphalidae) head jerk to knock aside the ichneumonid *Benjaminia euphydryadis* Viereck (Stamp 1982). *Heliothis virescens* larvae foul the bodies of the braconid *Toxoneuron* (formerly *Cardiochiles*) *nigriceps* (Viereck) by lunging and vomiting (Hays & Vinson 1971).

Bodyguards can lower parasitism. Ants tend groups such as soft scales, aphids, and mealybugs to obtain honeydew, reducing parasitism by aggression and disruption of parasitoid behaviors (Gross 1993). The caterpillar *Jalmenus evagoras* Schmett, which feeds on Australian acacia trees, is parasitized less frequently on trees with ants (Pierce et al. 1987). Ant tending can be an important factor reducing success for some classical biological control programs. In some groups (Hemiptera, Membracidae, and Coleoptera), maternal guarding of egg masses or groups of nymphs protects offspring from parasitoids (Maeto & Kudo 1992, Gross 1993).

Killing immature parasitoids if attacked

Hosts, even after they have been discovered and parasitized, may be able to destroy immature parasitoids through **encapsulation**, a process in which blood cells adhere to immature parasitoids to make a capsule. Reactive molecules such as hydrogen peroxide released within the capsule kill the parasitoid (Nappi & Vass 1998). If all eggs are killed, the host survives. Parasitoids, however, have at least two strategies to circumvent encapsulation: evasion and countermeasures.

The evasion strategy

Some parasitoids avoid encapsulation by developing externally. Venom paralyzes the host and preserves it from decay, and parasitoid larvae feed externally like predators (Askew & Shaw 1986, Godfray 1994). External parasitism, however, is largely restricted to leaf- or stem-miners, borers, pupae in cocoons, or gall makers, where some physical structure keeps parasitoid larvae and hosts together.

In contrast, internal parasitism allows use of unconcealed hosts such as caterpillars, aphids, or mealybugs. Also, internal parasitism of larvae or nymphs permits hosts to grow before death. Internal parasitoids, however, risk encapsulation. Some species evade this hazard by attacking the host egg, which lacks an immune system, or by inserting eggs into ganglia, where encapsulating blood cells have no access (Hinks 1971, Godfray 1994), although this is not a complete strategy, as they must eventually leave the ganglion to develop. However, most internal parasitoids must physiologically engage and defeat encapsulation using a variety of countermeasures.

The countermeasures strategy

Internal parasitoids of larvae, nymphs, or adult insects must defeat host immune systems. Unlike mammals, insect immune systems lack specificity and do not produce antibodies capable of recognizing and binding to specific foreign antigens. Insect immune systems mount both cellular and serum responses, but the main defense against parasitoids is encapsulation by blood cells. This is a coordinated response of aggregation, adhesion, and flattening of hemocytes, resulting in the isolation of the parasitoid inside a cellular capsule, within which toxic reactive compounds are released and kill the parasitoid (Nappi 1973, Nappi & Vass 1998). Encapsulation is sometimes accompanied by deposition of a dark pigment called melanin, a process dependent on phenoloxidase activity. Factors affecting the strength and rapidity of encapsulation (Vinson 1990, Pathak 1993, Ratcliffe 1993) include host age, host and parasitoid strain, superparasitization, and temperature (Blumberg 1997).

Apart from encapsulation as a host defense mechanism, symbiotic bacteria, particularly *Hamiltonella defensa*, can confer resistance to parasitism in clones of some aphids (Oliver et al. 2003, 2005).

Countermeasures used by parasitoids to defeat encapsulation include host choice, saturation, polydnaviruses, venom, teratocytes, and anti-recognition devices such as special coatings on eggs. Examples include the following.

1. Some parasitoids oviposit in young hosts, which often are least effective in encapsulation (Debolt 1991).
2. Parasitoids may deposit supernumerary eggs in hosts that exhaust the supply of encapsulating blood cells (Blumberg & Luck 1990), leaving other eggs to survive.
3. Two families of wasps, the Braconidae and Ichneumonidae, use genes from viruses (*Polydnaviridae* and *Bracnoviridae*) to deactivate host encapsulation. These viruses are transmitted to hosts in calyx fluid injected during oviposition (Stoltz & Vinson 1979, Stoltz 1993). The viral genes, in some cases, destroy lamellocytes, one of the hemocytes important in encapsulation (Rizke & Rizki 1990, Davies & Siva-Jothy 1991). They also help regulate the host's physiology and development to favor the parasitoid (Whitfield 1990). Some researchers suggest that these viral genes are no longer part of an independent entity but now form an integral part of the parasitoid's genome (Federici 1991, Fleming & Summers 1991). Also, another group of viruses, the family Reoviridae, help suppress host defenses (Renault et al. 2005).
4. Venoms (Moreau & Guillot 2005) and other materials injected at oviposition can interfere with signaling pathways used to initiate encapsulation. *Leptopilina boulardi* Barbotin et al. introduces substances into *Drosophila melanogaster* Meigen that stimulate the serine proteinase inhibitor Serpin 27A, which negatively regulates phenoloxidase. Enhancement of Serpin 27A reduces phenoloxidase levels, preventing effective encapsulation (Nappi et al. 2005). Venoms also participate in the suppression of encapsulation in some host/parasitoid systems by inhibiting the physical spreading of hemocytes over the surface of the parasitoid egg or, in other cases, by directly killing such cells (Zhang et al. 2004).
5. Teratocytes are giant cells, often derived from the serosal membranes of parasitoid eggs, that have a variety of functions in promoting successful parasitism. These include providing nutrition to developing parasitoids (Qin et al. 1999) and reduction of encapsulation by inhibition of phenoloxidase activity (Bell et al. 2004).
6. Some tachinids evade encapsulation by physically breaking up the developing capsule.
7. Eggs of some hymenopteran parasitoids have coatings on the egg surface that are not recognized by the host immune system.

Additional defenses are certain to be found with study of more species.

REGULATING HOST PHYSIOLOGY

Successful internal parasitoids, in addition to defeating host defenses, must positively regulate hosts to obtain maximum resources and other benefits (Lawrence & Lanzrein 1993, Beckage & Gelman 2004). Regulation may include manipulating molting, feeding, reproduction, or movement. Parasitism may lengthen the feeding stage, induce extra larval stages or precocious metamorphosis, block molting (Jones 1985, Lawrence & Lanzrein 1993), or induce or break host diapause (Moore 1989). Parasitoid regulation of host physiology can help: (1) link host and parasitoid seasonal life histories, (2) correctly time parasitoid development, (3) place hosts in the stage needed for parasitoid growth, and (4) reallocate nutrients from host egg development to parasitoid growth.

Some parasitoids use cues about host diapause to regulate their own state (Schoonhoven 1962), so that they emerge when hosts are in stages suitable for oviposition. When the tachinid *Carcelia* sp. develops in a univoltine species, it enters diapause, but when the same parasitoid develops in a bivoltine species, it continues to develop, has another generation, and enters diapause with its host at the end of the second generation (Klomp 1958). Success of parasitoids introduced to new regions for biological control can be affected by the degree of host/parasitoid synchrony. This in turn is influenced by the diapause phenology of each species and their relation to each other. In Australia, the synchrony of adult tachinids (*T. giacomellii*) with their pentatomid hosts (*N. viridula*) is imperfect because of such complexities, affecting the outcome of this biological control project (Coombs 2004).

In other cases, parasitoids, rather than passively reacting to host conditions, actively control them. The gregarious parasitoid *Copidosoma truncatellum* (Dalman), for example, causes its host *Trichoplusia ni* (Hübner) to undergo an extra larval molt (Jones et al. 1982), thus lengthening its feeding period and increasing resources for the parasitoid's brood. Another parasitoid, *Chelonus* sp., causes *T. ni* to prematurely initiate metamorphosis. Parasitized larvae spin cocoons, but do not pupate (Jones 1985). This ensures that the protective structure of the cocoon is provided to the developing parasitoid before the host's death.

Parasitism may also partially or completely suppress egg maturation by the host in some species, such as parasitism of *Anasa tristis* (De Geer) by *Trichopoda pennipes* Fabricius (Beard 1940, Beckage 1985). This effect is believed to benefit the parasitoid by making nutrients available that would otherwise be sequestered in developing oöcytes (Hurd 1993). Suppression of host reproduction can increase the efficacy of a biological control agent by ending egg laying even before causing host death (Van Driesche & Gyrisco 1979).

PATCH-TIME ALLOCATION

Local areas where hosts have been discovered (patches) and attacked must eventually be abandoned so the parasitoid can search for new host patches. Knowing when to leave a host patch is an important part of parasitoid biology. It might seem that a parasitoid should remain on a plant (or other host patch) until all hosts have been found, but this becomes inefficient if other favorable patches remain to be discovered. The study of how animals evaluate resource patches and decide when to move on is called **optimal foraging**. Foraging behaviors of many animal groups have been investigated (MacArthur & Pianka 1966, Vet et al. 1991). In the 1960–1990 period, much research was done to determine what rules, cues, and processes govern parasitoid foraging (Godfray 1994, van Alphen & Jervis 1996). Here we summarize the influences that affect parasitoids after they start intensified local search on a host patch. At some point intensified search ends. It may end when parasitoids deplete their available eggs and leave to search for nectar or other foods to replenish energy stores. Or parasitoids may leave patches still having eggs to deposit. Why does that happen? What judgments does the parasitoid make about the patch and what stimuli are encountered that determine behavioral outcomes?

Simple models of foraging behavior

Historically, three search rules were proposed to describe when foragers should abandon a patch (van Alphen & Vet 1986): number expectation (Krebs 1973), time expectation (Gibb 1962), and giving-up time (Hassell & May 1974, Murdoch & Oaten 1975). Foragers that hunt with the expectation of encountering a *fixed number* of hosts should leave a patch after that number has been encountered, whether or not additional hosts were still available. Strand and Vinson (1982), for example, found that *T. nigriceps* always abandons tobacco (*Nicotiana tabacum* L.) foliage after one host larva is attacked. This worked because hosts were solitary and each patch therefore had at most one host. However, by itself, this strategy provides no mechanism for abandoning patches that contain no hosts, so additional factors must also affect parasitoid behavior. Foragers that hunt with a *fixed time* expectation would leave patches after that time has elapsed whether or not hosts had been encountered or additional hosts remained undiscovered. Such a strategy would explain the inversely density-dependent patterns of parasitism often seen in nature. Alternatively, foragers hunting with a *fixed giving-up time* would abandon a patch after a preset time had elapsed without encountering a suitable host. A later modification envisioned that if hosts were encountered, the clock could be reset, and the patch would be abandoned only when no new hosts could be found within this reset period. Whether any of these models, or some more complicated scheme, describes how any real parasitoid forages must be determined from observations in nature. But first, we should ask about the kinds of cues a parasitoid might encounter that would affect a parasitoid's behavior on a patch.

Factors influencing patch-time allocation

At least nine factors affect patch-time allocation (van Alphen & Jervis 1996): (1) a parasitoid's previous host contacts, (2) its egg load, (3) host kairomone concentration in the patch, (4) encounters with unparasitized hosts, (5) encounters with parasitized hosts, (6) timing of encounters, (7) encounters with the marks of other parasitoids, (8) encounters with other parasitoid individuals, and (9) superparasitism.

It is not possible to definitely state that each factor has a positive or negative impact on residence time of a parasitoid on a patch, because a factor's influence may differ within and among parasitoid species, and may depend on past experience or current circumstances of the individual. Some generalities, however, can be recognized. In the following section, positive means an influence likely to increase patch time, and negative means one likely to decrease patch time.

1. Previous contacts with the same host species (positive). Parasitoids with previous contact with a given host are likely to react more strongly (through conditioning) to a patch that contains the same host. This may prolong time spent on that patch. Van Alphen and van Harsel (1982) showed that foraging time of *Asobara tabida* Nees increased when presented with a host species to which it had been conditioned 24 h previously.

2. Egg load (positive at high levels). The number of mature eggs a parasitoid has at any given moment influences its tendency to search for hosts (Minkenberg et al. 1992). On discovering a patch, a parasitoid begins to oviposit, decreasing available eggs. Eventually, low egg loads permit parasitoids to be more strongly influenced by competing demands, such as the desire to replenish nutrient stores by feeding. For the aphelinid *A. lingnanensis*, females with few eggs deposited small clutches (Rosenheim & Rosen 1991).

3. Patch kairomone concentration (positive influence). The more kairomone (indicating host presence) a parasitoid finds on a patch, the longer it is likely to remain there. Waage (1978, 1979) found that the parasitoid *Venturia canescens* Gravenhorst increased its patch-time allocation in response to increased kairomone left in the media by larvae [*Plodia interpunctella* (Hübner)]. Dicke et al. (1985) showed a similar response for the parasitoid *L. heterotoma* to its host's kairomone even when no hosts were present.

4. Encounters with unparasitized hosts (positive influence). The object of parasitoid search is to find unparasitized hosts. Therefore, encounters with unparasitized hosts, except for solitary species that occur one to a patch, increase patch search time; for example, *V. canescens* (Waage 1979) and *A. tabida* (van Alphen & Galis 1983).

5. Encounters with parasitized hosts (assumed negative, but may be positive). Encounters with parasitized hosts generally decrease patch time (e.g. *V. canescens*, Waage 1979; *L. heterotoma*, van Lenteren 1991). However, in some parasitoid species, contact with parasitized hosts has no detrimental effect on search time on patches (*A. tabida*, van Alphen & Galis 1983), and may even increase search time if parasitized hosts have potential to be successfully superparasitized.

6. The timing of encounters (variable influence). The patch-time allocation model of Waage (1979) and van Alphen and Jervis (1996) assumes that parasitoids have a certain level of motivation to search for hosts when they find a host patch, based on past experience and the parasitoid's response to kairomones on the patch. This motivation wanes spontaneously over time, but can be increased or decreased based on influences encountered on the patch (see list above). The exact timing of such encounters, therefore, is important because long periods between positive stimuli may allow motivation to diminish to levels too low to retain the parasitoid (Figure 3.12). In contrast, the

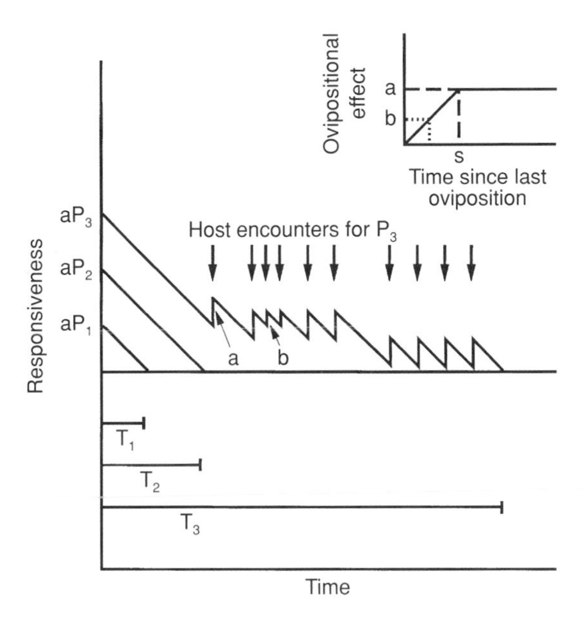

Figure 3.12 Models of retention times on patches for foraging parasitoids incorporate an innate tendency to stop responding to host kairomone over time, coupled with changes in the degree of responsiveness to kairomone due to encounters on the patch. Encounters that lead to oviposition increase retention while encounters with parasitized hosts may decrease motivation. T_1–T_3 are the duration of parasitoids 1, 2, 3 (P_1–P_3) on the patch (after Waage 1979). Reprinted from Van Driesche and Bellows (1996) with permission from Kluwer.

same string of events, differently timed, could produce a longer search time.

7. Encounters with marks of conspecific parasitoids (negative influence). Some parasitoids mark exploited host patches with pheromones that reduce search time of other females (or themselves) entering the patch later (Price 1970, Sheehan et al. 1993).

8. Encounters with other parasitoids (negative influence). Encounters on patches with conspecific adults may reduce foraging time (Hassell 1971, Beddington 1975).

9. Engaging in superparasitism (potentially a positive influence). Superparasitism is engaged in when already-parasitized hosts are encountered, so the influence of the two events is impossible to separate. However, for species that are competitive under conditions of superparasitism, encountering a previously parasitized host can be a positive influence, particularly if transit times to new patches are long or hosts are scarce (Waage 1986, van Dijken & Waage 1987, van Alphen 1988).

Behavioral mechanisms producing foraging patterns

Behaviors that retain parasitoids on a patch include: (1) shifts to walking in a looping or spiraling manner (with a consistent right or left bias) or a zigzag pattern (alternating right and left turns), in place of more straight-line motion, (2) moving less often or for shorter distances per movement, (3) departing from each resource item on the patch in a random direction, which can be caused by turning completely around several times on the resource item during its exploitation, and (4) reversing direction at patch boundaries when contact is lost with a kairomone widespread on the patch.

Behaviors that allow parasitoids to leave a patch include: (1) resumption of normal straight-line walking due to decay of resource-induced looping patterns and (2) failure to engage in direction reversal at patch edges (where contact with patch kairomones is lost) due to habituation to the kairomone.

Field studies of natural enemy foraging

Models and laboratory studies of foraging create hypotheses about how parasitoids might forage. However, field studies are required to validate theoretical models. Waage (1983) demonstrated parasitoid aggregation (*Diadegma* spp.) on high-density host patches under field conditions, a prediction of foraging models. Casas (1989), for the apple leafminer parasitoid *Sympiesis sericeicornis* Nees, showed that leafmines could be detected while the parasitoid was in flight adjacent to the leaf, but determining whether mines contained suitable hosts required landing. Sheehan and Shelton (1989) found that the braconid wasp *Diaeretiella rapae* (McIntosh) did not discover large patches of host plants (collards, *Brassica oleraceae* L.) faster than small patches, but was slower to leave large patches. The number of arrested parasitoids on a patch, therefore, was determined by decisions to leave patches, not factors affecting patch discovery. These studies and others (e.g. the study by Driessen & Hemerik 1992 of the time and egg budget of the vinegar fly parasitoid *L. clavipes*; the comparison by Völkl

1994 of the foraging behavior of *Aphidius rosae* Haliday at different spatial scales; and the examination by Heimpel et al. 1996 of the interactions between egg limitation and host quality on the dynamical behavior of a parasitoid) now allow comparisons between laboratory and field behaviors for particular parasitoids. Further work will refine our understanding of parasitoid foraging (Casas et al. 2004), but the principal components are now understood.

In the broadest terms, an understanding of the foraging decisions of an individual parasitoid will be driven by genetic factors (fixed differences among individuals), the degree of phenotypic plasticity in the species (variable differences among individuals that reflect past learning and other experiences), and the physiological status of the individual at the moment, relative to its needs for food, mates, or hosts (Lewis & Martin 1990, Lewis et al. 1990, van Alphen & Jervis 1996, Outreman et al. 2005, Wang & Keller 2005). Statistical analyses such as proportional hazards models have been used to integrate the complex factors influencing departure decisions (Burger et al. 2006).

Chapter 4

PREDATOR DIVERSITY AND ECOLOGY

Predators are species that have a life stage that kills and eats living animals for development, sustenance, and reproduction. Unlike parasitoids, insect predators are typically larger than their prey and require more than one prey item to complete development. Also, unlike almost all parasitoids, a number of insect predators are nocturnal. Predators are nearly universal, affecting all pests in all habitats to some degree. Insects are eaten by other insects, spiders, and birds and other vertebrates. Spider mites are eaten by thrips, beetles, and predatory mites; pest snails by predatory snails and birds. Juvenile predators use prey for growth, whereas adults use prey for maintenance and reproduction.

Intelligent manipulation of predator complexes for biological control in cropping systems requires knowledge of predator taxonomy and biology, specificity, and rates of predation. This chapter provides an overview of predator diversity and discusses those groups that have been important in biological control. For information on taxonomy and biology of predatory insects and mites see Clausen (1962), Arnett (1968), Hodek (1973, 1986), Hagen et al. (1976, 1999), Foelix (1982), Gerson and Smiley (1990), New (1992), Sabelis (1992), Dixon (2000), and Triplehorn and Johnson (2005).

NON-INSECT PREDATORS

Non-insect predators are found in several groups of invertebrates such as spiders, mites, and snails, and in groups of vertebrates, including birds, mammals, fish, reptiles, and amphibians.

Spiders

Spiders (Aranae) (Figure 4.1) are all predaceous (Foelix 1982). Spiders often show habitat specialization but rarely are specialized as to the prey species they consume. The potential importance of spider complexes in pest control is widely recognized (Clarke & Grant 1968, Mansour et al. 1980, Riechert & Lockley 1984, Nyffeler & Benz 1987, Bishop & Riechert 1990), but their actual significance in particular crops varies from substantial (e.g. rice in southeast Asia) to none (e.g. apples in Massachusetts, USA), depending on the target pest. Because spiders lack host specificity, they are not suited for introduction to new regions to control specific pests.

Figure 4.1 Wolf spiders (Lycosidae) do not build webs but rather actively pursue prey. Photograph courtesy of Jack Kelly Clark, University of California IPM Photo Library.

The appropriate manner to use spiders in biological control is therefore as local generalist predators to help retard population growth of diverse complexes of pests in crops. This can be achieved through conservation in crops of the local native spiders (Riechert & Lockley 1984). Features of spider biology that have important influences on their action as biological control agents include the ability of many species to colonize new areas through ballooning as spiderlings, the relatively high numbers of spiders per unit area of land, and their movements in and out of crops in response to temperature and moisture conditions (Riechert & Bishop 1990).

Mites

Some 27 mite families prey on or parasitize invertebrates, but only eight are important to biological control: Phytoseiidae, Stigmaeidae, Anystidae, Bdellidae, Cheyletidae, Hemisarcoptidae, Laelapidae, and Macrochelidae. Phytoseiidae are the most important and best known. Other families may become recognized as valuable as our knowledge increases (see Gerson & Smiley 1990, Gerson 1992).

Spider mites (Tetranychidae) became important crop pests after 1950 due to natural enemy destruction by pesticides. They can rapidly develop resistance to miticides. Biological control of spider mites depends on conserving their predators, especially phytoseiids (Hoy 1982, Gerson & Smiley 1990; Figure 4.2). Phytoseiid diet strongly influences their role in biological control; dietary groupings are proposed and discussed by McMurtry and Croft (1997). Phytoseiid conservation has been studied in many crops, including apples (Hoyt & Caltagirone 1971), grapes (*Vitis vinifera* L.) (Flaherty & Huffaker 1970), and strawberries (*Fragaria* × *ananassa* Duchesne) (Huffaker & Kennett 1956). Pesticide-resistant strains of a few species have been used to inoculate orchards (Croft & Barnes 1971; see Chapter 21 for details). An understanding of species-specific phytoseiid ecology is essential for successful use, including seasonal ecology, movement on and off crops, the role of surrounding vegetation, requirements for refuges to pass unfavorable seasons (Gilstrap 1988), and need for foods other than spider mites (see Chapter 22 for options).

Several phytoseiids are reared commercially for use against spider mites in greenhouses and on high-value outdoor crops such as strawberries (Huffaker & Kennett 1956, Overmeer 1985, De Klerk & Ramakers 1986). Phytoseiids have only occasionally been used as agents of classical biological control. One well developed example is that of the cassava green mite [*Mononychellus tanajoa* (Bondar)], which invaded Africa and caused important crop losses of cassava. This loss was significantly reduced by the release of a South American phytoseiid, *Typhlodromalus aripo* (De Leon) (Yaninek & Hanna 2003).

Figure 4.2 An adult mite, *Euseius tularensis* Congdon (Phytoseiidae), eating a citrus thrips [*Scirtothrips citri* (Mouton)] larva. Photograph courtesy of Jack Kelly Clark, University of California IPM Photo Library.

Snails

Predatory snails such as *Euglandia rosea* (Ferrusac) and *Rumina decollata* Risso have been used as classical biological agents against invasive plant-feeding snails. The introduction of *E. rosea* on Pacific islands (Laing & Hamai 1976) to control the giant African land snail, *Achatina fulica* Bowditch (a crop pest), however, has been an ecological disaster because this snail shows far too little prey specificity for use as a classical biological control agent. Its introduction has caused the local extinction of some non-target land snails of great cultural and scientific interest (Hadfield & Mountain 1981, Murray et al. 1988, Hadfield et al. 1993, Coote & Loève 2003). In contrast, introduction of the decollate snail to California, USA, apparently controlled the brown garden snail, *Helix aspersa* Müller, with no recorded harm to native snails (Fisher & Orth 1985).

Vertebrates

Many birds and small mammals feed on insects, but because of their broad diets most species are not safe for use as agents of classical biological control (Davis et al. 1976, Legner 1986, Harris 1990). Measures to conserve native birds and mammals can, however, sometimes increase mortality of pests in stable habitats such as forests (Bruns 1960, Nuessly & Goeden 1984, Crawford & Jennings 1989, Higashiura 1989, Zhi-Qiang Zhang 1992). However, there is little proof of the effectiveness of such agents for control of specific pests (Bellows et al. 1982, Campbell & Torgersen 1983, Torgersen et al. 1984, Atlegrim 1989).

Fish have been used effectively as biological control agents against mosquito larvae in small impoundments (Miura et al. 1984). The two most widely used species are poecilid top-feeding minnows: the mosquito fish (*Gambusia affinis* Baird & Girard) and the common guppy (*Poecilia reticulata* Peters) (Legner et al. 1974, Bay et al. 1976). Introductions of such mosquito fish, however, can damage native fish populations through competition or hybridization (Arthington & Lloyd 1989, Courtenay & Meffe 1989).

MAJOR GROUPS OF PREDATORY INSECTS

Predaceous insects of potential use in biological control are found in Dermaptera, Mantodea, Hemiptera, Thysanoptera, Coleoptera, Neuroptera, Hymenoptera, and Diptera (Hagen et al. 1976, Triplehorn & Johnson 2005), with Hemiptera, Coleoptera, Hymenoptera, and Diptera being most important. More than 30 families of insects are predaceous and, of these, the Anthocoridae, Nabidae, Reduviidae, Geocoridae, Carabidae, Coccinellidae, Nitidulidae (*sensu* Cybocephalidae), Staphylinidae, Chrysopidae, Formicidae, Cecidomyiidae, and Syrphidae are commonly important in crops. For information on taxonomy and biology of predatory insects and mites see Clausen (1962), Arnett (1968), Hodek (1973), Foelix (1982), Gerson and Smiley (1990), Hagen et al. (1999), and Triplehorn and Johnson (2005).

Predatory thrips (Thysanoptera)

Most thrips are phytophagous, and some species are pests of cultivated plants. Two families, however, contain

Figure 4.3 An adult *Franklinothrips* sp. thrips (Aeolothripidae). Photograph courtesy of Jack Kelly Clark, University of California IPM Photo Library.

predators: Aeolothripidae (Figure 4.3) *Franklinothrips orizabensis* Johansen, which feeds on thrips, mites, pollen, and lepidopteran eggs, and Phlaeothripidae, for example *Leptothrips mali* (Fitch), which feeds on mites.

Predatory bugs (Hemiptera)

There are many families of predaceous bugs. Various aquatic groups (Notonectidae, Pleidae, Naucoridae, Belostomatidae, Nepidae, Gerridae, Veliidae) include generalist predators that are probably important in suppressing mosquito larvae, aquatic snails, and pest insects on rice (Sjogren & Legner 1989). In crop fields and orchards, many families of predatory bugs influence pest abundance, including the following.

Anthocoridae

Minute pirate bugs are important predators of mites, thrips, aphids, and eggs and young larvae of pests such as European corn borer (*Ostrinia nubilalis* Hübner) (Coll & Bottrell 1991, 1992). Several *Orius* species are reared commercially to control thrips in greenhouses (Plate 4.1a; Gilkeson 1991). Some species have been moved to new locations, such as *Montandoniola moraguesi* (Putton) that was introduced into Hawaii, USA, for control of the invasive Cuban laurel thrips, *Gynaikothrips ficorum* (Marchal) (Clausen 1978).

Miridae

Many plant bugs are pests, but some predaceous species are valuable biological control agents (e.g. *Deraecoris* species in orchard crops) and a few have been imported into new regions. *Tytthus mundulus* (Breddin) was introduced to Hawaii and contributed to control of the sugarcane leafhopper, *Perkinsiella saccharicida* Kirkaldy (Clausen 1978). *Macrolophus caliginosus* Wagner is used to control whiteflies on greenhouse tomatoes in Europe (Avilla et al. 2004).

Geocoridae

Big-eyed bugs (*Geocoris* spp.) are significant predators of whitefly nymphs in cotton (Gravena & Sterling 1983) and of mites, thrips, and aphids in orchards.

Nabidae

Many nabids are predaceous and are most common on grass and herbaceous plants. Nabids feed on insect eggs, aphids, and other small, slow, or soft-bodied insects. *Nabus ferus* L. is a predator of the potato psyllid, *Paratrioza cockerelli* (Sulc) and the sugar beet leafhopper, *Circulifer tenellus* (Baker).

Predatory lacewings (Neuroptera)

Larvae of green lacewing (Chrysopidae) (Plate 4.1b) are predaceous on aphids, whiteflies, mealybugs, thrips, and eggs of various insects. Adults may or may not be predatory, depending on species. Several species are reared commercially, although their use is often not very effective because they are cannibalistic, expensive to rear, and have high food needs for survival after release (Hoddle & Robinson 2004; see Chapter 26). In outdoor crops, attack by generalist predators on augmentatively released lacewings lowers their ability to suppress pests such as aphids (Rosenheim et al. 1999). However, green lacewings probably contribute to conservation biological control in a number of crop systems, and therfore remain of interest (McEwen et al. 2001).

Predatory beetles (Coleoptera)

There are more than 300,000 beetle species in over 110 families. Many groups are important predators, especially the Coccinellidae, Carabidae, and Staphylinidae (Clausen 1962, Arnett 1968).

Coccinellidae

For reviews of the biology of coccinellids and their use in pest management see Hodek (1970, 1973). Obrycki and Kring (1998) discuss their use in biological control. Introduction of a ladybird beetle, *Rodolia cardinalis* (Mulsant), to control the cottony cushion scale (*Icerya purchasi* Maskell) (Plate 4.1c) in California in the 1880s initiated classical biological control because of the dramatic pest control achieved by this predator (Caltagirone & Doutt 1989). Introduction of the African *Hyperaspis pantherina* Fürsch to the island of St. Helena saved the endemic gumwood tree *Commidendrum robustum* (Roxb.) DC from extinction by suppressing the invasive scale *Orthezia insignis* Browne (Fowler 2004). Introductions of coccinellids against scale pests have worked more often than introductions against aphids (Clausen 1978, Dixon 2000). Some introduced coccinellids have become pests by forming large overwintering aggregations in houses (*Harmonia axyridis* Pallas) (Kovach 2004) or they have depressed densities of native coccinellids (*H. axyridis* and *Coccinella septempunctata* L.; Turnock et al. 2003; see Chapter 16).

Native coccinellids are predators of aphids, scales, eggs of various insects, spider mites, and other pests. Their conservation in crops can help suppress pests. In the USA, the native species *Coleomegilla maculata* (De Geer) is an important predator of eggs of various Lepidoptera and of the potato pest *Leptinotarsa decemlineata* (Say) (Hazzard et al. 1991).

Carabidae

Most ground beetles are generalist predators that live on or near the ground and feed mostly at night (Den Boer 1971, Thiele 1977, Den Boer et al. 1979, Erwin et al. 1979, Dajoz 2002). Some species climb plants in search of prey. Carabids are important predators in forage, cereal, and row crops (Hance & Gregoire-Wibo 1987). Agricultural practices that enhance carabids include strip rather than area-wide pesticide application (Carter 1987), retention of some weeds in crops, application of manure to increase organic matter (Purvis & Curry 1984), and planting strips of perennial grasses and raised mounds (beetle banks) in grain fields (Thomas et al. 1991, MacLeod et al. 2004).

A few carabids with specialized habits have been introduced for control of invasive pests, such as *Calosoma sycophanta* (L.) in North America for the control of the gypsy moth, *Lymantria dispar* (L.). Some carabids, such as *Scaphinotus* spp., feed on snails.

Staphylinidae

Most staphylinids are predaceous and some are important

predators of eggs and larvae of flies that breed in manure (Axtell 1981) and of species that attack the roots of young onions, cabbage, and broccoli (Read 1962).

Histeridae

Some histerids prey on manure-breeding flies. *Carcinops pumulio* (Erichson) is an important predator of *Musca domestica* L. eggs and larvae in poultry houses (Axtell 1981). In West Africa, introduction of *Teretrius nigrescens* (Lewis) (Plate 4.1d) controlled the larger grain borer, *Prostephanus truncatus* (Horn), an invasive pest of stored corn and cassava (Schneider et al. 2004).

Cleridae

Larvae and adults of most clerids are bark beetle predators; for example *Thanasimus* spp. are important predators of *Ips typographus* (L.) in central Europe (Mills & Schlup 1989).

Cybocephalidae

This group, sometimes included as part of the Nitidulidae, are predators on pests such as scales. Some species, such as *Cybocephalus* nr. *nipponicus* Endrody-Younga, have been introduced for classical biological control of invasive diaspidid scales (Van Driesche et al. 1998a).

Predatory flies (Diptera)

There are many predaceous fly families. The most important to biological control have been the Cecidomyiidae, Syrphidae, and Chamaemyiidae.

Cecidomyiidae

These flies are predaceous on aphids, scales, whiteflies, thrips, and mites (Barnes 1929). *Aphidoletes aphidimyza* (Rondani) is reared and sold for aphid control in greenhouses (Markkula et al. 1979, Meadow et al. 1985). Predaceous cecidomyiids are common predators of aphids in outdoor crops and potential exists to enhance their effectiveness using conservation biological control practices.

Syrphidae

Syrphids (Plates 4.1e and 4.1f) are important predators of aphids (Hagen and van den Bosch 1968), and some species have been introduced against invasive aphids.

Figure 4.4 Many ants (Formicidae) are predators of insects; here *Formica aerata* (Francoeur) is attacking a larva of the peach twig borer (*Anarsia lineatella* Zeller). Photograph courtesy of Jack Kelly Clark, University of California IPM Photo Library.

Chamaemyiidae

Larval chamaemyiids eat aphids, scales, adelgids, and mealybugs and are probably important in the natural control of some pest aphids; for example, *Leucopis* sp. nr. *albipuncta* Zetterstedt feeds on the apple pest *Aphis pomi* De Geer (Tracewski et al. 1984). Some have been introduced for control of invasive pests, such as *Leucopis obscura* Haliday introduced into Hawaii to control Eurasian pine adelgid, *Pineus pini* (Macquart) (Culliney et al. 1988).

Predatory ants (Hymenoptera, Formicidae)

Ant species include herbivores, scavengers, and predators (Hölldobler & Wilson 1990). All ants are social and the number of individuals per colony may be very large. Predaceous ants (Figure 4.4) can be a large source of non-specific mortality for insects. Ants are important in suppressing pests in forests and crops (Adlung 1966, Fillman & Sterling 1983, Way et al. 1989, Weseloh 1990, Perfecto 1991). Ants in citrus were manipulated by Chinese farmers for pest control 2000 years ago (Coulson et al. 1982), and colonies of green weaver ants continue to be managed in tropical plantations.

OVERVIEW OF PREDATOR BIOLOGY

Most predators cannot complete their life cycles on a single host, but must find, subdue, and consume a series of hosts to mature and develop eggs. Consequently, most predators require high prey densities and must have a mobile, highly efficient searching stage to locate prey. Spiders and predaceous mites are wingless but may be dispersed by wind. Predaceous insects have winged adults, which are more mobile than nymphs or larvae. Dispersing adult insects often have well developed senses of sight and olfaction that allow females to locate high-density prey patches. Some predators actively hunt and chase prey by searching foliage or soil visually or tactilely or by capturing prey in mid-air. Other groups such as crab spiders are ambush predators that wait in flowers and capure prey as they approach. Unlike many parasitoids, predators have nearly even sex ratios (50:50) because predators are never arrhenotokous and rarely parthenogenetic. In most instances, unmated female predators will either not lay eggs, or, if oviposition occurs, the infertile eggs do not hatch. Unlike all parasitoids, many predators are nocturnal or crepuscular (Doutt 1964, Pfannenstiel & Yeargan 2002).

Predators vary in the breadth of their prey ranges, from stenophagous species such as *Rodolia* beetles (coccinellids), whose larvae feed only on margarodid scales, to polyphagous groups such as lacewing (chrysopid) larvae that feed on aphids, caterpillars, mites, scales, thrips, and whiteflies. Most predators are somewhat restricted by prey body size, being able only to subdue prey smaller than themselves (Symondson et al. 2002). In some species, adults and larvae exploit similar prey species but attack different prey life stages because of size constraints. As immature Hemiptera, predatory mites, and spiders grow, they attack progressively larger prey. Also, some predators are habitat specialists, restricting their foraging to particular plant species or habitats.

In addition to prey, many predators consume foods of plant origin (Wackers et al. 2005). In some groups, diets change with life stage. Larvae of some lacewings and flies are predaceous, whereas adults are pollen or nectar feeders. In other groups, predators may exhibit some dietary flexibility in all life stages, consuming items such as sap, nectar, pollen, fungal spores, or honeydew when prey are scarce (Hagen et al. 1976, Symondson et al. 2002). A few groups, such as some mirids, suck plant sap when very young, but become predatory as they mature. At any age, however, such predators may revert to feeding on leaves or other plant parts when prey are unavailable.

Dietary requirements of predators affect their ability to suppress pests. Many predators must consume several prey before reproducing. Delayed reproduction often results in a slow numerical response to increasing prey populations and reduces the chances of acceptable control by some predators (Sabelis 1992). Furthermore, the functional responses of predators level off more quickly than do those of many parasitoids because predators become satiated from feeding, which results in lower attack rates per unit time spent searching and handling prey (Sabelis 1992). The functional response may be further modified if predators are distracted by alternative prey items that reduce attack rates on the pest. Finally, generalist predators with a broad host range may not show an aggregative or numerical response to any individual prey species, unless that species is dominant among all available prey (Symondson et al. 2002).

PREDATOR FORAGING BEHAVIOR

Predators search for prey over substantial distances in a variety of habitats. While searching, they may find many potential prey species, some patchily distributed, on a variety of plants. How then is prey finding to be efficient enough for population growth of the predator? Prey finding and use are affected by many factors, including: (1) volatile or tactile cues released by prey, the chemical and physical properties of the prey's host plant (Messina & Hanks 1998, De Clercq et al. 2000), (2) the sex of the predator, (3) the prey species attacked (Parajulee et al. 1994, Donnelly & Phillips 2001), (4) the spatial distribution of prey (Ryoo 1996), (5) predator behaviors such as searching capacity and arrestment on host patches (Ives et al. 1993, Neuenschwander & Ajuonu 1995), (6) discovery of alternative prey (Chesson 1989), and (7) prey defenses and habitat complexity (Hoddle 2003). All these factors can affect how effectively a predator finds prey and consequently how low it suppresses the pest density.

While both immature stages and adults of most predators are mobile enough to seek out their prey, adults often forage over longer distances. The non-flying immature stages must respond to more local cues (Hagen et al. 1976). Larval coccinellids track aphids by following volatile odors, and older larvae, which are more mobile, search more efficiently (Jamal & Brown 2001).

The ease of prey location depends on the predator's long- and short-distance searching efficiency, density, and spatial distributions of prey populations, the need for non-prey foods as part of the diet, and interactions with other members of the same or higher trophic level. Predators respond to a sequence of cues, starting with those that attract predators over long distances to prey habitats. Then, if suitable stimuli are encountered in the habitat, localized prey search results, leading to prey discovery, assessment, and use.

Habitat location

Finding the prey's habitat is usually done by reproductively mature females looking for oviposition sites. In some species, adult predators may emerge or break diapause already in a favorable crop or forest habitat and immediately begin to search for prey. Alternatively, predators that live in annual crops may need to move to find prey if last year's location is no longer suitable.

Three potential sources of long-distance cues exist: the habitat (e.g. plants), the prey itself, or chemicals released by pest-damaged plants. Undamaged plants composing the habitat may release large quantities of odor, but the odor is there whether or not prey are present. In contrast, odors from the bodies of prey, such as pheromones or odors from prey frass, are reliable indicators but are produced in small amounts that may not be detected easily. In some instances, natural enemy responses to volatile pest pheromones over long distances are both strong and reliable, and this behavioral trait can be used to monitor important predators of target pests. For example, the predator *Rhizophagous grandis* (Gyllenhal) (Coleoptera: Rhizophagidae) is attracted to traps baited with a kairomone produced by the bark beetle *Dendroctonus micans* Kug (Coleoptera: Scolytidae), which has led to improved population monitoring of both predator and pest (Aukema et al. 2000).

The third source of odors – those from plants actively being damaged by the herbivorous prey – is both reliable and produced in large quantity. For example, plants damaged by feeding spider mites are highly attractive to phytoseiid mites, which feed on spider mites (Sabelis & Van de Baan 1983, de Boer & Dicke 2005, Shimoda et al. 2005). Similarly, the predatory thrips *Scolothrips takahashii* Priesner, which is a specialist predator of spider mites, is attracted to bean plants damaged by *Tetranychus urticae* Koch. These thrips are not attracted to undamaged leaves, mechanically damaged leaves, or spider mites or their products, but do respond to damaged plants with spider mites. In field trials, bean plants with spider mites attracted mobile adult *S. takahashii*, but uninfested plants did not (Shimoda et al. 1997). Methyl salicylate, a compound in many blends of herbivore-induced plant volatiles, attracts predators, such as *Chrysopa* spp. (James 2006).

In some cases, predators may respond to mixtures of odors that include both herbivore-induced plant volatiles and volatiles from the prey itself. Volatiles released by disturbed aphids or by barley subjected to aphid feeding are both highly attractive to some coccinellids, whereas uninfested plants or undisturbed or non-feeding aphids are not. This suggests that aphid alarm pheromone ([E]-β-farnesene) functions in predator attraction (Ninkovic et al. 2001). Similarly, hydrophilid beetles that are generalist predators of banana weevil, *Cosmopolites sordidus* (Germar), are attracted to weevil-damaged banana pseudostems and attraction is stronger if weevil aggregation pheromones are also present (Tinzaara et al. 2005).

Understanding which plant compounds attract predators has led to the field testing of synthetic analogs such as methyl salicylate (MeSA) as lures to attract predators and increase their density in crops (James 2003, James & Price 2004). MeSA is a volatile form of salicylic acid, a plant signaling compound implicated in inducing plant resistance to pathogens and repelling some pest species (James & Price 2004). Controlled release of MeSA in hops and grapes resulted in four to six times more natural enemies than plots lacking MeSA dispensers. A variety of parasitoids and predators (e.g. Coleoptera: Coccinellidae; Diptera: Empidiidae, Syrphidae; Hemiptera: Anthocoridae, Geocoridae, Miridae; Hymenoptera: Braconidae; Neuroptera: Chrysopidae, Hemerobiidae) occurred in greater numbers in plots with MeSA compared to control blocks, and pest spider-mite densities were subsequently lower in MeSA-treated areas.

The role of plants in predator attraction has implications for conservation biological control. In some cases, non-crop plants may be an important source of predator-attracting compounds. In barley, weeds increased attraction of ladybird beetles, whether or not aphids were present, suggesting the value of retaining some non-crop plant diversity in crop fields (Ninkovic & Pettersson 2003). Conversely, herbivores feeding on novel crops may go undiscovered by local predators if these new plant species do not produce the critical attractive volatiles. In this situation, native pests may escape predation and become more damaging on the new crop plant (Grossman et al. 2005).

Prey finding

After predators arrive at a favorable prey habitat, they must locate prey. If initial inspection of the habitat leads to evidence of prey in the local area, the predator is likely to engage in **intensified local search**. The behaviors characteristic of intensified local search include more frequent turning, resulting in a sinuous (rather than straight-line) search path, and slower walking, which allows for a more thorough examination of leaf surfaces. Such behaviors can be triggered by prey frass (Wainhouse et al. 1991, Jones et al. 2004), prey materials such as wax or honeydew (van den Meiracker et al. 1990, Heidari & Copland 1993, Jhansi et al. 2000), volatile and non-volatile olfactory cues released by prey (Shonouda et al. 1998, Jamal & Brown 2001), vibrations from prey chewing (Pfannenstiel et al. 1995), or short-range visual detection of prey (Stubbs 1980).

The efficiency of localized search can be influenced by many factors, including host-plant architecture, the predator's hunger status, patch marking by conspecific predators, patch quality, and prey products (e.g. frass or honeydew). Plant architecture (i.e. plant height, leaf number, and leaf area) can affect attack rates by predators (Messina & Hanks 1998) because the more complex the plant's morphology becomes, the fewer prey will be found in a given amount of time (Hoddle 2003). This has been illustrated experimentally studying predator foraging efficiency on pea varieties with mutations for leaflessness, broad, rolled, or slender leaves (Kareiva & Sahakian 1990, Messina & Hanks 1998).

If, on the other hand, initial inspection of a newly discovered habitat fails to reveal any host cues, predators are more likely to engage in straight-line walking, which allows a larger amount of habitat to be examined. Such linear search patterns by predators occur when predators are seeking but not encountering prey. Simple experiments (Karieva & Perry 1989) and complicated modeling (Skirvin 2004) have demonstrated that foraging predators will search greater areas when canopy connectivity is high and linear movement is not interrupted by breaks in travel corridors. During this phase, plant architecture can influence search efficiency because the more divided and obstructed the plant foliage, the harder it will be for such search to continue. Features that enhance connectivity, such as leaf overlap between host plants, are favorable because such bridges allow efficient walking movement among plants (Kareiva & Perry 1989). In contrast, tangled vegetation or highly dissected plant structures may cause breaks in travel corridors that cannot be breached easily.

Hungry predators search less effectively because they walk more slowly, rest more frequently and for longer periods, and cover less distance compared to well fed predators (Henaut et al. 2002). Also, predator age can affect search: young, starved predators engage in extensive linear foraging earlier than older, similarly starved predators. This probably occurs because older predators have greater nutritional reserves. However, searching by older predators may also be influenced by the effects of learning associated with previously encountering and consuming prey (Lamine et al. 2005).

Prey acceptance

After a prey has been contacted, the age and experience of the predator, the size of the prey, and the prey's defensive actions can influence the success of attack. The chemical composition of the prey's cuticle may elicit biting or sucking by the predator (Hagen et al. 1976, Dixon 2000). The importance of surface chemistry to predators has been demonstrated by painting acceptable prey with cuticular preparations from unacceptable prey. In such experiments, predators rejected painted prey because the incorrect chemical search image was encountered (Dixon 2000). In many instances, the decision to attack may depend on rapid assessment of the relative risks (damage from prey defense) compared with the potential nutritional benefits of the species at hand.

Prey suitability

For any given predator, prey species will vary in their quality as food for survival or egg development. Potential prey species can be divided into three groups: (1) species that support both development and reproduction, (2) species that can be eaten, but do not support reproduction and contribute to lower fitness, and (3) unpalatable or noxious species that are not eaten (Dixon 2000). If predators consume too many prey from group 2 (that are substandard in quality), immature predators may fail to complete development or, if they do, the adults may be small, shorter lived, and lay fewer eggs. Conversely, high-quality prey result in shorter developmental times, less mortality of the immature stages, and larger females with enhanced fitness (Hoddle et al. 2001a). In some cases, specific prey may be sought to remedy a dietary deficiency the predator may be experiencing. For example, some vitamin deficiencies can enhance the responsiveness of some predatory mites to prey kairomones that signal the availability of essential elements. This modified prey selection response is lost when the essential dietary component that was missing is obtained. Predators can then switch to other more easily captured or preferred prey (Dicke & Groenveld 1986, Dicke et al. 1986).

PREDATORS AND PEST CONTROL

Because humans have long observed the effects of vertebrate predators, a general understanding of predator biology existed that was easily extended to invertebrate predators. Consequently, some of the earliest known human activities in biological control involved the deliberate use of generalist predatory insects, such as the manipulation of ants in citrus and date groves (DeBach & Rosen 1991). Some pest groups lack parasitoids, so predators may be their only effective natural enemies. This is the case for adelgids (for which predatory derodontid beetles and coccinellids are the key natural enemies) and phytophagous mites (which are preyed on by predatory mites, coccinellids, fly larvae, and thrips). Therefore, by necessity, predators with narrow prey breadth must be used in some biological control programs (Hagen et al. 1999).

Predators of arthropods can be divided into two broad categories: (1) generalist predators that provide substantial, but often unrecognized, natural control of many potential pests and can be enhanced through conservation biocontrol programs (see Chapter 22) or augmentative releases (see Chapters 25 and 26) and (2) specialized predators that, in addition to the above-mentioned uses, can be introduced to new locations as part of classical biological control programs (Hagen et al. 1976).

Generalist predators and natural control

Generalist predators are those that consume several kinds of prey separated by some pre-defined level of taxonomy. For example, a predator may be defined as a generalist if it feeds on prey in different families. A wide prey range can be beneficial because: (1) predators may attack multiple prey stages (e.g. eggs through adults), reducing the need for the predator to be closely synchronized with a particular pest life stage, (2) higher predator numbers can be maintained on alternative species, enabling rapid suppression of the pest should it suddenly increase, and (3) larger, more diverse predator complexes may be retained in annual cropping systems.

Estimates of the number of species range from over 500 in alfalfa (Pimentel & Wheeler 1973) to 1000 in cotton (Whitcomb & Bell 1964). Closer examination of sampling data, however, shows that relatively few of these species maintain persistent populations in crops (O'Neil 1984). Generalist predators that do breed in crops, however, are commonly found in many different crops, suggesting that these predators may share a common set of adaptations that enable them to be successful in the crop habitat (O'Neil & Wiedenmann 1987, O'Neil 1997). In soybeans, O'Neil (1984, 1988) and Wiedenmann and O'Neil (1992) showed that a stable group of predator species consistently maintained a low, but relatively constant rate of predation over a broad range in prey density (defined as prey per unit of leaf area). To do this, predators increase the area they search as the crop grows and the prey are diluted over an increasing leaf area. The relatively low attack rates of generalist predators suggest that they will only provide important pest suppression early in the annual crop's cycle, when pests are scarce (Wiedenmann et al. 1996).

Because food is often scarce, generalist predators exhibit several important life history trade-offs (Wiedenmann & O'Neil 1990, Legaspi & O'Neil 1993, 1994, Valicente & O'Neil 1995, Legaspi & Legaspi 1997), particularly between survival and development, and between fecundity and survivorship. Predators favor their survival at low prey densities by slowing their rate of development (Wiedenmann et al. 1996). Also, when prey are scarce, predators reduce their reproduction, which slows their rate of population increase. In addition, to sustain life when prey are absent, many predators feed on plants (Wiedenmann & O'Neil 1990, Legaspi & O'Neil 1994, Valicente & O'Neil 1995). In summary, generalist predators can remain in crop fields because they are not dependent on one prey type; they have a searching strategy that allows them to locate prey at low densities; and they exhibit life history trade-offs that allow them to sustain populations in crops with only low rates of predation.

Generalist predators in short-term crops

The transitory nature of annual crops and associated production practices (tillage, weed control, pesticide applications, harvesting, burning, fallow periods, and rotations) is widely believed to limit the number, diversity, and impact of predators (Hawkins et al. 1999, Bjorkman et al. 2004, Thorbek & Bilde 2004). If generalist predators are to be effective natural enemies in such rapidly changing environments they should: (1) be rapid colonizers able to keep pace with changes in pest populations, (2) be able to persist in the crop even when key pests are scarce, (3) have flexible feeding habits so as to rapidly exploit new food sources, and (4) have high reproductive and dispersal abilities and low competitive and interference capacity (Ehler & Miller 1978, Ehler 1990).

Favorable combinations of these attributes can allow generalist predators to control pests in some annual crops (Symondson et al. 2002). A literature analysis of manipulative cage and field experiments assessing the impacts of individual predator species or groups showed that in over 70% of cases predators (or complexes of predators) provided significant pest control. For example, a carabid beetle and lycosid spider complex controlled aphids in mid-season winter wheat production (Lang 2003), and, in another case, a complex of Hemiptera (geocorids and nabids) provided control of Colorado potato beetle (*L. decemlineata*) and aphids under certain conditions (Koss & Snyder 2005). Furthermore, manipulated populations of individual predator species have been shown to reduce crop damage or increase yields in 95% of experimental studies. Unmanipulated generalist predator complexes reduced pest populations in 79% of cases studied, and damage was reduced or yield increased in 65% (Symondson et al. 2002). For example, the combined impact of lycosid spider and carabid beetle predation on cucumber beetles significantly reduced beetle densities and increased yields of spring cucumbers (Snyder & Wise 2001).

In some instances, resident generalist predators may provide a strong defense against new invasive pests in short-term crops. For example, when the soybean aphid, *Aphis glycines* Matsumura, invaded the USA in 2000, existing coccinellids, anthocorids, and chamaemyiids greatly reduced the impact of this aphid (Fox et al. 2004).

Generalist predators in long-term crops

Perennial crops are less affected by destructive harvesting or tillage, which consequently favors natural enemy activity (Hawkins et al. 1999). Generalist predators can control both native and invasive pest arthropods, but their importance may be overlooked or underestimated because predation is not readily apparent and is difficult to quantify (Michaud 2002a). Nevertheless, generalist predators have provided either partial or significant control of pests such as mealybugs, scales, and spider mites in crops such as peaches (James 1990), grapes (James & Whitney 1993), citrus and avocados (Kennett et al. 1999), apples and almonds (AliNiazee & Croft 1999), and forests and shade trees (Dahlsten & Mills 1999, Paine & Millar 2002). Secondary pests that do not significantly damage the harvestable commodity have been controlled most successfully. Foliar feeders, for example, are more likely to be controlled to grower satisfaction by predators than are species that damage fruits. Pests with exposed life stages are typically more vulnerable to attack by generalist predators than are cryptic or concealed species (AliNiazee & Croft 1999).

Biological control of invasive pests is most likely to succeed in plantings of long-lived exotic plants because these non-native species often support a smaller set of herbivores compared to native plant communities. These simplified, more linear foodwebs allow introduced predators to operate with less interference from other predators. However, biological control programs in perennial crops can be disrupted by invasions of new pests that may either be poor targets for classical biological control (e.g. thrips or fruit-boring insects) or have damage thresholds that are too low to achieve by biological means (e.g. insects vectoring plant pathogens). For example, pesticide use in California avocado orchards was historically minimal because the important pests like greenhouse thrips, *Heliothrips haemorrhoidalis* (Bouché) (Thysanoptera: Thripidae), avocado brown mite, *Oligonychus punicae* (Hirst) (Acari: Tetranychidae), six-spotted mite, *Eotetranychus sexmaculatus* (Riley) (Acari: Tetranychidae), and omnivorous looper, *Sabulodes aegrotata* (Guenée) (Lepidoptera: Tortricidae), were adequately controlled by generalist predators (Fleschner 1954, Fleschner et al. 1955,

McMurtry 1992). However, subsequent invasions by new pest mites, thrips, and tingids have led to increased use of broad-spectrum, long-lasting pesticides. Avocado growers now rely less on natural pest suppression by predators and routinely use pesticides in areas of heavy pest pressure. The situation has arisen because the new exotic pests are difficult targets for classical biological control (e.g. thrips) and augmentative releases of commercially available native predators (i.e. predatory mite releases against spider mites; lacewing larvae and predatory thrips releases against thrips) either fail or are too expensive (Hoddle et al. 2002a, 2004, Hoddle & Robinson 2004).

However, in some cases resident guilds of native and exotic predators can provide rapid and important natural control of new invasive pests. For example, Asian citrus psyllid, *Diaphorina citri* Kuwayama (Hemiptera: Psyllidae), was attacked and substantially suppressed by several coccinellid species after it invaded Florida, USA (Michaud 2004).

Specialized predators in classical biological control

In many cases new invasive pests are not adequately controlled by pre-existing guilds of generalist predators. For example, although many local predators fed on red gum lerp psyllid, *Glycaspis brimblecombei* Moore (Hemiptera: Psyllidae), after it invaded California and established on eucalyptus, these predators failed to provide control (Erbilgin et al. 2004). Consequently, more specialized species, parasitoids in this case, had to be introduced. However, some groups such as adelgids lack parasitoids, so their control depends on the importation of specialized predators like derodontid beetles in the genus *Laricobius*.

About 12% of successful classical biological control programs have been due to predators, and predator introductions have been most effective against pests that are sessile, non-diapausing, and associated with stable perennial systems (Hagen et al. 1976). The most successful predators have been multivoltine species with non-diapausing and stenophagous adults that are efficient, long-lived hunters. Effective predator species tend to have population turnover rates that equal or exceed those of the prey populations (Hagen et al. 1976). Predators with narrow prey ranges may establish more readily in classical biological control programs compared to generalist predators, which may not compete successfully against a well established resident complex of indigenous predators. Generalist predators may also pose threats to desirable non-target species, such as other natural enemy species, through competition or intraguild predation (IGP).

Stenophagous predators have been extremely important in classical biological programs (e.g. *R. cardinalis* against *I. purchasi*) and in augmentative and inundative biological control (e.g. phytoseiid mites). *Rodolia cardinalis* has been used globally for the biological control of *I. purchasi* in agricultural settings (Caltagirone & Doutt 1989). Due to its high efficacy and limited prey range, *R. cardinalis* has even been used in the Galápagos Islands National Park, where *I. purchasi* endangers rare native plants. Specificity testing prior to release confirmed the narrow feeding range of this natural enemy, thereby clearing it for release in these unique and fragile islands (Causton 2004). Another relatively specific coccinellid, *H. pantherina*, has been used on the island of St. Helena in the south Atlantic against a South American scale, *O. insignis* (Hemiptera: Ortheziidae), which threatened the endangered endemic gumwood trees of the island. *Hyperaspis pantherina* almost never lays eggs in the absence of its prey, *O. insignis*, and over 90% of predator eggs are laid on female *O. insignis*, suggesting a very close, almost parasitoid-like relationship between predator and prey (Fowler 2004).

Phytoseiid mites have received intensive study as augmentative biological control agents of phytophagous mites and thrips on various annual and perennial crops. Phytoseiids have a diversity of lifestyles related to food utilization, which gives many members of this group relatively high prey specificity. Four general categories of phytoseiids are recognized (McMurtry & Croft 1997), as follows.

Type I phytoseiids are specialized predators of phytophagous *Tetranychus* species. They are represented by species of *Phytoseiulus*, especially *Phytoseiulus persimilis* Athias-Henriot, which is regularly used in annual outdoor and greenhouse crops to control *T. urticae* (McMurtry & Croft 1997).

Type II phytoseiids are selective predators of tetranychid mites that inhabit dense webs and are represented primarily by species of *Neoseiulus* and *Galendromus* (McMurtry & Croft 1997). Augmentative releases of *Neoseiulus californicus* (McGregor) and *Galendromus helveolus* (Chant) have successfully controlled persea mites, *Oligonychus perseae* Tuttle, Baker, and

Abbatiello, on avocados (Hoddle et al. 1999, Kerguelen & Hoddle 1999), although release rate, frequency, and timing are critical for control (Hoddle et al. 2000).

Type III phytoseiids are generalist predators that may show high fidelity to particular host plants (rendering them functionally more specific). This category has species in most phytoseiid genera. *Typhlodromalus aripo* is a Type III predator that, following its introduction to Africa, successfully controlled the cassava green mite, *M. tanajoa* (Gnanvossou et al. 2005). Other Type III phytoseiids feed on thrips, whiteflies, mealybugs, and scale crawlers, but these foods are usually less preferred than mites or pollen. Most Types III species have limited utility for augmentative releases against pests. An exception is *Neoseiulus (Amblyseius) cucumeris* (Oudemans), which is used for thrips control in greenhouses (McMurtry & Croft 1997).

Type IV phytoseiids are specialized pollen feeders that also feed on mites and thrips. This group is represented by one genus, *Euseius*, for which population increases depend more on pollen availability than prey abundance (McMurtry & Croft 1997). Consequently, populations of Type IV phytoseiids can be bolstered by supplying plant-based foods (see the section on phytophagy below). Significant impact on target prey may not always result (see following section), but these mites can be quite effective in some cases (James 1990).

EFFECTS OF ALTERNATIVE FOODS ON PREDATOR IMPACT

Suppression of a pest by a predator may be affected by other foods used by the predator. Specifically, a predator's ability to consume alternate prey or feed on plants when the target prey is scarce may alter the predator's impact.

Plant feeding by predators

Plant feeding allows many generalist predators to survive longer and maintain higher populations when prey are scarce. Consequently, plant-derived foods can be important to predators (Wäckers et al. 2005). However, a diet of plant food alone is often insufficient for growth of immature predators and reproduction of adults. Access to plant foods can reduce attacks by predators on each other in the absence of prey, but preferential consumption of plant foods may reduce attack rates on the target pest. Furthermore, this phytophagy may adversely affect harvestable crops.

Generalist predators like ambush bugs (Heteroptera: Phymatidae) can maintain themselves on nectar while they wait for prey (Yong 2003). Pollen can be an important food for coccinellids, increasing reproduction under field conditions (Lundgren et al. 2004). The avocado thrips predator *Franklinothrips orizabensis* Johansen (Thysanoptera: Aeolothripidae) readily feeds on avocado pollen and leaf sap, but doing so exclusively lowers thrips fitness (Hoddle 2003). Consumption of plant foods creates new opportunities for predator conservation. Intercropping maize in cassava, for example, allows the phytoseiid mite *T. aripo* to persist on maize pollen during periods of prey shortage (Onzo et al. 2005). Shelter belts of pollen-shedding plants can enhance populations of predatory mites on fruit trees and sustain predator mites while prey are scarce (Grout & Richards 1991a; Smith & Papacek 1991).

Effects of alternative foods on short-term predation rates may, however, be positive or negative. Adding pollen can reduce predator mite attack on whitefly nymphs (Nomikou et al. 2004). Conversely, in field experiments, geocorid predation on aphids increased when there were plenty of bean pods for predator feeding (Eubanks & Denno 2000). Pruning fruit trees to promote succulent vegetative growth (more suitable for predator feeding) can enhance predator mite populations (Grafton-Cardwell & Ouyang 1995), as can fertilization (Grafton-Cardwell & Ouyang 1996), or planting leguminous cover crops (Grafton-Cardwell et al. 1999). However, these foods may benefit the pest as well as the predator, potentially increasing crop damage (Grout & Richards 1990). In some cases plant feeding may directly scar or otherwise damage the crop. In apple orchards, feeding by the phytoseiid *Typhlodromus pyri* (Scheuten) can scar fruit (Sengonca et al. 2004), and predatory mirids used in greenhouses for whitefly control may damage tomatoes when prey are scarce (Lucas & Alomar 2002).

Alternative prey

Generalist predators sometimes switch between feeding on the target pest and alternative prey. Prey switching may reflect food preference, or the alternative prey may be easier to subdue, more nutritious, or temporarily more abundant. Alternative prey, from the human point of view, are species other than the primary pest, although some alternative prey could otherwise be pests. From the predator's point of view, alternative prey are supplemental food sources that can provide sustenance but might not support reproduction (Hodek & Honěk 1996, Soares et al. 2004).

Use of alternative prey can affect biological control in at least two ways: (1) biological control of the pest may improve if feeding on alternative prey leads to higher natural enemy fecundity or survival or (2) biological control may decrease if attack rates on the pest decrease due to preference for the alternative prey or detrimental impacts on predators from eating the alternative prey (Hazzard & Ferro 1991). In the first instance, the prey–prey interaction is negative as there is a symmetrical negative effect of each prey species on the other's density, an outcome termed **apparent competition** (Holt 1977). When apparent competition occurs, the presence of one prey species helps build predator populations that then increase their attack rate on the second prey, potentially improving biological control of the target pest (Holt 1977).

In the second case, prey–prey interactions are positive because the alternative prey draws or deflects attacks by predators, thereby reducing impact on the target pest by the predator (Holt 1977). This disrupts biological control. For example, eggs of *O. nubilalis* are eaten by a suite of generalist coccinellids, but egg predation declines when corn leaf aphids, *Rhopalosiphum maidis* (Fitch), and corn pollen are abundant (Musser & Shelton 2003). In contrast, aphid control by an assemblage of generalist predators (mostly carabids, staphylinids, and Hemiptera) in spring barley was improved in Sweden by the presence of alternative prey (dipterans, collembolans, and other herbivores). These alternative prey species either increased predator attraction to fields or enhanced predator reproduction, the strongest effect occurring early in the growing season (Östman 2004).

Theoretical models suggest that the presence of alternative prey will eventually increase overall biological control of the target pest by a predator that utilizes both prey species if the alternative prey has a strong positive effect on predator reproduction (Harmon & Andow 2004). This outcome is expected when predators are food-limited, and alternative prey are both abundant relative to the target prey and available for an extended period. Prolonged persistence of dense populations of alternative prey increases the likelihood of a decline in the target prey density. This results from shared predation and an increase in predator numbers owing to reproduction facilitated by a high food supply. In contrast, behavioral factors may reduce a predator's efficacy against a primary pest in the presence of alternative prey. This may occur, for example, if feeding on alternative prey satiates the predator or uses up foraging time. Reliable generalizations about the effects of alternative prey on target prey mortality cannot be readily made due to these conflicting influences.

INTERFERENCE OF GENERALIST PREDATORS WITH CLASSICAL BIOLOGICAL CONTROL AGENTS

Generalist predators may suppress or interfere with populations of arthropods released against weeds (Goeden & Louda 1976) or for arthropod biological control. In the case of a generalist predator attacking a weed biological control agent, no special descriptive term has been created, but the process is not rare. For example, the gorse spider mite, *Tetranychus lintearius* (Dufor), which was released in New Zealand in 1989 for control of gorse, *Ulex europeaus* L., failed to control the weed because a generalist predator, *Stethorus bifidus* (Kapur) (Coleoptera: Coccinellidae), suppressed the growth of *T. lintearius* populations (Peterson et al. 1994). In Oregon, USA, where *T. lintearius* was also established for the biological control of gorse, it was fed on by a complex of phytoseiid predatory mites released for augmentative control of spider mites (Pratt et al. 2003a), especially *P. persimilis*.

When generalist predators interfere with the action of predators released for arthropod biological control, the interaction is termed **intraguild predation** (IGP) because both species are in the same feeding guild (Rosenheim et al. 1995). Local native generalist predators can interfere with exotic predators or parasitoids released for control of invasive pest insects. For example, effectiveness of the encrytid wasp *Psyllaephagus bliteus* Riek, released in California for the biological control of the eucalyptus psyllid *G. brimblecombei*, has been reduced by anthocorid (Hemiptera) predation on parasitized psyllids (Erbilgin et al. 2004). Research interest in IGP is intended to determine whether interference among natural enemies reduces their impact on target pests, either in general terms or in particular cases. IGP may be unidirectional, where one natural

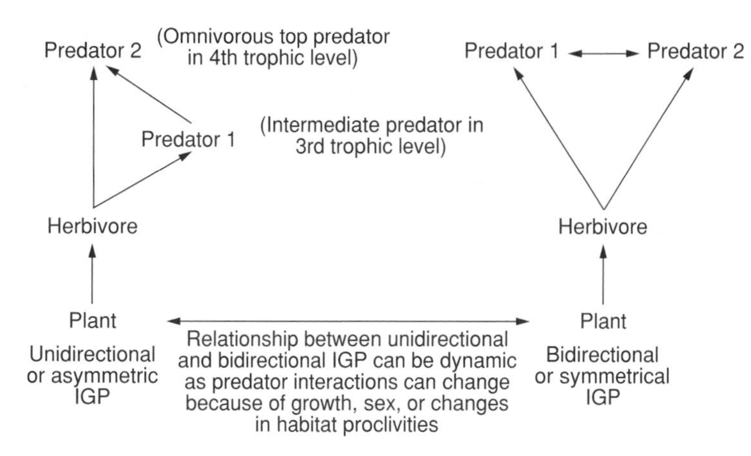

Figure 4.5 Trophic webs illustrating two types of IGP that can occur in biological control systems involving two upper-tropic-level organisms (exploiters) and their common herbivorous prey species (victims). Arrows indicate the direction of energy flow. Modified from Rosenheim et al. (1995).

enemy uses another for food, or bidirectional where each species uses the other for food (Figure 4.5). In both instances, the predators must share a common prey item, which results in competition.

IGP effects on parasitoids

Interference (IGP) among arthropod parasitoids and predators appears to be common in field situations. These interactions are being studied because they may affect the success of biological control projects. Predators affect parasitoids mainly by eating immature parasitoid larvae associated with parasitized pests (Rosenheim et al. 1995). This is an asymmetric interaction in which the predator always wins. This sort of interaction, regardless of its impact on the parasitoid, will complicate field measurement of parasitoid-caused mortality, requiring the use of marginal rate analysis in construction of life tables (Elkinton et al. 1992). Parasitization may even increase the odds that a predator can find and attack the prey. Parasitized aphid mummies are sessile and aggregated, making them particularly vulnerable to predation. Predation risk of aphid mummies increases if adjacent food attracts predators into the local vicinity, increasing the likelihood of discovery (Meyhöfer & Hindayana 2000). Behavioral changes experienced by parasitized gregarious sawfly larvae make them more likely to be attacked by predatory pentatomids (Tostowaryk 1971). However, parasitized hosts may become less preferred as prey as the immature parasitoid ages. For example, the coccinellid *R. cardinalis* readily attacks *I. purchasi* scales that have either eggs or young larvae of the parasitic fly *Cryptochaetum iceryae* (Williston) within them but will not attack scales with mature larvae or pupae of the fly (Quezada & DeBach 1973). Similarly, predatory hemipterans show increasing discrimination of parasitized lepidopteran eggs as parasitoids mature in the host (Brower & Press 1988).

The consequences to biological control of asymmetrical attack by an intraguild predator depend on the relative rate of exploitation of parasitized and unparasitized prey (Rosenhiem 1998). If generalist predators preferentially consume parasitized prey, they may reduce the efficacy of the parasitoid. Some aphids, for example, defend themselves from predators to some degree by kicking and moving when healthy, but parasitized aphids (mummies) do not (Snyder & Ives 2001). Predation by a hemipteran predator on parasitized Lepidoptera larvae caused populations of a stored product pest to nearly double due to disruption of biological control (Press et al. 1974). Alternatively, if predators consume parasitized and unparasitized prey in the ratios encountered (no preference) then predation will not affect the parasitoid's impact (Colfer & Rosenheim 2001, Snyder et al. 2004, Harvey & Eubanks 2005, McGregor & Gillespie 2005). Predatory mirids, for example, attack parasitized and healthy whitefly nymphs in greenhouses at rates that are dependent solely upon encounter frequency (McGregor & Gillespie 2005). Laboratory assessments of IGP can easily overestimate the degree of impact of IGP on biological control in a system and field trials may show no adverse impact of IGP even when laboratory studies suggest it may occur (Snyder et al. 2004).

IGP effects on predators

Some predators eat other predators, especially if they are smaller and hence easy to attack and consume. Predator attacks may be unidirectional (i.e. one predator species dominates another) or bidirectional (i.e. both predators attack each other, as can occur when adults of different species attack each other's immature stages). Both interactions appear common in agroecosystems (Rosenheim et al. 1995). The consequences for pest control can be neutral (Rosenheim et al. 1995), positive (Chang 1996), or negative (Rosenheim 2005). Neutral or beneficial consequences occur if the largest (dominant) predator efficiently exploits prey and preferentially feeds on prey rather than on the intermediate predators (Colfer & Rosenheim 2001). Larger predator complexes may even have synergistic effects if different predator species alter pest behavior in ways that make pests more vulnerable to natural enemy attack (Harvey & Eubanks 2005). However, negative effects on pest suppression may occur if the top predator preferentially preys on intermediate predators, especially if the top predator is less efficient at finding and killing the target pest than the intermediate predator that it suppresses (Colfer & Rosenheim 2001, Colfer et al. 2003, Rosenheim 2005). This sort of predator–predator interaction may prevent some seasonal inoculative releases of predators from establishing reproducing populations (Colfer et al. 2003) and consequently failure to provide pest control. For example, releases of lacewing immatures into cotton do not control whitefly populations because lacewing larvae or eggs are consumed by resident generalist predators such as minute pirate bugs (Anthocoridae) (Rosenheim et al. 1999).

Also, asymmetrical predation on native predators by an invasive predator can be an important factor in successful invasion of the exotic predator and displacement of native competitors. Declines of native coccinellids in parts of the USA have been associated with the invasion of larger, more aggressive, non-native coccinellids, such as *C. septempunctata* and *H. axyridis*, which actively attack native coccinellids even in the presence of aphid prey (see Chapter 16).

PREDATOR AND PREY DEFENSE STRATEGIES

Predators have many natural enemies and effective predators must both defeat prey defenses and protect their own life stages from attack. Coccinellids, for example, have over 100 parasitic insects, mites, and nematodes attacking them in addition to several entomopathogens (Hodek & Honěk 1996). Lacewings are common native generalist predators that demonstrate some forms of defense. They protect their eggs by placing them on long stalks that are not recognized as food by predators walking over the leaf surface and make it difficult for parasitoids to attack them (Canard & Volkovich 2001). Larvae of some lacewings camouflage themselves with pieces of prey, exuviae, or plant material held on their backs by hook-like appendages (Canard & Volkovich 2001). Lacewing larvae feeding on ant-tended Hemiptera may cover themselves with prey waxes as a chemical disguise to escape potential ant aggression (Szentkirályi 2001).

Chemical defenses are used by many predators. Lacewing eggs are often covered with oily protective substances, and lacewing larvae may expel defensive droplets from the anus at attackers. Some species of adult lacewings release repellant odors composed of tridecene and skatole from specialized prothoracic glands (Szentkirályi 2001). Coccinellids, when discovered and attacked, often feign death and may exude distasteful fluids from leg joints (Hodek & Honěk 1996). Chemical protection (distastefulness) is often conspicuously advertised with aposematic coloration (i.e. bright red and black patterns).

Predators that are not chemically protected may protect themselves when attacked by dropping from plants (Sato et al. 2005). Some species mimic the coloration of chemically protected species (Hodek & Honěk 1996). In other cases, predators may depend on escape through speed and agility (e.g. tiger beetles; Pearson & Vogler 2001) or use thick cuticles as a protective barrier (e.g. ground beetles; Sabelis 1992).

To feed, predators must defeat prey defenses. Prey may use many of the same defensive strategies discussed above: avoidance of detection (i.e. crypsis and camouflage), post-detection defenses (i.e. chemical and mechanical defenses, Müllerian mimicry), or deception (i.e. Batesian mimicry). Prey can reduce attack rates by predators in different ways as well. Group defense strategies such as high-density aggregations employed by colonial pest species can dilute predation risk or enhance the effectiveness of communicating potential danger with airborne pheromones that may reduce their per capita risk. For example, aphids use alarm pheromones that warn conspecifics of danger, which prompts walking or dropping from high-risk areas on the host plant. Aphids that are attacked can exude siphuncular waxes to impede predator movement, or use leg kicking to dislodge predators from plants (Dixon 2000). Some pest species may recruit bodyguards (e.g. ants) for protection from predators and provide nutritional rewards such as honeydew to the protective attendants for their services. In studying the biology of any particular predator, its prey's defenses must be considered, along with the predator's responses to these defenses. Knowledge of the limits of effectiveness of a predator's attack strategies will be useful in understanding its potential application in applied biological control programs.

Chapter 5

WEED BIOCONTROL AGENT DIVERSITY AND ECOLOGY

THE GOAL OF WEED BIOLOGICAL CONTROL

The goal of weed biological control is not to eradicate the weed but rather to reduce its vigor so that desirable plants can coexist. Weed biological control does not strive to duplicate the population-regulatory processes of the pest's native environment. When local natural enemies suppress a native plant, both specialists and generalists are involved. In contrast, weed biological control relies on the introduction of just the most specialized of a plant's natural enemies, whose impact is often increased because they are introduced without the specialized parasitoids or predators that attack them in their native range. Such introduced specialist herbivores can profoundly affect the abundance, productivity, and vigor of their food plant when the main factor limiting their populations is food supply. Under these circumstances, the biological control agent may attain densities in the recipient habitat that greatly exceed those in its native range. Desireable plants can then compete more successfully as productivity of the invasive weed declines. Competition then further suppresses the productivity and growth of the invasive plants.

This chapter discusses the kinds of agents used for classical biological control of weeds, which are mainly insects, mites, nematodes, and fungal pathogens. We also discuss non-specialized herbivorous fish that have been used for suppression of aquatic plants in some cases. The taxonomic diversity of potentially useful species is limited only by the diversity of invasive plants, but some groups have been used more often and more successfully than others.

TERMS AND PROCESSES

Natural enemies of plant pests are often classified according to their dietary breadth or host range in terms of the diversity of their host species. These are largely artificial constructs of what is really a continuum, but the concept is useful nonetheless. Some species clearly have very broad host ranges, while others have very narrow ones. The former are referred to as polyphagous or euryphagous, or simply as generalists. They exploit hosts from several different higher taxonomic categories (e.g. families, or orders and possibly even classes). The lobate lac scale [*Paratachardina lobata lobata* (Chamberlin)], for example, develops and reproduces on over 120 species of woody plants in 44 families (Howard et al. 2002). The grass carp (*Ctenopharyngedon idella* Val.), a fish species used as a biological control agent, feeds by browsing on a wide range of aquatic plants. Generalists theoretically utilize host species in proportion to their abundance, thus reducing plentiful species the most. However, the dietary range of a generalist is often constrained by physical or mechanical barriers (e.g. thorns, leaf texture, etc.). Generalists tend to avoid unpalatable species, shifting community composition towards fewer, less-palatable plant species. For example, in pond experiments grass carp selectively removed aquatic plants in order of preference but avoided *Myriophyllum spicatum* L. and *Potamogeton natans* L. These inedible species then increased in biomass and reached levels similar to the total biomass of all plant species in more diverse ungrazed ponds (Fowler & Robson 1978).

There are certain advantages to being a generalist, such as the ability to exploit alternative food sources, but there are also costs (Harper 1977). The efficient acquisition and digestion of food requires specialization. The most successful weed biological control introductions have involved highly host-specific exploiter species. Specialists usually have adaptations to overcome plant defense traits. Specialists frequently are members of taxonomic clades, each of which has diversified on species within a single plant group because of the group's shared phytochemistry. Such species can usually reliably be expected to be host-specific and, as such, are candidates for biological control programs (Andres et al. 1976).

The term **host species** refers to a plant on which the herbivore can complete its development, reproduce, and obtain other requisites for survival (e.g. shelter, moisture, enemy-free space, etc.). Such plant species are referred to as developmental hosts or complete hosts so as to distinguish them from food plants, which may be eaten but do not fully support the herbivore. Those that utilize a single plant species as a developmental host are referred to as monophagous. Monophagous species are the most desirable biological control agents because they pose minimal risk to other plant species. Stenophagous organisms typically utilize a few host species that are phylogenetically related (often in the same genus). When they are released in locations lacking plants closely related to the target weed, stenophagous herbivores exploit only the target and are functionally monophagous.

HERBIVORY AND HOST FINDING

Insects use plants in a variety of ways, with food being only the most obvious (Strong et al. 1984). Water in plant tissues, for instance, helps them avoid desiccation. Plants provide sites for oviposition and pupation. Some insects protect themselves from predation as well as desiccation by residing within plant tissues (leaves, stems, roots, bark), while others construct shelters from plant parts. Chemical compounds in plants are used by some insects as defensive secretions that discourage predators. For example, larvae of *Oxyops vitiosa* Pascoe cover themselves with oils from melaleuca foliage, which protects them from ants (Montgomery & Wheeler 2000, Wheeler et al. 2002, 2003).

Host plants present numerous challenges to plant feeders. Many have structures (thorns, urticating hairs, resinous glands, trichomes, etc.) that impede attachment to, ingestion of, or movement on the plant (Dussourd 1993). Physical barriers and qualitative defenses (toxins) may deter feeding, and quantitative defenses (digestibility-reducing compounds) can inhibit acquisition of adequate nourishment (Rhoades & Cates 1976). In addition, the low nutritive quality of most plants makes it difficult to obtain adequate nutrition for growth and development (White 1993).

Most plant-feeding insects use relatively few species as hosts. Sensory inputs, processed by the central nervous system, determine which to accept or reject (Bernays & Chapman 1994). The process of host acceptance involves a sequence of behaviors governed by external stimuli. Each behavior is triggered by a specific environmental cue, which must attain a minimal level (threshold) to induce the response. Acceptance at one step then enables progression to the next if the net stimulus is positive. Thus, for a plant to be a suitable host, the insect must (Bernays & Chapman 1994): (1) discern the plant's presence at long range and move towards it, (2) distinguish the plant at close range from a confusing array of other species and approach it, (3) find suitable sites on the plant for feeding and/or egg-laying, (4) be stimulated to taste the tissue, (5) be stimulated to ingest the tissue, continue feeding, and/or oviposit, (6) obtain (as an immature stage) adequate nourishment from the tissue to grow and develop, and (7) be able to mature sexually on the diet.

This process, in general, dictates that few of the plants encountered by any particular insect will serve as hosts. Plant chemical defenses further limit the number of acceptable species. Chemical defenses are metabolically expensive to overcome, so most insects restrict their diets to plants with similar defenses.

The damage produced by plant-feeding insects varies in its impact on the plant (see Chapter 20; Janzen 1979). Leaf feeding, for instance, reduces photosynthetic area, disrupts fluid and nutrient transport, induces desiccation of the leaf tissue, and opens the leaf to infection by opportunistic pathogens. It is seldom lethal, however, because most plants can recover from complete defoliation [although evergreen species, with metabolically expensive foliage (Thomas 2000) may succumb more readily than deciduous species]. Most plants compensate for lost tissue by producing new leaves so long as storage tissues and meristematic tissues remain undamaged. Repeated loss of photosynthetic tissues can, however, severely retard growth or even kill a plant when stored resources become depleted (Ohmart & Edwards 1991). Highly synchronized defoliation can also have serious implications, as in the case of *Lixus cardui* Olivier in which overwintering adults emerge in large numbers over a short period just as the thistle is about to bolt. Defoliated thistles suffer reduced growth and reproduction and early senescence (Briese et al. 2004). This loss of reserves affects the plant's ability to withstand stress from herbicides, drought, and frost. Also, even partial defoliation can inhibit flowering and thus cause population-level consequences through reduced seed production (Louda 1984).

Organisms that feed internally, particularly on meristematic tissues, often affect the plant more seriously than external defoliators. Larval or adult feeding inside leaves or stems may destroy the plant's ability to transport nutrients and fluid throughout the plant, causing desiccation, leaf curling, and wilting. Feeding in plant crowns frequently destroys newly forming leaves, reproductive organs, and vegetative propagules. Loss of meristems decreases the plant's ability to replace damaged tissue, subsequently reducing the overall productivity of the plant. Loss of reproductive structures and vegetative propagules can severely curtail plant population regrowth, which is particularly devastating to annual plants. Further, direct damage to storage organs impedes growth and recovery from other stresses. Gall formation creates an energy sink, thus depriving other plant structures of photosynthate, which can lead to reduced flowering. Galls also modify the architecture of the afflicted plant. Ultimately, the level of plant damage sustained is related to the number of phytophagous insect species it hosts, the per capita damage, and the densities they achieve.

Phytophagous insects can also vector plant diseases or facilitate entry of phytopathogens into the plant. Many sap-feeding insects transmit viral diseases capable of killing plants. Also, some sap-feeding insects are thought to inject toxic saliva into the wound, producing necrosis of surrounding tissues, such as with psyllids (Hodkinson 1974) and the sugarcane spittlebug (Hill 1975).

HERBIVORE GUILDS

Plant-feeding insects are sometimes classified as to whether they feed externally or internally. Those that live on and feed from the exterior of the plant are considered ectophages. Leafminers, gall insects, stem borers, and others that live and feed concealed within plant tissues are considered endophages.

Phytophagous insects exploit host plants in five general ways (Strong et al. 1984): (1) external feeding in which chewing mouthparts are used to bite plant tissue, most notably from the leaves, (2) external feeding in which piercing-sucking mouthparts penetrate the plant tissue then draw out cell contents or fluids from the vascular system, (3) external feeding in which rasping-sucking mouthparts abrade the plant surface then suck the fluids seeping from the wound, (4) internal feeding which creates burrows or mines within the plant tissue, and (5) creation of galls, where insects live and feed on hypertrophied plant tissue. Many phytophagous insects possess different feeding mechanisms at different life stages and could thus be included in more than one category.

Herbivores that exploit the same resource in a similar manner are often referred to as a feeding guild (Crawley 1983, Price 1997). Leaf feeders, for example, may be divided into the pit-feeding guild, the strip-feeding guild, or the sap-feeding guild (Root 1973). Plants growing in their native environment are often fed upon by representatives from numerous guilds (high species packing), whereas invasive plant species usually have depauperate faunas with numerous vacant niches. For example, Briese (1989a) and Briese et al. (1994) compared the phytophagous insect fauna of *Onopordum* thistles between Europe, the native range, and Australia, where they are invasive. They noted a virtual absence of endophages in Australia whereas this guild comprised 54% of the European fauna.

GROUPS OF HERBIVORES AND PLANT PATHOGENS

Nearly all efforts at weed biological control have involved classical biological control based on the introduction of insects or plant pathogens from the plant's native range (Julien & Griffiths 1998). In a few cases, generalist fish, such as the grass carp, have been used to graze down the biomass of macrophytic water plants in a non-specific way. A few other vertebrate species, such as geese, goats, and sheep, have been used to remove plants from local, often fenced, areas (De Bruijn & Bork 2006). There have been a few attempts to utilize native, plant-feeding insects to control introduced weeds by augmenting naturally occurring populations (Frick & Quimby 1977, Frick & Chandler 1978, Sheldon & Creed 1995). Also, efforts have been made to develop bioherbicides for crop weeds using locally occurring pathogenic fungi but these have rarely been economically successful (see Chapter 24).

Insects and mites as plant biological control agents

Most herbivores released for weed control have been insects due to high species diversity, their size, high degree of host specialization, and their potential for rapid population growth (Andres et al. 1976). Insects that feed directly on the living tissues of plants are confined to nine orders: Collembola, Orthoptera, Phasmida, Hemiptera (including the former Homoptera), Thysanoptera, Coleoptera, Diptera, Hymenoptera, and Lepidoptera (Strong et al. 1984). Representatives of seven of these orders have been used in past weed biological control attempts: Collembola and Phasmida being the exceptions (Julien & Griffiths 1998). Lepidoptera and Coleoptera have contributed 76% of the 341 species used for weed control. Among these seven orders, biological control species have been drawn from some 57 insect families, eight of which account for about 65% of the introductions: Curculionidae (19%), Chrysomelidae (17%), Cerambycidae (4%), and Bruchidae (3%) (all Coleoptera); Pyralidae (8%), Tortricidae (4%), and Noctuidae (3%) (all Lepidoptera), and Tephritidae (7%) (Diptera) (Julien & Griffiths 1998). In addition to insects, mites in the families Galumnidae (Julien & Griffiths 1998), Eriophyidae (Goolsby et al. 2004a), and Tetranychidae (Hill & Stone 1985, Hill et al. 1991) have been used (Briese & Cullen 2001). Overall, biological control has been attempted against about 135 weed species in some 43 plant families. About half of the targeted weed species have been members of three families: Asteraceae, Cactaceae, and Mimosaceae. Few generalities about the biology of these insect families are possible, inasmuch as the species included within them are quite diverse in their habits. Broader descriptions of herbivorous insect families are given by CSIRO (1970), Arnett (1985), and Triplehorn and Johnson (2005). We have drawn heavily upon these sources in our following descriptions of the biologies and life histories of the various groups.

Chrysomelidae (Coleoptera)

This is a large diversified family with over 3700 species that have evolved specialized relationships with many kinds of plants, although food plants are only known for about a third of the described species (Jolivet & Verma 2002). Species within the family have been grouped into some 20 distinctive subfamilies. Most of those used as weed biological control agents are in the subfamilies Alticinae (flea beetles), Chrysomelinae, Cassidinae (tortoise beetles), Chlamisinae, Cryptocephalinae, Galerucinae, Hispinae, or Hylobinae.

Most species are phytophagous although some are detritivorous, coprophagous, carnivorous, oöphagous, nematophagous, entomophagous, or cannibalistic. Adult chrysomelids generally feed openly on foliage and flowers. They are not strong flyers so they are vulnerable to predation and parasitism (Jolivet & Verma 2002).

Jolivet and Verma (2002) note that most species are oviparous. Eggs may be laid on the food plant or scattered on the ground, singly or in masses. They may be covered with secretions, excreta, or other materials, or enclosed in a case (oötheca). Those that lay fewer eggs generally provide greater protection for them. Larvae may feed in the open on foliage, or they may be leafminers, stem borers, or root feeders. Some are aquatic, and some create galls (subfamily Sagrinae). Larvae of *Cryptocephalus* feed on bark and plant debris and form a protective case (scatoshell) that protects them from ants (Jolivet & Verma 2002). Free-living larvae possess various means of protection, including coatings of fecal material; chemical, behavioral, or structural defenses; and parental care or subsociality (Jolivet & Verma 2002). Pupae are exarate and sometimes protected by a cocoon. Naked pupae formed on foliage may be chemically defended, spiny, or aposomatic. The pupa of at least one species produces defensive sound emissions (Jolivet & Verma 2002).

Most chrysomelids are oligophagous, although some are polyphagous. Their typical life-history pattern involves feeding and oviposition on leaves, with pupation on the foliage or after dropping to the ground. However, many variations in this typical pattern occur (Jolivet & Verma 2002).

About 62 species of chrysomelids have been used in biological control programs, and 36 (58%) have established at least once. Twenty-one species (58% of those established) have successfully produced at least local control of some 13 targeted weed species. St. John's wort (*Hypericum perforatum* L.) was controlled in the western USA by *Chrysolina hyperici* (Forster) and *Chrysolina quadrigemina* (Suffrian) (Figure 5.1; see McCaffrey et al. 1995 for summary). The alligatorweed flea beetle, *Agasicles hygrophila* Selman and

Figure 5.1 Adult of the chrysomelid *Chrysolina quadrigemina* (Suffrian). Photograph courtesy of Jack Kelly Clark, University of California IPM Photo Library.

Vogt, successfully controlled alligatorweed, *Alternanthera philoxeroides* (Martius) Grisebach (Julien 1981, Buckingham 1996). *Senecio jacobaea* L. was controlled by two herbivores, one of which was the root-feeding chrysomelid *Longitarsus jacobaeae* (Waterhouse). *Calligrapha pantherina* Stål successfully controlled spinyhead sida (*Sida acuta* Burman) in northern Australia (Flanagan et al. 2000). Two species of *Galerucella* have been introduced to control purple loosestrife (*Lythrum salicaria* L.) in North America with promising results at many sites (Blossey et al. 1996, Dech & Nosko 2002, Landis et al. 2003). Several species of *Apthona* have controlled leafy spurge, *Euphorbia esula* L., over large areas in some habitats of the north central USA (Nowierski & Pemberton 2002, Hansen et al. 2004). *Diorhabda elongata* Brulle has begun to show effective control of saltcedar (*Tamarix* spp.) in some areas of the western USA (DeLoach & Carruthers 2004). The tortoise beetle *Gratiana boliviana* Spaeth (subfamily Cassidinae) has been released in Florida, USA, against tropical soda apple (*Solanum viarum* Dunal) (Medal et al. 2004).

Curculionidae (Coleoptera)

The nearly 50,000 species of weevils comprise numerous subfamilies, some of which been elevated to familial status within the superfamily Curculionoidea (e.g. Zimmerman 1994). Most species are readily recognizable by the long, slender projection that bears the mouthparts, commonly referred to as a snout. Mandibles at the end of this snout are used to remove or chew holes in plant tissue.

Because of the number and diversity of weevils, it is difficult to generalize about their biology and ecology. They occupy nearly all land regions of the world, from the driest deserts to the most humid tropics (Zimmerman 1994). Nearly all are plant feeders, although at least one species, *Ludovix fasciatus* (Gyllenhal), preys on eggs of grasshoppers in the genus *Cornops* (Bennett & Zwolfer 1968). Some larvae feed externally, although the majority feed internally. They eat tissues of virtually all parts of plants including roots, bark, sapwood, heartwood, stems, twigs, leaves, buds, flowers, pollen, seeds, fruits, and dead and dying plant material (Zimmerman 1994). Some species are aquatic or subaquatic, living completely underwater or in air-filled tissues of underwater plants.

Most weevils are good fliers although some are flightless, having reduced wings. Some undergo seasonal flightless periods when the indirect flight muscles deteriorate. Flight muscle degeneration–regeneration sometimes alternates with ovarian maturation and degeneration (Buckingham & Passoa 1985, Palrang & Grigarick 1993).

Most species insert their eggs into plant tissue or between leaves that have been glued together, although some are laid directly on the soil. Oviposition often takes place in a hole excavated by the adult female with her snout, although some use specialized caudal appendages or ovipositors for this purpose (Zimmerman 1994). The hole may or may not be covered with a plug of excrement or other material.

The cylindrical larvae are usually legless, whitish, and grublike. Some feed on foliage while exposed, some cover themselves with excrement, but most feed internally within the plant tissues. Pupation may take place in a cocoon attached to the plant, within a burrow in a plant, loose in the soil, or in a hardened cell composed of soil particles. Development from egg to adult occurs within days for some species, years for others (Zimmerman 1994).

About 68 species of curculionid (including apionids) have been used in biological control projects. Of these, 49 (72%) have established at least once. Among the established agents, 26 (53%) have produced at least local control in at least one area. The impacts of

Figure 5.2 Adult of the melaleuca weevil, *Oxyops vitiosa* Pascoe. Photograph courtesy of Steven Ausmus.

14 species (29%) are unknown, either because the introductions are too recent or because evaluations were not done. Only nine of the established species (18%) have been considered totally ineffective. Examples of weevils that have been effective include *Rhinocyllus conicus* (Frölich), which controlled nodding thistle, *Carduus nutans* L., in Canada (Harris 1984) and elsewhere; *Neohydronomus affinis* Hustache, which has controlled water lettuce (*Pistia stratiotes* L.) in several countries (Harley et al. 1984, Dray & Center 1992, Cilliers et al. 1996); *Neochetina eichhorniae* Warner and *Neochetina bruchi* Hustache, which have controlled waterhyacinth [*Eichhornia crassipes* (Mart.) Solms] in many countries (Center et al. 2002); and *Microlarinus lypriformis* (Wollaston), which together with *Microlarinus lareynii* (Jacquelin du Val), partially controlled puncture vine (*Tribulus terrestris* L.) in the south western USA and Hawaii, USA (Huffaker et al. 1983). The melaleuca leaf weevil, *O. vitiosa* (Figure 5.2), has drastically reduced the invasive potential of *Melaleuca quinquenervia* (Cav.) Blake, an invasive wetlands tree from Australia (Pratt et al. 2005). The Mediterranean sage root crown weevil, *Phrydiuchus tau* Warner, provides good control of *Salvia aethiopis* L. at many sites along the Pacific coast of the USA (Coombs & Wilson 2004). *Stenopelmus rufinasus* Gyllenhal has nearly eliminated the floating fern *Azolla filiculoides* Lamarck from South Africa (McConnachie et al. 2004). *Cyrtobagous salviniae* Calder and Sands has provided spectacular control of giant salvinia (*Salvinia molesta* D.S. Mitchell), another floating fern, in many countries (Julien et al. 2002), including the USA (Texas; P.W. Tipping, personal communication). Australian weevils in the genus *Melanterius* have contributed to the control of weedy *Acacia* species [and the closely related *Paraserianthes lophantha* (Willd.) Nielsen] in South Africa (Impson & Moran 2004). Three weevil species provide effective control of the invasive tree *Sesbania punicea* (Cav.) Benth. in South Africa (Hoffmann & Moran 1998). The root weevil *Cyphocleonus achates* (Fahraeus) has been reported as effectively controlling spotted knapweed (*Centaurea stoebe* L.) at sites in Montana (Story et al. 2006).

Cerambycidae and Buprestidae (Coleoptera)

Larvae of these families typically bore in woody stems and can have developmental periods of 2 years or more. The Cerambycidae are distinctive, more or less cylindrical beetles that usually have long antennae and deeply notched eyes. Adults are often brightly adorned, but they may also be cryptically colored. Larvae have smallish heads, are legless, whitish, and usually bore in the heartwood of trees, but some live in plant stems, or bore into roots or the wood of buildings. They are somewhat spindle-shaped, being fleshy and elongate, tapering from anterior to posterior. In tree-feeding species the adults lay their eggs in crevices in the bark or in holes created by the female. The larvae bore into the wood, making tunnels that are round in cross-section. Some attack living trees, but most prefer weakened or dying trees or branches, or freshly cut log. Some girdle twigs and then oviposit in the isolated portion.

Buprestid adults have short antennae and are dorso-ventrally flattened. Many species are metallic blue, black, green, or copper-colored. Adults are quite active during the daytime and are often found nectaring on flowers. Most fly when disturbed although some retract their legs and drop to the ground. The larvae are spindle-shaped but are dorso-ventrally flattened and with a wide expansion of the prothorax. Their galleries tend to be oval in cross-section when they bore in wood. They also burrow under bark, in roots, or in the stems of herbaceous plants. Some small species create galls, some girdle twigs, and some are leafminers.

At least 17 cerambycids and three buprestids have been used in biological control projects. All three buprestids established at least once and two have provided some level of control (Julien & Griffiths 1998). *Agrilus hyperici* (Creutzer) was released in the USA, Australia, Canada, and South Africa to control St. John's wort (*H. perforatum*) and contributes to control

Figure 5.3 Cerambycid larva in wood of melaleuca tree. Photograph courtesy of Matt Purcell, CSIRO.

in Idaho. *Lius poseidon* Nap was released on Koster's curse [*Clidemia hirta* (L.) D. Don] in Hawaii, but its impact is unclear. *Sphenoptera jugoslavica* Obenberger was released in the USA and Canada for control of diffuse knapweed (*Centaurea diffusa* Lamarck), but it also exploits other knapweed species.

Cerambycids (Figure 5.3) have successfully established field populations in 10 cases but only four species are regarded as effective (Julien & Griffiths 1998). *Alcidion cereicola* Fisher was released in Australia and South Africa for control of cacti in the genera *Harrisia* and *Cereus*. *Archlagocheirus funestus* (Thomson) was released in Hawaii, Australia, and South Africa to control cacti in the genus *Opuntia*. *Megacyllene mellyi* (Chevrolat) was released in Australia to control groundsel bush (*Baccharis halimifolia* L.). It established only locally but reduced weed densities by up to 50% (Julien & Griffiths 1998). *Plagiohammus spinipennis* (Thomson) was released in Hawaii, Guam, Palau, and South Africa to control *Lantana camara* L. It provides partial control in Hawaii in areas of high rainfall but is ineffective at drier locations.

Bruchidae (Coleoptera)

Conflicts of interest often arise regarding the control of invasive trees since they are likely to have economic uses. Many species of exotic acacia, for example, are used in southern Africa for firewood, but these species are also jeopardizing unique floristic areas (Impson & Moran 2004). In this situation, the focus has often been on the selection of biological control agents that would reduce plant reproduction without killing existing trees (Dennill & Donnelly 1991). Seed feeders seem ideally suited to this purpose and bruchid beetles are well-known, highly specific seed predators.

Bruchids are short, stout-bodied beetles with elytra that do not completely cover the tip of the abdomen. The body narrows towards the anterior end and the head bears a short, broad snout. Although they attack seeds of several plant families, they are most prevalent in the Leguminosae (*sensu latu*). Seed beetles usually lay single eggs on seeds or pods, although some species oviposit on flowers. The larvae burrow into the seed where they devour the endosperm. They may develop completely in a single seed or feed on many seeds within a pod. Some species prefer immature seeds, others prefer mature seeds, and some only attack seeds on the ground. Bruchids usually pupate within the seed and upon completing development cut round holes in the testa through which to emerge.

At least 12 species of bruchid have been used in attempts to control mostly leguminous weeds by reducing the plant's reproductive potential. Eleven species established successfully; one species (*Algarobius bottimeri* Kingsolver, released in South Africa on *Prosopis* spp.) failed to establish. Despite reports of high levels of seed mortality, these agents are generally regarded as ineffective. However, two species [*Acanthoscelides puniceus* Johnson and *Acanthoscelides quadridentatus* (Schaeffer)] introduced into Thailand have reportedly destroyed up to 80% of *Mimosa pigra* L. seeds in some areas. However, two species [*Algarobius prosopis* (LeConte) and *Neltumius arizonensis* Schaeffer] established in South Africa in an attempt to control mesquite (*Prosopis* spp.) also destroy up to 70% of the seed crop, but are viewed as ineffective. Grazing of the seeds by cattle and recruitment of native parasitoids are considered the main causes of their lack of impact (Impson et al. 1999).

Pyralidae (Lepidoptera)

Most of the following information is derived from Munroe (1972). This is the third largest family in the Lepidoptera. Moths are small- to moderate-sized, with a long, scaled porrect proboscis. They often appear to be triangular in shape when at rest. Many species are dull, but several are strikingly colored. The family is ubiquitous, occurring in most areas and habitats. Some are even aquatic. The egg choria are usually thin and the eggs are sometimes flattened and lens-shaped. The obtect pupae are often enclosed in silken cocoons, although those formed in plant tissue may be naked.

Larvae are usually cylindrical, with a well-formed head capsule, prolegs, and distinct setae. Larval feeding habits are quite varied and species may be foliage feeders, borers, or feed on stored products such as beeswax. They often feed within webs or leaves tied together with silk. Some are leafminers and some live within air-filled tissues of aquatic plants whereas others possess gills and are fully aquatic.

One of the most famous weed biological control agents is the phycitine species *Cactoblastis cactorum* (Bergroth), which successfully controlled prickly pear cacti (*Opuntia* spp.) in Australia. At least 26 pyralid species have been employed in biological control, but only half established field populations, and only six have contributed to suppression of their target weeds: *Arcola malloi* (Pastrana) on alligatorweed, *C. cactorum* on *Opuntia* spp., *Euclasta gigantalis* Viette on rubbervine [*Cryptostegia grandiflora* (Roxb.) R. Br.], *Niphograpta albiguttalis* (Warren) on waterhyacinth, *Salbia haemorrhoidalis* Guneé on lantana, and *Tucumania tapiacola* Dyar on *Opuntia aurantiaca* Lindley. Except for *C. cactorum* and *E. gigantalis* Viette, these moths have not been reported to significantly impact their target pests.

Arctiidae (Lepidoptera)

Tiger moths are often brightly colored with conspicuous stripes, bands, or spots. When at rest they hold their wings over their bodies in a tent-like fashion. Eggs, which are often laid in clusters, are usually hemispherical in shape with surface sculpturing. The larvae have dense, often colorful setae, arranged in clumps on verrucae. Some species have urticating hairs. Dense setae give larvae a fuzzy appearance, hence the common name for some as woolybears. Pupae form inside cocoons created mainly from larval setae and small amounts of silk. Feeding habits are varied; some are generalists, but some are highly host-specific. Most are external leaf feeders on herbaceous or woody plants, although some feed within seed pods and some on lichens. Four species have been used as biological control agents, three of which have suppressed their target weeds successfully. *Tyria jacobaeae* contributed to the suppression of tansy ragwort, *S. jacobaea*, in Oregon, USA (McEvoy & Cox 1991). However, its further redistribution in the USA is not recommended because it also attacks some native *Senecio* species. *Rhynchopalpus brunellus* Hampson provides partial control of Indian rhododendron (*Melastoma malabathricum* L.) in Hawaii. After initial failures and repeated releases of massive numbers, the arctiid moth *Pareuchaetes pseudoinsulata* Rego Barros finally established in South Africa and is now beginning to control the triffid plant, *Chromolaena odorata* B. King and H. Robinson (C. Zachariades, personal communication), much as it has done in several other countries (Julien & Griffiths 1998).

Dactylopiidae (Hemiptera, formerly Homoptera)

Cochineal insects (Figure 5.4) are related to scale insects and mealybugs. They are native to tropical and subtropical regions of the Americas, where they feed on *Opuntia* cacti. Females produce carminic acid as a defensive substance, which is an important natural red dye used in textiles, food, drink, and medicines. Cochineal insects are sessile, soft-bodied, and reside beneath a fluffy, waxy, white covering. They feed on the cactus juices, using piercing–sucking mouthparts to penetrate the surface of the host plant. The wingless females are much larger than the winged males. The males are winged and short-lived. Eggs are laid under the body of the female and they hatch within a few hours of being laid. First-instar female larvae, called crawlers, are very active, have long waxy filaments on the dorsum, and are wind dispersed (Moran et al. 1982). Crawlers settle on to a feeding site within a day or two and immediately begin secreting a protective

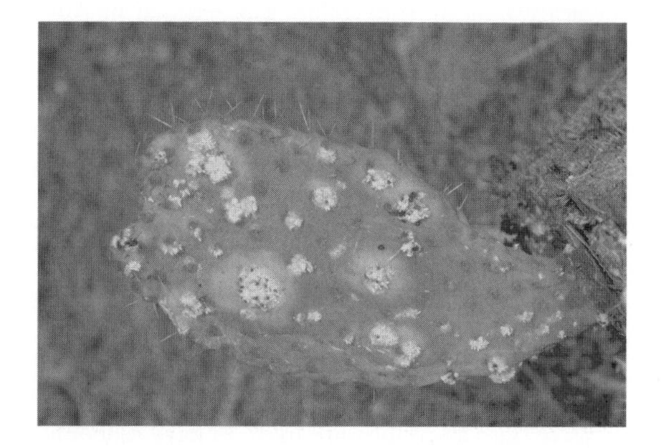

Figure 5.4 Colonies of *Dactylopius* sp. (Dactylopiidae) on a cactus pad. Photograph courtesy of Bob Richard, USDA-APHIS-PPQ.

covering. They then insert their mouthparts into the plant tissue and begin feeding. Their legs and antennae shrink and they remain at that location thereafter. The males leave the parent female, and fly off to locate a female with which to mate. Most of the above information is provided by Mann (1969). Guerra and Kosztarab (1992) review the biosytematics of the family.

Dactylopiids feed on cacti and while the number of species in the family is not large, perhaps only nine or 10 species, the family has played an important role in the successful biological control of several species of cacti. Four of six (or eight, depending upon the taxonomic interpretation) species of dactylopiids employed for biological control established and have provided control of their target cacti. *Dactylopius austrinus* De Lotto, for example, suppresses the cactus *Opuntia stricta* (Haworth) Haworth in Australia (Julien & Griffiths 1998) and South Africa (Hoffmann et al. 1998a, 1998b).

Tephritidae (Diptera)

These flies (Figure 5.5) are small- to medium-sized insects, most of which have banded wings. The maggot-like larvae tunnel in seed heads, form galls, or feed in fruits. A few are leafminers and at least one species lives in termite galleries. Females bear a heavily sclerotized ovipositor, which they use to insert eggs into living plant tissue.

Twenty-three species have been used in biological control projects, mainly species that feed in seed heads of thistles, knapweeds, and other plants or species that form galls. Seventeen species (74%) have established, but only seven species (41%) have contributed to the control of their target weeds, all of which are in the Asteraceae.

Figure 5.5 Adult of the seed head fly, *Urophora quadrifasciata* (Meigen). Photograph courtesy of Bob Richard, USDA-APHIS-PPQ.

Pteromalidae (Hymenoptera)

Gall wasps affect plants by inducing galls that divert nutrients from growth and reproduction. One species, *Trichilogaster acaciaelongifoliae* (Froggatt) (Plates 5.1a–5.1d) has controlled an invasive tree, *Acacia longifolia* (Andrews) Willdenow, in South Africa (Dennill & Donnelly 1991).

Acari

Only mites from the Eriophyidae, Tetranychidae, and Oribatidae (or Galumnidae) have been used in weed biological control. Briese and Cullen (2001) review the use of mites as plant biological control agents. The eriophyids are extremely small mites (about 0.15 mm in length) that feed on plant tissues (material below is from Kiefer et al. 1982). They are commonly known as gall, rust, bud, or blister mites and are extremely specialized plant feeders. As these names denote, some species cause galls, whereas others feed externally and discolor fruits or other plant parts. They are soft-bodied, spindle-shaped or vermiculiform, with two body regions and two pairs of legs. The life cycle of most species is simple, but some species that infest deciduous trees have a more complex alternating life cycle involving a morphologically distinct overwintering form of the female. Gall-formers cause hypertrophy of plant cells resulting in abnormal growth of leaf or bud tissue and other abnormalities. Symptoms of eriophyid injury vary by plant part and include the following: on buds, shoots, stems, and twigs there may be bud blisters, bud and twig rosettes and stunting, discolored buds and bud scales, enlarged buds, premature bud drop, galls, brooming, shoot, stem, and twig discoloration; on flowers there may be abnormal shape, blisters, discoloration, failing to open, galls, premature drop; and on fruits there may be abnormal shape, blisters, damaged seeds, discoloration, galls, hardening, premature drop; and leaves may show abnormal shape or distortion, blisters, discoloration, hair-like epidermal growth (erineum), galls, mosaic virus disease, stunting, webbing or coating, and russeting, bronzing, and withering.

Spider mites (Tetranychidae) are plump-bodied mites that form colonies in "webs" on the foliage of their host plant. The life cycle of tetranychids consists of egg, larva, protonymph, deutonymph, and adult stages. A quiescent stage occurs between each immature stage; the nymphochrysalis, deutochrysalis, and teleiochrysalis, respectively (van de Vrie et al. 1972). Mating takes place immediately upon female emergence. Fertilized eggs produce females whereas unfertilized eggs produce males. Rates of development of the immatures may be influenced by the quality of the foods as well as by environmental conditions (van de Vrie et al. 1972). Spider mites feed on plant juices by piercing the leaf epidermis with two sharp, slender, whip-like cheliceral stylets. The damage is first noticeable as small, diffuse, tan-colored patches on the leaf surface comprised of small, scratch-like stippling but extensive chlorosis and browning of the tissue may eventually result. Spider mites are commonly suppressed by generalist phytoseiid mites.

Species of galumnoid mites occur in a wide variety of habitats including moss, forest litter, and rotting wood (Krantz 1978), but they rarely feed on leaves of living plants (Walter & Proctor 1999). At least two species do feed on live plant tissue. Cordo and DeLoach (1976) describe the biology and life history of the waterhyacinth mite (*Orthogalumna terebrantis* Wallwork). Species in the genus *Hydrozetes* burrow and feed in the thalli of duckweeds (*Lemna* spp.) (Walter & Proctor 1999).

Only five species of mites have been released for weed biological control (Briese & Cullen 2001). All have established in at least one region and all five have contributed to the control of the target weeds (Julien & Griffiths 1998, Olckers & Hill 1999, Briese & Cullen 2001, Coombs et al. 2004). Three of the five are in the family Eriophyidae. The potential biological control uses of this family are reviewed by Gerson and Smiley (1990), who note that eriophyids, while slow acting, are often highly specific as to the hosts on which they feed. All three eriophyid species that have been released have suppressed their target weeds: *Aceria malherbae* Nuzzaci against bindweeds (*Convolvulus* spp.), *A. hyperici* against St. John's wort (*H. perforatum*), and *Eriophyes chondrillae* (Canestrini) against rush skeltonweed (*Chondrilla juncea* L.). An eriophyid mite (*Floracarus perrepae* Knihinicki and Boczek; Plates 5.1e and 5.1f) has been proposed for release against Old World climbing fern [*Lygodium microphyllum* (Cav.) R. Br.] in Florida (Goolsby et al. 2004a). *Cecidophyes rouhollahi* Kraemer was approved for release in Canada against cleavers (*Galium aparine* L. and *Galium spurium* L.) (Sobhian et al. 2004) but did not establish in Alberta, probably due to insufficient cold-hardiness (A.S. McClay, personal communication). One spider mite (*Tetranychus lintearius* Dufour) has been released, to control gorse (*Ulex europaeus* L.), but its effectiveness was reduced by generalist phytoseiids. The oribatid *O. terebrantis*, which was probably accidentally released in the USA, has been deliberately released in several other countries against waterhyacinth with little effect, although it reportedly causes severe damage to the weed in South Africa (Hill & Cilliers 1999).

Fungal pathogens as plant biological control agents

Interest in the use of phytopathogens as weed biological control agents developed after 1970, based on several precedent-setting projects in which introduced pathogens controlled invasive weeds. In 1971 and thereafter, strains of rust fungus *Puccinia chondrillina* Bubak and Sydow were moved from Europe to Australia, where they controlled two of three genetic forms of skeleton weed, *C. juncea*, a pest in wheat fields (Hasan & Wapshere 1973, Hasan 1981). The plant pathogens of interest for classical biological control of invasive plants are the rusts and smuts.

Rusts (Order Uredinales)

Many species of rust are highly host-specific pathogens of vascular plants. Rusts, named for their red-colored, air-borne urediniospores, are obligate parasites. Spores for use in releases must therefore be produced on live plants. Because a rust is likely to affect only a few species, sometimes only a single species, many rusts are excellent candidates for classical biological control. Eleven of the 18 cases of successful introductions of fungi against adventive weeds listed by Julien and Griffiths (1998) are rusts. The most important of these have been the control of skeletonweed (*C. juncea*) by *P. chondrillina*, blackberries and relatives (*Rubus* spp.) with *Phragmidium violaceum* (Schultz) Winter, and *Acacia saligna* (Labillardiére) Wendland by *Uromycladium tepperianum* (Saccado) McAlpine.

Additional projects have focused on the use of rusts. *Puccinia myrsiphylli* (Thuem.) Wint. has been released into Australia against bridal creeper [*Asparagus asparagoides* (L.) Druce], a pest of natural areas (Kleinjan et al. 2004). It has strongly affected the target plant at release sites (Morin et al. 2002). Also, new strains of blackberry rust (*P. violaceum*) are being released in Australia to suppress non-cultivated invasive blackberries (*Rubus* spp.) (L. Morin, personal communication; Bruzzese 1995, Evans et al. 2004).

Smuts (Order Ustilaginales)

Many smuts are obligate pathogens of vascular plants. Many smut fungi infect host plants systemically; such infections weaken plants and may disrupt seed production. Spores are dark in color and easily dispersed by air. Smuts show high levels of host specificity and are good candidates for weed biological control. The white smut pathogen *Entyloma ageratinae* Barreto and Evans was introduced into Hawaii, where it successfully controlled its target, hamakua pamakani [mistflower, *Aegeratina riparia* (Regel) King and Robinson] (Trujillo 1985).

Fish as plant biological control agents

At least 30 species of fish have been investigated for biological control of aquatic plants (van Zon 1977). Generalist feeders in the Cyprinidae (carp), Cichlidae, and Osphronemidae have been used for non-specific weed control in irrigation ditches or ponds, where elimination or near elimination of all macrophytes is desired. The potential for such fish to cause damage to native plants and fish is high. Each introduction must be considered very carefully, taking into account the potential for subsequent spread to other water bodies by flood or casual relocation by people. Although many species have been considered, in reality only the grass carp has been used widely and on a large scale (van der Zweerde 1990). In some instances, sterile hybrids or sterile triploids are used to minimize the risk of establishing breeding populations of introduced fish.

Chapter 6

ARTHROPOD PATHOGEN DIVERSITY AND ECOLOGY

Arthropod pathogens include bacteria, viruses, fungi, nematodes, and protozoa (Brady 1981, Miller et al. 1983, Maramorosch & Sherman 1985, Moore et al. 1987, Burge 1988, Tanada & Kaya 1993). Protozoa, however, have little importance in biological control, being mostly debilitating rather than lethal to their hosts. Microsporidia, formerly considered protozoa and now placed in the fungi, have potential importance as classical biological control agents but mostly are of concern as damaging possible infections in colonies of arthropod biological control agents.

Pathogens are an important part of natural control. Spontaneous epizootics of pathogens sometimes occur in pest populations (Fuxa & Tanada 1987), as for example the viral and fungal epidemics that periodically decimate gypsy moth larvae [*Lymatria dispar* (L.)] in North America (Gillock & Hain 2001/2002). Use of pathogens in classical or inoculative augmentative biological control has included programs against the rhinoceros beetle [*Oryctes rhinoceros* (L.)] on coconut on Pacific islands by an introduced *Oryctes* virus, and control of a spruce sawfly [*Gilpinia hercyniae* (Hartig)] in Canada by a baculovirus and of a *Sirex* woodwasp by a nematode *Deladenus* (*Beddingia*) *siricidicola* (Bedding) in Australia. While still uncommon, instances where arthropod pathogens are used in classical biological control may increase in the future.

Most research on the use of pathogens for biological control, however, has focused on efforts to formulate micro-organisms for site-specific application as biopesticides (Cherwonogrodzky 1980, Federici 1999, 2007). Here we consider the biology of key pathogen groups and factors affecting their transmission dynamics as factors in natural control. In Chapters 23 and 24 the potential of arthropod pathogens as biopesticides is discussed.

BACTERIAL PATHOGENS OF ARTHROPODS

Of the various pathogen groups, bacteria have been most successfully brought into commercial use. Bacteria are amenable to such use because several important species can be grown in fermentation media and do not require expensive rearing methods. Most emphasis has been placed on *Bacillus thuringiensis* Berliner, which has at least 65 known subspecies and many thousands of isolates. *Bacillus thuringiensis* is a complex of subspecies commonly found in habitats such as soil, leaf litter, insect feces, and insect guts (Federici 2007). Some *B. thuringiensis* products contain both live bacteria and associated toxic proteins; others contain only the bacterium's toxins. Interest in *B. thuringiensis*, combined with developments in molecular biology, led to the production of transgenic crops (especially cotton and corn) that express enough Bt toxins to protect plants from key pests. Although bacteria as biopesticides have remained a niche product, transgenic *Bt* plants have transformed pest control in some crops (see Chapter 21). For a brief historical account of the development of Bt crops, see Federici (2005).

While many species of bacteria can cause disease in arthropods, those that do not form resting spores (such as species of *Pseudomonas*, *Aerobacter*, *Cloaca*, or *Serratia*) usually cause disease only when the host is physiologically stressed. However, one species – *Serratia entomophila* Grimmont, Jackson, Ageron, and Noonan, the causative agent of amber disease – has been developed as a biopesticide and is marketed in New Zealand for control of the pasture pest scarab *Costelytra zealandica* (White) under the name Invade (Jackson 1990). Granular formulations of this bacterium remain active in inoculated soils for up to 5 months (O'Callaghan & Gerard 2005).

Spore-forming bacteria can more easily infect healthy hosts, following ingestion of spores. Species such as *B. thuringiensis*, *Bacillus sphaericus* Neide, and *Paenibacillus popilliae* (Dutky) (formerly in *Bacillus*) (Pettersson et al. 1999) are the pathogens, in order of decreasing importance, that have most often been investigated for possible use as bioinsecticides.

Paenibacillus popilliae is a pathogen of the Japanese beetle, *Popillia japonica* Newman, and other turf scarabs. Infections by this pathogen are referred to as milky spore disease because of the milky color of the hemolymph of diseased hosts. Despite the importance of the pests targeted with this pathogen, *P. popilliae* has largely failed as a commercial biopesticide (although it is still commercially available) because it does not readily produce spores when grown in fermentation media (Lüthy 1986). Since spores are the stage used in biopesticide products, this has prevented inexpensive commercial production. Inefficiency of mass production, combined with a low level of efficacy after application, has suppressed interest in this pathogen, reflected by the fact that only 14 research articles were located in the CAB International database under this name for 1999–2004.

The second species, *B. sphaericus*, is of interest because it kills mosquito larvae, as does one subspecies of *B. thuringiensis* (*B.t. israelensis*; see below; Singer 1990, Baumann et al. 1991, Charles et al. 1996). This species can be produced by fermentation and its insecticidal activity is due to crystalline toxins that are released when the insect digests spores it has taken in with its food. The host range of this bacterium is limited to a few genera of mosquitoes (Wraight et al. 1981, Singer 1987, Osborne et al. 1990). Genes coding for the toxin have been identified and transferred to other bacteria (Baumann et al. 1987, 1988, Baumann & Baumann 1989). Genes from *B. sphaericus* have been used to produce recombinant organisms expressing toxins of both this species and *B. thuringiensis* (Park

Figure 6.1 Indianmeal moth caterpillars (*Plodia interpunctella* Hübner) killed (dark) by the bacterium *Bacillus thuringiensis* Berliner, contrasting with a healthy one (white). Photograph courtesy of Jack Kelly Clark, University of California IPM Photo Library.

et al. 2003, 2005). *Bacillus sphaericus*, either alone or in combination with other materials, remains of interest for control of *Culex* spp. and mosquitoes that breed in polluted water. A commercial product (VectoLex) is being tested (Shililu et al. 2003, Brown et al. 2004). Work is focused on: (1) searching for more-lethal strains, (2) development of cheaper rearing methods, to lower production costs (Poopathi et al. 2003), (3) management of resistance development by mosquitoes (Park et al. 2005), and (4) field testing of formulated products. Overall, this pathogen appears to have a potential niche as a mosquitocide and research on its development continues, as illustrated by the 225 articles recovered from CAB International for 1999–2004.

Bacillus thuringiensis is the most extensively marketed bacterial pathogen of arthropods (Figures 6.1 and 6.2) (Beegle & Yamamoto 1992, Entwistle et al. 1993, Whalon & Wingerd 2003). Perhaps as many as 50,000 isolates have been collected, from which 65 serotypes, based on flagellar antigens, have been recognized and given subspecies names. Most of these serotypes affect caterpillars, and some, for example *B.t. kurstaki*, have been used against various lepidopteran pests of fruits, vegetables, and forests. The subspecies *israelensis* is effective against fly larvae, including mosquitoes, blackflies, sewage flies, and fungus gnats (de Barjac 1978, van Essen & Hembree 1980, Mulla et al. 1982). The subspecies *B.t. tenebrionis* infects chrysomelid beetles, such as the Colorado potato beetle, *Leptinotarsa decemlineata* (Say) (Herrnstadt et al. 1987).

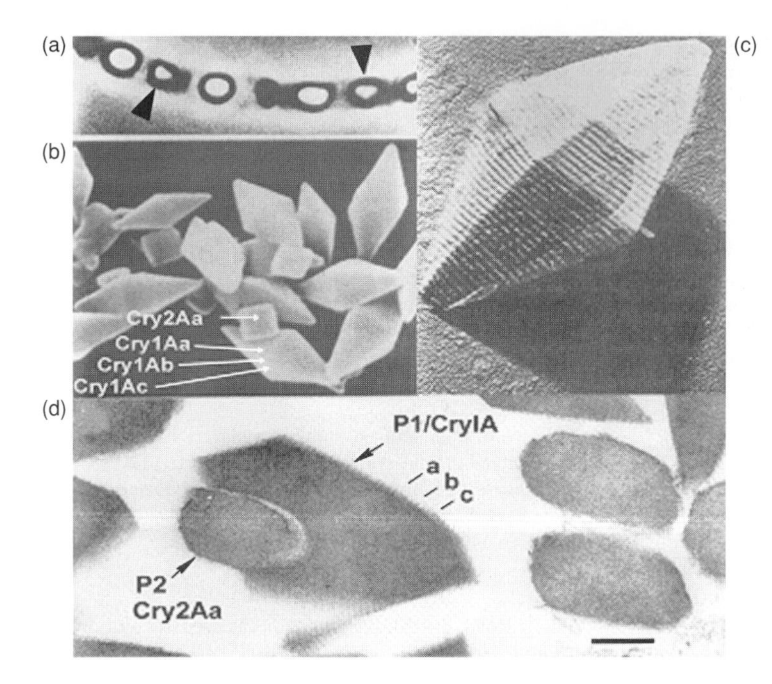

Figure 6.2 Micrographs of *Bacillus thuringiensis* cells and toxins: (a) cells with spores; (b) purified Cry1 and Cry2 crystals; (c) enlarged view of Cry1-type protein crystal; and (d) view of embedded cuboidal Cry2A crystal (P2) in the bipyramidal crystal (P1); scale bar, 200 nm. Reprinted with permission of Springer, with modification of caption, from Federici (2007); (c) micrograph by C.L. Hannay.

Insects that ingest *B. thuringiensis* spores are killed by the combined effects of poisoning by toxins and bacterial multiplication. Biopesticides containing this pathogen are important in organic farming and integrated pest management because of their compatibility with parasitoids and predators. Total use, however, remains minor compared to conventional insecticides in crops (Whalon & Wingerd 2003).

Bt genes, however, have been used to produce pest-resistant crop varieties (Vaeck et al. 1987), and grower adoption of these has been extensive. In 2006 approximately 50% of all cotton and corn grown in the USA were Bt varieties. Similar high rates of use also occur in some other countries (James 2002, Shelton et al. 2002). When plant-expressed Bt toxins control key pests, pesticide use decreases, allowing much higher survival of natural enemies in the crop (Dively & Rose 2003, Naranjo & Ellsworth 2003; see Chapter 21). Resistance to Bt toxins, delivered either as biopesticide sprays or transgenic plants, is possible (Tabashnik et al. 1990), and monitoring and management to delay resistance are important aspects of the use of the toxins of this pathogen. Broader social issues have also been raised concerning use of Bt crop plants (Gray 2004).

VIRAL PATHOGENS OF ARTHROPODS

Of the various families of insect viruses (Entwistle 1983, Moore et al. 1987, Tanada & Kaya 1993), only the Baculoviridae (Granados & Federici 1986; with one exception) are important as biopesticides or causes of natural epizootics (Figures 6.3 and 6.4). Baculoviruses usually kill their hosts and are only known to infect insects (Payne 1986). This family contains the nucleopolyhedroviruses (NPV) and granuloviruses (GV). The nonoccluded viruses (i.e. *Oryctes* virus), formerly placed in the Baculoviridae, are now unclassified (Jackson et al. 2005). For information on molecular aspects of the baculovirus infection cycle and organization of the baculovirus genome see Blissard and Rohrmann (1990).

One role for baculoviruses in biological control is that of a natural pathogen, causing periodic cycles of disease. Such pathogens might be local native species or introduced viruses targeted at invasive species. Fuxa (1990) lists 15 cases in which baculoviruses were introduced and established successfully against invasive pest insects. The level of control, however, is rarely high unless virus levels are augmented artificially. A few virus introductions have, however, controlled their

Figure 6.3 Larva of silver-spotted tiger moth [*Lophocampa argentata* (Pack.)] killed by a baculovirus, seen hanging in the head-down position that facilitates contamination of foliage by virus from disintegrating cadaver. Photograph courtesy of Jack Kelly Clark, University of California IPM Photo Library.

target pests. A nucleopolyhedrovirus of the invasive pest sawfly *G. hercyniae* permanently suppressed the pest after the virus was accidentally introduced into eastern Canada (Balch & Bird 1944). More deliberately, the intentional introduction of a non-occluded virus of the coconut beetle [*Oryctes rhinoceros* (L.)] suppressed the pest on coconut palms for nearly 4 years on south Pacific islands, but requires ongoing management for continued efficacy (Zelazny et al. 1990, Mohan & Pillai 1993).

Baculoviruses can also be formulated as biopesticides. However, since all viruses are obligate parasites, they must be reared in living insects or insect cell cultures. Consequently, few viruses have been successful as commercial products because production costs are high and product use is limited by high host specificity. In the absence of profitable products from private business, some viral biopesticides have been produced at public expense. In Brazil, government support has allowed the development of the nucleopolyhedrosis virus of the soybean defoliator *Anticarsia gemmatalis* Hübner (Moscardi 1983, 1999), and this biopesticide has been adopted by some soybean farmers (Corrêa-Ferreira et al. 2000).

FUNGAL PATHOGENS OF ARTHROPODS

Fungi can be classical biological control agents, biopesticides, or part of natural control through the

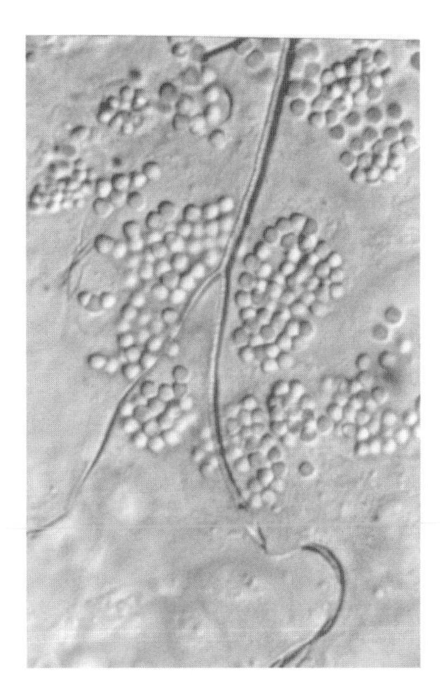

Figure 6.4 Micrograph of nucleopolyhedrovirus in hypodermis tissue of beet armyworm, *Spodoptera exigua* (Hübner). Photograph courtesy of J.V. Maddox; reprinted from Van Driesche and Bellows (1996) with permission from Kluwer.

epidemics they periodically cause in some arthropods (Goh et al. 1989, Carruthers & Hural 1990). They play little role in conservation biological control because manipulations to create fungal epidemics on demand in specific locations are not generally practical or available.

Cases of successful classical biological insect control using introduced fungi have been rare. The fungus *Zoophthora radicans* (Brefeld) Batko from Israel was introduced to Australia to aid in the suppression of the aphid *Therioaphis trifolii* (Monell) f. *maculata* (Milner et al. 1982). The accidental introduction of the Asian fungus *Entomophaga maimaiga* Humber, Shimazu et Soper into the northeastern USA caused high mortality of gypsy moth larvae (Webb et al. 1999) and is believed to have kept this pest under control, at least in New England (USA), since 1990.

Most research with fungi as biological control agents has focused on efforts to develop them as biopesticides (Ferron 1978, Gillespie 1988, Bateman & Chapple 2001, Bateman 2004). Successful development of mycoinsecticides has been frustrated by narrow host ranges and poor germination of conidia after application (Moore & Prior 1993). A product for control of African desert locust (*Schistocerca gregaria* Forskal) and other pest grasshoppers has been developed based on *Metarhizium flavoviride* Gams (=*M. anisopliae* var. *acridum*). This product has been promoted by international aid groups as a more environmentally safe solution to locusts in Africa and has also been tested for use in Asia and South America, with high levels of grasshopper mortality in field trials (Li et al. 2000, Magalhães et al. 2000, Zhang et al. 2000). An Australian *Metarhizium* species (Green Guard®) is currently registered for use in Australia against locusts (Lawrence 2006).

In some cases, mycopesticides, once applied, act like classical biological control agents, reproducing at high enough levels to continue to cause mortality at significant levels for several years without repeated application. This is the case, for example, with *Beauveria brongniartii* (Saccardo) Petch, which is applied in Swiss grasslands and orchards to control the cockchafer *Melolontha melolontha* L. This fungus has been detected in soil 14 years after application (Enkerli et al. 2004), which is believed to contribute to this product's success in controlling the pest (Zelger 1996).

Over 400 species of fungi that infect insects have been recognized (Hall & Papierok 1982). Their taxonomy is covered in Brady (1981) and McCoy et al. (1988) and their biology, pathology, and use in pest control in Steinhaus (1963), Müller-Kögler (1965), Ferron (1978), Burges (1981a), McCoy et al. (1988), Tanada and Kaya (1993), and Khetan (2001). Most attention has focused on about 20 species (Zimmermann 1986) in 12 genera (Roberts & Wraight 1986). These include *Lagenidium* (now considered not a true fungus, but a member of the Kingdom Straminipila) and *Entomophaga, Neozygites, Entomophthora, Erynia, Aschersonia, Verticillium, Nomuraea, Hirsutella, Metarhizium, Beauveria,* and *Paecilomyces* (all true fungi, in the Kingdom Eumycota).

Lagenidium (Kingdom Straminpila)

Members of this genus infect mosquito larvae and do not require an alternate host to complete a life cycle. *Lagenidium giganteum* Couch is registered as a pest-control product in the USA.

Entomophaga, Entomophthora, Neozygites, and *Erynia*

Fungi in these groups (all Entomophthoraceae) are important as naturally occurring pathogens but do not sporulate well on fermentation media and are not used as biopesticides. Hosts of these groups include caterpillars, beetles, aphids, and mites. For taxonomy and biology information see MacLeod (1963), Waterhouse (1973), Remaudière and Keller (1980), Humber (1981), Ben-Ze'ev et al. (1981), and Wolf (1988).

Imperfect Fungi (=Deuteromycota)

Species of *Aschersonia, Verticillium, Nomuraea, Hirsutella, Metarhizium, Beauveria,* and *Paecilomyces* belong to the Imperfect Fungi. This is an artificial group of species whose sexual forms (the basis for fungal classification) either have not yet been found or for other reasons cannot be confidently placed in the other fungal groups. *Hirsutella thompsonii* Fisher is a well-studied pathogen of eriophyid rust mites (McCoy 1981). *Beauveria bassiana* (Balsamo) Vuillemin has a wide host range (Figure 6.5) and is currently registered as a pesticide in the USA (de Hoog 1972). *Beauveria brongniartii* is registered for use in Switzerland against scarabs. *Metarhizium* species have been developed for control of locusts. *Paecilomyces, Verticillium,* and *Aschersonia* species have been studied as pathogens of whiteflies, aphids, and scales.

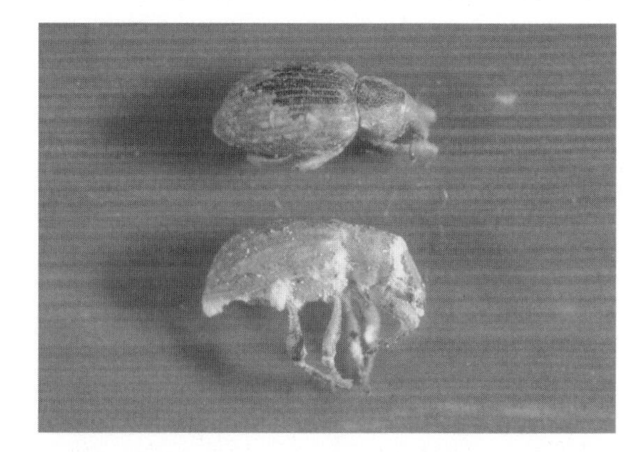

Figure 6.5 Adult rice weevil [*Sitophilus oryzae* (L.)], seen in side view, with hyphae of *Beauveria bassiana* (Balsamo) Vuillemin emerging from the cadaver. Photograph courtesy of Jack Kelly Clark, University of California IPM Photo Library.

NEMATODES ATTACKING ARTHROPODS

Of the 30 or more families of nematodes, nine have potential for insect biological control. Two cases exist in which introduced nematodes have suppressed an invasive insect. In Australia, the phaenopsitylenchid *D. siricidicola* introduced from New Zealand provided effective control of the European wood wasp, *Sirex noctilio* (Fabricius), a major pest in pine plantations (Bedding 1984). In Florida, USA, the steinernematid *Steinernema scapterisci* Nguyen and Smart was imported for control of an invasive *Scapteriscus* mole cricket in turf (Parkman et al. 1993, 1996).

Apart from the above cases, nearly all interest in nematodes for biological control of insects has been in the commercial production of steinernematid and heterorhabditid nematodes for use as biopesticides (Gaugler & Kaya 1990, Kaya 1993, Grewal et al. 2005, Adams et al. 2006). These nematodes harbor symbiotic bacteria able to kill the host rapidly (Kaya 1985, Burnell & Stock 2000).

Figure 6.6 *Steinernema carpocapsae* (Weiser) nematodes emerging from a host cadaver in water. Photograph courtesy of Jack Kelly Clark, University of California IPM Photo Library.

Steinernematidae and Heterorhabditidae

Many species of *Steinernema* and *Heterorhabditis* (Figures 6.6 and 6.7) have been commercialized as biopesticides (Gaugler & Kaya 1990, Kaya 1993, Kaya & Gaugler 1993, Tanada & Kaya 1993, Bullock et al. 1999, Koppenhöfer & Fuzy 2003). These nematode families have been used as commercial pest-control agents because they have the following attributes (Poinar 1986):

- a wide host range,
- an ability to kill the host within 48 h,
- a capacity for growth on artificial media,
- a durable infective stage capable of being stored,
- a lack of host resistance,
- apparent safety to the environment.

These nematodes invade hosts through natural openings (mouth, spiracles, anus) or wounds and penetrate into the hemocoel. Bacteria in the genera *Xenorhabdus* or *Photorhabdus* are released and kill the host quickly. Nematodes then develop saprophytically in the cadaver. See Lewis et al. (2006) for a review of the ecology and behavior of these nematodes in relation to their use for pest control, and Grewal et al. (2006) for information on nematode chemoreception and nematode biology in relation to heat and dryness. Gaugler and Kaya (1990) and Kaya and Gaugler (1993) provide information on rearing these nematodes and using them for pest control. These nematodes are only effective in moist environments such as soil or wet foliage in tropical climates. Heterorhabditid species turn host cadavers bright red (Figure 6.8). Commercial markets for some species have been established and large-scale production systems developed (Kaya 1985, Gaugler & Kaya 1990).

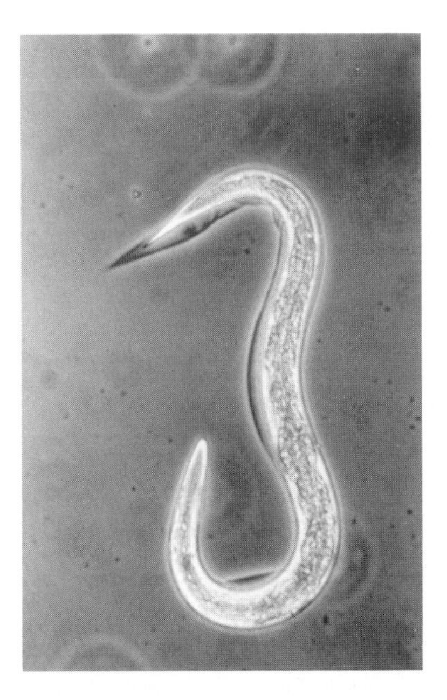

Figure 6.7 A close-up of a single *Steinernema* sp. nematode. Photograph courtesy of R. Gaugler; reprinted from Van Driesche and Bellows (1996) with permission from Kluwer.

Figure 6.8 Scarabaeid grubs infected with *Heterorhabditis* sp. nematodes turn a characteristic red color (right), in contrast to the cream color of uninfested grubs (left). Photograph courtesy of Jack Kelly Clark, University of California IPM Photo Library.

Phaenopsitylenchidae

The nematode *D. siricidicola* was introduced from New Zealand into Australia where it contributed substantially to the suppression of a major pest of conifer plantations, the European wood wasp, *S. noctilio* (Bedding 1984). The nematode infects larvae, but does not kill the host. It later invades the ovaries of the adult wood wasp, destroying the eggs. The wasp, however, continues to oviposit, with the result that nematodes rather than eggs are deposited in new trees, spreading the nematode.

GENERALIZED ARTHROPOD PATHOGEN LIFE CYCLE

To understand the value of any pathogen as part of the natural control affecting a pest, one has to understand the pathogen's biology. To complete their life cycles successfully, most pathogens must contact a host, gain entrance to the host's body, reproduce within one or more host tissues, and emit some life stage that subsequently contacts and infects new hosts. How any particular pathogen does these things will strongly influence which kinds of hosts it infects and how much impact it will have on the host's average density. Here we discuss these processes and compare them among different pathogen groups. When used as biopesticides, some aspects of a pathogen's biology, such as efficiency of transmission, become less important.

Host contact

Most arthropod pathogens lack a mobile stage (except nematodes and water molds, such as *Lagenidium* spp.). Therefore, host contact depends on chance encounters with hosts by spores or some other infective stage that is moved by wind, rain, or other organisms. The efficiency of contact between a pathogen and its hosts is determined by the spatial patterns of the infective stage and that of the host, and the survival of the infective stage over time. The occlusion bodies of nucleopolyhedroviruses from the cadavers of diseased gypsy moth larvae (*L. dispar*), for example, are released when host cadavers rupture. Virus occlusion bodies are initially concentrated near the site of host death, but later become distributed over nearby foliage (especially foliage directly beneath host cadavers) by rain (Woods & Elkinton 1987). Similarly, wind redistributes fungal conidia, which are initially concentrated near host cadavers, to new locations throughout the habitat.

The dispersal of pathogens among a group of hosts is called **horizontal transmission** (Figure 6.9). A few pathogens are transmitted between generations of hosts from mother to offspring (**vertical transmission**; Figure 6.10), a process that eliminates the need to contact new hosts randomly. Some pathogens are even able to actively seek out hosts. Some entomopathogenic nematodes use chemical cues such as CO_2 and host feces to detect hosts (Ishibashi & Kondo 1990) and then move toward them by swimming in the water between soil particles. Similarly, the motile zoospores of aquatic species of *Lagenidium* actively swim toward either hosts (in response to chemicals emitted by hosts) or light (which brings them to the water surface where mosquito larvae occur; Carruthers & Soper 1987).

Host penetration

Once a pathogen has contacted its host, it must penetrate the host's body and reach the susceptible tissues. The arthropod cuticle provides protection from many pathogens. Most bacteria and viruses cannot cross the external cuticle and must enter arthropods through the thinner wall of the midgut after being ingested. Consumption of food that is contaminated with pathogens is a major route of contagion for chewing arthropods. Sucking arthropods, in contrast, escape exposure to such contamination by feeding on internal

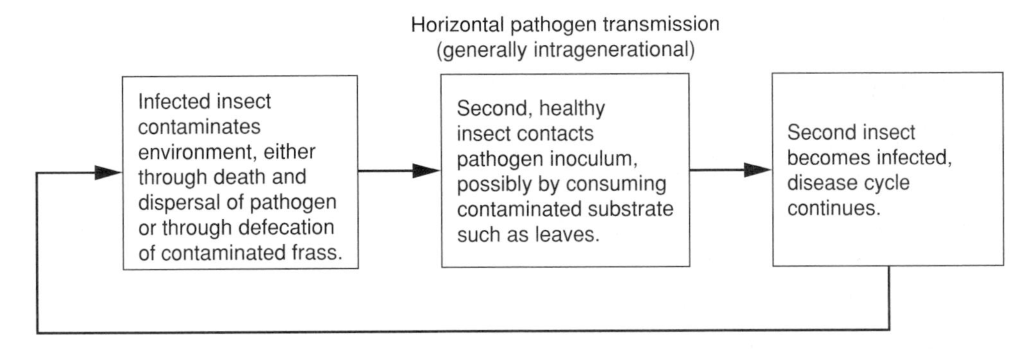

Figure 6.9 Horizontal pathogen transmission is between members of the same generation, usually from physical contact with cadavers or feces of infected individuals. Courtesy of J.V. Maddox; reprinted from Van Driesche and Bellows (1996) with permission from Kluwer.

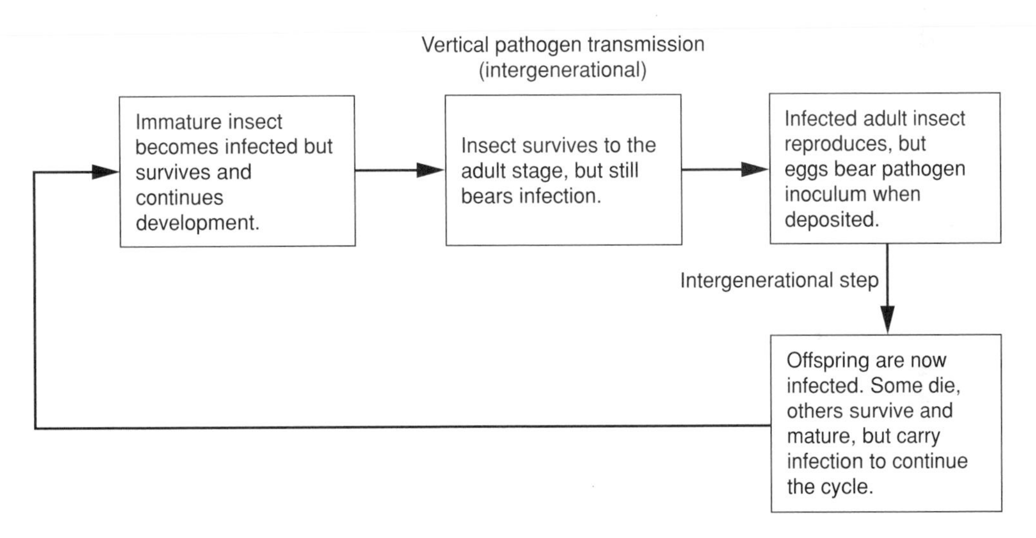

Figure 6.10 Vertical transmission of pathogens is between members of two succeeding generations, usually from mother to offspring via the egg. Courtesy of J.V. Maddox; reprinted from Van Driesche and Bellows (1996) with permission from Kluwer.

plant fluids, which are relatively free of entomopathogenic microbes. As a consequence, sucking insects such as aphids are less affected by pathogens such as bacteria and viruses that must enter hosts by ingestion.

In contrast, some nematodes and fungi are better at penetrating the insect integument. Steinernematid and heterorhabditid nematodes can enter hosts through wounds or spiracles using mechanical pressure and enzymes. Heterorhabditids can also cut the integument with a tooth-like structure. *Deladenus* nematodes use a stylet to enter the host. Fungi use special structures called penetration hyphae to exert mechanical pressure on the cuticle, together with the production of enzymes capable of digesting the chitin in the cuticle (Figure 6.11).

Reproduction in the host

Once a pathogen has penetrated its host, the pathogen must reproduce in a susceptible tissue. Some pathogens can reproduce in virtually all tissues, but others require specific tissues. The non-occluded *Oryctes* virus, for example, reproduces principally in fat body and midgut epithelium. The range of tissues a pathogen can infect influences the number of infective stages of the

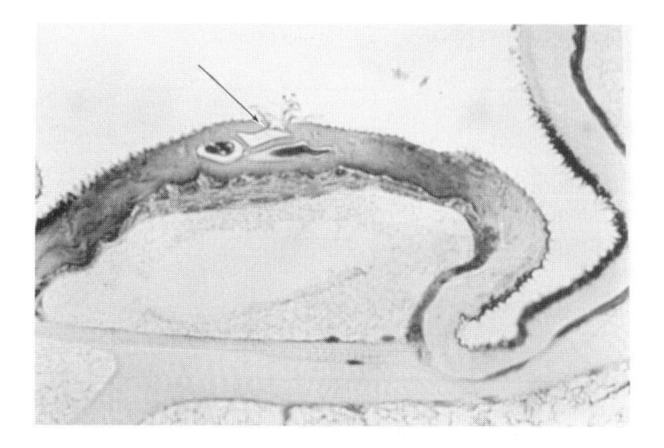

Figure 6.11 Micrograph of the penetration of cuticle of the pine sawfly, *Diprion similus* (Hartig), by a fungal hypha (arrow). Photograph courtesy of M.G. Klein from Klein and Coppel (1973) *Annals of the Entomological Society of America* **66**:1178–80; reprinted from Van Driesche and Bellows (1996) with permission from Kluwer.

pathogen that can be produced per host. Pathogens infecting all tissues may be more economical to rear than pathogens causing more selective infections.

Some pathogens that are obligatory parasites of living cells, like baculoviruses, only reproduce while the host is still alive. In contrast, steinernematid and heterorhabditid nematodes are largely saprophytes and most of their reproduction occurs after hosts are killed by associated symbiotic bacteria. Consequently, nematodes are able to use most host tissue for reproduction.

Escaping the dead host

To complete its life cycle, a pathogen must escape from the old host and find new ones. In the special case of vertical transmission (from parent to offspring), the pathogen contacts new hosts when the mother contaminates her own eggs. Usually, however, the pathogen must leave the old dead host, enter the larger environment, and in some way contact a new host. If a pathogen kills its host, the pathogen's offspring can escape the cadaver when it decomposes, as occurs when virus-killed caterpillars "melt" and fall apart. The offspring of entomopathogenic fungi (conidia) escape cadavers of dead hosts when special hyphae (conidiophores) grow through the cadaver's cuticle and produce aerially dispersed conidia. Release of conidia by some species of fungi is passive, but others discharge conidia eruptively. Nematodes can leave hosts in several ways, depending on the nematode group. For steinernematids and heterorhabditids, juveniles or adults can swim away from host cadavers in the water between soil particles. In other groups, nematodes may be dispersed through the host's reproductive tract during oviposition attempts of infected hosts.

Pathogen reservoirs and resting stages

Following the release of infectious pathogens back into the environment, the continuity of the pathogen's population depends on contacting new hosts. Because the host's presence in time and space may be patchy and unpredictable, pathogens require adaptations both for dispersal and persistence. Dispersal in the environment is largely accomplished by wind and rain, with host contact being largely a matter of chance. Random encounters with new hosts are more likely if hosts are aggregated. Insects such as whiteflies, aphids, and caterpillars or other insects undergoing high-density population outbreaks are especially favorable for disease transmission. Insects being reared in colonies in laboratory or commercial settings, unless reared individually, are also especially susceptible to the propagation of disease because of close proximity and high density.

When hosts are scarce in time or space, pathogen survival requires that the pathogen have some durable stage that can persist for fairly long periods. This increases the odds that some pathogens will eventually contact hosts. Spores of *Bacillus* species and occlusion bodies of baculoviruses are examples of durable pathogen stages. These stages end up in soil, where they persist. Rain can splash soil on to foliage, which moves some spores or virus back on to foliage, resulting in the chance that new hosts will ingest pathogens.

EPIDEMIOLOGY: WHAT LEADS TO DISEASE OUTBREAKS?

Epizootics are outbreaks of disease in an animal population and are part of natural control. Epizootics of baculoviruses and entomopathogenic fungi are common, whereas epizootics of bacteria such as *B. thuringiensis* are rare. The likelihood that an epizootic will occur is influenced by characteristics of both the host and the pathogen, the host population density and distribution, and environmental conditions such as temperature, rainfall, and humidity. The study of how such factors affect disease outbreaks is termed **epizootiology** [see Fuxa & Tanada (1987) for a discussion of epizootiology of insect diseases]. Below we consider what features of the host, the pathogen, or the environment lead to epizootics. However, when pathogens are used as biopesticides, applied where needed in large quantities, natural dynamics are largely superseded by the artificially imposed conditions and thus even some pathogens with poor transmission dynamics (such as *B. thuringiensis*) may be useful as biopesticides.

Host features that influence disease rates

Among the host factors that can affect the development of an epizootic are host density, spatial distribution, health of hosts, age, molting status, and behavior. Because the dispersal stages of pathogens, such as fungal conidia or viral occlusion bodies, decrease in abundance as the cube of the distance from the nearest previously infected host, contact rates with new hosts are highest when hosts are close together. Disease transmission is increased when insects occur in reproductive colonies (like aphids) or groups (like tent caterpillars), or have significantly aggregated spatial distributions (like whiteflies). For chewing insects such as caterpillars, horizontal transmission is facilitated by contact with feces or fragments of host cadavers, which is most likely at high caterpillar densities, such as in gypsy moth outbreaks.

The health of hosts also affects pathogen transmission because hosts stressed by other pathogens, poor nutrition, or adverse physical conditions are often less resistant to infection. Diseased individuals may also increase the dispersal of the pathogen by unusual behaviors. Infected individuals frequently die in relatively high positions on their food plant or habitat. Some caterpillars infected with virus migrate upward (perhaps as a response to starvation) and die at the tips of branches, a behavior that positions the cadaver to contaminate foliage lower down as the cadaver disintegrates.

Similarly, age and molting status affect susceptibility to infection. Young caterpillars are often more susceptible to *B. thuringiensis* and viruses. Newly molted insects, in which the cuticle is still rather thin, are more susceptible to fungi. Conversely, molting may prevent infection in some insects, such as aphids, if condia are shed on the cast off cuticle before penetration of the host.

Pathogen features that influence disease rates

Pathogen characteristics that influence disease rates include infectivity, virulence, production of toxins, nature of the pathogen life cycle, and the density, distribution, and persistence of the pathogen's dispersal stage. Pathogen genotype influences infectivity and virulence to a given host. Infectivity is the ability of the pathogen to penetrate the host's body and virulence is the ability, once inside the host, to cause disease. Pathotypes vary significantly with regard to which host species can be attacked successfully. In fungi, strains may vary in the level of enzymes produced by penetration hyphae, changing their infectivity to the host. In *B. thuringiensis*, isolates differ in the kinds and quantities of the toxins they produce. These differences in toxins determine which groups of hosts are susceptible to lethal infections by particular *B. thuringiensis* isolates.

Pathogen life cycles vary from simple to highly complex, with some requiring alternate hosts. Complex life cycles may limit pathogen transmission if alternate hosts or special conditions are available in only some habitats or periods. The requirement for copepods or ostracods as alternate hosts by Straminipila fungi in the genus *Coelomomyces*, for example, means that continuous reproduction of this pathogen following artificial application is only possible if these hosts are present (Tanada & Kaya 1993).

The density, distribution, and persistence of a pathogen's infective stages are important in determining both the normal rate of a disease and the frequency and intensity of epizootics. The presence of the infective pathogen stage is insufficient to cause epizootics in the absence of favorable environmental conditions. However, abundant, persistent sources of infective pathogen stages in the habitat favor the occurrence of epizootics.

Spread of a given pathogen in the habitat will depend on the nature of the release mechanism from the host. Wind-blown fungal conidia are likely to be more widely dispersed than viruses liberated by liquefaction of host cadavers with local contamination of foliage in the drip zone below cadavers. Persistence of the infective stage of a pathogen will be strongly influenced by the stage's tolerance of damaging physical factors, particularly ultraviolet light, high temperatures, and dryness. Some microhabitats, especially soil and protected spaces such as bark crevices, provide physical conditions that are more favorable to pathogen survival. Host contact with these zones, or movement of material from them to areas where hosts feed, will be important influences on rates of infection.

Environmental factors that influence disease rates

Temperature, humidity, desiccation, light, and soil characteristics all influence disease outbreaks (Benz 1987). Effects of temperature on disease rates are complex. Temperature changes can directly affect either the pathogen or the host, but the effect on disease rate can only be understood by also considering the impact of different temperatures on the host's behavior, growth, and movement. The route of entry of the pathogen can affect this process. For organisms in which ingestion of contaminated food is the principal route of entrance, infections can only be acquired at temperatures that permit hosts to feed. For fungi, which enter hosts through the integument, infections may be acquired at temperatures below those at which the hosts feed, if temperatures are favorable for fungal spore germination and hyphal growth.

Humidity, free water, and desiccating conditions are important in some situations. High humidity levels generally favor outbreaks of fungi, promoting both the germination of existing condia and formation of new conidia on cadavers. High humidity and soil moisture also favor nematode epizootics. Bacterial and viral disease rates are less influenced by these factors. Rain has relatively little direct effect on disease rates and does not wash significant amounts of infective pathogen stages from plant surfaces (Benz 1987). Desiccation, in contrast, is an important mortality factor for many pathogens, including nematodes and bacteria, and many pathogens have special stages adapted to withstand desiccation. These include the occlusion bodies of baculoviruses, the spores of some bacteria (*Bacillus*), the resting spores of fungi, and the eggs and juvenile resting stages of some nematodes.

The deleterious effect of sunlight, especially ultraviolet light, on baculoviruses is well known. Baculoviruses deposited on upper leaf surfaces exposed to sunlight are typically inactivated in a short period, ranging from a few hours to a few days. Fungal spores are also sensitive to light, but the conidia of many species are protected by light-absorbing pigments. Soil, because it is often moist and dark, is a favorable location for the survival of resting stages of bacteria, baculoviruses, and fungi. Soil pH and organic content can influence the rate of degradation of pathogens, as does the species composition and abundance of soil micro-organisms. Thus soil management used in agriculture can influence the rates of disease in crop fields (see Chapter 22 on conservation biological control).

Part 3

INVASIONS: WHY BIOLOGICAL CONTROL IS NEEDED

Chapter 7

THE INVASION CRISIS

URGENCY OF THE INVASION CRISIS

Governments and societies need to understand the principles of classical biological control and support its application financially if we are to respond intelligently to the invasive species crisis. We consider invasive species to be any non-indigenous species that is established where it did not evolve and that is physically separated from its area of origin by a geographic barrier. For our purposes, a species is invasive whether or not it is damaging (see Pyšek et al. 2004 for discussion of terminology in relation to invasive plants). Most invasive species are harmless but others are highly damaging, either to economic interests or to natural ecosystems. Despite efforts to check the spread of invasive pests, damaging new insects, plants, and pathogens continue to spread.

Invasive species can destroy crops or kill native plants or animals over large areas. Emerald ash borer (*Agrilus planipennis* Fairmaire), from China, infests 8000 km^2 in Michigan and has killed 6–8 million ash trees. It will likely destroy millions, even billions, of ash trees throughout North America unless checked with biological control agents. The invasion of North America by one Eurasian bivalve – zebra mussel (*Dreissena polymorpha* Phallas) – has imposed economic costs on water users (factories, waste-water treatment companies, or water-supply companies) that run into the billions of dollars annually. It is also likely to drive as many as 50 native pearly mussels to extinction. A hybrid marine alga [*Caulerpa taxifolia* (Vahl) C. Agardh] bred for aquaria is blanketing the Mediterranean sea bed with a toxic algal carpet likely to affect fish and other sea life in drastic, yet unclear ways.

Even plants and animals deliberately imported for beneficial uses can become pests. Kudzu [*Pueraria montana* (Lour.) Merr. var. *lobata* (Willd.) Maesen and Almeida], brought to the USA from Japan in 1876 and promoted for control of soil erosion, is now a dense mat smothering flowers, shrubs, and trees across 2.8 million ha (Britton et al. 2002). The European starling (*Sturnus vulgaris* L.), introduced to New York City, USA, in the 1890s for frivolous reasons, now accounts for one in every five wild birds in North America. Starling competition for nest cavities suppressed native blue birds (*Sialis sialis* L.), which only recovered due to a massive nest-box construction program.

A thousand examples, each painful, many bizarre, others banal, could be cited of invasive species' damage, of species brought in accidentally or deliberately for economic gain with no thought of future consequences. Throughout evolutionary time, isolation of species by habitat separation and geographic barriers (separate continents, mountain ranges, oceans, lakes) has allowed selection and divergence to create a beautiful and bewildering array of plants and animals. Human beings are now randomly mixing the world's species, bypassing natural barriers, transporting any species anywhere for any purpose. The results are often ugly, ecologically disastrous, and costly.

So, what can be done? Prevention comes first, and better regulatory policies, more thoroughly implemented, could greatly reduce the influx of damaging species (Hedley 2004, Baker et al. 2005). Preventing species introductions is, however, technically difficult. Also, political interest in preventation programs is diminished by trade interests and the fact that most introduced species are of little importance. European meadow flowers growing along North American roadsides cause no problems and are a minor part of the local flora in disturbed areas. Open societies, free trade, and biosecurity are difficult to blend. People want novel plants; businesses want to sell whatever is profitable; governments want international trade with few restraints to promote economic growth. With such desires, prevention will be at best a marginal success and no cure at all after invaders establish. Rather, each effort – education, inspection, eradication of colonizing populations, and biological control of widely established ones – has its part to play.

For species with a clear potential to cause significant damage, eradication through chemical or mechanical means should be attempted immediately after initial detection, if biologically feasible. Damage to the Mediterranean sea bed by *Caulerpa* alga was so clear that its detection in California, USA, spurred an immediate government effort to eradicate it, using divers to inject bleach under tarps on the sea bed placed over algae. Sometimes, however, the threat posed by an invasive species is unknown or the species is not detected before it has spread over a considerable area. When ash trees in Michigan started dying from borers, the insect responsible was not recognized as an invader, but thought to be a similar native species. When emerald ash borer was understood to be an invader, it was too late for eradication since it had already spread over thousands of square kilometers. Eradication of invasive plants with ranges exceeding 1000 ha is rarely economically feasible (Rejmánek & Pitcairn 2002). Eradication of cryptic, hard-to-detect, small flying species with such large ranges is impossible.

Chemical and mechanical controls can reduce invasive species in small areas, but usually cannot protect extensive large natural areas because such controls become too expensive, disruptive, and polluting when applied to large areas. Only classical biological control has the right features (self-spreading, permanent, self-reproducing, high specificity) to solve such problems. In Chapter 8, options for control of invasive species are discussed and compared with classical biological control, which itself is covered in Chapters 11 and 12. In this chapter we will first develop basic concepts about invasive species and discuss their origins, biology, and impacts.

CASE HISTORIES OF FOUR HIGH-IMPACT INVADERS

Caulerpa taxifolia: "killer alga" of the Mediterranean

The poisonous alga *C. taxifolia* never lived in the Mediterranean Sea, but in 1984 a 1-m^2 patch was found directly beneath the cliff where the Oceanographic Museum of Monaco sits. Then, it could easily have been eradicated, but no action was taken. By 2001, 1 m^2 had become 80 km^2 of infested sea bottom, along 190 km of coast, and the alga was spreading rapidly (see www.pbs.org/wgbh/nova/algae/chronology.html for details on the chronology of the spread). Eradication was no longer an option. The delay and denial that prevented eradication allowed boats to spread the pest around the Mediterranean (Meinesz 2004). Dense algal meadows developed over sandy bottom habitats (Plate 7.1a), increasing structural complexity but adding little to local food webs since the alga is toxic to all but the most specialized herbivores. What this vegetational change will mean for native biodiversity or commercial fisheries is still unclear. Mostly, the research has not yet been done. Initial results have found that toxins released by the alga in the water appear to have suppressed some organisms (Bartoli & Boudouresque 1997), but others have increased (Relini et al. 1998). Fishermen speak of plummeting catches of commercial species. These findings are likely just the beginning of efforts to determine the impacts of this invader on the Mediterranean Sea's ecosystem.

Where did this invasive alga come from? DNA analyses show the invading population came from plants distributed by the aquarium trade (recall the location of the initial population just below Monaco's national aquarium). Surveys have tentatively identified the plant's origin as Moreton Bay, Australia, where it is a native species (Jousson et al. 2000, Meusnier et al. 2002, Schaffelke et al. 2002, Murphy & Schaffelke 2003). This alga poses an invasion threat around the world, and consequently the USA has banned its commercial importation. When the alga was detected off the Californian coast in 2000, the state moved aggressively to eradicate the small patches then present, using bleach injected under tarps placed over plants on the sea bed (Withgott 2002, Williams & Schroeder 2003). Eradication from Californian waters has been successful, but the plant still exists around the world in thousands of aquaria, each a potential source for future invasions.

Can anything be done to lower the density of this alga in the Mediterranean? Few species have been found that eat this toxic plant, apart from sea slugs (ascoglossan mollusks) (Figure 7.1; Thibaut & Meinsez 2000). Unlike terrestrial plants, which typically are attacked by scores or hundreds of species of arthropods (which provide ample opportunities to find a safe, effective biological control agent), the number of herbivores

Figure 7.1 Ascoglossan sea slugs (here, *Elysia subornata*) are among the few groups of herbivores able to eat the toxic alga *Caulerpa taxifolia* (Vahl) C. Agardh. Photograph courtesy of Alexandre Meinesz, University of Nice.

eating marine algae is extremely limited and so far none has been found that has high impact, is specific to *C. taxifolia*, and is adapted to the cool waters of the Mediterranean. Further surveys are needed to see whether any such herbivores or pathogens exist in the native range of the alga.

Brown tree snake destroys Guam's forest birds

Silent Spring by Rachel Carson took its title from an image of forests without birds (no birds to sing, hence a silent spring): a future she feared the forests of her native Maine would suffer due to indiscriminate use of the pesticide DDT (widely applied against mosquitoes and forest insects in the 1950s and 1960s). Some bird populations were indeed suppressed and some even extirpated from parts of the USA by DDT, but 50 years later, these bird species have recovered. Banning the compound in the 1970s allowed ospreys and herons, which had been locally suppressed by DDT but were still present, to naturally increase. This natural recuperation, coupled with active restoration programs for bald eagles and peregrine falcons, which had disappeared from the region, led to a full recovery, after residues disappeared. Rachel Carson's fears, however, have quietly come to pass on a distant Pacific island called Guam. The forests of this US military base have gone silent as virtually all of its native forest birds have disappeared. Even introduced urban land birds are gone! Pesticides were not the culprit, nor hunting, nor loss of habitat. The silencing of these forests was caused by the brown tree snake (*Boiga irregularis* Fitzinger) (Plate 7.1b; Jaffe 1994, Rodda et al. 1997, 1999, Fritts & Rodda 1998), a non-native invader from northern Australia and New Guinea (Figure 7.2).

The snake reached Guam, an island with no native tree snakes, during the 1950s on military planes. It found the birds and lizards of Guam to be easy and abundant prey. By 1985, this snake had reached densities of 100 per hectare (Fritts & Rodda 1998) and progressively, native forest birds (Plate 7.1c) disappeared. Bats and reptiles were also affected. Currently in most forest areas only three native vertebrates – all lizards – still survive. Several introduced skinks or geckos provided alternative prey that allowed the snake to remain high even as the native birds disappeared (Fritts & Rodda 1998). Unlike DDT, whose harm could be ended by legislative action, the brown tree snake is a self-replicating biological pollutant that does not dissipate with time. While some decline in brown tree snake densities may now be occurring (due to depletion of it prey base), it poses a high risk of expanding to new islands.

Economically, the brown tree snake has also been devastating to Guam. Its habit of climbing on wires and entering electrical boxes causes over 200 outages per year, costing over US$4.5 million (Fritts et al. 2002). Because Guam is a major air transportation hub for the Pacific basin, the presence of high densities of this snake on Guam greatly increases the risk (otherwise rather negligible) that the snake will invade Hawaii, USA, or countless other snake-free Pacific islands, causing new ecological and economic impact with each new jump. Trapping, poison baiting, and installation of snake-proof fences have been used to create snake-free areas around airports and cargo storage areas. Dogs have been trained to detect snakes at airports in cargo or wrapped on wheels of airplanes. But dogs have detected only two-thirds of all snakes in staged tests. Several snakes have made it to Hawaii and

Figure 7.2 The native range of the brown tree snake (*Boiga irregularis* Fitzinger). N.S.W., New South Wales; N.T., Northern Territory; P.N.G., Papua New Guinea; QLD, Queensland. Figure courtesy of G. Rodda, USGS; reprinted from Rodda et al. (1999) with permission from Cornell University Press.

have been detected within snake-control areas around airports.

However, to save Guam's birds, large forest areas must be cleared of snakes (Engeman & Vince 2001). Perimeter trapping can suppress, but not eradicate, snakes in forests stands as large as 18 ha if sustained for 5–6 months (Engeman et al. 2000). To remove snakes from large, remote forests, better systems are needed. Air-dropping of mice carcasses laced with snake-killing poison seems to hold promise and is being investigated (Shivik et al. 2002). But none of these solutions will be permanent because eradication is not achieved. Bird-restoration programs based on chemical or mechanical snake control will fail if suppression efforts are not maintained. How can this snake be suppressed permanently? Biological control has been traditionally focused on suppressing weeds and pest insects. That experience is not helpful for this case. What little has been achieved in suppressing pest vertebrates has been done with pathogens. So far surveys in Asia for pathogens potentially useful against brown tree snake have been disappointing (Telford 1999, Caudell et al. 2002, Jakes et al. 2003). Currently, biological control options seem unavailable (Colvin et al. 2005). Meanwhile, the birds of Guam – those that have survived in zoos – wait for the time to go back to their native forests.

An Asian adelgid destroys hemlocks in the eastern USA

Hemlock woolly adelgid (Plate 7.1d), *Adelges tsugae* Annand (Hemiptera: Adelgidae), is an exotic insect from Asia that invaded eastern North America and is killing large numbers of hemlock trees (Figure 7.3; McClure 1987, 1996). Infested trees can die in as little as 4 years (McClure 1991). First collected in Virginia in 1951 on planted hemlocks (Stoetzel 2002), this adelgid is spread by birds and now occurs from North Carolina to New England, USA (USDA-FS 2004).

In some instances, native predators or parasitoids have been able to feed on and suppress new invasive pests. However, in the case of hemlock woolly adelgid, surveys in Connecticut (McClure 1987, Montgomery & Lyon 1996) and in North Carolina and Virginia (Wallace & Hain 2000) have shown that local natural enemies are ineffective. While predacious Cecidomyiidae, Syrphidae, and Chrysopidae were found associated with the pest, their densities were too low to reduce its populations. Because existing adelgid natural enemies in the eastern USA have little ability to suppress the pest, a project of classical biological control was begun to introduce predators from other areas, including lady bird beetles from the pest's native range

Figure 7.3 Eastern hemlock trees killed by hemlock woolly adelgid (*Adelges tsugae* Annand). Photograph courtesy of William M. Ciesla, www.Forestryimages.org.

Figure 7.4 Dense infestations of kudzu, *Pueraria montana* (Lour) Merr. var. *lobata* (Willd.) Maesen & Almeida, smother native flowers, trees and other vegetation. Photograph courtesy of Kerry Britton, www.Forestryimages.org.

in Japan [*Sasajiscymnus* (=*Pseudoscymnus*) *tsugae* (Sasaji & McClure)] (McClure 1995) and areas in China (*Scymnus camptodromus* Yu et Liu, *Scymnus sinuanodulus* Yu et Yao, and *Scymnus ningshanensis* Yu et Yao) (Plate 7.1e; Montgomery et al. 2000, Yu et al. 2000, Yu 2001) and derodontid beetles in the genus *Laricobius* (all adelgid specialists; Plate 7.1f) from the western USA (*Laricobius nigrinus* Fender) (Zilahi-Balogh et al. 2003a, 2003b) and Japan (*Laricobius* n. sp.).

To not pursue biological control of this pest would allow it to spread across the range of eastern hemlock, degrading the whole hemlock-dependent community. Studies in the Delaware Water Gap (Pennsylvania, USA) showed that 20% of the park's hemlocks had been killed by this pest and 60% were in decline (Evans 2004). Loss of hemlock affects native species dependent on the cool habitat generated by hemlock stands such as the solitary and red-eyed vireos, black-throated green warbler, blackburnian warbler, and ovenbird (Young et al. 1998), brook trout (*Salvelinus fontinalis* Mitchill), various salamanders, and certain mosses and flowering plants. Hemlock-dominated stream stretches were two and a half times more likely to have brook trout than hardwood-dominated stretches, and under hemlock, trout were twice as abundant (Evans et al. 1996, Snyder et al. 1998).

Kudzu smothers wildflowers of the southeastern USA

The introduction of kudzu (*P. montana*) seemed like a good idea: it grew rapidly, covering the eroded soils on farms in the southeastern USA affected by the drought of the 1930s. It even made good cattle feed, so the US Soil Conservation Service paid farmers to plant kudzu on 486,000 ha. Seventy-three million plants were produced for this use by special nurseries (Tabor & Susott 1941). Today, with kudzu making 2.8 million ha unproductive for either human use or nature perhaps (Everest et al. 1991), the plant is no longer seen as a savior of threatened soil. Fortunately there is little spread by seed. However, the plant is very tenacious and able to spread via runners, leading to thick mats that smoother other vegetation, including mature trees (Figure 7.4). Little native plant diversity survives such competition.

THE EXTENT OF HARMFUL IMPACT BY INVADERS

Measures of impact

How bad is the invasive species problem? One measure is simply the percentage of species in a local fauna or flora that are not native. For example, 27% of plant species in Florida, USA, are non-indigenous (925/3448) (Gordon 1998). Similar calculations can be made for any group (clams, crawdads, insects, mammals, etc.). Assuming that such invaders are not just mere rarities in the invaded habitats, a rising percentage of invasive species in the community is indeed cause for concern. However, such an approach can be misleading, because it does not account for the abundance or damage associated with particular invasive species. A fuller accounting of impacts considers the damage from individual invasive species. Millions of hectares of kudzu smothering whole plant communities is a high-impact invader, but chicory (*Cichorium intybus* L., a European plant that occurs at low densities alongside American roads) is not. The impact of a transformer plant (*sensu* Pyšek et al. 2004), like Australian paperbark trees [*Melaleuca quinquenervia* (Cavier) Blake], on the Florida Everglades is only captured by understanding its ability to transform sawgrass marshes into swamp forests (Versfeld and van Wilgen 1986, Vitousek 1986, Turner et al. 1998).

Given that invaders vary hugely in their effects, one way to understand invasive species' significance is encyclopedic local knowledge. Simberloff et al. (1997) compiled such information for various plants and animals in Florida. In the USA, state or regional exotic-pest-plant groups have created regional lists of invasive plants, categorized by level of threat. This approach focuses attention on the species likely to achieve the greatest geographic range or be most damaging to local native species or communities. Analogous efforts for invasive animals would be helpful.

Thoughts not captured in the above discussion lurk in the words time lag and synergy. Although compiling lists of low-impact invasive species may seem a waste of resources, it may have value in spotting emerging threats. Some species invade explosively, rapidly becoming damaging, fast-spreading pests. But some do not. For some species long periods of time are needed before populations have enough propagules or the right set of circumstances to explode over the landscape (Crooks 2005). *Mimosa pigra* L. was introduced near Darwin, Australia, in about 1891 (Miller & Lonsdale 1987). It remained a minor weed for nearly a century until the 1970s when the area experienced unusually heavy rains that dispersed the plant's water-borne seeds throughout the Adelaide River floodplain. Previous overgrazing of the area by feral water buffaloes had disturbed the soil, providing excellent germination sites. Within 10 years *M. pigra* thickets covered 45,000 ha, with major infestations in Kakadu National Park, a World Heritage Site (Londsdale et al. 1988).

Another feature of invasions not captured by an explicit focus on high-impact invaders is synergy: the ability of some invaders to facilitate the population growth and spread of others. In Hawaii, pigs, strawberry guava, mosquitoes, and bird malaria are synergistic. Pigs eat guava fruits and spread seeds deep into native forest. With more guava, there are more pigs, which form bigger wallows that hold water in forest habitats and allow mosquitoes to breed. Pulling mosquitoes deeper into forests brings avian malaria into contact with more native forest birds, which die due to lack of resistance to this non-native disease. Collectively, pigs, guava, mosquitoes, and avian malaria have effects far beyond what any one of them alone would have (Simberloff & Von Holle 1999, Van Driesche & Van Driesche 2000). Combinations of invaders can generate increasing impacts, leading to "invasional meltdown" of native communities. Lower-altitude habitats in Hawaii now contain few native birds or plants due to just such a process.

How many bad apples in the bushel? The tens rule

Given that invasive species vary, what are the odds that any new invader will be an ecological or economic disaster? The tens rule is a gross generalization that asserts that about 10% of imported species establish feral populations and that 10% of those feral species will become damaging (economically or ecologically; Williamson 1996, pp. 31–43). One of the original datasets supporting this rule was for British plants. Of 1642 widely planted exotic plants, 210 became established in nature (12.8%) and 14 became severe pests (6.7%) (Williamson 1993). The Mediterranean Sea has 85 species of established exotic macrophytic plants. Of these nine (10.6%) are considered pests, having taken over the roles of keystone species or become economically harmful (Boudouresque & Verlaque 2002). In a review of invasive species in the USA, it was found that across a range of taxa, between 4 and 18% of non-indigenous species that establish go on to become high-impact pests (U.S. Congress, Office of Technology Assessment 1993, fig. 2.2).

Many groups fit the tens rule. Some that do not include crops, biological control agents, and birds or mammals on oceanic islands. Many crops are well adapted to living outside cultivation (e.g. wild apples in North America, mulberry trees, figs in some climates, fennel, asparagus, etc.), and have establishment rates of 20–30%, but generally are not regarded as pests (human bias?), with some exceptions such as stands of fennel on Santa Cruz, one of the Channel Islands of

California, where restoration of native vegetation is being attempted (USEPA 2001). Birds and mammals on oceanic islands also exceed the tens rule. In the Hawaiian Islands over 50% of introduced birds have become established (Williamson 1993). The rate of establishment for mammals on oceanic islands approaches 100% (see data for Ireland and Newfoundland in Williamson 1993). Some areas like Hawaii seem to be "over-invaded" (McGregor 1973) and may be at greater risk than suggested by the tens rule. Rates of establishment and impact for insects released as biological control agents are also higher than expected, precisely because this is the sought-after effect of selected, not random, species. Rates of 36 and 37% have been recorded for establishment and impact when both weed and insect biological control agents are combined (see table 2.6 in Williamson 1993 with data from Lawton 1990 and Hawkins & Gross 1992).

Trends in invasion rates and the effects of free trade agreements

Are things getting worse or is the invasion rate more or less constant? Locally one can answer this question, but globally the data are too hard to compile. Locally, for example, things in the Galápagos Islands are getting worse as more people move to the islands and bring their favorite species (Mauchamp 1997). Large movements of people between regions always bring invasions. European colonization of Australia, New Zealand, Hawaii, and the Americas set in motion thousands of species invasions, some deliberately, some accidentally. By 1900, governmental restrictions on movement of plants were imposed in the USA and elsewhere to slow insect and plant-pathogen invasions. Invasions stimulated by colonization continue. Indonesian mass migration to the island of New Guinea and Brazilian agricultural settlement of the western Amazon are very recent examples.

International trade is a major vector of species to new regions. Trade is increasing globally, with goods being moved faster, further, and in larger quantities. Governments inspect items in trade to attempt to exclude invasive pests. The US Department of Agriculture (USDA) Animal and Plant Health Inspection Service (APHIS) inspects cargo at ports and also attempts eradication of newly detected populations of threatening invasive species. Work et al. (2005) estimated that 42 new insect species established in the USA between 1997 and 2001 due to four pathways based on cargo in trade. As trade has increased, however, the inspector's job has become more difficult, with more to inspect and less time to do it. Only 1 or 2% of items are actually checked. Cargo containers, the standard shipping method, means that to check any goods, containers must be separated out and opened, a time-consuming and costly process. Invasions of the USA in the 1990s by such high-impact pests as emerald ash borer and Asian long-horned beetle [*Anoplophora glabripennis* (Motschulsky)] suggest that inspection is very imperfect.

HOW DO INVASIVE SPECIES GET TO NEW PLACES?

Natural dispersal

Some invaders reach new areas through natural dispersal. This process has shaped the world's biota over evolutionary time. Obviously the plants and animals present on oceanic islands when humans first found them arrived there by themselves. The cattle egret (*Bubulcus ibis*) reached South America in 1877, presumably by flight. Sugarcane smut (*Ustilago scitaminea* Sydow) reached Australia in 1998, presumably as spores blown from Indonesia. Naturally arriving species are not necessarily benign to the communities they invade. They can be damaging. However, the rate of natural invasion is dramatically lower than the rate of human-assisted invasions. This difference, not in kind but in rate, is the root of the current invasion crisis.

Hitchhikers and stowaways

Apart from biological control agents, insects are rarely deliberately imported. Most species are moved unintentionally on plants or cargo (see Sailer 1983 for a history of US insect invasions). Plant importations can lead to insect and pathogen invasions. Cassava mealybug likely reached Africa on imported planting material. Other insects have moved in wooden packing material or other goods. Asian long-horned beetle and emerald ash borer invaded the USA from China as larvae or pupae in crates or pallets made of untreated wood.

In nearly all countries, it is understood that such hitchhiking invasive species should be kept out if possible. Prevention of such introductions is therefore a matter of how much society is willing to pay or forego in trade to properly control the vectoring commodities. A century ago, it was generally recognized that moving soil along with plants made pest movement easy and detection nearly impossible. Consequently, moving untreated soil with plants was prohibited. Similarly, untreated logs with intact bark are an excellent means to move pathogens, borers, and bark beetles and their importation is now banned by many countries.

Businesses that import species to sell

Some invasive species are valuable plants or animals that were imported for commercial use. Many plant species, for example, are moved between biogeographic regions for use as crops, forestry trees, or ornamentals. Many imported plants have caused economic or ecological damage. Cacti, native to the Americas, were brought to Australia by early settlers. Cacti were well adapted to the arid climate and free of pests. They spread and eventually infested nearly 24 million ha, half so densely that the land had no economic value (DeBach 1974). An important added feature of plant invasions is that they often benefit from widespread planting (causing high propagule pressure); for example, suburban yards foster plant invasion of surrounding natural habitats by providing abundant seed sources.

Invasive animals are imported by the pet and aquarium trades, which constantly search for novel things to sell. Freshwater exotic aquaria fish are widely produced in outdoor ponds in Florida, from which large numbers of individuals periodically escape in time of floods. This has led to the establishment of at least 31 species in local waters (Courtenay 1997). Many terrestrial vertebrates have also become established via the pet trade, including various birds, lizards, frogs, and even monkeys (Stiling 1989).

There are few legal controls on the sale of groups of organisms popular with the plant or pet trade industries. Importers do not have to prove that new species are safe and not likely to become invasive. Only a few known culprits are excluded; the rest get the benefit of the doubt.

Farmed plants and animals

Farmers, foresters, and ranchers at times import new species for commercial production. Crop plants have been moved around the world, but even when invasive are usually viewed as benign. Only in extreme cases, such as strawberry guava in Hawaii, are invasive food plants viewed as pests. Demand for importation of new crop species can increase when immigrant groups seek to produce their traditional crops in new locations. In the USA, for example, demand is growing by Asian communities for importation and production of water spinach (*Ipomoea aquatica* Forsk), even though this crop is already known to be invasive in the southern USA (see www.iisgcp.org/EXOTICSP/waterspinach.htm#origin).

Foresters routinely move tree species among biogeographic regions. Northern-hemisphere conifers such as pines, cypress, or fir have been widely planted in southern-hemisphere countries, which lack similar softwoods. Vast plantations of *Pinus* have been established in Chile, New Zealand, Australia, and South Africa. Species of eucalyptus (from Australia) are widely planted in South America and Africa. In the southern hemisphere, imported trees are invading native grasslands and forests. Yet, commercial foresters feel justified in planting any tree anywhere if it is profitable to do so.

Common farm animals (pigs, cattle, goats, rabbits, and sheep) were widely released on to mammal-free oceanic islands in the age of sail (Chapuis et al. 1994, Desender et al. 1999). Farm-animal liberations, usually concurrent with rat and cat invasions (Atkinson 1985, Veitch 1985), have been a major cause of extinction of endemic plants and birds on oceanic islands. Even on continents the new species brought in by animal farmers have had serious effects. Escaped American mink (*Mustela vison* Schreb.) are now affecting numbers of water birds in Europe (Ferreras & MacDonald 1999). South American nutria (*Myocastor coypus* Molina) are damaging coastal wetlands in the eastern USA. Aquaculturists move shrimp, bivalves, and fish, which may either become pests in their own right, or harbor pathogens able to infect related native species (Kuris & Culver 1999, Anderson & Whitlatch 2003).

Government-supported releases

Government, by virtue of its control of many resources and ability to set the rules for species movement, exerts a powerful influence over species invasions. Many species invasions are planned and supported by governments. In Australia, public acclimatization societies were formed to "Euroform" the continent by establishing familiar trees, ornamental plants, fish, game, and other species that immigrants associated with home. In the USA, soil-conservation agencies introduced plants like kudzu to heal eroded land and grasses such as Lehmann lovegrass (*Eragrostis lehmanniana* Nees) to increase forage for cattle on public grazing lands (Anable et al. 1992). Game fish like rainbow trout have been widely introduced by public fish and game agencies into rivers and lakes in the USA and elsewhere, often damaging native fish and amphibians (Knapp & Matthews 2000). Game birds like the ring-necked pheasant (*Phasianus colchicus* L.) and chukar [*Alectoris chukar* (Gray)] were introduced to the western USA to provide additional hunting opportunities. In countries such as the UK, Australia, New Zealand, and South Africa, public policies have promoted forestry based on plantations of exotic trees. Planting of non-native trees over large areas retards the restoration of native forests and harms native plants and wildlife (see Richardson 1998 for a review).

Governments also conduct classical biological control introductions to suppress pests. If done well, this sort of introduction is part of the solution of the invasive species problem. But if the policies and procedures guiding the choice of pests selected as targets for control and the agents considered acceptable to introduce are not based on ecological principles, biological control introductions can also become damaging invasive species (Johnson & Stiling 1998, Goodsell & Kats 1999, Boettner et al. 2000, Kovach 2004).

Smuggled species and their associated organisms

One additional source of invasive species is smuggled items. Instances appear to have occurred in which plant material was smuggled into countries because it would not be permitted via official channels. Smuggling of avocado seedlings from Mexico into the USA has occurred, for example, because of shortages of seedlings from US sources. Such trees can easily vector foliage pests. Similarly, ethnic groups wanting to bring in forms of citrus not available in the USA might bring in plants infected with citrus greening disease, which could potentially destroy the US citrus industry. In Hawaii, people continue to smuggle in their pet snakes despite a $200,000 fine if they are caught (Kraus & Cravalho 2001, Kraus 2003).

WHY DO SOME INVASIONS SUCCEED BUT OTHERS FAIL?

The success or failure of individual invasions can turn on many features and predicting outcomes is not easy. Factors usually believed to favor invasions include: (1) high propagule pressure, (2) low biotic resistance, and (3) disturbance.

Propagule pressure

Propagule means any seed, body part, or individual that can start an invasive population. For plants, propagules usually are seeds or plant fragments. For animals, propagules would be individuals or colonies, of adults or immatures. Propagule pressure is the simple idea that increasing the number of propagules released increases the odds that the species will establish, especially if repeated releases are made. Propagule durability is also very important. If propagules remain viable for long periods, seed banks develop that allow a species to survive bad periods and repopulate when conditions are favorable. In addition, species with easily dispersed propagules are more likely to be effective invaders. For plants, ease of seed dispersal depends greatly on seed morphology. For species using animals to disperse seeds, the presence or absence of a good seed disperser can play a crucial role.

For naturally invading species, the above features are set by their biology (how many seeds are produced, how do they disperse, etc.). In other cases, human activity sets both propagule number of distribution. Suburban homes built in forested areas, for example, provide multiple locations from which shrubs or other plants are free to disperse into the wild. The planting of large numbers of ornamental plants increases the propagule pressure of commonly used species, causing our gardens to become staging areas for species invasions of surrounding areas.

Biotic resistance

After arrival, invaders must experience positive population growth if their numbers and range are going to increase. Otherwise the initial group will die out. Positive population growth requires that death rates be lower than reproductive rates. Biotic resistance is the concept that some places are more favorable than others to an invading species because of fewer deaths from herbivores, predators, or pathogens. For plants, biotic resistance would also include competition from other plants for limited resources or space, which reduced growth and seed production. Sea birds initially colonizing new islands do better on predator-free islands compared with ones with rats.

Habitat disturbance

"Disturbance prepares the seedbed." Disturbance is most easily visualized in relation to plant invaders. For some kinds of plants, disturbed soil, where local species have been eliminated, lowers the impact of competition on seedling survival of the invader, making establishment easier. Disturbance may be caused by animal grazing (Merlin & Juvik 1992), fire (Milberg & Lamont 1995), mechanical action of rivers (Hood & Naiman 2000), human actions, or storms. Habitat disturbance may also lower predation rates. On Christmas Island, for example, red land crabs (*Gecarcoidea natalis* Pocock) are a key source of mortality for the invasive giant African snail (Lake & O'Dowd 1991). Logging lowered crab densities, making such disturbed areas more prone to snail invasion than intact rain forests.

INVADER ECOLOGY AND IMPACT

Some effects of invaders can be photographed: a brown tree snake swallowing the eggs of a Guam flycatcher would be worthy of *National Geographic*. Other impacts – like hemlock woolly adelgid's gradual killing of its host trees – only become visible after many years. Linking the adelgid to stands of dead hemlock is feasible, if a bit indirect. But who would tie the decline of a native butterfly (*Pieris napi oleracea* Harris) with the invasion an exotic pest butterfly [*Pieris rapae* (L.)], without careful teasing out of the invisible link of shared parasitism (Benson et al. 2003)? Even harder links to make occur when invasive species change habitat characteristics in ways that send native species populations into long, slow declines as their habitats become too dry, or burn too often, or have too much nitrogen in the soil.

Direct kill

Invasive insects and plant pathogens can be selective, killing most of a few favored hosts, but allowing the remaining community members to adjust as best they can. The invasive fungus that destroyed the American chestnut [*Castanea dentata* (Marsham) Borkjasuer] affected a single species. Other tree species, mainly oaks, filled in the gaps. Direct kill by more generalist predators can cut a wider path. The introduction of red foxes to Australia reduced the abundance of at least 11 medium-sized marsupials (Kinnear et al. 2002).

Competition for space or resources

Invasive plants may outgrow native plants, dispossessing them of access to soil and light. Some invasive species may directly smother native species, such as skunk vine (*Paederia foetida* L.), which drapes itself thickly over trees in Florida's hardwood hammocks (Pemberton & Pratt 2002). Other invasive plants simply increase their ground coverage to the detriment of native species, as when purple loosestrife (*Lythrum salicaria* L.) replaces cattails (*Typha* sp.) in freshwater marshes (Blossey 2002). Even some animals, mainly species with low mobility, can be dispossessed of their living space. Dense zebra mussel encrustations severely affect pearly mussels (Unionidae), filtering food and fouling native mussels' valves.

Changing food webs

Every trophic relationship, such as A eats B, is embedded in a broader food web (see Chapter 9). In some cases, an invader's actions can change large portions of the community food web, greatly increasing the invader's impact. For example, when Nile perch (*Lates niloticus* L.), a large predatory fish, was released into Lake Victoria in East Africa, the food web underwent a massive contraction, with perhaps as many as 200 native fish species disappearing (Goldschmidt 1996, Seehausen et al. 1997) and most of the food energy being redirected into Nile perch and two lesser predators. Interestingly, some evidence suggests that species feared extinct may not all be gone, merely greatly reduced in number. Furthermore it seems that overfishing of Nile perch is allowing some fish species to partially recover (Balirwa et al. 2003). Invasive plants can also dramatically alter community food webs, by dominating the producer level. Bitou bush

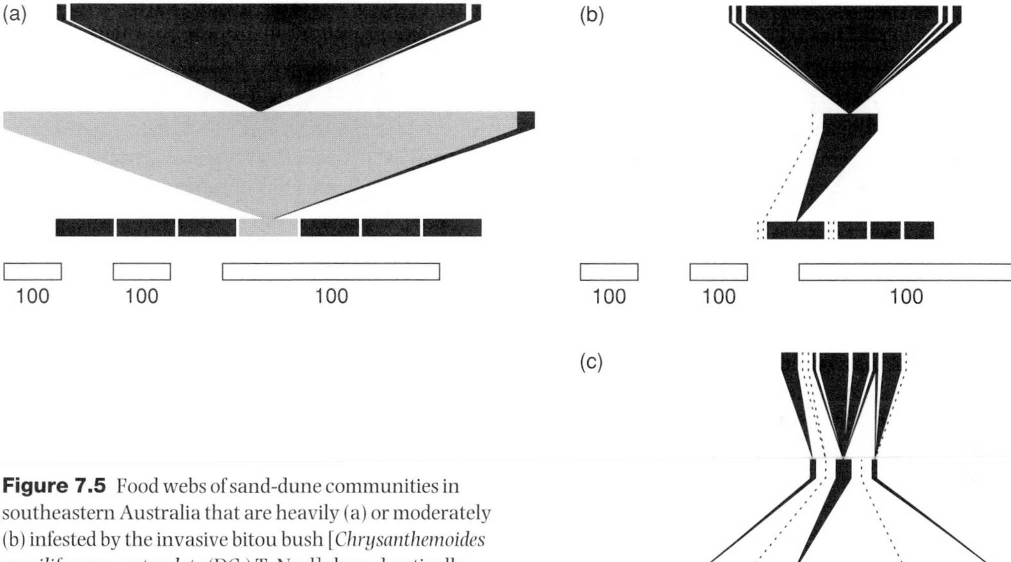

Figure 7.5 Food webs of sand-dune communities in southeastern Australia that are heavily (a) or moderately (b) infested by the invasive bitou bush [*Chrysanthemoides monilifera* ssp. *rotundata* (DC.) T. Norl] show drastically reduced species diversity compared to the same communities free of this weed (c). Reproduced with permission from Willis and Memmott (2005).

[*Chrysanthemoides monilifera* ssp. *rotundata* (DC.) T. Norl], an invasive plant of sand-dune communities in southeastern Australia, greatly lowers herbivore and parasitoid level diversity in invaded communities (Willis & Memmott 2005; Figure 7.5).

Changing the physical features of the habitat

Invaders can also change invaded habitats physically and chemically. For example, beavers (*Castor canadensis* Kuhl) convert cold-water streams into warm, still-water habitats. Species able to physically define a habitat are sometimes called ecosystem engineers (Crooks 2002). Such species can modify habitats in a variety of ways, including: (1) increasing fire frequency or intensity in grasslands (D'Antonio & Vitousek 1992), (2) lowering water tables (Neill 1983, Vitousek 1986), (3) increasing soil salinization (Kloot 1983), and (4) enhancing nitrogen in sterile soils (Vitousek 1990, Ley & D'Antonio 1998, Hughes & Denslow 2005).

Chapter 8

WAYS TO SUPPRESS INVASIVE SPECIES

The first response to the crisis of invasive species should be to slow the rate of invasion by implementing policies and practices aimed at prevention. However, prevention sometimes fails, and thus monitoring to detect newly established invaders is important. Early detection may make eradication feasible. If prevention and eradication fail, active controls will be needed, including: (1) habitat management, (2) pesticides, (3) mechanical tools, and (4) biological control. Several factors affect which options are best for particular cases, including the extent of the infested area, cost of control, and whether the need is for permanent or temporary suppression. If the goal is to permanently suppress an invasive species over an entire landscape, biological control is the most practical method. In farmers' fields or small nature preserves, mechanical or chemical control may be feasible.

PREVENTION: HEADING OFF NEW INVASIONS THROUGH SOUND POLICY

Prevention begins with sound policies that minimize the risk of invasion (Van Driesche & Van Driesche 2001). Prediction of which species are likely to become damaging invaders is a valuable first step. This requires a broad knowledge of the taxonomy and biology of various groups of organisms and specific information about the invasiveness of particular species in other regions. For species predicted to be invasive, the risk of introduction can be assessed by **invasion pathway analysis**, the study of how specific invaders move geographically. If the potential key vectoring processes are understood, methods to limit unwanted introductions can be devised. When risk from a process is very low, it may be more effective to tax the vectoring activities rather than prohibit them and use the proceeds to eradicate or control the invasions that occur (Hayes 1998).

Predicting which species could be high-impact invaders

The life-history characteristics of a species and the degree of its invasiveness elsewhere are indicative of its potential for further invasions. More use needs to be made by governments of such information. Several principles can guide the process.

Apply the same standards to all groups of organisms being introduced

Risks associated with the introduction of exotic plants have historically been dramatically underestimated and in most countries there is relatively little emphasis on determining the invasive potential of introduction of new plant species. Conversely, most people assume insect introductions, even those of biological control agents, will be damaging. Different groups of organisms are regulated, if at all, under different laws for different purposes. Plant introductions are regulated mainly to prevent the introduction of insects and plant pathogens. In the USA (and most other countries) plants are assumed to pose no risks, unless they are on a tiny list of noxious weeds. In part this is because people enjoy plants and assume they are beneficial. Also, it may reflect unwillingness of governments to interfere with the commercial trade in exotic plants. Few countries other than Australia and New Zealand require pre-introduction analysis of the potential for new plants to become invasive.

Be guided by experiences in other countries

Species that are invasive anywhere are more likely to become pests if introduced to new regions with similar climates (National Research Council 2002). Parts of South Africa, Australia, Chile, California (USA), and the area around the Mediterranean Sea all have similar climates. Thus, a species invasive in any one, should be assumed to be a risk in the others as well. For example, the European plant *Hypericum perforatum* L. has become an invasive pest in Australia, California, South Africa, Chile, New Zealand, and Hawaii, USA (Julien & Griffiths 1998): all regions with areas of Mediterranean-type climate.

Study the invasion potential of species in valuable groups

If the economic value of a plant group is high, the invasion potentials of individual species in the group should be determined. Such detailed knowledge can allow a group's benefits to be enjoyed while avoiding some risks through preferential use of the least invasive species. For example, exotic pines are important for plantation forestry in the southern hemisphere because there are few native conifers with commercial properties. Whereas use of native trees should be favored, as long as the forestry use of exotic trees continues, it is valuable to know which species in commonly used genera like *Pinus* and *Eucalyptus* are most invasive. Studies in South Africa (Richardson 1998) have shown that, for exotic pines, propagule pressure correlated well with invasive risk potential, with the most invasive species being those, such as *Pinus greggii* Englemn., that mature early, set many, light seeds, and do so at short intervals.

Avoid species with a structural competitive advantage or against which biological control is not feasible

Some kinds of organisms are more likely to be highly damaging or impossible to control and these species need to be recognized and avoided scrupulously. Among these are vines, floating aquatic plants, grasses, and plant pathogens. There are many examples of damaging invaders in these groups [vines: Asiatic bittersweet, *Celastrus orbiculatus* Thunb.; skunk vine, *Paederia foetida* L.; Old World climbing fern, *Lygodium microphyllum* (Cav.) R. Br.; kudzu, *Pueraria montana* (Lour) Merr. var. *lobata* (Willd.) Maesen & Almeida; floating plants: waterhyacinth, *Eichhornia crassipes* (Mart.) Solms.; giant salvinia, *Salvinia molesta* D. S. Mitchell; red water fern, *Azolla filiculoides* Lamarck; water lettuce, *Pistia stratiotes* L.; alligatorweed, *Alternanthera philoxeroides* (Mart.) Griseb.]. The ability of these plants to float on the water or to climb on native trees enables them to pre-empt access to light, thus favoring them in competition with native plants.

Groups such as grasses and plant pathogens are of special concern because there seems to be nothing that can be done about them if they become damaging invaders. No examples exist of successful biological control of grasses, although efforts against a few species are underway. Similarly, we should be very concerned about invasive plant pathogens because the diseases they cause [e.g. chestnut blight due to *Cryphonectria parasitica* (Murr.) Barr.; dogwood anthracnose by *Discula destructiva* Redlin; butternut canker by *Sirococcus clavigignenti-juglandacearum* Nair, Kostichka, & Kuntz; and Dutch elm disease by *Ophiostoma ulmi* (Buisn.) Nannf.] have decimated important forest trees in North America and classical biological control cannot control such pests.

Pathway analysis: study of how invasions spread

Prevention efforts can be made more efficient by focusing on vectoring processes rather than particular species. Plants, soil, ballast water, hull fouling, and wooden packing material are important means for moving some invasive species.

Plants

Some insects that attack plants are attached to them or present inside stems or fruits. Such insects move easily with their host plants. San Jose scale [*Quadraspidiotus perniciosus* (Comstock)] was moved around the world on apple nursery stock. By 1900, so many pest insects had reached the Americas, New Zealand, Australia, and South Africa on imported plants that these countries enacted laws requiring that plants be inspected and certified as insect-free before importation. These measures were adopted because many of the imported pests did serious damage to agriculture, forestry, and horticulture. Success in excluding herbivorous insects varies with the inspection rigor and trade volume of a given country.

In contrast, reducing the risk that movement of plants will introduce pathogens of related native plants has been less successful. In part, this is because these are microscopic organisms that are difficult to detect, their potential impact is often unknown, and sampling requires more expertise and time to determine risk. But also, the ecological basis of the problem is more complex. With insects, for the most part, the goal was to detect known pests. In contrast, fungi or bacteria on foreign plants may be harmless to those species but turn out to be lethal to related native plants. Detection of such new-association pathogens whose lethality is not yet even suspected is not possible by the means used to reduce plants as vectors for invasive insects. Rather, societies need to actively study organisms that might move between plant species and become pathogens and then control their invasions by limiting the importation of the plants likely to be vectors. Such work is rarely done. Instead, such connections are usually determined only after damaging invasions have occurred, mainly for the purpose of limiting further spread. Attempts to control the spread of sudden oak death fungus [*Phytophthora ramorum* (S. Werres, A.W.A.M. de Cock & W.A. Man in't Veld)] from California to the rest of the USA is an example (see the USDA sudden oak death website, www.aphis.usda.gov/ppq/ispm/pramorum/). While not proven, it is believed that this pathogen was imported on *Rhododendron* plants (Martin & Tooley 2003, Rizzo & Garbelotto 2003) to California, where it now has infected and killed native oaks with a subsequent change in forest tree compostion. Federal authorities are trying to prevent its dissemination throughout the North America through quarantines on plants that host the pathogen. However, on some hosts this pathogen is non-lethal and asymptomatic, making detection nearly impossible.

Soil

Mixtures of living soil were often shipped internationally before 1900 when plants were moved to new countries. This practice was banned in the USA shortly after 1900 because insects and pathogens are both common and undetectable in untreated soil. To stop the movement of soils containing living organisms, shippers could eliminate soil (ship plants with bare roots), heat-treat it, or fumigate it with pesticides to kill insects and pathogens.

Ballast water

Unlike soil, the importance of ship ballast water (Figure 8.1) as a means of spreading invasive species was not recognized legally until recently. It has long been known that ballast water holds exotic species that can establish after being discharged in new regions.

Figure 8.1 Discharge of ballast water from ocean-going vessels when in port is a major route of invasion for aquatic species. Photograph courtesy of Dave Smith.

But it was not until the zebra mussel disaster that the seriousness of such invasions impressed itself on the USA government. Zebra mussel (*Dreissena polymorpha* Pallas) was found in 1986 near Detroit, Michigan, USA (Schloesser 1995). It spread rapidly, sometimes reaching 700,000 mussels/m^2 (Schloesser 1995). Companies with water-intake and -outfall pipes must now actively clean their pipes chemically or mechanically. More importantly, this mussel is a severe competitor with native pearly mussels, many of which were already threatened (Ricciardi et al. 1996, Martel et al. 2001). Zebra mussels reduce the food concentration in the water column and foul the valves of native mussels, preventing them from closing properly. The potential damage from a zebra mussel invasion of North America was recognized as early as 1921 and the likely vectoring mechanism (larvae in ship ballast water) by 1981 (Schloesser 1995). New legislation passed in the USA now requires that ships manage their ballast water to reduce the transport of invasive species, either by chemically treating the water or exchanging water in mid ocean so that no foreign freshwater ballast is brought into US lakes or rivers.

Hull fouling

Similar to ballast water, hull fouling of ships has great potential to spread non-native marine species over great distances. As ships enter ports, changes in salinity and water temperature induce spawning of hitchhiking organisms (Minchin & Gollasch 2003). Thus, hitchhikers have the potential to spawn in any or all ports visited by their host vessel. Antifouling coatings painted on to ship hulls intended to minimize fouling often deteriorate with time. In addition, many large-vessel hulls are designed with recesses in the hull that provide refuge from turbulent water flow for mussels, barnacles, polychaete worms, and crustaceans (Coutts et al. 2003). On 186 vessels inspected in the North Sea, exotic species made up 96% of the hull-fouling organisms, and 19 posed a high risk for establishment (Gollasch 2002).

Wooden packing material

Wooden crates and pallets used to ship goods from China were the pathway in the 1990s for the invasion of the USA by two highly damaging forest pests: Asian long-horned beetle [*Anoplophora glabripennis* (Motschulsky)] and emerald ash borer (*Agrilus planipennis* Fairmaire). The former is confined to a few small infestations and may be succumbing to eradication efforts. The emerald ash borer (Figures 8.2 and 8.3), however, was not detected before occupying thousands of square kilometers in Michigan. Although its eradication is being attempted by massive cutting of ash trees (Figure 8.4), this is unlikely to succeed. These invasions illustrate that untreated wooden packing materials pose a high risk for invasion of pests of native trees.

Figure 8.2 Adult of the emerald ash borer (*Agrilus planipennis* Fairmaire), a borer from China that has killed over 6 million ash trees in the central USA and Canada. Photograph courtesy of Deb McCullough, USDA Forest Service.

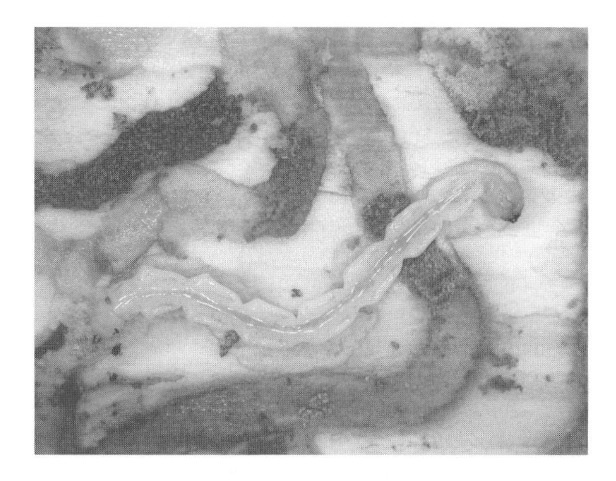

Figure 8.3 Larva of the emerald ash borer (*Agrilus planipennis* Fairmaire). Photograph courtesy of Deb McCullough, USDA Forest Service.

Figure 8.4 Eradication of the emerald ash borer (*Agrilus planipennis* Fairmaire) was attempted in the USA based on cutting all ash trees within a half mile (0.8 km) of any infested ash trees discovered in surveys. Photograph courtesy of Deb McCullough, USDA Forest Service.

New measures requiring heat or pesticide treatment of wooden packing materials are being implemented.

ERADICATION BASED ON EARLY DETECTION

When prevention fails, invasive species arrive at new locations. Inspection of cargo at international borders offers some chance to intercept and exclude arriving pests. The odds of successful detection, however, are low because less than 5% of goods are inspected.

When detection fails, invaders may establish. If incipient populations are found early, eradication should be attempted for highly damaging species using chemical or mechanical methods. Early detection and aggressive mechanical control swiftly eradicated the South African sabellid polychaete worm (*Terebrasabella heterouncinata* Fitzhugh and Rouse) in California. This worm arrived on South African abalone imported for mariculture (Kuris & Culver 1999) and was first detected infecting native *Tegula* snails near the outflow from an abalone-rearing facility. Eradication was achieved by hand removal of *Tegula* snails to densities too low to sustain sabellid transmission. New contamination was prevented by filtering the facility's waste water and halting the dumping of shell debris in the intertidal zone. Similarly, the black-striped mussel, *Mytilopsis sallei* (Recluz), was eradicated from Darwin Bay (Northern Territory, Australia) by treating the infested marina with high concentrations of bleach and copper sulfate (Bax 1999).

Eradication, however, becomes less feasible as the size of the infested area increases or after wide dispersal of the invader's propagules has occurred. At that point, the objective of the control program is likely to focus on slowing the spread of the invader by preventing additional areas from becoming infested, rather than eradication. At that point, eradication efforts should be abandoned in favor of containment or use of suppression tactics such as biological control.

INVADERS THAT DO NO HARM

Most invasive species do not become pests. If non-native species are not economically important and do not strongly affect native species or communities, they should be ignored even though some conservationists find them objectionable on principle as biological pollutants. Resources, however, are not available to attempt to control all non-native species. For some invasive species, an initial outbreak may be followed by decline to non-pest levels (e.g. McKillup et al. 1988). Brown citrus aphids [*Toxoptera citricida* (Kirkaldy)], for example, appear to be suppressed in Puerto Rico and Florida, USA, to levels not likely to spread tristeza virus (a critical disease of citrus) by existing generalist aphid predators (Michaud 1999, Michaud & Browning 1999). In such cases, control efforts are not required. Natural enemy importations should be reserved for use against species not declining spontaneously and which pose quantifiable environmental and economic threats that justify program initiation.

CONTROL OF INVASIVE PESTS IN NATURAL AREAS

For high-impact invaders in natural areas, control options include habitat management, chemical and mechanical control, and introduction of natural enemies. Each method has advantages and disadvantages that should be considered when choosing the best approach for particular problems. Sometimes chemical or mechanical approaches can be combined with biological control programs, especially against long-lived woody plants. For example, felling is being used with the application of herbicide to stumps in the Florida Everglades, USA, to clear existing stands of the invasive tree *Melaleuca quinquenervia* (Cavier) Blake and prevent resprouting (Figures 8.5 and 8.6). Concurrently, exotic insects are being released to reduce seed production, kill seedlings, and suppress growth of saplings and stump sprouts.

Figure 8.5 Cutting of pole size or large melaleuca plants [*Melaleuca quinquenervia* (Cavier) Blake], an invasive Australian tree in the Florida Everglades, is done to speed removal of existing plants, with suppression of seeds and new seedlings provided by biological control agents. Photograph courtesy of Ted Center, USDA-ARS.

Figure 8.6 Herbicides are also used to kill large melaleuca plants [*Melaleuca quinquenervia* (Cavier) Blake] in the Florida Everglades, and when sprayed on cut stumps, to prevent stump sprouting. Photograph courtesy of Steve Ausmus.

Habitat management

Mismanagement of land or water can sometimes cause exotic plants or insects to proliferate. Solving such problems begins with improvement of basic management practices. Overgrazing, for example, may confer a competitive advantage on unpalatable exotic species, causing them to increase. If this is the root cause of an invasive plant problem, altering the grazing regime, not use of biological control, should be the first thing considered. For example, pursuing biological control of native *Opuntia* cacti on the Caribbean island of Nevis in the 1950s (Simmons & Bennett 1966) was an error because the dense stands of cacti (some native, some introduced) in pastures were due to overgrazing. Changing cattle management, combined with some herbicide use, could have solved the problem.

Chemical and mechanical controls

Against plants

Mechanical and chemical controls work well for temporary suppression, sometimes even eradication, of many invasive plants, especially larger species on a limited area. Herbicides, for example, were used to clear gorse (*Ulex europaeus* L.) from pastures being replanted with koa trees to expand habitat for Hawaiian birds. The area requiring treatment was small (about 800 ha) and treatment had a permanent effect because gorse does not regrow under koa trees (Van Driesche & Van Driesche 2000).

Chemical and mechanical controls can be implemented immediately when and where needed, making them ideal for "weeding" small preserves where unique floral or other species need to be protected quickly on a limited area. Potential concerns in managing treated areas include whether the invasive weed will regrow after treatment and, if so, how quickly; whether treated areas will be invaded by new weeds; whether emerging native vegetation will be competitive, and how this competition will be affected by other factors such as removal of exotic vertebrates. On Santa Cruz Island, California, shooting and trapping were used to remove sheep and pigs, which allowed native vegetation to regrow. However, in some parts of the island, native seed banks were depleted and regrowth was dominated by the exotic herb sweet fennel (*Foeniculum vulgare* Miller), which required herbicide treatment, together with replanting of native vegetation. At such locations, dense stands of fennel developed, covering several thousand acres (see The Nature Conservancy website on island fox recovery plan at http://nature.org/wherewework/northamerica/states/california/features/sci_recovery.html.

Whether there is a role for chemical or mechanical control against a pest plant needs to be determined on a case-by-case basis, taking into account one's management goals, available resources, and the biology of the plant (see Cronk & Fuller 1995 and Myers & Bazely 2003). Some groups of plants, such as grasses or plants with deep or persistent root systems, or those able to regenerate from fragments, will be especially difficult to control and may not be suitable targets.

Against insects

Insects that infest large natural areas can seldom be eradicated by mechanical or chemical means. Suppression, however, is sometimes possible. Traps have been used to suppress Africanized honey bees along the leading edge of their invasion into the USA from Mexico, and bait sprays, together with releases of sterile males, are used against the Mediterranean fruit fly [*Ceratitis capitata* (Wiedemann)] in California (Anon 1988, Carey 1992). Small outlying infestations of the gypsy moth [*Lymantria diapar* (L.)] in the western USA have been eradicated by aerial pesticide treatment of infested forests with insect growth regulators or *Bacillus thuringiensis* Berliner (Dreistadt & Dahlsten 1989). Amdro® (hydramethylnone) baits were used successfully in the Galápagos Islands to remove the little fire ant (*Wasmania auropunctata* Roger) from Marchena Island (21 ha) (Causton et al. 2005).

Against vertebrates

Poison baits, traps, fencing, and shooting can be used to suppress or eradicate invasive mammals and other vertebrates. Inside fenced nature reserves, large mammals can be eradicated, for example pig removal inside forest-bird reserves in Hawaii. On small islands, poisons have been used to remove cats, rats, mice, and rabbits. Round Island in the Indian Ocean, once a lush tropical forest habitat, was degraded to barren slopes with only remnant vegetation following the introduction of goats and rabbits. Partial restoration has been achieved with the use of poison baits against rabbits (North et al. 1994) and shooting of goats. Native palms and some reptiles are now recovering (Bullock et al. 2002). Rodents have been eliminated with poison baits on small oceanic islands of New Zealand (Taylor & Thomas 1993), California (Jones et al. 2005), and British Columbia, Canada (Taylor et al. 2000) to protect rare birds or allow their reintroduction. The techniques were developed on small islands, but have been adapted successfully for use on increasingly larger islands. Eradication of vertebrates is becoming increasingly feasible (Veitch & Clout 2002, Lorvelec & Pascal 2005).

Biological control

Suppressing invasive species by importing specialized natural enemies from their native ranges is an old idea that began for the control of crop pests and later was extended to address pests of natural areas. In 1855, Asa Fitch in the USA suggested importing parasitoids of the European wheat midge [*Sitodiplosis mosellana* (Géhin)]. In 1863, a non-native cochineal insect, *Dactylopius ceylonicus* (Green), was moved within India to suppress cacti (Goeden 1978). In 1884, *Cotesia glomerata* (L.) from Europe was established in North America against *Pieris rapae* (L.) (Clausen 1978). In 1888, the ladybird beetle *Rodolia cardinalis* (Mulsant) was imported from Australia to California, where it suppressed cottony cushion scale, *Icerya purchasi* Maskell, a major pest of citrus. This was the precedent that demonstrated the effectiveness of the method (DeBach & Rosen 1991).

More than 100 species of invasive insects and 40 weeds have been controlled permanently by introductions of natural enemies (Clausen 1978, Cameron et al. 1989, Greathead & Greathead 1992, Julien & Griffiths 1998, Waterhouse 1998, Waterhouse & Sands 2001, Mason & Huber 2002). Biological control through introduction of natural enemies is permanent and self-spreading (see Chapters 11–13). Once agents are established, they can reproduce by themselves and spread over large areas with minimal human assistance and persist year after year without any further cost. This means that pests in natural areas can be dealt with using this approach. In contrast, chemical or mechanical controls are often either too costly or polluting for widespread, repeated use. Potential risks to native species from introduced natural enemies of invasive pests must be predicted (Chapters 17 and 18) and judged to be acceptable before particular natural enemies are released. If this is done and if projects have sound ecological justifications, the classical biological control is environmentally safe.

FACTORS AFFECTING CONTROL IN NATURAL AREAS

The main factors affecting the choice of control method against an invasive pest are the size of the infestation to be suppressed, cost, and social agreement on the pest status of the species. Mechanical and chemical control are commonly used in small nature preserves because they can be implemented quickly with good effect against local problems. Control efforts are under the immediate control of the preserve manager and can use volunteer labor, which is free and helps educate the public about invasive species impacts. Hand weeding a small preserve simultaneously of several important invasive plants would be a common example of such an approach. Costs of such work may run to a few hundred dollars per hectare, which is affordable at a small scale (5–10 ha) by private groups and with government support can be implemented on areas up a few thousand hectares. Such costs are not usually sustainable, however, if the goal is to clear hundreds of thousands or millions of hectares of an infestation. Also, such approaches rarely work against invasive insects and, if attempted, the needed insecticides are likely to be polluting and damaging to native species. One of the largest mechanical/chemical weed-clearance projects is the clearance of melaleuca trees from the Florida Everglades (see Chapter 12) by cutting and herbiciding. In South Africa the Working for Water Project is employing tens of thousands of workers to clear invasive pines and other trees to restore water flows. This project also allows the government to provide much-needed jobs, which has increased its social popularity.

In contrast, biological control can effectively control an invasive pest over an entire landscape regardless of size. Indeed, the larger the infestation, the more appropriate the use of this method. Biological control has high start-up costs, requiring long-term financial support for its implementation. Start-up costs are directed at basic research to understand the ecology of the pest in its native range, locate and study its natural enemies, select and import those likely be specific, measure their host ranges, and finally release and evaluate them. Projects often can take decades and costs can run to millions of dollars unless the target pest is well known and has been controlled previously in other countries. In such repeat control projects, control in new areas will be quicker and cheaper, being limited mainly to the cost of any further testing of the host range that might be needed and costs to establish natural enemy colonies of known agents and establish them in the field.

Because biological control agents will spread to their ecological limits, they cannot be confined to particular properties based on ownership or country. Consequently, there must be broad agreement that the target species is a pest whose reduction is desired over the whole ecological region (e.g. in the USA, agents are likely to spread to Canada or Mexico). Any conflicts between social or political groups that see the target pest differently must be resolved before the release of biological control agents. This problem is not usually an issue in chemical and mechanical control projects, which are easily confined with particular boundaries.

CONTROL OF INVASIVE SPECIES IN CROPS

Farmers and foresters also have problems with invasive species. However, unlike the problems discussed earlier in this chapter, these problems do not necessarily need to be solved over the whole region, but just in the farmer's own fields. Crop pests may have little or no impact on natural areas because of the difference in vegetation, although this may not always be true, as in the case of cottony cushion scale, a citrus pest that attacks native plants in the Galápagos National Park (Causton 2004). Because crop pests cost farmers money from lost production, they are willing to spend money on their control. This means that in addition to government-run programs of classical biological control, there are other options in crops that are not feasible in natural areas. These options include use of methods not covered by this book (pesticides, cultural controls, use of resistant plants, traps, or manipulation of insect behaviors) as well as additional forms of biological control such as: (1) manipulations to preserve or enhance natural enemies, (2) applications of pathogens as pesticides, or (3) release of parasitoids or predators that have been reared by commercial businesses and sold to the farmer. These approaches can be used against both exotic and native pests in farms and tree plantations, as will be discussed later in this book.

Part 4

NATURAL ENEMY INTRODUCTIONS: THEORY AND PRACTICE

Chapter 9

INTERACTION WEBS AS THE CONCEPTUAL FRAMEWORK FOR CLASSICAL BIOLOGICAL CONTROL

Species invasions and introductions of biological control agents take place within ecological communities whose composition can strongly affect outcomes of both processes. Components of communities can retard or facilitate invasions, provide spontaneous biological control of invaders, or make introduced biological control agents fail through biotic resistance. Similarly, prediction of non-target risks from biological control introductions proceeds from an inventory of the possibly affected species in the communities where the biological control agent is released or spreads. Although the main factors affecting the success of insect populations are trophic effects and resource constraints, success of invasive plants and their biological control agents is also affected by competition between the target plant and other plant species in the community. Community ecology, therefore, is an integral part of planning and understanding programs of biological control.

TERMINOLOGY

Trophic pyramids of who eats whom are called **food chains**. In classical insect biological control, the agents in the upper trophic level (in relation to the pest) are the parasitoids and predators that attack the pest. In weed biocontrol, herbivorous insects are the trophic level of interest. Species in the next level up (hyperparasitoids for insect projects and parasitoids for weed projects) are undesirable forces that are eliminated in quarantine during importation. Local (native or exotic) parasitoids or hyperparasitoids may, however, attack the biological control agent in the recipient country after release. Whereas generally this does not entirely negate the impact of introduced natural enemies, in some cases, especially for weed biocontrol agents, it may.

Since few species of herbivores, parasitoids, or predators are strictly monophagous, each occupies a place in several cross-linked food chains and the full array of these linkages form the **food web**. Figure 9.1 gives an example of a food web for an introduced plant (red gum), invasive pest insect (red gum lerp pysllid), and its introduced natural enemies in California, USA.

Every species is embedded in a food web; **natural control** is the total mortality imposed on a species by all the consumers at higher tropic levels within the species' food web. Natural control, which may be weak or strong, often restrains the density of native species to low levels. For exotic species that exist at pest densities, natural control in the invaded region, by definition, is insufficient. The goal of classical biological control

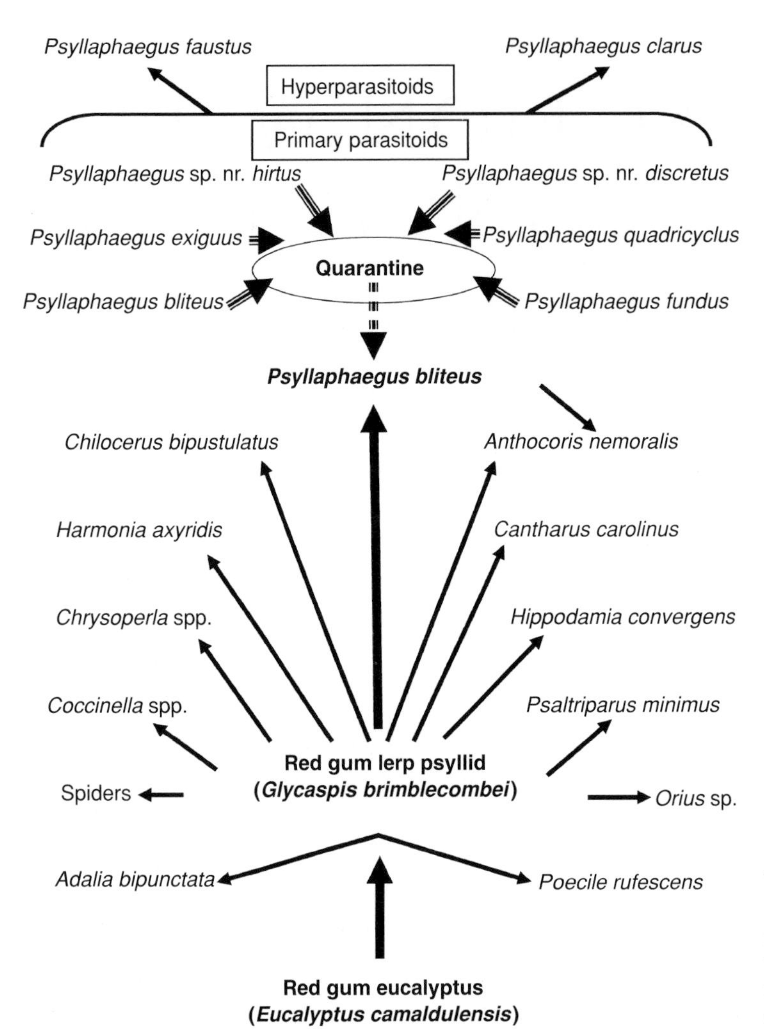

Figure 9.1 The food-web context of arthropod biological control is illustrated by that of the red gum lerp pysllid (*Glycaspis brimblecombei* Moore) and its parasitoids on eucalyptus in California, where all major components are introduced species. Drawing courtesy of Mark Hoddle.

is to increase natural control by adding new, more effective species of antagonists.

Species may also limited by their competitors. This is rare for herbivorous insects, but **interspecific competition** is a major force affecting plant densities. The strength of plant–plant competition commonly affects both invader success and the impact of weed biological control agents. The invasion of Hawaiian forests by *Miconia calvescens* DC., a Central American understory tree, is the result of both escape from its natural enemies (Killgore et al. 1999, Seixas et al. 2004) and its greater shade tolerance compared to native Hawaiian plants (Baruch et al. 2000). For plants, damage from herbivores and pathogens and competition with other plants for resources are both strong determinants of number and biomass (Polis & Winemiller 1996) and are linked by trade-offs (Blossey & Notzold 1995, Blossey & Kamil 1996). Blossey and Kamil (1996) used experimental comparisons of plant genotypes in the native and invaded ranges of purple loosestrife (*Lythrum salicaria* L.) to show that its invasion of North America may have involved both escape from natural enemies, and, in this natural enemy-free environment, selection for plant genotypes that allocated more resources to competitive abilities (vegetative growth) at the expense of herbivore defense. When both trophic and competitive relationships must be considered, the framework is called an **interaction web** (Wootton 1994).

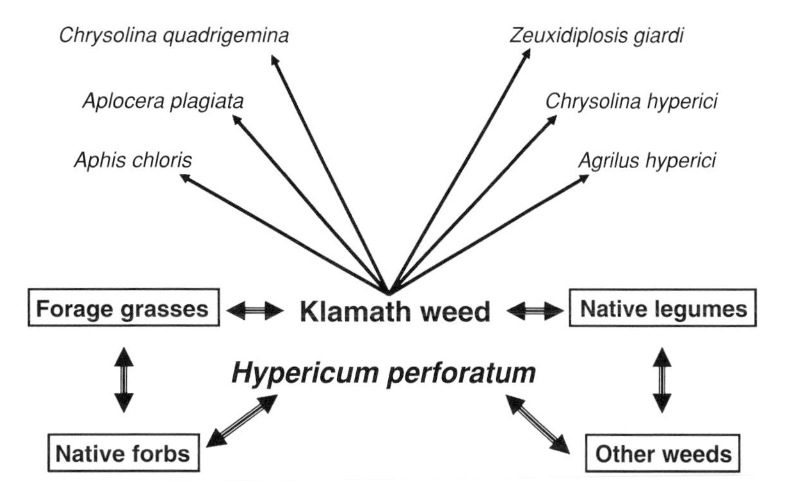

Figure 9.2 The food-web context of weed biological control is illustrated by that of St. John's wort (*Hypericum perforatum* L.) in California. Drawing courtesy of Mark Hoddle.

In general terms, restraints on the growth of native or invasive species may come from either the upper or lower trophic levels, giving rise to the terms **top-down** or **bottom-up limitation**. Top-down influences are antagonistic or feeding relationships (predation, parasitism, herbivory, infection), in which a population is attacked or consumed in some way by a species in an upper trophic level. Examples would include a caterpillar feeding on a plant, a parasitic insect attacking the caterpillar, a bird eating the parasitic insect, or a fungal pathogen attacking any of these organisms.

Bottom-up influences act in the opposite direction within the food web; that is, from the lower trophic levels. Gall-forming insects, for example, usually require the target plant structure to be at a precise developmental stage for successful attack. Arroyo willow gall sawfly [*Euura lasiolepis* (Smith)], for example, can only initiate attack between bud burst and shoot elongation; otherwise tissues become too tough. The supply of shoots at this growth stage limits gall sawfly density more than attack by natural enemies on immature gallmakers (Craig et al. 1986). Similarly, densities of plants may be set by the adequacy and extent of resources (proper soil, rainfall, the thermal environment), while predators may find their population density set by prey availability.

In addition to top-down and bottom-up influences are competitive interactions that occur among species at the same trophic level for some limiting resource (light, space, or nutrients; Figure 9.2).

Which of these influences is most important in setting animal and plant population sizes has been subjected to extensive experimentation. Combinations of influences in which none of the above forces alone is strong enough to set a species' typical density exist and may be common. Top-down and bottom-up forces may act simultaneously, with bottom-up forces setting the stage upon which top-down forces act (Stiling & Rossi 1997). Also, in addition to varying over time, the relative balance of top-down and bottom-up forces may vary spatially due to environmental features of the habitat. In salt marsh communities of the northeast Atlantic coast, for example, herbivorous planthoppers (*Prokelisia* spp.) are more strongly influenced by plant quality and vegetational complexity in the low marsh (subject to greater tidal immersion, which limits spider density), whereas in the high marsh, top-down predation by spiders becomes more important (Denno et al. 2005). Rather than being static relationships, the relative importance of different influences may change as circumstances in the community shift (such as new species invasions, natural enemy introductions, or climate change).

FORCES SETTING PLANT POPULATION DENSITY

Plants are commonly affected by competition for water, light, or nutrients, which is shown by increases in plant size after mixed species stands are thinned. Increases in size often lead to increased reproduction, which in turn may result in increased plant density in later generations (Harper 1977, Solbrig 1981). The stage for competition is the physical habitat, which may provide few or abundant resources. In barren habitats, scarcity of physical resources limits plants directly and there is little plant–plant competition (e.g. Oksanen et al. 1981, 1996). As habitats improve and primary production rises, competition comes into play, but herbivory still may not be important. As productivity increases further, herbivores can also be supported and herbivory may become an important influence on plant density. In very productive natural systems, greater productivity supports enemies of herbivores, which may suppress herbivores so that plant–plant competition again becomes important.

Herbivores that affect plants may be either generalists or specialists, native or exotic. Biological control of weeds generally is concerned with the effects of specialized, invertebrate herbivores that have coevolved with the weed in its native range. Moving plants as seeds to distant locations separates them from attack by such coevolved specialized invertebrate herbivores. In the invaded region, many exotic plants may be suppressed by some combination of limited resources, plant competition, and attacks by local generalist native herbivores, both invertebrates and vertebrates. Indeed, it appears that native, vertebrate, generalist herbivores often prevent exotic plants from becoming pests, but this effect is lost when native vertebrates are replaced by exotic vertebrates, to which the exotic plants (but not the local native ones) may have evolved defenses (Parker et al. 2006).

However, some plants will not be controlled by native generalist herbivores, which, together with escape from specialized natural enemies, will allow these plants to increase in density and become environmental or economic weeds. This is true especially for plants that are toxic or unpalatable or not reachable by vertebrate grazers. Loss of invertebrate natural enemies is illustrated by insects found on the invasive plant purple loosestrife (*L. salicaria*). In North America, the invaded area, only 59 phytophagous species have been recorded on this plant, and none causes appreciable damage (Hight 1990). In contrast in its native range in Europe, this plant typically occurs at low density in association with over 100 species of plant-feeding insects (Batra et al. 1986), which attack all parts of the plant. Although most of these herbivores have limited impacts, some strongly damage the plant. This has been demonstrated by dramatic decreases in plant biomass, seed set, and abundance in North America after these important insects were introduced (Blossey & Schat 1997, Nötzold et al. 1998, Stamm Katovich 1999, Landis et al. 2003, Piper et al. 2004). *Galerucella* spp. chrysomelid beetles, released into stands of purple loosestrife in the USA, defoliated plants, which led to reduced plant size, reduced seed set, and lower plant density over several years. Loosestrife declines were followed by increased growth of other plants in the community, demonstrating the decline of purple loosestrife's competitive edge due to damage from herbivores (Corrigan et al. 1998, Nötzold et al. 1998, Landis et al. 2003, Hunt-Joshi et al. 2004).

Biological control practitioners should assume that competition from other plants is likely to be part of the mechanism by which introduced herbivores reduce the density of many invasive plants, together with stresses from climatic and edaphic factors (Center et al. 2005). Some cases exist in which a single herbivore species has provided complete control of an introduced plant [e.g. *Salvinia molesta* D.S. Mitchell and *Azolla filiculoides* Lamarck by *Cyrtobagous salivinae* Calder and Sands and *Stenopelmus rufinasus* Gyllenhal, respectively (Thomas & Room 1986, Hill 1999)], and in other cases success clearly required the joint action of several herbivore species acting together [e.g. *Sesbania punicea* (Cav.) Benth., controlled by *Trichapion lativentre* (Bèguin-Billecocq), *Rhyssomatus marginatus* Fåhraeus, and *Neodiplogrammus quadrivittatus* (Olivier) acting together].

FORCES SETTING INSECT POPULATION DENSITY

Interspecific competition does exist among herbivorous insects, particularly among species of scales or other Hemiptera that share the same food plant. McClure (1980), for example, demonstrated negative effects of competition between two high-density species of hemlock scales invasive in North America. Intraspecific competition may be more common, particularly for invasive species that occur at high densities (e.g. McClure 1979). However, insect–insect competition does not beneficially affect insect biological control in any manner analogous to what occurs in weed biological control.

Some insect species' populations can be limited by bottom-up effects, particularly if food plants restrict oviposition success. When plants are highly defended or when susceptible structures are present either only briefly or at unpredictable times, insect population growth may be constrained by lack of needed host plants of proper quality, as mentioned above for the arroyo willow gall sawfly.

In contrast to the above forces, many groups of plant-feeding insects are limited specialized parasitoids and predators. This is the reason why insects such as scales, aphids, and mealybugs routinely explode to high densities when they escape their natural enemies by invading new regions. The many instances in which invasive insect populations have been reduced dramatically following introduction of their natural enemies in programs of classical biological control show the importance of population regulation by the upper trophic level for insects. Ash whitefly [*Siphoninus phillyreae* (Halliday)] (Bellows et al. 1992a), winter moth [*Operophtera brumata* (L.)] (Embree 1966), larch sawfly [*Pristiphora erichsonii* (Hartig)] (Ives 1976), and California red scale [*Aonidiella aurantii* (Maskell)] (DeBach et al. 1971), for example, were reduced by four, two, three, and one orders of magnitude, respectively, following natural enemy introductions against these species in locations where these were invasive.

Groups of insects that have immature stages that occur deep within protective media such as soil or plant tissues may, however, be less accessible to natural enemies. Gall insects, root-feeding species, or insects that tunnel in plants (borers or fruit-infesting species) suffer less mortality from natural enemies than exposed feeders (Gross 1991, Cornell & Hawkins 1995, Hawkins et al. 1997) and there are fewer cases in which such pests have been suppressed by introduced natural enemies (Gross 1991).

Suppression of insect populations by generalist natural enemies may also occur. Continued suppression of winter moth in Canada after the introduction of the specialist parasitoid *Cyzenis albicans* (Fallén) is due at least in part to predation by generalist carabid beetles preying on unparasitized winter moth pupae in the soil (Roland 1994). Similarly, two generalist coccinellids and larvae of the sryphid fly *Pseudodorus clavatus* (F.) were the primary agents responsible for limiting the growth of brown citrus aphid [*Toxoptera citricida* (Kirkaldy)] colonies in Puerto Rico and Florida, USA (Michaud 1999, 2003, Michaud & Browning 1999).

There are even a few cases in which introduced pathogens have suppressed insect populations: (1) the fungus *Entomophaga maimaiga* Humber, Shimazu and Soper, which can prevent outbreaks of the gypsy moth [*Lymantria diapar* (L.)] in the northeastern USA (Webb et al. 1999, Gillock & Hain 2001/2002) and (2) an *Oryctes* virus, which suppresses rhinoceros beetle [*Oryctes rhinoceros* (L.)] in coconut palms in western Samoa (Bedford 1986). However, these two examples are unusual and in general pathogens have rarely been shown to restrict insect populations within narrow density bounds. Baculovirus epidemics, for example, often only arise after pests have reached high densities, probably due to poor virus transmission at low densities.

PREDICTIONS ABOUT PESTS BASED ON FOOD WEBS

Native arthropods

In natural (i.e. non-farm, non-forest plantation) systems, we should expect that natural control will act to limit the density of many native herbivorous insects. In such systems the actions of natural enemies are typically complex and take place within food webs with many linkages (Hawkins et al. 1997). However, human actions may lead to the loss of natural control. For example, crops may be bred that lack defenses against insects. Crops grown in large uniform stands, especially perennial crops, may favor pest population increase by eliminating host finding by the pest. In plantations, reduction in associated vegetation (compared to natural forests) may lower the availability of alternative hosts and floral resources needed by natural enemies, causing even some native species to become serious pests. Also, such plantations may consist of introduced plants favorable to a local native herbivore, but not favorable to the local natural enemies. This may result in loss of natural control because local natural enemies are absent or ineffective within the plantation; outbreaks of the native insect *Oxydia trychiata* (Guénee) (Lep.: Geometridae) in exotic pine plantations in Colombia (Bustillo & Drooz 1977) are an example of this process.

In some cases, natural control may act to some degree but be inadequate for human purposes if even low densities of the pest cause unacceptable losses. Pests that directly attack high-value products, such as fruits, fall into this category. In apple production for fresh fruit only about 1% infestation by fruit pests like apple maggot [*Rhagoletis pomonella* (Walsh)] or codling moth [*Cydia pomonella* (L.)] can be tolerated. Natural control does not reach this level of pest suppression. Similarly, insects that vector plant pathogens are rarely suppressed to an acceptable level through natural control because disease transmission by a few infected insects can quickly lead to large economic losses.

Based on these considerations, we can expect the following for native herbivorous arthropods.

1. In natural plant communities most herbivores will not be common enough to damage plants severely.
2. Loss of natural control will be a common consequence of intensive farming or plantation forestry.
3. In farm fields, indirect pests (ones that attack a part of the plant that is not directly harvested and sold) such as mites, leafminers, scales, or mealybugs are more likely to be amenable to biological control than direct pests (species that attack the marketed item) such as fruit feeders.
4. Organic farming methods are more likely to suppress native pests (although not all of them) than introduced pests (which probably will lack effective natural enemies unless they have been the successful target of a previous classical biological control program).
5. Herbivores whose larvae feed where there are few natural enemies (deep in soil or plant tissues) are least likely to be suppressed by natural control, unless other life stages are more exposed.
6. Disruption of natural control by use of pesticides is likely to occur and may be remedied by changing pesticide use patterns.

Exotic plants and arthropods

Most invasive species are not considered pests. In part this is because they attack non-economic plants, or, if they are plants, they remain in disturbed areas and do not invade natural areas. Also, some species simply fail to attain damaging densities due to local generalist natural enemies (**biotic resistance**), in combination with effects of local climate and resource limitations (e.g. Gruner 2005). Such biotic resistance may exert its effect concurrent with the invasion (such that the species is never registered as a pest), or it may happen with a lag, after the invader has increased to pest density. For example, in Australia, populations of the invasive millipede *Ommatoiulus moreletii* (Lucas) declined after an initial period of high density due to the attack by a native rhabditid nematode (McKillup et al. 1988). Because it may not be immediately clear whether high densities associated with a new invader will persist, it is important to allow enough time to pass before initiating a classical biological control program against a new invasive species in order to see if the local natural enemies are capable of suppressing the pest (Michaud 2003). Also, in some cases, invasive species populations are later controlled by the invasion of its own specialized natural enemies from its native range (a process sometimes called **fortuitous biological control**). For example, San José scale [*Quadraspidiotus perniciosus* (Comstock)] has spread around the world on tree fruit planting material from its original native range in the Russian Far East, but one of its specialized parasitoids, *Encarsia perniciosi* (Tower), has spread along with it (Flanders 1960), partly suppressing the invading scale in new regions.

However, if an invasive species has persisted for many years at damaging levels, it will rarely be controlled spontaneously by natural enemies (although this does occasionally occur), and the introduction of specialized natural enemies from the pest's native range will be required. Chapter 10 discusses population theory as a basis for classical biological control concepts.

Chapter 10

THE ROLE OF POPULATION ECOLOGY AND POPULATION MODELS IN BIOLOGICAL CONTROL

JOSEPH ELKINTON

BASIC CONCEPTS

The science of population ecology provides the conceptual and theoretical framework within which the applied discipline of biological control is practiced. Biological control workers use concepts from population ecology to predict the effectiveness of agents considered for release or to evaluate the effectiveness of agents that have been released. Some biological control workers use population models to aid in this process. Here we review the basic concepts of population ecology and consider the various classes of models that have been used.

A fundamental property of the population dynamics of all species is that the number or density of individuals will increase at an ever-increasing rate when conditions are favorable. The simplest example of such growth is illustrated by the replication of single-celled organisms. A bacterium might divide every hour, so that a colony that began with one individual would grow to 2, 4, 8, 16, . . . 2^t, where t is the number of hours or replications. With insects and many other organisms, the rate of replication with each generation is potentially much faster because each individual produces many offspring in each generation instead of two. Mathematically, we refer to this process as **geometric growth** and the general equation is:

$$N_{t+1} = \lambda N_t = N_0 \lambda^{t+1} \qquad (10.1)$$

where N_t and N_{t+1} are the size of the population in generations t and $t + 1$ respectively, λ is the multiplication rate per generation, and N_0 is the initial population size at $t = 0$.

For organisms that reproduce continuously, the same process is expressed with the following **exponential growth** equation (eqn. 10.2; Figure 10.1):

$$\frac{dN}{dt} = rN \quad \text{or} \quad N_t = N_0 e^{rt} \qquad (10.2)$$

where N is the population size or density, dN/dt is the growth rate (the change in density per unit time), N_0 and N_t are defined as above, e is the base of Naperian logarithms, and the constant r is the instantaneous per capita rate of increase. When the birth and death rates

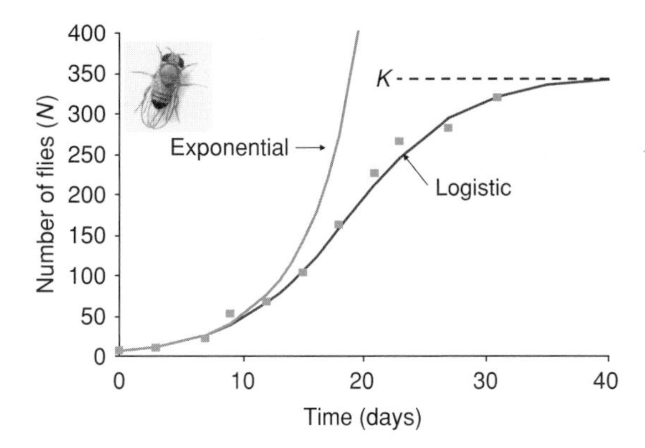

Figure 10.1 Fit of the logistic model to a laboratory population of fruit flies by Pearl (1927) and the estimated exponential rate of growth of this population if the effects of resource limitation were removed.

are equal, $r = 0$ and the population ceases to grow. When the death rate exceeds the birth rate, r is negative and the population declines.

It is obvious that no population can continue growing indefinitely; sooner or later, it will reach a density above which individuals can no longer obtain the resources they need to survive. This density is known as the **carrying capacity** of the environment. For different species in different habitats, the carrying capacity will be determined by competition for particular resources. For desert plants, water is typically the limiting resource. For many animals, food supply determines the carrying capacity. As a population expands toward the carrying capacity, the rate of growth slows down. This process is typically represented by the logistic equation (eqn. 10.3; Figure 10.1), which was first applied to population growth by Verhulst (1838) and later used independently by Pearl and Reed (1920):

$$\frac{dN}{dt} = rN - \frac{rN^2}{K} \qquad (10.3)$$

Here K is the carrying capacity and r and N are defined as above. The first term (rN) represents exponential growth. The effect of the second term (rN^2/K), often called environmental resistance, increases as N becomes large. As N approaches K, the rate of growth (dN/dt) approaches zero.

There are a number of assumptions inherent in the use of the logistic equation to represent population growth. The first of these is that population density will approach and then remain stable about the carrying capacity (K), unless otherwise disturbed. In actuality, most populations fluctuate in density, even populations kept in the laboratory under constant environmental conditions. Another assumption is that the shape of the curve is symmetrical above and below the mid-point. In fact, few population systems, even in the laboratory, follow the exact trajectory predicted by the logistic model. The importance of the logistic equation is its contribution to theoretical ecology. It captures the most basic processes of population dynamics: exponential growth and the effects of factors that limit growth. Variations of the logistic equation have been explored by many individuals; indeed, it is the foundation of a large body of work in theoretical population ecology. Lotka (1925) and Volterra (1926) extended the logistic model to describe competition between species and predator–prey interactions. These models have been widely adapted to model the impact of biological control agents on their host populations, which we discuss below. May (1974, 1976) used a discrete-time version of the logistic to demonstrate the possible existence of deterministic chaos in the dynamics of natural populations. This work suggested that some of the erratic fluctuations in density that characterize most populations was caused not by the influence of random factors such as variable weather conditions but by the inherent mathematical properties of population growth and the limits to growth, including the impact of natural enemies that are embodied in the logistic equation. Other applications of the logistic include models of food webs and interactions between many species in a community. Thus theoretical models of this type have played an important role in population ecology.

Although no one doubts that competition for resources confers an upper limit on the growth of all populations, it seems clear that many populations of animals and plants persist at densities far below any obvious carrying capacity determined by resource availability. Biological control is based on the assumption that natural enemies can reduce the target pest populations to such low densities and that many of the most important pest species are invasive organisms that have escaped the natural enemies that keep them at densities below the carrying capacity in their county of origin. Classical biological control seeks to reunite the pests with these natural enemies.

Density dependence

The low densities characteristic of most species fluctuate within a fairly narrow range of values. For a population to remain at constant density, the birth + immigration rate must equal the death + emigration rate. Each individual must on average replace itself with one surviving offspring. Indeed, for any species to persist over evolutionary time, the average rate of gain must equal the average rate of loss, although these quantities may vary considerably from year to year. Organisms that experience high mortality compensate by producing lots of young. For this fundamental reason, most population ecologists believe that most populations are stabilized by factors that are **density dependent**. Such factors influence birth rates or death rates in a way that varies systematically with density, such that populations converge to densities at which birth rates and death rates are equal and the density is at equilibrium. Such factors act as a negative feedback system that is analogous to regulation of room temperature by a thermostat. If densities rise above the equilibrium value, the death rate exceeds the birth rate and the population returns to equilibrium. If densities fall below the equilibrium value, the birth rate exceeds the death rate and the density rises.

These concepts were introduced to ecology by Howard and Fiske (1911), who were engaged in the importation of parasitoids of gypsy moth, *Lymantria dispar* L., to North America, one of the first large biological control projects. They believed that populations cannot persist for long unless they contain at least one density-dependent factor that causes the average fecundity to balance the average mortality. The term density dependent was coined by Smith (1935), another early proponent of biological control.

Predators, pathogens, and parasitoids often cause mortality to their hosts that is density dependent. The proportion or percentage of the population killed by these factors varies systematically with density. An increase in the proportion dying with increasing density is called **positive density dependence**; a decrease in the proportion dying with increasing density is termed **negative** or **inverse density dependence**. A mortality factor is **density independent** when the proportion killed varies in ways that are unrelated to population density. Many abiotic factors, such as mortality caused by subfreezing temperatures, act in a way that is independent of density. Although many insect population ecologists focus on sources of mortality, density-dependent changes in fecundity may also serve to stabilize densities or lead to changes in population growth as density increases. Certainly, competition for resources is a density-dependent process that will stabilize a population at the carrying capacity if causes of mortality do not intervene at lower densities.

Functional and numerical responses

Density-dependent predation or parasitism may arise from two different sources: the numerical response and the functional response (Solomon 1949). The **numerical response** is an increase in the density or number of predators or parasitoids in response to increasing prey density. The numerical response can arise from increased reproduction or survival of predator or parasitoid offspring induced by increases in prey availability, or it can arise at a local scale, from an aggregative response whereby predators and parasitoids are attracted to sites with high densities of prey.

The **functional response** is an increase in the number of prey taken per predator or parasitoid at increasing prey density. Important contributions to the understanding of the functional response were made by Holling (1959, 1965). In laboratory experiments, Holling presented individual predators with different numbers of prey. He showed that the number of prey consumed over a specified time interval increased with the number of prey available, but at a decreasing rate towards an upper maximum (Figure 10.2a). This effect is caused by an upper limit in the predator's capacity for consumption and by the increasing proportion of time devoted to handling the large number of prey at the expense of time spent searching for prey. Above this limit, further increases in prey density will not cause higher consumption. The proportion of prey consumed plotted against prey density declines steadily (Figure 10.2b), illustrating that the functional response is inherently inversely density dependent. Without a numerical response, predators and parasitoids are unlikely to stabilize a host population. Further work by Holling showed that under some important conditions, the functional response can lead to positive density-dependent predation. Whenever increases in prey density result in some change in the foraging behavior of the predator or parasitoid, such that they forage more efficiently or concentrate their efforts on the particular prey species, the number taken will accelerate with increasing host density (Figure 10.2c), and the proportion taken will increase (Figure 10.2d) over the lowest range of prey densities. Holling termed this a type III functional response in contrast to type II, which is the continuous decline in proportion taken evident

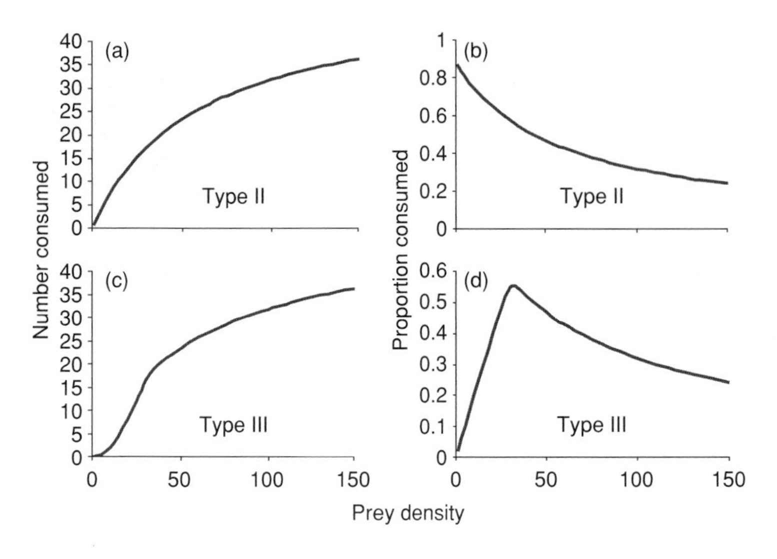

Figure 10.2 Number of prey consumed per predator and the corresponding proportion consumed for a type II (a, b) and type III (c, d) functional response (see text for details). Adapted from Holling 1965.

whenever there is no change in foraging behavior in response to changes in prey density (Figures 10.2a and 10.2b). Holling (1965) demonstrated a type III response for shrews foraging for sawfly pupae. He envisioned type III responses to be characteristic of vertebrate predators, which have a relatively high capacity for learning and behavioral change. However, the type III functional response has subsequently also been demonstrated in many insect predators and parasitoids (e.g. Hassell & Comins 1977).

Specialist and generalist natural enemies

Specialist or **monophagous** natural enemies are those that attack a single host species. **Oligophagous** natural enemies restrict their attacks to a closely related group of species. **Generalist** or **polyphagous** natural enemies attack a wide range of host species. The distinction is important because generalists and specialists typically respond very differently to changes in host density. Specialists are most likely to exhibit a numerical response to changes in density of their prey because they depend on no other food sources and their seasonal development is closely linked with that of their prey. Generalists may exhibit little or no numerical response, because they depend on many types of prey and may shift from one prey to another, depending on which species are available. In fact, it is very common for many natural enemies, especially generalists, to exhibit inverse density dependence, wherein mortality declines as prey density increases and thus cannot stabilize prey densities. Such predators may, nevertheless, play an important role in suppressing prey density even though the resulting densities are unstable or are stabilized by other factors. In annual crop systems, for example, both the plant resource and their insect pests are ephemeral and long-term stability is not particularly important.

Complex density dependence

Many systems exhibit **complex density dependence**, such as when mortality from particular natural enemies may switch from positive to negative density dependence as host density increases. For example, bird predation on forest-dwelling caterpillars may be density dependent at the lowest density but then shifts to inverse density dependence as the densities exceed the capacities of the predators to respond numerically and the functional response approaches the upper limit of prey consumption (Figure 10.3; Mook 1963, Krebs 2005). Under such conditions, the prey densities may escape into an outbreak phase, which is characteristic of a few species. Outbreak populations are typically subject to a different suite of density-dependent factors, such as viral diseases and starvation, which only become major sources of mortality when densities are high. These factors can maintain populations at a high-density equilibrium, but more frequently they cause the collapse of populations back down to a

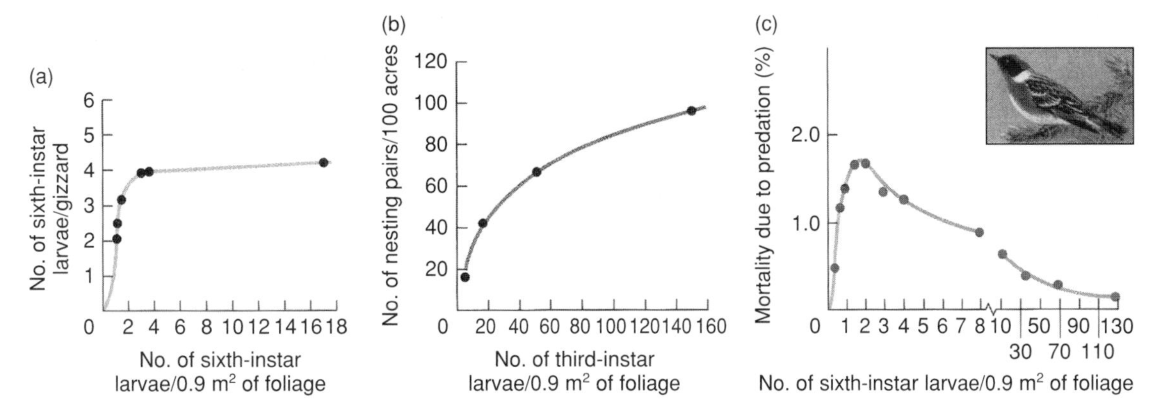

Figure 10.3 Functional response (a) and numerical response (b) of bay-breasted warbler (*Dendroica castanea*) to various densities of spruce budworm *Choristoneura fumiferana* (Clemens) and the combined impact (c) on the percentage of larvae consumed. Data from Mook 1963; reprinted with permission from Krebs (2005).

low-density, endemic phase. In contrast, generalist predators that might consume the majority of prey individuals at low density are likely to consume a tiny fraction of the population at high density, even though they are attacking the same or higher numbers of prey individuals at these high densities.

Southwood and Comins (1976) proposed a "synoptic model" as a general feature of insects that occasionally go into an outbreak phase (Figure 10.4). Earlier expressions of this idea can be found in the writings of Morris (1963) and Campbell (1975). The model is depicted by plotting R_0, the net reproductive rate, against density. In the resulting figure higher mortality produces a **natural enemy ravine** (also called a predator pit; see Chapter 27) at low density at which the population is maintained at equilibrium ($R_0 = 1$) by natural enemies. The natural enemy ravine separates two "ridges," one at high and one at low density, where mortality is lower and population densities increase. At very high density, other mortality factors such as starvation and disease cause the populations to collapse. At the extremes of low density, an **Allee effect** (Allee 1931) comes into play, caused by the failure of individuals to find mates and reproduce. Populations in this range decline inexorably to extinction. Such low densities are infrequent in most natural populations but Allee effects have been widely proposed as one reason why new invaders often go extinct and do not get established and why biological control agents often fail to establish if they are released in inadequate numbers (Hopper & Roush 1993, Liebhold & Bascompte 2003).

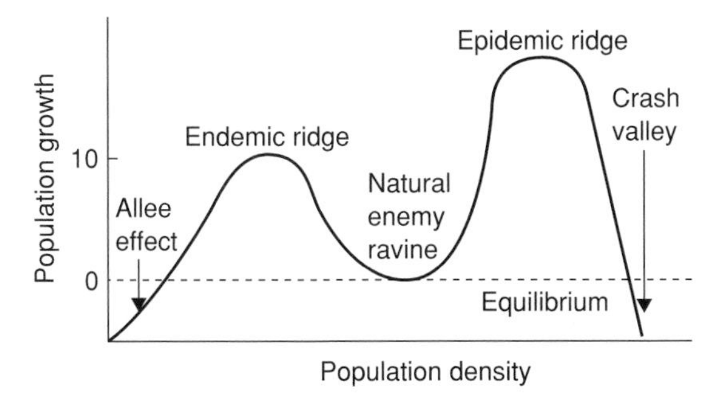

Figure 10.4 The synoptic model of complex density dependence. Adapted from Southwood and Comins (1976); redrawn from Elkinton (2003) with permission from Elsevier.

The model of Southwood and Comins (1976) exemplifies a **multiple-equilibrium system**, meaning that the population is regulated and potentially stabilized at more than one density. Morris (1963) proposed such a model for spruce budworm *Choristoneura fumiferana* (Clemens) and the mechanism he envisioned is illustrated in Figure 10.3c. Predation by generalist predators is positively density dependent at low density but inversely density dependent at higher density. We should note, however, that the maximum mortality caused by bay-breasted warbler, *Dendroica castanea*, in Figure 10.3c is 2% and this bird is but one of a large suite of natural enemies attacking the budworm. Subsequent analysis of the population dynamics of spruce budworm by Royama (1984) specifically rejected the multiple equilibrium model for this system. So it is important to understand that the Southwood and Commins model is a plausible, but not necessarily universal, general description of outbreak systems.

Southwood and Comins's (1976) conceptual model crystallizes our thinking about one attribute of biological control agents that has long been obvious to practitioners of biological control. Some agents are effective at suppressing or maintaining low host densities (in the natural enemy ravine), whereas others are effective only at high density and may be responsible for terminating outbreak densities. In the gypsy moth system for example, low host densities are thought to be maintained primarily by predation by small mammals (Liebhold et al. 2000), whereas outbreak densities that cause defoliation are invariably reduced by a virus disease (Doane 1976). It is typical for baculovirus disease epidemics among insects to be associated with high densities. This arises from the mode of transmission. Insects become infected when they ingest viral particles given off by dead or dying hosts, an event that occurs rarely at low density but with very high frequency at high density. Parasitoids also vary in their effectiveness with host density. Many host/parasitoid systems are maintained indefinitely at low host density by parasitoids that have good searching ability; they can locate hosts and cause high levels of mortality when hosts are very sparse. Agents with this ability are considered outstanding candidates for biological control. Other parasitoids, however, are more effective at outbreak densities. For example, the tachinid parasitoid *Cyzenis albicans* (Fallén) is widely cited as a major biological control success in controlling winter moth *Operophtera brumata* (L.), invasions of Nova Scotia (Embree 1960, 1965) and British Columbia (Roland & Embree 1995), Canada. It causes high mortality and the decline of outbreak densities but much lower parasitism in endemic densities, which are largely maintained by beetle predators (Roland 1994, Roland & Embree 1995). A major reason for this is that *C. albicans* lays microtype eggs on the foliage surface which must be consumed by the host for it to become infected and *C. albicans* is attracted to defoliated trees where it lays its eggs. When winter moth densities decline to lower levels they no longer cause significant defoliation and *C. albicans* is either unable to find appropriate foliage for oviposition or else squanders its eggs on leaves damaged by other species.

All of the early proponents of biological control (Nicholson 1957, DeBach 1964a; Huffaker & Messenger 1964) had no doubt that density dependence was a key feature of successful biological control systems. The degree to which natural enemy attacks need to be density dependent has never been completely resolved. Murdoch et al. (1995) considered several case studies of the most successful biological control projects in history and concluded that almost none of them showed convincing evidence of density dependence. These examples included *C. albicans* controlling winter moth *O. brumata* in Canada and *Aphytis* parasitoids controlling California red scale [*Aonidiella aurantii* (Maskell)]. The latter system has been studied intensively by Murdoch and colleagues (recently summarized by Murdoch et al. 2005), and yet no evidence for density dependence has ever been detected. We revisit both examples below.

The failure in many systems to demonstrate density-dependent mortality caused by successful biological control agents raised the possibility that the persistence of many successful biological control agents was primarily a **metapopulation** process. The term metapopulation, coined by Levins (1969), refers to the idea that natural populations of most species consist of many subpopulations that are partially isolated from one another, and that dispersal of individuals between the subpopulations occurs at a limited rate. It has long been suggested that natural enemies may indeed drive their hosts and consequently themselves, if they are specialists, to extinction within these subpopulations. Provided that this happened asynchronously among subpopulations, then dispersal and emigration of both hosts and natural enemies between subpopulations could recolonize the extinct subpopulations, and the metapopulation as a whole might persist indefinitely. Nicholson and Bailey (1935) invoked this idea to explain the persistence of host/parasitoid systems in the face of the extinctions predicted by their model. Andrewartha and Birch (1954) used this idea to explain the persistence of many species in the absence of density-dependent processes. A variety of recent investigators have attempted to model metapopulation processes (see Hanski 1989 for review) and have shown that they can indeed cause prolonged, but not indefinite, population persistence in the absence of density-dependent stabilizing factors.

Most field studies of biological control agents acquire measures of percentage parasitism or percentage mortality from replicated study plots for one or more host generations. Plots of mortality against host density in these studies reveal **spatial density dependence**, which is variation in mortality with density among populations in different locations. In contrast, theoretical work on models of population systems focus on **temporal density dependence**, which represents variation over host generations at a single location. It is important to understand that spatial density dependence may or may not lead inevitably to temporal density dependence. It all depends on the details of the functional and numerical responses that result in the density dependence of natural enemy attacks. For example, Gould et al. (1990) showed that mortality of gypsy moths caused by the generalist tachinid parasitoid, *Compsilura concinnata* (Meigen), increased dramatically with gypsy moth density in a series of experimental populations created on 1-ha plots with different densities at several locations in the same year (Figure 10.5a). The density-dependent response was evidently a behavioral one by the fly. It was not at all clear the extent to which such responses would occur in studies where density varied temporally instead of spatially. Only in the latter studies would the reproductive response of the fly to changes in gypsy moth density be measured. In the case of *C. concinnata* a numerical response across generations of gypsy moth is highly constrained due to its multivoltine nature which requires that it complete three or four late-summer generations on other hosts. Indeed, a 10-year study of parasitism in naturally occurring populations of gypsy moth (Williams et al. 1992) revealed no evidence of temporal density dependence and far lower levels of parasitism by *C. concinnata* (Figure 10.5b). A major part of the difference between the two studies was one of spatial scale. Outbreaks of gypsy moth tend to occur synchronously on the spatial scale of many square kilometers. The ability of this fly to regulate low densities of gypsy moth is thus unresolved. It is possible that it may indeed play a stabilizing role in suppressing incipient outbreaks that occur on the spatial scale no larger than a few hectares. A variety of other studies have shown that density dependence is detectable at some spatial scales but not others, notably the work of Heads and Lawton (1983) on holly leafminer (*Phytomyza ilicis* Curtis).

Detection of density dependence

Some of the difficulties in detecting density dependence in natural populations are statistical in nature. Population data that have been analyzed for density dependence consist of two fundamental types. In some systems we have data on percentage parasitism or percentage mortality, as well as measures of host density. In other systems the only data available are the numbers of hosts present in successive generations. Data of the latter type are known as **time-series data** and a variety of tests have been proposed to detect the existence of density-dependent processes in them. If populations are regulated by density-dependent factors, then densities should tend to increase at low

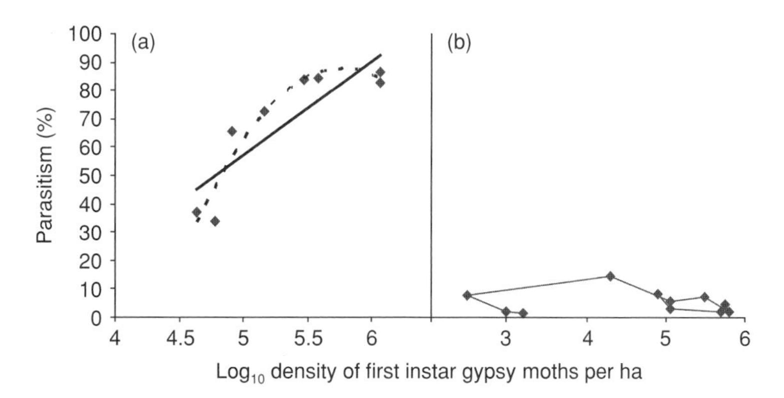

Figure 10.5 (a) Percentage mortality of gypsy moth [*Lymantria dispar* (L.)] caused by the parasitic fly *Compsilura concinnata* (Meigen) for a series of experimental populations with different densities in the same year (from Gould et al. 1990); (b) a time series of percentage mortality caused by *C. concinnata* in a 10-year study of gypsy moths in naturally occurring populations (from Williams et al. 1992). The solid line in (b) connects consecutive generations. Gypsy moth densities were converted from egg masses per hectare in the original to first instars per hectare by assuming each egg mass yielded 300 larvae. Both figures redrawn from Elkinton (2000) with permission from Elsevier.

density and decrease at high density. Thus plots of population change (R), where $R = \log \frac{N_{t+1}}{N_t}$, against log N_t should have a negative slope in regulated populations (e.g. Smith 1961). However, various investigators showed that regression analyses applied to plots of this type were biased and would typically have a negative slope, suggesting positive density dependence even when none existed (Watt 1964, Eberhardt 1970, Royama 1992). Subsequent authors proposed more sophisticated statistical tests that involved bootstrap or related techniques (Pollard et al. 1987, Dennis & Taper 1994). These tests also have statistical limitations. First, they lack statistical power. One needs 20–30 generations of data to reliably find density dependence when it exists (Solow & Steele 1990, Dennis & Taper 1994). Data-sets that long are rare in ecology. Second, the ubiquitous presence of measurement error can bias these tests (Shenk et al. 1998, Freckleton et al. 2006).

The problems of detecting density dependence when the data consist of percentage mortality in particular life stages and host density over successive generations are less severe. However, the data points are not independent from one generation to the next, so that standard least-squares regression is not valid when applied to data of this kind. Vickery (1991) offers a resampling solution similar to that of Pollard et al. (1987) for time-series density data as discussed above.

Time lags in density-dependent responses are common in population systems. For example, it is typical for a predator or parasitoid to respond numerically to changes in density of its host, but this response typically lags behind that of its host by at least one generation. The result is that peak predator density and hence peak mortality of the host will occur after the host has declined dramatically from peak density. Plots of mortality against density may reveal no positive relation between the two, even if it is clear that the predator is regulating its host. Such responses are known as **delayed density dependence**. A different set of techniques has been developed to detect it. The first of these techniques were graphical in nature (Hassell & Huffaker 1969, Varley et al. 1973). Turchin (1990) and Royama (1992) proposed statistical tests based on the analytic methods of Box and Jenkins (1976) that have had wide application in econometrics and the physical sciences. Turchin (1990) used this method to show that time-series data on 10 of 14 species of forest Lepidoptera in Europe had significant delayed density dependence. Other authors have pointed out statistical limitations of these procedures (e.g. Williams & Liebhold 1995). Ultimately, detecting density dependence correctly when it exists or avoiding the false demonstration of density dependence when it does not exist remains a challenge in many population systems.

POPULATION MODELS

Lotka–Volterra models

A major focus of population ecology has been to develop population models to study the effects of natural enemies on their prey. Much insight has been gained from simple mathematical expressions that relate prey density to that of changes in density of specialist predators or parasitoids. One approach, pioneered by A.J. Lotka (1925) and V. Volterra (1926) entailed a simple modification to the logistic equation by adding a term that represents prey consumption on densities of both host and predator:

$$\frac{dN}{dt} = r_1 N - k_1 PN \qquad (10.4)$$

$$\frac{dP}{dt} = -r_2 P + k_2 PN$$

where N and P are the respective densities of host and predator and the rates of population growth are given by dN/dt and dP/dt. In the first equation, the first term on the right represents exponential growth of the host ($r_1 N$) in the absence of the predator, whereas in the second equation, the first term on the right represents exponential decline in the predator ($-r_2 P$) in the absence of the host. The second term on the right in each equation represents the effects of predation, which is determined by the encounter rate of host and predator and is proportional to P^*N. The model predicts a **predator–prey oscillation** (Figure 10.6a) characterized by delayed density dependence. The changes in density of the predator, or parasitoid, lag behind those of its host. The highest rates of attack on the prey, or host, occur at peak predator density, which occur after the host population density has declined. Laboratory studies of predator–prey interaction frequently show such predator–prey oscillations, as in the well known study of Utida (1957) with a parasitoid of the azuki bean weevil, *Callosobruchus chinensis* (L.)

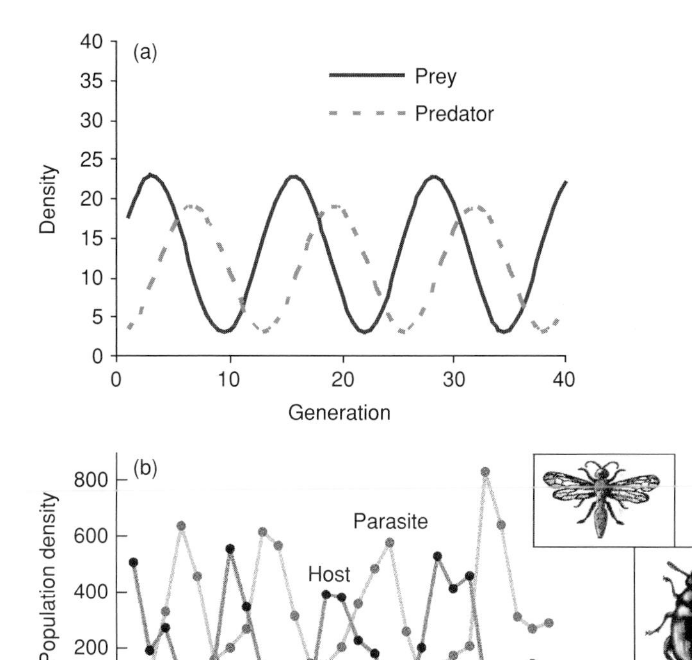

Figure 10.6 (a) Predator–prey oscillation as predicted by a Lotka–Volterra model (after Elkinton 2003); (b) host–parasitoid oscillation of the azuki bean weevil in laboratory culture (after Utida 1957, reprinted with permission from Krebs 2005).

(Figure 10.6b). The implications for biological control were that we should expect host populations to fluctuate following establishment of a biological control agent, rather than necessarily persisting at a constant equilibrium density.

Host–pathogen models

Anderson and May (1980, 1981) developed models for the interaction of hosts and pathogens analogous to those of Lotka and Volterra for predator and prey. In their formulation hosts exist in two states: susceptible S and infected I. They presented a whole series of models. The one below (model G of Anderson & May 1981 following the notation used by Dwyer 1991) is for insect pathogens which have a free-living transmission stage P such as the spores of fungal pathogens or the occlusion bodies of insect baculoviruses.

$$\frac{dS}{dt} = r(S+I) - bS - vSP \qquad (10.5)$$

Rate of change of susceptible individuals = reproduction-non-disease deaths – transmission

$$\frac{dI}{dt} = vSP - (\alpha + b)I \qquad (10.6)$$

Rate of change of infected individuals = transmission – death of infected individuals

$$\frac{dP}{dt} = \lambda I - \mu P - v(S+I)P \qquad (10.7)$$

Rate of change of pathogens in the environment = release from infected individuals – pathogen decay – consumption of pathogens by the host

Here S is the density (or number) of susceptible hosts, I is the density of infected hosts, and P is the density of free-living pathogens outside the host. The model expresses the instantaneous rates of change of these three variables. The per capita rate parameters are as follows: v is the transmission constant (essentially the encounter rate of host and pathogen), r is the reproductive rate of the host, b is the non-pathogen-induced death rate, α is the pathogen-induced death rate, λ is the number of pathogen particles (progenies) produced by a cadaver of an infected host, and μ is the decay rate of the pathogen outside the host. Terms in the original model representing host recovery from infection are omitted because recovery is considered negligible in most forest insect/baculovirus associations.

Anderson and May applied their models to the European larch budmoth, *Zeiraphera diniana* Guenée, an insect that defoliates larch forests periodically in the European Alps. They estimated model parameters from the literature and found that their model predicted oscillations in host density that closely matched those recorded in the field by Auer (1968). Regular oscillations of density have long been reported for a number of forest insects. Anderson and May's pioneering work suggested that pathogens might be responsible for these oscillations. Furthermore, they showed that there exists a host density threshold below which infections can no longer be sustained in the host population and the persistence of the pathogen depends upon its ability to survive in the environment outside of the host. Thus, even though the prevalence of the pathogen declines to zero in the low-density phase of the host, nevertheless the pathogen alone was responsible for the dynamic behavior involving cycles of outbreaks (at least in the model). It is important to note, however, that various studies on larch budmoth have suggested that other factors, including effects of defoliation on host-plant quality (Benz 1974), may cause the population cycles and that some outbreaks of budmoth have not been accompanied by virus epizootics (Baltensweiler & Fischlin 1988). This example illustrates the important point that a model may successfully mimic a particular set of field data but that does not imply that the model is necessarily a correct description of the system.

The Anderson–May models have been extended by many subsequent investigators to other host–pathogen systems and the models we adapted include additional factors and details of host–pathogen biology. For example, Briggs and Godfray (1995) added stage structure (e.g. larvae, pupae, adults) to Anderson–May models and investigated the behavior of several alternative versions, including those in which only the larval stage is susceptible and transmission may or may not occur until after the death of the infected hosts. Their models exhibited complex dynamics, including the occurrence of cycles with durations equal to or less than the developmental time of the host. Briggs and Godfray (1996) explored the behavior of models where the host is regulated at a low-density equilibrium by some other factor but occasionally escapes into an outbreak phase which is regulated by the pathogen. Dwyer et al. (2004) applied a similar model to the gypsy moth system.

Nicholson–Bailey models

A different class of models appropriate for parasitoids and host populations with discrete generations was initiated by Thompson (1924) and Nicholson and Bailey (1935). These models were difference equations, in contrast to the differential equations of the Lotka–Volterra type. The general form of the model expresses host or prey density N in generation $t+1$ as:

$$N_{t+1} = \lambda N_t f(N_t, P_t) \qquad (10.8)$$

where λ is the rate of increase per generation of the host in the absence of parasitism and $f(N_t, P_t)$ is the proportion of hosts escaping parasitism in the previous generation (t). Similarly, the number of parasitoids in the next generation P_{t+1} is given by:

$$P_{t+1} = cN_t[1 - f(N_t, P_t)] \qquad (10.9)$$

where $1 - f(N_t, P_t)$ is the proportion of hosts attacked by parasitoids in generation t and c is the number of surviving parasite progeny produced per parasitized host. The notation $f(N_t, P_t)$ stands for any function of N_t and P_t. Variations in the model involve incorporating different factors into $f(N_t, P_t)$. The simplest version for $f(N_t, P_t)$ proposed originally by Thompson (1924), and in a different form by Nicholson and Bailey (1935), assumes that all hosts are equally likely to be attacked and that parasitoids search at random, such that the proportion that escape is given by the zero term of the Poisson distribution. In other words, parasitoid attacks or ovipositions are distributed at random among available hosts, including those already parasitized, and thus the proportion escaping attacks is given by $f(N_t, P_t) = e^{-aPt}$. The parameter a is a measure of parasitoid search efficiency. This model predicts that hosts and parasitoid will experience density oscillations of ever-increasing amplitude until both go extinct. As indicated above, Nicholson and Bailey invoked metapopulation processes as a possible explanation of the long-term persistence of such systems, but Nicholson (1957) himself was a major proponent of the universal existence of density-dependent stabilizing factors in natural populations. Beginning in the 1960s, M.P. Hassell, R.M. May, and colleagues began an exploration, extending over several decades, of the various factors that would stabilize host–parasitoid interactions in models of this type. These factors included **mutual interference of parasitoids**, **patchiness of hosts or parasitoid attacks**, and **variation in host susceptibility**.

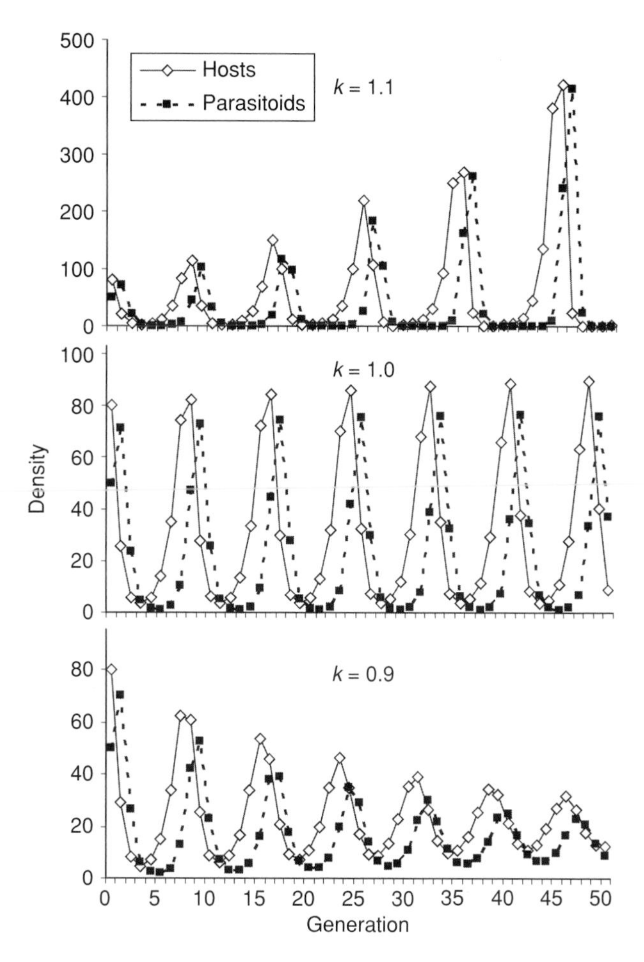

Figure 10.7 Stabilizing effects of parasitoid aggregation independent of host density in the model of May (1978). When aggregation is low ($k > 1.0$) the model predicts ever-increasing oscillations resulting in eventual extinction of both host and parasitoid. When aggregation is high ($k < 1.0$) model converges on a stable equilibrium.

Aggregation of parasitoid attacks may be density dependent – that is, directed toward patches of high host density (Hassell & May 1973) – or it may be unrelated to or independent of patches of high host density (May 1978); both kinds of aggregation will stabilize Nicholson–Bailey models (Pacala & Hassell 1991). Reviews of this literature can be found in Hassell (2000). The model proposed by May (1978) for aggregation of parasitoid attacks independent of host density illustrates this class of models nicely:

$$N_{t+1} = N_t \lambda \left(1 + \frac{aP_t}{k}\right)^{-k} \qquad (10.10)$$

Here host survival $f(N, P) = \left(1 + \frac{aP_t}{k}\right)^{-k}$ is the zero term of the negative binomial distribution, which is used widely to represent aggregated patterns of dispersion. The parameter k is a measure of clumping or aggregation, in this case of parasitoid attacks. May (1978) showed that values of $k < 1$ (high aggregation) produced damped oscillations that converged on an equilibrium density whereas values of $k > 1$ (low aggregation) produced divergent oscillations that resulted in eventual extinction of both host and parasitoid, as in the original Nicholson–Bailey model (Figure 10.7).

Application to biological control of winter moth

Most of the work with Nicholson–Bailey class models has focused on exploring various theoretical ideas such as the general features causing overall stability. However, Hassell (1980) used only a slightly more complex version of eqn. 10.10 to explicitly model one of the most striking biological control successes of all time as mentioned above, namely the control of winter moth, *O. brumata*, in Nova Scotia by the introduction of the tachinid parasitoid *C. albicans* (Embree 1960, 1965). The model was made possible by the detailed life-table studies of winter moth conducted by Varley and Gradwell (1968) and Varley et al. (1973) in England and of outbreak populations of winter moth by Embree (1965) before and after the population was permanently suppressed by the release of *C. albicans*. This life-table information allowed Hassell to partition the overall generational rate of increase λ into components due to fecundity (f) and survival s of successive life stages due to the action of all other mortality factors: $\lambda = fs_1s_2 \ldots s_n$. In some cases, these life-stage survivals were functions of host density. To model the impact of parasitism by *C. albicans*, Hassell used the same negative binomial term as in eqn. 10.10 and he estimated the clumping parameter k from field data on numbers of *C. albicans* larvae per caterpillar in data collected at Varley and Gradwell's field site in England. He found that k increased with host density, so in his model k was not a constant as in May (1978) but a function of host density. Other changes included adding a functional response to the estimates of search efficiency parameter a. For further details see Hassell (1980). The resulting model did a remarkable job of estimating the observed temporal pattern of winter moth decline and parasitism increase in Nova Scotia (Figure 10.8), as documented by Embree (1965, 1966, 1971), as well as the resulting low-density equilibrium and periodic fluctuations in density of winter moth. Follow-up work by Roland (1994) explored the role of pupal predators in maintaining Canadian populations of winter moth at low density in a manner entirely similar to that documented by Varley and Gradwell in England. Roland (1988) hypothesized that the presence of *C. albicans* helped provide a food source for pupal predators in the winter months and enhanced predator populations that then had a greater impact on winter moth pupae in the summer and accounted for the increase in winter moth predation that Roland's analysis suggested helped cause the collapse of high-density winter moth populations in North America. As indicated above, just because a model accurately predicts the overall dynamics of a population system in a given set of data does not imply that the model is necessarily a complete or correct account of how the system works. Alternative models based on other sources of mortality might perform just as well. Thus although model construction is an important component of any attempt to understand the behavior of a system, any model should be greeted with a healthy skepticism.

Desirable attributes of biological control agents

Beddington et al. (1978) attempted to summarize what had been learned from the many theoretical studies of model host/parasitoid systems that could be useful to biological control practitioners. They reviewed the various models in the Nicholson–Bailey family to determine which features result in a low value of the predicted equilibrium density in the presence of the parasitoid expressed as a proportion of the carrying capacity of the host in the absence of the parasitoid. In other words, they focused on the attributes that produced the largest reduction in density in proportion to the density without the parasitoid. They concluded that the most important properties of effective biological control agents are high search efficiency and high aggregative ability (patchiness). These conclusions applied to specialist but not polyphagous parasitoids.

Murdoch and Stewart-Oaten (1989) further analyzed the stabilizing role of aggregation in Nicholson–Bailey models (Hassell & May 1973, May 1978, Pacala & Hassell 1991). They also constructed analogous Lotka–Volterra equations that allowed for continuous redistribution of parasitoids to regions of high host density during the life span of both host and parasitoid. The Nicholson–Bailey models only allow redistribution at the beginning of each host generation, whereas most parasitoids can respond continuously to changes in host density as they occur during the life span of both species. They confirmed the conclusions of Hassell and May (1973) that parasitoid aggregation stabilizes Nicholson–Bailey models by reducing parasitoid efficiency. However, they noted that there is a trade-off between density-dependent parasitoid aggregation and equilibrium host density. Specifically, as density-dependent aggregation (μ) increases, stability increases

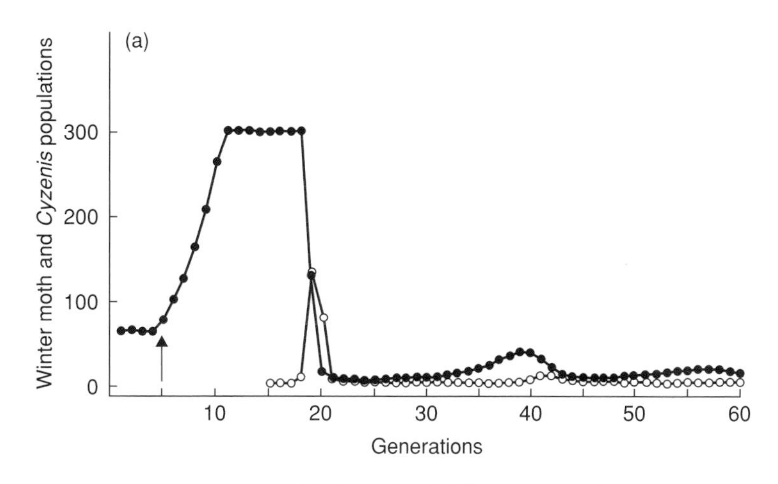

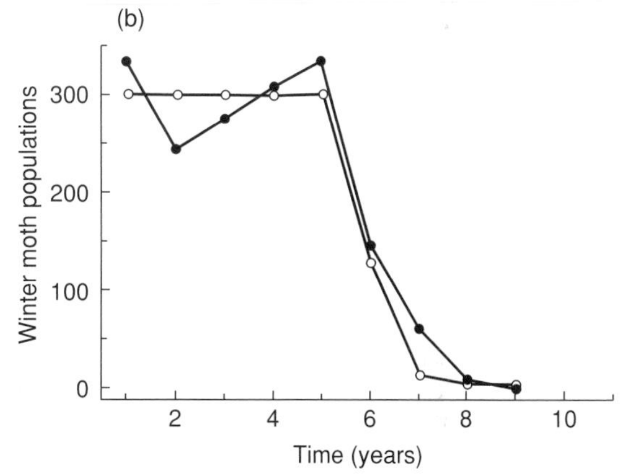

Figure 10.8 Application of modified version of model of May (1978) by Hassell (1980) to predict the dynamics of *Cyzenis albicans* (Fallén) attacks on winter moth in Nova Scotia. (a) Simulated density of winter moth (●) and *C. albicans* (○) populations following introduction as in Nova Scotia. The arrow represents the switch between model parameters characteristic of low-density populations in England to those characteristic of outbreak populations in Nova Scotia; (b) comparison of winter moth densities predicted by the model (○) with those observed by Embree (1966) in Nova Scotia (●) during the period (1958–63) when *C. albicans* was establishing in the population. Reproduced from Hassell (1980) with permission.

(Figure 10.9), but so does the equilibrium density of the host. The implication was that theoretical ecologists should pay more attention to the equilibrium density of the host and less to stability because the former was more important in terms of practical effect. Murdoch et al. (2003) argue that the trade-off between stability and equilibrium host density was a general feature of many related models of host–parasitoid interaction. They showed further that with analogous Lotka–Volterra equations which allow continual re-assortment of parasitoids to patches of high host density the effects of parasitoid aggregation were very different from results with Nicholson–Bailey models. Density-dependent parasitoid aggregation usually destabilizes the system. Density-independent parasitoid aggregation (patchiness, as in May 1978) has no effect on the stability of the system. Several authors (e.g. Godfray & Pacala 1992) criticize the mathematical details of Murdoch and Stewart-Oaten (1989), but follow-up modeling in parasitoid aggregation both within and between generations (Rohani et al. 1994) confirmed these general conclusions. These were: (1) that density-dependent aggregation within a generation destroys the stabilizing influence of between-generation aggregation and thus has little overall effect on stability and (2) that density-independent aggregation within a generation has no effect on the ability of between-generation aggregation to stabilize a system.

Godfray and Waage (1991) used a simple time-lagged continuous-time model (in the Lotka–Volterra class) to help select the best of two encyrtid parasitoids as candidates for introduction to West Africa for control of mango mealy bug, *Rastrococcus invadens* Williams, a major invasive pest of mango and citrus. Their model indicated that the species *Gyranusoidea tebygi* Noyes would achieve a much lower host

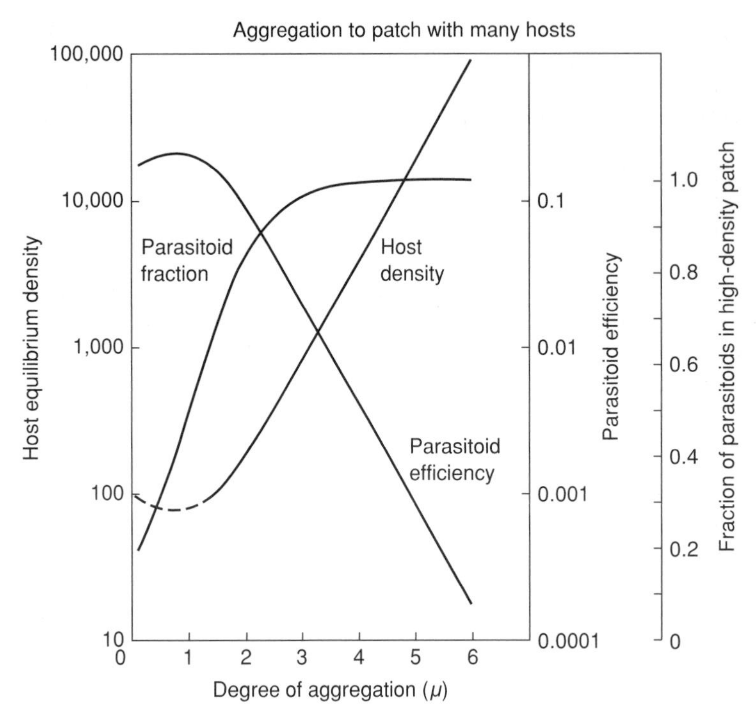

Figure 10.9 Trade-off between stability and equilibrium host density as a function of parasitoid aggregation to patches of high host density in the model of Hassell and May (1973), from Murdoch and Stewart-Oaten (1989).

equilibrium density than the other species (*Anagyrus* sp.) over a range of possible model parameter values (search efficiencies) so they concluded it would be best to focus rearing and release efforts on *G. tebygi*. The model results came after this species had already been released, but they illustrated the possible use of such models to help biological control workers decide which of several promising agents to focus on.

Models of spatial spread

Thus far we have discussed models that treat each population as an isolated unit. We have already pointed out that most species consist of metapopulations and that dispersal is a major component of the dynamics of all populations. One of the first attempts to model the spread of populations was by Skellum (1952), who added terms to a Lotka–Volterra-type model to represent dispersal in addition to population growth. Skellum fit this model to data on the European spread of the muskrat, an invasive species from North America. Subsequent researchers have used modifications of the Skellum model to represent spread of released natural enemies in host populations. For example, Dwyer and Elkinton (1995) combine an Anderson–May-type host–pathogen model with a Skellum-type formulation of dispersal to model the spread of a baculovirus released from a point source into a disease-free population of gypsy moths. Dwyer et al. (1998) extended this model to analyze the spread of the fungal pathogen of gypsy moth that was introduced accidentally to North America and spread across the landscape beginning in 1989 (Hajek et al. 1990a). Many other researchers have attempted to model spread of natural enemies. For example, Harrison (1997) used a dispersal model of the western tussock moth, *Orgyia vetusta* (Boisduval), and its parasitoids in the coastal scrub habitat of California, USA, to show that it is the dispersal of tachinids from patches of high host density that prevents outbreaks of the poorly dispersing tussock moth from expanding.

Complex simulations

The models described so far are relatively simple. They contain a small number of parameters or variables and leave out much of the known biology of the host and its natural enemies. Theoretical ecologists focus on such models because they can be analyzed by a variety of mathematical tools and can be used to address questions of general ecological significance. They hope that the models capture the essential features of the systems that they represent. Applied ecologists, in contrast, are often drawn to more complex models, because they wish to understand the complex interplay of environmental and biotic variables that accounts for the density fluctuations of particular species of interest. With modern computers, there is virtually no limit to the complexity that can be built into such models, but this does not mean that the resulting simulations will necessarily be useful or revealing. Many highly complex models of major pest systems were constructed in the 1970s and 1980s, when high-speed computers first became widely available, but most were abandoned because they were based on hundreds of parameter estimates that far exceeded the knowledge available about the interactions of many natural enemies and the environmental factors that influence the dynamics of these population systems. Typically they failed to accurately predict the behavior of the systems they represented, and they were often too complicated to understand. An example especially familiar to the author was the Gypsy Moth Life System model constructed by the US Forest Service (Sheehan 1989, Sharov & Colbert 1994) with input from many colleagues conducting gypsy moth research. The model predicted the growth of forest stands and gypsy moth populations under the influence of a complex of natural enemies that included 10 introduced parasitoids established in North America from one of the earliest and largest biological control efforts. Despite many decades of research, the impact of these parasitoids on the system is very poorly understood. The model included these parasitoids along with disease agents and native predators, of which there are many species. The problem was that almost nothing is known about the population dynamics of any of these species and much of the model was based on guesswork. Not surprisingly, few gypsy moth workers had much faith in the predictions of the model and it was never evaluated or tested (Sharov 1996). Instead, far more insight into the dynamics of the gypsy moth system came with the development of far simpler models (e.g. Wilder et al. 1994, Dwyer et al. 2004). Highly complex models constructed for many other major pest systems also ended with models that were abandoned. There has been relatively little discussion of the lessons that were learned from these massive modeling efforts (but see Liebhold 1994, Logan 1994, Sharov 1996). As a result of these failures many population ecologists and biological control workers became disillusioned with modeling as a viable approach to understanding biological control.

Applications: *Aphytis* and California red scale

Despite these early failures, efforts to construct detailed population models have continued with various biological control projects. Theoretical ecologists who construct models of biological control systems now understand the need to strike a balance between model complexity and simplicity. The most useful models are those that include only enough complexity to capture the essence of the population system under study. It is often the case that only by constructing models of intermediate complexity can the reasons for success of particular biological control programs and the dynamical behavior of host populations under biological control be understood. We have already illustrated how Hassell (1980) used a very simple model to describe the biological control of winter moth by *C. albicans*. We now illustrate this process with another of the most successful and most heavily researched system in biological control history, the California red scale, *A. aurantii*, on citrus. The parasitoid *Aphytis melinus* DeBach was introduced to California from India in 1957 (DeBach & Sundby 1963). It rapidly displaced the previously established parasitoid *Aphytis lignanensis* Compere, particularly in the more arid inland sites. This system has been studied in detail by W.D. Murdoch and colleagues for several decades. Various models, both simple and complex, have been constructed, culminating in a model (Murdoch et al. 2005) that explains the low-density stability of the host–parasitoid interaction. The system is one of the simplest imaginable systems of biological control, involving only a single specialist parasitoid *A. melinus* and its host. No other predators and parasitoids play a significant role in pest control. The parasitoid maintains what appear to be very stable host densities several orders of magnitude below that of its carrying capacity. It maintains this stability on various spatial scales including that of the individual tree, so stability does not arise as a metapopulation process with locally unstable subpopulations. Despite its simplicity, the cause of the evident stability in the system has proven very difficult to understand. Various studies have failed

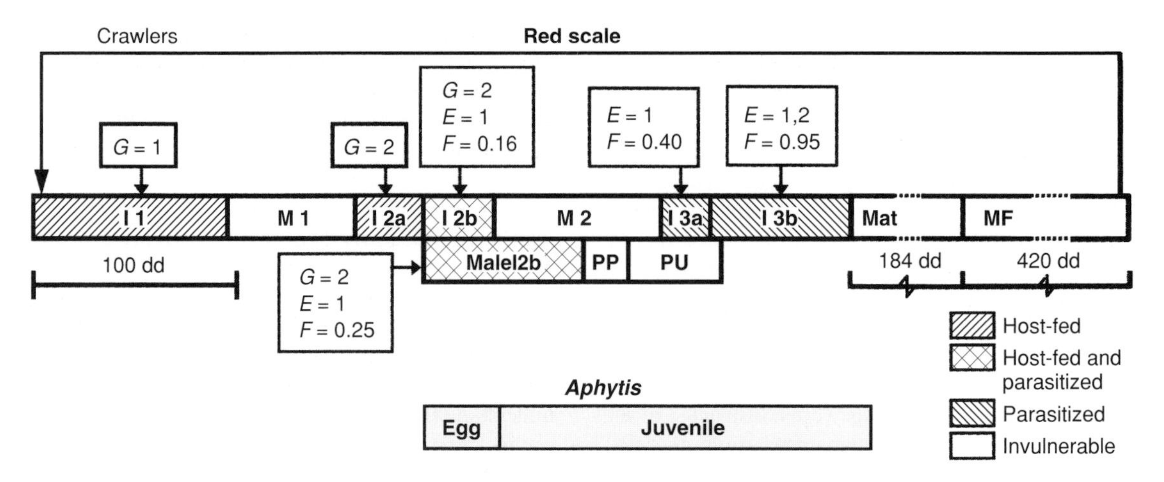

Figure 10.10 Red scale [*Aonidiella aurantii* (Maskell)] and *Aphytis melinus* DeBach life stages represented in the model of Murdoch et al. (2005). Width of stage indicates duration (degree-days, dd). G, gain in egg-equivalents from a meal; E, number of eggs laid; F, fraction that are female; I, instar; M, molt; males and females are distinguishable after I2a. Invulnerable adult females: Mat, mature females; MF, crawler-producers. PP, PU, pre-pupal and pupal males, respectively.

to demonstrate temporal density-dependent parasitism by *A. melinus* in this system at any spatial scale. Simple models invariably predict unstable oscillations of scale density and extinction of the parasitoid. Various studies have examined and eliminated possible causes of this stability. For example, Reeve and Murdoch (1985, 1986) tested and refuted the idea that stability was caused by a partial refuge from parasitism at the center of the tree.

Murdoch et al. (2005) constructed a detailed model that incorporated much of what they had learned from several decades of research on this system and was elaborated from earlier model versions (Murdoch et al. 1985, 1987, 1996). The model is stage-structured, meaning that it represented the development and parasitism of successive life stages of host and parasitoid. These different life stages are shown in Figure 10.10. The model consisted of a series of differential equations representing rates change of the different life stages of both host and parasitoid. Implementation of the model in the computer involved daily updating of each state variable (density of life stages) based on physiological time steps or day-degrees. The model closely predicted the outcome of a manipulative experiment that involved creating outbreaks of scale insects in individual trees and documenting the resulting response of *A. melinus* and its effect on scale density (Figure 10.11). Murdoch et al. (2005) then manipulated model parameters to understand model sensitivity to various factors that accounted for the low-density stability of the system. Chief among these were the existence of a long-lived adult stage that is invulnerable to the parasitoid and a rapid parasitoid development time relative to that of the host.

This example represents one of relatively few detailed simulations that both predict the outcome of particular host–parasitoid manipulations in field experiments and allow researchers to understand the various factors that account for the observed dynamics of the system. The model was made possible because of decades of biological research on various aspects of the system and its relative simplicity involving just a single host and a dominant parasitoid. Few other biological control systems have such simplicity. This is a major reason why model construction and simulations have played a relatively modest role to date in most biological control projects. Nevertheless the example illustrates the potential that models have to elucidate system dynamics.

This chapter does not discuss models of biological control of weeds. The impact of biological control agents on plants is fundamentally different from the impact on arthropods. The agent rarely kills the entire plant and has a variable impact on different plant parts.

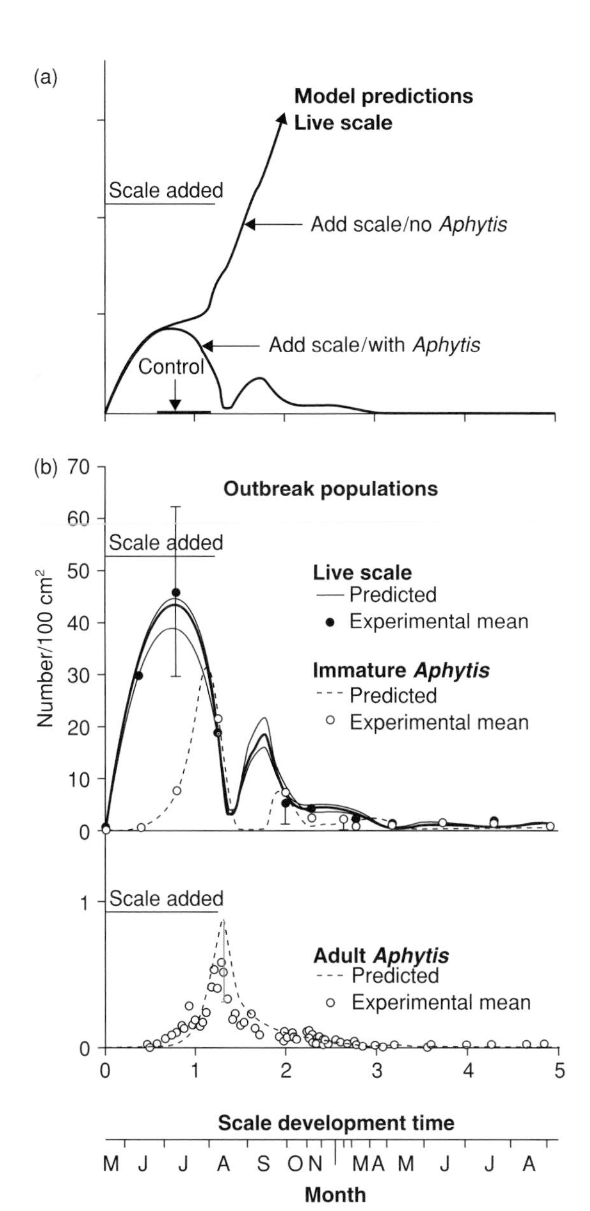

Figure 10.11 Model of interaction of *Aphytis melinus* DeBach with red scale, *Aonidiella aurantii* (Maskell), and results of field augmentation of red scale crawler reported in Murdoch et al. (2005). (a) Density of live scale predicted by the model. Scale eventually increase exponentially in absence of *Aphytis*, but return to control density with *Aphytis* present. (b) Top: thick lines, model prediction; thin lines, highest (*Aphytis* lag shorter) and lowest (immature scale death rates (*opposite*) higher) densities predicted when parameters increased or decreased, individually, by 10%. Vertical lines, range of live scale densities (four experimental trees) on dates when prediction is furthest from observed mean. Bottom: vertical line, range for date closest to the peak when counts were made in all four trees. The model parameters were estimated independently of the experimental data. Figure from Murdoch et al. (2005) with permission.

Thus the models constructed to represent this process are fundamentally different from those used for insect populations. For a recent example of a model-based analysis of a biological control of weeds see the work of Shea et al. (2006) on the management of nodding thistle *Carduus nutans* L. in Australia.

There are many other models that have been constructed for particular biological control systems. It is not possible or necessary to review them here. This purpose of this chapter has been to expose readers to some of the major classes of simple models that form the foundation of more complex models aimed at simulating particular host/natural enemy systems. Examples of models that have been developed for biological control include the work of Gutierrez and colleagues on the biological control of Cassava mealy bug, *Phenacoccus manihoti* Matile-Ferrero, by the encyrtid *Apoanagyrus* (formerly given as *Epidinocarsis*) *lopezi* (De Santis) in Africa (Gutierrez et al. 1988). Another example is a model by Barlow et al. (1996) of the impact of the pathogen *Sphecophaga vesparum vesparum* (Curtis) that has been released in New Zealand as a biological control agent for the introduced yellowjackets *Vespula vulgaris* (L.) and *Vespula germanica* (F.).

In conclusion, all biological control workers share a common heritage of basic concepts in population ecology, such as exponential and logistic population growth, density dependence and functional and numerical responses. These concepts provide a framework that allows biological control scientists to think clearly about the projects that they work on. Construction of simple theoretical models of host–parasitoid and predator–prey interactions have enabled ecologists to understand the basic dynamics we can expect from such systems. Work with these models has produced an understanding of some of the attributes we seek in effective biological control agents. Many of these same attributes, however, such as high search efficiency and rapid population growth relative to the host, arise from common sense and practical experience in biological control. There are a number of examples of successful application of population models to actual biological control systems, but those examples were often constructed long after the biological control agents were released. Successful model development depends on field and laboratory data that may take many years to acquire. Many early attempts to construct complex simulations failed because of too much complexity or because they required data that far exceeded what was available. Nevertheless, the examples cited in this chapter illustrate what is possible and that it is only by constructing population models that we can understand interplay of hosts, natural enemies, and environmental factors that result in the observed dynamics of populations and the success or failure of biological control agents. The examples discussed are relatively few because we typically lack enough data from the field to allow adequate model construction.

Chapter 11

CLASSICAL BIOLOGICAL CONTROL

INTRODUCTION

Introduction of natural enemies as a form of biological control includes: (1) **classical biological control**, in which the targeted pest is an invasive species and the introduced natural enemies are species from its native range and (2) **new-association biological control**, in which there is no previous evolutionary association between the target pest and the introduced natural enemies. Some targets of new-association projects are native pests. Others are invasive species whose origin is unknown or whose associated natural enemies are insufficient to suppress the pest populations.

CLASSICAL BIOLOGICAL CONTROL

Justification for classical biological control

The *ecological justification* for classical biological control is that the high density of many invasive species is caused by their escape from specialized natural enemies found in their native range, which were left behind in the invasion process. Biological control is an applied ecological process that reassociates pests with their missing natural enemies by importing them. Since many invasive species, at high densities, harm the communities they invade (see Chapter 7), their suppression is ecologically beneficial to a broad range of native species. Thus, classical biological control of hemlock woolly adelgid (*Adelges tsugae* Annand) is justified as a means to preserve stands of native trees and the other native species dependent on hemlock stands as habitat. Similarly, biological control of old world climbing fern [*Lygodium microphyllum* (Cav.) R. Br.] is justified because it is the only practical way to prevent the destruction of the tropical hammock tree island communities of the Everglades National Park and adjacent wetland areas in Florida, USA.

The *economic justification* for classical biological control is that it is often more feasible, more efficient, and less environmentally damaging than approaches such as pesticides, augmentative releases of reared natural enemies, and habitat manipulations for suppression of invasive species over large areas. When pesticides are used, for example, control is temporary, merely solving the pest problem for a single cropping season. Consequently, pesticides must be reapplied annually. Classical biological control solves pest problems permanently (see below) and thus avoids issues of pesticide pollution and the yearly costs associated with either augmentative or conservation biological control.

History and success rates

Several thousand introductions of natural enemies (agent-by-country combinations) have been made worldwide for classical biological control of arthropods or weeds since the method's beginnings in the 1880s (Clausen 1978, Luck 1981, Greathead 1986b; Greathead & Greathead 1992, Julien & Griffiths 1998). Such summaries as cited here have been used in meta-analyses to compare rates of establishment and control associated with different groups of natural enemies or pests (Hall & Ehler 1979, Hall et al. 1980, Hokkanen & Pimentel 1984, Julien et al. 1984, Greathead 1986a; Waage 1990, Hoffmann 1996). From these summaries, it had been calculated that *60% of all projects have a positive effect*. In 17% of all projects natural

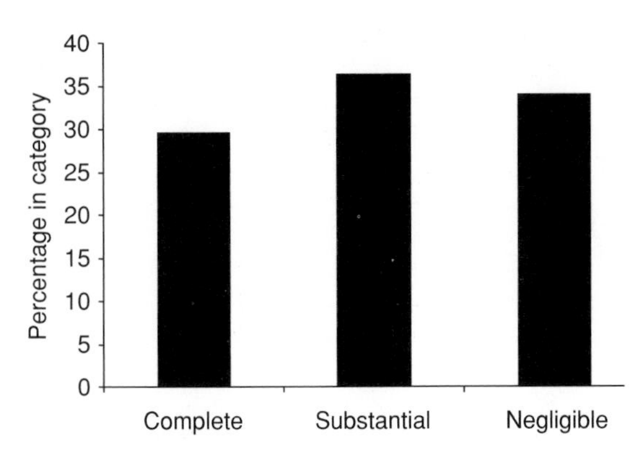

Figure 11.1 Average success of weed biological control projects from South Africa and Hawaii in three different categories. Complete means a high level of control with no need for additional practices; substantial means sufficient control that use of other control tactics is significantly reduced; negligible means that there was no effect. Data from McFadyen (1998), p. 379.

enemy introductions result in complete control (invasive species no longer considered a pest) and in 43% of projects, the pest is substantially or partly controlled, reducing its ecological damage or reducing the amount of pesticide needed to provide complete control. A similar rate of success (66% for complete plus substantial) has been calculated for weed biocontrol projects (McFadyen 1998; Figure 11.1).

In many cases, considerable time must pass before the benefits of introductions can be measured. In New Zealand, 83% of biological weed control projects were found to have achieved partial or full control of the target pest, provided that enough time had passed to see the projects' effects (Fowler et al. 2000).

Key features of classical biological control

Permanency

Unlike augmentative biological control (where the goal is only to protect one season's crop; see Chapters 25 and 26), classical biological control agents are chosen to have the capacity to permanently establish and spread. Such permanency means that classical biological control solutions, once achieved, require no further action in subsequent years. This allows problems to be addressed for which no one is willing to pay repeatedly for control year after year. It requires government support for an extended implementation period (5–10 years for insect targets, 5–20 for weed targets) to select a target pest and then find, screen, release, and evaluate promising natural enemies. Classical biological control projects may be rapid or slow in achieving their goal, but if successful they all have an end point beyond which the control continues indefinitely. The permanency of classical biological control eliminates the pollution that might otherwise follow from annual applications of pesticides.

Spread to ecological limits of agent

Populations of effective classical biological control agents spread naturally to new areas until they reach their ecological limits or encounter a geographic barrier. The braconid *Peristenus digoneutis* Loan was released as a new-association agent against the tarnished plant bug [*Lygus lineolaris* (Palisot de Beauvois)] in eastern Pennsylvania and northern New Jersey (USA) and first recovered in 1984 (Day 1996, Day et al. 1998). It established and spread north and east through New England and New York into Canada, but did not spread south. Continued research eventually showed that this species does not survive the warmer winters south of about 40°N latitude (Day et al. 1998). This climatic barrier sets the southern geographic limit of this species.

Because spread of natural enemies to their ecological limits is normal, to ensure safety of classical biological control projects, researchers must anticipate as accurately as is possible what the eventual range of the natural enemy is likely to be (see Chapter 15 for discussion of the prediction of geographic ranges of invading species). Such predictions have two uses: (1) predicting where, geographically, a particular agent has potential to contribute to suppression of the target pest and (2) identifying what regions the natural enemy will invade and hence what native species the agent is likely to contact after release. This information guides choice of species that should be included in the species test lists during host range testing.

Goolsby et al. (2000a) compared the climate of the native range of the melaleuca gall fly (*Fergusonina turneri* Taylor) in Australia to the climate of south Florida where this insect was being considered for release. They predicted, based on this comparison, that this insect should be able to establish throughout the Florida range of melaleuca. In contrast, Stewart et al. (1999) predicted that the effectiveness of the alligatorweed flea beetle (*Agasicles hygrophila* Selman and Vogt) would be very limited in New Zealand because of less than optimal climate in most of the country.

When trying to predict which native species will have their ranges invaded by a new agent, it should not be assumed that the natural enemy's range will be exactly the same as that of the target pest. Natural enemies that are able to exploit other hosts may eventually have ranges that are larger than those of their target pest. For example, *Cactoblastis cactorum* (Bergroth), a pyralid moth introduced for control of a few weedy or invasive species of *Opuntia* cacti in the Caribbean will very likely have a final range that encompasses all of the Caribbean, the coastal strip from Florida through Texas, USA, and much of Mexico (Zimmermann et al. 2001). Similarly, the weevil *Rhinocyllus conicus* (Frölich), introduced for control of nodding thistle (*Carduus nutans* L.) and two other species in the same genus, invaded areas of the USA, such as the Nebraska sandhills, where native thistles were present but the invasive pests were not (Louda et al. 2005).

Potential for high level of control

Pest suppression that can be achieved through classical biological control of pest arthropods ranges from little (<20%) to modest (50%) to spectacular (99.99%). In some cases, the consequences of a project were recorded simply as an increase in yield of a crop on which a pest had been suppressed. Control of cottony cushion scale in California, USA, in the 1880s, for example, allowed yield of market-quality citrus to increase 200% after pest suppression (DeBach 1964a). Control of olive scale, *Parlatoria oleae* (Colvée), in California allowed the cull rate to drop from 43% before the project (1956–8) to 0.3% by 1966 after two effective parasitoids had been established (DeBach et al. 1976). Control of cassava green mite [*Mononychellus tanajoa* (Bondar)] in Africa in the 1990s by the introduced phytoseiid *Typhlodromalus aripo* De Leon increased root production by a third (J.S. Yaninek, personal communication; Echendu & Hanna 2000). In other cases, reductions in pest density of over 99% have been measured directly for various scales, mealybugs, whiteflies, Lepidoptera, and other pests (van den Bosch et al. 1970, Beddington et al. 1978, Summy et al. 1983, Bellows et al. 1992a; Bellows 1993; see Figure 11.2 for other examples).

Determining efficacy of weed biological control agents is more complex than for insect biological control projects. There is no simple method that can be used since agents may variously affect plant number, biomass, or reproduction. Amounts of tissue removed or damaged is not necessarily a good measure because some tissues are vital to the plant whereas others are not. For example, a plant may be able to sustain the loss of large quantities of leaf tissue, but a small amount of damage to meristematic tissue can be lethal. Damage in some cases may cause very little harm to the parent plant but result in the nearly complete cessation of seed production. Control may be complete in one area but poor in another. These points are well illustrated by outcomes of biological control projects directed against waterhyacinth in Africa. In the last 10 years, the weevils *Neochetina eichhorniae* Warner and *Neochetina bruchi* Hustache have provided control levels from 5 to 100% in West Africa. In East Africa, the biomasss of a 15,000-ha infestation on Lake Victoria was reduced by 70–80% in 3–4 years. In many parts of South Africa, these weevils have not been effective.

Speed of impact on pests

Natural enemies are introduced in small numbers relative to the pest, so reproduction of the natural enemy through a series of generations (commonly between 6 and ten) is nearly always necessary before the pest's density begins to decline. Pest declines often start first at release sites. Regional declines in the pest, logically, take longer. The percentage of damaging infestations by the Asian armored scale *Unaspis euonymi* (Comstock) on euonymus plants fell dramatically at release sites within 1–2 years (Van Driesche et al. 1998b) after release of the coccinellid *Chilocorus kuvanae* (Silvestri) in Massachusetts, USA. Statewide declines took longer. Releases were begun in 1988; by 1994, the predator had dispersed across the whole state and occurred on 26% of the plants in the landscape with heavy scale infestation, but the percentage of plants with heavy scale infestations had not changed from pre-release surveys. When the state was resurveyed in 2002, the predator was found on 43% of heavily infested shrubs and had caused a 35% decline in the proportion of plants with damaging scale populations (Van Driesche & Nunn 2003).

The time required for visible impacts of herbivorous insects on target plant populations varies from as little as a year to decades. Control has been most rapid for floating ferns in the genera *Salvinia* and *Azolla* (Room et al. 1981, Room 1990, Hill & Cilliers 1999, Cilliers et al.

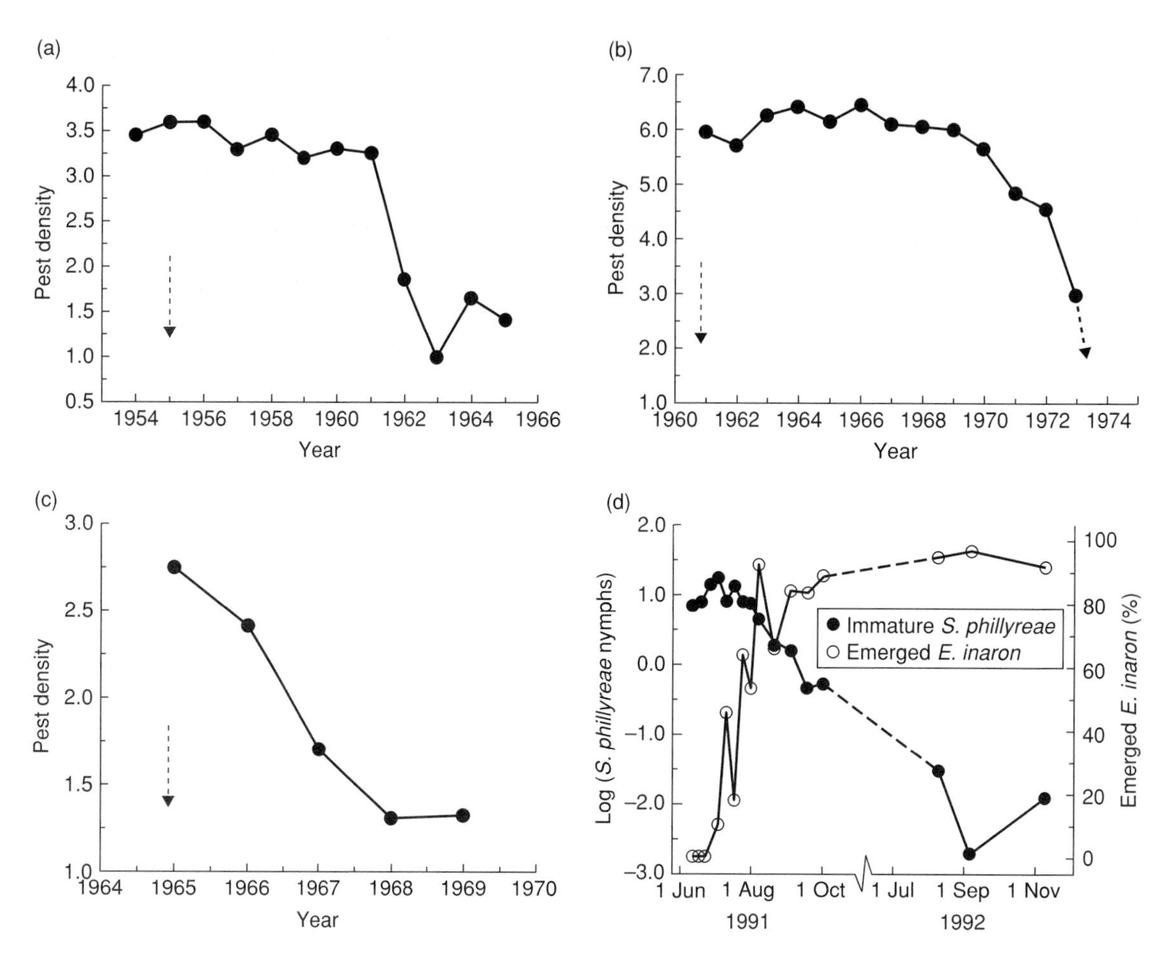

Figure 11.2 Classical biological control can reduce densities of invasive insects by one to four orders of magnitude (99–99.999% control), as shown by changes in pest density for (a) winter moth *Operophtera brumata* (L.) by *Cyzenis albicans* (Fallén) (after Embree 1966); (b) larch sawfly [*Pristiphora erichsonii* (Hartig)] by *Olesicampe benefactor* Hinz (after Ives 1976); (c) California red scale [*Aonidiella aurantii* (Maskell)] by *Aphytis melinus* DeBach (after DeBach et al. 1971); and (d) ash whitefly [*Siphoninus phillyreae* (Haliday)] by *Encarsia inaron* (Walter) (after Bellows et al. 1992a). Reprinted from Van Driesche and Bellows (1996) with permission from Kluwer; original sources are in the references cited.

2003, McConnachie et al. 2004), which have rapid biomass turnover, simple structural geometry, and vulnerable life histories (often reproducing only by vegetative means). After the weevil *Stenopelmus rufinasus* Gyllenhal was introduced into South Africa for control of *Azolla filiculoides* Lamarck, some weed mats disappeared within 2 months and most were gone within a year (Hill 1999). Likewise, releases in Texas, USA of the weevil *Cyrtobabous salviniae* Calder and Sands on giant salvinia (*Salvinia molesta* D.S. Mitchell) reduced coverage and biomass by as much as 99% within 21 months (P.W. Tipping, personal communication). For long-lived woody plants, control may require more time, even after the insect populations have had time to increase to damaging levels. The chrysomelid *Diorhabda elongata deserticola* Chen introduced in 1999 to control saltcedar shrubs (*Tamarix ramosissima* Ledeb., *Tamarix chinensis* Lour., *Tamarix parviflora* DC, *Tamarix canariensis* Willd., and hybrids) in riparian habitats of the North American southwest took about 3 years to build up to population levels needed to cause local defoliation (DeLoach et al. 2004), but because trees have abundant carbohydrate reserves they are able to regrow from roots and dormant buds. Plant death will require repetition of the defoliation process, with smaller, less competitive plants returning after each cycle.

Safety compared to chemical control

In North America, in the 1960s and 1970s, pesticides caused widespread significant harm (see Chapter 21). Widespread use of chlorinated hydrocarbons such as DDT, chlordane, dieldrin, and heptachlor produced toxic residues in food chains that poisoned many kinds of wildlife, directly killing songbirds and destroying eagle, hawk, falcon, pelican, and heron populations by thinning the shells of their eggs, leaving parents with no offspring. As environmentally damaging chlorinated compounds were phased out in favor of less residual but more acutely toxic organophosphates and carbamates, potential for human poisonings among applicators and farm workers (Figure 11.3) increased alarmingly. Against this background, the advantages of biological control were obvious: use of natural enemies to suppress pests instead of pesticides spares wildlife (birds and mammals), reduces the toll of injuries from accidental poisoning, and reduces pesticide residues in food. Consequently, biological control in all its forms was embraced in the 1960s as an environmentally friendly, "green" technology.

In the 1980s, Howarth (1983, 1991) shattered this uncritical consensus by pointing out cases in which classical biological control introductions had apparently damaged native non-target species (see Chapter 16). Since then, many authors have expanded knowledge of both the potential for and actual past occurrence of such impacts (Clarke et al. 1984, Turner et al. 1987, Delfosse 1990, Diehl & McEvoy 1990, Miller 1990, Simberloff & Stiling 1996, Duan et al. 1997, Louda et al. 1997, Pemberton 2000, Blossey et al. 2001a, Munro & Henderson 2002, Pearson & Callaway 2003, Henneman & Memmott 2004, Ortega et al. 2004, Johnson et al. 2005). Collectively, such new information and the discussions it stimulated led to a better understanding of potential risks of biological control of arthropods. This happened concurrently with a general increase in the level of care biological control practitioners brought to their projects (see Chapter 17 on host range estimation). Extensive focus on past damage, however, today has the potential to obscure the great benefit of biological control and to overlook its ability to incorporate higher standards of environmental safety into its procedures. Compared to the toxic and polluting effects of many pesticides, pest suppression through introduction of natural enemies has an excellent safety record. Although some early problems caused by pesticides have been eliminated, new ones may have arisen, such as disruption of normal embryogenesis in amphibians (Figure 11.4).

Figure 11.3 Farm workers, such as these people harvesting strawberries, have significant potential exposure to pesticide residues while working in the crop. Photograph courtesy of Helen Vegal.

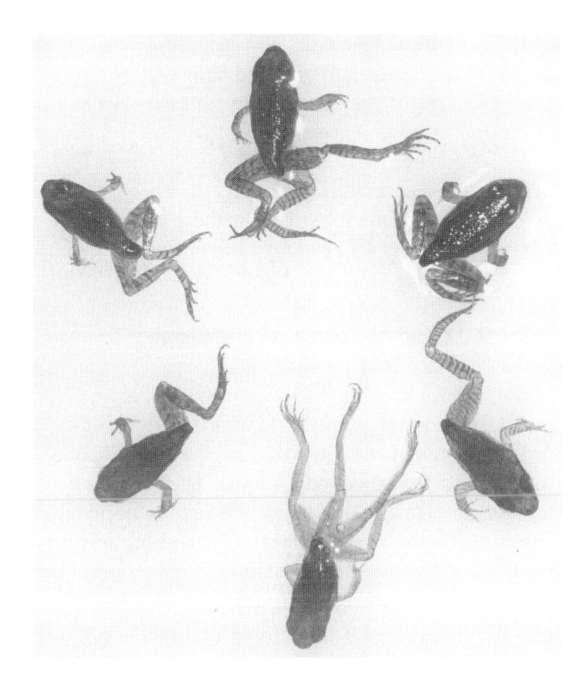

Figure 11.4 Deformations in frogs are suspected, in part, of being caused by exposure to residues of crop herbicides. Photograph courtesy of Joseph Kiesecker.

Classical biological control clearly poses no threats to people, domestic animals, or most plants (Pemberton 2000). If care is taken to introduce only specialized natural enemies, effects on non-target native insects or plants can be largely forecast and will be limited to closely related species. Use of natural enemies is especially sound as a means of pest control in natural areas, where there is no one to pay for other more costly methods of pest suppression, and in crop fields in resource-poor countries where farmers cannot afford to buy pesticides or other tools to control pests.

Selected case histories of classical biological control

There are hundreds of species to which the technique of natural enemy introduction has been applied successfully, including arthropods, aquatic plants, terrestrial plants, and even a few vertebrates. We present here the details of several insect and weed control programs to illustrate: (1) proper selection and identification of the target pest, (2) importance of adequate host-range estimation, (3) non-target studies, (4) the complexity of some projects, and (5) necessary follow-up evaluations. Further information on these issues is developed in the subsequent section that discusses the sequential steps typical of most classical biological control projects.

Spotted knapweed in western North America

Spotted knapweed, *Centaurea stoebe* ssp. *micranthos* (= *C. maculosa* L.), is one of a complex of Eurasian knapweed species that have invaded North American rangelands (Figure 11.5). It is a biennial or short-lived perennial that spreads mainly by seed (Watson & Renney 1974, Powell et al. 1994). It infests about 1.2 million hectares in the USA (Story et al. 2004a) and Canada. The weed, being drought resistant, poor forage, and allelopathic, displaces most other plants and often forms monocultures (Watson & Renney 1974, Harris & Myers 1984, Bais et al. 2003). It thus alters ecosystem structure and function by diminishing biodiversity, allowing increased soil erosion, and reducing forage for wildlife and livestock (Story et al. 2004a).

A biological control project was initiated in 1961 with surveys in Europe (Schroeder 1985, Müller-Schärer & Schroeder 1993), which detected 34 insect species, two mites, and two fungi on *C. stoebe* ssp. *maculosa*, 20 of which were candidates for use as biological control agents (Schroeder 1985). In the early 1970s, the gall-forming tephritids *Urophora affinis* Frauenfeld and *Urophora quadrifasciata* (Meigen) were released in Canada (Harris 1980a, Harris & Myers 1984, Müller-Schärer & Schroeder 1993), the latter on *Centaurea diffusa* Lamark. An additional 11 insect species were released by 1992. All established initially, but one (*Pterolonche inspersa* Staudinger) later disappeared (Story et al. 2004b).

Figure 11.5 A dense stand of spotted knapweed, *Centaurea stoebe* ssp. *micranthos* (= *C. maculosa* L.), which has greatly reduced both the economic and ecological value of many North American grasslands. Photograph courtesy of Jim Story.

Figure 11.6 The tephritid fly *Urophora affinis* Frauenfeld is a biological control agent introduced into North America against spotted knapweed, *Centaurea stoebe* ssp. *micranthos* (= *C. maculosa* L.). Photograph courtesy of Robert D. Richard, www.Forestryimages.org.

In 1973, *U. affinis* (Figure 11.6) was released in Montana, USA, where it established and dispersed quickly (Story & Anderson 1978). *Urophora quadrifasciata* was not released in the USA but dispersed from Canadian sources into Montana where it was found in

Figure 11.7 A galled seed head of spotted knapweed, *Centaurea stoebe* ssp. *micranthos* (= *C. maculosa* L.) induced by *Urophora* spp. of tephritid flies. Photograph courtesy of Jim Story.

Figure 11.8 The mouse *Peromyscus maniculatus* Wagner is able to use larvae of *Urophora* spp. tephritid flies in knapweed [*Centaurea stoebe* ssp. *micranthos* (= *C. maculosa* L.)] galls as a supplemental food. Photograph courtesy of Milo Burcham.

1981 (Story 1985). Both tephritids lay eggs in open knapweed inflorescences. Larvae of both flies induce galls (Figure 11.7) in which they feed (Harris 1980a, 1996). These flies reduced seed production of *C. stoebe* ssp. *maculosa* up to 95%, but the weed continued to spread and so the need for additional agents was recognized (Harris 1980b, Maddox 1982, Müller-Schärer & Schroeder 1993).

Pearson et al. (2000) asserted that the gall flies, which were extremely abundant on spotted knapweed in Montana, had little effect on the weed but were an important supplemental food for deer mice (*Peromyscus maniculatus* Wagner) (Figure 11.8). This food allowed mice to reproduce earlier and develop larger populations at sites with dense knapweed infestations, affecting the local food web (Figure 11.9). Pearson et al. (2000) failed, however, to discriminate between the effects of the weed itself and the effects of the gall fly as the ultimate cause of the mouse concentrations inasmuch as neither their study nor that of Pearson and Callaway (2006) compared mice densities in knapweed stands of similar density with and without gall flies (Smith 2006). Rather, their study compared mice numbers between sites with high and low knapweed infestations (Pearson & Callaway 2006). Pearson and Callaway (2003) speculated that elevated mouse densities posed a risk to human health by increasing environmental levels of the hanta virus, but they provided no evidence of increased disease in the local human population (Smith 2006). Pearson and Callaway (2006) have, however, shown that densities

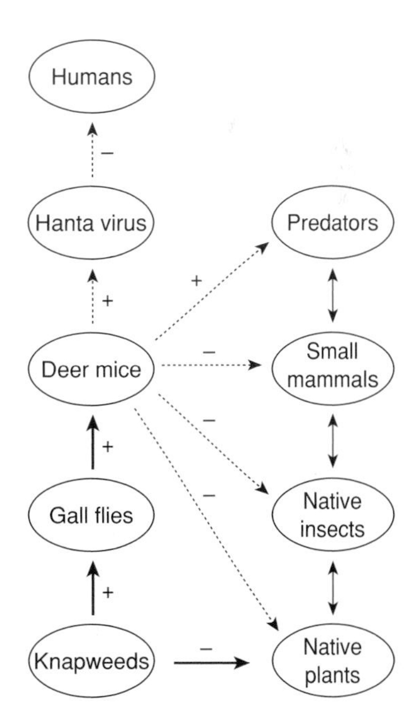

Figure 11.9 A food web showing the directions of interactions that link the knapweed gall fly to human hanta virus levels via a stimulation of mouse reproduction when mice are provided with galled knapweed heads as food. Reprinted from Pearson and Callaway (2003).

of mice infected with hanta virus are approximately twice as great at sites with high knapweed infestations (with *Urophora* galls) as sites with low knapweed densities.

These studies illustrate the possibility of food-web effects arising from a biological control agent that increases its own density but fails to reduce that of its target weed. These studies, however, do not place the observed mouse densities in the larger context of the density variation normally found in deer mice populations among years and habitats in the region. In that context, is a doubling of mice density unusual? Nonetheless, this outcome has heightened the awareness of the need to consider the odds of effectiveness of a biological control candidate during its screening process (McClay & Balciunas 2005) and possible consequences if control is not achieved.

However, changes are still occurring in this system. Additional agents [especially the root weevil *Cyphocleonus achates* (Fåhraeus)] (Figure 11.10) are proving effective. Story et al. (2006) documented declines in spotted knapweed densities of 77 and 99% at two sites in western Montana over an 11-year period associated with the root weevil *C. achates* (see also Corn et al. 2006). Such lowered knapweed densities would not support elevated deer mice populations (Pearson & Callaway 2006). It is significant to note that plant densities were only reduced with the combined attack of gall flies, reducing seed production, and root-feeding weevils, causing higher rates of plant death (J.M. Story, personal communication). This contradicts the view of Pearson et al. (2000) that the gall flies play no role in knapweed control (i.e. they are an ineffective agent) but rather suggests that the gall fly is a necessary but not sufficient agent by itself. The efficacy of the combination of the gall fly and root weevil illustrates the concept of cumulative impact by multiple biological weed control agents.

Figure 11.10 The weevil *Cyphocleonus achates* (Fabricius), a root-feeding biological control agent released against spotted knapweed [*Centaurea stoebe* ssp. *micranthos* (= *C. maculosa* L.)]. Photograph courtesy of Jim Story.

Acacia saligna in South Africa

This tree is a major environmental pest in Cape fynbos, a hot spot of plant diversity the size of Portugal that contains 8600 plant species, 5600 of which are endemic (Cowling & Richardson 1995). All of tropical Africa, an area 235 times larger, contains only 3.5 times as many species (Cowling & Richardson 1995). Alien plants have invaded the fynbos and threaten many endemic plants and their associated insects (Richardson et al. 1992). The fynbos, being nutrient-poor, is especially vulnerable to nitrogen-fixing plants, such as *Acacia* species, that alter nutrient cycling (Yelenik et al. 2004). Port Jackson willow [*Acacia saligna* (Labill.) Wendl.], is the most threatening invader in the region, forming dense thickets and displacing native species (Morris 1999, Henderson 2001).

Acacia saligna is native to southwestern Australia (Henderson 2001). While conducting surveys there for biological control agents, Stefan Neser noticed that the plant was severely attacked by a gall-forming rust fungus, *Uromycladium tepperianum* (Sacc.) McAlp (Morris 1991). Although this fungus was known from several species of *Acacia* (van den Berg 1977), the possibility of host-specific genotypes was considered (Morris 1991). Their existence was ultimately confirmed by Morris (1987), who found that 20 *Acacia* species and four *Albizia* species were unaffected when inoculated with *U. tepperianum* teliospores from *A. saligna*. This narrow host range enabled the fungus to be released in South Africa in 1987. Although effects were expected to be slow (Morris 1987), the rust spread rapidly, producing up to 5000 galls on large trees (Morris 1999). As few as between one and five galls killed seedlings and saplings; on older trees several hundred galls were required (Morris 1999). It was thought that the galling did not directly kill the trees but predisposed them to other stresses, particularly drought, ultimately reducing densities by 90–95% (Morris 1999).

This highly successful project demonstrated the feasibility of using biological means to control large woody trees and the need for in-depth study of candidate agents. It also reconfirmed the value of plant pathogens for classical biological control and provided another example of the usefulness of gall-forming organisms. As with many other projects, it also showed the value of sublethal effects that stress plants, causing them to die from other causes. Finally it demonstrates the absolutely critical role of biological control in conservation of native plants threatened by highly competitive invasive species. No other form of control had any serious potential for protecting the fynbos plant region.

Pink hibiscus mealybug in the Caribbean

This project, conducted in the Caribbean in the 1990s, illustrates the continuing need for biological control in its most traditional form to control pest Hemiptera as they spread to new areas on plants in international trade. Pink hibiscus mealybug (*Maconellicoccus hirsutus* Green) invaded Grenada in 1993 (Kairo et al. 2000, Michaud 2003), infesting shoots, flowers, and fruits of many plants, most importantly ornamental hibiscus (*Hibiscus rosa-sinensis* L.), soursop (*Annona muricata* L.), cotton (*Gossypium hirsutum* L.), cocoa (*Theobroma cacao* L.), and citrus (*Citrus* spp.) (Cock 2003, Gautam 2003). The mealybug reached high densities and spread rapidly to other islands and adjacent mainland areas. Mealybugs caused immediate losses to the tourist industry by reducing the beauty of ornamental plants around hotels. Losses also occurred in several major crops and inter-island trade was affected through in-effective quarantines enacted to control spread. Grenada, and Trinidad and Tobago, suffered an estimated $US10–18 million in losses the first year (Michaud 2003). In contrast, Puerto Rico, where effective parasitoids were introduced almost immediately after an invasive population was discovered, suffered no economic losses.

Control of this pest was greatly facilitated by previous successful control of the same species in Egypt in the 1920s (Clausen 1978), where it had invaded, presumably from India. Natural enemies previously introduced to Egypt – the coccinellid *Cryptolaemus montrouzieri* Mulsant and the encyrtid parasitoid *Anagyrus kamali* Moursi – were also released in the Caribbean, as well as another encyrtid parasitoid, *Gyranusoidea indica* Shafee, Alam and Agarwal. The coccinellid had little or no effect, even though it did establish. Costly augmentative releases of *C. montrouzieri* were somewhat useful as a stop-gap measure to reduce extremely high populations on limited areas of high-value plants but could not provide areawide control. Control by the parasitoids, especially *A. kamali*, however, was rapid and complete (Kairo et al. 2000).

A new aspect associated with this project that was not part of the earlier work in Egypt was a concern for possible effects on non-target mealybugs. To gauge the specificity of *A. kamali*, nine mealybugs were assessed (Sagarra et al. 2001). Of these, *A. kamali* oviposited in two species, but failed to develop. This parasitoid was therefore judged to be relatively specific and beneficial. In contrast, *C. montrouzieri* is a known generalist mealybug predator. The failure of the Grenadian government to appreciate the difference between these two agents illustrates that the desire to protect non-target insects is not widespread. However, once the effectiveness of the parasitoids had been demonstrated, use in new areas was limited to them.

This project clearly shows that familiar pest groups continue to invade new regions, creating new important problems. Classical biological control has the ability to respond rapidly to such invasions, provided institutions with the required scientific staffing and funding are maintained by governments and have the legal mandate to intervene at the earliest phase of such invasions, when projects can be most effective in preventing damage.

Chestnut gall wasp in Japan

The chestnut gall wasp (*Dryocosmus kuriphilus* Yasumatsu) (Figure 11.11) was introduced to Japan from China during World War II, and became a critical pest

Figure 11.11 The invasive chestnut gall wasp (*Dryocosmus kuriphilus* Yasumatsu) is a severe pest of chestnut production in Japan when not suppressed by biological control agents. Photograph courtesy of Seiichi Moriya.

Figure 11.12 Galls of *Dryocosmus kuriphilus* Yasumatsu on chestnut in Japan. Photograph courtesy of Seiichi Moriya.

in chestnut orchards by galling buds (Figure 11.12), which reduces nut formation. This pest was virtually uncontrollable with pesticides because gall wasp larvae are protected inside plant tissues. A resistant variety of the chestnut was introduced into cultivation and provided control in the 1950s, but by 1960, galls began to be found on this variety as well, suggesting that a new form of the gall wasp had developed. Approximately 40% of the shoots were galled in the early 1980s, before biological control was attempted (Figure 11.13; Moriya et al. 2003). A biological control project was started after the pest was discovered in China and a new torymid parasitoid, *Torymus sinensis* Kamijo, was recovered. This species was introduced to Japan (Moriya et al. 1989). Reliable identification of *T. sinensis* required molecular markers because of the existence of a similar, but ineffective, Japanese parasitoid (*Torymus beneficus* Yasumatsu and Kamijo) (Yara 2005). Following the introduction of *T. sinensis*, the level of shoot galling decreased, reaching 3% by 1992, far below the economic injury level of 30%.

Eucalyptus pests in California

Since 1850, over 90 species of eucalyptus have been imported as seed from Australia into California for a wide range of ornamental and other uses. For more than a century, these species remained virtually pest-free. However, beginning in the 1980s, a series of eucalyptus-feeding insects invaded California (Figure 11.14),

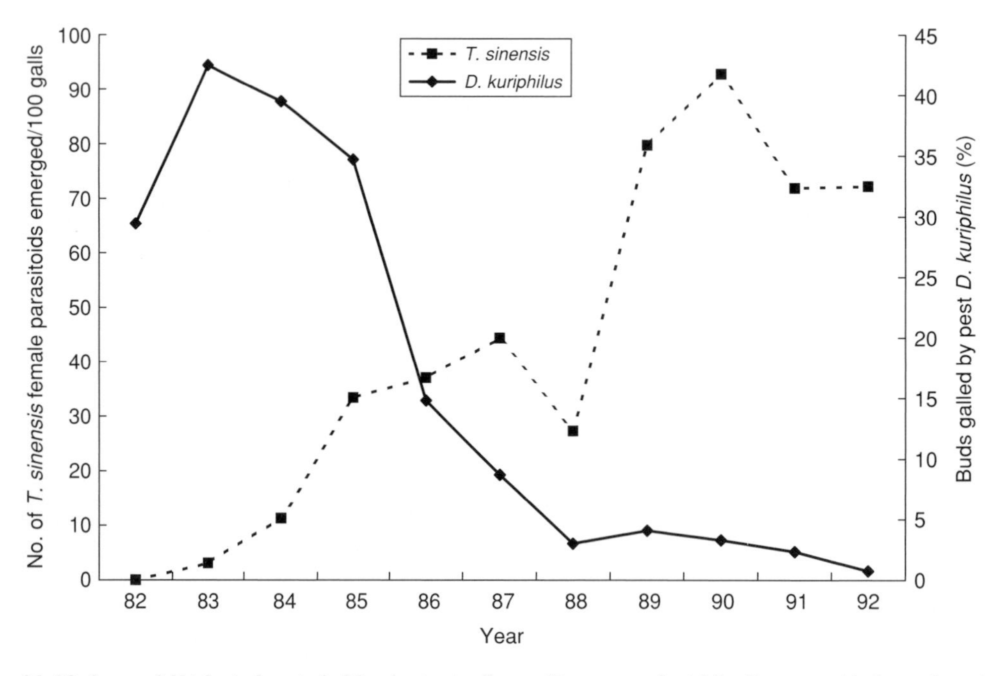

Figure 11.13 Successful biological control of the chestnut gall wasp (*Dryocosmus kuriphilus* Yasumatsu) in Japan from the introduction of the Chinese torymid parasitoid, *Torymus sinensis* Kamijo. Redrawn with permission from Moriya et al. (1989).

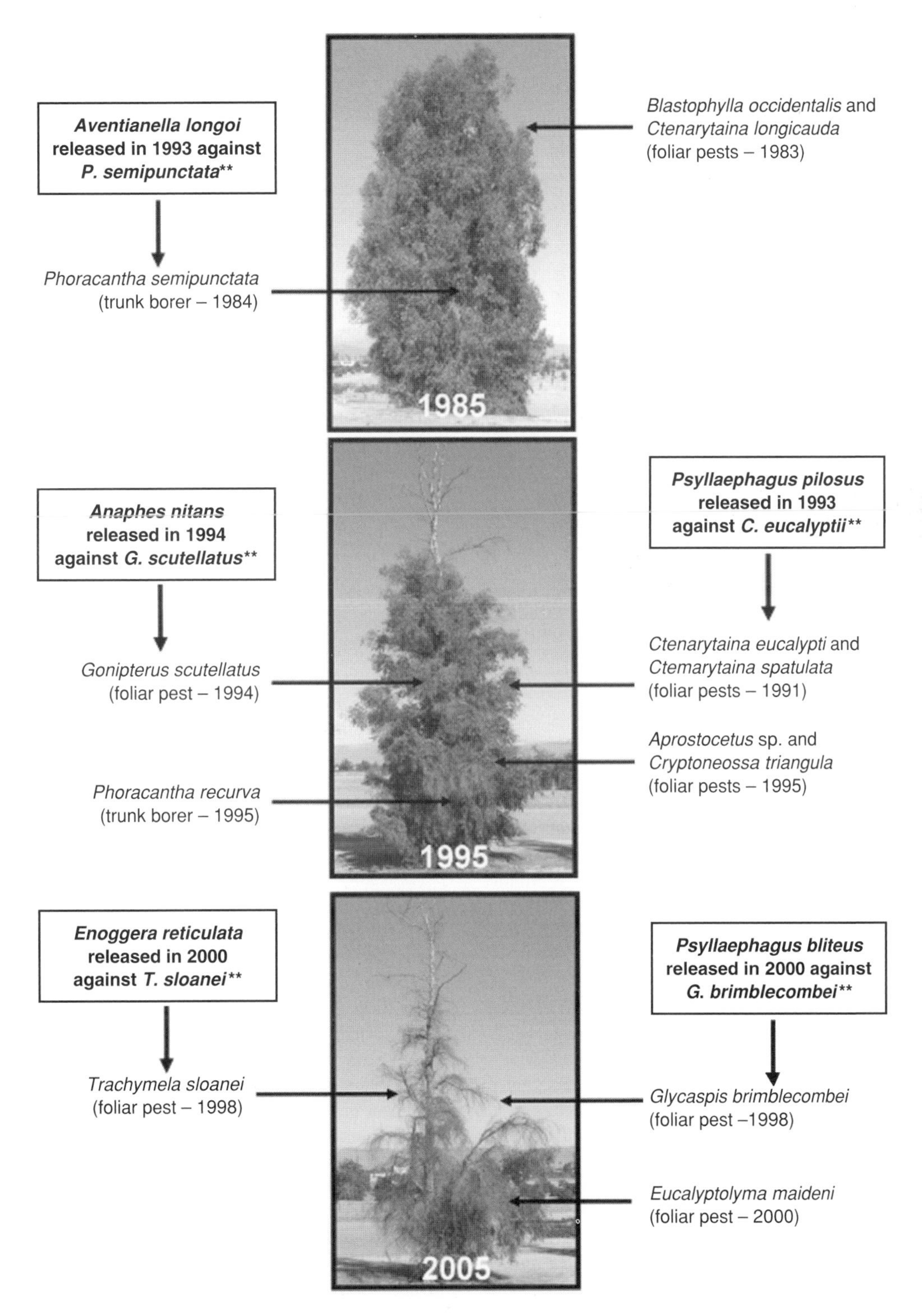

Figure 11.14 *Eucalyptus* species native to Australia have been grown in California since around 1850. These important landscape trees were not subject to significant herbivory by coevolved phytophages in California for more than 100 years. By 1985, three eucalyptus-feeding insects had established in California, by 1995 another six species had established, and by 2005 three new herbivores had established. Four of these pest species have been the subject of classical biological control and ** indicates successful projects where pest densities were reduced to non-economic levels, and * indicates projects that failed to provide adequate control of the target. Drawing courtesy of Mark Hoddle.

Figure 11.15 A *Eucalyptus* tree damaged by the eucalyptus borer *Phoracantha semipunctata* (Fabricius), a species invasive in California. Photograph courtesy of Jack Kelly Clark, University of California IPM Photo Library.

damaging and killing eucalyptus. The first invasion was of the borer *Phoracantha semipunctata* (Fabricius) (Coleoptera: Cerambycidae) (Figure 11.15), which was first detected in California in 1984 (Paine & Millar 2002). Subsequently, 15 other species of eucalypt herbivores have reached the state, including another *Phoracantha* (in 1995), the leaf-feeding weevil *Gonipterus scutellatus* Gyllenhal (in 1994), the chrysomelid *Trachymela sloanei* Blackburn (in 1998), and at least six psyllids, including blue gum psyllid (*Ctenarytaina eucalypti* Maskell) and two lerp psyllids (*Glycaspis brimblecombei* Moore and *Eucalyptolyma maideni* Froggatt) (Paine & Millar 2002).

Parasitoids from Australia have been introduced against four of these invaders. The borer *P. semipunctata* has been completely controlled by the egg parasitoid *Aventianella longoi* Siscaro (Mymaridae) (Hanks et al. 1995, 1996). The snout beetle *G. scutellatus* (Hanks et al. 2000) and the psyllids *C. eucalypti* and *G. brimblecombei* (Hodkinson 1999, Dahlsten et al. 2005) have also been suppressed by introduced parasitoids. Introductions are planned or in progress against the second borer species, the leaf beetle, and the lerp pysllids.

These events show that the high productivity that exotic plants often have in introduced ranges can be abruptly lost when pests from the plants' native ranges invade. Declines in productivity may be especially steep if the introduced range is physically marginal for the plant, a limitation that may be tolerated in the absence of herbivores, but unsupportable once these invade. The magnitude of refuge-enhanced productivity at risk of loss from invasions is high. For example, consider the case of rubber tree plantations in southeast Asia. Plantation production of rubber trees (*Hevea brasiliensis* Muell. Arg.) is not biologically feasible in the Amazon (its native range) because of associated native pests. Were these to invade the refuge for the plant created by transporting it in the early twentieth century to southeast Asia, cheap rubber might cease to exist, with massive impacts on the world's motorized economy. A similar collapse of citrus production in its Florida refuge may now be starting with the invasion from Asia (the native home of citrus) of the uncontrollable disease citrus greening.

Another lesson from work on eucalyptus pests in California is that borers, once considered an unlikely target for classical biological control, may be feasible in a number of cases. The invasion of emerald ash borer (*Agrilus planipennis* Fairmaire; Coleoptera: Buprestidae) into the north central USA, where it is killing millions of ash (*Fraxinus* spp.) trees (Anon 2004, Herms et al. 2004) illustrates the critical importance of this precedent.

Larger grain borer in Africa

The larger grain borer, *Prostephanus truncatus* (Horn) (Coleoptera: Bostrichidae), was accidentally introduced into Africa in the 1970s in shipments of maize grain from developed countries sent as food aid. In Africa, dried maize grain is stored on farms (Figure 11.16) and is a staple food for millions of people. Storage conditions do not permit insect exclusion and the larger grain borer, which can feed on dried maize grain, quickly became a major pest in farm-stored maize and cassava, causing losses of up to 30% (Borgemeister et al. 1997). To reduce losses, predators of the larger grain borer were sought in Central America, its native range, and the histerid beetle *Teretrius* (formerly *Teretriosoma*) *nigrescens* (Lewis) was introduced to both East and West Africa in the early 1990s. One complexity affecting this project is that the pest is also able to feed on dead wood and thus reservoir pest populations exist in woodlands. Beetles from

Figure 11.16 Corn for on-farm consumption in Africa is stored in simple structures that are accessible to the larger grain borer, *Prostephanus truncatus* (Horn) (Coleoptera: Bostrichidae), an introduced pest capable of destroying a large part of the stored grain. Photograph courtesy of K. Hell, IITA.

Figure 11.17 Pheromone traps can be used to monitor local levels of both the larger grain borer, *Prostephanus truncatus* (Horn), and its introduced predator, the histerid *Teretrius nigrescens* (Lewis). Photograph courtesy of K. Hell, IITA.

woodland populations can migrate to new maize stores as they are created. The predaceous histerid finds its prey through attraction to the pest's aggregation pheromone. Consequently, numbers of predators and prey caught in traps (Figure 11.17) baited with this pheromone could be used to monitor *T. nigrescens'* establishment and spread and to measure changes in pest abundance over time (Figure 11.18).

In West Africa, catches in pheromone traps documented the predator's establishment and rapid spread following its release in 1992. Increasing numbers of predators in traps were associated with decreasing numbers of the pest (Borgemeister et al. 1997). Surveys in 1995–7 in Benin showed a sharp reduction in numbers of first-generation grain borers, and farm surveys showed a decrease in the infestation rate and loss (Borgemeister et al. 1997). In East Africa, Hill et al. (2003) found an 80% drop in numbers of pest beetles breeding in Kenyan woodlands compared to before the predator's introduction. This was a critical finding, suggesting that areawide reduction in the larger grain beetle in its natural reservoir was possible, despite modeling results that suggested that this predator would not be effective because of its low growth rate relative to that of the predator (Holst & Meikle 2003). Declines of the larger grain borer have been documented in both Togo and Benin in West Africa and in the areas with the predator for the longest period (southern Togo and Benin) grain losses during storage have dropped to levels equivalent to those before the pest's invasion (Schneider et al. 2004).

This project is important because it again demonstrates the critical role biological control can play in protecting the food supplies of rural people. It demonstrates the use of pheromone traps as a tool for monitoring progress in such projects. It also shows the interplay between a pest reservoir (here, dead wood in forests) and pest populations on the critical resource (here, farm-stored maize), and the limits of modeling in predicting field outcomes when populations interact over a complex landscape.

Step-by-step description of the process

All classical biological control projects move through similar steps (Van Driesche & Bellows 1993): (1) choosing appropriate targets and generating support, (2) obtaining correct pest identification, (3) surveys of the pest's natural enemies in the invaded area, (4) identifying the pest's native range, (5) collecting natural enemies in targeted locations, (6) judging the potential of candidate natural enemies to suppress the pest, (7) creating colonies of natural enemies in quarantine, (8) estimating each natural enemy's host range, (9) petitioning for release, (10) release and establishment, (11) assessing impacts on the pest and non-target species, and (12) assessing the program's completeness and economic value.

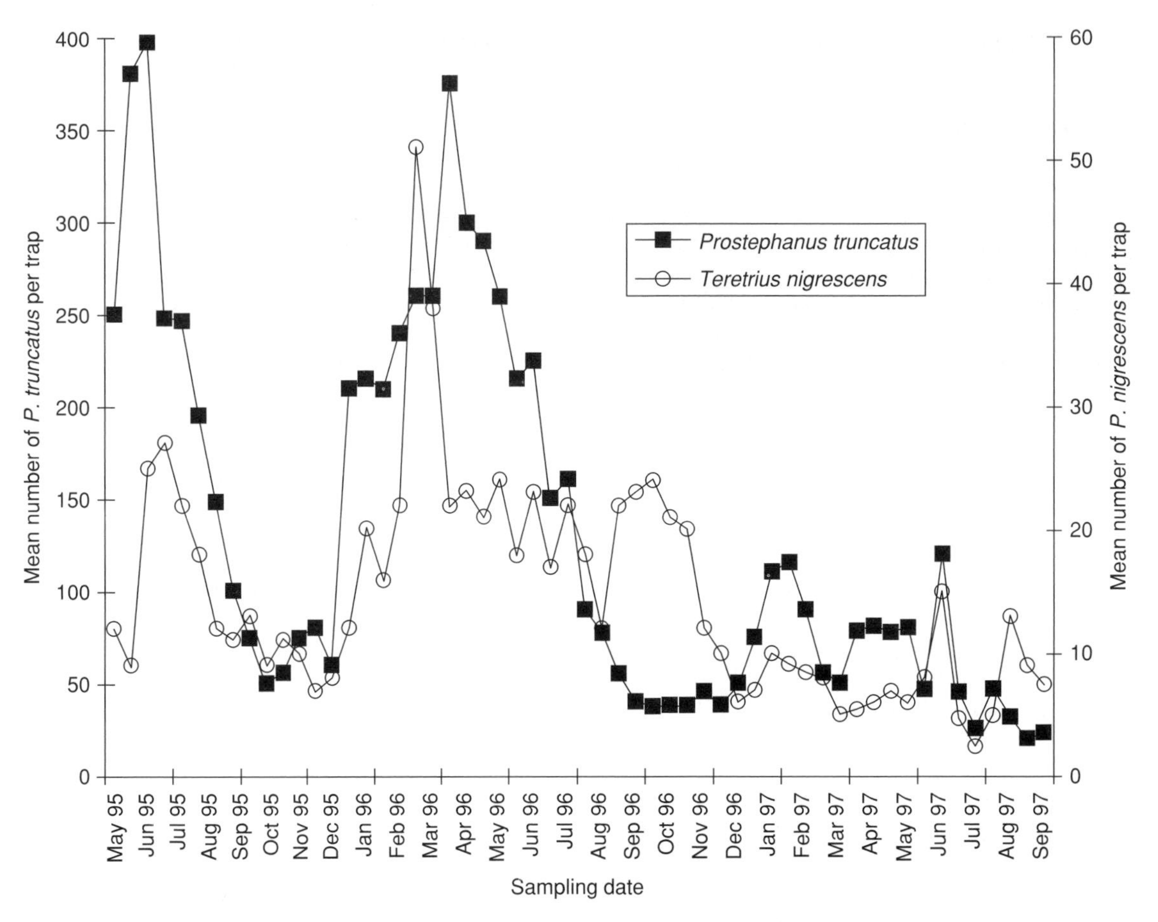

Figure 11.18 Captures in pheromone traps of the larger grain borer, *Prostephanus truncatus* (Horn), and its introduced predator, the histerid *Teretrius nigrescens* (Lewis), show progressively lower seasonal peaks of pest captures in the southern Guinea savanna following the increase in abundance of the predator in the second year (1996). Redrawn with permission from Schneider et al. (2004).

Step 1: choosing appropriate targets and generating support

Pests selected as targets should be important, either economically or ecologically, because staffing and funding used on one project are unavailable for another. The impacts of the invasive species should be measured before starting the project (e.g. Ross et al. 2003, Brown et al. 2006). Native plants and all but a few native insects are inappropriate as targets. Since invasive species sometimes are controlled by local or self-spreading agents (Simberloff & Gibbons 2004), pests selected as targets should be species that have persisted as pests for several years or more [e.g. Vercher et al. (2005) in Spain showed that local parasitoids of the citrus leafminer (*Phyllocnistis citrella*) Stainton failed to provide control 7 years after the invasion]. Ideally target species should be ones that occur at low densities in their native ranges.

There should be broad social agreement on the need for suppression of the selected pests, with no unresolved conflicts of interest between groups. For example, in Australia, the proposal to control the range weed *Echium plantagineum* L. was opposed by beekeepers who considered it bee pasture (Cullen & Delfosse 1985). In South Africa many tree species introduced for forestry and agroforestry have become invasive. These species have commercial value and many are used as firewood by local people. To resolve such conflicts, governments may need to establish courts of arbitration to decide what is in the best interest of society.

Finally, the historical record of classical biological control can shed light on the odds of success, which can also be taken into account in deciding whether a given species is going to be an easy or difficult target. Mealybugs and armored scales would be examples of easy targets, because many species in these groups have been controlled successfully through introductions of natural enemies. In contrast, grasses and soil-inhabiting larvae of scarab beetles have never been controlled with classical biological control.

Step 2: obtaining correct pest identification

Projects start by obtaining an authoritative identification of the pest species, perhaps also including a molecular characterization of the invasive population to provide a match to the population from which the invader originated. This latter process can be very useful if the pest's known distribution is extremely wide. For example, this approach showed that the invasive population of old world climbing fern (*L. microphyllum*) in southern Florida matched populations in northern Queensland, Australia (Goolsby et al. 2004b), which therefore was the likely source of the Florida population.

After the pest's identity is known, the available information on its distribution, biology, host range, pest status, and its natural enemies and those of its close relatives can be compiled. General knowledge of related species is useful because near-relatives or similar-appearing species may occur in the area to be searched and each of these species may have specific natural enemies. For example, in searching for natural enemies of the cassava mealybug (*Phenacoccus manihoti* Matile-Ferrero) in South America, entomologists had to distinguish this species from its closest relative, *Phenacoccus herreni* Williams. Both species have similar parasitoid species, some of which are shared, but the species that was key to success was unique to *P. manihoti* (Neunschwander 2003).

When a pest is an undescribed species, its most closely related relatives must be determined to guide the collection of natural enemies. Otherwise time may be lost collecting natural enemies from the wrong species. The weed *S. molesta* was originally misidentified as *Salvinia auriculata* Aublet and as a consequence the search for effective natural enemies was misdirected to Trinidad and Guyana, where *Cyrtobagous singularis* Hustache, was collected but which proved ineffective. In the 1970s, after the pest was recognized as a new species (later described as *S. molesta*), matching of specimens to herbarium records showed its native range to be southern Brazil. A search for natural enemies there seemed to suggest that the available natural enemies consisted of the same three species (a grasshopper and a moth, along with the weevil) that had previously been found on *S. auriculata* (Forno & Bourne 1984). At first, the Brazillian population of the weevil was thought to be a local race of the previously encountered species. However, when the Brazilian weevil was released in Australia, it proved extremely effective and more detailed taxonomic studies showed that it was, in fact, a new species, later named *C. salviniae* (Room et al. 1981, Calder & Sands 1985, Moran 1992). This weevil has since been released in many other countries with equal effect. This project served to highlight the critical importance of taxonomy in biological control.

Step 3: surveys of the pest's natural enemies in the invaded area

To avoid introducing natural enemies that are already present or cannot be distinguished from those already present, the pest in the invaded region should be surveyed and its natural enemies inventoried. In some cases, molecular markers may have to be developed to ensure that such species can be told apart from any new species to be introduced. For example, in the project against the B strain of the sweetpotato whitefly, *Bemisia tabaci* (Gennadius), in the USA, populations of various aphelinid wasps in the genera *Eretmocerus* and *Encarsia* were introduced from many countries. Molecular markers were used to identify each population and to distinguish them from the native parasitoids already present in the area (Goolsby et al. 1998, 1999).

Step 4: identifying the pest's native range

To collect natural enemies for a classical biological control project, foreign populations of the target pest have to be located. The native range of a pest might be inferred from: (1) records of occurrences of the pest or its relatives, (2) communication with scientists where the pest is believed to be present, (3) examination of specimens in collections of world museums (e.g. the British

Museum of Natural History, the Smithsonian SEL) and regional institutions in the presumed native range, (4) study of genetic variation in populations of the pest from different locations, and (5) actual surveys in potential locations. Based on the results of such efforts, the likelihood of any given area being the native range of the pest has to be assessed (see Chapter 13 for suggestions on methods).

Step 5: collecting natural enemies in targeted locations

After certain areas have been chosen for surveys, collections are made (see Chapter 13). Depending on the level of accessibility and availability of local universities or research stations, foreign collecting is either done through short trips made by scientists from the country importing the natural enemy species, by hiring local scientists, or and deploying staff to the collecting region for extended periods of time. The third approach is more effective, as quick trips to a region often do not allow enough sites to be examined at enough times of the year to find all natural enemies of interest. However, each approach can work.

Things to take into account in collecting natural enemies include the following (see Chapter 13 for details).

1. Transportation, health, and safety of the person making the collections must be taken into account given circumstances in the proposed collecting areas.
2. Necessary permits should be secured, both to export the natural enemies from the area of collection and to import them to a quarantine laboratory in the receiving country.
3. The receiving quarantine laboratory should be provided with the hosts or plants needed to rear the newly collected natural enemies.
4. Adequate provisions must be made for rapid shipment, taking into account all regulations and procedures, which may be ill defined and changeable.
5. Searches need to be made that include various seasons, elevations, and climates, as natural enemies may vary. Searches also need to sample all the pest's life stages or parts (for plants).
6. There must be separate handling of potential biotypes or cryptic species, keeping collections separated by location and host. Natural enemy populations vary genetically and this variation may be important to attributes such as habitat choice, host preference, physiology, life-cycle parameters, behavior, or host specificity. This variation is a resource to be recognized and managed in the importation process. Introduction of new biotypes of a natural enemy from different locations has proven crucial to success in several past projects. In 1959, *Trioxys pallidus* Haliday from France was introduced into California to control the walnut aphid [*Chromaphis juglandicola* (Kaltenbach)], but this population established only in southern California. In 1968 a population of the same parasitoid from Iran, a region with a similar climate, was imported and was more successful (van den Bosch et al. 1970). Biotypes can also be important in the case of plant pathogens. Molecular tools now make recognition of biotype differences easier than ever before (Roehrdanz et al. 1993, Antonlin et al. 1996, Legaspi et al. 1996, Alvarez & Hoy 2002, Vink et al. 2003, Kankare et al. 2005; see Chapter 15).

Step 6: judging the potential of candidate natural enemies to suppress the pest

It has been suggested that it is inefficient for biological control projects to introduce more than one or two "best" species (e.g. Ehler 1995). This argument is predicated on the assumption that it is possible to judge how populations newly brought together will quantitatively affect each other in a new environment over a period of years (e.g. as tried by Godfray & Waage 1991, Mills 2005). If efficacy could be predicted, it would also have the benefit of avoiding the use of species that establish but fail to suppress the host (and hence remain abundant themselves). Avoidance of such species has been suggested as an important means to avoid unwanted indirect effects of biological control agents (McClay & Balciunas 2005). However, there are many example of projects for which the introduction of several agents has led to excellent control (Huffaker & Kennett 1969), and in some instances multiple agents have clearly been shown to be essential (Hoffmann & Moran 1998).

As a practical matter, predictions needed to choose a best agent would have to be based on either laboratory data or information gained from the native range. Predictions based on laboratory assessments are limited by their inability to assess such biological factors as agent dispersal, responses to climate, and effects of alternate hosts (Messenger 1971, Eikenbary & Rogers 1974, Mohyuddin et al. 1981, Legner 1986). In addition, release from attack by hyperparasitoids, cleptoparasitoids, or other predators (present in the pest's native range) may make predictions of performance in the new location uncertain. Given the well known principle that new properties emerge at new levels of organization, it is hardly surprising that attributes of individuals measured in cages are poor predictors of performance of populations in the field.

A different approach to choosing natural enemies to introduce may be to look for "vacant attack niches" in the life system of the pest in the invaded area compared with that in the native range. For the codling moth, *Cydia pomonella* (L.), Mills (2005) used a stage-structured model of its life cycle in the recipient area (Califonia) and the native range (central Asia) to identify natural enemies causing high levels of mortality in Asia to stages with low mortality in California. Selection criteria for choosing species to introduce were that the new parasitoids should not show antagonistic interactions with existing parasitoids, should cause at least 30% mortality in the native range, and should attack stages (second instars and the cocoon) lacking natural enemies in California.

In practice, natural enemies often are discovered or approved for release sequentially and thus the question of how many and in what order to release species often is replaced by what species are available first, followed by quitting when the pest is controlled. A separate question of great importance is when to stop work on a species that has some promise but for which flaws are discovered prior to release, or which fails to show any suppression of the host after release. In general, before any particular species is released, two questions need to be answered positively: (1) is it plausible that the natural enemy *might* be effective and (2) is it safe? (for the latter, see Chapters 17 and 18).

Plausible species are those that share certain attributes thought to be favorable for success and do not have any obvious features that would make them unfit or unsafe for the intended use. In terms of parasitoids and predators, Coppel and Mertins (1977) proposed a list of such desirable attributes: ecological match to the host's habitat, temporal synchronization, density responsiveness, high rate of reproduction, high searching capacity, high dispersal capacity, host specificity, compatibility with host physiology, simple food requirements, and freedom from hyperparasitism. Godfray and Waage (1991) describe how preliminary observations on life-history characteristics may be combined using mathematical models of population interactions. Of particular importance is the emerging view that successful parasitoids, for example, are those with faster rates of population growth than the pests against which they are released (high generation-time ratios, GTRs; Kindlmann & Dixon 1999, Barlow et al. 2003). For a discussion of the theory of classical biological control, see Chapter 10.

Finally, studies on candidate natural enemies and their hosts in the native country can provide insight into the potential impact of specific natural enemies, in particular where a given species might be most successful in terms of factors such as elevation, climate, and habitat.

Step 7: creating colonies of natural enemies in quarantine

Good practice and the laws of many countries require that natural enemies collected in foreign locations be shipped to quarantine laboratories so that no harmful organisms are accidentally introduced. Quarantines are buildings especially designed for containment of organisms, in which the imported materials can be inspected safely (see Chapter 13). Quarantine design and operating procedures in the USA are reviewed by Coulson et al. (1991). The United Nation's Food and Agriculture Organization has published guidelines for quarantine procedures suitable for use during introductions of biological control agents (Anon 1992).

Quarantine laboratories provide a place where all undesired organisms found in shipments by accident can be removed and destroyed and the desired organisms – the biological control agents – preserved and used to create rearing colonies. Consequently, to fulfill their mission, quarantine laboratories must provide an environment conducive to successfully breeding natural enemies (and their hosts), while at the same time preventing their escape.

Colonies of candidate natural enemies must consist of only a single species of natural enemy and its rearing host, prey, or food plant. The natural enemy colony must be demonstrated to be free of hyperparasitoids (for parasitoids), parasitoids (for weed biological control agents), and pathogens. The natural enemy must be shown, by breeding in the quarantine laboratory, to be able to complete its life cycle on the target pest.

Without a well-run quarantine laboratory, desirable natural enemies may be lost and foreign collecting efforts thus wasted. It may be necessary to maintain colonies of natural enemies for several generations to study their biology and host specificity before a release petition can be prepared. During this period, all materials supporting the colonies must be managed to prevent invasions by unwanted species. If plants form part of the support needed to rear the natural enemy, invasion of unwanted herbivores (aphids, thrips, mites, etc.) must be prevented or suppressed.

If the desired natural enemy is a pathogen, the objectives of establishing and maintaining a pure colony of the organism remain the same as described above. However, the actual quarantine procedures used are those of a microbiological laboratory, including special filtering of incoming and exhaust air supplies, containment and treatment of all waste water, and special culturing conditions to minimize contamination of microbial cultures (Melching et al. 1983).

Step 8: estimating each natural enemy's host range

Before a pure culture of a new natural enemy can be considered acceptable for release into the environment, information must be collected that allows its likely host range to be estimated, relative to the fauna or flora of the areas to which the agent might eventually spread (for details, see Chapter 17). Information comes from previous host records in the literature, observations of what species are attacked by the agent in the area of its origin, and laboratory host-range testing in quarantine or, sometimes, as field tests in the country of the natural enemy's origin.

For herbivorous insects and plant pathogens, host-range estimation has been a routine part of classical biological control for more than 75 years. Initially, such testing focused on testing crops, ornamentals, and other valuable plants to ensure that herbivore or pathogen introductions would not introduce a new plant pest. In the 1970s, with the advent of endangered species laws, protection of threatened native plants was added as an objective for testing plans. Currently, the goal is to forecast the host range and avoid any important impacts on any native plant species, endangered or not (Zwölfer & Harris 1971, Frick 1974, Wapshere 1974a, 1989, Woodburn 1993).

For parasitic and predaceous arthropods, host range testing was not originally required, as the concern of governments was to protect plants, not native insects. Since the early 1990s, however, a consensus has emerged among biological control scientists that host-range estimation should be a standard part of all classical biological control projects (Van Driesche & Hoddle 1997). Methods to make such estimates for predaceous or parasitic insects are being developed (Van Driesche & Reardon 2004, Bigler et al. 2006), but such testing is not yet required in most countries.

Step 9: petitioning for release

For most countries, the decision to release a new biological control agent into the environment, with the intent that it establish, is regulated by law. Although details vary by country, such laws should seek to ensure that no important damage to non-target native species happens, but, if some damage is inevitable, that it is judged acceptable before release in view of the important harm done by the pest whose control is being sought.

Step 10: release and establishment

Field establishment of new natural enemies is a crucial step. Establishment is usually defined as the presence of a breeding population of the natural enemy 1 year after the last release. Establishment of the natural enemy is assessed by sampling, either directly for the released agent or indirectly (for parasitoids) by collecting hosts and rearing to detect parasitism. The historical record shows that 34% of attempts to colonize natural enemies succeed (Hall & Ehler 1979). The probability that new agents will become established in the field can be increased by careful attention to a series of ecological, technical, human, and financial considerations (Beirne 1984, Van Driesche 1993), which are discussed in Chapter 19.

Step 11: assessing impacts on the pest and non-target species

Measurement of change in the target pest's density is basic to classical biocontrol programs (see Chapter 20 for methods). If feasible, pest densities should be measured in control plots before natural enemies are released, as such pre-release information is valuable in establishing the pest density baseline to which future densities are compared. If this is not feasible or if the impact of a previously released agent is to be assessed, other approaches are needed. Evaluations done as the biological control project unfolds provide guidance on agent effectiveness, allowing mass rearing to support future releases to concentrate on the best species. For example, an assessment of the impacts of the natural enemies released in western North America against tansy ragwort *Senecio jacobaea* L. – the flea beetle *Longitarsus jacobaeae* (Waterhouse) and the arctiid moth *Tyria jacobaeae* L.– showed that the flea beetle was the more important agent (McEvoy et al. 1993), allowing work in new areas to concentrate on that species.

Evaluation includes assessment of impacts on non-target species. Field sampling should concentrate on whatever non-target species seemed most likely to be affected based on the pre-release risk assessment. For example, athel pine [*Tamarix aphylla* (L.) Karst.] is an exotic but valuable shade tree in Mexico that has been predicted to be at risk of attack from the saltcedar biological control agent, *D. elongata* (Chrysomelidae), released in Texas. Athel therefore is the obvious species to monitor for unwanted effects from this release.

Step 12: assessing the program's completeness and economic value

When a project ends, its completeness should be assessed. Has the pest been reduced adequately in all invaded areas, what economic or ecological benefits resulted, and what, if any, harm occurred to non-target species? A benefit/cost ratio for the project should be calculated to provide to government agencies to whom requests for support of new projects must be justified. If the pest was not suppressed to the desired level, the program's objectives should be reconsidered in light of what has been discovered. In particular, the need for different natural enemies should be addressed, especially if natural enemies were discovered during the exploration phase of the program that were not introduced. If it appears that additional natural enemy species or biotypes can be introduced successfully, then this should be done and the impact of these new species evaluated. If no further natural enemies are known, additional exploration may be needed.

Overall classical biological control is an economically very productive social investment. Australian projects have had an average benefit/cost ratio of 10.6:1, with a maximum exceeding 100:1 (Tisdell 1990). In crops, cumulative benefits increase yearly because of the absence of pest damage in each subsequent production season and lower pesticide use. Classical biological control is particularly valuable in protecting agriculture in developing countries where for many farmers pesticides are too costly or unsafe (e.g. Herren & Neuenschwander 1991, Zeddies et al. 2001). Benefits from control of pests of natural areas can be ecological or economic. Monetary values of ecological improvement are more complex to measure. Economic benefits vary according to how the pest disrupted human activities, including things such as loss of water supply, reduced transportation opportunities, reduced fisheries, etc. (Thomas & Room 1986, Bangsund et al. 1999).

NEW-ASSOCIATION BIOLOGICAL CONTROL

In addition to importing natural enemies from the home range of an invasive pest (classical biological control), importation can be used in at least two other ways in which new combinations of natural enemies and pests are brought together.

1. Some important pests are native species. Pimentel (1963) suggested that such native pests might be controllable with parasites and predators collected from relatives of the pest present in other biogeographic regions. He introduced the phrase new association, which now refers to the use of one organism for the biological control of another with which the biological control agent has had no previous evolutionary connection. For such projects, the potential source of natural enemies would be closely related (same genus or tribe) species or those that are ecologically similar to the target pest but found in separate biogeographic areas (other continents) with similar climates.
2. In other cases, a pest may be invasive, but its origin unknown. For example, the gracillariid moth *Cameraria ohridella* Deschka and Dimic is a moth of unknown origin that was first observed in Macedonia in the 1970s. It is now invasive throughout much of Europe as a high-density pest leafminer on horse chestnut trees (*Aesculus hippocastanum* L.) (Kenis et al. 2005). A European native range for the pest seems doubtful due to its recent spread and the fact that the genus is from the Americas and Asia. A potential host switch may have occurred, which if true complicates matters since surveys only on horse chestnut might fail to find the species. Plans to collect natural enemies from species taxonomically related to the invader are being considered.

Examples of new-association biological control

A number of precedents demonstrate that new-association natural enemies can, at least in some instances, suppress organisms with which they had never had previous evolutionary contact:

1. An early famous example was the work of Tothill et al. (1930), who used a tachinid fly to suppress a devastating coconut pest, the zygaenid moth *Levuana iridescens* B-B., in Fiji in the 1920s. Researchers at the time viewed the pest as invasive, but this point has been argued. Regardless of the pest's source, the origin (if not Fiji) was never determined and the natural enemy ultimately controlling the pest [the tachinid *Bessa remota* (Aldrich)] was imported from a different zygaenid [*Brachartona catoxantha* (Hampson)].
2. The native geometrid moth *Oxydia trychiata* (Guenée) became a defoliator in Colombia tree plantations after exotic pines were planted. It was later suppressed by the introduction of the North American egg parasitoid *Telenomus alsophilae* Viereck, which had never before been associated with *O. trychiata* (Bustillo & Drooz 1977, Drooz et al. 1977).
3. The European rabbit, *Oryctolagus cuniculus* (L.), was suppressed in Australia with a virus from a species of rabbit from South America (Fenner & Ratcliffe 1965).
4. The sugarcane borer, *Diatraea saccharalis* (F.), was suppressed in Barbados with *Cotesia flavipes* (Cameron), a braconid wasp from India associated with other species of borers in graminaceous plants (Alam et al. 1971).
5. Tarnished plant bug (*L. lineolaris*) density in alfalfa has been reduced by 60% (Day 1996) by the braconid *P. digoneutis*, a parasitoid of *Lygus* native to Europe. This parasitoid was selected based on study of the host/parasitoid associations found in Europe (Kuhlmann & Mason 2003).
6. Invasive prickly pear cacti in Australia [*Opuntia stricta* (Haworth) Haworth var. *dillenii* (Ker Gawler) L. Benson and *O. stricta* (Haworth) Haworth var. *stricta* (Haworth) Haworth], which originated from the Gulf Coast of North America or the Caribbean (Mann 1970, Julien & Griffiths 1998), were controlled by the South American pyralid moth *C. cactorum* from Argentina, where it attacked other *Opuntia* (Dodd 1940).

Constraints and failures

However, in many new host/parasitoid combinations, the target host may be physiologically unsuitable or unattractive to parasitoids taken from related species. Predicting which new hosts will be susceptible to a parasitoid or herbivore's attack is not feasible on the basis of theory. Rather, each natural enemy/pest combination has to be tested. This approach was used by Ngi-Song et al. (1999) to assess which species of stem borers (*Chilo* and *Diatraea* spp.) supported successful development of various *Apanteles* and *Cotesia* parasitoids of interest when hosts and parasitoids were combined in novel combinations. Five of the parasitoid species studied showed lower attack rates on novel hosts. In two of 17 novel combinations, no hosts were parasitized. In seven combinations, broods were small or died due to host unsuitability. Unsuitable hosts were those in which parasitoids were unable to defeat host encapsulation processes and hence were killed as eggs or young larvae (e.g. Alleyne & Wiedenmann 2001). In another case, parasitoids reared from coniferous-feeding geometrids in Europe for possible use against the eastern hemlock looper [*Lambdina fiscellaria fiscellaria* (Guenée)] in Newfoundland were unable to develop successfully in this new host (West & Kenis 1997).

For herbivores, unsuitable host plants will be those that do not have chemical cues that induce herbivore oviposition, or in other instances, species that are nutritionally inadequate for development of the immature herbivore (see Chapter 12).

Success rates for new-association programs

Clearly some new host/parasitoid combinations exist that are suppressive of the host. Pimentel (1963) argued that new associations should be even more suppressive on average than when natural enemies have coevolved with their hosts. This was believed to be so because antagonists with no prior close relationship to a species would not have been subjected to any evolutionary pressures towards the attenuation of the agent's virulence (Hokkanen & Pimentel 1984, 1989). However, success rates do not appear to actually differ between classical biological control and new-association projects (Goeden & Kok 1986, Schroeder & Goeden 1986, Waage 1990).

Are new-association projects ethical?

Targeting native plants

In the past, some scientists argued that economically undesirable native plants [such as species of *Prosopis* (mesquite) in southwestern US grasslands] were appropriate targets for natural enemy introductions

(De Loach 1980, 1985). Pemberton (2002), however, argues that native plants are never acceptable as targets for natural enemy importations because: (1) native weeds are likely be abundant species on which many other native species would depend, (2) biological control agents would spread to parks and nature reserves where native plants would be viewed as part of the flora, regardless of any economic problems they might cause elsewere, (3) new-association natural enemies would likely have broader host ranges, increasing risks to non-target native species, and (4) projects against native plants, at least in some countries, would be politically unacceptable, would not be permitted to proceed to the release stage, and hence would be a waste of resources.

Targeting native insects

Some native insects are important pests that have been successfully reduced in density by natural enemy importations [e.g. the fruit pest *L. lineolaris* by *P. digoneutis* from Europe (Day 1996) and the sugarcane borer *D. saccharalis* by *C. flavipes* (Alam et al. 1971)]. Other pests of native trees have been suggested as potential targets, including the white pine weevil [*Pissodes strobi* (Peck)], which kills leaders of regenerating white pines (*Pinus strobus* L.), greatly reducing their timber value (Mills & Fischer 1986, Kenis & Mills 1994), as well as spruce budworm [*Choristoneura fumiferana* (Clemens)] and spruce budmoth (*Zeiraphera canadensis* Mutuura and Freeman) (Mills 1983, 1993). These projects, however, have not gone further than identifying potential candidate natural enemies from European congeners of the pest. Control of *Lygus* species is the only new-association project in the USA against insects that is actively being carried out. Whether or not new projects would be socially acceptable is not clear.

Targeting exotic insects or plants with new-association natural enemies

Some exotic insects and weeds have been controlled through the introduction of natural enemies from other host species (e.g. *Levuana* moth in Fiji and *Opuntia* cacti in Australia, as mentioned above). These projects are ecologically justified because the resulting reduction in density of the pest moves the invaded community back towards to its pre-invasion state. The fact that the natural enemies lack previous evolutionary association with the target, given they have been demonstrated to have adequate host specificity for safety in the recipient country, is not an ethical issue.

Sources of potential natural enemies

For new-association projects, sources of natural enemies are not necessarily obvious. The general approach is to search on congeneric species, or on less-related species that have similar life histories or ecology, in regions with similar climate. For insect targets, it also is useful to collect natural enemies from hosts on the plants on which the target pest feeds. When Tothill was unable to locate the native range of the coconut moth, he collected other moths in the same family found on coconut in a broad geographic region with similar climate. In other projects, the feeding habit and plant association of the pest insects, rather than taxonomic affinity *per se*, was the basis for finding new-association parasitoids. Thus, when *C. flavipes* was identified as a parasitoid for sugarcane stem borer, it was because it attacking other borers in species of large-stemmed grasses, which were not necessarily closely related to sugarcane borer (Alam et al. 1971). A drawback to this approach is that such species will, by definition, have broader host ranges than classical biological control agents.

Potential risks of new-association introductions

If biological control introductions are directed against exotic species of unknown origin, the risks are the same as for any classical biological control project. Projects against native species should be limited to those against native insects, as opposed to undesirable plants. In such cases, the only additional risk posed by such projects, over those of a similar classical biological control project, would be that the specialized native parasitoids or predators of the target pest that might be reduced in density or suffer range reductions if their native host is controlled. Day (2005) found that after the suppression of the tarnished plant bug (*L. lineolaris*, native to USA), by the introduced European parasitoid *P. digoneutis*, the native parasitoid *Peristenus pallipes* (Curtis) remained present in the target habitat (alfalfa fields), with some change in abundance on the target pest (from 9 to 2% parasitism). However, there was little or no change of *P. pallipes*' abundance on its other hosts, such as

Adelphocoris lineolatus (Goeze) (changing from 12 to 10% parasitism) and *Leptopterna dolabrata* (L.) (increasing from 17 to 21%).

SUMMARY

Classical biological control is the most effective and valuable form of biological control. It allows permanent solutions to be developed for invasive pests, either economically important crop pests or ecological pests in natural habitats. No other form of biological control permits cost-effective, permanent solutions to such problems. Classical biological control contributes to human health through the reduction in pesticide use, a issue of continuing importance. Impacts of classical biological on non-target species have occurred, but potential exists to modify the process of selecting agents for introductions so as to reduce such risks. New-association biological control is a variant on classical biological control that is appropriate for some arthropod targets, particularly invasive pests whose areas of origin cannot be identified, and, in a few cases, some native species that cause large economic losses.

Chapter 12

WEED BIOLOGICAL CONTROL

DIFFERENCES AND SIMILARITIES BETWEEN WEED AND ARTHROPOD PROGRAMS

Weed biological control follows the same progression as for arthropod pests (see Chapter 11). It involves the selection, importation, and establishment of specialist herbivores or phytopathogens into a new environment. Sites are usually inoculated with few individuals and control depends upon the ability of the agent to increase and attain critical population levels. These populations, upon establishment, become self-sustaining, inflict damage to the target, then disperse to new stands of the weed, and ultimately attain a long-term balance with it. Successful control can produce dramatic vegetational changes (Plate 12.1), causing monotypic weed stands to be replaced with more diverse native vegetation. Most programs have evaluated the impact of the introduced agents on the target weed. In addition, post-release investigation of non-target damage and food-web effects now receive increased attention.

Weed biological control is based on both lethal effects (somewhat rare) and the cumulative stress from non-lethal impacts. Plant-feeding insects and phytopathogens alter plant reproduction, competitive ability, growth rates, seedling recruitment, and many other aspects of weed biology. Knowledge of plant physiology, plant ecology, weed science, plant taxonomy, phylogenetics, and other fields of botany are important for weed biological control programs. Herbivorous insects can affect the susceptibility of weeds to phytopathogens as well, so insect–pathogen interactions can become important. Effects of weed biological control agents may be subtle impacts that accumulate over long time periods, making evaluate difficult, especially if well defined baseline plant information is not available. Thus, pre-release data are often sought to establish the status of the weed population before implementation of biological control. The time required to conduct a weed biological control project is therefore greater than for arthropod projects because of this and the greater emphasis on determination of the host range. It is not unusual for weed projects to require 20 years or more for completion (Harris 1985).

Insect biological control projects usually involve the introduction of one or a few biological control agents. In contrast, weed biological control more commonly involves the introduction of a complex of agents. For example, twice as many weed biological control projects (40%) as ones aimed at insects (21%) employed three or more agents (Denoth et al. 2002). Single agents may provide complete control for architecturally simple plants (such as *Azolla filiculoides* Lamarck and *Salvinia molesta* Mitchell) (Hill 1999, Cilliers et al. 2003) that do not reproduce sexually, but multiple agents are more likely to be needed for widely distributed, genetically variable, architecturally complex species with multifaceted modes of reproduction. An extreme case is that of *Lantana camara* L., against which nearly 40 insect species have been released in over 30 countries. This woody shrub comprises an extremely variable genetic complex with over 600 cultivars, some of which do not exist naturally (Winder & Harley 1983, Baars & Neser 1999, Day & Neser 2000, Day et al. 2003). *Lantana* produces numerous bird-dispersed seeds (Thaman 1974) and hybridizes readily. Its foliage and seeds are toxic, it can be spiny, and it invades a wide variety of habitats. After slashing and burning, there is extensive growth of new plants from suckering (Greathead 1968). Although progress has been made in some areas (e.g. Hawaii, USA), complete control of this weed has not been achieved anywhere.

WHY PLANTS BECOME INVASIVE

Non-native plants that are introduced into new areas often become larger, more robust, more numerous, and produce more flowers and seeds than in their native ranges (Siemann & Rogers 2001, Stastny et al. 2005). These attributes may enable them to outcompete native plants and invade natural communities, sometimes to the near exclusion of native plants. Three mechanisms driving plant invasion have been proposed: the enemy release, biotic resistance, and increased resource availability hypotheses.

The **enemy release hypothesis** postulates that introduced plants experience less attack by herbivores and other natural enemies, allowing them to increase in density and to expand in distribution. Carpenter and Cappuccino (2005) have shown a positive, albeit weak, correlation between plant invasiveness and lack of herbivory. Others have further postulated that such reduced herbivory lessens the need for plant defenses. This **optimal defense theory** suggests that there is an allocational trade-off between traits for herbivore resistance and those for plant growth. Proponents of this theory assert that defensive attributes are lost or reduced through natural selection when they are no longer needed. This putatively allows allocation of a greater proportion of photosynthate to growth and reproduction, leading to increased competitive ability (Blossey & Notzold 1995, Bossdorf et al. 2005). This may, however, make the invader more vulnerable to herbivory. Others, however, argue that introduced species may not lose defenses (Genton et al. 2005, Stastny et al. 2005) or may even evolve increased resistance to herbivory (e.g. Leger & Forister 2005), presumably due to an enhanced ability to capture sufficient resources to meet all allocation needs. This may be further complicated by induced plant responses to herbivory (Karban & Myers 1989), which may increase or decrease plant resistance (Tuomi et al. 1984, Williams & Myers 1984). For *Senecio jacobaea* L., invasive populations have decreased investment in defensive compounds directed against specialist herbivores (species not present in the invaded range), but conversely have increased levels of compounds directed against generalist herbivores (Joshi & Vrieling 2005). A further postulate of this hypothesis would be that after specialized natural enemies are reassociated with a population of an invasive plant, that population should begin to reinvest in the defensive compounds needed to defend against the specialist herbivores. Evidence has been found in North American herbaria of reacquisition of toxic furanocoumarins by the invasive European plant *Pastinaca sativa* L. after the subsequent invasion of its specialized herbivore, the parsnip webworm [*Depressaria pastinacella* (Duponchel)] (Zangerl & Berenbaum 2005).

The **biotic** (or **ecological**) **resistance hypothesis**, first proposed by Elton (1958), suggests that invasive plants would fail to establish or proliferate due to their interactions with native species, especially competitors and generalist herbivores (Maron & Vilà 2001, Levine et al. 2004).

The increased **resource availability hypothesis** suggests that a plant community becomes more susceptible to invasion whenever the amount of unused resources increases (Davis et al. 2000). Blumenthal (2005) merged the enemy release and increased resource hypothesis as the resource-enemy release hypothesis, which asserts that these two mechanisms act together to allow or prevent invasion. Clearly, simple answers will not suffice and synthesis is needed to better understand these mechanisms. Shea and Chesson (2002) have moved in this direction with their concept of "niche opportunity," which recognized three temporally fluctuating factors that contribute to an invader's growth rate: resources, natural enemies, and the physical environment. They equate low niche opportunity with biotic resistance and suggest that more diverse communities are less susceptible to invasion.

Classical weed biological control seeks to nullify the benefits of enemy release by introducing natural enemies, while at the same time recognizing that this alone may not provide adequate control of a particular invasive plant. For plants that have become invasive because of natural enemy release, especially those invasive plants that have reduced defensive attributes, biological control can be highly effective. However, in most cases, biological control must be supplemented with other control approaches (Hoffmann 1996). Blumenthal (2005) noted that "successful management of plant invasions may require both biological control, which aims to reduce the effects of natural enemy release by introducing natural enemies from an invader's native range, and methods aimed at limiting or reducing resource availability." In many, if not most cases, the solution involves integrated management that employs biological control as a basic strategy. It should be noted, however, that biological control can also be an effective solution even when the ultimate cause of the weed problem relates to something other than release from natural enemies.

SELECTING SUITABLE TARGETS FOR WEED BIOLOGICAL CONTROL

There are many ways to prioritize plants targeted for biological control. Highest priorities may be assigned to the most damaging weeds, to projects that would be least expensive to undertake (e.g. projects that have been successful elsewhere), those most easily done (e.g. surveys most readily done in friendly countries with willing collaborators), those that lack alternatives (biological control as the last resort), those most amenable to biological control (target susceptibility), those most likely to succeed (enhancement of success rates), or those that are most environmentally acceptable (safety to non-targets). Unfortunately, biological control is often viewed as the method of last resort and projects are initiated based on political expediency or funding opportunities. Peschken and McClay (1995) have developed a scoring system to assist in target selection that integrates a variety of these factors. This system, however, places the greatest emphasis on the economic importance of the weed with little concern for the number of related native plant species (an indicator of potential for non-target effects). Likewise, good targets with few native relatives might not be selected if they were primarily environmental weeds causing little economic loss. The Peschken and McClay system, however, provides a basis for a revised scheme placing more emphasis on environmental damage and protection of native species.

CONFLICTS OF INTEREST IN WEED BIOLOGICAL CONTROL

The politics surrounding weed biological control become complicated when the targeted plant is not universally regarded as undesirable (Tisdell et al. 1984). Cattlemen in Australia, for example, refer to "Patterson's curse" when discussing *Echium plantagineum* L., whereas beekeepers know it as "Salvation Jane". Obviously, ranchers favored control of this inferior pasture browse that contains potentially toxic compounds, whereas the beekeepers value it as a nectar source for honey production (Piggin 1982). Resolution of this conflict required the intervention of the courts and ultimately the passage of the Australian Biological Control Act, which now weighs individual interests against the national good (Delfosse 1985). The introduction of an exotic biological control agent is generally irreversible, the agents being impossible to eradicate after establishment (Harris 1988), so all viewpoints merit consideration. This dictates conservative actions and exhaustive inquiry (DeBach 1964b) and it must be determined that biological control serves the public interest. Any conflicts must be resolved before initiating a project. The process of initiating and conducting weed biological control is highly regulated in the USA, at both the state and federal levels, under plant protection laws that protect agricultural interests by preventing the introduction of plant pests. Some have argued, however, that the US system, by not providing adequate opportunities for public input and disclosure, is not well suited to the identification and resolution of conflicts of interest (Miller & Aplet 2005). A more aggressive model of conflict resolution exists in New Zealand, based on legislation explicitly governing the introduction of new organisms.

FAUNAL INVENTORIES: FINDING POTENTIAL WEED BIOLOGICAL CONTROL AGENTS

Choosing areas in which to conduct natural enemy surveys

Once a weed has been chosen as a target, it must be decided where to search for natural enemies. This requires delimiting the species' native range and determining where the invasive population originated. Such determinations may involve study of regional floras and other literature, examination of herbarium specimens and records, consultation with botanists, review of historical documents, climate matching, and genomic analyses. Many modern molecular techniques help in this process (Goolsby et al. 2006a; see also Chapter 15). The most promising search areas within the native range may be further defined by identifying ecoclimatic regimes that approximate those in the intended release area (see Chapter 14) (McFadyen 1991). Finally, it may be important to match plant genotypes to ensure that the correct genetic variant of the weed is surveyed and the optimal natural enemy biotype secured.

When selecting search areas for biological control agents of a particular weed, one often encounters the phrase "the area of origin" as being the most likely area in which to find herbivorous species that are specific to the target plant. This generally means the location where the plant species evolved although it sometimes is used to refer to the exact area within the weed's entire geographic range from which the weed was introduced. These two areas may or may not be the same. Present distributions are not always indicative of past evolutionary events so it may be difficult to define the evolutionary area of origin (McClay et al. 2004). Clues exist, however, as enumerated by Darlington (1957) and Cain (1943), that provide insight into the geographical history of a taxon. The two most useful of these are as follows: (1) The **center of diversity** of the species complex assumes that the origin will be where the most species in the group are found. Udvardy (1969) noted that a high ratio of endemics to widespread species is an indication of how long a particular taxonomic group has existed within an area. Such centers of diversity are where specialized herbivores would most likely have evolved (Wapshere 1974b). (2) The **degree of differentiation** clue states that there should be greater differences among populations of a species, among species within a genus, or genera within a family, in areas where the group has endured the longest. Molecular comparisons of genetic diversity among taxa allow such comparisons. The area of greatest haplotype diversity can be located for some pests and this used to infer areas of origin.

Conducting surveys

After a search area is identified, faunal inventories are done to compile as complete a list as possible of herbivorous species (usually insects or mites) and phytopathogens that exploit the plant. Species accumulation curves (also called rarefaction curves) can help determine how thoroughly an area has been surveyed and compare species richness among areas (Müller-Schärer et al. 1991). This involves plotting the number of species encountered against the number of individuals sampled, or some other measure of sampling effort (Krebs 1999, Heard & Pettit 2005). These data typically form a curve that levels off as the common species are collected and increased effort is needed to detect rarer species. The curve's asymptote estimates the total species in the community that are countable using the methods employed.

During surveys, hundreds of species may be enumerated. For example, 116 phytophagous insects were identified from kudzu [*Pueraria montana* var. *lobata* (Willd.) Maesen and S. Almeida] in China. More than 450 insect species were identified from *Melaleuca quinquenervia* (Cavier) Blake during surveys in Australia (Balciunas et al. 1994a) and nearly as many from *Mimosa pigra* L. in the Americas (Heard & Pettit 2005). Obviously, not all can be studied thoroughly, compared, and then ranked to pick the best candidate as is sometimes advocated (e.g. Myers 1985, Denoth et al. 2002), so prioritization and expert vetting is required. Opportunity often dictates which organisms undergo further evaluation. This choice may depend on the sequence of discovery, rarity or commonness of the organism, its range, seasonality, or hardiness, ease of rearing and developing research colonies, the time required for development (some wood-borers, for example, require 2 years to complete their life cycle), the investigator's knowledge of similar species, and a measure of intuition. It takes considerable time to develop a biological control agent as a candidate for release. So, rather than enumerating all possible species then conducting lengthy studies on every species in the list before selecting candidates, promising agents are usually quickly selected and studies begun on these few. These are not random selections as sometimes implied (Myers 1985), but rather thoughtful choices based on available information, direct observation, the experience and knowledge of the research group, and the practical realities encountered.

The concept of targeted agent selection, in which the biology of the targeted weed is compared to the mode of action of candidate agents, should play an important role in the selection process by identifying the type of agents needed (Briese 2006a). Surveys in both the native and invaded ranges of the weed show which kinds of natural enemies are missing in the invaded area, indicating which sort of species might usefully be introduced. Comparative studies of the population dynamics of the weed in native and invaded areas using population models can determine how the critical transitions or driving forces within the plant's life cycles may differ between these areas, further pointing to introductions that might be useful (Briese et al. 2002a, 2002b, 2006a, Jongejans et al. 2006). Such studies can indicate what traits of candidate agents are most likely to affect the weed's population dynamics. This information then guides the search for candidate agents (Briese 2006a). Obviously, this requires extensive study of both the weed and the candidate agents in both the native and adventive ranges, which necessitates a long-term commitment of resources. This approach is facilitated by availability of laboratories and staff in the area where the studies are needed (Goolsby et al. 2006a) but becomes much more difficult in remote or dangerous areas not amenable to frequent visits.

Opinion varies on whether common, widely distributed species or rare, sparsely distributed species make the best weed biological control agents. Common, widespread species are likely to tolerate a wider range of environmental conditions and their success on the host plant is apparent. These agents are often found early during surveys and are the first to be released, leading some to think that these are the most likely to succeed. Heard and Pettit (2005) noted that the weevil and the psyllid that have had an impact on *M. pigra* seed production in Florida were in fact widespread and abundant in the native range, whereas the rarer species have not had much impact.

On the other hand, plants are less likely to have adapted to damage caused by rare species, especially if suppression by natural enemies has caused their rarity. Such species, if released from their natural enemies through importation, are likely to increase in population size significantly. For example, the flower bud gall wasp *Trichilogaster acaciaelongifoliae* (Froggatt), which in South Africa controlled *Acacia longifolia* (Andr.) Willd., was rare and localized in its native Australia (Neser 1985) due to heavy parasitism and competition with larvae of a gracillarid moth. However, within two generations the gall wasp reduced seed production by 95–99%, decreased tree biomass, and increased tree mortality (Dennill 1985, Dennill et al. 1999). The success of this project (Dennill & Donnelly 1991) illustrates the potential value of rare species.

SAFETY: "WILL THOSE BUGS EAT MY ROSES?"

The most important consideration in selecting a biological control agent is the degree of risk to non-target plants in the recipient area. There are two components to this risk, the first being the adoption of non-target plants as complete hosts with attendant long-term, irreversible consequences and the second being spillover effects from temporary feeding on a non-target plant with short-term, localized, and reversible consequences (Briese & Walker 2002). Both of these possibilities should be considered during the testing process. Briese and Walker (2002) recommend separating the measurement of risk into three elements: phylogenetic relatedness, biogeographic overlap, and ecological similarity. Wapshere's (1974a) phylogenetic approach remains the central element in risk assessment, but this newer approach also gives weight to the predicted geographical distribution of the agent in the recipient region relative to distributions of non-target plants potentially at risk. Also, this newer scheme takes into consideration the degree of ecological similarity between the target species and a non-target species, with regard to specific requisites needed by the agent to survive and complete its life cycle.

Determination of host range is an exercise in risk assessment rather than a mechanism to ensure the absolute safety of a candidate agent (Briese & Walker 2002). There will always be risk, but the risk can be minimized. A test plant list should be developed before initiating host-range studies. Priority is given to phylogenetic lineages that are most closely allied to the target weed and progressively less emphasis is assigned to more distantly related taxa (Wapshere 1974a). This process requires knowledge of plant phylogeny and taxonomy, which has greatly improved with the advent of molecular systematics (Goolsby et al. 2006a). Briese and Walker (2002) suggest that ecology and biogeography be added as additional modifying filters to the above risk assessment. Other criteria often used to establish test plant lists include: (1) the known hosts of species closely related phylogenetically to the candidate biological control agent (especially congeneric species), (2) ecologically similar sympatric plant species that occur in the same habitat as the target weed, (3) economically important plant species, especially those grown in the same climatic zone as the target weed, and (4) distantly or unrelated plants with similar phytochemistry that could be attractive to the agent.

The safest targets are usually those with no native or economically important relatives in the area where control is needed (i.e. low or no phylogenetic or biogeographic overlap). Many plant species are included in host-range testing schemes for political rather than scientific reasons, which has led some to suggest that such testing, while politically comforting, adds little useful information and should be abandoned (Briese & Walker 2002, Briese 2003). Harris (1989), however, cautions against neglecting the political aspects of a biological control program. Others argue that phylogeny alone is not a sufficient criterion; that chemical similarity better predicts host use, so species with similar secondary chemistries are important to include in host-range assessments (Becerra 1997, Wheeler 2005).

Useful information about host ranges can also be derived from field observations or open-field tests in the country of origin that compare the occurrence of the candidate agent on the target plant with that on coexisting species, especially congeners or species of economic importance. All species of interest may not occur together, however, so desired comparisons are not always possible. To remedy this, non-target species can be interspersed with the target species in garden plots or placed at field sites to determine their potential use by the prospective agent (Clement & Cristofaro 1995, Uygur et al. 2005). This testing method produces valuable data on host range under natural conditions. However, the drawback comes from the fact that insect densities are difficult to regulate and may not attain the population levels comparable to those produced in non-native regions after having been freed of natural enemies. Also, all such open-field host-range tests are intrinsically choice tests by nature and thus pose some risk that low-ranked hosts may go unrecognized, yet be attacked if the agent disperses beyond the range of the target weed (see Chapter 17). Briese et al. (2002c) minimized this possibility by the use of a two-phase open-field test wherein the first phase allowed a choice between the host and non-target species. The target weed plants were then cut at the beginning of the second phase after the insects had colonized them which forced them to use the non-target species, emigrate, or starve. This approach was used to evaluate four candidates for control of *Heliotropium amplexicaule* Vahl. All four species fed only on the target weed and a closely related species of *Heliotropium* during the first phase. One of the candidate species, a thrips, disappeared quickly after the preferred host was removed while another, a leaf beetle, persisted for several days on the related plant before also disappearing. In contrast, the third species, another leaf beetle, rapidly colonized and fed on the related *Heliotropium*. These results demonstrated the host-selection behaviors of these insects under normal conditions as well as during the extreme circumstances that might occur after an agent has destroyed the target weed (Briese et al. 2002c). DeLoach (1976) used a similar approach to evaluate a weevil (*Neochetina bruchi* Hustache) for control of waterhyacinth. They planted non-target plants around a small pool containing waterhyacinth plants that were infested with the weevils then sprayed the waterhyacinth with an herbicide. They then monitored the non-target species to determine whether the weevils moved to them as their host plants died.

Building on the above types of information, most host-range data come from experiments done under controlled conditions (see Chapter 17). These bioassays challenge the prospective agent with various test plant species usually in a caged environment. Host selection is a process involving a sequence of linked behaviors (Heard 2000), so it is important to first determine when and at what stage host selection occurs (Wapshere 1989). For example, the female melaleuca gall fly (*Fergusonina turneri* Taylor) selects the larval host because the larvae cannot survive outside of the plant to move to another host. In this case, oviposition is the critical stage in host selection so there would be no need to test larval feeding. Other more mobile external feeders, such as grasshoppers, can readily move among plants so the point of host selection is less discrete. This type of information is needed to design tests in which candidates are allowed to choose either between or among non-target test species (choice minus control) or between or among test species and their normal host (choice with control), or are not given a choice and forced to either subsist on the test plant or perish (starvation tests). Results are usually discussed in terms of the candidate's performance on the test species relative to the normal host. Performance assessment invokes measurements of survival, feeding rates, residence time on the plant, amount of tissue consumed, growth and developmental rates, reproductive rates, and other life-history parameters. The resultant data must be considered in the aggregate to provide a complete picture of the suitability of test plant species as potential alternate hosts (van Klinken 2000).

PRE-RELEASE DETERMINATION OF EFFICACY

It has long been recognized that it would be desirable to predict the effectiveness of biological agents before introducing them. This would reduce the risk of negative effects (see Chapter 18 on indirect effects) and increase efficiency by not wasting resources in evaluating ineffective agents (McClay & Balciunas 2005). Harris (1973) developed a scoring system that was later modified by Goeden (1983) in an attempt to select effective insect agents. These systems suffer from a bias toward western US rangeland weeds and emphasize characteristics of the agents while ignoring equally important plant traits. Rather, what is called for is to predict how much damage a given agent can do to the weed and compare this with the amount of damage needed to control the weed (Harris 1985). Wapshere (1985) criticized the Harris and Goeden schemes and elaborated on an earlier proposed ecoclimatic approach (Wapshere 1970), which emphasizes the effects an agent is observed to have on a weed in its native range. However, observation of such effects during short periods do not always reveal the root causes of a weed's density or degree of damage.

In general terms, the effect of an agent depends on four factors: (1) the per capita damage inflicted, (2) the rate of increase of the agent's population, (3) the duration of attack, and (4) timing of attack. The first aspect is highly predictable from laboratory testing, the others less so. No one has yet successfully predicted the impact of a new biological control agent before its release. Nonetheless, since high per capita impact is a necessary if not sufficient condition for an effective agent, measurement of per capita damage before release is a good first step. Such information ensures that an agent approved for release at least has some potential to control the weed. Modeling approaches advocated by Briese (2006a), in conjunction with information on per capita impact, can help determine that the proper stage of the weed's life cycle is targeted by the agents chosen for release.

Because high per capita impact is necessary but not sufficient, effects on plant performance do not correlate directly to effects on plant population dynamics (Crawley 1989). Ultimately, the impact of an agent depends on a blending of the per capita level of damage it inflicts, how great a density the agent can attain after release, and the significance of the damage to the plant's population dynamics (Cullen 1995). Unfortunately, biological control outcomes are not generalizable and the impact of one insect on its host plant does not provide any means to predict the outcome of another agent on another plant, even if they bear some similarities. Each case is unique because climatic factors, competition, and predation, as well as other novel aspects of the environment, all vary. Rather than striving for the ability to predict efficacy across projects, **adaptive management** is a better goal, in which the objective is to learn as quickly as possible how the agents at hand affect the target species in the recipient environment and then use that information to guide the project (Shea et al. 2002). Such adaptive management also entails analyzing successful projects conducted in other areas against the same weed to determine how much of the previous success might be transferable to a new location.

HOW MANY AGENTS ARE NECESSARY FOR WEED CONTROL?

How many and what combinations of agents it takes to reduce the density of invasive plant species is a widely debated question. The answer, however, differs for each unique insect–plant association and is not generalizable. It is what the agents do to the plant and how their impacts interact, not the mere number of agents, that cause reductions in weed population densities (Hoffmann & Moran 1998).

Some ecologists argue that the number of biological control agents introduced against a given target weed should be severely minimized, either to be "more science-based" (Harris 1977), or to allay social concerns about unpredictable potential indirect effects (Denoth et al. 2002), or out of fear that interference among agents might lessen the total impact achieved (e.g. Crowe & Bourchier 2006). Other ecologists have articulated a contradictory concept called "cumulative stress" (Harris 1981, 1985), wherein several species acting together are more likely to surpass a damage threshold beyond which the plant is no longer able to tolerate additional stress and succumbs. This principle has, in fact, been demonstrated in the case of *Sesbania punicea* (Cav.) Benth. in South Africa. Hoffmann (1990) determined that control of this small tree would require 99.9% reduction of seed production. Three insects were introduced: *Trichapion lativentre* (Bèguin-Billecocq), which destroys buds; *Rhyssomatus marginatus* Fåhraeus, which feeds on seeds; and *Neodiplogrammus quadrivittatus* (Olivier), which bores into stems. *Trichapion lativentre* reduced seed set by up to 98% by attacking flower buds and *R. marginatus* destroyed up to 88% of the remaining seed crop (Hoffmann & Moran 1992). Together, they completely suppressed reproduction. Decline of existing stands, however, was most closely associated with presence of *N. quadrivittatus* (the borer), provided one or both of the other seed-suppressing insects was present (Hoffmann & Moran 1998). Collectively, these three agents reduced the weed to insignificance, which would not have occurred with one of them acting alone (Hoffmann & Moran

1998). However, other cases, such as the interaction between *Urophora affinis* Frauenfeld and the weevil *Larinus minutus* Gyllenhal, when attaching seeds heads of *Centaurea stoebe* L. ssp. *micranthos*, show that potential exists for biological control agents to act antagonistically to an extent that reduces rather than increases weed control (Crowe & Bourchier 2006). So, judgments must be made in each case if additional agents seem likely to be beneficial or not.

The value of multiple agents is obscured by some reports in the literature: Myers (1985), based on *post hoc* analysis of 26 successful projects listed by Julien (1982), concluded that, in 81% of the cases, success was attained from the introduction of a single insect species rather than by a cumulative stress from several. However, this was based on subjective assessments of how much each agent contributed, not experimental analyses of impact where treatments consisted of various numbers of agents. This analysis does not distinguish between a single agent being solely effective versus its being the dominant agent in a group of agents that collectively cause enough stress on the plant to provide control. Denoth et al. (2002), using an updated database (Julien & Griffiths 1998), recognized that rates of agent establishment did not differ between single-agent and multiple-agent projects, and that success increased (albeit, weakly statistically) as more agents were released.

The historical record of weed biological control, therefore, does not support either view strongly (one agent alone being effective compared with cumulative stress from multiple agents) and clearer answers are available for single cases, such as the *S. punicea* example discussed above (Hoffmann & Moran 1992), which strongly supports the cumulative stress mechanism. It should be noted that the order in which agents are released may be important. One agent might, for example, weaken the target plant in ways that increase its susceptibility to another agent. Or, conversely, one agent might deplete a critical resource needed by a second agent. In the first case, establishment of the subsequent agent would be facilitated; in the second case it would be inhibited.

RELEASE, ESTABLISHMENT, AND DISPERSAL

The many years of diligent work that lead up to the release of an agent are wasted if the agent fails to establish, so it is prudent to expend significant effort to maximize the likelihood of field colonization (see Chapter 19). For populations of plant-feeding insects to establish, they need favorable climate, suitable host plants, absence of severe competition, shelter, suitable soil type, and other physical factors (Sutherst 1991). Localized incipient populations may also die out due to uncommon random events like floods or fire (e.g. Hoffmann & Moran 1995), and, in such cases, further release effort could lead to a more successful outcome. There have been about 1200 attempts worldwide to establish biological weed control agents (Julien & Griffiths 1998, supplemented by multiple sources), an attempt being the release of one biological control agent against one weed in a single geographic area. Agents successfully established in 720 cases, whereas establishment clearly failed in 347 cases, and results were inconclusive for the remaining 133. Hence, one-third of all attempts for which the outcomes are known failed due to lack of establishment. Project success rates, then, might be drastically improved by expending more resources on this aspect of a program. This requires an extensive understanding of the ecological requirements of the agent being released.

When few individuals are available for release, a choice must be made between making a few larger releases or several smaller ones. The better choice depends upon characteristics of the agent as well as field conditions (see Grevstad 1999a). Minimum viable populations (MVP) of some species, such as the melaleuca weevil *Oxyops vitiosa* Pascoe, seem to be very low (Center et al. 2000), because they can establish populations with relatively few individuals, perhaps even a single gravid female. In these cases, it is wise to guard against the risks of local extinction by releasing the available insects over many sites. Other species, for which MVP levels are higher, may require the release of large numbers at a few locations (e.g. Memmott et al. 1998, Grevstad 1999b) and perhaps continual augmentation of the founder colonies with supplemental releases (e.g. Center et al. 1997a).

Numbers of individuals (but not their genetic diversity) can be increased by mass rearing, but this can be difficult when the agents have to be reared on plants. Another frequently used strategy is to first establish a nursery colony at a single field site to provide stock for dissemination to other areas. Cages may be used early in a release program to keep the individuals of the founder colonies together to maximize their chances of finding mates. Cages also screen out some predators and protect the agents and their plants from storms or other damage (Briese et al. 1996). Cages are removed after the population establishes and numbers have increased. Site or plant conditions can also sometimes be manipulated to increase the likelihood that the agent will establish or increase more rapidly. For example, reproduction in some insects is related to the nutritive value of the plant tissue (e.g. *Cyrtobagous salviniae* Calder and Sands). Tissue nitrogen levels may be increased by using fertilizers, thereby facilitating population increases (Room & Thomas 1985). Other species may need flushes of new plant growth (e.g. *O. vitiosa*), which can be induced by pruning or mowing.

Following establishment, it is useful to measure rates of dispersal to determine where to make additional releases. Disperal monitoring frequently employs transects radiating spoke-like from the release point in four cardinal directions. Observers follow the transect while continuously searching or looking at fixed points for the agent. Traps or sentinel plants can be positioned at monitoring points to facilitate detection. The dispersal distance averaged over all transects divided by the time after release (or after establishment) provides the population diffusion rate. Not all organisms disperse in a symmetrical, radial pattern though, so monitoring techniques may need to be modified as appropriate (e.g. Grevstad & Herzig 1997). More sophisticated designs that employ a spatial grid with regularly spaced sampling locations may also be employed. These are usually done using Geographic Information Systems (GIS) and overlaying the grids with maps of physical (lakes, streams, forests, soils, etc.) and environmental (wind direction, rainfall patterns, etc.) landscape features. Distributions are assessed periodically by repeatedly surveying quadrats from the grid. The data can then be analyzed using spatial modeling.

EVALUATION OF IMPACTS

Impacts of an introduced biological control agent on a target weed must be considered on at least three different levels: (1) on the performance of individual plants, (2) on plant populations at the local level, and (3) at the landscape or regional level. Some of these considerations are the same as for biological control of arthropod targets (see Chapter 20), but others, particularly some aspects of impact on performance of individuals, are distinctive.

Effects on the performance of individual plants may be measured in a laboratory or field setting and usually involve comparisons of plant reproduction (flowering and seed set), plant stature, and vegetative growth between treatments with or without the biological control agent. In some cases, densities of the biological control agent may be varied to determine critical damage thresholds. Insect densities may be regulated using enclosures, exclosures, or insecticides. Insecticide exclusion experiments are often the most valuable inasmuch as cages can be a confounding factor. Insecticides, however, generally fail to totally eliminate the agent's population and densities on the non-treated plants are nearly impossible to control. Treatments, then, are best regarded as low or high. These studies are vital to the recognition of the symptoms of bioagent-induced stress in natural field settings.

As noted above, showing that biological control agents affect plant performance is entirely different from demonstrating that they affect plant population dynamics (Crawley 1989). For this reason, studies on plant populations at the local level are needed. In order to be considered successful, it must be demonstrated that biological control reduces plant density or coverage or enables more efficient control by other means. Determination of population-level effects usually involves studies of recruitment, stand density, extent of coverage, and rates of expansion. Ideally, plant populations should be experimentally compared between sites where the agent has been released and similar control sites where it has not been released (McClay 1995). Baseline data, from which changes can be measured following establishment of biological control agents, are advantageous, so monitoring should begin even before releases are made. Relationships between plant traits can be elucidated so that allometric relationships can be established. These enable the estimation of variables that require destructive sampling from more easily obtained morphometric measures (e.g. biomass from plant height).

The geographical extent of agent impacts should next be determined by doing assessments at many sites. This often precludes collection of detailed data. An example is provided by a study of the hydrilla leafmining fly (*Hydrellia pakistanae* Deonier). Hydrilla in tanks was stocked with varying numbers of flies and it was found that the percentage of leaf whorls that were damaged provided an indicator of larval intensity. Damage to 60–70% of the leaf whorls was a threshold associated with biomass reduction. Monitoring during different times of the year at several locations revealed that damage levels rarely exceeded 15% of the leaf whorls, indicating that the flies were not causing sufficient damage to affect plant density. These data indicated a need for introduction of additional biological control agents (Wheeler & Center 2001).

The least detailed evaluation approach involves broad regional assessments of the range and extent of the weed, usually as total regional area and how this changes as agents exert their effects. This involves satellite imagery or aerial surveys or the concerted effort of large numbers of participants doing ground-based surveys. Such assessments are usually done by resource management agencies rather than individual research groups. These data, along with information from the previous evaluation phases, provide a complete picture of the impact of biological control at the landscape level.

NON-TARGET IMPACTS

Concern exists over potential non-target effects of introduced biological control agents. It is therefore important to test laboratory-derived host range predictions in the field. Potential non-target hosts should be identified at release sites and periodically checked for damage. Field plots containing test plants of non-target species can be established in weed-infested areas and then monitored on a regular basis. Or, common garden experiments may also be used. Or, the biological control agent may be released directly on to potential non-target plants and then observed to determine their residence time. Monitoring to detect potential non-target effects, or their absence, is now required as a condition of release permits in the USA (APHIS-PPQ Form 526).

WHEN IS A PROJECT SUCCESSFUL?

Success in weed biological control must be judged on a per-project, not per-agent, basis (McFadyen 2000). Success rates of individual agents are not important: indeed, it is to be expected that only some of the agents used in a control program will contribute to its success. Most authors use success to mean only complete success, in which no other control measures are required to reduce the weed density to non-problem levels. However, this neglects the importance of partially successful projects, which have value because less effort is required to control the weed, either because the density or extent of the weeds is reduced, or the weed is less able to reinvade cleared areas or slower to disperse (Hoffmann 1996). Even moderate amounts of stress can reduce the competitive ability of the plant and make it less invasive.

Successful biological control agents often act by preventing outbreaks, not by reducing populations that are already high. Therefore, in order to see biological control in action, it may be necessary to perceive the outbreaks that never happen. This is a difficult feat, at best, and explains why so many biological control projects are incompletely evaluated and even successful projects may be undervalued or forgotten. Thus, statistics on success rates should be viewed with suspicion and considered conservative, inasmuch as only obvious successes are reported. Furthermore, weed declines may occur incrementally over many years or even decades and may not be easily observed, especially when the observational baseline shifts over time, or personnel changes over decades interrupt collection of critical data. Success of projects should be assessed in terms of the project's own original objectives.

A project is economically successful when economic benefits of weed suppression exceed project costs. The harm done by weeds, however, is sometimes difficult to measure, as is the benefit brought about by biological agents. The benefits of biological control continue to accrue indefinitely over time, so that the cost/benefit ratio of a project increases annually after successful control. Furthermore, economic factors such as changing inflation rates, prices of agricultural produce, or returns on alternative investments make computing benefits complex (Room 1980). Page and Lacey (2006) conducted an economic analysis of 104 years of weed biological control projects in Australia. They found that the annual return over that period was AUS$95.3 million for an average annual outlay of just $4.3 million. The total return was estimated at about $10 billion, making it one of the most successful scientific programs in the nation's history.

Ecological success, the relevant measure for projects against weeds of natural areas, is not measured in economic terms, but rather the degree to which the invaded natural communities return to their pre-invasion state as the competitive effects of the invasive weed are reduced.

CONCLUSIONS

Classical weed biological control programs have become increasingly sophisticated, being comprised of numerous, disparate facets often involving multiple laboratories, foreign and domestic, as well as the efforts of numerous scientists with expertise in widely divergent fields. These increasing demands for comprehensive knowledge about both the target weed and the agents proposed for release make small programs of weed biological control difficult and potentially irresponsible. However, most biological control programs do not follow a linear sequence of steps, such as described in the literature (e.g. Wapshere et al. 1989). More often, they require adaptive management in which goals are set according to a general strategy but with a learning plan such that the prescribed steps can be altered as knowledge increases and uncertainty decreases (McFadyen 2000). The adaptive approach allows the efficient use of human and material resources, which are often limiting. Projects should be done in their entirety, including follow-up evaluations. Commitments should be sought at the outset to ensure that project evaluations are included as part of an overall funding package. Follow-up should include non-target and food-web effects as well as evaluations of the efficacy of the agents, both individually and collectively.

Part 5

TOOLS FOR CLASSICAL BIOLOGICAL CONTROL

Chapter 13

FOREIGN EXPLORATION

The following chapter presents information aimed at scientists actually conducting foreign exploration, together will background on design and operation of quarantine facilites to handle material collected. For further information on these activities, see Bartlett and van den Bosch (1964), Boldt and Drea (1980), Klingman and Coulson (1982), Schroeder and Goeden (1986), and Coulson and Soper (1989). Work in the native range during foreign exploration can help in both selecting potentially more effective natural enemies and more accurately understanding the host ranges of the agents encountered (Goolsby 2006a). Foreign explorers should familiarize themselves with such opportunities and devise a strategy for maximizing the value of collecting time spent in foreign exploration.

PLANNING AND CONDUCTING FOREIGN EXPLORATION

Selecting survey locations

Depending on how much is known about an invasive species, selecting natural enemy collection areas can be completely straightforward or extremely uncertain. Invasion of the western USA by ash whitefly, *Siphoninus phillyreae* (Halliday), was followed immediately by collections in Europe and the Middle East where the whitefly and its natural enemies were well known. In contrast, efforts to collect natural enemies of giant water fern (*Salvinia molesta* Mitchell), cassava mealybug (*Phenacoccus manihoti* Matile-Ferrer), beech scale (*Cryptococcus fagisuga* Lindinger), avocado thrips (*Scirtothrips perseae* Nakahara), and banana weevil [*Cosmopolites sordidus* (Germar)] were all thwarted, at least initially, because the pests were either new to science or their area of origin was uncertain.

For invasive plants, the species itself may be well known but may have an extremely large range, with unclear genetic relatedness among geographic populations, making it difficult to select priority areas for natural enemy collection. Also, the invasive population of a pest may include several species or hybrids derived in complex ways from several parent species. For example, saltcedars (*Tamarix* spp.) invasive in the southwestern USA consist of four species and their hybrids (DeLoach et al. 2003). The parent species of the pest populations have a collective native range that extends from North Africa to China (Milbrath & DeLoach 2006). Molecular analyses (see Chapter 15) can unravel such relationships and help identify which locations might be the best to search for natural enemies.

To deal with difficult cases, several sources of information can be helpful, including literature on the pest (or its relatives), professional overseas contacts, climate matching (see Chapter 14), and, for stenophagous insect pests, the biogeography of the host plant. If the pest is initially known from many widespread areas, molecular tools can be used to determine which location is most probably the source of the invasive population (e.g. Williams *et al.* 1994, Biron et al. 2000, Gaskin 2003, Goolsby 2004). The population of hemlock woolly adelgid (*Adelges tsugae* Annand) invasive in the eastern USA, for example, might potentially have come from at least three areas (western USA, Japan, and China), but molecular analyses showed conclusively the origin to have been Japan (Havill et al. 2006). Such molecular comparisons can also shed light on whether a pest infestation derives from one or several independent sources (e.g. Carter et al. 1996), which indicates that separate natural enemy collecting might be desirable for the subpopulations of the pest in the invaded area.

When the invasive pest is unknown outside of the invaded area, finding the native range will depend on surveys, guided by taxonomic, biogeographical, and climatic inferences. Such surveys might explore areas where relatives of the pest are known to occur or where the host plant evolved (for invasive insects). Once populations of the pest have been located, their genetic diversity can be measured to determine where it is greatest, which is likely to be the area of origin. Areas of origin are also predicted to be those where the pest's natural enemies are most diverse. These and other concepts are useful for generating hypotheses about a pest's origin, but none provides an infallible mechanism for locating the native range. Below we discuss applications of these concepts.

Where the closest relatives of the pest species occur

If a pest is unknown apart from its invasive range, its native range might be where the largest numbers of congeners occur and, specifically where the most closely related species are found. Horse-chestnut leafminer (*Cameraria ohridella* Dschka and Dimic), an invasive species in Europe, is unknown elsewhere. Europe is not seen as its native range because of its recent eruptive spread there and its lack of specific parasitoids. Since other members of *Cameraria* are from the Americas and Asia, not Europe, those areas are seen as potential areas of origin (Kenis et al. 2005).

Where the pest's host plants evolved

For insects with high host-plant specificity, the center of evolution of those plants, if known, may be the area of origin of the insect. For example, some citrus pests such as certain scales are believed to have originated in southeast China, the area of origin of the genus *Citrus*. Host specific avocado-feeding insects are likely to have evolved in Central America (Hoddle et al. 2002b). Or, if the history of plant movement is known, this may provide evidence as to where an insect population came from. For example, the coffee leafminer, *Leucoptera coffeella* (Guérin-Méneville), an important pest in the Americas, is suspected of being an invader, but its point of origin is unknown. Clues to its origin include the little-noticed presence of this insect on coffee in Madagascar and the island of Réunion, the presence of several native coffee species on Réunion, and the arrival of *Coffea arabica* var. *bourbon* to the Americas by way of Réunion. Collectively these facts point to Réunion as possibly the native range of this moth (Green 1984).

The native range is where the pest species shows greatest genetic diversity

The torymid gall wasp *Megastigmus transvaalensis* Hussey, in Africa, feeds on plants in the genus *Rhus*, but in the rest of the world is associated with *Schinus*. Scheffer and Grissell (2003) analyzed genetic variation in a mitochondrial cytochrome oxidase sequence and found extensive variation in African populations but no variation elsewhere. They concluded the insect was of African origin. Gwiazdowski et al. (2006) used the same approach to assess likelihood of a western European compared with a southeastern Europe/western Asian origin for an eriococcid scale (*C. fagisuga*) that is an invasive forest pest in North America.

Where the pest's natural enemies are most diverse

If a pest is known from several areas and genetic analyses do not separate derived from ancestral locations, diversity of the associated natural enemy fauna might be helpful. Pschorn-Walcher (1963) in working with various sawflies suggested that the area of origin would be associated with large complexes of specialized parasitoids. He suggested that Central Europe was the likely home of the sawfly *Pristiphora erichsonii* (Hartig), where its parasitoid complex is large and distinctive, not the UK, where the sawfly has relatively few parasitoids. Similarly, Kfir (1998) used this concept to argue that South Africa, not Europe, was likely to be the native home of the crucifer-feeding diamond back moth [*Plutella xylostella* (L.)], which had previously been thought to be from Europe because that is the area of origin of the cultivated *Brassica* crucifers this pest commonly attacks.

Planning a foreign collecting trip

Planning a foreign exploration trip starts with summarization of all available relevant information, including pertinent taxonomic information, identities of relevant specimens in museum collections, and notes from previous collecting trips in the search areas. This information, together with recent correspondence with collaborators, is used to determine the most suitable seasons and locations for the natural enemy surveys. In addition to picking locations to search, planning for a foreign collection trip must address obtaining necessary permits and visas, finding a competent collector and collaborator in the areas to be searched, and assembling the needed equipment.

Permits

The first step in organizing a foreign collecting trip is to obtain the necessary permits from the countries to be visited to collect and export live material, and also the necessary permits to import the collected material into the collector's home country. Importation planning requires developing an agreement with an authorized quarantine facility for receiving and processing imported material. Export of dead specimens collected for study or as voucher specimens is also regulated in some countries. Since laws will vary by country and may change with time, biological control workers should seek up-to-date local information by inquiring directly with a colleague. The collector must provide the quarantine facility copies of the importation permits, expected dates of shipments, arrangements for customs clearance of shipped material, and clear plans for the establishment of colonies of the imported natural enemies. In support of development of these colonies, arrangements with the receiving quarantine laboratory must be made so that necessary host material is available for use in rearing.

Permission for importation to quarantine does not imply permission for release into the environment. Typically, any plausible biological control agent may be imported into quarantine for study. Release from quarantine requires the development of adequate data to assess potential risks of the agent to the local fauna or flora (see Chapter 17).

Qualifications for an explorer/collector

People collecting natural enemies in foreign countries must be good planners, highly flexible and adaptable, and rugged travelers. They must have a broad knowledge of the target pest, its potential natural enemies, and possible host plants. Foreign collecting may require work in areas that are difficult to reach and lack efficient infrastructure and services. A local bilingual assistant familiar with local customs is often necessary for gaining secure access to potential collecting sites. Collaborations with local research institutions often increase the effectiveness of surveys. Surveys may also be subcontracted to organizations specialized in natural enemy survey and collection, such as the USDA-ARS overseas biological control laboratories, CABI Bioscience (UK), CSIRO (Australia), or other national or regional agencies with appropriate expertise.

Equipment

Arrangements should be made before departure for all needed equipment (Table 13.1). In cases where work is being conducted in conjunction with a laboratory in the foreign country, some items such as microscopes may be available there. **Collecting equipment** may include plant-harvesting or soil-digging tools; sweep nets and beat trays; rearing cages; sample storage vials or bags; maps and travel guides, notebooks, and global positioning system (GPS) devices to record collection locations; and ice chests or other insulated storage containers to prevent overheating of collected material. Digital cameras are essential for recording the condition of collection locations and for recording initial identities of targets, candidate natural enemies, and host plants. **Identification and handling**

Table 13.1 List of equipment for exploration for natural enemies

Collection equipment
Sweep nets, maps, camera
Spades, shovels, trowels, pruning shears, saws, gloves
Collecting bags (paper, plastic), ice chest or portable refrigerator
Identification and handling equipment
Microscope and light, magnifying glass and lenses
Reference texts for local floras, etc.
Small vials for isolating specimens, labels
Silica tubes for collecting specimens for DNA analysis
Cards for labeling groups of vials
Record notebook or forms, scissors, ruler
Insect pins, forceps (fine and large), alcohol, honey
Packaging and shipping supplies
Shipping boxes (cardboard or wood)
Styrofoam or insulated outer containers
Cold or hot packs
Address labels, quarantine labels
Copies of shipping permits, tape, string

equipment may include forceps, fine camel-hair paint brushes, scalpels, razor blades, scissors, plant pruners, probes, microscopes, hand lenses, optivisors, lights, silica tubes for DNA samples, vials or other containers either to preserve natural enemies for future DNA analyses or to house collected natural enemies for colony establishment, parafilm for sealing vials, honey for feeding natural enemies (if allowed by the host country), Petri dishes with prepared media for inoculation with pathogens, and literature pertaining to identification of host plants, hosts insects, and natural enemies. **Packaging and shipping supplies** will include primary external containers, internal insulated containers, frozen gel packs, wrapping materials, cushioning material, shipping boxes, tape, shipping labels, transparent label envelopes, and permits.

Collecting specimens and recording field data

During surveys, as much material as possible should be collected. Local scientists or their graduate students can be helpful in locating native habitats or agricultural plantings suitable for collecting. Field notes should be kept that include dates of all collections; names of contacts, villages, farms, natural areas, or national parks visited (with GPS coordinates, elevation, and digital pictures of sites); notes on seasonal weather patterns at collection sites, habitat types, plant communities, host plant species, and host species located, as well as natural enemies found. Collections should be made in areas not subject to pesticide application, avoiding commercial farms (which are likely to be relatively sterile) in favor of organic farms, untreated backyard plants, and botanic or public gardens.

Since collection sites may be far from airports, it may be necessary to keep material alive while traveling for many days. Active insect stages should be fed appropriate food, provided with water, and kept cool. Humidity in the collection containers must be in the 40–75% range to avoid killing by dehydration. Similarly, high humidities that could promote growth of fungal pathogens and free water, in which insect may drown, must be avoided. Wherever possible, field-collected insects should be carried in insulated ice chests, which can be kept cool by recharging them with ice. Alternatively, a more sophisticated type of chest exists that uses battery power or DC current to cool the chest, eliminating worries about overheating or water damage (from melting ice). Proper labeling of the collected samples is very important as it is the label that accompanies the material to the quarantine facility; the collector should not rely on field notes alone, but provide all the important information with the specimens. Each collection should be labeled with a unique accession number that can be used to track the collection from field, to quarantine, and eventual release as a biological control agent.

Adequate time should be allowed at the end of each day to sort, label, and process material for storage or shipment; this may exceed the time spent collecting. Often it is advisable to stay 2 days in each location, to collect in both morning and evening, when insect activity is greatest. Access to a local laboratory facility may provide working space and facilities, but in general the explorer should be prepared to process collections in a hotel room or similar accommodation.

Collected material should be shipped as frequently as possible to ensure that insects reach the quarantine facility in the importing country alive. More frequent, smaller shipments are advantageous because any one shipment may be lost due to delays or errors. Shipments should be made at the beginning of the work week to avoid unnecessary shipping delays over weekends. Collections made at the end of a trip should be hand carried to the importing country, if permitted. Allowance must be made for any inspections required by the exporting country prior to final authorization to ship. The most durable life stage of the natural enemy, if available, should be selected for shipment. Such stages would include insect pupae, diapausing stages, or eggs. For pathogens, host cadavers, spores, or fungal hyphal colonies in agar might be stages chosen for shipment.

SHIPPING NATURAL ENEMIES

Shipment of live natural enemies (Figure 13.1) is a critical part of most classical biological control programs (Bartlett and van den Bosch 1964, Boldt & Drea 1980, Bellows & Legner 1993) and many losses occur at this step. Shipment should be made by an express shipping service, air freight, or priority airmail, and collectors will likely have to return to major cities to access such services. If possible, select the day of shipment so that packages arrive in the home country early in the working week to facilitate rapid handling of the package and avoidance of the delays common on weekends. Personnel at the port of entry and quarantine facility

(a)

(b)

(c)

Figure 13.1 A container for international shipment of natural enemies. (a, b) Note the use of Styrofoam insulation and pack of artificial ice (in center) for cooling (the foam top of the box has been removed for the photograph). (c) Assembled package and associated shipping labels. Photographs courtesy of USDA/BIRL, (a) M. Heppner, (b) S.R. Bauer, and (c) R.M. Hendrickson; reprinted from Van Driesche and Bellows (1996) with permission from Kluwer.

should be notified by e-mail, fax, or phone of the details for each shipment (name of the carrier, routing information, time of arrival, and airway bill number). The external packaging material should bear the necessary permits and address labels to facilitate recognition and handling by customs and agricultural inspection personnel to avoid delays at the port of entry, a common source of death to shipped natural enemies.

Some general recommendations for shipping insects include the following.

1. Ship by the fastest route available, using a well-thought-out plan for rapid clearance of customs and reshipment to quarantine.
2. Ship materials in insulated ice chests or Styrofoam containers fitted snuggly inside cardboard boxes.
3. Avoid fresh plastic materials (which may give off toxic fumes) and gelatin capsules (which soften in high humidity).
4. Seal dishes and vials well (parafilm is useful); use a larger number of small containers rather than a few larger ones, and do not overfill packages with material.
5. Label dishes or vials with accession number, collecting date and location, and, if available, the name or group of the agent (which must match the permit).
6. Avoid overheating *en route* by placing frozen gel packs inside packages.
7. Avoid condensation in packages by trimming plant material to a minimum, using cloth not plastic bags, and ventilated vials or Petri dishes, and adding absorbent materials inside vials.
8. For herbivorous insects, provide ample plant material as food. Use of Zip-loc bags inflated with air works well for eggs and early life stages that are susceptible to being crushed.
9. Avoid excessive dryness, if necessary, by adding a Petri dish filled with a saturated salt solution and then sealed with semi-permeable membrane (such as Opsite Wound®, available from medical suppliers) (Hendrickson et al. 1987).

10. Pack natural enemies collected from soil in moistened excelsior or sphagnum moss (if allowed by the permit conditions).
11. Provide shipped adults with a moist, absorbent material, such as paper toweling, as a resting substrate.
12. Feed adult parasitoids by placing small spots of honey placed on the inside of the containers. Provide adult moths or flies access to sugar water, Gatorade®, or honey, either on the inside of the glass containers or on cotton wicks or sponges (note: importation of honey may be prohibited in some countries, such as Australia and New Zealand, but can be purchased on arrival). Moistened raisins may also be used (Bartlett & van den Bosch 1964). Pollen may be provided to feed predaceous mites.
13. Limit the growth of saprophagous organisms on diseased arthropods or plant tissue by either transferring pathogens to artificial media prior to shipping or dividing samples of pathogens into several small lots to limit contamination arising from any particular specimen.

OPERATING A QUARANTINE LABORATORY

A quarantine facility is designed to be a highly secure area in which shipments of organisms from abroad can be opened, contaminants excluded, and the desired natural enemies reared while their safety for release in the receiving country is determined.

Design and equipment

Rooms for handling beneficial arthropods must be sealed from the outside environment through special construction details, control of entrance through multiple doors, and screening of air-exchange ducts (Leppla & Ashley 1978). Personnel should wear laboratory coats and shoe covers while in the facility, remove them when exiting, and leave them in the facility. Rooms designed for use with pathogens require additional precautions because of the size of the organisms under study (see Melching et al. 1983 and Watson & Sackston 1985). Pathogen-handling areas must be sealed, and air recirculated through a double set of filters able to remove particles down to 0.5 μm in order to remove airborne fungal spores and bacteria. Exhaust air must pass through a third, deep-bed filter before being vented. Air pressure within the facility must be lower than the outside atmosphere to prevent unanticipated air exchange from inside the facility to the outside. Work spaces within the quarantine area are typically divided into smaller cubicles to limit contamination among study areas.

Identification of imported arthropods typically requires a binocular dissecting microscope (10–120×) with a high-quality, fiber-optic light. A compound microscope is necessary to screen cultures of beneficial arthropods for entomopathogens and to identify imported pathogens. Taxonomic literature needed for identification of introduced species should be available in quarantine or accessible through internet connections. Other necessary equipment in quarantine laboratories includes: (1) rearing cages or containers to separate and house natural enemies, (2) a water source and other facilities to prepare media for culturing micro-organisms, (3) autoclaves or furnaces to sterilize or burn unwanted contaminants or shipping materials, (4) areas for maintenance of plants, (5) refrigerators or cold rooms to maintain organisms in a diapause state, including a freezer for killing arthropods, (6) growth chambers for holding natural enemies at desired culturing temperatures, (7) carbon dioxide to anesthetize arthropods, and (8) various tools, ranging from hammers and screwdrivers to repair and adjust cages, to forceps and probes for handling minute arthropods.

Personnel and operating procedures

Sound and comprehensive administration and operating policies are basic to the safe and effective operation of a secure quarantine laboratory. There should be a single quarantine officer responsible for all aspects of security for a facility. Consolidation of responsibility increases security, ensuring that each organism is properly handled and that adequate records of all organisms received, shipped, or otherwise processed are kept. The quarantine officer should be familiar with the regulations and laws governing the conduct of the facility, develop and maintain contact with the regulatory personnel at the local ports of entry through which shipments arrive, process all incoming shipments, maintain the necessary records, and oversee the functioning of the facility.

The quarantine officer is also responsible for maintaining contact with all the people likely to be involved in the collection and shipping of natural enemies to the quarantine facility. Quarantine staff should be familiar with the concepts and practices of biological control and how they apply to the specific projects and organisms with which they are involved. Project scientists conducting quarantine research should have knowledge of the taxonomy and life histories of targeted pests and possible natural enemies. Such information can be critical in planning host specificity studies and in verifying the nature of the relationship between a presumed primary natural enemy and its host.

All workers in the quarantine facility should be familiar with the physical operation of the containment facility and its equipment, particularly ventilation, power, and utilities, and with function of the autoclave or other sterilizer. Names and telephone numbers of maintenance and repair personnel to be contacted in cases of needed repairs should be available for normal workdays as well as weekends and holidays. Janitorial needs within the quarantine facility should be handled by quarantine personnel. Equipment and facility service or repair personnel must be apprised of the importance of quarantine security and should be accompanied by quarantine personnel when working in the security area. Visiting scientists and regulatory personnel should also be accompanied in the facility. Casual visits by individuals or groups should not be permitted in the containment area.

Fire, earthquakes, vandalism, and illness may interrupt quarantine operations. Instructions should be posted on entryways advising emergency personnel which entry and exit procedures will be least likely to breach quarantine security. Telephone numbers for contacting the quarantine officer and assistant, during both business and off-duty hours, should be posted at the entrance to the facility.

MANAGING INSECT COLONIES IN QUARANTINE

After natural enemies have been collected and shipped to a quarantine laboratory, sustainable colonies of the agents must be created. This requires: (1) successful recovery of natural enemies from overseas shipments, (2) maintenance of host material for rearing, and (3) successful reproduction of the agent under quarantine conditions.

Processing overseas shipments

Shipments of live organisms should be opened inside a transfer box (an observation cage with a glass top and closed sides and fitted with one or two cloth sleeves through which the material can be manipulated). This precaution allows for the safe initial separation of any potential contaminants from the natural enemies being shipped. Live organisms are collected into glass vials; packaging materials are heat treated in a dry oven or autoclave and discarded. The live organisms are then screened taxonomically; those known to be undesirable (hyperparasitoids or unwanted phytophagous arthropods) are killed and preserved in 75–95% ethanol as vouchers. If ultra-low freezers are available, frozen samples should also be stored to preserve material for molecular studies. Potentially beneficial organisms are separated by species, host plant, and collection locality, observed for mating, and placed in isolation cages with the appropriate host for propagation. Each colony in quarantine should be assigned a specific number that would refer to the source of the original accession of material.

If shipments consist of plant pathogens, precautions against unauthorized release are those of a microbiological laboratory, and include special air-filtering requirements as discussed above, and sterilization of water and soil supplies (both entering and exiting the quarantine laboratory), typically by autoclave, before discharging them from the area. Personnel must shower before leaving the laboratory, leaving their work garments inside the facility.

Managing natural enemy and host colonies

The establishment of laboratory cultures of candidate natural enemies and necessary hosts or host plants, free from contamination, is a major objective of quarantine operations. Some quarantine cultures, particularly those maintained on plants, may be subject to infestation by pest arthropods such as aphids, whiteflies, thrips, mealybugs, and mites. The control of these organisms must be undertaken with care, and a particular objective should be avoiding the use of pesticides where possible. If possible, plants used to propagate natural enemies (or their hosts) should be grown from seed in a quarantine greenhouse. If field-harvested or purchased plants are used (as with woody species that are needed in some scale projects), they should be treated with a non-residual pesticide and held in a separate area to verify they are pest free before use. Requirements for pesticide persistence and selectivity will depend on the pest to be killed and the natural enemy to be reared. Release of commercial natural enemies such as predaceous mites to control spider mites or thrips may be useful when pesticides cannot be used. However, plants should be screened carefully for generalist predators, should these have been used, before placing plants in colonies.

Quarantine cultures can also be subject to infestation by parasitoids (for phytophagous natural enemies), hyperparasitoids, parasites (such as mites or nematodes), and pathogens. Cross-contamination between colonies (especially parasitoids) being reared in quarantine may occur. Care by quarantine personnel is critical in identifying such problems early, isolating the affected cultures, and eliminating the unwanted organisms. Molecular bar-coding of parental material at the time of importation can be used to ensure that subsequent generations of the colony remain pure (Goolsby et al. 1998). Elimination of pathogens from arthropod cultures is based on a mixture of destroying infected materials, together with sterilizing and frequently changing rearing containers (Etzel et al. 1981). Surface sterilization of eggs, for example by briefly dipping in 10% solution of bleach (sodium hypochlorite solution) in water, may also be successful (Briese & Milner 1986).

Protecting imported pathogen lines from contamination requires provision for isolation of microbial cultures. Where microbes are grown in artificial media, isolation may be accomplished by maintaining cultures in sealed containers, in different growth chambers, or in different rooms. Where cultures are maintained on live hosts, provision must be made for the maintenance of uninfected host material (either plants or arthropods) outside the quarantine facility. Where particular biotypes of pathogens or arthropods are being studied, tools to reliably identify strains must be available, using molecular markers (see Chapter 15) or other methods. Representative specimens should be kept in cryogenic storage for comparison with later introductions or recoveries, or to evaluate isolated lines for genetic drift or contamination.

Before natural enemies can be cleared for release from quarantine, adequate host specificity must be demonstrated to indicate safety to the biota of the release country. Information for release petitions can come from several sources (literature records, field surveys in the area of origin, and laboratory host-range tests). Colonies of natural enemies reared in quarantine are used to perform laboratory host-range tests while the agent is still in quarantine. In Chapter 17 we discuss how to conduct these tests. Concerns that the quarantine laboratory must address are: (1) how to rear the candidate natural enemies in large, reliable numbers for many generations, (2) how to ensure that the culture retains its integrity (free from contaminants or invasions by similar species), and (3) how to retain the genetic features of the original strain without selection for adaptation to laboratory conditions.

DEVELOPING PETITIONS FOR RELEASE INTO THE ENVIRONMENT

For weed biocontrol agents

To petition for release of new plant-feeding agents, a formal process must be followed in the USA and in other developed countries engaged in weed biological control. Once colonies of the weed control agent have been established in the quarantine laboratory, interactions begin with a government committee charged with overseeing the development and evaluation of host-range data and the resulting risk/benefit assessment. This interagency review committee (Technical Advisory Group, TAG) must approve the list of non-target species to be included in host-range testing to ensure completeness. Results, as they become available, must be submitted to the TAG with an analysis. If the TAG is satisfied that the agent in not likely to pose any significant danger to native plants, it is recommended to APHIS (or similar regulatory agency in other countries) to be approved for release. The acceptable degree of host-range breadth for a new agent is not fixed, but may vary from one program or area to another, even for the same target or natural enemy species depending on the circumstances in the release area. Generally, natural enemies are cleared for release if the release will not put desirable or valuable species at risk, either in the area of release or in areas within the likely natural spread of the agent. However, incidental feeding on non-target plants is not usually a prohibitive impediment to release, and natural enemy species need not be strictly monophagous to be acceptable.

For biocontrol agents of arthropod pests

For parasitic and predaceous insects, some countries (New Zealand, Australia) have formal legal requirements that require host range testing, and stipulate a procedure to be followed. The USA, however, does not have any such requirement at this time. Rather, the process occurs in two steps: first the USDA-APHIS-Plant Protection and Quarantine (PPQ) must be petitioned to make a finding that the species to be released is not a plant pest in the context of the law. This is generally the case for most parasitic and predaceous insects. Second, in the USA, an Environmental Assessment is written by the researcher requesting the release, describing the host range of the agent relative to the native insects in the release area (unless such an assessment already exists, as is the case for certain genera that are commonly used in insect biological control). The Environmental Assessment further provides an analysis of the relative risk/benefit consequences of the proposed release. USDA-APHIS sends the petition for release to the North American Plant Protection Organization (NAPPO) Biological Control Panel for review and recommendation. The NAPPO review is sent to anonymous reviewers and then a recommendation is returned to the USDA-APHIS. Petitions for release of agents in Canada or Mexico are also sent to NAPPO and recommendations forwarded to all three countries' representatives. If a review of this assessment yields a finding of no significant impact, the release may proceed.

Balancing estimated risks and benefits

When assessing the suitability of a candidate natural enemy for release to the environment, the risk of attack on non-target species must be compared to expected damage to the environment if the pest remains uncontrolled. This is done by first using the estimate of the agent's host range and knowledge of the local fauna or flora to estimate the degree of risk an introduction might entail. This is then compared with the ecological or economic benefits being sought. Putting these two together allows the project scientists to describe the likely cost/benefit ratio for the proposed introductions. The goal is only to introduce agents when there is real need for the pest to be suppressed and when the agent appears reasonably specific. But, in the last step, the judgment about the estimated risks and projected benefits becomes a political decision made on behalf of the community by its government. This process should be an open process that solicits public input and acts with continued consultation with conservation biologists.

Chapter 14

CLIMATE MATCHING

An important tool in classical biological control projects is the matching of climates between invaded areas and the native range of the invasive species or other collection areas for natural enemies. This allows the search for natural enemies to be directed towards areas with the best climatic match, which should produce natural enemies with the best chance of establishing in the recipient country after release (Bartlett & van den Bosch 1964, González & Gilstrap 1992, Hoelmer & Kirk 2005). This is a tool best used during the initial planning of a project to guide exploration activities.

To establish viable populations, insects require climatic conditions suitable for reproduction and development. Temperature, in particular, is one of the key climatic factors affecting establishment and spread, along with rainfall amounts, rainfall patterns, soil moisture and pH, and photoperiod. Successfully establishing agents must withstand local climatic extremes (e.g. excessive cold or wet) and exploit favorable intermediate conditions for development and population growth.

With this in mind, this chapter has three objectives.

1. To look at the application of **climate matching** between the pest or natural enemy's donor range (i.e. where the organisms are from) and that of the intended receiving range (where they have invaded or are to be released) for the purpose of determining where in the donor range would be best to search for natural enemies that are well adapted for the receiving range.
2. To discuss the use of **inductive modeling** to infer from the climate of a species' home range (a) whether it could spread to any given area of interest, (b) whether an established population is likely to spread further, and, if so, (c) whether the range expansions will be transient or permanent. Potential spread and establishment of permanent pest populations is of special interest because it has implications as to how risks of natural enemy introductions to non-target native species should be managed. Such predictions bring into better focus what non-target species the introduced natural enemies might come in contact with as the pest spreads, provided the introduced natural enemies are also adapted to the climates of the areas into which the pest is spreading.
3. To illustrate the use of climate data in **deductive modeling**, which is the application of laboratory-derived population demographic statistics and degree-day estimates for development of natural enemies. Computer programs are then configured appropriately with pertinent weather-station data to provide estimates of the strength of natural enemy population growth in areas where spread is expected, to determine where incursion is likely, and what the expected intensity of impact on target pest populations could be.

CLIMATE MATCHING

Climate matching before conducting expensive foreign exploration in a vast home range and before starting importations and time-consuming safety evaluations of agents that may not be well matched climatically can increase the likelihood of selecting more suitable agents (Goolsby et al. 2005a, Hoelmer & Kirk 2005). Such focusing of the search area through climatic matching is often necessary because many target pests have extremely large geographic ranges, which encompass many climatic and ecological zones (noting, however, pests that are damaging in their native range will have better known, larger ranges, whereas species that are not pests in their native ranges will have poorly known distributions, and perhaps falsely smaller ranges). To narrow a pest's known range down to a tractable sized region where natural enemy searches can be focused, it can be useful to first determine which parts of the pest's home range best match the invaded range where natural enemy releases are desired.

Climatic mismatch between areas surveyed in the home range and the receiving range is presumed to be an important factor limiting establishment and impact of natural enemies, and has most likely been a cause of failure in some biological control programs (Bartlett & van den Bosch 1964, Beirne 1975). For example, natural enemies of St. John's wort (*Hypericum perforatum* L.) released in Australia were initially collected in England in the 1920s and 1930s. Only one of the five agents from England established in Australia. In contrast, five of six agents collected later in southern France, in areas with a Mediterranean climate that better matched the release areas, established, including the most successful agent, the chrysomelid *Chrysolina quadrigemina* (Suffrian). This is one of earliest examples of improving natural enemy establishment rates through carefully matching the climate of the home range with that of the intended introduced range (Syrett et al. 2000).

Determination of how similar climatic conditions are between selected locations within the home range and the receiving range can be done using climate-matching software (e.g. CLIMEX, bioSIM, BIOCLIM, DOMAIN, and HABITAT; Baker 2002). These programs have been developed to identify and map areas of the world with similar climates using historical weather records from numerous locations worldwide. The programs allow weighting of specific environmental factors (e.g. rainfall and temperature) if desired when estimating the degree of the climatic match and when mapping potential species distributions. Alternatively, computer programs can be used to link weather data to information on how climate can affect a given species' phenology or distribution (i.e. reactions to heat stress or cold/wet stress, etc.; Hoddle 2004, Hoelmer & Kirk 2005). CLIMEX is a program used commonly for these types of analysis and was designed with biological control applications in mind (Sutherst & Maywald 1985, Sutherst et al. 2004). CLIMEX will be used to illustrate important points in this chapter where relevant.

Interpreting output from CLIMEX is based on maps produced by the program. Dots on output maps can be set to represent a variety of possible climatic variables (e.g. average temperature, average maximum temperature, relative humidity, etc., or an index produced by a combination of these and other variables). The larger the dot at a specific location, the better the average prevailing climatic conditions at that location for the species of interest (Figure 14.1).

Retrospective climate-matching analyses for aphelinid parasitoids released in the USA for biological control of the silverleaf whitefly, *Bemisia argentifolii* Bellows and Perring (Hemiptera: Aleyrodidae), demonstrates the importance of climate matching as an indicator for predicting species establishment (Goolsby et al. 2005a). Percentage climate matches between areas can be determined using the Match Climates function in CLIMEX. This function can be used to compare the average climatic conditions (e.g. maximum and minimum temperatures, total rainfall, rainfall pattern, relative humidity, soil moisture, and combinations of these factors) in the natural enemy's area of origin to those of the introduced range. Component indices can range from 0 to 100, with 100 being an exact match between the two locations for the parameter of interest. In the Goolsby et al. (2005a) study, the average climatic match index value for parasitoids establishing in a new area was approximately 75%, whereas for parasitoids failing to establish the average climatic match index was approximately 67% (data averaged from table 2 in Goolsby et al. 2005a). However, high climatic match indices do not guarantee establishment, and some parasitoid species with climatic match indices of 80% failed to establish, indicating that factors other than climate can be very important in affecting establishment once suitable agents with good climatic tolerance are identified (Goolsby et al. 2005a). Impact on the target was greatest when the natural enemy with a close climatic match exhibited a narrow host range and high attack rates (Goolsby et al. 2005a). The clear results of these analyses led to the recommendation of a specific parasitoid species, *Eretmocerus hayati* Zolnerowich and Rose (Hymen-optera: Aphelinidae) from Pakistan, from a large pool of candidate species, as the top priority for evaluation for release in Australia for control of *B. argentifolii* in cotton-growing areas (Goolsby et al. 2005a). *Eretmocerus hayati* is now established in many locations in Queensland, Australia, and is spreading rapidly (Goolsby 2007).

The effectiveness of climate matching in predicting the establishment of natural enemy strains collected from different locations within the home range is being assessed retrospectively with molecular techniques (Iline & Phillips 2004). In some instances climate matching of an area within the home range of a natural enemy has not accurately predicted performance of the natural enemy in the introduced range where climate was similar. When these failures occur, it may be useful

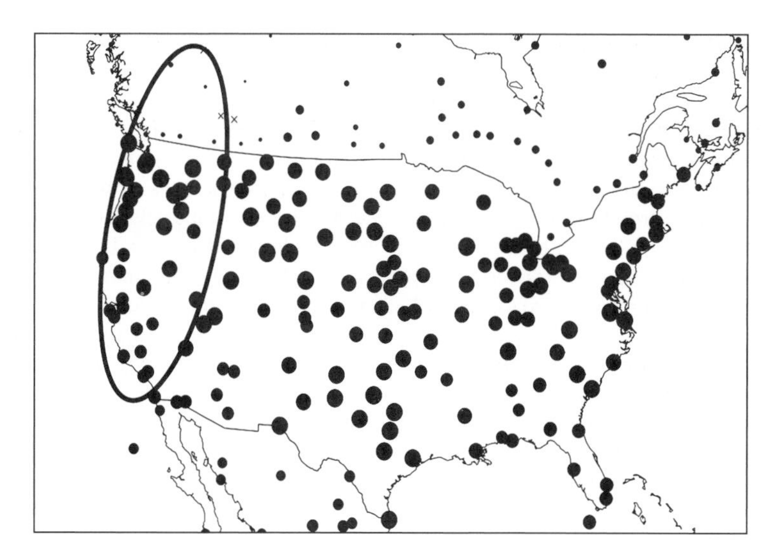

Figure 14.1 The CLIMEX Match Climates function illustrates how similar the climatic averages in Auckland, New Zealand, are to locations in North America. The level of similarity is given by the Match Index parameter, which is an average of up to seven component indices, including minimum and maximum temperatures, rainfall (quantity and seasonality), relative humidity, and soil moisture. The larger the black dot on the map the closer the average climatic match between Auckland and locales in North America. The ellipse delineates the home range of a hypothetical pest native to North America that has established in Auckland. CLIMEX suggests that average year-round climatic conditions in the pest's home range that are most similar to Auckland are in the western and southwestern coastal areas of the known range, and foreign exploration for climatically pre-adapted natural enemies for potential release in New Zealand should be initiated here and not in the northern extremes of the range where the dots are smallest. Climate-matching exercises of this nature may have important practical utility for foreign exploration and natural enemy selection. Map drawn by M. Hoddle.

to investigate other factors unfavorable to natural enemies. Some factors that might prevent establishment of natural enemies that are climatically well matched include attack by generalist predators such as ants, lack of genetic diversity (in uniparental species) needed for post-release adaptation, long-term inbreeding in the laboratory before release, and fluctuations in relative humidity (van Klinken et al. 2003).

The use of a scientifically based process to choose arthropod natural enemy species for preliminary host-range screening is important because first, host specificity testing in quarantine is time consuming, difficult, and expensive. Ranking of candidate natural enemies can expedite evaluations and reduce reliance on release of larger numbers of species about which less is known. Money is saved and success rates for biological control both in terms of establishment and impact may increase. Second, accumulation of successful applications of these techniques supports development of biological control theory.

INDUCTIVE MODELING: PREDICTING SPREAD AND INCURSION SUCCESS

In many instances, little detailed information is available on the climatic responses or the developmental and reproductive biology at various temperatures of the target pest or its natural enemies. Despite this impediment, it is possible to make accurate estimates of how an organism will respond to average prevailing climatic conditions in a new area through the application of inductive modeling, also referred to as inverse or inferential modeling (Sutherst & Maywald 2005). This is done by inferring an organism's climatic responses to conditions based on its distribution in the home range and extrapolating those responses to an invaded range. This can be done very simply in CLIMEX; a climatic template of the home range is chosen from a default menu (e.g. a Mediterranean or subtropical climate template) that is most representative of climatic conditions in the home range. Climatic parameters affecting organismal responses defining the selected template are tweaked until distribution maps result that most closely resemble the known home range of the pest or natural enemy. The assumption here is that the parameter settings defining the climatic responses of the organism of interest are then close estimates of the real parameters affecting its distribution. CLIMEX and other programs have no knowledge of the impact of host-plant availability, interspecific competition, natural enemy activity, etc. on the distribution of the species in its home range. Modeling programs only use weather-station data to describe the resulting distribution of the organisms of interest. Consequently, climatic conditions are assumed to be primarily responsible for the observed distribution that defines the home range.

The glassy-winged sharpshooter, *Homalodisca coagulata* (Say) (Hemiptera: Cicadellidae), is an important pest in California, USA, because it vectors a pathogenic bacteria, *Xylella fastidiosa* Wells et al., which kills a variety of ornamental and agricultural species (e.g. grapes and almonds) by clogging the xylem and preventing water conduction. This pest is native to the southeast USA and northeast Mexico and invaded California in the late 1980s. After a substantial lag period, *H. coagulata* populations exploded, and it began to move rapidly northwards out of southern California causing substantial economic damage by vectoring *X. fastidiosa* in vineyards. Although no information on the effect of different temperatures on developmental and reproductive biology are available, inferential modeling was conducted to infer *H. coagulata*'s climatic limits from its known home range (Figure 14.2) and then this model was applied to determine its potential new range in California and globally (Figure 14.3). This pest has subsequently invaded French Polynesia, Hawaii (USA), and Easter Island as predicted by the deductive model (Hoddle 2004). Such an approach helps to alert biological control practitioners to the possible geographic range natural enemies might be required to operate in, and also provides suggestions as to other areas where the pest may occur naturally but gone unrecorded. For example, the Yucatan Peninsula and Caribbean may yield *H. coagulata* populations with previously unknown parasitoid complexes that could be useful in biological control projects against this pest.

Because climatic response data are typically missing for many important pests (and their natural enemies), there is a need for increased use of inductive modeling to estimate the risks posed by such pests. Such model predictions can provide general estimates of the risk that pests will establish and thus threaten agricultural enterprises or natural areas under current and potential future climates (i.e. changes due to global warming) (Sutherst & Maywald 2005). Inductive modeling has been used to assess the risk of various exotic insect pests invading new areas (MacLeod et al. 2002, Vera et al. 2002, Hoddle 2004, Sutherst & Maywald 2005), global spread of important plant diseases (Paul et al. 2005), risk assessment for establishment and range expansion of transgenic predatory mites (McDermott & Hoy 1997), expected range of weed natural enemies in introduced areas (Mo et al. 2000), and edaphic and climatic factors limiting the spread of pestiferous soil mites (Robinson & Hoffmann 2002). With an ever-increasing number of publications using climate modeling software, especially CLIMEX, to investigate

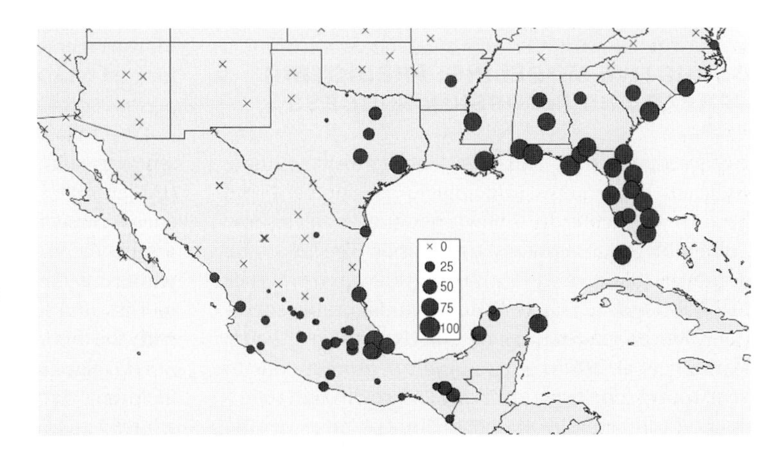

Figure 14.2 Map of the distribution of *Homalodisca coagulata* in its home range generated in CLIMEX by inductive modeling. Model parameters in the Temperate Template were adjusted iteratively until the observed distribution was realized. Large black dots indicate high climatic suitability for *H. coagulata* and crosses indicate unsuitable areas. For full details on how the model was prepared see Hoddle (2004).

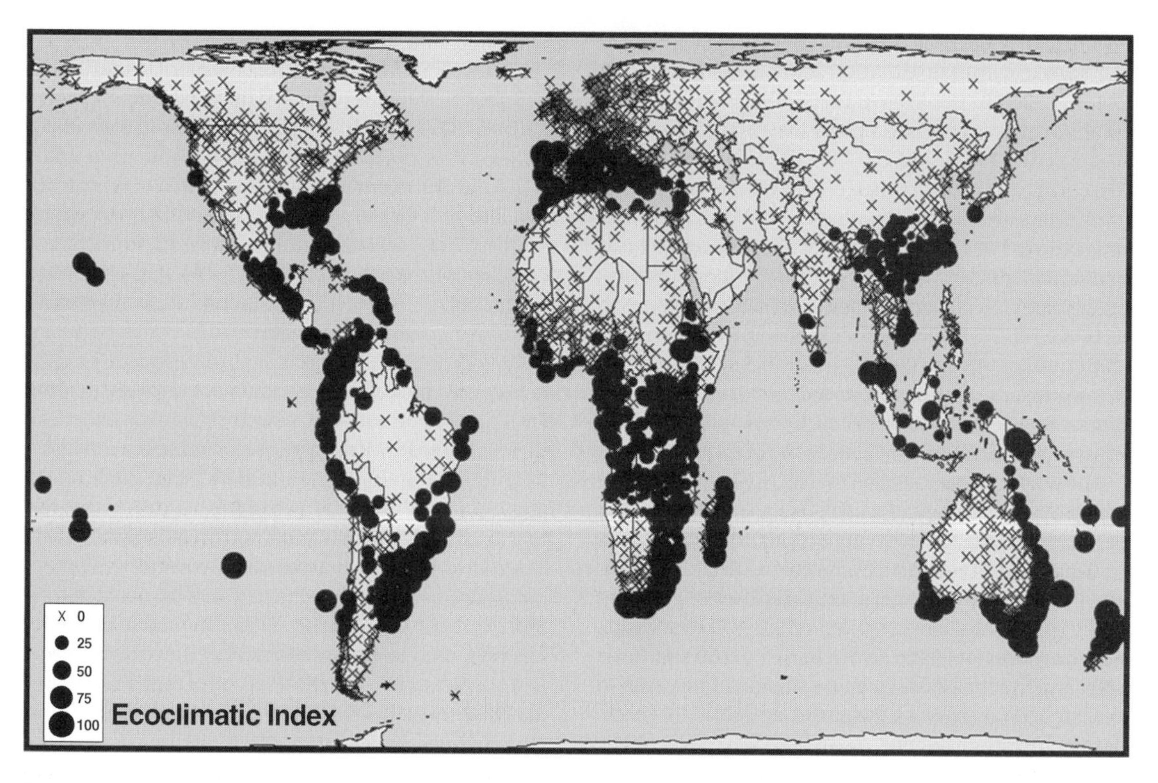

Figure 14.3 Predicted global distribution of *Homalodisca coagulata* from inductive modeling. Crosses indicate unsuitable areas for *H. coagulata*, and black dots indicate areas of varying climatic suitability. After Hoddle 2004.

climate-related hypotheses about pest and natural enemy spread and impact, global collaboration and information sharing across international research communities via the worldwide web is being advocated (Sutherst et al. 2000).

DEDUCTIVE MODELING: PREDICTING SPREAD AND INCURSION SUCCESS

Accurate prediction of an insect's ability to accumulate sufficient degree-days to complete development and begin reproduction in a new area may indicate how vulnerable that region is to invasion by an exotic organism (Sutherst 2000, Baker 2002), and whether incursion will be transient due to unfavorable conditions for prolonged periods (Jarvis & Baker 2001, Hatherly et al. 2005) or potentially permanent due to year-round conditions favorable for growth and reproduction (Sutherst 2000, Baker 2002). Consequently, survival is influenced not only by critical temperatures (i.e. upper and lower lethal temperature thresholds), but also the length of time of the exposure. All arthropods die when exposed to prolonged periods of excessive heat, unless they have unique adaptations to deal with this stress. To aid prediction of overwintering survivorship in inhabited areas, most research has focused on the ability of insects to survive prolonged periods of cold.

Two reliable predictors of overwintering survival when used in combination are $LTime_{50}$ and $LTemp_{50}$, the length of exposure time or temperature experienced, respectively, at which 50% of the experimental insects die (Leather et al. 1993). The outdoor overwintering survivorship of some species of pest thrips, moths, and whiteflies in the UK correlates closely with the amount of time spent at −5°C (Bale & Walters 2001). Similar studies for biological control agents, including predatory bugs, mites, coccinellids, and whitefly parasitoids, have demonstrated a strong correlation between $LTime_{50}$ in the laboratory at 5°C and overwintering survival in the UK (Hatherly et al. 2005). These types of data for natural enemies can be used to assess the invasion and establishment risk in new areas and may be an important component of environmental risk assessment studies before release of natural enemies where permanent establishment is not desired (i.e. augmentative control; Hatherly et al. 2005). The utility of this work has been demonstrated by research on the phytoseiid mite *Neoseiulus californicus* (McGregor). This is a commercially available species that is used in Europe and elsewhere for control of pest mites in greenhouses, and it has unexpectedly established populations outside of protected environments (Hart et al. 2002). In cold climates, subtropical or tropical species unable to survive short periods of exposure to cold can survive in greenhouses or other protected environments (e.g. arboreta). Some pest thrips [e.g. *Thrips palmi* Karny and *Frankliniella occidentalis* (Pergande)] can overwinter in European greenhouses. These source populations can invade outdoor field crops each spring and summer, and if temperatures are sufficiently warm over a long-enough period, rapid population growth may result causing economic damage (Morse & Hoddle 2006). For pests and natural enemies with obligatory diapause, knowledge of temperature and day-length cues that initiate and terminate the resting stage are required to accurately determine when life cycles start and end in a given area.

For any insect population to survive in an area, it must not only be able to tolerate prevailing extremes of hot and cold, but also accumulate enough thermal units to complete immature development and function adequately as an adult. To determine whether temperatures in a given area remain above a critical minimum threshold long enough to enable complete development, an estimate of the number of degree-days required for maturation is needed. The degree-day model is based on empirical observations relating developmental rate to temperature, and that over most of this relationship a linear interaction occurs (Campbell et al. 1974). As the extremes of cold and heat tolerance are approached, the relationship becomes curvilinear (Lactin et al. 1995). Applying results from degree-day analyses derived from constant temperatures is fraught with difficulties when assessing real-world conditions that experience unpredictable fluctuating temperatures (Baker 2002). In addition to controlling development, temperature influences other physiological processes that are critical to the survival of natural enemy populations. For instance, temperatures that provide the fastest development of immature parasitoid stages may result in lower survivorship rates and reduced fecundity of progeny (Pilkington & Hoddle 2006). Incorporation of more detailed biological data (e.g. degree-day requirements) into programs like CLIMEX can potentially overcome inaccurate predictions about natural enemy establishment and impact when only matching climate parameters in the home range to the introduced range are used to estimate success (van Klinken et al. 2003).

Despite potential drawbacks, degree-day models have been very useful in determining pest and natural enemy phenology in the field and for assisting decisions on whether pest-control interventions are necessary. In this instance, weather-station data are used to assess the degree-day accumulation for the pest or natural enemy, and online programs are available for specific pests (see www.ipm.ucdavis.edu/WEATHER/ddretrieve.html). However, temperature stations are often sparsely distributed, may not be near cropping systems, can be affected by unrepresentative microclimate effects, or lack sufficient years of data for meaningful analyses. Inconsistencies such as these can be alleviated to some extent by interpolating weather data across a range of weather stations in the area of interest. Interpolated weather data can be combined with phenology or demographic models and analyzed with GIS software to generate colored maps that display various estimates of, for example, the number of generations in a given area, or net reproductive output (Plate 14.1).

CONCLUSIONS

Climate is a major factor affecting the establishment and reproductive success of invading species (i.e. pests and deliberately introduced natural enemies) in new areas. Biological control practitioners have long agreed that similarity between climates in the donor range and introduced range should be considered carefully and areas with greatest similarity selected for prospecting for natural enemies. Despite the assumed importance of climate matching, very few empirical evaluations have been conducted to explore this underlying hypothesis explicitly. However, this situation appears to be changing and the limited number of retrospective analyses that have been conducted tentatively support the importance of climatic matching and establishment and impact success of natural enemies. Computer modeling combined with biological and ecological data for natural enemies and target pests is greatly assisting investigations of the influence of climate on population growth and geographic spread of organisms. It should be remembered that climate may not be the only factor affecting establishment and spread of an organism. Host availability, overwintering sites, resident competitors, and generalist natural enemies, for example, will all interact to various degrees with climate to affect establishment success, proliferation, spread, and impact.

Chapter 15

MOLECULAR TOOLS

RICHARD STOUTHAMER

Rapid developments in molecular biology have made many new techniques (Figure 15.1) available for the genetic characterization of populations of animals and plants. Molecular ecology has led to many insights into the ecology and genetics of populations, species, and higher taxa. These techniques can help answer questions of importance to biological control programs. For instance, from which area of the native range does an invasive species originate? Is this target pest parasitized by parasitoid species A or B? Are these two morphologically similar populations of natural enemies in fact different species?

The greatest impact of molecular methods on biological control so far has been improved species and biotype recognition. Many natural enemies are exceedingly small, with only a limited morphological character set useful for their identification, and the application of molecular techniques has substantially simplified the identification of some of these species. Incorrectly identified natural enemy species have led to failed biological control programs (Gordh 1977). For example, the augmentative release of incorrectly identified species of *Trichogramma* in some instances has led to a reduction of natural control by resident *Trichogramma* species (Stouthamer et al. 2000a).

Figure 15.1 Extracting samples for DNA sequencing. Photograph courtesy of M. Hoddle.

DNA sequences are also used extensively as additional characters to determine the phylogenetic relationships between different taxa. Well supported phylogenies for a natural enemy and its target pest can be very helpful in predicting life-history characters of related species and this can help with selection of the most promising and host-specific natural enemies (Briese & Walker 2002).

Another application of these methods is to determine the area of origin of an invasive population. Delineation of a smaller area within a vast home range may make it possible to collect natural enemies that have coevolved with the invasive population. Natural enemies adapted to the source pest population may be better synchronized with the pest and consequently control the pest more efficiently in the new range (Goolsby et al. 2006b).

This chapter is not meant to be an exhaustive overview of the molecular methods that may be used in biological control. Instead, I present an overview of the most commonly used molecular markers, including an explanation as to what each method is, how a method could be applied to a biological control project, and how these tools have been used to answer questions related to biological control. In the second part of the chapter a short overview is given of the most appropriate technique for answering a set of questions that may be relevant for biological control projects. Most of the examples given stem from biological control of arthropods; however, many of these techniques are similarly applicable to biological control of weeds. Little attention is given in this chapter to the methods that are used to analyze some of the more advanced applications of molecular markers. These analyses are critical and often are only applicable if certain conditions are met, and new students of this field are advised to carefully read the most recent literature and if possible consult with population geneticists before undertaking advanced analyses of population genetics. For technical details about analytical DNA techniques see Hoy (1994) or Hoelzel (1998). For an overview of the general principles of molecular ecology, consult publications by Avise (2004), Beebee and Rowe (2004), and Freeland (2005).

TYPES OF MOLECULAR DATA

Molecular data can be classified in several ways. (1) Data may be either nucleotide sequences or DNA or protein fragments of varying molecular masses that form visible bands in different positions on appropriate gels. (2) DNA used in analyses may be from the nucleus, which represents inheritance from both parents in most cases, or may be DNA from organelles such as mitochondria or chloroplasts, inherited only through the female line, or from endosymbionts like *Wolbachia*, also inherited through the mother. (3) Genetic material may be drawn from single- or multiple-copy sources. For example, genes coding for ribosomal RNA (rRNA) are present as many copies in a cell. Such multi-copy genes have the advantage that more template DNA is present in a cell and consequently such genes should be easier to amplify in the polymerase chain reaction (PCR). Single-copy genes (one copy of the gene per gamete) are generally the genes that code for proteins.

Fragment analysis

Isozymes

Isozymes are different enzymes that catalyse similar types of reactions in the cell. Often these isozymes are related to each other because they originated through gene duplication. For our purposes allozyme analysis is most relevant when allozymes are different allelic forms from the same enzyme-coding locus; that is, a particular isozyme. During the 1970s and 1980s, the use of allozyme markers was common. Two methods are used to separate different allozymes: (1) gel electrophoresis and (2) isoelectric focusing. In gel electrophoresis, the different allozymes are separated by pulling them through a gel matrix. The speed of allozyme movement is determined by the size of the protein and the way it is folded. In isoelectric focusing, a gradient of isoelectric points is created in a buffer on top of a membrane. Each allozyme variant will accumulate at the position of its isoelectic point (i.e. the position in the gel at which the protein has no net electrical charge) along the gradient. Once the enzymes have been separated they are made visible by using indicator dyes. These dyes will change color in the presence of the appropriate substrates and cofactors for the particular isozyme.

How do you find allozymes to analyze?

A large number of different isozymes are present in insects. To find the isozymes that show the appropriate level of variation many different ones can be tested. An overview of recipes and techniques for many of the different isozymes is given in Richardson et al. (1986). For application of enzyme electrophoresis, the insect specimens need to have been preserved in such a way that their proteins have not been degraded. This means that either freshly killed individuals need to be used or the specimens need to have been frozen shortly after they have been killed. Individuals are then homogenized in a buffer and the resulting solution is loaded into a gel for starch electrophoresis or placed on top of a membrane if isoelectric focusing is used. Details of these methods can be found in Unruh et al. (1983) and Kazmer (1991). Enzyme electrophoresis as outlined briefly above is not used much anymore for population studies, having been superseded by PCR-based molecular methods. PCR methods have proven to be more practical, mainly because of the ease with which DNA can be preserved for later analysis (i.e. field-collected specimens can be killed and preserved in 95–100% ethanol and kept cool). Even though protein electrophoresis techniques have several disadvantages compared to PCR-based studies, allozymes have an important advantage in that the same protocols can be used to determine the genetic make-up of many different species.

What are allozymes used for in biological control?

Allozyme electrophoresis can be used for species/

biotype recognition, population genetic studies, analysis of the gut contents of predators to determine which prey species have been eaten, or to ascertain whether particular host insects are parasitized.

Examples of allozymes being used in biological control

Unruh et al. (1983) used starch gel electrophoresis to show the effects of prolonged mass rearing on genetic variation in natural enemies used for biological control. As a model system, they examined laboratory cultures of *Aphidius ervi* Haliday (Hymenoptera: Braconidae) and, based on the decline in heterozygosity as measured with starch gel electrophoresis, they concluded that even in cultures where 100 females were used as breeding stock in each generation, four of the eight loci that initially each had two alleles became fixed for one or the other of the allelic forms. The smaller the number of females used in each generation, the faster allelic diversity was lost, suggesting that the quality and genetic diversity of natural enemies being produced in successive generations was diminishing despite attempts at lessening adverse effects from inbreeding.

Kazmer and Luck (1995) used allozymes manipulated by isoelectric focusing as a marker to experimentally measure the effect of parasitoid size on host-finding ability in the field. The egg parasitoid *Trichogramma pretiosum* Riley (Hymenoptera: Trichogrammatidae) was used as their experimental natural enemy species. The size of these parasitoids is determined largely by the size of the rearing host (i.e. the egg in which they developed). Parasitoids reared on *Sitotroga* eggs are small whereas those reared on larger *Heliothis* eggs are substantially bigger and presumably more fit. Kazmer and Luck (1995) first determined the genetic make-up of the population of *T. pretiosum* already present in the field and found two alleles of the enzyme phosphoglucomutase (PGM). In laboratory cultures of *T. pretiosum*, two additional PGM variants (A and B) were present. These two allozyme variants were crossed into a line collected from the tomato fields where the authors' *T. pretiosum* studies were carried out. By repeated backcrossing, two new lines of *T. pretiosum* were created that were almost genetically identical to the line from the tomato fields, except that one line was homozygous for the allozyme PGM-A and the other line for allozyme PGM-B. Next, individuals of these marked lines were released together in the field, comparing either small PGM-A individuals with large PGM-B individuals or the reverse. The relative performance of large and small wasps could then be tested directly by determining how many artificially deployed host eggs that were parasitized resulted in offspring containing either marker, which indicated whether the mother was a large or small wasp. Kazmer and Luck (1995) showed that larger *T. pretiosum* were better at finding host eggs in the field, confirming that parasitoid size is a good indicator of fitness.

In several studies, allozymes have been used as markers to distinguish species or subpopulations of minute insects (Pintureau 1990, 1993, Pinto et al. 1992, Ram et al. 1995, Burks & Pinto 2002, Iline & Phillips 2004). Allozymes have also been studied to determine the identity of food eaten by various predators (Vennila & Easwaramoorthy 1997, Greenstone 1999, Harwood & Obrycki 2005). Finally, allozyme markers have been used to determine the presence of immature mymarid parasitoid larvae inside the eggs of the glassy-winged sharpshooter, *Homalodisca coagulata* (Say) (Hemiptera: Cicadellidae) (Byrne & Toscano 2006).

Randomly amplified polymorphic DNA (RAPD) markers

What are RAPD markers?

RAPD markers are markers that can be obtained by PCR using RAPD primers. RAPD primers are generally only 10 base pairs (bp) long (in normal PCR reactions, primers are around 20–30 bp long) and in each PCR reaction only a single RAPD primer is used, which functions as both the forward and the reverse primer. Because the primer is only 10 bp long, there may be many places in the genome of an organism where the primer can bind. The RAPD PCR will only give a PCR product if the two binding places (one on the forward strand and one on the reverse strand of DNA) in the genome are close enough together that within one PCR cycle the section of DNA between the two primers can be polymerized. It takes time for the polymerase enzyme to read and copy the DNA. If the two priming sites are far away from each other the polymerase may not be able to copy the whole length of the DNA between the two priming sites. For instance, if the two priming sites, one on the forward strand and one on the reverse strand, are less than 2000 bp away from each other, a PCR product will be formed. But if the distance is

10,000 bp, no exponential amplification of the 10,000 bp will take place. RAPDs are dominant markers, meaning that the marker is either present or not. Consequently, individuals that are homozygous or heterozygous for the marker will be indistinguishable from each other.

How do you find RAPDs?

RAPD primers can be readily bought in kits from specialized suppliers or they can be ordered separately. For general use, purchasing the commercial kits is the best and cheapest approach. To determine which of the primers will work on your natural enemy or pest of interest, many different primers will need to be tested until primers are found that show the level of variation that is needed for the problem to be studied (see below). Use of RAPD primers generally results in several different DNA sequences being amplified. When this PCR product is visualized following gel electrophoresis, several bands of different sizes will be visible. For some applications it is desirable that there are a lot of differences between individuals in this banding pattern, whereas for other applications less variation is better. For example, high variation is best for paternity analysis because if several different males could be the father of a particular offspring, then the paternity is most easily determined if the potential fathers differ substantially in their RAPD fingerprints. However, if one wants to use RAPDs to distinguish between two closely related species, it would be better if all individuals within each species show the same banding pattern, and for these patterns to differ between the species. Once appropriate, effective primers are identified for the intended task, one may need to order additional primers with the required specific sequences from a commercial supplier. Because RAPD primers are very short, it is important to optimize the PCR conditions to obtain consistent results. RAPD PCR can be optimized for conditions in a specific laboratory, but there often appears to be a problem with the portability of an optimized RAPD PCR protocol (van Belkum et al. 1995). Consequently, reaction conditions that work well in one laboratory do not necessarily work well in other laboratories, and this is one of the reasons why this method has fallen out of favor with many researchers.

A method that has many of the advantages of RAPDs but does have portability is the use of amplified fragment length polymorphisms (AFLPs; Vos et al. 1995). In short, this method is based on first cutting the DNA of a specimen into fragments using specific restriction enzymes, followed by attaching DNA linkers to the fragments. This is followed by a step where the fragments with linkers attached to them are amplified by PCR using primers that bind to the linkers. Subsequently this mixture is used as the template for later PCRs, now using primers that consist of the linker sequence but with additional bases added on the 3′ end. The product of these PCRs is electrophoresed on a gel and the banding pattern can be analysed. AFLPs are used extensively in mapping studies of genomes; however, they can also be used for population studies. Compared to some other methods, the protocol used for this method requires a large number of steps in the laboratory, making it less attractive.

What are RAPDs used for?

RAPD PCR is used as a fingerprinting technique for determining the paternity of offspring, to differentiate between species or biotypes, and for the genetic mapping of traits. RAPD PCR is a particularly popular technique for the recognition of fungal biotypes (Dodd & Stewart 2003, Dodd et al. 2004, Pujol et al. 2005, Zhou et al. 2005).

Examples of RAPDs being used in biological control

In a very careful study Kazmer et al. (1995) showed some of the pitfalls of using RAPD markers and this paper is recommended reading for anybody considering their use. The goal of Kazmer et al. (1995) was to use RAPD markers to distinguish between several closely related strains of the aphid parasitoid *Aphelinus asychis* Walker (Hymenoptera: Aphelinidae). They studied the repeatability of the pattern of RAPD bands in replicates of the same DNA sample and found that 25% of the gel bands generated in a PCR were not found in another replicate from the same PCR. This problem was further compounded because the hybrid offspring of two parasitoid lines sometimes did not show the expected banding pattern, and gels contained bands of slightly different sizes.

DNA fingerprinting based on RAPD PCR is often used to confirm that individuals or strains belong to a particular species or biotype. The B biotype of *Bemisia tabaci* Gennadius (Hemiptera: Aleyrodidae), for example, can be recognized using RAPD PCR (De

Barro & Driver 1997). Whereas RAPDs can often be used for such purposes, sometimes the variability within the species or biotypes of interest makes it difficult to assign individuals to the correct group. This problem was encountered by Gozlan et al. (1997) when they tried to use RAPD markers to distinguish between geographic populations of an *Orius* species (Hemiptera: Anthocoridae). The goal of their research was to find markers that could be used to assign individuals as belonging to these distinct geographic populations to determine the success of a released population of *Orius* relative to the local population of the same species. But they found that the RAPD patterns were too variable within geographic populations to be useful.

RAPD markers were used successfully by Edwards and Hoy (1995) to follow the fate of a laboratory-selected, insecticide-tolerant population compared with a field population of a parasitoid of the walnut aphid [*Chromaphis juglandicola* (Kaltenbach)]. Using the frequency of the different RAPD markers in both populations they used a statistically discriminant function to distinguish the populations from each other and assign individuals to either the laboratory or field population. Next, they allowed these two populations to interbreed in the laboratory and they followed over time the genetic composition of the interbreeding population in cages where insecticides were either applied or not. Their results showed that irrespective of the insecticide treatment the markers associated with the insecticide-resistant line persisted much better in the laboratory than the markers associated with the field-collected line. This led to the conclusion that the insecticide-tolerant selected line had adapted to laboratory conditions.

Inter-simple sequence repeat (ISSR) markers

What are ISSR markers?

ISSR markers are related to RAPD markers in that they are the result of PCR amplification of unknown parts of the genome of the study organism. ISSR patterns can be obtained by amplifying DNA from organisms using commercially available ISSR primers. ISSR primers consist of a series of dinucleotide repeats followed by two non-repeated bases. For example, an ISSR primer could be CTCTCTCTCTCTTG [or $(CT)_6TG$]. Often the last two bases (i.e. TG) are degenerate, which means that during the manufacturing of the primer on the same position in the primer two or more bases are possible. The degeneracy of a position in the sequence is indicated with the following letters: Y means C or T, and R means A or G. For instance, the degenerate primer $(CT)_6RT$ will consist of a mixture of the primers $(CT)_6AT$ and $(CT)_6GT$. The advantage of using degenerate primers is that they allow annealing to and amplification of a variety of related DNA sequences.

ISSR primers make use of microsatellite DNA sequences (see below) scattered throughout the genome of organisms. Because there are generally many different hypervariable microsatellite sequences in the genome of organisms, the ISSR makers are likely to amplify many different DNA regions. ISSR primers are treated as dominant markers, so a particular band is either present or lacking. The fact that at a single locus several alleles may all result in a band of somewhat different sizes is commonly ignored.

How do you find ISSRs?

ISSR primers can be bought in sets from commercial companies. Similar to RAPD primers, extensive optimization of these primers is needed for them to work in a reliable and repeatable fashion.

What are ISSRs used for?

They can be used for the same experimental purposes as RAPD primers.

Examples of ISSRs being used in biological control

ISSRs have been used mainly in the study of invasive weed populations to determine their level of genetic variability and potential hybridization with native species. Ash et al. (2004) used ISSRs to study the invasive lance-leaved waterplantain in Australia. Their analysis of the ISSR markers showed that the different populations of this plant in different areas were likely due to separate importations of this weed and that it is probable that seeds of this plant have been transported between infested areas. The fact that there have been two genetically distinct importations of this weed (most likely from different areas in the plants home range) may have important implications for future biological control efforts against this weed using plant pathogenic fungi. Tests to determine the efficacy of fungi against the target weed should include specimens from both populations and any hybrids, as pathogen efficacy may vary among genotypes.

ISSR and RAPD primers have been used to identify and distinguish biotypes of Russian thistle (*Salsola tragus* L.) found in California (Sobhian et al. 2003). Both markers gave the same results, showing that there were two biotypes (A and B) and that in field tests in Uzbekistan a potential biological control agent (a gall midge) was capable of attacking both biotypes, although A was preferred.

In New Zealand, the alfalfa pest *Sitona discoideus* Gyllenhal (Coleoptera: Curculionidae) is controlled successfully by the parasitoid *Microctonus aethiopoides* Loan (Hymenoptera: Braconidae). The population of this parasitoid used in New Zealand most likely originated from Morocco but was imported from a population previously established in Australia (Phillips et al. 2002). A second alfalfa pest, *Sitona lepidus* Gyllenhal (Coleoptera: Curculionidae), was later discovered in New Zealand that was not parasitized by *M. aethiopoides*. In Europe, several biotypes of *M. aethiopoides* are known that can successfully parasitize both species of *Sitona*. From experiments it became clear that the New Zealand population of *M. aethiopoides* from Morocco was not capable of successfully reproducing on the European *S. lepidus*, whereas the French strain of *M. aethiopoides* could reproduce on both the European and New Zealand *S. lepidus*. Using ISSR markers, the French and the New Zealand (Moroccan) populations of *M. aethiopoides* were compared and the authors concluded that there were genetic differences between them. These differences were sufficiently minor that these populations would not be considered different species, but still the populations differed in ways that mattered to their use as biological control agents. In a follow-up study Vink et al. (2003) used phylogenetic methods to determine the relationships among several *M. aethiopoides* populations and determined that there was a clear difference between parasitoid specimens collected from *S. discoideus* and *S. lepidus*.

ISSR primers have also been used to differentiate between species and populations of several *Gonatocerus* species (Hymenoptera: Mymaridae) used in the classical biological control of *H. coagulata* (de Leon et al. 2004, de Leon & Jones 2005).

When different populations are compared with ISSRs or microsatellites, it is important to make sure that field-collected individuals are compared and not individuals from a laboratory culture, which may have been started with a limited number of individuals that do not include the full genetic variation in the field population. In addition, for parasitoids that lay their eggs in hosts that are clustered, or that are gregarious, it is important to note that all the offspring of a single host cluster are most likely the offspring of a single mated female and will be very similar genetically. To properly sample species with these characteristics, samples of single individuals should be taken from many clusters (e.g. egg masses).

Individuals originating from laboratory cultures are by necessity very related, and when two laboratory cultures of a species are compared one is not accurately sampling the amount of genetic variation that is present in the whole population. Such a comparison is likely to mistakenly find that most variation is between subpopulations (cultures) and very little genetic variation is present within cultures, leading to the incorrect conclusion that the cultures represent distinct biotypes or subspecies.

Microsatellites

What are microsatellites?

Microsatellites, simple sequence repeats, or short tandem repeats (STRs) are all different names for the same type of marker. Microsatellites consist of tandemly repeated DNA sequences, where the repeat unit consists of only 1–6 bp and the whole repetitive region spans less than 150 bp. The locus of a microsatellite marker can have many different alleles, each with a different number of repeats. The many different alleles at a microsatellite locus are thought to come about by replication slippage of DNA strands containing many repeat units (Schlotterer 2000). Microsatellites are typically neutral and co-dominant, making them very useful as molecular markers in population studies.

How do you find microsatellites?

Finding microsatellites for a particular organism can be an involved, time-consuming, and expensive process (Zane et al. 2002). Generally, genomic DNA is extracted from the organism and cut into shorter pieces (≈500 bp) using different restriction enzymes. Next, these DNA fragments are ligated directly into plasmids. Commonly, before ligation the DNA fragments will be enriched for DNA strands containing microsatellite DNA sequences by using selective hybridization to pull out the DNA fragments containing microsatellite repeats. Subsequently, the plasmids containing the DNA fragment are used to transform bacteria and the transformed bacteria are isolated using selective media that favor bacteria containing an insert. If non-enriched DNA is used for the transformation, then the bacterial colonies containing inserts need to be screened to determine which ones contain a microsatellite insert. Next, those plasmids containing an insert are sequenced. The sequences that contain both microsatellite repeats and the DNA sequences flanking the repeat can be used for designing PCR primers. The PCR primers are then designed and tested on different individuals of the species of interest to determine whether they indeed amplify a product of the expected size. Primers that amplify a microsatellite locus need to be tested to determine: (1) that the microsatellite locus is polymorphic (i.e. there are several alleles at the microsatellite locus) and (2) that no null alleles are present. Null alleles are cases where the primers are unable to amplify the microsatellite locus (Selkoe & Toonen 2006). Mutations in the priming regions for the microsatellite locus are another cause for the failure of the primers to amplify. Null alleles should be avoided because many of the statistical methods used to analyze microsatellite DNA assume that within populations alleles will be in Hardy–Weinberg equilibrium. If null alleles are present then individuals that are heterozygous for one of the null alleles will be scored as homozygous (Selkoe & Toonen 2006).

What are microsatellites used for?

Microsatellites are used to answer such questions as: from which population does this individual originate? What are the genetic relationships between individuals? What is the mating structure of a population?

Examples of microsatellites being used in biological control

Microsatellites have not been commonly used in biological control studies. A survey of the literature shows that microsatellite primers have been developed for many organisms of biological control interest (Bon et al. 2005, Brede & Beebee 2005, Slotta et al. 2005, Williams et al. 2005, Lozier et al. 2006). Microsatellites have been used to determine the likely source of invasive species populations (Bohonak et al. 2001, Tsutsui et al. 2001, Augustinos et al. 2002, Facon et al. 2003, Baliraine et al. 2004, Hufbauer et al. 2004, Clarke et al. 2005, Grapputo et al. 2005). However, very few studies have as yet been published on using microsatellites for addressing other questions of importance to biological control.

Two studies have used microsatellite markers to determine the population genetic structure of aphid parasitoids. Baker et al. (2003) studied *Diaeretiella rapae* MacIntosh (Hymenoptera: Aphidiidae), a parasitoid that has been distributed worldwide, including Australia, for various aphid biological control projects. Genetic diversity of *D. rapae* in Western Australia was low, indicative of severe genetic bottlenecks during the importation and colonization process. The authors speculate on the implications of this low genetic diversity on the potential ability of this *D. rapae* population in controlling the expected invasion of the Russian wheat aphid, *Diuraphis noxia* (Kurdjumov) (Hemiptera: Aphididae), given that not all populations of *D. rapae* are able to control this pest.

Hufbauer et al. (2004) used mitochondrial DNA and microsatellites to reconstruct the history of the introduction to the USA of the parasitoid *A. ervi* (Hymenoptera: Braconidae), which was intended to control the pea aphid, *Acyrthosiphon pisum* (Harris) (Hemiptera: Aphididae). Approximately 1000 parasitoid pupae were imported from France in 1959 and reared for a number of generations before they were released in the field. When comparing the present US population with French and Hungarian *A. ervi* populations, it became clear that during its introduction this species experienced a mild genetic bottleneck. Also, the US populations have undergone some post-release differentiation since geographic separation and genetic distinctness were positively correlated among US populations.

Microsatellite markers have also been used to distinguish different clones of the species *Trichogramma cacoeciae* Marchal (Hymenoptera: Trichogrammatidae). This egg parasitoid is a completely parthenogenetic species, consisting of a number of different clones (Vavre et al. 2004, Pizzol et al. 2005). Microsatellite markers can be used to test the relative performance of different clones of this species against pests of interest.

Gene sequences

There are three types of gene sequence that I will

consider: (1) DNA sequences coding for proteins, (2) rRNA sequences, and (3) mitochondrial genes.

DNA sequences of genes coding for proteins

DNA sequences are transcribed to form the primary RNA transcipt, which is subsequently processed before it becomes the mature messenger RNA. During the processing, stretches of sequence that do not code for amino acids are removed. These non-coding stretches of DNA are called introns, and the stretches of DNA that code for proteins that end up in the messenger RNA are called exons. Messenger RNA is subsequently translated into a protein on the ribosomes. Many protein-coding genes only have a single copy in the genome of the organism.

How do you find DNA sequences? For many protein-coding genes, primers have been developed (Brower & Desalle 1994). A comparison of the utility of several protein-coding genes for phylogenetic purposes in insects is given by Danforth et al. (2005). Many primers for protein-coding genes are discussed in the references cited in this paper.

What are DNA sequences used for? The most common use of DNA sequences is to determine the phylogenetic relationships among different species. For phylogenetic studies using DNA sequences it is essential to be able to align the DNA sequences with high confidence. Using DNA sequences coding for proteins simplifies the alignment problem.

Examples of DNA sequences being used in biological control No studies of direct importance to biological control using DNA sequences of protein-coding genes have been published. In general, nuclear protein-coding genes are not used much in population studies, because of their low level of variation. It is much simpler to use microsatellites for these purposes. Different-sized introns in protein-coding genes have been used in studies to determine the origin of several fruit fly invasions (Villablanca et al. 1998), using what has been termed exon-primed intron-crossing (EPIC) PCR. In these PCR reactions, the intron is amplified and because introns have higher mutation rates than exons, within-population variation in intron alleles can be substantial. In some species there is a substantial size variation in introns (Gasperi et al. 2002). Intron sequences are potentially very useful for determining the origins of populations of invasive species that have limited genetic variation because of bottlenecking during the invasion process. In many fruit fly species, the genetic variation of many markers is very low, because the population has gone through sequential genetic bottlenecks when an invasive population with reduced genetic variation in one area forms the source population of secondary invasion.

rRNA sequences

Ribosomes are the structures within cells where messenger RNA is translated into protein. Ribosomes in insects consist of three parts, named 5.8S, 18S, and 28S. Genes coding for these parts occur in repeat units in the genome, and each repeat unit consists of the gene for 18S rRNA, a spacer ITS1, a gene for 5.8S rRNA, a second spacer ITS2, and the gene for 28S rRNA followed by the intergenic spacer. While the sequences of the ribosomal genes are very conserved between species, the sequences for the spacer regions can vary substantially even between closely related species. Although there are many copies of the ribosomal repeat per nuclear genome, the sequences of the 18S, 5.8S, and 28S are generally all identical within an individual including the 28S D2, which is highly conserved and used as a very conservative identifier of species. Different D2 sequences indicate different species, but the same D2 sequence does not guarantee that two individuals are conspecific. However, for the spacer regions ITS1 and ITS2, there may be several different sequences within an individual. Although the differences generally are small, it precludes the direct sequencing of the ITS regions. Copies of the ITS regions often differ in the number of microsatellite repeats found in the ITS sequences. This difference causes the different sequences to differ by a number of base pairs and the sequencer will read different bases at the same position in the sequence. Consequently ITS PCR products need to be cloned before sequences can be obtained (for an example see Stouthamer et al. 1999).

How do I find rRNA sequences? Because the ribosomal gene sequences are very conserved, primers located in these conserved area can be used to amplify the DNA from many different organisms. For an overview of how to obtain rRNA sequences and for lists of primer sequences see Gillespie et al. (2005) and Hillis and Dixon (1991).

What are rRNA sequences used for? DNA sequences of the rRNA coding genes are used for phylogenetic studies. Because the sequences of the 18S, 5.8S, and 28S evolve very slowly they are often used to determine the higher-level classification of insects, such as relationships among orders. However, certain areas in the rRNA-coding genes are useful at lower taxonomic levels. These sequences include the various extension regions of the 28S rRNA (Gillespie et al. 2005). The ITS regions are not easily used in phylogenetic studies because their alignment is uncertain. ITS regions, however, are used extensively for the recognition of species, especially cryptic species or biotypes. Many different applications using ITS spacers are found in the biological control literature.

Examples of the use of ribosomal sequences in biological control Sequences of both the D2 and ITS spacers have been used to identify species that lack clear morphological features that can separate entities. The D2 extension region of the 28S rRNA is very useful for determining whether two different individuals belong to the same species, but the differences between closely related species are small. D2 has been used for the identification of many different species of the very important parasitoid genus *Encarsia* (Hymenoptera: Aphelinidae) (Babcock & Heraty 2000, Schmidt et al. 2001, Manzari et al. 2002, Pavis et al. 2003).

ITS1 and ITS2 sequences have been used in many different biological control projects for species recognition. ITS regions of closely related species often differ not only in DNA sequence but also in their size (i.e. the number of base pairs). These two features make the ITS regions very suitable for cheap and reliable species recognition. For instance, in the genus *Trichogramma*, many species can be differentiated simply by the size of the PCR product following amplification with ITS2 primers (Stouthamer et al. 1999). Different species, with similar-sized PCR products, can often be distinguished by the pattern of restriction fragments (i.e. restriction-fragment length polymorphism, RFLP) after digesting the PCR product with different restriction enzymes. Size of the PCR product and the restriction fragments following digestion with restriction enzymes has been used to produce keys to identify the species of *Trichogramma* (Silva et al. 1999, Stouthamer et al. 1999, Ciociola et al. 2001, Pinto et al. 2002).

If only a few species are present in a particular area, then primers can be designed that amplify only the DNA of a single species. Davies et al. (2006) used this method in Australia to distinguish between the two *Trichogramma* species found in cotton, constructing species-specific primers. These primers were developed for those parts of the ITS2 sequence where the sequence of both species differs. The primers were then tested to verify that they only amplified the DNA of the intended species. In addition, primers were also constructed in such a way that the size of the species-specific PCR product differed between species. The species identity of an unknown individual belonging either to species A or B could be determined by a multiplex PCR reaction. In this reaction, a general ITS2 forward primer and the two species-specific reverse primers were added to each PCR reaction to identify the species. One of the advantages of such an approach is that it results in a positive species identification. The specimen being analyzed is either species A or species B. If no PCR product is obtained, then the unknown individual may belong to another species, which can then be identified by simply amplifying and sequencing the complete ITS2.

Species-specific primers based on ITS sequences are also used to determine whether hosts are parastitized, and if so by which parasitoid. Zhu et al. (2000) designed specific primers against two common parasitoids of the Russian wheat aphid. Using these specific primers they were able to: (1) identify adults of the two parasitoids to species and (2) determine the species of parasitoid in a parasitized host by extracting parasitoid DNA from parasitized aphids. This method was so sensitive that they were able to detect parasitoid larvae inside aphid hosts that were only 1/1000 the size of an adult parasitoid and 1 day after an aphid was parasitized the parasitoid could be detected. Many other studies have used similar approaches (Greenstone 2006).

Mitochondrial genes

Mitochondrial genes differ from genes located on the nuclear chromosomes in that they have purely maternal transmission. This means that all offspring inherit all their mitochondria from the mother; the father's mitochondria are not transmitted. The genome of mitochondria is small compared to the nuclear genome.

How do you find mitochondrial genes? Many different primers are commercially available for identifying mitochondrial genes. Packages can be bought that contain different primer sets for several different mitochondrial genes (Simon et al. 1994).

What are mitochondrial genes used for? Over the last few years, the mitochondrial cytochrome oxidase (COXI or COI) gene has been used for identification purposes in projects known as species barcoding. The idea behind barcoding is to sequence the COI gene of as many different species as possible and then use the COI gene sequence to identify unknown specimens from analyzed sequences from catalogued or named species. This approach allows unknown specimens (both adult and larval stages) to be characterized and assigned an identification or name tag. Many papers have been published that oppose the idea of barcoding life for several different reasons. In some cases, the COI gene of two closely related species does not differ, yet the species are recognized as being different (Moritz & Cicero 2004, Hurst & Jiggins 2005). Also, the statistics used to delineate undescribed species with COI genes have been criticized (Will & Rubinoff 2004). Despite these shortcomings, this method does appear to have its utility and can be particularly suitable in combination with the sequencing of additional genes or when biological and morphological characteristics are also studied to supplement COI data.

Mitochondrial DNA sequences have been used most commonly in phylogeography, which is the study of processes governing geographic distributions of genealogical lineages, especially those within and among closely related species (Avise 2000). Such analytical phylogeographic methods can be of substantial use to determine the origin of invasive species or introduced natural enemies.

Examples of mitochondrial genes being used in biological control projects Species barcoding using COI genes has not yet been used much in biological control. However, it is thought to have a great potential to help identify potential invasive species (Scheffer et al. 2006) and natural enemies (Greenstone et al. 2005). Greenstone et al. (2005) determined the COI sequences for many carabid beetles and spiders found in crop fields. They used these sequences to create species-specific primers that allowed them to identify all the studied species regardless of the life stage collected. They point out that the density of larval stages of these species is often higher in the field than the number of adults, and the impact of immature stages on pest control is only rarely studied, in part, because of identification difficulties. Agustí et al. (2003) used the COI sequence of the pear psylla [*Cacopsylla pyricola* (Förster)] (Hemiptera: Psyllidae) to design species-specific primers to detect *C. pyricola* DNA in predator guts. After 8 h, all predators that had eaten between one and five psyllids still tested positive for the target pest. Perdikis et al. (2003) used mitochondrial DNA sequences to distinguish between two closely related predatory hemipterans encountered in field studies. Finally, Borghuis et al. (2004) used PCR-RFLP analyses of mitochondrial COII to distinguish between two closely related parasitoid species *Trichogramma minutum* Riley and *Trichogramma platneri* Nagarkatti (both Hymenoptera: Trichogrammatidae). These two species could not be distinguished using their ITS2 sequences alone (Stouthamer et al. 2000b).

Mitochondrial DNA sequences have been used in many studies to determine the origin of an invasive species. In the native range of an invasive pest often a clear association exists between particular mitochondrial sequences and a subarea of the total native range. This association between sequence pattern and geographic location can be used to determine the origin of an invasion. An example is the study by Havill et al. (2006) to determine the origin of the hemlock woolly adelgid, *Adelges tsugae* Annand (Hemiptera: Adelgidae), which has invaded eastern North America. Based on the sequences of the COI gene, the invasive population could be traced to Japan. The results of this analysis meant that both mainland China and Taiwan could be excluded as sources for the eastern North American population. Similar studies have pinpointed the source for *Phylloxera* invasions worldwide (Downie 2002), the pumpkin fruitfly [*Bactrocera depressa* (Shiraki)] in Japan (Mun et al. 2003), and the Chinese mitten crab (*Eriocheir sinensis* H. Milne Edwards) in North America (Haenfling et al. 2002). Hufbauer et al. (2004) used several molecular techniques to study the origin of the North American *A. ervi* parasitoid populations. This species was introduced from western Europe in 1957 and Hufbauer et al. (2004) showed that most North American *A. ervi* indeed have mitochondrial sequences in common with the western European and Middle Eastern *A. ervi*, confirming their conjectural area of origin. In the Pacific northwest (USA), a second mitochondrial sequence was found that was most simlar to sequences found in *A. ervi* from Japan, indicating that most likely a second introduction took place there.

IMPORTANT BIOLOGICAL CONTROL ISSUES THAT MOLECULAR TECHNIQUES CAN ADDRESS

Species identification

In biological control programs, the identity of the pest or of potential natural enemies may be unknown or only poorly understood relative to other similar taxa. Under such circumstances the work of a competent systematist is needed to resolve critical taxonomic questions. However, with the wide availability of molecular methods and their relative simplicity and ease of use, it is now feasible for biological control workers to characterize taxa of interest, thereby allowing a systematist to later identify the species. This approach means that projects will not stall because of taxonomic uncertainties, which is often referred to as the taxonomic impediment in biological control. Most attention has been given to the barcoding method, where the sequence of the COI gene is used to identify species even if a species name is unavailable or true identity is uncertain. Although this method has disadvantages, which have been discussed above, an initial characterization of a taxon through barcoding can be useful. Determining the COI sequence is relatively straightforward, general primers are commercially available that work on many different species, and the resulting PCR product can be direct sequenced so that the cost of characterizing a specimen is low. One of the major problems from the applied point of view is that, within a species, several and sometimes substantially different COI DNA sequences exist. In Hymenoptera, the D2 region of the 28S rRNA appears to be very good for species recognition. Within a species there appears to be very little variation in this sequence, whereas between species there are differences which helps to make identification easier (Heraty 2004). The PCR reaction using the D2 primers works very reliably, and the PCR product can also be direct sequenced. Finally, the sequence of the ITS2 is used in a number of genera for species identification (Stouthamer et al. 1999, 2000b, Alvarez & Hoy 2002). ITS2 is a sequence that can be easily amplified using published primers that work on a wide variety of organisms. The disadvantage of ITS is that it is a multi-copy gene that has within-individual variation. This makes it necessary to clone the DNA before sequencing thus making this technique a more expensive process than sequencing the 28S D2 rRNA or the COI gene.

Differentiation between species

Another problem where molecular methods can play a vital role in biological control is in differentiating between species. Sometimes the species that can be expected in a field sample are known, but their species identity may be difficult to determine. This situation can occur if:
• only a single sex of the species can be identified morphologically;
• immature stages cannot be allocated to a particular species;
• eggs or larval stages of parasitoids inside their host would need to be kept alive for a long rearing period before the identifiable adult stage emerges;
• morphological identification requires extensive and time-consuming specimen preparation and the services of a limited number of taxonomic experts.

Method 1

The cheapest method for species recognition is to develop a PCR protocol that will amplify different-sized products from the different species of interest. This can be accomplished by amplifying a gene region where the species of interest differ in the size of their PCR products. The size of the internally transcribed spacers (ITS1 and ITS2) of the ribosomal repeat often differs between species. The species can be distinguished by amplifying the ITS region and determining the size of the product on a gel. Any other gene where a particular primer set results in a product with a species-specific size could be used as an identification tool.

Method 2

Sometimes two or more species can be distinguished by the species-specific restriction pattern of a particular gene. In this case, the PCR products of the selected gene for the species of interest are identical, but a restriction enzyme cuts the PCR products for each species into different-sized fragments. Gel electrophoresis can be performed on this restricted DNA and species identity for unknown specimens can be assigned.

Method 3

The final method involves developing species-specific primers that amplify a PCR product of a different size for each species. To develop this technique, a single gene is typically used where the species differ substantially in the sequence of this gene. The ribosomal spacers, ITS1 and ITS2, are very suitable for this purpose. To use this technique, the DNA sequence needs to be known for each species; these sequences are then aligned to determine the conserved parts (i.e. the parts of the sequence that are present in all of the species) and the variable parts (i.e. those sequences only found in a single species). Next, for each species, a reverse (or forward) primer is developed that binds to the DNA of only that species and can be used with the general forward (or reverse) primer for that gene region. The primer sets for each species are first checked to make sure that they amplify the gene region on that species. Next they are tested on the other species to make sure that the primers do not amplify the DNA of those species as well. Once it has been established that the species-specific primers indeed work on the species of interest, they can then be tested in a multiplex PCR reaction to determine their effectiveness for identification. In a multiplex PCR reaction, several primers are used; in this case it would be the general forward primer and the species-specific reverse primers. If a specimen belongs to species A in the multiplex PCR the species-specific product for species A will be amplified.

Where did the invasive species originate?

The markers of choice for determining the origin of invading populations are generally mitochondrial DNA sequences. Often in the area where a species originates, different mitochondrial types (i.e. sequences) are found at a higher frequency, or perhaps are restricted exclusively to a particular subarea within the native range. By comparing the mitochondrial type of the invading population to samples of the pest from throughout the native range, some areas of the native range can be excluded as source populations because genetic matches are not made. Following this first cut, a more precise determination of the source population may be obtained using microsatellite DNA markers. Mitochondrial sequences can be very informative if the invasion originated from the native range of the pest; however, for some of the worldwide pests, new areas are often colonized from areas that had been colonized earlier. With each subsequent invasion genetic variation is lost and very few mitochondrial types remain. Under such circumstances it becomes impossible to use the mitochondrial sequences to determine the origin of secondary and tertiary invasions (Bohonak et al. 2001), but the analysis of microsatellite data may shed light on the invasion process.

Determining what a predator ate in the field

Prey items present in the gut of a predator can be identified using PCR for a time period of around 1–2 days. The DNA that is present in the insect gut generally degrades with time. Consequently, it is important to use a gene region that occurs in many copies in the genome of the prey item (ribosomal genes, mitochondrial genes) and to design prey species-specific primers that amplify a relatively short PCR product. The reason for this is that when DNA is degraded, it is repeatedly cut into shorter strands. For the PCR to work, both priming sites must be present on a contiguous stretch of DNA, which is more likely for short sequences. When the prey species-specific primers have been designed and tested on the prey species from which they originate, they then should be tested to make sure that they are indeed species specific and do not amplify the predators own DNA. Once specificity has been determined, the prey-specific primers are tested with predators that have fed on various numbers of the prey species: (1) to ascertain that the primers work on a consumed prey and (2) to determine the relationship between detectability of the prey item in the gut and time since feeding. A review of all the possible methods to measure predation using molecular techniques is given by Symondson (2002).

Which strain of a natural enemy is more effective?

Microsatellite markers are excellent tools to determine the relative importance of augmentative natural enemy releases in already existing populations. Because microsatellites are so variable, many different alleles of a particular microsatellite locus will be present in field populations. For example, assume that in a natural enemy population, for locus A and locus B, three different alleles exist, each with a frequency of 0.33. Assuming that the population is in Hardy–Weinberg equilibrium, the $A_1A_1B_1B_1$ genotype in the field will occur at a frequency of one individual in 81. If a strain is created in the laboratory that is $A_1A_1B_1B_1$, such a strain can then be used to determine the effect of augmentative releases in the field. If we release parasitoids that are $A_1A_1B_1B_1$ as females that had mated with their brothers (as is common with many hymenopterous natural enemies), then we can determine the effect of augmentative releases on the parasitization of hosts by collecting hosts from the field and determining the frequency of $A_1A_1B_1B_1$ offspring emerging from these hosts. If the release did not have any influence then we would expect only one in 81 individuals to be $A_1A_1B_1B_1$. However, if a substantially higher frequency of $A_1A_1B_1B_1$ individuals emerges from parasitized hosts after the release, the increased numbers of this genotype then reflect the offspring of the augmentative release. Because so many different genotypes can be reared in the laboratory that occur at a low frequency in the field, several marked populations can be released in succession and after each release it will be obvious which individuals are the offspring of the augmentative release. Such studies can be used to determine the optimal timing and number of natural enemies for augmentative releases. This approach has already been used to determine the importance of the size of natural enemies on their efficacy as biological control agents (Kazmer & Luck 1995).

Sometimes several strains of a natural enemy have been collected and could be released for the classical biological control of a pest. The question comes up, which one may be more suitable for the control of the pest population? Often it is possible to use neutral markers (i.e. ISSRs, microsatellites, etc.) to distinguish between the different populations of natural enemies. However, if these different populations are able to interbreed then the neutral markers are not very useful in determining which of these populations does better in the long term, because the association between the neutral marker and the population origin will be lost quickly. In general it will only be possible to test the short-term performance of the different lines as described in the example above for the relative performance in the augmentative releases.

Is the pest or natural enemy infected with symbionts?

Over the last 20 years it has become obvious that many insects are infected with symbionts. It has been estimated that as many as 76% of all insect species are infected by *Wolbachia* (Jeyaprakash & Hoy 2000). In many cases these symbionts are required by the host for its survival and reproduction. However, many insects are also infected with secondary symbionts that are not required for the normal functioning of the insect. Some populations are polymorphic for infection with secondary symbionts. Secondary symbionts may have unusual effects on their hosts. For example, some secondary symbionts of aphids confer resistance to parasitism (Oliver et al. 2003, 2005).

Symbionts that are classified as reproductive parasites are also extremely common and can cause crossing incompatibility between infected and uninfected individuals. It is particularly important to know the infection status of different populations when they are mixed either in the laboratory for mass production or when they are released in the field where a population is already established (Mochiah et al. 2002). Mixing infected and uninfected populations can result in the depression of the population growth of the natural enemy. Several different reproductive parasites (e.g. *Wolbachia*, *Cardinium*, and *Rickettsia*) are known and all can be easily detected by PCR (Weeks et al. 2003, Zchori-Fein & Perlman 2004, Hagimori et al. 2006). Some other reproductive parasites manipulate the offspring sex ratio produced by their host. For example, in many Hymenoptera complete parthenogenesis (thelytoky) is caused by an infection with a *Wolbachia*, *Cardinium*, or *Rickettsia* species (Stouthamer et al. 1990, Zchori-Fein et al. 2001, Hagimori et al. 2006). Such infections may be beneficial for biological control applications because the infected female parasitoids produce only daughters, thereby promoting more rapid population growth and pest suppression. Thelytoky also overcomes problems associated with finding mates at low densities, which may result in the failure of natural enemies to persist in a given area (Stouthamer 1993). Finally, there are a number of reproductive parasites that cause the death of male offspring produced by an infected female. Such infections are commonly found in Coccinellidae (Hurst & Jiggins 2000). It is probably best to remove such male-killing infections from natural enemy populations before their release.

CONCLUSIONS

Many new molecular tools have been developed in the last 20 years that have practical application for biological control for species identification, determining pest areas of origin, studying the efficacy of natural enemy biotypes, and determining the magnitude of non-target impacts and habitat infiltration. These tools can help improve the efficacy of biological control, reduce risks from non-target impacts, and enhance understanding of how genetic structures of pests and natural enemy populations affect pest regulation and stability.

Several areas are still underexplored with these techniques. For instance, can we go back to old "failed" biological control projects in which the natural enemy was established but did not provide economic control of the pest? We can determine whether the natural enemy population had experienced strong genetic bottlenecks which may have impaired the ability of the population to grow and adapt to the local conditions. We can use several of the markers discussed above to determine the level of genetic variation present in the population. If genetically impoverished populations of natural enemies are detected, additional genetic variation may be imported to enhance the biological control efficiency of these already established populations. We can expect benefits from introducing additional genetic variation. One of the most dramatic examples is the work by Spielman and Frankham (1992). In experiments to determine how addition of genetic variation affected small, partially inbred *Drosophila melanogaster* (Diptera: Drosophilidae) populations, they showed that the addition of a single male *D. melanogaster* immigrant to these populations doubled the relative reproductive fitness of these populations. Additional examples of changes following the introduction of genetically divergent individuals in populations are given in Tallmon et al. (2004).

With our increased ability to pinpoint the origin of our invasive pests it now is also possible to test hypotheses about the best areas within the native range of a pest to collect natural enemies. In addition, in augmentative biological control it is now possible using genetically marked lines to determine the optimal timing and numbers of natural enemies for release. Do such natural enemy releases really contribute to control? Similarly, we can expand on the work of Kazmer and Luck (1995) and determine which natural enemy traits are important for their performance in the field. Is it better to rear larger, more expensive, natural enemies or is the number that is released more important? Is it better for the field performance of parasitoids to release them after they have just emerged and are not experienced with hosts, or does giving them oviposition experience improve their field performance? Does feeding parasitoids with honey before release improve their biological control performance? How important is the genetic make-up of a line for performance and impact in biological control? We often assume genetic make-up is important; however, this has never been tested empirically in a biological control system (Hopper et al. 1993). Releasing different genetic lines of the same species, recognizable by different genetic markers, can answer these types of fundamental questions.

Our ability to easily distinguish species of minute parasitoids now makes it possible to precisely test their relative performance in inundative biological control. For example, with *Trichogramma*, several species can be released simultaneously in the same field plot. In the past it would have been very difficult to determine the species identity of parasitoids emerging from hosts, but now with our ability to easily identify individuals such tests can be done with relatively little effort with high confidence in the results.

In biological control, selective breeding of natural enemies for improved traits associated with their biological control performance has not been pursued to any great extent. Notable exceptions include selecting natural enemies for pesticide resistance (Hoy 1990), for improved offspring sex ratio (Wilkes 1947), and enhanced temperature tolerance (White et al. 1970). In the past it was difficult to distinguish different genetic lines from each other and quantifying the relative performance of these lines by necessity involved testing each different line in separate plots to avoid identification problems. Using some of the molecular markers discussed above it is now possible to test differently selected lines of the natural enemies in the same field plot.

Part 6

SAFETY

Chapter 16

NON-TARGET IMPACTS OF BIOLOGICAL CONTROL AGENTS

BIOLOGICAL CONTROL AS AN EVOLVING TECHNOLOGY

The safety of classical biological control to non-target species has been debated widely (Simberloff & Stiling 1996, Follett et al. 2000, Lynch & Thomas 2000, Louda et al. 2003a). The historical record shows that technology, ideas, and values affecting biological control have all changed over time. Early practitioners were not professionals and in the earliest cases had no training or qualifications for the task and acted in pursuit of their own, rather than the public's, interests. Consequently, a clear trend exists from early rather damaging impacts associated with amateurs moving vertebrates, to better and safer projects as a professional class of biological control practitioners developed. Some practitioners serve agricultural interests, whereas others work to resolve problems caused by invasive environmental pests in wild areas.

This progression is illustrated here by showing how the goals, attitudes, knowledge, and techniques employed for biological control have changed from the 1800s to the present. The stories used to move the discussion along are arranged in roughly chronological order, to gain a perspective on this historical dynamic. We focus in part on how reasons have changed for making natural enemy introductions. Early efforts were strictly economic, aiming to reduce damage to crops or forests from invasive insects or weeds. Later, this broadened and some projects were conducted solely to protect a native species or ecosystem from the harm by invasive species. Similarly, the degree of knowledge employed in such efforts has increased dramatically, from almost none in the earliest days (nineteenth century) to highly sophisticated molecular studies that are parts of many projects in the twenty-first century. Efforts by plantation owners or acclimatization societies in the nineteenth century were based on folklore and "common knowledge," which often was nothing more than knowing that a certain predator ate the pest. As biological control efforts developed from private actions to government-supported programs, the depth of knowledge about natural enemy biology, behavior, ecology, and genetics increased dramatically.

We also focus on the importance of who makes an introduction. Risk has declined greatly as this has shifted away from private plantation owners, to government scientists hired to serve the needs of particular groups, to teams of scientists seeking to balance ecological and economic interests for the benefit of the whole society. Biological control projects are now largely run either directly by government agencies or by international groups (such as CABI BioScience) working for countries. Countries, however, vary in available scientific expertise and resources. Countries also vary in their attitudes about how much risk to non-target native species is acceptable, given economic pressures imposed by a pest, in the context of the nation's wealth and food security. Small countries retain rights to make choices about what species they feel they need to introduce; however, these choices affect all the countries in the biogeographical region and so regional coordination is critical.

Overall, this chapter's goal is to illustrate biological control as an evolving human activity that has at times been misused but which is capable of being wielded with great precision when good policy and sufficient resources are applied.

THE AMATEUR TO EARLY SCIENTIFIC PERIOD (1800–1920)

It was no accident that classical biological control began in temperate areas colonized by Europeans. Such locations (e.g. Canada, the USA, Australia, New Zealand, South Africa, and others) were the sites of frequent and damaging pest invasions as a by-product of the movement of European agriculture, people, and goods to colonies with similar climates. Invasion routes have since become more complex, reflecting modern international trade, but historically it was in these colonies that the desire to combat pests with biological introductions first arose. Newly established plantations of exotic plants must have been relatively pest-free at first, but this would have changed as pests of the crops reached these new regions. Avocado and eucalyptus, for example, were both virtually pest-free in California, USA, for their first hundred years, but then were attacked by a series of new invasive pests. Biological control was born in an effort to combat this. The earliest pests were often vertebrates deliberately brought by the settlers themselves (e.g. rabbits) or that accompanied them as stowaways (e.g. rats). Being large and obvious, these sorts of pests were of earliest concern to planters, who often reacted to them based on their own personal notions of how things worked in their home country. (Perhaps such as "Rabbits? Foxes take care of that sort of problem," or "Rats? When I was stationed in India, I saw mongooses do wonders against rats.") Acting on such simple notions, in a unregulated environment where governments had not yet thought that it was their business to tell people what animals or plants they might move about the globe, planters turned to vertebrate introductions in the hope that they would solve their problems.

During this period, a couple of other overriding attitudes helped shape events. First, there was little or no concern about the effects of introduced predators on native wildlife. This was just beginning to be observed and thought about. Concern for such impacts had little or no influence on events in the nineteenth century. A second important view was that plants (being useful as crops, timber, or ornamentals) were uniformly assumed to be beneficial unless painful experiences had proven otherwise (we are still struggling with this attitude today) and, by contrast, insects that fed on plants were uniformly felt to be pests or at least of little importance.

So, with that background, we discuss four projects to illustrate the status of things during this period and note the first steps towards change. We start with some vertebrate introductions by private persons or groups (mongooses in the Caribbean to control rats) or governments (cane toad in Australia). We then step backward in time and switch to invertebrate introductions to discuss the precedent-setting case of the successful control of cottony cushion scale on California citrus. As our final example for this period, we discuss the first "super project" of biological control, the attempt to suppress the gypsy moth in New England (USA).

1872: mongoose in the Caribbean for rats in sugarcane

Sugar estates were developed in the Caribbean during the 1600s and 1700s to exploit profits from the high demand for sugar in Europe. European ships brought new species of rats (first *Rattus norvegicus* and later *Rattus rattus*) to many parts of the Caribbean, including Jamaica. Very early, ship rats became abundant and destroyed up to a quarter of the annual sugar crop (Roots 1976). In 1872, a Jamaican sugar planter, Mr W.B. Espeut, imported small Indian mongooses (*Herpestes auropunctatus*) (Figure 16.1) from Nepal and released them on his plantation. Rat populations were reduced, so other planters purchased animals from Mr Espeut for release in additional areas, including Puerto Rico (1877), Barbados (1878), St. Croix (1884), and Cuba (1886), later followed by many other areas (see Thulin et al. 2006 for a history of introductions). In Hawaii, USA (where this same mongoose had previously been introduced), an examination of 356 mongoose feces showed that 52% contained only the remains of rats and mice, while the rest included insects as well (Pemberton 1925). However, in Trinidad, the diet of mongooses also included various birds, lizards, snakes, frogs, and toads (William 1918). Ground-nesting birds seemed especially likely to be attacked. In the Caribbean, the mongoose is blamed for extermination of the burrowing owls of Antigua and Marie Galante and the Jamaican nighthawk, among others

Figure 16.1 The introduction of the small Indian mongoose (*Herpestes auropunctatus*) was unscientific and highly damaging to native wildlife. Photograph courtesy of Rick Taylor, Borderland Tours.

Figure 16.2 The marine toad [*Bufo marinus* (L.)] is a poisonous species whose introduction has been highly damaging to predators wherever it has been released. Photograph courtesy of Don Sands, CSIRO.

(Lever 1994). Eight lizards are also believed to have been driven to extinction in the Caribbean because of this mongoose, including ground lizards in the genus *Ameiva* and the *Mabuya* skinks (these skinks, however, survive elsewhere on mongoose-free islands). Here, we do not give a full account of the havoc caused by this predator. Lever (1994) sums it up saying that "the mongoose in the West Indies has helped to endanger or exterminate more species of mammals, birds, and reptiles within a limited area than any other animal deliberately introduced by man anywhere in the world."

Although this example is consistently held up as an example of biological control gone bad, it is important to note that the early introductions of mongoose in many areas were not made by biologists based on scientific information, but rather the private action of sugarcane planters, based not on science about the diet and biology of this predator but rather a layperson's general knowledge and attitudes.

1935: cane toad in Australia for gray-backed grub in sugarcane

The marine toad, *Bufo marinus* (L.) (known as the cane toad) (Figure 16.2), is a generalist predator of insects that caught the attention of sugarcane growers in the Caribbean in the 1800s. Native to Surinam in South America, it was moved repeatedly by private sugarcane growers because of the belief that it reduced grubs attacking sugarcane. In quick succession, the toad was moved from Surinam to Martinique to Barbados, to Jamaica, and so on. This was later followed by its introduction by professional entomologists to Hawaii and, in 1935, to Queensland, Australia. Before this time, some information had been obtained (Dexter 1932) that cane grubs were at least part (25%) of the diet of marine toads living in sugarcane fields in Puerto Rico. On this basis, the importation to Australia in 1935 was a project of the Bureau of Sugarcane Experiment Stations in Queensland, in contrast to the nineteenth-century releases in the Caribbean, which were private actions by individual sugarcane planters. The target pest in Queensland was a scarab, the grey-backed cane beetle [*Dermolepida albohirtum* (Waterhouse)].

However, the introduction of this toad to Australia was not based on much real scientific consideration of its likely benefits, and its risks were not even considered. The most obvious of these risks stems from its toxicity (due to a toxin called bufotenine produced in the parotid glands). Naïve individual predators that ate marine toads often died. In Queensland, evidence of the impacts of this introduction are not well defined (because of lack of information on native species before its introduction), but the toad is believed to have contributed to the decline of the quoll (*Dasyurus* sp.), a marsupial "cat," as well as various native snakes (Lever

1994). It has had no impact at all on the target cane grub (Waterhouse & Sands 2001). Viewed globally, Lever (1994) considers the movement of this toad to be the most destructive of all amphibian introductions, largely because of the effects of its toxin on native species.

For our purposes, it is important to note that, unlike the mongoose introductions in the Caribbean, the movement of the marine toad to Australia was done with the support of a government agency. Why government failed to protect the common good in this case is an important question. The agency involved likely saw its constituency as the sugarcane growers. Perhaps the agency was not aware of potential risks from the cane toad, although by 1935 they could have learned of impacts in the Caribbean if they had sought the information. Or perhaps they did not see it as their role to be concerned about anything other than reducing pest problems for the industry they were created to serve. This failure points out the need for a broad review of all biological control introductions (and indeed all exotic species introductions), because a narrow review by just the group affected by the target pest (or by their closely associated servant government agencies) may miss or not care about other important considerations.

1886: vedalia beetle success and ladybird madness

The control of the cottony cushion scale (*Icerya purchasi* Maskell) in California by an Australian coccinellid, *Rodolia cardinalis* Mulsant, is pointed to repeatedly as the beginning of the era of "scientific" biological control. Although this is partly true, the creation of scientific biological control at that time was still in its infancy. The cottony cushion scale invaded California in about 1868, probably on *Acacia* imported as an ornamental, at about the time when the citrus industry was developing in the new state. By 1887, this scale was dramatically reducing citrus yields and growers were seeking help (summarized from Caltagirone & Doutt 1989).

Unlike the sugarcane growers discussed above, the California citrus growers in the 1880s looked to their state government for help rather than taking actions by themselves. This stemmed from two factors. (1) California had by that time enacted a plant quarantine law (to prevent pest introductions) that would have legally limited the scope of such private action. (2) Also, finding natural enemies of a pest scale would have required specialized knowledge of insect taxonomy and biogeography. By this time, the science of entomology had developed enough so that it was recognized that the pest scale was an invasive species from Australasia. With USDA funding, a California state delegation to a trade exposition in Melbourne was used as a pretext to also send an entomologist, Albert Koebele, to Australia to investigate the kinds of natural enemies of *I. purchasi* found there (Caltagirone & Doutt 1989). Ultimately, both a fly [*Cryptochetum iceryae* (Williston)] and larvae of an unidentified coccinellid (later named *R. cardinalis*) were found and send to California. Both established quickly and immediately controlled the pest. Both the pest and ladybird beetle were highly visible, so growers easily understood the process.

Two events followed from the success of this project. One was that the state of California in 1923 charged the University of California to engage in research to allow for further use of biological control. This mandate to both do research and conduct projects on biological control on behalf of California agriculture institutionalized the discipline and gave it a steady source of funds, scientific staff (at first mostly taxonomists but later also ecologists), and laboratories. This stimulated a rapid expansion of biological control.

Another effect of the success of *R. cardinalis*, however, was less laudable and points to the low degree of understanding that went into this project, which indeed succeeded in large measure because of chance. By good luck, the pest scale was a margarodid scale. For this precise group, ladybird beetle species in the genus *Rodalia* are true specialists. This specialization underpins both the success of *R. cardinalis* in controlling the target *Icerya* species and accounts for the virtual complete lack of non-target harm from the project. That these points were not understood at the time is illustrated by the so-called ladybird fantasy that gripped the state following this first success. During 1891–2, Koebele continued to collect and ship other coccinellids from Australia, New Zealand, New Caledonia, and Fiji, in the hope of repeating the success of the cottony cushion scale project against other scales and mealybugs in California citrus. Over 40,000 individuals of some 40 species were sent to California (Caltagirone & Doutt 1989). None controlled any of the intended pests; only four even established. The fact that so much was expected of them indicates just how thin actual knowledge about natural enemies, ladybirds, biological control, and population dynamics was at the time. By 1912, this craze had died out (for lack of any follow-up successes) and a professor at the University of California, Harry Smith, was appointed to give guidance to biological control efforts in California and to initiate scientific studies of how natural enemies really worked. *This was the true beginning of the scientific period of biological control.*

The California ladybird fantasy of the early twentieth century died out, but the same expectations have re-emerged periodically elsewhere. USDA introductions of various exotic coccinellids continued throughout the 1960–90s. Some exotic coccinellids were targeted to specific pests but in other cases the target was nothing more than a pest category (aphids for *Coccinella septempunctata* L.). Below we will discuss two of these cases: *C. septempunctata* and *Harmonia axyridis* Pallas.

Figure 16.3 The tachinid fly, *Compsilura concinnata* (Meigen), is a polyphagous parasitoid that causes high levels of mortality to such non-target insects as the giant silkworm moths. Photograph courtesy of Michael Thomas.

1905–11: brown-tailed and gypsy moths – shotgun or super project?

In the mid-to-late 1800s, two species of defoliating moths in the family Lymantriidae invaded North America. The gypsy moth, *Lymantria dispar* (L.), was brought to the USA from Europe in about 1869 by an amateur entomologist wishing to breed this species with native silk worms to benefit a local silk industry, and the insect escaped. Brown-tailed moth, *Nygmia phaeorrhoea* (Donovahan), another European species, invaded Massachusetts on its own in about 1897. Both caused extensive defoliations of deciduous trees, but the brown-tailed moth was also a public health hazard due to skin and lung irritations (capable of causing death) from contact with its hairs. A combined project targeting both species was mounted by the USDA from 1905 to 1911 (with additional work later). This was a very large project with extensive funding and many scientists. Work was conducted in Europe, Russia, and Japan. Because the pests had large native ranges and because it was not apparent that any single agent or small group of species controlled their numbers in these areas, the decision was made to introduce a long list of species, essentially any non-hyperparasitic species found associated with the pests in surveys that could be obtained in adequate numbers to propagate. Some 40–80 species (accounts vary) of parasitoids were released against the gypsy moth, as were several species of predaceous insects. Of these parasitoids, at least 10 established and six have become common. Approximately 20 species of parasitoids were released against the brown-tailed moth and eight established. These species provided, at best, partial control of the gypsy moth (Waters et al. 1976, Dahlsten & Mills 1999), but complete control of the brown-tailed moth (Waters et al. 1976). Subsequently, the exotic fungus *Entomophaga maimaiga* Humber, Shimazu & R.S. Soper, completed the biological control of gypsy moth (Webb et al. 1999), at least in New England, where significant outbreaks have not occurred since the early 1980s. The tachinid *Compsilura concinnata* (Meigen) proved to be an important parasitoid of both gypsy moths (Liebhold & Elkinton 1989, Gould et al. 1990) and the brown-tailed moth.

Compsilura concinnata (Figure 16.3), a generalist parasitoid of caterpillars, has, however, also become a major mortality factor for some of the largest, most colorful of North America's native silkworm moths. It was known at the time of its introduction to attack many host species, but native insects were not seen as important and the mission at the time was to protect the forest trees from defoliation. Within 13 years of this species' 1906 introduction, some entomologists raised concerns over attacks by *C. concinnata* on the large and beautiful native giant silkworms (Culver 1919). However, the wide host range of the fly was seen by others as desirable, allowing it to maintain higher numbers even when gypsy moth numbers were low. Releases of *C. concinnata* continued to be made in new parts of the USA as late as 1986, despite such concerns, which were later shown to be well founded (Boettner et al. 2000).

This project illustrates that, on the technical side, biological control was making rapid progress in areas such as parasitoid taxonomy, biology and rearing. Conversely, it also shows that ideas concerning the population dynamics behind classical biological control and safety to non-target species were still in their infancy. The assumption, for example, that it would be necessary to introduce a very large number of species of parasitoids found associated with the target pest in its native range was not justified. We now have many examples where control of invasive insects has been achieved by just one or two specialized species. The project also illustrates that societal values (plants are good, insects are unimportant) influenced decisions about biological control introduction. Specifically, the polyphagous nature of *C. concinnata* was known at the time of its first release, but this fact was not seen as the detriment that it is now recognized to be.

A DEVELOPING SCIENCE MAKES SOME MISTAKES (1920–70)

National governments asserted regulatory authority over importations of beneficial insects during this period, with projects carried out by either government scientists or ones working for agricultural commodity associations with government oversight of importation. The study of natural enemy biology increasing became the basis for agent selection. Standards and goals, particularly in defining what outcomes were desired, continued to evolve. Society was clearly "plant-centric" early in the era, inasmuch that plant protection (of crops, forests, and sometimes a highly valued non-economic plant) was the goal. Damage to native plants *per se* was not a social consideration and was considered acceptable, although this reversed by the end of the period. In contrast, the same care was not extended to native insects until the 1990s, nearly 25 years later. The technical ability to forecast host ranges of weed biological control agents also began to be developed during this period. Owing to a lag in social awareness, this period was both an era when biological control was seen as a wholly green technology (although the term was not used in that period) while at the same time field studies demonstrating damage from past projects were being carried out.

In this section we analyze several biological control projects that have been discussed widely as examples of non-target impacts of classical biological control: (1) the coconut moth's control in Fiji (which has been variously represented as a major success against an invader or extinction of a native species), (2) the release of a highly damaging predaceous snail on Pacific islands, (3) the release of tachinids in Hawaii for stink bug control, (4) a thistle feeder [*Rhinocyllus conicus* (Frölich)] that has attacked native thistles, and (5) the cactus moth *Cactoblastis cactorum* (Bergroth), which was imprudently released in the Caribbean without consideration of likely impacts on non-target cacti, many of which were already known as hosts for this moth.

1925: coconut moth in Fiji – extinction of a native or control of an invader?

In 1924, defoliation of coconuts in Fiji was rampant and the economy and culture of this island group was at risk of collapse. The cause of the defoliation was feeding by caterpillars of a small blue moth, *Levuana iridescens* Bethune-Baker (Zygaenidae) (Figure 16.4), which was first recorded in 1877. Because of the lack of any earlier records of such damage, and other reasons, the moth was presumed to be an invasive species when biological control was attempted (Tothill et al. 1930). A search for its native home, presumed

Figure 16.4 The coconut moth (*Levuana iridescens* Bethune-Baker) was a devastating pest of coconut in Fiji that was controlled successfully by the introduction of the tachinid *Bessa remota* (Aldrich). Photograph courtesy of M. Hoddle.

to be in the little-explored island-continent of New Guinea, was unsuccessful and eventually a tachinid fly, *Bessa remota* (Aldrich), was collected from a related coconut pest, the zygaenid *Artona catoxantha* Hampson. This zygaenid was found mining coconut leaves in Batu Gajah, Federated States of Malaya, where it was heavily parasitized. The tachinid fly was taken to Fiji, cultured, and released, causing high rates of mortality to the pest, which later became extremely rare or disappeared.

The disappearance of this moth has been interpreted by Robinson (1975) to mean this species went extinct on Fiji. Further, the species was represented by Robinson as a native species, despite the arguments of Tothill to the contrary. Howarth (1991) repeated this claim as an indication of an island extinction caused by a biological control introduction. But this interpretation has several problems (Kuris 2003, Hoddle 2006). First, arguments exist that the coconut moth was not a native but rather an invader to Fiji, based on: (1) absence of any recorded defoliating outbreaks of coconut before 1877, (2) the observed spread of the moth within the Fijian group, which would not be happening if it were native unless its host plant was being planted in new locations, and (3) the apparent absence of parasitism. Assuming it was an invader, coconut moth presumably existed elsewhere in southeast Asia and may still do so, even though levels have been too low to detect. Second, there is reason to believe that even on Fiji, the coconut moth remains present, albeit at extremely low levels. Rather than being driven to extinction by 1929 as claimed by Howarth (1991), the species was recorded in both 1941 (Sands 1997) and 1956 (Paine 1994), although it is certainly extremely rare if still present. Lack of subsequent records may be due to either confusion with another coconut pest that caused similar damage or the departure of British entomologists after Fijian independence. Currently, one of us (Hoddle) is attempting to re-collect levuana moth in Fiji. Another consequence of this project that seems more clearly to represent unintended harm to non-target native insects is the apparent disappearance from Fiji (but not globally) of the zygaenid *Heteropan dolens* Druce (Robinson 1975).

1950s–80s: predatory snails in the Pacific

The herbivorous giant African snail, *Achatina fulica* Bowditch, was deliberately spread for its edibility to many Pacific islands during the twentieth century. This species became a pest in gardens at various locations, creating a demand for its control. In response, the predatory snail *Euglandia rosea* (Ferrusac) (Figure 16.5) was introduced from Florida (USA) into Hawaii (and subsequently from there to other areas) as a predator of *A. fulica*. *Euglandia rosea* is a generalist predatory snail that locates its prey by following slime trails of snails (Clifford et al. 2003). It can consume *A. fulica*, but it prefers smaller prey species (Cook 1989). Land snails such as *Achatinella* spp. in Hawaii and *Partula* snails on Moorea have undergone species radiation and are classic examples of island evolution. As such, their conservation and scientific value is very high. The prey range of *E. rosea* was not studied before its introduction to these islands, but subsequently has been found to be quite broad. Many native snail species declined drastically or went extinct after its introduction (e.g. *Partula* spp. on Moorea, Murray et al. 1988; *Achatinella mustelina* in Oahu, Hawaii, Hadfield & Mountain 1981, Hadfield et al. 1993; as well as some aquatic snails, Kinzie 1992). Despite little evidence that *E. rosea* was an effective predator of the target pest and warnings by biologists that it would harm native snails, introductions of *E. rosea* to new locations continued (Civeyrel & Simberloff 1996, Cowie 2001). In the Society Islands, of 61 original endemic tree snail species only five remain (Coote & Loève 2003). Although a number of factors have contributed to the loss of these native snails, predation by *E. rosea* is the dominant reason.

Figure 16.5 The predatory snail *Euglandia rosea* (Ferrusac). Photograph courtesy of Ken Hayes, University of Hawaii, USA.

Why this introduction was made is perplexing. The need was rather limited. The conservation resource potentially at risk was fairly obvious. Evidence that the predator was effective was meager. Inertia seems the most obvious answer, together with a fairly low level of biological knowledge actually used in the decision-making processes in the various locations. Local agricultural officials seem to have agreed to the introductions for agricultural reasons without consultation with conservation biologists. Awareness of the risks and of the pointlessness of this approach seems to have increased. A 1998 leaflet from the South Pacific Commission recommended against any further releases. CABI BioScience was consulted when Western Samoa sought help in the 1990s from the United Nations Food and Agriculture Organization (FAO) with its *A. fulica* problem. CABI BioScience advised against the introduction of *E. rosea* to the islands. This case illustrates that new information about effectiveness and risks of past biological control projects does not always come to the attention of local decision-makers in a timely way. Rather, some political units empowered to make such decisions may be handicapped by lack of people qualified to judge such issues. Such officials may accept simplistic recommendations or simply imitate what other regional entities have done.

1962–3: parasitoids of *Nezara viridula* in Hawaii

The pest stink bug *N. viridula* L., of Mediterranean or Ethiopian origin, invaded Hawaii in 1961. This species is a major pest of many fruit, nut, and vegetable crops (Waterhouse & Norris 1987). In an attempt to lower the pest's density, the nymphal/adult parasitoid *Trichopoda pilipes* (F.) (Diptera: Tachinidae) and several populations or species under the name *Trissolcus basalis* Woolaston (Hymenoptera: Scelionidae), an egg parasitoid, were released and became established (Davis 1964). The herbivorous native scuttellerid *Coleotichus blackburniae* White (the koa bug) and some species of predatory pentatomid bugs (*Oechalia* spp.) have since declined in abundance (F. Howarth, personal communication). *Coleotichus blackburniae* was known to be physiologically suitable as a host for both of the above parasitoids (Davis 1964) and was used as a substitute host for rearing the tachinid. Eggs of *T. pilipes* occur on pinned koa bug specimens in local museum collections (Follett et al. 2000).

A retrospective study (Johnson et al. 2005) was conducted to determine whether these parasitoids attacked the koa bug at rates high enough to reduce its populations. Work at 24 sites (mostly on the big island, Hawaii) found that egg parasitism by *T. basalis* was low and confined to elevations below 500 m on a single introduced plant (*Acacia confusa* Merrill). The highest parasitism rate was 26%, but this occurred at only one site, while nine sites showed no parasitism by *T. basalis* and three sites were in the 3–9% range. In contrast, egg predation by an invasive spider was high (average 34%, range 4–88%). Tachinid parasitism of adult koa bugs was nearly zero at 21 of 24 sites, but increased significantly at three sites with high koa bug densities, reaching 70% among females and 50% among fifth-instar nymphs.

Do these data indicate high or low impact? Some have suggested that the dynamics of the bug and fly, both highly dispersive, are such that bug populations escape parasitism for a while but eventually are highly affected at the local level when colonies are eventually discovered by the fly (F. Howarth, personal communication). If so, this demonstrates the difficulty of reconstructing an interaction when the condition of the non-target species before the introduction can no longer be observed.

What other lessons can be learned from these outcomes? First, the parasitoids were released in a sensitive area (an oceanic island with high levels of endemism) without any consideration of impact on native insects, which seemed likely to occur. This would not be acceptable today. Also, in this system, the species of insects needed for conducting the host-range tests were taxonomically well known and readily available. This would have made host-range testing relatively easy. Finally, and most importantly, it points directly at the issue of what level of protection should non-target inverterbrates receive. Legally, there is no guidance on this last point in most locations, even today.

1968–9: *Rhinocyllus conicus* for thistle control in North America

The weevil *R. conicus* (Figure 16.6) was introduced from France to North America during 1968–9 for the control of the invasive nodding thistle, *Carduus nutans* L. (a species complex) (Julien & Griffiths 1998, Gassmann & Kok 2002). Nodding thistle was first reported in the USA in 1953 in Pennsylvania but

Figure 16.6 The thistle seed-head-feeding weevil *Rhinocyllus conicus* (Frölich) has both controlled its target pest and damaged non-target thistle populations. Photograph courtesy of Loke Kok, www.Forestryimages.org.

spread rapidly and established damaging infestations in over 42 states by the 1970s.

There are no native North American thistles in the genus *Carduus*, but there are several North American genera in the same subtribe, the Carduinae, specifically *Cirsium*, of which North America has about 100 native species. Field host records in Europe exist for *R. conicus* that show that species in four genera in the Carduinae are used as hosts (*Carduus*, *Cirsium*, *Sylibum*, and *Onopordum*). Host-range testing for this species was based on screening of agricultural crops (like artichokes), horticultural species, and some European thistles, including *Cirsium* species (Zwölfer & Harris 1984). Since the horticultural or crop species were not attacked and because the potential use of North American native thistles by the agent was deemed either unlikely or unimportant (all thistles being considered pests), release was approved. The target weed was controlled successfully (Gassmann & Kok 2002, Roduner et al. 2003). However, larval feeding in seed heads of over 20 native thistles has been observed (Turner et al. 1987, Louda et al. 1997). The impact of *R. conicus* on one of these species has been studied in detail (Louda 1998). Seed loss was shown to affect Platte thistle (*Cirsium canescens* Nuttall) populations, which are seed-limited (Louda & Potvin 1995, Rose et al. 2005).

There are two lessons from this case. First, the initial selection of the agent was flawed (by current social perspectives) because it was based on the social judgment that native thistles were not a resource worthy of protection. It was fairly clear from the data available at the time of introduction that native thistles might be attacked, but this information was ignored by the scientists in charge. Social values relative to native plants have changed since the 1960s. Now it is generally held that all native plants merit protection, not just economically important species. In this sense, this project is reflective of attitudes that no longer exist in many countries, nor among most scientists working on weed biological control.

Second, this case raises an important point about the interpretation of host-range test data. Early host-range assessors found this weevil showed a preference for the target plant and they predicted that this preference would limit its impact on non-target species. That proved not to be the case, since significant attack on some native thistles did occur. At first it was presumed that this was due to a change in host preference. However, a host-range reassessment using insects from Platte thistle found that the weevil's preferences had not changed (Arnett & Louda 2002), but rather that field outcomes were caused by weevils encountering this low-ranked host in the absence of its preferred host.

1957: *Cactoblastis cactorum* in the Caribbean

Cacti have been moved out of their native ranges in the Americas to dry areas around the world. In Australia and South Africa, species of *Opuntia*, such as *Opuntia stricta* Haworth and others, have became invasive in wild lands. In some areas, dense cactus patches spread over millions of hectares, rendering them useless economically and reducing their ecological value. In Australia, a government commission was created in the 1920s to seek a biological control solution for a 20-million-ha infestation of *O. stricta*. Surveys of herbivores associated with the cactus in its native range (Argentina) led to the identification of over 50 species. Larvae of one of these, the pyralid moth *C. cactorum* (Figure 16.7), provided dramatic and rapid control after introduction to Australia. Larval feeding opened pads to a bacterial disease that caused the plants to die within a few years (Dodd 1940). This introduction is generally recognized as one of the more valuable, and safe, projects in the history of weed biological control.

However, in 1957, in a separate project, this moth was introduced by the Caribbean nation of Nevis (with

Figure 16.7 Larvae of *Cactoblastis cactorum* (Bergroth). Photograph courtesy of Ted Center, USDA-ARS.

subsequent introductions in 1960 to Montserrat and Antigua) in response to native *Opuntia* spp. cactus infestations in pastures that had developed as a result of overgrazing (Simmonds & Bennett 1966). Several decades later, *C. cactorum* invaded Florida and subsequently spread northward along the coast (Johnson & Stiling 1998). In Florida, the endangered cactus *Consolea* (formerly *Opuntia*) *corallicola* Small is being attacked (Stiling et al. 2004). A greater threat is the potential for attack on the much larger *Opuntia* flora of Mexico (Zimmermann et al. 2001), some of which are also economically important. Far from a success, this application of *C. cactorum* has proven to be an embarrassment and a potential economic and ecological disaster, all for an easy solution to a minor problem that most likely could have been corrected by lower grazing rates and herbicide applications to pest cactus stands.

In neither of these introductions was the host range of the moth assessed specifically, but it was understood at the time that the moth fed widely on many species of prickly pear cacti. For Australia, this was sufficient information to demonstrate safety to native plants because there are no native cacti in Australia. Any cactus populations in Australia would be exotic species planted as ornamentals or feral offspring of such plants. Therefore, native Australian plants were not at risk. However, the same was not the case for the Caribbean introductions. Quite the opposite: the Caribbean borders on the heartland of the native distributions of literally hundreds of species of *Opuntia*. In that context, safety of *C. cactorum* would require extensive testing of native species, because it would have to be assumed that the moth would eventually spread throughout the islands, Florida and into Mexico and would only be safe if it were species-specific (which it is not). In fact, the introduction was made for the purpose of controlling some native cacti, despite the fact that many *Opuntia* species are grown as crops in Mexico (for edible cactus pads and as host plants for cochineal insects used as dye stuffs). Some partial, after-the-fact host-range testing (using species from Florida) found that *C. cactorum* accepted all tested species of *Opuntia* for oviposition and for larval feeding (Johnson & Stiling 1996). The contrast between these programs illustrates that the degree of knowledge about an organism's host range necessary to ensure safety is geographically dependent.

BROADENING PERSPECTIVES (1970–90)

Biological control was initially seen during this period as a green technology that allowed pesticide use to be reduced. However, by the end of the period, concern over pesticides had decreased (due to regulatory cancellation of the most offending materials and development of safer products), while concern over non-target effects of classical biological control increased greatly. This was due to new information from study of older projects and the momentum that new ideas frequently gain in science. A more detailed understanding of the impacts on native species of some past natural enemy releases was developed during this period through research on selected systems where impact was suspected. Here we discuss the cases of: (1) two coccinellids whose introduction increased concern over non-target effects through competition, one of which became a minor nuisance pest itself, (2) two weevil parasitoids that have different host range widths, well predicted by laboratory testing, and (3) two recent weed biocontrol projects that reflect the high level of care that projects currently take to ensure the absence of both trophic and indirect impacts.

Released 1957–8 or invaded 1973/1988? continued confusion with coccinellids

As classical biological control agents, coccinellid species vary from very effective to useless. Highly specific species such as *R. cardinalis* against cottony cushion scale often control their target hosts with minimal potential for unwanted effects. However, fascination with coccinellids has induced agencies to engage in coccinellid introductions in cases where the degree of host specificity was low, the need vague rather than specific, and where unwanted side effects should have been anticipated (Strong & Pemberton 2000). The introductions of *H. axyridis* and *C. septempunctata* illustrate several of these issues. In each of these cases, a decision was made in the 1950s (*C. septempunctata*) or 1960s (*H. axyridis*) to introduce the species, but recoveries were never made following releases, which were considered to have failed. Then, decades later (1970s for *C. septempunctata* and 1980s for *H. axyridis*) the species showed up, in both cases near a port city. The inference was made that the beetles were spontaneous invaders. Support for this assumption included a very limited geographic distribution near a port when first detected, coupled with the similar pattern of invasion for five other species of ladybird beetles that were never purposefully released in the eastern USA (Day et al. 1994). Following their invasions, these ladybird beetles were embraced by the USDA and quickly redistributed as biocontrol agents to many locations, leading each to become widespread and dominant in its habitat (*C. septempunctata* in meadows and row crops, and *H. axyridis* in orchards and forests).

Coccinella septempunctata was first released in the USA in 1957–8 but despite release of 150,000 laboratory reared beetles in 10 states and one Canadian province, establishment was never detected (Schaefer et al. 1987). It was found in New Jersey during 1973, possibly having entered through nearby ports. Subsequently, this population was redistributed widely as a biological control agent (Angalet et al. 1979). The concern over *C. septempunctata*'s presence has not been reductions of non-target prey, although some potential risk may exist for immature stages of rare butterflies (Schellhorn et al. 2005). Rather, the concern has been over potential competitive displacement of native coccinellids in the same feeding guild. After establishment, *C. septempunctata* became the dominant ladybird beetle in various habitats in the USA and Canada, including stands of *Phragmites* reeds in coastal New Jersey (Angalet et al. 1979), alfalfa in the northeastern USA (Day et al. 1994) and Manitoba (Turnock et al. 2003), apple orchards in West Virginia (Brown & Miller 1998), and potato fields in Maine (Alyokhin & Sewell 2004).

Harmonia axyridis was first detected during 1988 in Louisiana and is believed to have entered through the port of New Orleans (Day et al. 1994). This establishment by accidental invasion followed previous failures to deliberately establish the species during 1978–81, mainly on pecans in Georgia (Tedders & Schaefer 1994). Subsequently, *H. axyridis* became the dominant coccinellid in pecan orchards, where it lowered the spring density of the two pest aphids on pecan from 100 to two per leaf (Tedders & Schaefer 1994). Several other pest insects have been reduced in abundance by this predator (Koch 2003). The relative abundances of native ladybirds in apples (Brown & Miller 1998), citrus (Michaud 2002b), and other crops (Colunga-Garcia & Gage 1998) have declined. While the abundance of native species on these crops is not of concern, their overall decline would be. That, however, is difficult to determine. In addition, this ladybird's presence in winter in homes in the northern USA and its presence on clusters of wine grapes at harvest (resulting in off flavor in wine) have made it a minor pest (Kovach 2004).

1982/1991: *Microctonus* weevil parasitoids in New Zealand

The release of two weevil parasitoids in New Zealand illustrates the ability of host-range testing to reduce non-target parasitism of native insects by identifying species with narrow host ranges. In this case, two closely related parasitoids, *Microctonus aethiopoides* Loan and *Microctonus hyperodae* Loan, were released in different decades and subjected to different levels of host-range testing. *Microctonus aethiopoides* was introduced into New Zealand in 1982 for control of *Sitona discoideus* Gyllenhal, which feeds on alfalfa, whereas *M. hyperodae* was released against the grass-feeding *Listronotus bonariensis* (Kuschel) in 1991. Only cursory host-range testing preceded release of *M. aethiopoides*, as it was well known as a parasitoid of the target pest. However, for *M. hyperodae*, extensive testing showed that it had a fairly narrow host range (Goldson et al. 1992). Both species have either already controlled their target pests (Goldson et al. 1993 for *M. aethiopoides*) or show strong likelihood of doing so (McNeill et al. 2002 for *M. hyperodae*).

Post-release studies showed that in the laboratory *M. aethiopoides* parasitized 14 of 19 non-target species offered (74%) and in the field attacked 33% of non-target host species sampled. In contrast, in the laboratory *M. hyperodae* parasitized 23% (7/31) of non-target species offered, but in the field attacked only 6% (3/48) of species sampled (Barratt 2004). In addition, parasitism by *M. aethiopoides* was detected in non-agricultural habitats (subalpine meadows) and parasitism of some non-target weevils was as high as that of the target host (Barratt et al. 1997).

The lessons from this case are that parasitoids effective against their target pests may still be oligophagous and use some species of non-target native insects as hosts. The ability of more rigorous host-range testing to help select those with the narrowest host ranges is also illustrated by the example. Finally, this case again shows that parasitoids can disperse out of agricultural fields and interact with native species in other habitats.

Tamarix and *Melaleuca* insects: responding carefully to potential complications

Weed biocontrol projects conducted in the 1990s and the following decade illustrate the lengths to which current projects go to avoid impacts on native species. To cite two, we mention the work against saltcedars in the southwestern USA and against melaleuca in the Florida Everglades (USA).

Tamarix spp. are Eurasian desert shrubs that were introduced to the USA into California at the end of the nineteenth century as ornamentals and to stabilize sand dunes along railroad lines. *Tamarix ramosissima* Ledeb. (and two other species or hybrids) became highly invasive along desert rivers and, because of deep rooting and poor regulation of water loss, caused the water-table levels to drop. Intense competition and drier soils transformed saltcedar-infested riparian areas, which become poor or unsuitable habitat for most native plants. Saltcedars are major environmental weeds that infest the highest-quality desert habitats and damage native plant communities over extensive areas. Saltcedars are also taxonomically distant from native North American plants, making it easier to obtain agents with the necessary level of host specificity. Extensive surveys in Europe and Asia detected a large guild of herbivorous insects associated with saltcedars, with species in at least 25 genera of insects (DeLoach et al. 1996). At least 300 insect species are specific to the plant genus. Fifteen species were tested for host specificity in laboratories in various parts of the native range and six species were sent to a quarantine laboratory in Texas for further study. The top candidate for introduction that emerged from this work was the chrysomelid defoliating beetle *Diorhabda elongata* Brulle *deserticola* Chen (Lewis et al. 2003a). Host specificity studies with 58 species of plants revealed it to be highly host specific to *Tamarix* species (DeLoach et al. 2003). The native plant genus of greatest concern was *Frankenia*, but *D. elongata* larvae rarely survived (less than 1.6%) on species in this genus. A risk analysis showed that *D. elongata* would not threaten any of the three species of *Frankenia* in the USA (Lewis et al. 2003b).

Two concerns did emerge with this project. First, one species of introduced *Tamarix* [*Tamarix aphylla* (L.)] has value as an ornamental, chiefly in Mexico, and is likely to be used as a minor host. Second, invasive stands of *Tamarix* have been adopted as nesting sites by an endangered bird, the southwestern willow flycatcher (*Empidonax traillii extimus*) (Dudley & DeLoach 2004) because its normal nesting trees, cottonwoods, have been displaced by saltcedar. As a result, the USDA and the US Fish and Wildlife Service entered into extensive discussions on the importance of this potential risk and how it might be mitigated. The conclusion was that the risk was small because the *Tamarix* plants were not likely to all die quickly and there would be sufficient time to manage the conversion of vegetation away from saltcedar to native cottonwoods. Early beetle releases were to be made in areas distant from known nesting areas, and cottonwoods were to be replanted where needed in advance of loss of *Tamarix* as nest sites. Initial releases of this beetle have been made and early indications are that microclimatic matching of source populations to release areas will be needed to obtain establishment and promote high impact. Populations from Fukang, China and Chilik, Kazakhstan were found to be able to reproduce and overwinter successfully at sites north of 38°N, but south of that point these populations enter diapause prematurely and fail to establish (Lewis et al. 2003a). New populations from Crete are being considered for those areas.

Paperbark tree [*Melaleuca quinquenervia* (Cavier) Blake] is an Australian swamp forest tree planted as an ornamental in Florida. It invaded the Everglades in about 1900 and by the 1980s posed a severe threat to this ecosystem. A recovery project integrating stem cutting and herbicide treatment of foliage and stumps (to rapidly remove larger plants) and release of biological control agents (to reduce seed production and lower seedling survival) is underway (Rayamajhi et al. 2002). Ten agents have been evaluated in Australia, of which five have been introduced to US quarantine for further study. Of those, three have been approved and released: a weevil, *Oxyops vitiosa* Pascoe; a pysllid, *Boreioglycaspis melaleucae* Moore; and a gall fly, *Fergusonina turneri* Taylor. Herbivory by the weevil causes tip die-back and also affects photosynthate allocation, with defoliated trees drastically reducing flowering and seed production (Pratt et al. 2005). In addition, the weevil, especially in conjunction with a naturally invasive rust fungus (*Puccinia psidii* G. Wint.), inhibits sucker growth on stumps (M.B. Rayamajhi, personal communication). The psyllid decreases the growth and survival of seedlings, reduces photosynthetic capacity of the leaves, and causes the leaves to drop prematurely (Franks et al. 2006, Morath et al. 2006). A common garden study has shown that saplings protected from such herbivory using systemic insecticide grow rapidly and flower prolifically whereas those not so protected hardly grow at all and produce almost no flowers (P.W. Tipping, personal communication). Tree densities in mature stands in Florida have been reduced by 85%, mostly due to the loss of the smaller, suppressed trees in the understory. Canopy coverage has also been reduced by about 70%, which has allowed light to penetrate to the forest floor and native species to re-establish (M.B. Rayamajhi, personal communication). Management agencies now have more time to remove existing stands because follow-up treatments are less necessary. In addition, areas that have been cleared are less likely to be reinfested from surrounding, unmanaged stands (T. Center, personal observations).

Another agent considered for use against melaleuca illustrates the care currently used in biological control projects: the melaleuca defoliating sawfly (*Lophyrotoma zonalis* Gagné) is a highly destructive agent that was of interest because larvae burrow beneath the papery bark of the tree to pupate. It was therefore an excellent candidate for wetter areas where agents that pupate in soil might not survive. This insect was found to be extremely host specific, with larvae only developing on three bottlebrushes (*Callistemon*) closely related to the target weed (Buckingham 2001). However, its introduction was withheld by project scientists who recognized during the testing program that toxic octapeptides (lophyrotomin and pergidin) existed in the larvae of this species (Burrows & Balciunas 1997, Oelrichs et al. 1999). Consumption of large quantities of a related sawfly have caused cattle deaths in Australia (Dadswell et al. 1985), so toxicity testing was initiated to determine if these peptides posed a risk to domestic animals or wildlife (Buckingham 1998). Cooperators at the USDA-ARS Poisonous Plant Research Laboratory in Login, Utah, USA, force-fed freeze-dried larvae to mice, which suffered no ill effects. Large pre-pupae and larvae were also offered to red wing blackbirds (*Agelaius phoeniceus*) at the USDA-APHIS Denver Research Laboratory, Gainesville, Florida. Most birds rejected the larvae but two ate them and later regurgitated without further adverse effects. Larvae were then freeze-dried and added to a dry diet. The birds ate the amended diet without harm. Thus, it seemed that risks to livestock or wildlife would be minimal and the host range was clearly acceptable.

However, project scientists were concerned about possible risks to migrating birds. Florida lies in the pathway of the Atlantic Flyway, a major North American migratory route. Birds that make landfall often arrive depleted and hungry after long flights from South or Central America. It was plausible that while in this condition these migrants, upon encountering masses of sawflies, might gorge themselves and thereby ingest a toxic dose of the octapeptides, which they might not tolerate in this frail state. Furthermore, little was known about potential effects on other insectivorous predators such as frogs and lizards, or how these toxins might accumulate in the food web. The scientists therefore decided independently to seek other agents and to not risk the release of the sawfly, despite a positive opinion from the Technical Advisory Group and the loss of several years of research (T. Center, personal communication). These examples illustrate current projects led by responsible biologists that attempt to foresee both conventional trophic effects and indirect non-target effects that agents might cause.

CURRENT PRACTICE AND CONCERNS

Here we sum up and describe the broad trends in non-target effects. We are interested in knowing whether they are increasing, stable, or decreasing, for both plant and insect targets. Is new information on old projects driving our view of the level of risk and, if so, does that overstate the degree of risk? Is fear of potential non-target impacts by biological control agents exacerbating impacts of invasive species by delaying or preventing new projects? What are current national or international standards for assessing risk for natural enemy introductions? Is the pesticide problem – one of the original drivers for promoting use of biological control – really solved or are there still important impacts that make further reductions in pesticide use desirable? Is biological control truly a green technology and can conservation and biological control groups develop a better mutual understanding to enhance their common cause in reducing impacts of invasive species?

Recognition and frequency of non-target impacts

The recognition of non-target impacts of classical biological control developed in two steps: (1) "would it happen?" and (2) "would it matter?" That some impacts were likely to occur was long known, but in many cases, as least for insect biological control, attack by a parasitoid or predator on native species as alternative hosts was considered a desirable feature rather than a liability. Some parasitoids of the gypsy moth (*L. dispar*), for example, like the tachinid *C. concinnata*, were known prior to introduction to be polyphagous, but this was not seen as a reason to abandon their release, as the goal was to protect trees, not insects. Similarly, that the weevil *R. conicus* was likely to feed on native thistles was anticipated and indeed documented by others long before Louda's detailed work on Platte thistle populations, but was not a concern because thistles were all grouped indiscriminately as weeds.

Concern over non-target impacts on plants developed before concern for impacts on native insects. Early reviews (e.g. Harris 1988, 1990) stressed that weed biocontrol agents did not cause plant extinctions and indeed only occasionally fed on non-target plants. A critical development in thinking on this topic was the review by Howarth (1991) that focused attention on the potential for harm to non-target species by classical biological control agents (of both weed and insect pests). The 1990s led to increased research on selected cases and writing of review articles on the topic (e.g. Cruttwell-McFadyen 1998, Pemberton 2000, Louda et al. 2003a). Pemberton (2000) analyzed the impacts of weed biological control agents in the USA (including Hawaii) and the Caribbean. For 111 (of 112) insects, three fungi, one mite, and one nematode that were established successfully, the only non-target plants fed on were those in the same genus as the target weed or other species that had been attacked in host-range tests (and thus predicted to be in the host range). Only one of these 117 agents attacked a plant that was not congeneric with the target weed or with plants accepted in host-range tests. This implies that direct attack on plants by weed biological control agents is very predictable using current host-range screening methods. This implies that in cases such as saltcedar and melaleuca, in which there are no native plants in the recipient country in the same genus, current practices will accurately identify any non-target plants at risk from attack. Cases in which native plants do exist in the same genus as the target weed will require more extensive evaluation of potential risks to native congeners before agents can be released. Cases such as *R. conicus* (e.g. Louda et al. 2003a, 2003b) do not indicate that prediction methods are flawed, but rather that their predictions were not taken seriously.

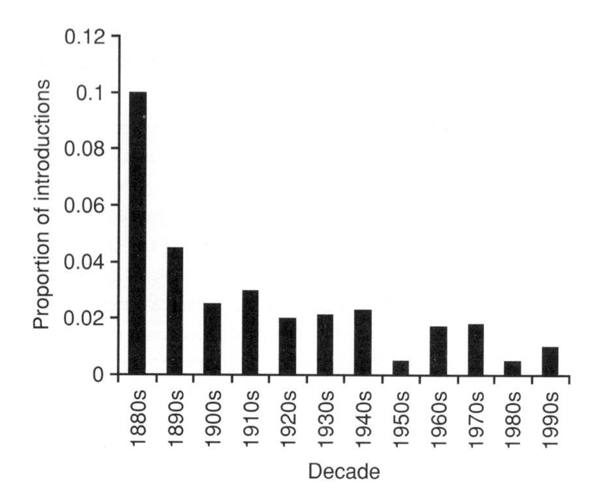

Figure 16.8 The proportion of biological control introductions that have harmed non-target species has declined historically, indicating that classical biological control has become increasingly safe. The analysis was based on ≈5800 unique agent×location introductions. Redrawn from Lynch et al. (2002), with permission from CABI Publishing.

With respect to projects of insect biological control, the process is less advanced. However, reviews (Lynch & Thomas 2000, van Lenteren et al. 2006a) have found that the rate of impacts with important population-level consequences, based on evidence in the literature, is low and most likely decreasing (Figure 16.8).

Laws and standards to reduce harm to native species

The introduction of herbivorous insects in the USA (and many other countries) has been prohibited for nearly a century by plant quarantine acts. USDA-APHIS has used this authority to protect plants from harm by introduced weed biological control agents.

Which plants receive protection has, however, evolved over time. Initially (c.1920–70), plants of concern were primarily important crops, forest trees, or ornamentals. In the 1970s, with the passage of an endangered species law, protection was extended to any officially listed species. By the 1980s, biological control practitioners and the committee charged with reviewing importation petitions (the Technical Advisory Group, TAG) took the position that all native plants (threatened or not) should be protected from significant damage by weed biological control agents. Impacts on exotic species of ornamental plants, as well as minor or transitory damage to any non-endangered native plants, were acceptable provided such harm was less important than the potential benefit of controlling the pest.

Legal authority for regulating the importation of parasitoids and predaceous insects for insect biological control in the USA is less clear. A clear need exists in the USA for a new law defining procedures and authorities, and establishing methods for assessing and balancing risks and benefits of potential projects (Messing & Wright 2006). In some countries (such as Australia and New Zealand), specific laws regulating biological control have been passed to provide for consistent standards and clear processes, but this has not yet been done in the USA. Which non-target insects would have to be protected and how stringent that protection should be are also unclear. There are only a few insects that are either economically important for production of products, such as honey or silk, or are legally classified as endangered. Previously introduced biological control agents, however, are one group of insects with clear economic importance. Risks to weed biological control agents that are closely related to targeted pest insects should be evaluated in assessing the agent's specificity.

Broadly, in the absence of anything more specific, the standard for evaluating proposed introductions of insect biological control agents is the risk/benefit perspective implicit in environmental protection acts. If projects provide a net economic or ecological benefit, some damage to non-target species is acceptable. When risks and benefits are both ecological, they can be directly compared. When benefits are economic but risks are ecological, comparisons are difficult. A need exists for a designated governmental body to act as the final arbitrator of whether a proposed introduction has a net benefit to society. Currently, only Australia and New Zealand have such systems.

International standards for the importation of biological control agents exist that can serve as non-binding guidance to countries lacking their own specific laws. In North America, these include the NAPPO (North American Plant Protection Organization) standards #12 (for entomophagous agents) and #7 (for phytophagous agents) (Anon 2000, 2001). In addition, the FAO of the United Nations has promulgated a Code of Conduct covering the release of exotic biological control agents (Anon 1997a). Standards for importing natural enemies into European countries have been reviewed and standardized (Bigler et al. 2005).

Applying host-range testing to candidate biological control agents

The key to keeping unacceptable impacts on non-target species to a minimum in the future will be application of host-range testing to new agents and public review of the evidence before release. A system to evaluate host specificity of weed biological control agents proposed for introduction is well established (in the USA reviews are conducted by the TAG). No comparable system is in place in the USA for review of parasitoids or predators but some have suggested that a similar TAG-like approach should be developed (Strong & Pemberton 2000; see Chapter 17 for the methodology of host-range testing).

Attempts should also be made to anticipate harmful indirect effects (see Chapter 18; Messing et al. 2006), especially if potential for such harm is suggested by the agent's ecology in its native ecosystem or by its basic biology (e.g. possession of toxins or other features likely to cause problems). However, methods for identifying the potential for such indirect effects are still being developed (Messing et al. 2006). Indeed, the potential for indirect effects is present for any species introduction (not just of biological control agents) and most large-scale human actions. In general, the mere potential for such effects, unless a clear and demonstrable threat, should not be an impediment to making needed introductions to combat invasive species in a timely way. Future discussions of non-target risks (see Bigler et al. 2006) will need to go deeper than mere use of a non-target species, to a consideration of impact (population depression or range reduction), which has so far been studied in only a few cases.

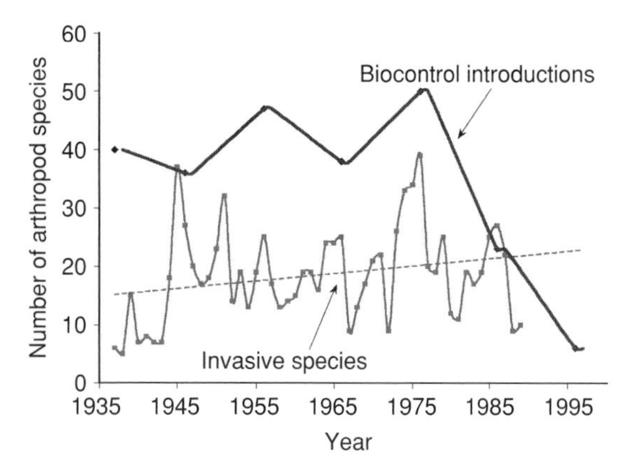

Figure 16.9 Owing to public perception of biological control as a risky process, rates of natural enemy introduction into Hawaii have declined sharply since about 1975, while rates of pest invasions have continued to increase, creating an increasingly unmet need for pest control. After Messing and Wright (2006).

Why not just say "no" to classical biological control?

One might conclude that concerns over risks of non-target impacts from biological control introductions could most effectively be dealt with by just ending future biological control introductions. In some regions such as Hawaii, the number of releases of new biological control agents has declined (Messing & Wright 2006) (Figure 16.9). This is unfortunate because many invasive species have serious harmful impacts on native species and need to be managed. If classical biological control is not used, such damage is likely to continue because chemical and mechanical controls are rarely effective over whole landscapes due to costs, pollution, and disruption (see Chapter 8). Decisions about environmental protection should weigh the damage from the invader against the typically much smaller risks of the biological control agent.

Increased pesticide use is not desirable. Biological control was emphasized in the 1960s and 1970s because problems from persistent pesticides were judged to be too serious to be allowed to continue. These included pesticide residues in food, water, human breast milk, arctic mammals, and various predaceous birds. A number of birds (e.g. eagles, hawks, herons, and ospreys) were regionally suppressed in numbers, some to the point of regional extermination. The most offending pesticides (e.g. DDT, chlordane, dieldrin, etc.) have been prohibited by laws in many countries and some new, safer pesticides registered. However, many significant problems remain that make a further reduction in pesticide use desirable. The two most important of these problems are damage to amphibians and disruption of mammalian (including human) hormone systems. Whereas amphibian declines have not been clearly linked to pesticides and are definitely tied to multiple causes, pesticides do appear to be part of the problem (Ankley et al. 1998, Kiesecker 2002). Finally, and perhaps most frightening, some pesticides mimic human hormones (specifically estrogen) at mere parts per billion, leading to various harmful effects on reproduction, including lowered sperm counts and feminization of males (Colborn et al. 1997, Schettler et al. 1999, Krimsky 2000, Bustos-Obregon 2001, Palanza & vom Saal 2002, Saiyed et al. 2003). For all of these reasons, turning away from biological control, which in effect would increase pesticide use, is undesirable.

"RE-GREENING" BIOLOGICAL CONTROL

Many conservationists responsible for specific preserves seek to control invasive species in relatively small areas, using mechanical or chemical methods. Biological control scientists are usually focused on correcting invasive species problems over the whole landscape. Interchange between these two groups has been inadequate. Many preserve managers have been exposed only to the negative characterization of biological control as part of the invasive species problem, rather than its most effective remedy. There is a need to make biological control better understood by conservation biologists and the general public. This will require increased precision and predictability of biological control introductions, coupled with emphasis on projects with ecological objectives, and ample public scrutiny.

Chapter 17

PREDICTING NATURAL ENEMY HOST RANGES

Given that a decision has been made to invest resources in estimating the host range of a specific candidate biological control agent, we need to know technically how to do this well. Sources of information useful in estimating a species' host range include: (1) literature records, (2) surveys in the native range, (3) tests in quarantine laboratories, and (4) field experiments in the native range. Here we describe how each of these sources contributes to estimation of the likely host ranges of parasitoids, predaceous arthropods, herbivorous arthropods, and pathogens.

LITERATURE RECORDS

Early in any project, researchers are likely to inventory the known natural enemies of the target pest, as reflected in published literature or attached to specimens in collections. Thereafter, the literature host records for those species can been assembled, giving some initial impression of which natural enemy of the pest shows specificity. Information in computerized databases (see especially CAB Abstracts and Agricola) omit material prior to 1971, when such computerization was begun. Earlier literature (back, generally, to at least 1900 or earlier if necessary) must be collated by hand from abstracting journals such as *Abstracts of Applied Entomology* or still older works on regional natural history or taxonomy of particular groups. Museum collections in countries where surveys are to be conducted are important sources of information, as specimens of natural enemies may include rearing or feeding information.

Use of information from literature records must take into account several potential issues affecting the meaning or quality of the literature records, including: (1) biotypes and symbionts, (2) errors, (3) negative information, and (4) host ranges of relatives.

Biotypes and symbionts

A general problem with information from literature records is that either the target pest or the natural enemy species of interest may consist of a series of biologically separated populations that have been mistakenly lumped together because of morphological similarity. For example, Old World climbing fern populations from different parts of Queensland differ as to whether the mite *Floracarus perrepae* Knihinicki and Boczek can attack the plant (Goolsby et al. 2006b). Similarly, populations of the same genetic make-up may differ in possessing or lacking symbionts conferring resistance to some parasitoids. For example, some strains of the bacterium *Hamiltonella defensa* make pea aphids [*Acyrthosiphon pisum* (Harris)] resistant to the branconid *Aphidius ervi* Haliday (Oliver et al. 2005).

Similarly, a natural enemy species may exist as regionally differentiated populations that differ in their host ranges. For example, molecular analyses have shown that the braconid *Microctonus aethiopoides* Loan, used for control of various forage weevils, consists of at least two biotypes, one (Moroccan) associated with *Sitona discoideus* Gyllenhal and the other (European) with *Sitona lepidus* Gyllenhal and species of *Hypera* (Vink et al. 2003). Because the European strain was parthenogenetic, both strains could be employed in New Zealand, against different pests, without genetic crossing (Goldson et al. 2005). Likewise, there are two biotypes of the encyrtid parasitoid *Comperiella bifasciata*

Howard, each adapted to parasitize only one of two closely related scales. The yellow scale biotype of *C. bifasciata* successfully parasitizes yellow scale, *Aonidiella citrina* (Coquillet), but does not develop on red scale, *Aonidiella aurantii* (Maskell) (Brewer 1971), whereas the red scale biotype does the reverse (Smith et al. 1997).

There are three important conclusions from the existence of biotypes. First, genetic markers must be developed to recognize the exact form of any agent employed so that it can be differentiated from similar appearing species present in the release area. Second, projects should not dismiss a species as a prospective natural enemy simply because the literature suggests a wide host range. Its host range needs to be determined if such a species seems to be potentially effective, to see if the literature might not reflect a complex of biotypes rather than the true host range of one single population. For example, a dolichopodid fly that damages waterhyacinth in South America was of interest as a control for this weed in the USA and South Africa. However, what was presumed to be a single species occurred on several other plants in the family Pontederiaceae, which discounted its potential value. Careful taxonomic study, however, revealed the presence of a complex of at least nine species, five of which feed on waterhyacinth (Bickel & Hernandez 2004). At least two of these, *Thrypticus truncatus* Bickel and Hernandez and *Thyrpticus sagittatus* Bickel and Hernandez, seem quite host-specific and are now regarded as potential biological control agents. Third, because biotypes might exist within a natural enemy species, a project should avoid the mistake of assessing the host range of one population and then collecting individuals for release from another population or set of populations. For example, release of the melaleuca weevil [*Oxyops vitiosa* (Pascoe)] was restricted to insects collected at just one location because those from other locations appeared slightly different (Madeira et al. 2001).

Errors

Researchers and political administrators evaluating release petitions should be aware that the scientific literature frequently includes some erroneous records, because the host (or target plant) or the parasitoid (or herbivorous insect) was misidentified. If a certain natural enemy is associated with a given species only by a single report, it should be given less credence than relationships documented multiple times. For example, when the petition was submitted for release of the weevil *O. vitiosa*, mention was made of museum specimens collected from two inland locations outside the normal range of the host plant *Melaleuca quinquenervia* (Cavier) Blake. A reviewer saw this as evidence of a broader host range than indicated and recommended against the release of this valuable agent on the basis of what was later determined to be an erroneous report in the literature.

Negative data

Another way to use the literature is to identify species that have been in extensive contact with the natural enemy of interest but are not recorded in the literature as hosts (De Nardo & Hopper 2004). Both native species in the donor area that are related to species of concern in recipient areas and non-native species that may have invaded or been introduced into the donor area can be of interest. For example, American plants imported into Europe as ornamentals are likely to be in contact with herbivorous insects under consideration for introduction to North America. Lack of attack in Europe on such American plants suggests they would not be attacked in America if the agent were introduced. Balciunas et al. (1994b), for example, took advantage of the fact that their laboratory in Townsville, Australia, was some distance from naturally occurring *M. quinquenervia* stands. However, the target plant as well as many of the test plants of interest existed as ornamentals in a local parking lot. This afforded them the opportunity to monitor for the presence of *O. vitiosa* on these plants on a regular basis. They observed an average of 158 eggs/tree, 108 larvae/tree, and eight adults/tree on the target plant but virtually nothing on any of the other 19 species of Myrtaceae present, which were non-target species theoretically at some risk.

Host ranges of congeners

Do the host ranges of the congeners of a candidate biological control agent provide information on the agent's likely host range? For parasitoids, congener host ranges are not very useful since many genera contain species of both wide and narrow host ranges.

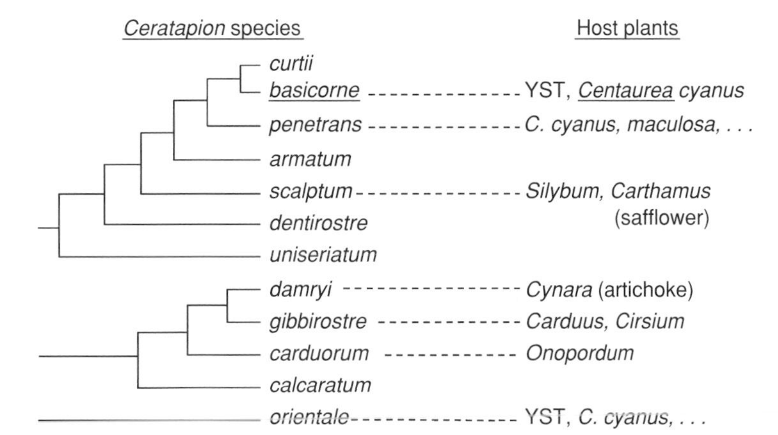

Figure 17.1 Information on the host ranges of close relatives of a potential biological control agent can add to the understanding of an agent's likely host range. Here, the host ranges of weevils in the genus *Ceratapion* are displayed on a phylogenetic tree, to place the candidate biological control agent, *Ceratapian basicorne* (Illiger) in better context. YST, yellow starthistle. Image courtesy of Lincoln Smith, USDA-ARS.

For example, the tachinid genus *Trichopoda* contains *Trichopoda giacomellii* (Blanchard), a narrowly specific species, and *Trichopoda pennipes* (Fabricius), a species with a significantly wider host range (Huffaker & Messenger 1976, Liljesthrom 1980). In contrast, the host ranges of an herbivore's congeners may be more informative (e.g. White & Korneyev 1989, Zwölfer & Brandl 1989). For example, the weevil *Ceratapion basicorne* (Illiger) is of interest as a biological control agent of yellow starthistle (*Centaurea solstitialis* L.) in the USA. As part of that effort the host ranges of this weevil's congeners are being noted and placed in a phylogenetic context (Smith 2007) (Figure 17.1).

SURVEYS IN THE NATIVE RANGE

Surveys in the native range are typically structured to discover natural enemies associated with the target pest. However, after doing such surveys and choosing a candidate for introduction, one can further survey the donor area to determine the host range of the agent. Although this does not indicate whether a particular species in the proposed recipient country might be attacked, it does provide information about the width of the host range in the donor area. For example, an Australian weevil of interest for biological control of the aquatic weed *Hydrilla verticillata* (L.f.) Royle fed on 16 other plant species and laid eggs in 11 species in the laboratory. However, field surveys conducted in the native range on the plants showed the host range was much narrower (Balciunas et al. 1996).

Field surveys in the donor region can also indicate the habitats in which the agent is found. For example, surveys in Europe of mirid bug parasitoids have indicated in which habitats the braconids *Peristenus digoneutis* Loan and *Peristenus stygicus* Loan (species being introduced to the USA) forage (Kuhlmann & Mason 2003).

Field surveys in the native range can be combined with laboratory host-range tests to validate the efficacy of such tests by subjecting the non-target species in the donor country to such laboratory testing. Work performed in Europe on eight species of mirids selected based on phylogenetic considerations (number of branches in the family cladogram from the normal host), together with spatial and temporal overlap between the normal host and the other test species, showed that the laboratory tests overestimated host ranges and attack rates, compared to results of field surveys with the same species (Haye et al. 2005).

LABORATORY TESTING TO ESTIMATE HOST RANGES

After natural enemies have been imported into quarantine in the country where release is intended, these species must be assessed against various native or economically important plants or host insects to predict their likely host range. Methods for doing this are well developed for weed biocontrol agents and are under development for agents directed at pest arthropods (see Van Driesche & Reardon 2004, Babendreier et al.

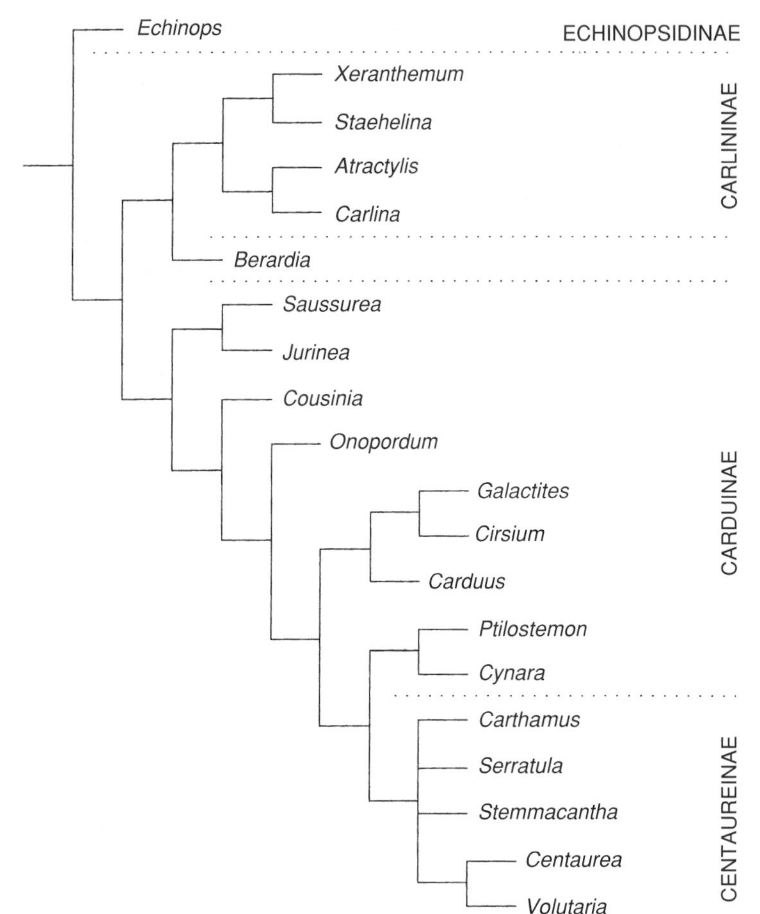

Figure 17.2 A phylogenetic tree of plant genera in four subfamilies, showing their genetic distance from the genus *Centaurea*, in which the target pest yellow starthistle (*Centaurea solstitialis* L.) is located. Using such trees, expanded to show species within genera, one can use the number of branching events to select non-target test species that have varying degrees of relatedness to the target pest with greater precision. Image courtesy of Lincoln Smith, USDA-ARS.

2005, New 2005). This work is done in quarantine and the resultant data are used to make the decision whether or not to approve release of the agent into the environment. There are several aspects in this process, including: (1) choosing the test list of species to be examined, (2) picking which agent responses to measure, as dictated in part by the agent's biology, (3) maintaining standard test animals and conditions, (4) choosing a particular hierarchy of test designs, and (5) interpreting the results.

Constructing the test species list

For early (pre-1960) weed biological control programs, test lists were viewed as lists of species of interest (mostly crops) to which the safety of the herbivore to be introduced for biological control had to be demonstrated. No attempt was made to define the fundamental host range (*sensu* van Klinken 2000) of the herbivore, but rather merely to assure safety to a set of specific plants. Two undesirable consequences were associated with this procedure. First, some plants on the test list were so unlikely to be attacked that testing was a waste of time and scientific staffing. Second, potential risks to plants of no economic importance were not considered.

As early as the 1960s it was recognized that plants at increasing taxonomic distance from a target weed were likely to be at decreasing risk of attack by a weed biological control agent. This occurs because the fundamental nature of a plant clade's secondary compounds is often preserved over evolutionary time as the plant clade diversifies. Concurrently, specialized herbivores able to feed on the plants track this diversification with their own evolution (Cornell & Hawkins 2003). Taxonomic relatedness to the target plant was therefore an early tool used to select for test plants most likely to be at risk, a process that came to be called the centrifugal method (Wapshere 1974a). However, unrelated but chemically similar plants sometimes exist and these may also be at risk (e.g. Wheeler 2005).

As this perspective was embraced, the goal of the testing procedure shifted from assessing safety to an *ad hoc* species list to defining the real limits of the agent's host range (termed the **fundamental host range**). Before the 1990s, test species were selected by choosing representatives from each of the categories of increasing size (genus, tribe, subfamily, family) in the taxonomic hierarchy. See Kuhlman et al. (2006a) for review of selection criteria for test species.

Since the 1990s, with the advent of molecular tools, there has been an explosion of studies presenting phylogenetic trees of plant groups based on base-pair sequences of various genes (Briese 2005, 2006b). Because these phylogenetic trees now exist in a great many groups, it is often possible to select test species based on the number of branching events (in the cladistic sense) that separate the target pest from potential test species. Species are thus selected from groups one, two, three, or four branching events removed from the target, rather than membership in the same genus, tribe, subfamily, etc., as done previously (e.g. Figure 17.2). It is, however, important to note that most branches in a cladogram have low statistical significance and that the number of branching events has no absolute meaning. The same node number can denote different amounts of genetic distance under several situations: (1) in trees based on complete compared with partial sampling of the taxa in the group, (2) in speciose compared with less speciose groups, and (3) whether or not subspecies populations are included as entities in the tree. Thus, this tool provides advice on choosing test plants but is not necessarily authoritative. The same approach can also be used to help interpret patterns of host use (Figure 17.3).

Phylogenetic trees of the species closely related to the target pest are less common for insect biological control projects than for plants, so they would not be available as a tool to select test species in many projects. In such cases, the researcher may want to create trees for the tribe or subfamily in which the target pest resides. If that is not possible, test species would have to be selected based on placement in hierarchical taxonomic categories, selecting species from the pest's genus, tribe, subfamily, family, and order (Wapshere 1974a). G.E. Heimpel (personal communication) suggests that sequences of *COI* gene (or other useful genes) could then be used as means of quantifying the degree of relatedness between the pest and each test species. It sometimes may be necessary to add additional test species if the genetic distances of one's first selections prove smaller than was supposed based on their taxonomic placement. For analysis after host-range testing has been completed, the acceptance and/or suitability of each test species could then be graphed against each species' genetic distance from the target pest to determine whether host suitability declines sharply or more gradually beyond some prescribed genetic distance.

This approach can also be used, in principle, to select the species of non-target insects to use as test species in defining the fundamental host ranges of parasitoids (Haye et al. 2005). However, at this time fewer phylogenetic trees are available for insects compared to plants. The same approach would also be recommended for choosing species of hosts to assess the host range of pathogens (of plants or of insects).

Selection of test species must consider both protection of native non-target species related to the target pest and introduced biological control agents that might be in the agent's host range (Kuhlmann et al. 2006a). For example, plans to introduce parasitoids of cabbage seedpod weevil [*Ceutorhynchus obstrictus* (Marsham)] into North America had to consider the native weevils related to the pest and 11 exotic species in the genus already used or proposed for use as weed biological control agents (Kuhlmann et al. 2006b).

Picking measurable responses for insects

The strategy for assessing the host range of a biocontrol agent will depend on how it finds, assesses, and attacks the pest. In most cases, hosts are chosen by the adult insect. Commonly measured responses include the following.

Oviposition preferences of adult natural enemies

This response is meaningful for a wide range of herbivorous, predatory, and parasitic insects. It is believed to be the limiting stage in host selection for many herbivores

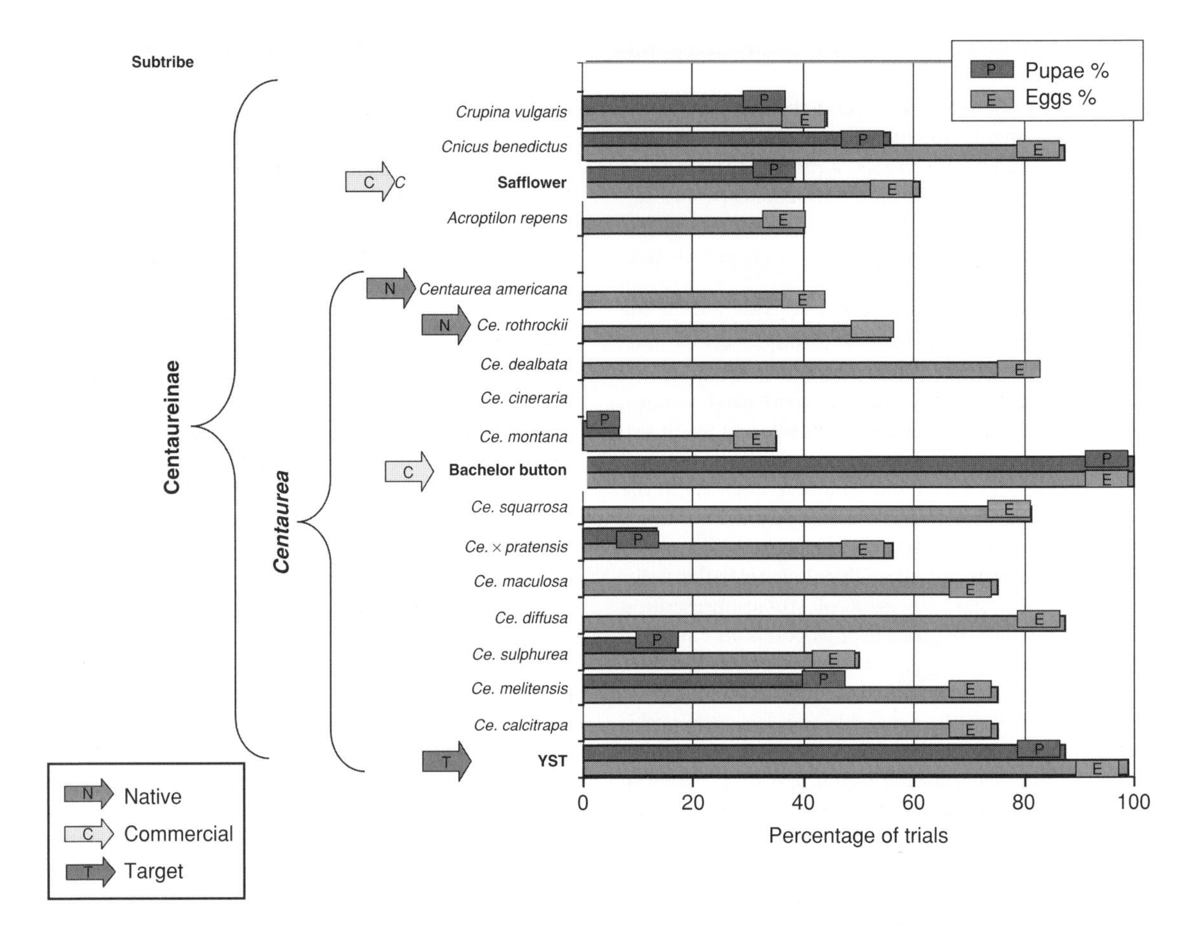

Figure 17.3 Placing responses to non-target species on to a phylogentic diagram can assist in data interpretation. Here we see the percentage of events leading to oviposition and to progeny pupation on various test species when these have been challenged by *Ceratopian basicorne* (Illiger), a potential biological control agent for yellow starthistle (*Centaurea solstitialis* L.). YST, yellow starthistle. Image courtesy of Lincoln Smith, USDA-ARS.

(except those with wandering larvae). In contrast, parasitoids may oviposit in more species than their larvae can develop in, at least for internal parasitoids whose hosts have active immune systems. For predators, measuring of oviposition preferences will be meaningful only if egg laying is closely associated with prey.

Feeding preference of adults or larvae

Feeding preference is a meaningful parameter for any life stage that feeds and is mobile enough to make a choice. Adults of some holometabolous herbivorous insects (moths, flies) may not feed on their larvae's host plant, while adults of some species in other groups such as Coleoptera or Hemiptera may do so. If both adults and larvae feed, each should be tested as their food choices may differ. This is true for both herbivorous insects and predators. For larvae, feeding preferences tests are meaningful for external-feeding, mobile larvae, which can make choices. Neonate and older larvae should be tested separately, as older larvae (with stronger mouthparts) may be able to eat some species that young larvae cannot. Feeding preference has no meaning for internal parasitoids or endophytic feeders like leafminers, whose feeding sites are determined by the oviposition choices of their mothers. In such cases, growth and development are the meaningful things to assess, within the range of hosts accepted for oviposition.

Larval growth and development

For all species, but especially internal feeders, the proportion of larvae that can successfully complete their development through to pupation when fed on a given host is a very significant measure of host suitability.

Oogenesis and continuation

A follow-on feature from host suitability for larval growth and development is to measure whether insects that mature on a given host are able to develop normal egg complements, based either on just larval-acquired resources from the host, or this in addition to further adult feeding on the same host (depending on the kind of agent). For parasitoids, it is also important to see whether progeny reared on a host maintain a normal sex ratio. A still further extension of this sort of suitability testing is a continuation test, in which the objective is to determine whether the host can support a series of generations of the agent with no loss of fecundity or survivorship.

Picking measurable responses for pathogens

Testing of plant pathogens focuses on infectivity, based on artificial placement of the inoculum (such as spores) on the susceptible tissues under physical conditions (temperature, relative humidity) known to promote infection in the target pest. Outcomes are measured in terms of the frequency and severity of any infections that result. Also, the course of the infection and the degree of its impact on the plant must be measured. The same approach is used to assess host ranges of arthropod pathogens. The major difference from assessment of plant pathogens is that for insect pathogens the only common outcome is death, whereas for plants other outcomes such as reduced growth, deformation, failure to set seed, etc., could be outcomes of infection.

Attributes of test animals that must be standardized or controlled

Several attributes of test animals can affect their willingness or ability to engage in the behaviors (feeding, oviposition) being measured in host-range estimation tests. These include age, hunger status, mating status, previous contact with the target pest, rearing history, and biotype. A species' fecundity will often vary with an individual's age. Young parasitoids of many species, for example, need time to mature eggs before they exhibit interest in potential hosts. During this period (or later, after a bout of oviposition), an agent's behavior may favor feeding over oviposition if carbohydrate reserves are depleted. For parasitoids, both mated and unmated individuals are capable of attacking hosts, but their choices may differ. Similarly, it is well established that previous contacts of a parasitoid with a host species can condition it to prefer the familiar host more than new species. Such conditioning may extend to preference for the natal host as well. Finally an agent's biotype will shape its host range. To obtain repeatable results, all of the above factors must be considered and as many brought to standard conditions as needed for a particular system.

Types of test design and their interpretation

The testing goal is to define the fundamental host range of the agent (the genetically determined limits to preference and performance) in order to predict field host specificity (Sheppard 1999, Spafford Jacob & Briese 2003, Sheppard et al. 2005, van Lenteren et al. 2006b). Test designs are no-choice, choice (in several variations), continuation, oogenesis, and open-field tests.

No-choice tests

In this design, the agent is confined with one test species at a time. No-choice larval feeding tests are called starvation tests as insects must eat the test plant (or prey) or starve. No-choice oviposition tests are run with adults. This design was the earliest approach for testing weed agents and is currently also used with insect parasitoids or predators. In the 1970s and 1980s, no-choice tests lost favor to preference tests (choice design) for assessing weed agents (as being more natural) but in the 1990s, the no-choice design was again emphasized to increase detection of low-ranked, non-target hosts (Briese 1989b, 2005, Thompson & Habeck 1989, Turner et al. 1990, Adair & Scott 1993, 1997, Woodburn 1993, Turner 1994, Balciunas et al. 1996, Peschken et al. 1997, Scott & Yeoh 1998). No-choice tests are most likely to detect whether a species is within the fundamental host range of the agent being assessed. Time-dependent effects (Browne & Withers 2002) can be detected by confining insects for increasing periods on the test plant. Plants or host insects that are accepted, either immediately or after a moderate period of deprivation, are regarded as physiological hosts. The status of hosts that are accepted only following prolonged deprivation is debatable because in nature insects would likely continue to search for more acceptable hosts. No-choice tests are time-consuming as each species requires a full and separate regimen of testing. Positive controls (exposure to the target pest) are needed to verify that any negative findings with non-target test species can safely be interpreted as rejected and are not due to physiological inability of the individuals tested to feed or oviposit. Positive controls are obtained either by simultaneously testing other individuals from the same rearing batch, or by subsequently exposing the individuals used in the main test to the target pest immediately afterwards (this latter is better termed a sequential-choice test). The accuracy of no-choice laboratory tests for prediction of the fundamental host range is being assessed by comparing such laboratory data with patterns of attack measured in the field with the same test species (Briese 2005, Haye et al. 2005). Real progress is likely in this area in the next decade.

Choice tests

Here, test species are presented to the agent in groups. There are several variations of this design. As commonly used, the agent is presented simultaneously, in the same cage, with the target pest and several non-target species. Choices tests have been used in two different circumstances. They may be used early in a screening program to rapidly exclude as non-hosts as many species as possible, by incorporating many species into tests. Lack of attack on a species was interpreted as implying that an unattacked test species was either a non-host or such a low-ranked host that important attack would not occur in nature. Choice tests may also be used after a series of no-choice tests to re-examine non-target species that received minor attack in no-choice tests. Frequently, attack rates on these non-target species would be lower or zero in the presence of the target pest due to preference. Lack of attack, however, when interpreted as indicative of non-host status (instead of just a low-ranked host), risks false-negative results.

Sequential-choice tests resolve this problem of preference masking low-rank status. Here, the pest and non-target species are presented one after another (A, B, A, B, A; where A is the pest and B one or several non-target species). This approach allows each species to be considered separately, but still provides a positive control for each individual agent tested. However, this sequence design has the disadvantage that the agent is exposed first to the pest species, which may condition the agent, heightening its preference for the pest. An alternative design is B, A, long break, B, A, which solves this problem, if time between exposures (long break) is sufficient to dissipate conditioning effects.

A third variation is called choice-test-minus-control, in which agents from a common source selected for use in a test are assigned at random to either: (1) a cage with many non-target test species of plants or (2) a cage with only the target pest (for example, see Heard et al. 2005). The second cage serves as a positive control on the physiological readiness of the test insects. The first cage rapidly screens a set of non-target plants, without the agents being distracted by the presence of the target plant (presumably the highest-ranking host). This design, however, while better than a choice test that includes the target weed, may still miss low-ranking hosts if they are ignored in the presence of a much higher-ranked host, even if that host is a non-target plant. This problem may be further addressed by repeating the test, iteratively removing the species receiving the most attack in the previous cycle until the lowest-ranked species has been assessed.

Continuation and oogenesis tests

Important effects on a non-target host are unlikely if the agent cannot maintain its population solely on that species. (Without such an ability, non-target effects would be reduced to spillover impacts of individuals immigrating from the target pest, as during episodes of high density on the target pest during the initial control cycle.) Oogenesis tests determine whether the agent can develop eggs when fed only the test species. Continuation tests determine whether a test species can feed and reproduce on the test species for several generations without reductions in fecundity, survivorship, or population size (Day 1999). Buckingham et al. (1989), for example, found that a population of the fly *Hydrellia pakistanae* Deonier died out within eight generations if reared exclusively on the non-target pondweed *Potamogeton crispus* L. In some cases, *de facto* continuation tests in another country may provide information valuable in assessing risk. In South Africa, the mirid *Eccritotarus catarinensis* (Carvalho) was released for waterhyacinth [*Eichhornia crassipes* (Mart.) Solms-Laub]. In laboratory assays it fed on pickerelweed (*Pontederia cordata* L.), a non-native invader in South Africa. The mirid failed to establish persistent breeding populations on pickerelweed, both in cages and at sites where waterhyacinth with populations of the mirid were adjacent to pickerelweed (Coetzee et al. 2003). These results constitute a field continuation test and show that, if the bug were to be introduced to the USA (where waterhyacinth is an invasive pest but pickerelweed is a native plant), it would be unlikely to establish itself on pickerelweed.

Open-field tests

Open-field tests have mainly been used with herbivores because of greater ability to manipulate the test species. Tests are run outdoors, either in a common garden plot of artificial composition or in natural stands of the weed into which potted non-target plants are added (Clement & Sobhian 1991, Briese et al. 1995, Clement & Cristofaro 1995, Briese 1999). The agent is either present as a natural population or is released artificially. The outcome scored is usually the number of eggs laid on each test plant. The open-field test was developed at the end of the 1980s (e.g. Clement & Sobhian 1991) on the assumption that removing test plants and insects from cages and letting them interact in an open space would eliminate errors that occurred when test insects were denied the option to leave the test arena. (In cages, with emigration denied, oviposition sometimes occurs on plants believed to not really be hosts, or even on the cage itself.)

Use of open-field tests requires that the test species of interest be present in the donor country or that permission be obtained to import them. This is often possible with plants (which may be present already through importation, or may safely be imported for testing and then destroyed without escape or reproduction). However this is never possible for insects, because the native insects of interest to the recipient country could be potential invaders in the donor region. For this reason, open-field tests are rarely used with insect biological control. If used, congeneric forms from the donor country are assessed as surrogates for the recipient country's native species. This was done, for example, by Porter et al. (1995), who exposed a series of local species of ants in Brazil to phorid parasitoids. By this means, data were obtained suggesting that these flies were host-specific at least to the genus level.

Open-field tests are choice tests and thus may overlook low-ranked hosts. A partial solution is to use a two-phase open-field test (Briese et al. 2002c). Steps in the such a test are: (1) creating a common garden plot containing the target weed and various non-target test plants, (2) allowing the candidate biological control agents to colonize the plot, (3) taking data on the agents' feeding and oviposition, and then (4) killing the target weed plants. This forces the agents to switch and accept lower-ranked hosts, emigrate, or die. When this approach was used by Briese et al. (2002c) for four candidate species attacking the weed *Heliotropium amplexicaule* Vahl, a pest in Australia, it was found that three agents either left or died, but one (an undescribed flea beetle, *Longitarsus* sp.), switched to feeding on the non-target species *Heliotropium arborescens* L.

INTERPRETATION OF TESTS

Choice and no-choice tests may sometimes produce opposite results (Table 17.1). Among the possible causes of such inconsistency are: (1) host preference, (2) time-dependent change, and (3) excitation of the central nervous system. In addition, any cage tests may be influenced by confinement itself if the biology of the species is distorted by confinement.

Host preference

Natural enemies, particularly parasitoids, often change their degree of responsiveness to a host following an initial contact with the species. Contact with a familiar host (i.e. a species normally attacked) enhances responsiveness to that species in subsequent contacts. Experienced females typically respond quicker and more strongly to the normal host than do naïve parasitoids (see review by Withers & Barton Browne 2004). Experience with a host can come from previous contact with the odor of a familiar host or the substrate (plant)–host complex. It may also be caused by experience gained from the rearing host, particularly if the parasitoid emerges from some item such as a cocoon or cadaver associated with the rearing host. With unfamiliar (i.e. novel) hosts, experience can also change responsiveness, either increasing or decreasing later response intensity. In a host-range test, both influences may operate and be hard to distinguish, but in general the control for this issue is to use naïve agents in tests.

Table 17.1 Interpretation of cases when results of choice and no-choice tests do not agree

	Choice test (negative result)	Choice test (positive result)
No-choice test (negative result)	Case I: test species is outside the host range.	Case II: test species is outside the host range and a positive result in choice test is likely due to CNS stimulation by other test plants.
No-choice test (positive result immediately)	Case IV-A: the species is inside the physiological host range and may be accepted in the field if encountered alone, or ignored if encountered in the presence of a more preferred host.	Case III: test species is inside the host range.
No-choice test (positive result after several days of deprivation)	Case IV-B: test species is outside the host range and the positive result in no-choice test is due to starvation, which is likely to promote dispersal, not feeding, under field conditions.	

A negative result means non-acceptance, and a positive result means acceptance. CNS, central nervous system.

The most constricting fact preventing this in some cases is the inevitability of some contact with the rearing host in many systems. If this is the target pest, this may distort the choices exhibited by the agent in favor of the rearing host. If an agent cannot be reared except on the target species, the best approach is to try to control (through dissection or removal of pupae from the rearing environment) contact of new adults with hosts (see Monge & Cortesoro 1996). Effects of rearing hosts mediated with pre-imaginal experience (that of immature stages) seems to be rare.

For herbivorous insects, preference for a high-ranked host in a choice test may cause a low-ranked host to receive no oviposition or feeding, making it appear erroneously as a non-host.

Time-dependent change

The response of female insects to cues associated with oviposition sites changes as the time since last host contact increases (Barton Browne & Withers 2002). As the period of host deprivation for an agent increases, the insect is increasingly likely to accept a less preferred host species for feeding or oviposition. In feeding tests, host deprivation means increasing hunger, often with the result that starved individuals feed on hosts that might be ignored by insects with more moderate hunger levels. In no-choice tests, for example, the test duration may be short or long and test duration may influence results. To detect such effects, a series of daily observations may be useful to see whether a test species is accepted immediately, or only after prolonged periods.

With respect to oviposition, host deprivation acts through its effects on the insect's egg load. If insects start the test with an initial high egg load, the passage of time without contact with the usual host may simply increase the likelihood that other less familiar or less favored hosts will be accepted. If, however, egg load declines over time due to egg resorption or egg dumping (oviposition in random locations) then responsiveness to test species may also decline. Some agents may be dissected to directly observe egg loads to determine whether they remain stable or decrease during a host-deprived period.

Excitation of the central nervous system

In some cases, if insects are simultaneously exposed to both the normal host and a novel host, the novel host may receive ovipositions because contact with the normal host has released oviposition behavior. For example, Field and Darby (1991) found that the parasitoid *Sphecophaga vesparum* (Curtis) (Ichneumonidae) oviposited in cells of a non-target wasp (*Ropalidia plebeiana* Richards) when these were artificially placed within 10 cm of brood of the target wasp (*Vespula* spp.), but no *R. plebeiana* brood were attacked when presented alone in a no-choice test. The occurrence of such events may be suspected if the host range increases in choice tests compared to no-choice tests.

Confinement effects

It is widely accepted that for most herbivorous insects (and likely for predaceous and parasitic ones as well), a series of behaviors leads to host location and acceptance by a foraging female (Vet et al. 1995). In most laboratory host-range tests, the small size and composition of the test cages prevent at least the early steps in such behavior sequences. This may lead some hosts to be attacked artificially in laboratory tests if early, discriminating steps are skipped. The likelihood and importance of such an event must be considered on a case-by-case basis and will be influenced by the biology and dispersal powers of the agent being studied.

EXAMPLES OF HOST-RANGE ESTIMATION

Dipteran parasitoids: phorid flies attacking fire ants

The red imported fire ant (*Solenopsis invicta* Burden) invaded the USA in the 1930s (Lennartz 1973) and now occupies over 120 million ha from Texas to Virginia (Callcott & Collins 1996). It reaches densities of 1800–3500 ants/m^2 (Macom & Porter 1996), causing a wide array of economic and ecological damage, including displacement of native fire ants. A classical biological control program has been inititated against this pest based on the observation that in its native range in Argentina, densities are only 10–20% as high (Porter et al. 1997). At least 20 species of phorid flies attack this pest in its native range, but are absent in the USA. Host ranges of several of these phorids (*Pseudacteon* spp.) were evaluated in support of their introduction to the USA (Porter & Gilbert 2004). The native insects most closely related to the target pest in the USA are *Solenopsis geminata* (Fabricius) and *Solenopsis xyloni* (MacCook). Other native ants in this genus in the USA either occur in habitats that are too dry to support the invasive fire ant or are species whose head size is not large enough to support development of the *Pseudacteon* flies being introduced (whose larvae mature in the host's head capsule). To be potentially suitable as a host for these flies, ants must have head widths between 0.4 and 1.6 mm.

Assessment of safety of the phorids proposed for introduction began with open-field tests in South America that compared attack rates on *S. invicta* with those on ants in other genera and subfamilies (Porter et al. 1995). This was followed by field exposures in Brazil of *S. invicta* compared with *S. geminata*, since both species occur there (Porter 1998). These field tests confirmed the existing published literature that the *Pseudacteon* spp. being studied exploited only *Solenopsis* species and preferred *S. invicta* over *S. geminata*.

The next phase of the assessment was based on laboratory assessments run in US quarantine. No-choice tests examined rates of attack behaviors and parasitization on *S. invicta* compared with *S. geminata* and *S. xyloni*. Results showed that *Pseudacteon tricuspis* and *Pseudacteon litoralis* rarely engaged in attack behaviors against these native fire ants and never parasitized them (Porter & Gilbert 2004).

Sequential choice tests for *Pseudacteon curvatus* and *Pseudacteon obtusus* (Table 17.2) measured preference between the target pest and the native fire ants because both of these flies attacked some native fire ants in no-choice tests. For *P. curvatus*, 75–85% of the female flies preferred the imported fire ant over either native fire ant (Porter 2000, Vazquez et al. 2004). Even *P. curvatus* flies reared in the laboratory on *S. xyloni* retained a strong preference for *S. invicta*. Similarly, over 95% of *P. obtusus* flies preferred the invasive species over the native fire ants. These data were interpreted to mean that even where target and non-target ants co-occurred, little attack on the native fire ants was likely.

Tests were also conducted to assess any potential for *P. curvatus* to become a nuisance species by looking for attraction to items such as ripe fruit or raw meat, carrion, or dung. Of more than 50 items tested, none was attractive to any *Pseudacteon* species (Porter & Gilbert 2004).

Post-release field studies with *P. tricuspis* (the first species to be released) confirmed the lack of attraction of this species to mounds of the native fire ant *S. geminata*, to trays with workers of *S. geminata* workers, or any of 14 other species of ants in 12 genera (Porter & Gilbert 2004). Post-release field tests with *P. curvatus* found that a few flies were attracted to *S. geminata*, but no oviposition was observed and attraction to

Table 17.2 Results of sequential choice tests (target, non-target, target) for several phorid flies of the genus *Pseudacteon* being considered for importation against the imported fire ant (*Solenopsis invicta* Burden) in the USA, with comparison to the non-target native species *Solenopsis geminata* (Fabricius)

Fly species (genus *Pseudacteon*)	No. attacking flies and attack rate per fly		
	Time 1: *S. invicta*	Time 2: *S. geminata*	Time 3: *S. invicta*
P. litoralis	23/23 2.33 attacks/fly	2/23 0.34 attacks/fly	20/21 1.11 attacks/fly
P. wasmanni	18/18 3.21 attacks/fly	2/18 3.1 attacks/fly	8/13 3.0 attacks/fly
P. tricuspis	25/25 1.91 attacks/fly	1/25 0.04 attacks/fly	15/21 1.17 attacks/fly
P. curvatus	20/20 1.53 attacks/fly	13/20 0.75 attacks/fly	–

Data from Gilbert and Morrison (1997).

S. geminata was only one-twentieth of the rate of attraction to *S. invicta*.

In summary, this group of parasitoids was predicted based on field tests in the native range and laboratory tests in quarantine to attack only *Solenopsis* species and to show a nearly complete preference for the invasive fire ant compared to the native species in the same genus. These predictions were later verified by post-release field tests.

Hymenopteran parasitoid of the pink hibiscus mealybug

Following the invasion of the Caribbean by the pink hibiscus mealybug, *Maconellicoccus hirsutus* (Green), in 1992, a classical biological control program was organized on behalf of the region by CABI BioScience. The project proposed to introduce the encrytid *Anagyrus kamali* Moursi, which had previously controlled the mealybug in Egypt (Kamal 1951). This species is a primary, solitary endoparasitoid of mealybugs, known from four genera (*Pseudococcus*, *Ferrisia*, *Nipaecoccus*, and *Planococcoides*) (Cross & Noyes 1995). It also kills mealybugs by host feeding. Species of the Anagyrini (the tribe to which *A. kamali* belongs) generally attack either one or few closely related mealybug species (Cross & Noyes 1995).

To assess the breadth of the host range of this parasitoid, Sagarra et al. (2001) examined the suitability of eight non-target species of mealybugs common in the Caribbean (specifically in Trinidad) in choice and no-choice tests. Test species were *Planococcus citri* (Risso), *Planococcus halli* Ezzat and McConnel, *Dysmicoccus brevipes* (Cockerell), *Pseudococcus elisae* Borchsenius, *Saccharococcus sacchari* (Cockerell), *Puto barberii* (Cockerell), *Nipaecoccus nipae* (Newstead), and *Plotococcus neotropicus* (Williams and Granara de Willink). Of these, the parasitoid probed three species (*P. citri*, *P. halli*, and *P. elisae*), but laid eggs in only *P. citri* and *P. halli*. In no-choice tests, 24 and 18% as many individuals of these non-target species were attacked compared to the target pest. Parasitoid immature stages, however, failed to mature in the non-target hosts. Thus, of the nine mealybugs considered, only the target pest was an actual host of *A. kamali*, which was released and controlled the target pest throughout the region.

Predaceous derodontid beetle feeding on hemlock woolly adelgid

The hemlock woolly adelgid, *Adelges tsugae* Annand, is a serious invasive pest of eastern [*Tsuga canadensis* (L.) Carrière] and Carolina hemlock (*Tusga caroliniana* Engelmann) (McClure 1991) for which a biological control was mounted because local natural enemies did not prevent tree mortality. A search for specialized predators was undertaken because adelgids lack

Table 17.3 Oviposition of *Laricobius nigrinus* Fender on various potential prey (adelgids, aphids, and scales) under both choice and no-choice conditions in comparison with the oviposition on the target prey, *Adelges tsugae* Annand

Test species	Oviposition of *L. nigrinus* (no. eggs/female in 3 days)		
	No-choice test	Choice test, on target	Choice test, on non-target
Pest (*A. tsugae*)	**12.2**	–	–
Adelges abietis	0.7	**7.6**	0.4
Adelges piceae	3.1	**10.1**	1.8
Pineus strobi	7.9	**12.3**	2.3
Cinara pilicornis	0.2	**12.4**	0
Myzus persicae	0.0	**9.8**	0
Chionaspis pinifoliae	0.1	**17.5**	0

Data from Zilahi-Balogh et al. (2002).

parasitoids. *Laricobius* spp. (Derodontidae) are specialized feeders on adelgids (Lawrence 1989). *Laricobius nigrinus* Fender, native to the western USA, where it is associated with *A. tsugae* (L.M. Humble, Canadian Forest Service, unpublished data), was imported and its prey range evaluated. The suitability of six species as potential prey was examined in comparison to the target host (Zilahi-Balogh et al. 2002). The test list consisted of two congeneric species [*Adelges piceae* (Ratzeburg) and *Adelges abietis* (L.)] and one other adelgid [*Pineus strobi* (Hartig)], all of which are conifer feeders. Less similar potential prey offered were two aphids [*Cinara pilicornis* (Hartig) and *Myzus persicae* (Sulzer)] and an armored scale [*Chionaspis pinifoliae* (Fitch)]. Of the two aphids, *C. pilicornis* feeds on conifers, whereas *M. persicae* does not. The armored scale offered feeds on pine.

In no-choice oviposition tests (Table 17.3), *L. nigrinus* laid eggs in association with all the test species offered except the aphid *(M. persicae)*. In choice tests, oviposition was observed only near the three non-target adelgids. In a longer assay (3 days) *L. nigrinus* laid 51% as many eggs on *A. abietis*, 43% on *P. strobi*, and 14% on *A. piceae* as on the target pest. On the target pest, 17% of eggs laid survived to produce adult beetles, but none successfully matured on any of the other test species (Table 17.4). In summary, the predator *L. nigrinus* is specific among the species tested to the target pest. Although some

Table 17.4 Development and survival of immature stages of *Laricobius nigrinus* Fender on various Hemiptera offered as prey, in comparison with the target prey, *Adelges tsugae* Annand

Test species	Reached fourth larval instar	Pupated	Emerged as adult
Pest (*A. tsugae*)	**58%**	**19%**	**17%**
Adelges piceae	11%	0%	0%
Pineus strobi	7%	0%	0%
Adelges abietis	0%	0%	0%
Cinara pilicornis	0%	0%	0%
Chionaspis pinifoliae	0%	0%	0%

Data from Zilahi-Balogh et al. (2002).

eggs were deposited near other prey species, these did not survive.

Predaceous coccinellid beetle and cottony cushion scale in the Galápagos

The cottony cushion scale (*Icerya purchasi* Maskell) is a polyphagous margarodid (Hale 1970) that has invaded 15 of the Galápagos Islands (Causton 2004), where it damages 62 native or endemic plants, of which six are endangered. The Park Authority commissioned a study of the proposed introduction of the specialized coccinellid *Rodolia cardinalis* Mulsant to suppress the invader. This coccinellid is believed to be native to Australia and has been released in over 60 countries for control of cottony cushion scale and has frequently provided effective control.

To assess potential risks of this predator to Galápagos insects, the host range of *R. cardinalis* relative to Galápagos insects was investigated. Both adult and larval host ranges were assessed. Because the desired test insects often required plants endemic to specific islands, a quarantine laboratory was built to conduct the study. Also, since most of the test insects could not be reared, they were collected in the field. This presented a complication that in some cases individuals tested were later found to have been parasitized and test results had to be discarded. Finally, because of drought conditions, some desired test species could not be located in adequate numbers. As a substitute in some cases, invasive insects belonging to important test groups (families) were used as substitutes. The final test list included the Galápagos' only native margarodid (*Margarodes similis* Morrison), any species listed in the literature as prey of any *Rodolia* species (or if not available, related species), and any Galápagos species morphologically similar to *I. purchasi* or likely to live in close proximity to *R. cardinalis*.

Twenty specific prey records for *R. cardinalis* were located, which suggested the prey range included the Margodidae, Pseudococcidae, Diaspididae, Dactylopiidae, and perhaps Aphididae. Given this breadth, 14 Galápagos Coccoidea and three aphids were considered as potential prey. In addition, several native predators were included to look for intraguild predation. Tests with *R. cardinalis* larvae were conducted with 16 species, from nine families. Feeding occurred only on the native margarodid *M. similis* and only after it emerged from its protective waxy cyst. Larvae, however, failed to complete development on this prey, dying within 1 week. On all other hosts, larvae died within 1–2 days.

Adults of *R. cardinalis* were tested using both naïve individuals and ones with prior feeding experience on the target pest. Six species (from five families) were tested with naïve beetles and eight species (from six families) were tested with conditioned beetles. Both conditioned and naïve beetles fed on *M. similis* that had emerged from cysts. However, adult *R. cardinalis* beetles were not able to break open the waxy cysts of *M. similis* and did not dig into the soil, where this root-feeding margarodid lives. No feeding was observed, directly or indirectly, on any of the other test species. No test species, including *M. similis*, stimulated oviposition by *R. cardinalis*. With the exception of two mealybugs, survival times of adult *R. cardinalis* were no greater on non-target prey than on water alone.

In summary, it was concluded that this species posed no threat to native insects of the park and it was released. Evaluations of its impact on the cottony cushion scale and recovery of the affected native have yet to be done.

Herbivorous insect: a gall insect of melaleuca

Of the many galls on *M. quinquenervia* in Australia, one type is caused by the invasion of the stem apex by a host-specific fly (*Fergusonina turneri* Taylor; Diptera: Fergusoninidae) and a mutualistic nematode (*Fergusobia quinquenerviae* Davies and Giblin-Davis). The nematodes are carried by the female flies and are simultaneously deposited with the fly eggs in a susceptible bud. The nematodes immediately begin to induce gall formation while eclosion of the fly larvae is delayed. Nutritious gall tissue is available for the larvae by the time the fly eggs hatch, and feeding by the fly larvae further enhances gall development.

Molecular studies have shown that these organisms have speciated within the Myrtaceae and each species pair has evolved a close dependence on one another and on a single host-plant species (Giblin-Davis et al. 2003, Davies & Giblin-Davis 2004, Scheffer et al. 2004, Taylor 2004). Thus, this mutualistic combination seems ideal for biological control of *M. quinquenervia* inasmuch as galling of the stem tips halts the indeterminate growth of the stem apex thus precluding flower and seed production on the affected axis.

This would likely reduce the enormous regeneration potential of *M. quinquenervia*, which is responsible for its success as an invasive weed.

The testing strategy for these two agents involved determining whether or not *F. turneri* would: (1) deposit eggs and nematodes on test plant buds, or attempt to do so, (2) choose buds of non-target plant species when melaleuca buds were unavailable, or (3) complete development on non-target species. Testing focused on oviposition which is the critical stage in host selection for these species. The susceptible stage of bud development was determined and test plants were pruned to induce bud formation. Flies were caged on the stems when the buds reached the appropriate stage. Several types of tests were done: (1) no-choice oviposition tests on cut stems, (2) multi-choice oviposition tests with and without melaleuca, using cut stems, (3) two-choice or four-choice oviposition tests on cuttings, (4) no-choice development tests on entire plants and on branches of potted plants, and (5) two-choice development tests on potted plants. Eight native myrtaceous species as well as closely related ornamentals were tested in Florida under quarantine conditions. In addition, a few non-related species were tested mainly for oviposition, which is the critical stage in host selection. Galls were produced only on *M. quinquenervia* as predicted by the field surveys, so permission for released was requested and granted.

Herbivorous mite: on Old World climbing fern

The Old World climbing fern (*Lygodium microphyllum*) is a pernicious invader in the Florida Everglades National Park, in the USA, that has the potential to affect this critical ecosystem drastically. It is particularly damaging to Everglades tree islands, which harbor most of the biodiversity in the region. Many tree islands are now thickly covered with this rampant plant, which has resulted in drastic changes in the structure and composition of the natural communities. *Lygodium microphyllum* occupies a broad range throughout the Old World tropics. One of the potential biological control agents of interest is the eriophyiid mite *F. perrepae*, which feeds on the leaflets of the fern, causing the edges to roll up and develop into galls. The mite has apparently developed local lineages, so it became important to identify the origin of the Florida plant so as to study the correct strain of the mite. It was discovered through DNA analysis that the population in Florida likely originated in northern Queensland or Papua New Guinea (Goolsby et al. 2006b). Mites collected from the Cape York Peninsula thrived on plant material from Florida, whereas they fared poorly on ferns from southern Queensland, and vice versa (Goolsby et al. 2006b).

Host testing of *F. perrepae* emphasized *Lygodium* species from North America and the Neotropics as well as fern species native to the southeastern USA. The mite is a minute, soft-bodied organism that was difficult to handle so Goolsby and colleagues (Goolsby et al. 2004b, 2005b, 2005c) developed a unique system for determining host range. Spores were germinated and the sporeling ferns were placed in small, thimble-sized pots. Ten mites were carefully transferred to an individual sporeling leaflet using a single eyelash. The young sporeling tissue was optimum for the development of the leaf rolls so these were used for the initial no-choice screening. *Floracarus perrepae* developed normally on the Florida genotype of *L. microphyllum*. There was also modest development on the North American native *Lygodium palmatum* (Bernahardi) Swartz. However, lethal minimum temperature and cold stress tests revealed that the mite would not likely establish in the more northern areas where *L. palmatum* was found. There was also minor development on six other fern species but leaf rolls were induced only on *Lygodium* species, with full rolls common only on *L. microphyllum*. These six non-target plant species that exhibited development as sporelings were retested as more mature plants. *Floracarus perrepae* developed only on *L. microphyllum* in both no-choice and choice tests with these plants. Goolsby et al. (2004b, 2005b, 2005c) concluded that *F. perrepae* was specific to *L. microphyllum* and posed little or no risk to native or cultivated ferns in North and South America. This species has now been approved for release in the USA.

RISK ASSESSMENT

The conclusion of an assessment of the host range of a new agent and of any indirect effects that might be evident is to conduct a risk assessment concerning the potential costs and benefits of its release in a particular recipient country or region, being further guided by ethical behavior (Delfosse 2005). One outcome might be to immediately reject release of the agent based on an obviously overly broad host range or significant attack rates on valuable test species (e.g. Cristofaro et al. 1998, Heard et al. 1998, Haye et al. 2006). Barring this, the gains and losses associated with the action must be compared.

Risk assessment (see Wan & Harris 1997, Andersen et al. 2005, Dhileepan et al. 2005, Wright et al. 2005, van Lenteren & Loomans 2006, and Bigler & Kölliker-Ott 2006 for examples and principles) starts by indentifying any risks implied by the test data, taking into account any mitigating factors of geography, climate, or other matters that might act to change risk in the field. This risk must then be balanced against the harm caused by the invasive species, now and projected for the future given any like spread or cumulative or synergistic impacts. Relative risks of these two events are then compared to determine whether the release would likely be a net improvement in public good, in terms of both economic and ecological outcomes. This process must address who benefits and who suffers and whether any risks are unacceptable. In general, risks of biological control agents should be judged by similar standards applied to other categories of exotic species introductions.

Chapter 18

AVOIDING INDIRECT NON-TARGET IMPACTS

Direct non-target effects of biological control are those in which the biological control agent attacks a native species, as would occur, for example, when a weed biological control insect ate a native plant or a released parasitoid attacked a native insect. **Indirect non-target effects**, in contrast, arise when the agent influences relationships among species within the food web of the target pest (Holt & Hochberg 2001, Pearson & Callaway 2005). Indirect effects are not predictable based on estimation of the biological control agent's host range. Instead, anticipation of indirect effects requires an understanding of how the new species will interact with other species in the community in which it will be established (for a review of potential prediction methods see Messing et al. 2006).

In this chapter, we first discuss the types of indirect interaction predicted by theory. Second, we ask whether an agent's efficacy can be predicted because, theoretically, important indirect effects only occur if the density of the natural enemy remains high for prolonged periods of time (an outcome not associated with successful biological control, but rather with agents that multiply but do not suppress the target pest). Finally, we discuss how, and to what extent, biological control projects should be required to predict and avoid indirect effects.

KINDS OF POTENTIAL INDIRECT EFFECTS

Three types of indirect effect have been described that might affect the consequences of a biological agent's introduction: (1) ecological replacement, (2) compensatory responses, and (3) food-web interactions (Pearson & Callaway 2005).

Ecological replacement

Ecological replacement occurs when an introduced pest replaces a native prey or plant species as the food or shelter for a native animal species. Invasive plants, for example, while generally harmful competitors for native plants (and hence also damaging to these plants' dependent specialized herbivores), may become habitat or food for native animals. In such cases, a successful biological control program against the invasive species on which the native one depends would remove essential food or shelter from the native species.

In New Zealand, some endangered weta (*Deinacrida* spp.) utilize invasive gorse stands (*Ulex europaeus* L.) (Stronge et al. 1997) as habitat. Stands of this thorny shrub protect weta from predation by introduced rats. Native birds also sometimes benefit from introduced plants. In the southwestern USA, a classical biological control project against invasive saltcedars (*Tamarix* sp.) led to concerns for an endangered bird [southwestern willow flycatcher, *Empidonax traillii extimus* (Phillips)] that nests in *Tamarix* stands. Formerly this bird nested in riparian cottonwood trees, which were displaced by *Tamarix*. To forestall any potential lack of nest sites, a plan has been developed to begin the biological control efforts away from the bird's nesting areas and to start replanting cottonwood trees to provide nest sites that can be used as *Tamarix* declines.

To avoid this sort of ecological replacement effect, one must ask whether any native species have become highly dependent on the proposed target pest for food or habitat. Dependency, not mere use, is a key issue. If the original resources used by the native species before the pest invasion remain abundant, then reduction of the invasive species that serves as an additional host will not be fundamentally damaging. Surveys of weed stands or insect pest populations can be done in the recipient area before the biological control project is initiated to identify significant use by native species.

A further complication occurs when a native exploiter population consumes but does not control an invader. In such cases, the invasive species may be a positive influence on a particular native species (the predator) but a negative influence on other native species that might suffer more predation from the larger populations of this native predator.

Compensatory responses

The concern here is whether the attack of a biological control agent could have the counterintuitive effect of making an invasive weed more, not less, competitive with native plants. This outcome is theoretically possible because some plants do respond to defoliation by increasing their growth or reproduction (e.g. Wan et al. 2003). However, no clear examples have documented such an outcome due to the introduction of a biological control agent of an invasive plant.

Food-web interactions

Invasive species and native species may share exploiters, which may be introduced biological control agents. For example, in North America the introduced parasitoid *Cotesia glomerata* (L.) attacks the invasive butterfly *Pieris rapae* (L.) and native species such as *Pieris napi oleracea* Harris (Benson et al. 2003). Resulting reductions in the native species superficially look like competition, but are really food-web-mediated impacts termed **apparent competition** (Figure 18.1) caused by the parasitoid. The pest butterfly supports high densities of the parasitoid that then attack the native buttefly. See van Veen et al. (2006) for a review of apparent competition.

Figure 18.1 When a pest and non-target species share a common introduced natural enemy, the interaction is termed apparent competition because there appears to be a negative effect of the pest directly on the non-target species. Redrawn from Lynch et al. (2002), with permission from CABI Publishing.

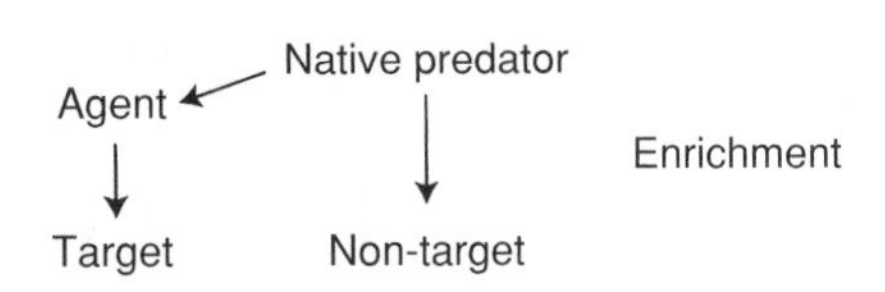

Figure 18.2 When the food or host supply of a native natural enemy is expanded because it can exploit an introduced natural enemy through intraguild predation, the native predator's impacts on its normal prey may increase or decrease. This condition is termed enrichment, because from the native predator's point of view, the food supply has been increased. Redrawn from Lynch et al. (2002), with permission from CABI Publishing.

Other food-web-mediated interactions include **food-web enrichment** (Figure 18.2), in which an introduced natural enemy becomes a resource for native organisms, allowing their populations, and hence their impacts, to increase. For example, the tephritid *Mesoclanis polana* Munro is a biological weed control agent that has been adopted as a host by native parasitoids in Australian bush infested by *M. polana*'s target, the weed *Chrysanthemoides monilifera* ssp. *rotundata* (L.) (Willis & Memmott 2005). (See also the spotted knapweed/mouse example discussed in Chapter 11.)

An important food-web effect that some natural enemies might have is direct toxicity to organisms that eat them. The melaleuca sawfly, *Lophyrotoma zonalis* Gagné, was among the agents studied in Florida, USA, for possible introduction against the weed tree melalueca. However, a review of the risk posed by certain toxins in the larvae of this sawfly found that risk, while low, was not acceptable and the agent was not introduced. Biological control researchers feared that starving songbirds, upon arriving to Florida after spring migration, might be poisoned if they gorged themselves on toxic larvae, even though laboratory tests with healthy birds eating the same food were unaffected (for details, see Chapter 16). Toxins are an identifiable feature of a natural enemy whose importance can be evaluated.

CAN RISK OF INDIRECT IMPACTS BE REDUCED BY PREDICTING NATURAL ENEMY EFFICACY?

Community ecology theory predicts that new species with dense populations are more likely to cause indirect impacts on local food webs than are rare species (Holt & Hochberg 2001). This suggests that successful biocontrol agents pose little such risk (because after pest numbers are reduced, agent numbers also decline). It has been suggested that greater risk may be associated with partially effective agents that do establish and become common and then remain abundant because they fail to suppress the pest's density. Therefore, to the extent that the likely efficacy of an agent can be predicted, when several agents are simultaneously available for release, it would be least risky to proceed with the species predicted to be most effective, wait to see if it suppresses the pest and then proceed to other species only if necessary.

However, focusing on efficacy as a tool to reduce risks of indirect effects can pose several problems. First, it is very difficult to predict field efficacy from laboratory studies. Second, agents are discovered, screened for safety, and approved for release on independent time lines. Rarely is a full set of information on all potential agents available at one time. Even for cases in which extensive knowledge is available on all potential agents, including indications of which is likely to be most effective, the above strategy would necessitate holding some species in reserve for years. Would these species be held in laboratory colonies where they would likely lose quality? Or would they be recollected later in the field, which may not be easy to do or may require another round of host testing since the new collection might differ from the original consignment? Third, clearly documented cases exist in which several agents working together have suppressed the target pest while none did so separately. This is especially so with weed control agents. In such a case, there is no best agent to identify. Finally, some sorting of agents already happens during natural enemy surveys, such that most damaging agents are often discovered earlier and chosen for additional work earlier.

Prediction of the efficacy of a biological control agent from laboratory data is inherently difficult because doing so requires that we know how one population (the agent) will interact with another population (the pest) in an outdoor environment that is likely to be different from where the agent was collected. Effects to predict natural enemy efficacy of necessity are somewhat different for agents directed against pest insects compared with weeds. Indicators of agent efficacy have long been discussed for insect targets (Turnbull & Chant 1961). Furthermore, some informal predictions about the probability of agent efficacy occur in every biological control project because most consider the degree of climatic match, host suitability, and favorable biology shown by the candidate agents. A great many attributes have been suggested as indicative of a "good" biological control agent, including short generation time (relative to the pest; Kindlmann & Dixon 1999), high fecundity, good searching ability, and a positive density-dependent response (of the agent's population, and hence not easily measured in the laboratory; see Chapter 10). This type of information has been used to exclude seemingly poor prospects.

One feature of importance that has been used in this manner is the relative size of the intrinsic rates of increase of the pest and natural enemy populations (Kindlmann & Dixon 1999). Agents directed against insect pests that they can out-reproduce will be favored in their numerical response, a feature identified as important by insect biological control theory (see Chapter 10). Froud and Stevens (1997) cited the higher rates of increase of the parasitoid *Thripobius semiluteus* Boucek relative to the invasive thrips *Heliothrips haemorrhoidalis* (Bouche) as a feature suggesting that the importation of this parasitoid into New Zealand might be effective. However, while perhaps a necessary condition, this feature alone is not sufficient for efficacy as poor searching ability or disproportional mortality of the agent (relative to the pest) during a climatically unfavorable season, or a period when hosts persist in an unattackable stage, can easily render such a natural enemy ineffective. For example, *Gonatocerus ashmeadi* Girault in Tahiti (a location with a continuously benign climate) rapidly controlled the glassy-winged sharpshooter [*Homalodisca coagulata* (Say)] (Grandgirard et al. 2006), but has not been able to do so in California, USA, largely because of seasonal interruptions in host availability (Hoddle, unpub.). Also, when agents and pests are actually interacting, the reproductive rate of the pest will be lowered by the presence of the natural enemy.

In insect biological control, the concern over prediction of agent efficacy focused on possible interference if several agents were released simultaneously. Would the released agents reduce the potential impact of the "best" species or would the more effective species dominate such that total pest suppression was maximized? Huffaker and Kennett (1969) analyzed several cases of biological control and concluded that multiple species release as a strategy did not lower pest control. However, they were not concerned at the time, neither with non-target effects on other insects nor with costs from assessing host ranges of multiple species. Had these factors been issues, the "best" release strategy might have been a more structured one, with releases spread out over time, beginning with the single species believed to have the greatest potential to suppress the pest. Ehler (1995) took this latter approach in choosing which species to release against obscure scale [*Melanaspis obscura* (Comstock)] on oaks in California. Of the 11 parasitoid species found on the scale in Texas, USA (part of the native range), four species accounted for more than 90% of all parasitism, but one of these was a hyperparasitoid. From the pool of 11 species, three were excluded because they were hyperparasitoids, four were excluded because they were undescribed, pending further information on their biologies, and one was excluded because it already occurred in California. This left three species for further consideration. The most specialized of these three [*Coccophagoides fuscipennis* (Girault)] was the least abundant and was not chosen for introduction. One species was dropped for consideration because it could not be reared readily in the laboratory. Releases were ultimately made of the remaining species, *Encarsia aurantii* (Howard), which was the most abundant, being an exotic invasive parasitoid that had come to dominate the obscure scale parasitoids complex in Texas. This species ultimately successfully controlled the pest in California. This approach was an attempt to reduce a complex of species through elimination and identify the species that was best for introduction. The choice made in this case may have limited the risk of indirect effects, but may have increased the risk of direct effects because the most host-specific species available was not chosen and the introduced species had a wide host range. This illustrates some of the practical and theoretical trade-offs inherent in agent selection for insect biological control.

Another approach to the problem of predicting efficacy is to assess natural enemy performance in laboratory cages before release and then measure how well the results predict field impacts. This approach is feasible for pests that rear well on small plants amenable to caging and in systems such as whiteflies in which both hosts and parasitoid have multiple generations of short duration. Goolsby et al. (2005a) compared post-release success with pre-release cage impacts of various parasitoid species and populations released into the western USA for control of *Bemisia tabaci* biotype "B" Gennadius. They concluded that climatic match and pre-release evaluations were predictive of post-release success.

Modeling of parasitoids and their hosts has also been explored as a means of predicting which parasitoids or combinations might be more valuable for introduction (Pedersen & Mills 2004). Godfray and Waage (1991) used this approach to do an after-the-fact prediction of which of two parasitoids available for release in West Africa to control mango mealybug (*Rastrococcus invadens* Williams) would have been most effective. The prediction, however, was never fully tested as the first species to be released controlled the pest (Bokonon-Ganta & Neuenschwander 1995). Mills (2005) took a broader view and used a stage-structured model of the codling moth [*Cydia pomonella* (L.)] in California to assess which life stage would be most vulnerable to additional parasitoid species in terms of impact on the pest population's rate of increase (r_m). This turned out to be the second larval instar or the cocoon as the desirable target stages. Criteria for parasitoid selection then were species that attacked one of these host stages, caused greater than 30% parasitism in the pest's native range, and had no potential for antagonistic interactions with other parasitoid species. Based on these criteria, Mills suggested that *Mastrus ridibundus* (Gravenhorst) was the most promising species for introduction. Release of this species appears to have caused high parasitism of codling moth cocoons (up to 70%) and some decline in damage in walnut orchards in California (Mills 2005).

In contrast to efforts to identify effective parasitoids or predators for insect biological control, evaluation of the likely efficacy of new weed biological control agents has focused largely on pre-release assessment of per capita impact of the agents. For weed biocontrol agents, McClay and Balciunas (2005) suggested that impact = range × abundance × per capita impact. They suggested that we can measure these attributes by scoring fecundity, voltinism, and host-plant suitability as predictors of abundance, and climatic match as a predictor for range. One approach considered has been to use artificial damage to plants (imposed by the research in the laboratory) to determine what sorts of damage most strongly affect plants. A general argument has been made to use this approach in advance of host specificity testing (see Raghu & Dhileepan 2005 for discussion of several cases where this approach has been followed). While of potential value, cases have been recognized in which mechanical damage does not simulate the effects of insect damage (e.g. Schat & Blossey 2005).

A per capita impact laboratory evaluation for the tephritid *Parafreutreta regalis* Munro, being released against Cape-ivy (*Delairea odorata* Lemaire), showed that this fly had important effects on the plant's performance under the test conditions (Balciunas & Smith 2006). We argue, however, that representing such per capita impact assessments as true predictors of field impact is misleading. Indeed, while urging the merits of early assessment of an agent's per capita impact, McClay and Balciunas (2005) acknowledge that local parasitoids and predators of introduced herbivores might change the expected abundance of the weed biocontrol agent in unpredictable ways. They rate prediction of post-release abundance (based on pre-release laboratory tests) as "very difficult." They assert, however, that the per capita effect would be relatively easy to score in the laboratory, or in some cases might be assessed in the country of origin using manipulative field tests. Balciunas and Burrows (1993) used insecticides to try to assess the impact of Australian insects on saplings of *Melaleuca quinquenervia* (Cavier) Blake. Goolsby et al. (2004a) used an acaricide exclusion test to assess the impact of the mite *Floracarus perrepae* Knihinicki and Boczek on *Lygodium microphyllum* (Cav.) R. Br. in Australia. An agent that does not have high per capita impact under the ideal conditions of laboratory testing will likely be an ineffective species in the field, unless it reaches extremely high densities, which can happen. Conversely, some agents that do have high per capita impacts in laboratory tests may still fail to be effective in the field for reasons such as poor adaptation to the local climate or attack by local natural enemies. To date, there do not seem to be any cases of selection of agents based on such predictions, followed by assessment of field outcomes relative to predictions. With time, the power of this approach to predict field efficacy will become clearer as more such pre- and post-release data-sets become available.

Part 7

MEASURING NATURAL ENEMY IMPACTS ON PESTS

Chapter 19

FIELD COLONIZATION OF NATURAL ENEMIES

Field establishment of natural enemies is a critical step in classical biological control because without establishment, there is no chance for spread and impact. Releases may fail for many reasons, some of which are related to the agent, some to the recipient site or community, and some to the techniques used. Too few agents may be released or the release may be managed badly (Beirne 1985, Hågvar 1991). Beirne (1975) found that higher establishment rates for parasitoids and predators in Canada were associated with large releases at semi-isolated, ecologically simple sites. No evidence was found that mass rearing increased establishment rates. However, a mass-rearing colony facilitates releases at a larger number of sites, which can accelerate the impact of programs directed against pests with large geographic ranges, and can allow projects to survive setbacks from chance events. Releases may also fail if the recipient community lacks some essential biotic component such as a required overwintering host or if local natural enemies attack the released agent at high rates.

LIMITATIONS FROM THE AGENT OR RECIPIENT COMMUNITY

Biological inadequacies of the agent or a mismatch between it and the recipient community can be significant causes of failure to establish. These include: (1) agents unable to survive the local climate, (2) parasitoids or predators that have an inadequate preference for attacking the target pest on its food plant, (3) weed control agents that acquire suppressive native natural enemies from the recipient community, (4) agents poorly synchronized with the phenology of the target in the new range, or (5) agents that lack an essential alternate host.

When possible, biological limitations of poor agents should be recognized and avoided by choosing better-adapted species or biotypes. Problems rooted in the recipient community, however, such as attack on the natural enemy by local predators, parasitoids, or hyperparasitoids, are predictable only generally and cannot be avoided.

Adaptation to the climate and seasonality of the recipient country

To survive in a new area, an agent must be able to survive the physical extremes of heat and cold, and wet and dryness where released. Also, agents must respond appropriately to the total environment by: (1) emerging in synchrony with the attackable host or plant stage and (2) entering diapause, if required, at an appropriate time. In general, introductions are more successful if natural enemies come from donor areas with climates similar to the recipient area (Messenger et al. 1976), although some agents have successfully transferred between very dissimilar climates (e.g. Bustillo & Drooz 1977). There are few actual studies of the importance of donor area climate as a predictor of establishment and contrary examples exist: for two insects from Argentina released into Australia against mesquite (*Prosopis* spp.), the climate of the collection location did not predict establishment success well for at least one of the agents (a gelechiid moth, *Evippe* sp #1), which became widely established but developed the highest populations at locations significantly warmer than its source area (van Klinken et al. 2003).

Climatic factors assumed to be important to establishment include extremes of temperature and humidity, effects of seasonal rainfall patterns on host and host-plant availability, and photoperiod. Initial assumptions, however, may be misleading and fail to correctly identify which aspect of climate actually restricts an organism's establishment. When the tortoise beetle *Gratiana spadicea* (Klug) failed to establish at some high-altitude sites in South Africa, cold winters were blamed. However, studies later showed the limiting factor to be low humidity (<57% relative humidity), which was damaging to the beetle's eggs (Byrne et al. 2002).

Climatic maps or computerized meteorological data can be used to map similarities between regions to help direct foreign collecting to areas with climates similar to the intended release areas (Yaninek & Bellotti 1987; see Chapter 14). Direct field studies, however, of rates of attack of the natural enemy on the pest at several locations that vary in their climate can reveal important information about the likely amplitude of the climatic ecological tolerance possessed by the agent. Goolsby et al. (2005b), for example, by studying the eriophyid mite *Floracarus perrepae* Knihinicki and Boczek in Australia, New Caledonia, and India, were able to predict that climate in the intended recipient location (southern Florida, USA) would not be an impediment to establishment.

Agents that tolerate a region's physical climate may still fail if local climate induces poor synchrony with the critical stage of its host or if the agent is not stimulated to enter diapause at the right time. For example, a population of the braconid wasp *Cotesia rubecula* (Marshall) collected in British Columbia, Canada, enters diapause whenever day length falls below 15–16 h (Nealis 1985). This induces diapause by the end of August, which is reasonable, given the imminent onset of a wet, cold fall. When this strain was moved to Missouri, USA (≈12° of latitude further south; Puttler et al. 1970), sensitivity to this day length caused the parasitoid to enter diapause in early September, when average temperature was more than 15°C. It is now recognized that survival of this parasitoid is low if exposed to such temperatures while in diapause. As a consequence, establishment in Missouri failed. Another population, collected in Beijing, China, was later released in Massachusetts, USA. These locations are within 2° of latitude and the parasitoid established readily (Van Driesche & Nunn 2002).

Climate may cause a potentially effective agent to fail if synchrony with the host is impaired. In New Zealand, the introduced gorse seed weevil [*Apion ulicis* (Forster)] failed to exert maximum impact on the target plant (*Ulex europaeus* L.) because reproductive diapause caused poor synchrony with gorse seeds. In New Zealand, the weevil emerged after most of the spring seed crop and was only available to attack summer seeds. This mismatch occurred because the plant in the new habitat set seed twice a year, rather than once as in Europe. While not actually preventing establishment, this mismatch reduced agent efficacy significantly (Cowley 1983). Similarly, *Rhinocyllus conicus* (Frölich) established less well on *Carduus acanthoides* L. than on *Carduus nutans* L. because of poor synchronization between *C. acanthoides* flowering and beetle oviposition (Surles & Kok 1977).

Finally, the climatic tolerance of the agent and those of the target pest may only partially overlap, such that an agent may not be suitable for some locations where the invasive species is a pest. The weevil *Perapion antiquum* (Gyllenhal), for example, while effective against *Emex australis* Steinheil in Hawaii, is not useful in Australia because the areas where this weevil would likely establish are physically distant from and climatically dissimilar to *E. australis* problem areas, to which the agent is poorly adapted (Scott 1992).

Inability to parasitize the target pest on its typical host plant

Plant features such as chemical composition, leaf texture, pubescence, and plant architecture can affect the ability of parasitoids and predators to attack otherwise suitable hosts (e.g. Elsey 1974, Keller 1987). If an agent is collected from the target pest's principal host plant, host-plant suitability is likely assured. If, however, the parasitoid or predator is collected from the target on a different plant, then problems may arise if the donor plant and recipient-area plant differently affect agent foraging or immature stage survival. For example, the parasitoid *Habrolepis rouxi* Compere is able to attack and mature well in California red scale, *Aonidiella aurantii* (Maskell), on citrus plants but the same insect feeding on sago palm (*Cycas revoluta* Thunb.) causes 100% mortality to the parasitoid's immature stages (Smith 1957).

Degree of attack by local natural enemies

Natural enemies released in biological control programs

may themselves be attacked by local species. For example, cocoons of *C. rubecula* [a braconid released against imported cabbageworm, *Pieris rapae* (L.)] are attacked by hyperparasitoids in Virginia, USA, and this may have contributed to its failure to establish there permanently (McDonald & Kok 1992).

Herbivorous insects released against weeds may be attacked by generalist parasitoids and predators present in the recipient region, a process that has been referred to as **biotic interference**, which is a component of biotic resistance (Goeden & Louda 1976). Examples include: (1) attack on lantana gall fly (*Eutreta xanthochaeta* Aldrich) galls by *Diachasmimorpha tryoni* (Cameron), a parasitoid introduced to control fruit-attacking tephritids (Duan et al. 1998), (2) attack on the loosestrife beetle *Galerucella calmariensis* L. by the mirid bug *Plagiognathus politus* Uhler (Hunt-Joshi et al. 2005), and (3) attack on the rush skeletonweed gall midge (*Cystiphora schmidti* Rubsaamen) by the pteromalid parasitoid *Mesopolobus* sp. in Washington state (USA) (Wehling & Piper 1988).

Whether or not attacks are trivial or render the biocontrol agent ineffective varies greatly. Among weed biocontrol agents released in South Africa (Hill & Hulley 1995), 40 of 62 species were attacked to some degree by native parasitoids. Agents that were poorly concealed endophytes (such as leafminers) were more frequently attacked than were exposed feeders. Introduced leafminers and spider mites typically attract generalist parasitoids and predators, such as the gorse spider mite (*Tetranychus lintearius* Dufour), which was attacked after release in Oregon, USA, by various phytoseiids, including *Phytoseiulus persimilis* Athias-Henriot (Pratt et al. 2003a). However, the degree to which failure to establish is due to such attack is unclear, because this interaction is a fleeting event that is rarely the focus of research.

Lack of essential alternate hosts in the recipient community

Some physically favorable recipient locations may lack essential biotic components for establishment of a new species. The eulophid parasitoid *Pediobius foveolatus* (Crawford) cannot overwinter in the target pest (*Epilachna varivestis* Mulsant, Coccinellidae) because it requires a species that overwinters as a larva (not as an adult, as does *E. varivestis*; Schaefer et al. 1983). Since no host with this biology occurs in North America, *P. foveolatus* did not establish.

MANAGING RELEASE SITES

Small populations of natural enemies are vulnerable to disturbance and chance events. To minimize potential disruption, release sites should be chosen to provide the natural enemy with sufficient insect hosts or food plants and be managed to protect the site from pesticides, fire, flood, or deliberate destruction. Site selection criteria will be less important if a great many releases can be made because the loss of a few sites will be insignificant. Releases should be made at sites spanning the range of local climates and habitats occupied by the pest to discover the kind of locations to which the natural enemy is best adapted.

Augmentation of host populations, if needed, may be accomplished by release of host arthropods from laboratory cultures or, for weed agents, by seeding or fertilization (Room & Thomas 1985). Management for other purposes may be underway at release sites, such as burning of grasslands. In such cases, it will be important to discover how such practices might affect agent establishment or persistence (Fellows & Newton 1999). Release sites should not be sprayed with pesticides and should be left unharvested if harvest would destroy the plot. In the case of weed agents, release sites should not be mowed or sprayed with insecticides or herbicides, unless herbicide applications help the agent attack the plant. If the critical habitat is a short-term crop, a series of crop plantings spaced over time can stabilize the availability of the crop (and hence the pest) over a longer period. Secure release sites with minimal uncontrolled public access should be chosen to minimize physical disturbance. Clear agreements describing the site's management should be worked out with the site owner or custodian.

QUALITY OF THE RELEASE

The quality of a biological control release can be affected by: (1) the number of individuals released, (2) their genetic diversity, health, nutrition, and mating status, (3) previous conditioning to the target host, (4) adequate protection during transport, and (5) appropriate choice of the life stage released.

Number released

An agent is most likely to establish if large numbers are released, at many sites, in several sequential years

(Beirne 1975, Memmott et al. 1998, 2005, Grevstad 1999b, Clark et al. 2001). For some species, releasing more insects per site is not better, provided a necessary minimum has been released (Center et al. 2000). Larger release numbers per site, however, may shorten the time it takes for the agent to reach levels causing visible impacts on the target pest (e.g. De Clerck-Floate et al. 2005). In the absence of specific information, release of several hundred individuals per site is probably reasonable. Once specific experience has been gained with a particular natural enemy, a minimum release number per site may become clear.

Fitness and health of stock

The natural enemies used for a release need to be in good health at the time of release, free of pathogen infections, well fed, already mated (if adults are released), and have a broad representation of the genetic characteristics of the original field population from which the culture was sourced.

The genetic fitness of the natural enemies actually released significantly affects the outcome of a release (Hopper et al. 1993). Hufbauer and Roderick (2005) review ways in which microevolution affects success and safety of biological control. Several potential problems exist, most importantly: (1) founder effects, (2) drift, (3) inbreeding depression, and (4) selection for laboratory conditions (Roush 1990a). **Founder effects** refer to failure of the initial collection to include an adequate representation of the genetic variation in the species. Evidence that this has affected biological control outcomes is rare, but certainly might occur. Molecular analyses of haplotype diversity between populations of biological control agents in donor and recipient locations now makes it possible to quantify such founder effects (Hufbauer et al. 2004). **Drift** refers to loss of variation while in culture due to random processes leading to loss of some alleles. This is mainly a concern when colony sizes are very low (<100 individuals). **Inbreeding** and **selection for adaptation to laboratory conditions**, however, are frequent, ongoing events of concern during laboratory rearing of a natural enemy. Genetic deterioration may occur when agents are reared for several generations in the laboratory (as is typically necessary for host specificity testing) (Center et al. 2006) (Figure 19.1).

Laboratory rearing selects for survival in an artificial environment. Inbreeding, while in general undesirable, can be used as a tool to prevent such adaptation. Because isolines maintained as separate rearing colonies have less genetic diversity, they will be less responsive to selection for laboratory conditions. But collectively, a group of such colonies still preserves all the genetic diversity from the original founding colony. An added benefit of many separate rearing lineages is improved disease control, because contamination is likely to be limited to just part of the colony. In general, it is advantageous to release populations into field sites as rapidly as possible, but unfortunately is now rarely possible. To retain diversity in laboratory cultures, they should be as large as possible and should offer as natural an environment as possible, including the necessity to disperse, locate mates, and find the host. Genetic selection may continue after release, as new populations are selected for the environment in the recipient country, possibly leading to improved performance over time (Hopper et al. 1993).

Healthy individuals are essential for successful establishment. Cultures should be maintained with optimal numbers of hosts to promote agent health. Natural enemies should be reared on the most preferred stages of their host to ensure that offspring are not stunted, which can reduce longevity and fecundity. Adults should be offered water and, for many species, a carbohydrate source such as honey before release. Mating before release is very desirable. Large cages and natural light may be needed to stimulate courtship and mating for some species of natural enemies.

Conditioning to target host

Many insects show increased responsiveness to their host after an initial contact with it. Consequently, individuals used for releases should be given opportunities to feed or oviposit on the pest. For many organisms this will happen naturally in the rearing culture. For organisms reared on alternative hosts, exposure to the target host can be arranged in the laboratory before release. Entomopathogens reared in artificial media may lose pathogenicity to the target host, which can be restored if the pathogen is cultured for a generation in the target host immediately before release.

Protection during transportation to release site

During transportation to the release site, agents should be housed in insulated containers to prevent

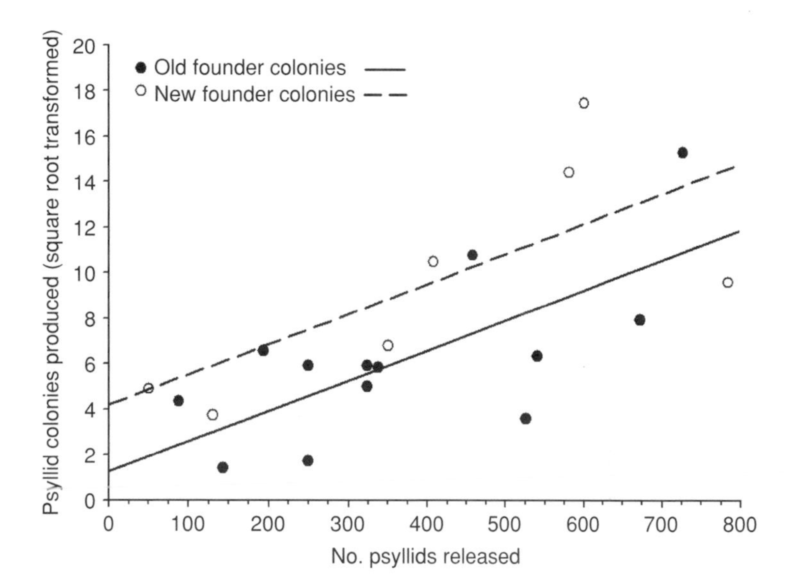

Figure 19.1 Evidence for some loss of genetic quality is seen in the slower rate of population growth of populations of the melaleuca psyllid (*Boreioglycaspis melaleucae* Moore) after their release in Florida, USA, if insects were taken from an older laboratory colony, compared with a recently established one. After Center et al. (2006).

overheating. If transporting or shipping requires more than a few hours, water and possibly food must also be provided. Avoid excessively low humidity during shipment. Releases should take place, if possible, in the early morning or evening to avoid extremes of temperature. Timing open field releases at dusk may inhibit overdispersal of species that are robust fliers. Natural enemies should be released on to plants in sheltered positions. Releases should not be made immediately after rainfall (when foliage is wet) or when storms are threatening.

Choice of life stage used for release

Several life stages may be suitable for release and advantages vary by species (Van Driesche 1993). Adults can immediately attack the pest, but highly mobile species might overdisperse their progeny making it difficult for offspring to find mates upon emergence (Allee effects; see below). Immature stages may be a more durable, abundant product of some mass-rearing programs. However, because of limited mobility and defensive capacity, immature stages are at risk of dying from predation or other causes before they mature and reproduce. For the coccinellid *Chilocorus nigritus* (Fabricius), Hattingh and Samways (1991) found that establishment success was greatest with adult beetles, followed by older larvae, and younger larvae. Release of eggs failed to produce establishment.

For parasitoids, release of laboratory-reared, parasitized hosts, is another option. This approach is particularly valuable for groups with delicate adults such as egg parasitoids; Moorehead and Maltby (1970) describe field release of eggs parasitized by the mymarid *Anaphes flavipes* (Förster). In some cases, it may be possible to collect field-parasitized hosts in sufficient numbers to use for redistribution to new locations, as was the case for larvae of cereal leaf beetle, *Oulema melanopus* (L.), parasitized by *Tetrastichus julis* (Walker) (Dysart et al. 1973). In other projects, parasitized mealybugs, whiteflies, or other pests have been used effectively to redistribute key natural enemies. Care should be taken to evaluate the condition of such collections, however, to verify that diseased individuals or hyperparasitoids are not also being redistributed. Parasitoids may also be released as colonies on plants bearing parasitized hosts. This allows natural enemies to emerge over time, providing a continual inoculation of adults into the environment.

Plant and insect pathogens can be released by dispersing the infective stage on a susceptible stage of the pest. Mechanical blowers, for example, were used to apply spores of the rust fungi *Puccinia chondrillina* Bubak and Sydow onto skeleton weed (*Chondrilla juncea* L.) plants (Watson 1991). For insect targets, pathogens have only occasionally been used as agents of classical biological control. When they are employed, infective stages can be applied directly if the pest occurs as accessible colonies, or in some cases infected hosts may be released to carry the pathogen into the field population. The *Oryctes* virus of coconut beetle [*Oryctes rhinoceros* (L.)], for example, was inoculated into field populations in western Samoa by feeding virus solutions to adult beetles, which then were liberated in the field. Infected beetles brought the virus into contact with larvae at the communal breeding sites in rotting palm trunks, where oviposition by many females occurs (Waterhouse & Norris 1987).

CAGING OR OTHER RELEASE METHODS

Arthropods may either be released into cages (Figure 19.2) or liberated freely into the environment (Figure 19.3). Depending on the details of the biology of the particular natural enemy, either approach may have some advantages. Cages have the advantage of preventing excessive rapid dispersal of the released individuals, and providing temporary protection from predators. Failure to form a breeding population due to overdispersal is a concern in some species and this has been termed the **Allee effect** (Allee et al. 1949, Hopper & Roush 1993). See Taylor and Hastings (2005) for a summary of literature on how this process influences biological invasions.

Hosts within cages also provide a sampling point that can be checked later to see whether the natural enemies reproduced. When cages are used, they should be large enough to enclose an excess number of hosts

Figure 19.2 Use of cages for establishment of the dipteran weed biological control agent *Hydrellia pakistanae* Deonier, released in Florida against the invasive aquatic plant *Hydrilla verticillata* (L. f.) Royle. Photograph courtesy of Ted Center, USDA-ARS.

Figure 19.3 Free release of the mymarid *Gonatocerus ashmeadi* Girault against the glassy-winged sharpshooter, *Homalodisca coagulata* Say, in Tahiti. Photograph courtesy of Julie Grandgirard and Jerome Petit.

(with respect to the potential reproduction of the females placed in the cage). Cages should be able to withstand wind, rain, inquisitive animals, or other conditions that are likely to be present at the release site. Cages usually should be removed a few days after the individuals have been released to prevent over-exploitation of the resource and free any surviving agents. Cages are used in weed agent releases in a very similar manner to that for parasitoids (Briese et al. 1996). For example, cages were used to attempt to obtain establishment of *Spodoptera pectinicornis* Hampson (Lepidoptera: Noctuidae), released in the USA against waterlettuce (*Pistia stratiotes* L.) after free releases had failed, likely because of predation and overdispersal. Very large cages can also be used to establish field insectaries from which agents can be conveniently harvested for further release.

When open releases are used, released insects should be placed where there are adequate populations of the target, in a stage susceptible to attack and when weather conditions are favorable. If the natural enemy is a predator or herbivore able to feed on the target in a variety of ages or life stages, timing is less likely to be an impediment to success. In contrast, for parasitoids, or herbivorous insects such as seed-head feeders that attack only a specific plant stage, releases must be timed more carefully so as to coincide with the necessary host or plant stages. In general, correct timing of releases can be best assured by directly sampling the host population to confirm the presence of suitable stages, which requires a clear understanding of what stages are preferred for attack by the agent.

When releases must be made over large areas, mechanical release systems may be helpful. In areas with limited access by road, airplanes can be used to drop release packages designed so that natural enemies escape successfully after impact. *Apoanagyrus* (formerly given as *Epidinocarsis*) *lopezi* (De Santis), for example, was released against cassava mealybug in roadless parts of tropical Africa from airplanes by dropping vials containing adult wasps, which were able to escape after vials reached the ground (Herren et al. 1987).

PERSISTENCE AND CONFIRMATION

The colonization of an agent may require repeated attempts, with variations on the approaches used, before successful establishment is achieved. Provisions should be made to make many releases and to repeat releases if necessary. Creativity should be used to explore the best colonization methods for the species at hand. After an effective method for establishing a particular species has been discovered, establishment at other locations can be achieved by repeating the previously successful method.

After releases of a natural enemy have been made, follow-up surveys are needed to detect its reproduction, spread, and impact. Sampling can be done using several approaches. If no other similar natural enemy is present in the system, as may be the case for the first natural enemy introduced into a pest population, simple visual inspection in the field (or examination of specimens reared in the laboratory from samples collected at the release site) may be sufficient to confirm establishment. Adults reared from samples or directly in the field can then be compared to voucher specimens to confirm identification, with assistance from an appropriate taxonomist.

Molecular tools can help confirm establishment, particularly if: (1) the agent is very similar to other species that occur on the same host in the region or (2) if detections are based on diapausing immature stages (such as parasitoid larvae) that would require long rearing periods before adults suitable for indentification could be obtained. Recoveries of *Peristenus* braconids released against *Lygus* plant bugs in the USA were assessed with molecular markers to avoid the necessity of a 10-month rearing process, during which many hosts in the sample often were lost to other causes (Erlandson et al. 2003, Ashfaq et al. 2004).

A newly released species can be tentatively considered established if it is detected over a period of at least 2 years. However, lack of detection in this time period is not conclusive evidence of failure, as in some cases the first recoveries of a released agent may not occur for several years. Only after concerted efforts at establishment in all available environments have failed should the conclusion be drawn that a species is not likely to establish in a particular region. Much can be learned by investigating factors impeding establishment in the field so these mistakes are not repeated with other agents.

Chapter 20

NATURAL ENEMY EVALUATION

Evaluating outcomes is important to all biological control programs. For augmentative and biopesticidal biocontrol, measurement of change in the pest's density or biomass after application may be all that is necessary. For conservation biological control, measurement of natural enemy/pest ratios during the crop cycle may be important to guide integrated pest management (IPM) decision-making. For classical biological control programs, evaluation efforts are needed to measure changes in the target pest's abundance or biomass, determine population-level mechanisms behind such changes, and monitor non-target species to look for unwanted impacts.

In this chapter we discuss: (1) natural enemy surveys in crops, (2) pre-release surveys of natural enemies in a recipient country, (3) post-release surveys to detect establishment and spread of new agents, (4) survey of non-target species to detect potential harmful impacts, (5) measurement of population impacts on the target pest, (6) separation of component mortalities of a natural enemy complex, and (7) economic evaluation of classical biocontrol programs.

NATURAL ENEMY SURVEYS IN CROPS

To use information about natural enemies in IPM crop protection systems, farmers or their IPM consultants must: (1) know which natural enemy species significantly affect the key pests of the crop, (2) have reliable sampling methods to measure their abundance, and (3) have models or tools that predict short-term impacts of the natural enemies on pest densities.

Identifying key natural enemies in a crop

To effectively manipulate or conserve natural enemies in a crop, pest managers must know which species really matter. Identifying key natural enemies begins with surveys (e.g. cassava in South America, Bellotti et al. 1987; apples in the northeastern USA, Maier 1994; maize in east Africa, van den Berg 1993; bananas in Indonesia, Abera et al. 2006). Surveys done when pests are at low densities may be more indicative of key species than surveys of outbreaking populations, which may attract additional species not involved in outbreak prevention. A variety of collection methods (pitfall traps, sweep netting, leaf sampling, etc.) can be used to capture predaceous arthropods in crops or collect pests from which to rear parasitoids. Potential predators found in surveys can be confirmed as actual predators either by: (1) directly observing predation in the field, (2) offering the pest to putative predators in laboratory tests (being alert for false-positive results due to the artificiality of the cage), or (3) detecting markers from the pest in field-collected predators using such techniques as enzyme-linked immunosorbent assay (ELISA; antigen–antibody) or DNA markers (see Chen et al. 2000, Hoogendoorn & Heimpel 2001, Harwood et al. 2004, and Chapter 15).

Measuring natural enemy abundance

To use information about natural enemies in IPM decision-making, the densities of the key species will have to be measured and correlated to the current pest density in the crop. As with all sampling efforts, consideration needs to be given to what level of sample precision is required and what approach will give the desired rates with minimal sampling effort (e.g. Gyenge et al. 1997). A common approach is to directly count predators or parasitized hosts and use this information to calculate predator/prey ratios or percentage parasitism values. Predator/prey ratios are commonly used to monitor biological control of pest mites in apples, grapes, and strawberries (Pasqualini & Malavolta 1985, Nyrop 1988). Similarly, the ratio of parasitized to unparasitized *Helicoverpa* moth eggs has been used to monitor pest pressure in processing tomatoes (Hoffmann et al. 1991). Coccinellid beetle density can be monitored using sweep netting or timed visual searches (Elliot et al. 1991).

Traps can also be used to monitor the density of some natural enemies. *Aphytis* spp. parasitoids of red scale, *Aonidiella aurantii* (Maskell), in South Africa, for example, can be monitored either with traps baited with scale pheromone or with visual (yellow) traps (Samways 1988, Grout & Richards 1991b). Parasitoids of fruit flies have been monitored by placing fruit in sticky-coated wire cages (Nishida & Napompeth 1974). Aggregation and sex pheromones of natural enemies can also be used as lures (Lewis et al. 1971).

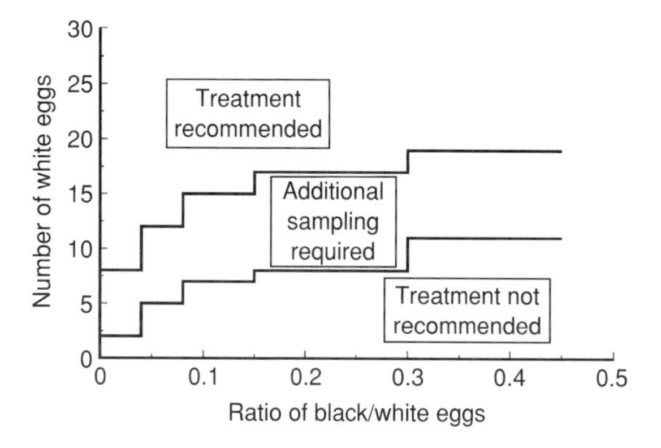

Figure 20.1 A sequential sampling plan applied to ratios of parasitized (black) and unparasitized (white) *Helicoverpa zea* (Boddie) eggs in processing tomatoes, to determine whether biological control is being effective or if pesticides need to be applied. After Hoffmann et al. (1991), reprinted from Van Driesche and Bellows (1996), with permission from Kluwer.

Forecasting pest suppression by natural enemies

To change IPM decisions based on measurements of natural enemy abundance requires the ability to forecast the impact of the natural enemy on pest density. One approach is simply to change the current estimate of pest density by recognizing that any pests that are parasitized or infected should be deleted from the current count. This modification is justified if the sampled stage is not the stage that directly causes the damage that has to be suppressed. For example, the tomato fruitworm, *Helicoverpa zea* (Boddie), is counted in the egg stage, but the damaging stage is the larva; therefore, any parasitized eggs should not be included in estimates of pest numbers. Similarly, counts in peppermint (*Mentha piperita* L.) of the variegated cutworm, *Peridroma saucia* (Hübner), can be used to modify the spray threshold for this species in Oregon, USA, because parasitism occurs in early instars, but damage is largely due to feeding of older larvae (Coop & Berry 1986).

More generally, counts of natural enemies or natural enemy/pest ratios can be used to modify projections of pest population growth (Nyrop 1988). For example, in vineyards in the Crimea, Ukraine, ratios of one predator mite [*Metaseiulus occidentalis* (Nesbitt)] per 25 phytophagous mites [*Eotetranychus pruni* (Oudemans)] were associated with phytophagous mite populations that did not increase to economically damaging levels (Gaponyuk & Asriev 1986). Similarly, sticky-trap catches of adult leafminers (*Liriomyza trifolii* Burgess and *Liriomyza sativae* Blanchard) and their parasitoids in watermelon (*Citrullus vulgaris* Schrader) in Hawaii, USA, enabled a 3-week forecast of future mine numbers (Robin & Mitchell 1987). Ratios of black (parasitized) and white (presumed healthy) eggs of *H. zea*, together with counts of white eggs per leaf, are used in processing tomatoes in California, USA, in a sequential sampling process to make decisions about the need to apply pesticides (Hoffmann et al. 1991) (Figure 20.1).

PRE-RELEASE SURVEYS IN THE NATIVE RANGE FOR CLASSICAL BIOLOGICAL CONTROL

The natural enemies that already exist in the area of intended introduction must be known before introducing new species against an invasive pest to ensure that newly introduced species are needed and recognizable. Methods used to collect natural enemies in surveys vary with the kind of natural enemy. Survey methods for parasitoids and predators are the same as for conducting surveys in crops, as discussed above. Herbivorous insects or mites in weed biological control pre-project surveys are most often found by visually searching plants at many locations throughout the growing season. This typically involves a careful search of all the parts of the plants, both externally and internally (for stem borers or leafminers), including excavation of the root system to search for root feeders or root borers. Laboratory feeding tests can confirm feeding and development by a given species on the target plant. Pathogens may be cultured from diseased hosts, infected plant tissue, or soil samples with appropriate media. Nematodes can be recovered from either bodies of infected hosts or soil samples, incubating them with larvae of the greater wax moth, *Galleria mellonella* (L.), to obtain a new infection.

Workers preparing for introductions of insects to attack invasive thistles in California for example, conducted extensive surveys to document native or self-introduced agents already associated with thistles in the state (Goeden & Ricker 1968, Goeden 1971, 1974). Other examples of such surveys include work in Florida, USA, to document herbivores associated with melaleuca trees (Costello et al. 2003) and surveys in South Africa to document native herbivores associated with various species of *Solanum* (Olckers & Hulley 1995). Pre-project surveys must be thorough, covering enough dates and locations to provide a full accounting of the species associated with the target pest. This body of information and preserved specimens (for DNA-marker work if that should prove necessary) is later compared to lists of candidate natural enemies that may be discovered during natural enemy surveys in the pest's native range.

Pre-project survey data are used for each candidate natural enemy to identify the most similar species already present. Means must then be devised to separate any species pairs similar enough to cause confusion, by either direct morphological comparisons or molecular analyses (Chen et al. 2002, Greenstone et al. 2005; see Chapter 15).

POST-RELEASE SURVEYS TO DETECT ESTABLISHMENT AND SPREAD OF NEW AGENTS

Methods for obtaining establishment of newly released natural enemies are discussed in Chapter 19. Surveys are required after releases are made to see whether the agent has persisted, established, and spread. The usual approach for parasitoids is to collect samples of the target host, which are then reared, dissected, or subjected to allozyme (Greenstone 2006) or DNA (Prinsloo et al. 2002) analysis. Foliage of the crop may be searched or collection methods such as sweep netting, beat sheeting, or trapping can be used to detect predators visually. Sampling for adult herbivorous insects is similar to that for predators. Damage alone may be the basis for detection if the feeding of a weed control agent is distinctive, as was the case with the weevil *Mogulones cruciger* Herbst attacking houndstongue (*Cynoglossum officinale* L.). Damage may serve to focus intensified efforts on plants where the insect is most likely to occur (De Clerck-Floate et al. 2005). Immature stages of internal plant feeders (stem borers, root borers, or seed-head feeders) are detected by dissection of plant parts or holding plant parts such as seed pods or galls for insect emergence. Clear glass vials attached to darkened chambers containing the plant material can trap emerging insects. Negative data do not necessarily mean that an introduction has failed.

An agent's range expansion can be measured after establishment to estimate rate of spread and habitat preferences, and whether these are affected by the agent's biology or habitat variables. Such sampling can also show how the agent is interacting with other species, especially other biological control agents of the same pest. Spread of a new agent can be assessed by sampling the host, using the techniques already mentioned, at increasing distances from a release point. For agents released against crop pests, non-crop host plants may also need to be sampled, if they exist. In some cases, traps can be used to measure the spread of a natural enemy. Sticky traps, for example, were used in Florida to monitor the distribution and abundance of *Encarsia opulenta* (Silvestri) and *Amitus hesperidum* Silvestri, parasitoids of the citrus blackfly, *Aleurocanthus woglumi* Ashby, following their introduction (Nguyen et al. 1983).

The rate of natural spread of an insect can be affected by the vegetation through which it must move. Two *Aphthona* chrysomelid species released against leafy spurge (*Euphorbia esula* L.) were studied in Alberta, Canada (Jonsen et al. 2001) and movement of one species was found to be faster through grass-dominated stands than in shrubby vegetation. Dispersal studies can also help predict how fast an infested region will be colonized. Studies in Florida with the weevil *Oxyops vitiosa* (Pascoe) predicted that from 135 release locations, the weevil would reach half of all melaleuca stands by June 2008, taking into account the effect of stand fragmentation on rate of dispersal (Pratt et al. 2003b). Such information can be used to adjust release plans. Dispersal surveys can also reveal an agent's interactions with other natural enemies. Surveys in

Virginia of spotted knapweed (*Centaurea maculosa* Lamarck) gall flies (*Urophora affinis* [Frauenfeld] and *Urophora quadrifasciata* [Meigen]) showed that although *U. quadrifasciata* arrived later, it quickly exceeded *U. affinis* in importance, suggesting it was the more effective species (Mays & Kok 2003).

In addition to natural spread, natural enemies may be moved unintentionally by human activity, such as movement of infested plants. This rate may be much greater than natural dispersal rates and may dominate the pattern of natural enemy spread for some species.

POST-RELEASE MONITORING FOR NON-TARGET IMPACTS

All classical biological control projects should assess possible impacts of newly released agents on non-target species (Barratt et al. 2006). Pre-release studies typically identify which local species, if any, might be marginally at risk of such attack and checking those species is the logical approach to post-release survey work. Day (2005), for example, assessed parasitism rates in non-target mirids in the northeastern USA of two introduced European euphorine braconids, *Peristenus digoneutis* Loan [released against *Lygus lineolaris* (Palisot), a native species] and *Peristenus conradi* Marsh [released against *Adelphocoris lineolatus* (Goeze), an invasive species]. He did so while also assessing these parasitoids' impacts on their target pests and found that there was no impact on the non-target invasive mirid *Leptopterna dolabrata* (L.), nor was *Peristenus pallipes* (Curtis), a native parasitoid of the target pest (*L. lineolaris*), eliminated.

Similar non-target impact surveys may be run in support of augmentative biological control releases. In Switzerland, consequences of releases of the egg parasitoid *Trichogramma brassicae* Bezdenko for control of European corn borer (*Ostrinia nubilalis* Hübner) in maize were assessed to see whether *T. brassicae* might affect the alternate hosts of the corn borer's native tachinid parasitoid (*Lydella thompsoni* Herting). Surveys in the habitats of the two principal alternative hosts (the noctuid *Archanara geminipuncta* Haworth and the crambid *Chilo phragmitellus* Hübner) showed, however, that the eggs of these species were either hidden or not attractive and thus were not parasitized by *T. brassicae* under field conditions (Kuske et al. 2004).

Non-target impacts of introduced pathogens can be assessed by direct host surveys during epizootics of the introduced pathogen. Hajek et al. (1996) showed that the exotic fungal pathogen *Entomophaga maimaiga* Humber, Shimazu & R. S. Soper of gypsy moth [*Lymantria dispar* (L.)] infected only two non-target caterpillars of 1511 specimens collected during epizootic field conditions. These two specimens, one each of two species, represented less than 1% of the total number collected of either species.

Potential effects of new weed biological control agents can be assessed by post-release surveys or experiments, focusing on any native or otherwise important species fed on during host range testing. For the saltcedar chrysomelid *Diorhabda elongata* Brulle, possible attack on plants in the genus *Frankenia* was assessed in the southwestern USA by planting *Frankenia* at sites where *D. elongata* populations were defoliating saltcedar (Dudley & Kazmer 2005). Surveys were conducted in South Africa, on non-target *Solanum* plants adjacent to sites where the lacebug *Gargaphia decoris* Drake was numerous on its target plants, *Solanum mauritianum* Scopoli and *Cestrum intermedium* Sendt. Only negligible feeding was found on any of two native or three exotic *Solanum* species (Olckers & Lotter 2004). Paynter et al. (2004) found that 16 of 20 weed biological control agents released in New Zealand were host-specific under field conditions, while two others attacked native plants to a minor degree and the last two attacked exotic plants. Center et al. (2007) found no effects by the melaleuca psyllid on 18 non-target species predicted before release to be suboptimal or non-hosts during laboratory host-range testing.

Separate from the above use of non-target impact surveys within classical biological control projects, independent retrospective surveys of the non-target impacts of biological control agents have also been done (e.g. Nafus 1993, Johnson & Stiling 1996, Duan et al. 1997, 1998, Louda 1998, Boettner et al. 2000, Barron et al. 2003, Benson et al. 2003, Johnson et al. 2005, among others; see Chapter 16 for discussion of this topic).

MEASUREMENT OF IMPACTS ON THE PEST

Classical biological control, in contrast to pesticides, is intended to have a prolonged, often permanent, effect on pest populations. Quantitative evaluation of these effects is an essential part of the process that provides guidance as to what worked, wholly or partially, and what did not. Evaluation of weed and insect biological control projects differ in important ways.

Agent impacts for insect targets typically involve increased pest mortality, or, in a few cases, decreased fecundity (e.g. Van Driesche & Gyrisco 1979), leading to a lower pest density, which remains stable over time. Three questions are of interest for insect and mites targets: (1) whether natural enemies are reducing pest/host populations, (2) if so, how this is being done, and (3) what the individual contribution of each natural enemy is to the overall mortality that can be measured in a pest's population.

The influences of natural enemies are much more varied for weed targets. Rarely is outright plant death the immediate outcome. Rather, natural enemies reduce plant performance by affecting growth and reproduction, reducing photosynthesis, disrupting water or nutrient conduction, nutrient storage, worsening effects of competition with other plants, or lowering tolerance to abiotic stresses such as drought or low soil fertility. Cumulatively these impacts lower the standing biomass, reduce coverage (the area of soil or water occupied by the plant), lower the output of seeds or other reproductive parts, deplete the seed bank, and lower the ability of defoliated plants to regrow or compete with other plants, etc.

Different parameters are therefore measured in weed compared with insect biological control and different conceptual frameworks are required to integrate and model population-level consequences of natural enemy impacts. Hence, in the following discussion, we consider insect and plant biological control impact evaluations separately.

Evaluating parasitoids and predators for arthropod biocontrol

The most direct means to assess the impact of natural enemies on the density of an insect or mite population is the use of the manipulative (or experimental) approach pioneered by Paul DeBach (see Luck et al. 1988, 1999 for reviews). This method is based on the direct comparison of pest density and natural enemy abundance in populations of the pest with and without the natural enemy of interest.

The use of this method requires the ability to find, establish, or create plots both with and without the natural enemy to be evaluated. This is done by manipulating the time and place of release of a newly introduced natural enemy or by using cages or selective insecticides to exclude an already widely established natural enemy from some areas. In classical biological control projects, this can be done either: (1) by comparing pest densities before releasing the agent to the densities after the agent has established (**before-and-after design**; **temporal design**) or (2) by comparing pest density in control and release plots that are spatially separated (**spatial design**). When the natural enemy of interest is already widespread, plots with and without the natural enemy can be created by exclusion, either with pesticides (**insecticide check method**) or with cages that protect a small pest population from attack by the widespread natural enemy (**cage-exclusion design**).

When the necessary plots with and without the natural enemy for the experimental evaluation approach cannot be created, **life tables** may be constructed and used to form inferences about the importance of particular natural enemies (Varley & Gradwell 1970, 1971, Manly 1977, 1989, Bellows et al. 1992b, Bellows & Van Driesche 1999). Life tables allow mortality from one natural enemy to be compared with other sources of mortality acting on the pest, and they allow the contribution to population regulation by a given natural enemy to be assessed. To construct life tables for insects, estimates must be obtained of the numbers entering each stage in the pest's life cycle, the fertility of the adult stage, and the numbers dying in each stage from specific sources of mortality, including the agent, to be evaluated. Typically data are needed for a series of generations.

Approach 1: field experiments for evaluating insect biological control

Before-and-after design If the introduction of a new natural enemy has not yet occurred, plots can be established and sampled for several pest generations to generate pre-release baseline data on pest density (the before estimate). These density values can later be compared to the pest density and survival rates in the plots after the natural enemy has been released, established, and given time to increase in number. This approach was used, for example, by Gould et al. (1992a, 1992b) to evaluate the effect of the parasitoid *Encarsia inaron* (Walker) on the ash whitefly, *Siphoninus phillyreae* (Halliday), in California, and by Borgemeister et al. (1997) to assess the impact of an introduced histerid predator beetle on a pest of stored grain in West

Africa. Because some plots will inevitably be lost over the course of such a multi-year evaluation, a fairly large number of plots (eight to 10 per treatment) should be established. This design works best with sedentary insects that have many generations each year, because differences develop faster and local densities are largely influenced by local processes. For pests such as Lepidoptera whose adults disperse long distances, local densities of immature stages in plots may be strongly influenced from year to year by adult movement, making it more difficult to measure natural enemy impact.

Spatial design When time is not available to collect pre-release data at a series of sites, the plots with and without the natural enemy can be created by establishing a set of study sites, releasing the new agent at some (chosen at random), and reserving the others as no-release controls (e.g. Van Driesche & Gyrisco 1979, Van Driesche et al. 1998b, Morrison & Porter 2005). Because successful natural enemies spread, some control sites may be invaded by the natural enemy (Van Driesche et al. 1998b, Morrison & Porter 2005). To compensate for this, control sites should be located as far as possible from release sites without changing basic geographic, climatic, or ecological conditions. There is no clear answer to the question of how far is far enough between control and release sites because the dispersal powers of a new natural enemy are unlikely to be known initially. Between 5 and 15 km is a reasonable figure, but longer distances may be desirable in some cases. For, example, Morrison and Porter (2005), in setting up plots to evaluate the phorid fly *Pseudacteon tricuspis* Borgmeier (attacking fire ants), originally positioned the control plots about 20 km from the release plots on the assumption that natural enemy spread would not exceed 3–4 km/year. In fact, spread exceeded 15–30 km/year and these sites were invaded before enough time had elapsed for the pest density to be affected. New control plots were established 70 km from release plots in an effort to maintain parasitoid-free sites for the test.

Exclusion design Cages have been employed extensively to evaluate resident natural enemies by excluding them from plots, plants, or plant parts infested with the pest (DeBach et al. 1976). Tests can last for several generations for pests such as scales, whiteflies, and aphids that are able to complete their whole life cycle inside small cages. For larger or more mobile species, studies may be more limited in scope, evaluating patterns of mortality among cohorts of immatures of a single generation, as was done for tree borers by Mendel et al. (1984).

The classic design for exclusion tests, pioneered by DeBach, consists of three treatments: a closed-cage, an open-cage, and a no-cage treatment derived by sampling the unmanipulated population (DeBach & Huffaker 1971, Knutson & Gilstrap 1989). The open cage is intended to have the same microclimate as the closed cage but still allow natural enemies to reach the pest. When pest densities and survival are similar between the open cages and the uncaged treatments, this suggests that there are no important cage effects. Differences between the closed-cage and open-cage treatments can then be considered to reflect the effect of the natural enemies. DeBach & Huffaker (1971) used open and closed leaf cages to measure the effect of parasitoids (*Aphytis* spp.) on California red scale, *A. aurantii*, on leaves of English ivy, *Hedera helix* L. Large cages were used by Neuenschwander et al. (1986) to evaluate the effect of *Apoanagyrus lopezi* (De Santis) on the cassava mealybug, *Phenacoccus manihoti* Matile-Ferrero (Figure 20.2). The plant area enclosed can vary from parts of leaves enclosed by leaf cages of 1–2 cm in diameter (Chandler et al. 1988), to sleeve cages on tree branches (Prasad 1989), or bucket cages over clumps of cereal crop plants (Rice & Wilde 1988), to field cages (1–3 m on an edge) over patches of crops such as alfalfa (Frazer et al. 1981, O'Neil & Stimac 1988a) or whole trees (Faeth & Simberloff 1981, Campbell & Torgersen 1983).

Potential cage effects include increased temperature (leading to faster pest developmental rates), restriction of pest and/or agent movement (leading to a local concentration inside the cage), and higher humidities (potentially raising rates of fungal disease). Temperatures and humidity inside cages should be monitored using small recording devices and compared to outside temperatures. Cage designs or locations should be modified if differences are found. Restricted movement of neonates is of concern only in multi-generation studies of sedentary forms such as scales. The importance of this issue needs to be determined for each species of interest by comparisons between open and closed systems. Potential enhancement of humidity in cages can be measured directly with data-logging equipment. Another important consideration is the thorough removal of all natural enemies from the closed-cage treatment. If possible, cages should be installed over

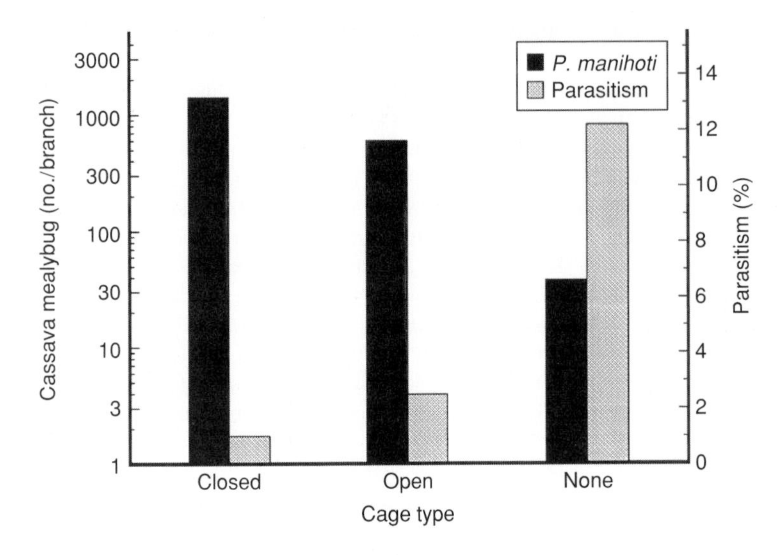

Figure 20.2 Numbers of the cassava mealybug, *Phenacoccus manihoti* Matille-Ferrero, and percentage parasitism by *Apoanagyrus lopezi* (De Santis) on branches of cassava covered with a closed sleeve cage or an open cage, or not covered with a cage. After Neuenschwander et al. (1986), reprinted from Van Driesche and Bellows (1996), with permission from Kluwer.

patches of pests when they are free of natural enemies. Otherwise it may be necessary to remove existing natural enemies from cages through trapping or the application of a short-residual insecticide.

Cages can also be used to partition the impacts of a complex of natural enemies into those attributable to individual species or groups by using cages that selectively exclude one group. Rice and Wilde (1988), for example, in a study of greenbug, *Schizaphis graminum* (Rondani), on wheat and sorghum used two mesh sizes, a small mesh that excluded both predators and parasitoids and a large mesh that only excluded the larger predators, allowing their separate impacts to be measured.

The use of barriers provides another approach for creation of natural enemy-free pest populations. For ground-dwelling predators that do not readily disperse through flight, strips of plastic or metal flashing can be used to isolate patches of a crop, such as wheat or alfalfa, from the rest of a larger field. These patches can then be cleared of the ground-dwelling natural enemies (such as carabids) by using a short-residual pesticide or through intensive trapping. This method has been used successfully to measure the effect of ground-dwelling predators on cereal aphids in spring barley in Sweden (Chiverton 1987) and on wheat in the UK (Winder 1990).

Insecticide check design Another way to exclude a natural enemy from a plot is to apply a pesticide that is toxic to the natural enemy and yet relatively harmless to the pest. Such sprayed plots can then be compared to nearby unsprayed plots that contain the natural enemy. The validity of this design depends on three conditions being met: (1) the chemical must be sufficiently toxic to the key natural enemies so that they are greatly reduced, (2) the chemical must cause very little injury to the pest, and (3) the chemical must not change the pest's fecundity, either directly or indirectly by inducing chemical changes in the pest's host plant. Furthermore, because entire natural enemy complexes are likely to be reduced by the chemicals applied, the impacts observed in the untreated plot must be attributed to the total complex that is excluded. The insecticide check method allows relatively large areas to be used as exclusion plots and allows study of groups that do not complete their life cycles well in small cages (Brown & Goyer 1982, Stam & Elmosa 1990). Also, dispersal of winged adults, scale crawlers, or other life stages is not restricted, as would occur in cages.

Selective chemicals have been used to conduct natural enemy exclusion experiments with scales (DeBach & Huffaker 1971), mites (Braun et al. 1989), aphids (Milne & Bishop 1987), thrips (Tanigoshi et al. 1985), and mealybugs (Neuenschwander et al. 1986) (Figure 20.3).

Laboratory tests must be conducted to validate each of the method's key assumptions when conducting natural enemy exclusion experiments based on the use of selective pesticides, laboratory tests must be conducted to validate each of the method's key assumptions. First, the relative toxicity of a series of potentially selective

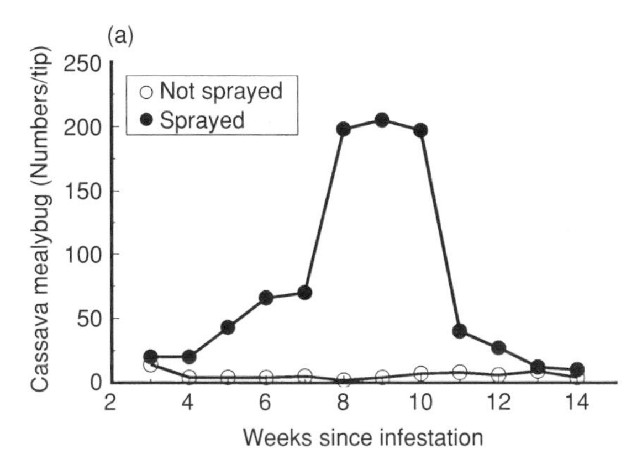

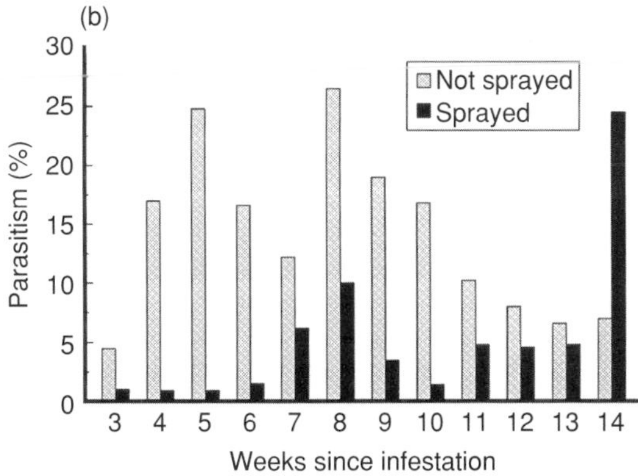

Figure 20.3 Selective chemicals may be used to demonstrate the effect of natural enemies. Densities (a) of cassava mealybug, *Phenacoccus manihoti* Matille-Ferrero, are higher and percentage parasitism (b) by *Apoanagyrus lopezi* (De Santis) is lower in sprayed compared with unsprayed plantings of cassava. After Neuenschwander et al. (1986), reprinted from Van Driesche and Bellows (1996), with permission from Kluwer.

pesticides must be measured for both the pest and the natural enemies to be excluded. From these results, a chemical must be identified that is low in toxicity to all stages of the pest and highly toxic to at least one life stage of the natural enemy (e.g. Braun et al. 1987a). The selective pesticides must then be further screened to see if they affect the pest's fecundity. Pesticide stimulation of pest fecundity has been noted for one or more chemicals for aphids (Lowery & Sears 1986), thrips (Morse & Zareh 1991), spider mites (Boykin & Campbell 1982), and planthoppers (Chelliah et al. 1980). Braun et al.'s (1987b) study of permethrin for use in cassava to suppress phytoseiids attacking the phytophagous mite *Mononychellus progresivus* Doreste in Colombia provides a good example of how to organize a program of laboratory tests to identify selective pesticides for use in natural enemy exclusion experiments. Finally, field data on the natural enemy's density in the sprayed plot must be collected to demonstrate that the technique was effective.

Approach 2: life tables for evaluating impacts of natural enemies of arthropods

Life tables can be used to organize information about mortality affecting an arthropod population. They are useful when the impact of natural enemies can't be directly measured or the importance of several sources of mortality needs to be compared. Insect life tables are divided into rows that correspond to life stages (such as eggs, small larvae, large larvae, pupae, adults) and into columns that summarize numbers entering (l_x) and dying (d_x) in each stage (summed over the whole generation), with causes of mortality separated as fully as possible (see Table 20.1 for sample life table).

Each life table reflects a single discrete generation of the pest. For species with overlapping generations, life tables can be constructed to summarize events for some time step. When information on the pest's fecundity is available, the effect of a natural enemy can be expressed in terms of its effect on the pest's population growth rate (R_0). Mortality rates can be analyzed to see whether particular factors are density-dependent (see Chapter 10) when a series of life tables is available (constructed for populations separated either spatially or temporally). The following discussion of life tables is divided into: (1) concepts and terms, (2) collecting data to build life tables, and (3) inferences from life tables.

Concepts and terms Understanding life tables requires understanding: (1) density compared with total numbers entering a stage, (2) recruitment to and from a stage, (3) apparent mortality, *k*-values and marginal attack rates, (4) population growth rate, and (5) key-factor analysis.

1. Total number entering a stage compared with density. The number of animals present on individual sample dates is termed **sample density** and this is the most commonly collected type of population data. Density is the net balance for the generation of all gains to the stage minus all losses through emigration, death,

Table 20.1 A sample life table, using data for *Pieris rapae* (L.) (after Van Driesche and Bellows, 1988)

Stage	Factor	Stage		Factor d_x	Marginal attack rate*	Apparent mortality		Real mortality		*k*-value factor
		l_x	d_x			Stage q_x	Factor q_x	Stage d_x/l_o	Factor d_x/l_o	
Egg		10.6690	0.1280			0.0120		0.0120		
	Infertility†			0.1280	0.0120		0.0120		0.0120	0.0052
Larvae		10.5410	10.5139			0.9974		0.9855		
	C. glomerata			4.6607	0.8675		0.4422		0.4368	0.8777
	Predation			5.8532	0.9806		0.5553		0.5486	1.7122
Pupae		0.0271	0.0084			0.3100		0.9855		
	Predation			0.0084	0.3100		0.3100		0.0008	0.1612
Adults		0.0187								
Fertility†		356.0								
Sex ratio†		0.5								
F_1 progeny		3.3285								
R_o (F_1/P_1)		0.3120								

For definitions of column headings, see Bellows *et al.* (1992b).
*Marginal rates are calculated by using equations in Elkinton *et al.* (1992) for the case of a parasitoid and a predator where the predator is always credited for the death of a host attacked by both agents, i.e. where $c = 1$.
†Values from Norris (1935).
Reprinted from Van Driesche and Bellows (1996).

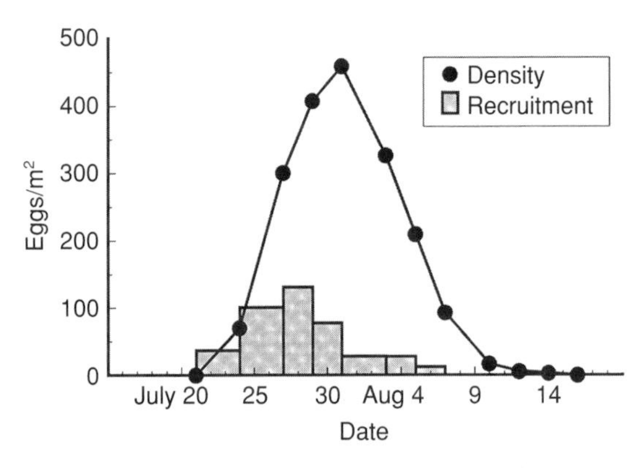

Figure 20.4 Density and recruitment measure different things. Life tables call for total recruitment per generation in their l_x columns, not the density on any particular date. Here density and daily recruitment of Colorado potato beetle [*Leptinotarsa decemlineata* (Say)] eggs are compared for one generation. After Van Driesche et al. (1989), reprinted from Van Driesche and Bellows (1996), with permission from Kluwer.

or molting up to the sample date. Life tables constructed from density data utilize samples collected for each stage of the pest from when it first appears until it is no longer present. Density data, being normally distributed, form bell-shaped curves (Figure 20.4), but neither the peak value nor the sum of all sample values equals the l_x value called for in life tables (Van Driesche 1983). Density data must be analyzed by **stage frequency analysis** methods to obtain an estimate of l_x (see below).

2. Rates of recruitment and loss. Because density data are complex parameters generated by multiple processes of gain and loss, more reliable estimates of l_x or d_x can be obtained by directly measuring the number of animals entering or dying in a particular stage over short periods of time. When such measures are repeated over the whole period over which gains or losses are occurring, total gain (termed **recruitment**) or loss can be summed directly from the raw data without complex analyses that invoke unrealistic assumptions (as in stage frequency analyses). See Figure 20.5 for a conceptual model of recruitment and loss-based sampling for life-table construction.

Recruitment occurs during each life stage. Daily per capita oviposition multiplied by the density of live ovipositing females, for example, would be the daily recruitment for an egg population. Analogous recruitment to each subsequent stage would occur due to development and molting. Daily predation (as animals

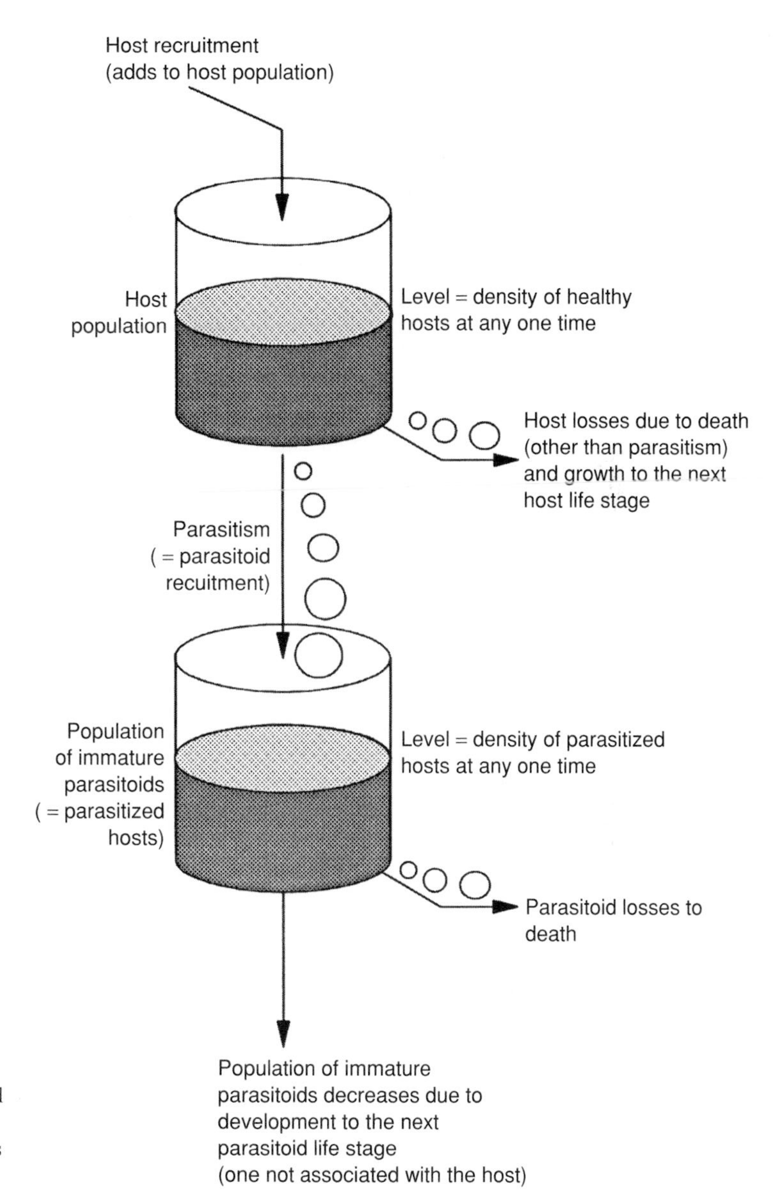

Figure 20.5 A conceptual model of recruitment to a host population, together with losses, resulting in the moment-to-moment host density, with linkage to the population of parasitized hosts (whose density is similarly determined by recruitments, as parasitoid ovipositions in hosts, and by losses). Reprinted from Van Driesche and Bellows (1996), with permission from Kluwer.

eaten per sample unit, not percentage eaten), summed over the whole period during which a stage is available to be attacked estimates d_x for predation for the stage sampled.

3. Apparent mortality, *k*-values, and marginal attack rates. The usual way to estimate mortality rates within a life table is d_x/l_x, where d_x is the number of animals observed to die within a given stage from a particular mortality factor and l_x is the number that entered the stage over the course of the generation. This is called **apparent mortality**, which may also be expressed for computational convenience as ***k*-values**, where $k = -\log(1 - \text{apparent mortality})$. *k*-Values are additive, such that the *K*-value for total mortality is given by $K = k_1 + k_2 + \ldots k_n$, where each component is the value for a particular source of mortality.

When only one source of mortality affects a stage, apparent mortality provides an unambiguous estimate.

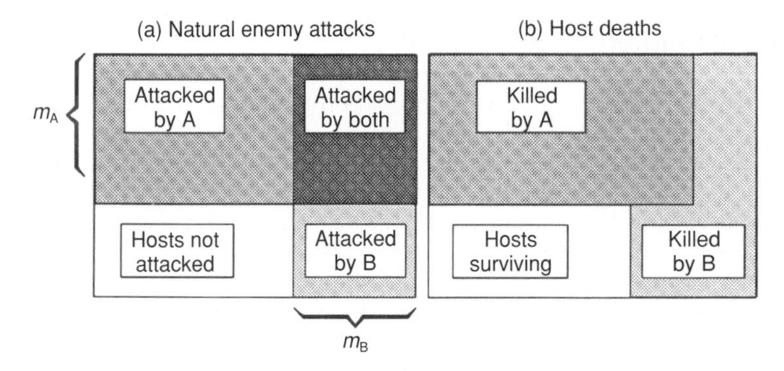

Figure 20.6 In populations subject to two (or more) contemporaneous mortality factors (here, factors A and B), the number attacked by each factor (a) will exceed the number killed by each (b) because some individuals will be attacked by both factors, but must necessarily die from only one. Quantifying the effect of each factor requires estimating the marginal attack rates (*m*) for each factor. After Elkinton et al. (1992), reprinted from Van Driesche and Bellows (1996), with permission from Kluwer.

However, if two or more sources of mortality act together within a stage, apparent mortality does not accurately reflect the strength of each agent because overlapping factors mask one another. Whereas some pest individuals will be affected by two or more sources of mortality, each will die from only one cause. In such cases, the underlying mortality rates behind the observed apparent mortality rates can be expressed as **marginal attack rates** (Royama 1981, Carey 1989, 1993, Elkinton et al. 1992). Marginal attack rates are defined as the level of mortality from an agent that would have occurred if the agent had acted alone (Figure 20.6). For some kinds of agents, such as parasitoids, this is equivalent to the number of hosts attacked (stung), even though some will later be killed by other agents (such as predators) instead of by parasitism. A general equation for calculating the marginal attack rate, *m*, for contemporaneous factors is given by Elkinton et al. (1992):

$$m_i = 1 - (1 - d)^{d_i/d}$$

where m_i is the marginal rate for factor i, d_i is the observed death rate from factor i, and d is the death rate from all causes combined. These calculations apply to a wide variety of cases (e.g. multiple parasitoids, parasitoids, and predators) and any number of contemporaneous factors. Slightly modified calculations provide marginal rates for other special cases (Elkinton et al. 1992).

4. Population growth rates. The impact of a new biocontrol agent can be summarized by computing the change it causes to the pest's population growth

Table 20.2 Hypothetical mortalities from two agents acting in separate life stages (k_1 acting on eggs, k_2 acting on larvae) over a series of generations (subjected to key-factor analysis in Fig. 20.7)

Generation	Number of eggs	Proportion mortality	Number of larvae	Proportion mortality	Number of adults	Feundity (eggs/female)	k_1	k_2	Total K
1	1000.00	0.985	15.00	0.400	9.00	100	1.82	0.22	2.05
2	900.00	0.985	13.50	0.800	2.70	100	1.82	0.70	2.52
3	270.00	0.975	6.75	0.140	5.81	100	1.60	0.07	1.67
4	580.50	0.980	11.61	0.060	10.91	100	1.70	0.03	1.73
5	1091.34	0.975	27.28	0.800	5.46	100	1.60	0.70	2.30
6	545.67	0.985	8.19	0.260	6.06	100	1.82	0.13	1.95
7	605.69	0.980	12.11	0.000	12.11	100	1.70	0.00	1.70
8	1211.39	0.980	24.23	0.320	16.47	100	1.70	0.17	1.87
9	1647.49	0.985	24.71	0.400	14.83	100	1.82	0.22	2.05
10	1482.74	0.980	29.65	0.600	11.86	100	1.70	0.40	2.10

Reprinted from Van Driesche and Bellows (1996) with permission from Kluwer.

rate. Population growth rates may be expressed either as the intergenerational **net rate of increase** (R_o) or the instantaneous **intrinsic rate of natural increase** (r_m) (Southwood 1978). R_o is the number of times a population increases or decreases from one generation to the next. Increasing populations have R_o values above 1, whereas values below 1 mean the population is decreasing by the proportion expressed (e.g. a value of 0.8 means a decrease to 8/10 of the original value). The instantaneous rate r_m is the rate of increase per unit of time (rather than per generation). Any value above 0 indicates an increasing population, whereas decreasing populations have values below 0. If paired life tables for populations with and without a natural enemy of interest are available, the difference in the two populations' growth rates is a direct and powerful measure of the impact of the natural enemy (see Van Driesche et al. 1994 for such an example).

5. Key-factor analysis. This is commonly misunderstood to be a means to identify the factor setting the typical density of the population. It is not. Rather it is a procedure used to identify which of several mortality factors observed for several generations of a pest has contributed most to the between-generation variation in total mortality (Morris 1959, Varley & Gradwell 1960). The common form of key-factor analysis (Varley & Gradwell 1960) is graphical. For a series of life tables spanning several generations, each mortality factor's strength is expressed as a series of *k*-values (one per factor per generation), together with the series of *k*-values for total mortality (the capital *K*-value). The mortality factor whose pattern looks most like that of total mortality is termed the **key factor**. Variations on this procedure have been developed that regress individual mortalities against total mortality, the factor with the greatest slope being the key factor (Podoler & Rogers 1975, Manly 1977).

Successful biological control agents need not be key factors and key factors do not necessarily set the pest's equilibrium density. This concept can be easily grasped through an example. Consider an insect population subject to two sources of mortality (Table 20.2), a common egg parasitoid that consistently kills 97–99% of all eggs in each generation and a fungal pathogen that kills from 0 to 80% of larvae in various years. The fungal disease is the key factor (see Figure 20.7) because it provides the most variation between years. However, the egg parasitoid provides, on a consistent basis, most of the mortality setting the average density and its removal would result in the greatest change in the insect's average density.

Collecting data to build life tables In life tables, rows represent developmental stages of the pest, or time periods in its development. Columns list numbers of pests that live to enter a given stage (l_x) and that die in a stage (d_x), and the division of these deaths into the numbers caused by each recognizable kind of mortality. Additional columns reformulate mortality as rates for ease of comparison, such as **apparent mortality** (based on decrease within each successive life stage, d_{xi}/l_{xi}), **real mortality** (based on decrease from first life stage, d_{xi}/l_{x0}), and **marginal attack rates** (accounting for contemporaneous attacks within a life stage by two or more factors). There are two principal methods

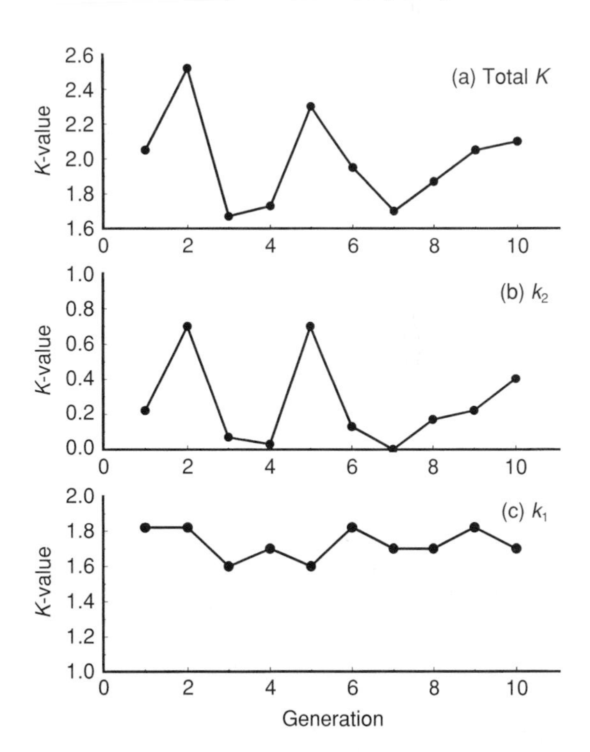

Figure 20.7 Graphical key-factor analysis for a hypothetical population depicted in Table 20.2. In this case, the key factor, by definition, is factor 2 (b) because this is the individual mortality factor that most closely matches the pattern of total mortality, *K* (a). However, factor 1 has a much greater average impact on the population and should it be added or removed would cause the greatest change in average pest density. Reprinted from Van Driesche and Bellows (1996), with permission from Kluwer.

used to obtain estimates for the l_x and d_x values required for construction of life tables: (1) stage frequency analysis and (2) direct measurement of recruitment.

1. Stage frequency analysis. Methods that attempt to obtain estimates of l_x or d_x from density data-sets are called stage frequency analysis methods and a series of methods has been published (Richards & Waloff 1954, Dempster 1956, Richards et al. 1960, Southwood & Jepson 1962, Kiritani & Nakasuji 1967, Manly 1974, 1976, 1977, 1989, Ruesink 1975, Bellows & Birley 1981, Bellows et al. 1982; see Southwood 1978 for reviews). Two methods, the graphical method of Southwood and Jepson (1962) (Bellows et al. 1989) and the second method of Richards and Waloff (Van Driesche et al. 1989), have been modified to allow the calculation of both the numbers of the pest entering a life stage and the numbers of the pest stage attacked by a specific natural enemy. Use of these methods generates estimates of the desired values, but these depend on a series of assumptions and the accuracy of stage frequency estimates is rarely determined independently.

2. Recruitment estimation. The alternative to stage frequency analysis is direct measurement of recruitment to each stage. Rather than measure, for example, the number of eggs present on each of a series of sample dates, the number of new eggs laid per time period can be measured over the entire period during which oviposition occurs. Summed across the whole ovipositional period, time-specific oviposition rates give total generational oviposition (l_x for the egg stage) as required

Table 20.3 Life table of the first generation of *Phyllonorycter crataegella* (Clemens) in an unsprayed apple orchard at Buckland, Massachusetts, USA (from Van Driesche and Taub, 1983)

Stage	Factor	Stage		Factor d_x	Marginal attack rate*	Apparent mortality		Real mortality		*k*-value	
		l_x	d_x			Stage q_x	Factor q_x	Stage d_x/l_o	Factor d_x/l_o	Stage	Factor
Egg		283	6			0.021		0.021		0.009	
	Infertility†			6	0.021		0.021		0.021		0.009
Sap larvae		277	63			0.227		0.223		0.112	
	Parasitism			35	0.134		0.026		0.123		0.062
	Residual			28	0.108		0.101		0.098		0.050
Tissue larvae		214	168			0.785		0.594		0.668	
	Parasitism			140	0.729		0.654		0.494		0.567
	Residual			28	0.206		0.130		0.098		0.100
Pupae†		46	0	0	0.000	0.000		0.000		0.000	
Adults		46									
Sex ratio†		0.5									
Fertility†		22									
F_1 progeny		505									
R_o		1.78									

*Marginal rates are calculated by using equations in Elkinton et al. (1992) for the case of a parasitoid and a predator where the predator is always credited for the death of a host attacked by both agents.

†Here, data for *Phyllonorycter blancardella* (Fabricius) (old name, *Lithocolletis blancardella*), a similar species, were used as an approximation for a similar species, where data were lacking. Data taken from Pottinger and LeRoux (1971).

Reprinted from Van Driesche and Bellows (1996) with permission from Kluwer.

for use in life tables. Simultaneous measurement of numbers dying from specific causes during each time interval allows a similar total estimate to be made for particular mortality factors. The ratio of these two values then measures d_x, the rate of mortality in the stage due to the factor (Van Driesche & Bellows 1988). Sampling methods to measure recruitment exist for a number of life stages and species, and the method can be applied wherever the biology of the species allows effective sampling approaches to be devised (Lopez & Van Driesche 1989, Van Driesche et al. 1990).

Inferences from life tables From one or a series of life tables several inferences can be made that shed light on the importance of particular natural enemies: (1) how much mortality the natural enemy causes relative to other factors, (2) whether the mortality from a new source is offset by changes in rates of mortality from other sources, and (3) whether the new factor has lowered the pest's population growth rate below replacement ($R_0 = 1$). Life tables for invasive insects can reveal the degree of attack by native natural enemies and differences in fecundity or survival rates, relative to those shown in the native range, provide a rational basis for developing integrated pest management strategies (e.g. Toepfer & Kuhlmann 2006).

Paired life tables When paired life tables are constructed from plots having and plots lacking a natural enemy, the impact of the natural enemy on population growth can be directly measured. This approach has been employed to assess the effect of introduced parasitoids of the citrus blackfly, *A. woglumi*, in Florida (Dowell et al. 1979), of the predator *Mesocyclops longisetus* Thibaud on immature stages of *Aedes aegypti* (L.) in tires in Yucatan, Mexico (Manrique-Saide et al. 1998), and of native parasitoids of the apple blotch leafminer, *Phyllonorycter crataegella* (Clemens), in Massachusetts, USA (Van Driesche & Taub 1983). In the *P. crataegella* case, life tables were constructed both for untreated plots, where parasitoids were present (Table 20.3), and for plots treated with pesticides, where parasitoids were nearly eliminated (Table 20.4). Life tables from the untreated plots indicated parasitism was a substantial source of mortality and that the net rate of increase (R_0) was 1.8. When parasitism was deleted from this life table (Table 20.5) R_0 increased to 7.7. In sprayed plots lacking parasitism (Table 20.4) an actual rate of increase of 9.41 was observed, similar to the hypothesized value.

Evaluating effects of weed biocontrol agents

Kinds of impacts measured

When assessing the effects of herbivorous insects or fungal pathogens on plants, a fundamentally different approach (compared to evaluation of insect biocontrol agents) is required because impacts on plant performance that potentially have population-level effects include a wide range of processes. McClay (1995) recommended attention to the following four elements when evaluating the impacts of biological control agents on weeds: (1) assess the weed, not the agent, (2) assess weed populations, not individual plants, (3) assess in the field, and (4) prove responsibility of the agent. The following parameters can be used to measure these impacts.

Death Plant death, unlike insect death, is not the usual criterion by which a weed control agent's impact is scored. In most cases, plant death, even when it occurs, may be a protracted process, with plant biomass declining as a "die-back" over considerable time. Instead of measuring plant deaths directly, changes in density, biomass, or coverage are observed over time. Story et al. (2006), for example, showed that spotted knapweed density at two sites declined by 99 and 77% after the root weevil *Cyphocleonus achates* (Fahraeus) increased dramatically at those sites.

Counts of plant deaths can more easily be recorded in small-plot garden experiments. Wenziker et al. (2003), while evaluating the crown weevil *Mortadelo horridus* Alonso-Zarazaga and Sanchez-Ruiz on thistles in garden plots in Australia, recorded 17 and 26% mortality of *Carduus pycnocephalus* L. and *Carduus tenuiflorus* Curtis plants, respectively. Likewise, Tomley and Evans (2004), by tagging individual rubbervine plants [*Cryptostegia grandiflora* (Roxburgh) R. Brown] determined that 75% of the plants died at a one field site due to the combined effects of environmental stress and an introduced rust [*Maravalia cryptostegiae* (Cummins) Ono], while at another more favorable site plants compensated for the associated defoliation.

For some purposes, plant organs (leaves, tillers, buds, capitula, etc.) may be viewed as populations that undergo natality and mortality (Harper 1977, 1981) and changes in their numbers may be more easily observed than change in number of whole plants. Waterhyacinth [*Eichhornia crassipes* (Mart.) Solms-Laub] mats, for example, consist primarily of large, conspicuous leaves that function as floats. These are

Table 20.4 Life table of the first generation of *Phyllonorycter crataegella* (Clemens) in a sprayed apple orchard at Buckland, Massachusetts, USA (from Van Driesche and Taub, 1983)

Stage	Factor	Stage		Factor d_x	Marginal attack rate*	Apparent mortality		Real mortality		k-value	
		l_x	d_x			Stage q_x	Factor q_x	Stage d_x/l_o	Factor d_x/l_o	Stage	Factor
Egg		433	9			0.021		0.021		0.009	
	Infertility†			6	0.021		0.021		0.021		0.009
Sap larvae		424	19			0.045		0.044		0.020	
	Parasitism			1	0.002		0.002		0.002		0.001
	Residual			18	0.043		0.042		0.041		0.019
Tissue larvae		405	34			0.084		0.079		0.038	
	Parasitism			17	0.043		0.041		0.039		0.019
	Residual			17	0.043		0.041		0.039		0.019
Pupae†		371	0	0	0.000	0.000		0.000		0.000	
Adults		371									
Sex ratio†		0.5									
Fertility†		22									
F_1 progeny		4072									
R_o		9.41									

*Marginal rates are calculated by using equations in Elkinton et al. (1992) for the case of a parasitoid and a predator where the predator is always credited for the death of a host attacked by both agents.

†Here, data for *Phyllonorycter blancardella* (Fabricius) (old name, *Lithocolletis blancardella*), a similar species, were used as an approximation for a similar species, where data were lacking. Data taken from Pottinger and LeRoux (1971).

sequentially and continuously produced on individual shoots, and plants must retain a constant leaf complement, with leaf production balancing senescence, or they will sink (Center & Van 1989). Center (1981, 1985) and Center and Van (1989) showed that biological control agents lowered leaf production rates relative to leaf senescence, and this was the mechanism behind the observed impact (sinking of plant mats).

Reduced growth Biological control agents sometimes affect plants in subtle ways, reducing growth even when there are no overt signs of plant deterioration. Detection of these effects may be complicated in natural settings due to the difficulty of establishing and maintaining satisfactory agent-free controls, but when done properly the ensuing data are often definitive (McClay 1995). Franks et al. (2006), using cages, were able to show under quasi-natural field conditions that the introduced psyllid *Boreioglycaspis melaleuca* Moore, but not the weevil *O. vitiosa*, reduced growth of *Melaleuca quinquenervia* (Cavier) Blake seedlings. Tomley and Evans (2004) tagged individual rubber-vine plants (*C. grandiflora*) to monitor the effects of the introduced rust fungus *M. cryptostegiae* on their growth. They documented a marked decrease in growth, especially in high rainfall areas. Likewise, Goolsby et al. (2004a), using potted plants, showed that the eriophyid mite *Floracarus perrepae* Knihinicki and Boczek had the potential to reduce growth of Old World climbing fern. Briese (1996) demonstrated the potential of a stem-boring weevil (*Lixus cardui* Olivier) to suppress the growth of thistles in the genus *Onopordum* by caging individual plants in a natural field population. Dennill (1985) compared growth of galled and ungalled stems of *Acacia longifolia* (Andr.) Willd. By doing so, he was

Table 20.5 Life table of the first generation of *Phyllonorycter crataegella* (Clemens) in an unsprayed apple orchard at Buckland, Massachusetts, USA, modified by deleting mortality from parasitism (from Van Driesche and Taub, 1983)

Stage	Factor	Stage		Factor d_x	Marginal attack rate*	Apparent mortality		Real mortality		k-value	
		l_x	d_x			Stage q_x	Factor q_x	Stage d_x/l_0	Factor d_x/l_0	Stage	Factor
Egg		283	6			0.021		0.021		0.009	
	Infertility†			6	0.021		0.021		0.021		0.009
Sap larvae		277	28			0.101		0.099		0.046	
	Parasitism			0	0.000		0.000		0.000		0.000
	Residual			28	0.101		0.101		0.099		0.046
Tissue larvae		249	51			0.206		0.180		0.100	
	Parasitism			0	0.000		0.000		0.000		0.000
	Residual			51	0.206		0.206		0.180		0.100
Pupae†		198	0	0	0.000	0.000		0.000		0.000	
Adults		198									
Sex ratio†		0.5									
Fertility†		22									
F_1 progeny		2178									
R_0		7.70									

*Marginal rates are calculated by using equations in Elkinton *et al.* (1992) for the case of a parasitoid and a predator where the predator is always credited for the death of a host attacked by both agents.

†Here, data for *Phyllonorycter blancardella* (Fabricius) (old name, *Lithocolletis blancardella*), a similar species, were used as an approximation for a similar species, where data were lacking. Data taken from Pottinger and LeRoux (1971).

Reprinted from Van Driesche and Bellows (1996) with permission from Kluwer.

able to determine that galling of buds by the wasp *Trichilogaster acaciaelongifoliae* Froggatt not only reduced seed production, but suppressed vegetative growth.

Change in biomass Biomass can be useful as a metric to assess the effect of herbivory, reflected either as loss of growth or tissue destruction. Also, a plant's biomass allocation among plant parts can reveal its responses to herbivory, such as compensatory growth or altered reproductive strategies. Change in biomass per unit of habitat area can often be readily measured, although if destructive sampling is used, this precludes repeated measurement of the same unit. Allometry, wherein non-destructive measurements of a correlated plant trait (such as height) are used to estimate biomass, can be used when there is a need to follow a plant's biomass over time (e.g. Van et al. 2000). Biomass integrates all aspects of plant biology and population dynamics. Waterhyacinth biomass per unit of water surface area, for example, integrates population density and plant stature into a single measure. Decreases in biomass may precede changes in percentage cover and thus be an earlier sign of impact by a biological control agent.

Change in percentage cover For plants, the proportion of ground or water surface that is covered by the invasive species is an easily measured, meaningful estimate of the pest population. It can be measured on a local level using hoops or quadrants placed on the surface, at intermediate scales through the use of global positioning system (GPS) devices to map changes in infestations, or at the landscape level with aerial photography or even satellite imagery.

Seed set In many plant biological control projects, some natural enemies are released to attack the seeds or other reproductive parts of the plant. Weevils and flies, for example, destroy developing seeds in the thistle seed heads. Their impact is directly measured by counting the percentage of seed heads infested, within each seed head size class, and then determining seed production in infested seed heads compared to uninfested ones. Damage to other parts of the plant can also result in reduced seed set. Briese (1996), for example, showed that attack by the stem-boring weevil *L. cardui* caused *Onopordum* thistles to produce fewer seed heads, which were smaller, suffered higher levels of abortion, and formed 80% fewer viable seeds. Pratt et al. (2005) estimated that undamaged melaleuca trees were 36 times more likely to flower than trees damaged by *O. vitiosa*. A single defoliation by this weevil caused an 80% reduction in the number of reproductive structures and herbivore-damaged trees produced 54% fewer seed capsules. In other cases, agents may change plant biomass or stature without reducing seed output (e.g. Hoffmann et al. 1998b).

Seed-bank size A follow-on measure to reduction of seed production is decline in the accumulated seed stores (seed bank) in the soil. Using insecticide exclusion experiments, Dhileepan (2001) found that only 3% of the *Parthenium hysterophorus* L. seedlings present in plots at the beginning of the growing season survived to produce flowers (and hence seeds) by the end of the season at one site due to exposure to biological control agents. This contrasted with 45% survival in plots where biological control agents were excluded. Exclusion of biocontrol insects resulted in a seven-fold increase in the soil seed bank the following season, whereas no increase in seed banks occurred in the biological control plots.

In the case of *A. longifolia*, prior to the introduction of the bud-galling wasp *T. acaciaelongifoliae*, seed densities in the soil reached 45,800 seeds/m^2 (Dennill & Donnelly 1991). These seeds are long-lived and are stimulated to germinate *en masse* by fires, which occur frequently in the floristically rich Cape fynbos biome of South Africa. Galling by the wasp reduced seed production nearly to zero at some sites (Dennill & Donnelly 1991), so that seed banks progressively became depleted. Similarly, *Mimosa pigra* L. produced seed banks in northern Australia that varied from 8500 to 12,000 seeds/m^2 (Lonsdale et al. 1988). The twig-boring moth *Neurostrota gunniella* Busck reduced seed rain by as much as 60%, reducing input to the seed bank (Lonsdale & Farrell 1998). Paynter (2005) showed that a stem-boring moth, *Carmenta mimosa* Eichlin and Passoa, reduced seed banks at floodplain sites from nearly 7000 seeds/m^2 to fewer than 3000 seeds/m^2.

Change in nutrient reserves Loss of stored nutrients reduces a damaged plant's ability to compete with other plants, to grow, and to set seed. Nutrients of perennial plants are stored in various organs, including roots and leaves (especially for conifers). Impacts of herbivory on nutrient stores can include: (1) reduction in stores through reduction in photosynthesis caused by defoliation and (2) direct damage to storage organs by insects that feed directly on roots or consume conifer needles. Assessment of impacts on nutrient reserves can be made by measuring storage organ size or chemically assessing quantities of starch or other reserves. Katovich et al. (1999), for example, showed that defoliation by *Galerucella* spp. reduces sucrose levels and starch reserves in roots and crowns of purple loosestrife (*Lythrum salicaria* L.).

Vascular system function Some herbivores disrupt water transport in plants by damaging the vascular system directly or by introducing pathogens that do so. The effect of such disruption may be death of branches or the whole plant. Measurement of this effect may be based on deaths of individual plants or flagging of branches.

Defoliation and lowered rates of photosynthesis Measurement of photosynthesis under field conditions is possible but technically more complex than some of the other measures of impact mentioned here. If reductions result from simple loss of photosynthetic area, it may be adequate to simply quantify defoliation as a percentage of available leaf area. However, plants can often compensate for the loss of some foliage so that defoliation doesn't necessarily result in a comparable reduction in photosynthesis. Also, galling or sucking insects can reduce photosynthesis without defoliation. Measurements of plant photosynthesis can be made for field plants, comparing damaged and normal plants of similar size. Florentine et al. (2005), for example, used gas-exchange measurements on leaves of *P. hysterophorus* plants to determine the effects of galling by the moth *Epiblema strenuata* Walker. They found that galling reduced net photosynthesis by 80–92%, depending on plant life stage. Doyle et al. (2002) determined that 10–30% leaf damage to *Hydrilla verticillata* (L. f.) Royle by the leafmining fly *Hydrellia pakistanae* Deonier caused a 30–40% reduction in photosynthesis, leaving barely enough to meet the daily respiratory requirement. Damage to 70–90% of the leaves reduced

(a)

(b)

Plate 3.1 (a) The tachinid *Erynniopsis antennata* Rondani is a parasitoid of the elm leaf beetle, *Pyrrhalta luteola* (Müller). (b) The encyrtid *Anagyrus kamali* Moursi.

(c)

(d)

Plate 3.1 *(Cont'd)* (c) The aphelinid *Aphytis melinus* DeBach attacking California red scale, *Aonidiella aurantii* (Maskell). (d) The trichogrammatid *Trichogramma pretiosum* Riley ovipositing in the egg of *Helicoverpa zea* (Boddie).

(e)

(f)

Plate 3.1 *(Cont'd)* (e) The mymarid *Gonatocerus ashmeadi* Girault. (f) The braconid *Aphidius colemani* Viereck. Photographs courtesy of (a, c–f) Jack Kelly Clark, University of California IPM Photo Library and (b) William Roltsch, California Department of Agriculture.

Plate 4.1 (a) *Orius tristicolor* (White) (Anthocoridae), a species used for augmentative control of thrips in greenhouse crops. (b) A chrysopid larva, a group predaceous on aphids and other pests.

(c)

(d)

Plate 4.1 *(Cont'd)* (c) An adult vedalia beetle [*Rodolia cardinalis* (Mulsant)] next to its prey, the cottony cushion scale (*Icerya purchasi* Maskell). (d) The histerid beetle *Teretrius nigrescens* (Lewis), a predator of the larger grain borer, *Prostephanus truncatus* (Horn), a pest of stored corn on subsistence farms in Africa.

(e)

(f)

Plate 4.1 *(Cont'd)* (e) An adult syrphid fly. (f) Syrphid larvae are predators of aphids. Photographs courtesy of (a–c, e, f) Jack Kelly Clark, University of California IPM Photo Library and (d) Georg Goergen, IITA.

Plate 5.1 (a) Galls of *Trichilogaster acaciaelongifoliae* (Froggatt) (Hymen.: Pteromalidae) on *Acacia longifolia* (Andrews); (b) adult gall wasp; (c) close-up of a gall; and (d) bisected gall showing larvae.

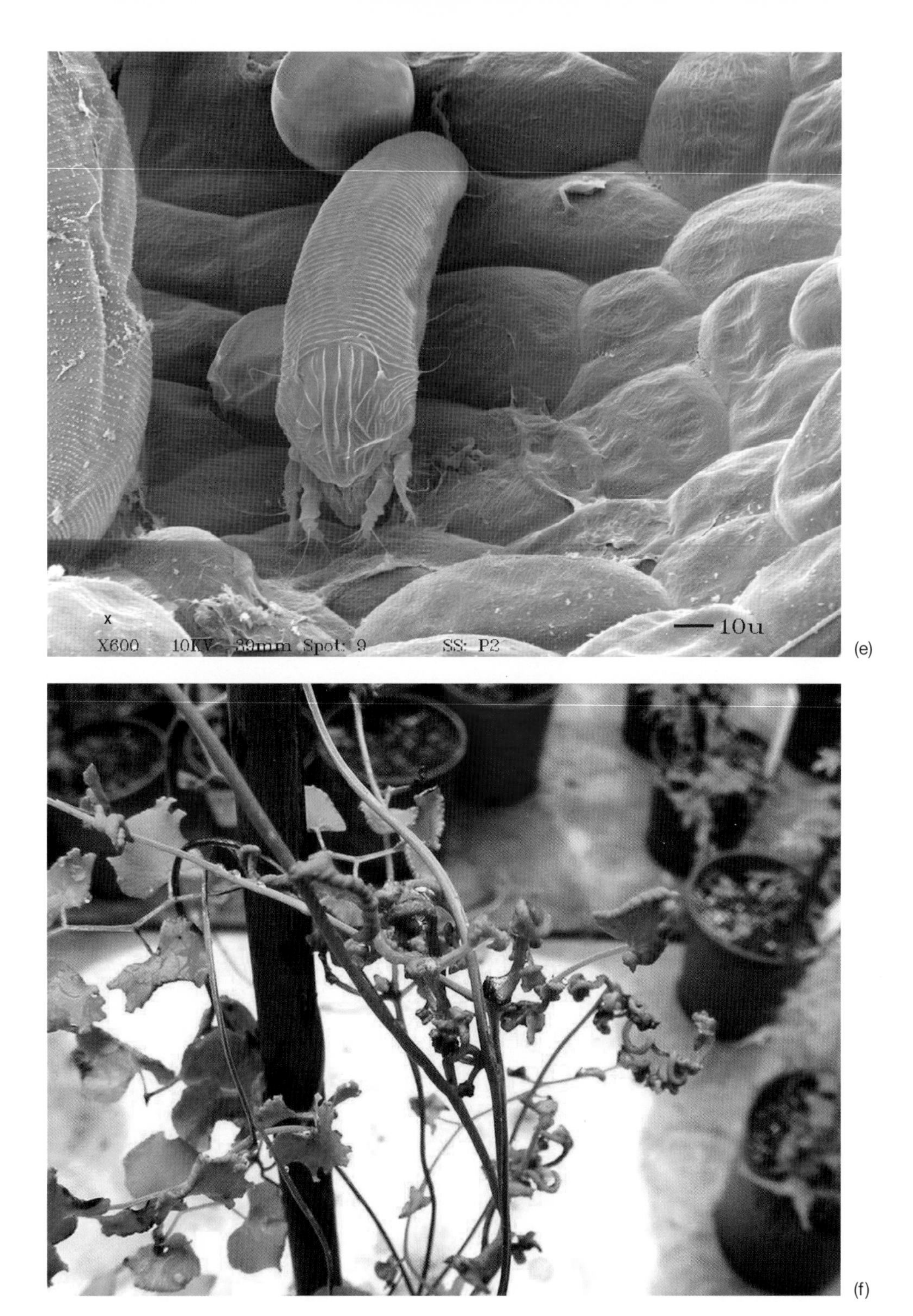

Plate 5.1 *(Cont'd)* (e) The eriophyiid rust mite *Floracarus perrepae* Knihinicki and Boczek, an herbivore associated with Old World climbing fern, *Lygodium microphyllum* (Cav.) R. Br.; and (f) damage from *F. perrepae*. Photographs courtesy of (a–d) S. Neser, PPRI and (e, f) John Goolsby, USDA-ARS.

(a)

(b)

Plate 7.1 (a) A stand of the toxic alga *Caulerpa taxifolia* (Vahl) C. Agardh covering the bottom of the Mediterranean Sea. (b) Brown tree snake (*Boiga irregularis* Fitzinger), an invasive predator that has decimated the forest birds of Guam.

(c)

(d)

Plate 7.1 *(Cont'd)* (c) The Guam Micronesian kingfisher (*Todirhamphus cinnamominus cinnamominus*) is one of the native birds of Guam decimated by the brown tree snake (*Boiga irregularis* Fitzinger). (d) Close-up of an adult hemlock woolly adelgid (*Adelges tsugae* Annand) and eggs.

(e)

(f)

Plate 7.1 *(Cont'd)* (e) *Scymnus camptodromus* Yu et Liu, a coccinellid predator from China of the hemlock woolly adelgid (*A. tsugae* Annand). (f) *Laricobius nigrinus* Fender, a derodontid predator from British Columbia (Canada) of the hemlock woolly adelgid (*A. tsugae* Annand). Photographs courtesy of (a) Alexandre Meinesz, University of Nice, (b) Christy Martin, CGAPS, Hawaii, USA, (c) W.D. Kesler, (d) Mike Montgomery, www.Forestryimages.org, (e) Dr Guoyue Yu, and (f) Rob Flowers.

(a)

(b)

Plate 12.1 (a, b) Rubber vine (*Cryptostegia grandifolia* R. Br.) is a severe pest of natural areas in tropical Queensland, Australia, smothering native vegetation (a); the rust *Maravalia cryptostegiae* (b), a pathogen imported from Madagascar, damages rubber vine heavily and is providing control (Vogler & Linday 2002).

(c)

(d)

Plate 12.1 *(Cont'd)* (c, d) Biological control of waterhyacinth [*Eichhornia crassipes* (Mart.) Solms-Laub] on Lake Victoria in Africa (Kisuma, Kenya) by introduced weevils provided dramatic control, changing solid mats in May 1999 (c) to open water by December of the same year (d).

Plate 12.1 *(Cont'd)* (e, f) The chrysomelid beetle *Diorhabda elongata* Brulle *deserticola* Chen, has been introduced in the southwestern USA for the biological control of *Tamarix* spp., which are Eurasian shrubs widely invasive in riparian areas. Here, (e) larvae are shown defoliating *Tamarix* shrubs.

(f)

Plate 12.1 *(Cont'd)* (e, f) The chrysomelid beetle *Diarhabda elongata* Brulle *deserticola* Chen, has been introduced in the southwestern USA for the biological control of *Tamarix* spp., which are Eurasian shrubs widely invasive in riparian areas. (f) The extent of defoliation is visible in an elevated view of a release site. Photographs courtesy of (a, b) Colin Wilson, (c, d) Mic Julien, CSIRO, and (e, f) Ray Caruthers, USDA-ARS.

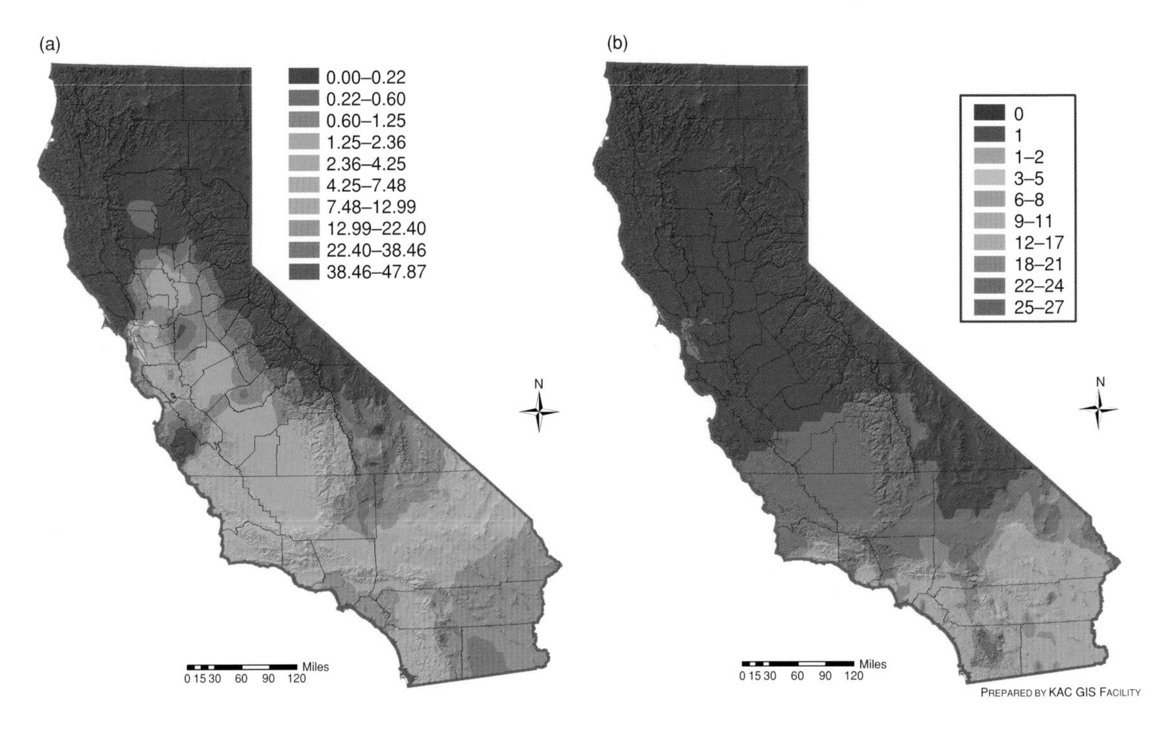

Plate 14.1 GIS mapping of the estimated life-table statistic, net reproductive rate (R_0) for the *Homalodisca coagulata* egg parasitoids (a) *Gonatocerus ashmeadi* Girault and (b) *Gonatocerus triguttatus* Girault in California. R_0 was determined in the laboratory for each parasitoid species across five experimental temperatures and the resulting relationship between R_0 and temperature was determined and modeled in GIS using data from 260 weather stations in California. The results are very striking: *G. ashmeadi* can be expected to infiltrate most of California and reproduce annually, whereas *G. triguttatus* may be severely restricted to localized regions in southern California. Maps drawn by M. Hoddle.

photosynthesis by about 60%, leaving insufficient photosynthate to meet respiratory demands, leading to the likely demise of the plants.

Increased susceptibility to pathogens Insect feeding may reduce plant performance by facilitating infections by pathogens. Charudattan et al. (1978) found that waterhyacinth infested with the weevil *N. eichhorniae* Warner and the mite *Orthogalumna terebrantis* Wallwork were more diseased than plants not infested with these arthropods. These diseases were caused by known pathogenic fungi as well as soft-rot bacteria that caused root and crown rots.

Increased susceptibility to physical stress In some cases, herbivory decreases plant tolerance to environmental stresses. *Eucalyptus* species grown in southern California, a region with summer drought (in contrast to the moister summers in the native ranges of many *Eucalyptus* species), showed little ability to tolerate both borer attack and summer drought, while being able to tolerate either one alone (Hanks et al. 1999). Assessment of such impacts can employ measures such as rates of plant death or percentage cover, in plots having both the insect plus the abiotic stress compared with plots with only the abiotic stress.

Decreased competitive ability Reduced ability to compete with other plants is a secondary measure of impact. This decrease can be caused by any or all of the changes described above. It can only be measured through experimental comparisons of plots having and lacking the natural enemy of interest, crossed with the presence or absence of competing vegetation. For example, McEvoy et al. (1993) found that interspecific plant competition combined with herbivory from the flea beetle [*Longitarsus jacobaeae* (Waterhouse)] inhibited the increase and spread of ragwort (*Senecio jacobaea* L.). Likewise, Center et al. (2005) showed that the weevils *Neochetina bruchi* Hustache and *N. eichhorniae* altered the competitive advantage of waterhyacinth (*E. crassipes*) over waterlettuce (*Pistia stratiotes* L.). A single waterhyacinth plant was competitively equal to 41 waterlettuce plants when the weevils were excluded, but the two species became competitively equal when the weevils were present.

Caging methods for use in weed control projects

Exclusion cages can be used effectively to study some short-term effects of weed biocontrol agents such as: (1) reduction in seed production on caged and uncaged inflorescences or (2) effects of galling on growth of branches or other structures on galled and ungalled plants. Gilreath and Smith (1988) used small cages on cactus pads to exclude parasitoids and predators of the beneficial herbivore *Dactylopius confusus* (Cockerell). This created high and low populations of this species and a 10-fold difference in the level of cactus-pad death from herbivory. Cages were used effectively to determine which of two agents (the cinnabar moth, *Tyria jacobaeae* L. or the ragwort flea beetle, *L. jacobaeae*) released against a grassland weed (tansy ragwort, *S. jacobaea*) in Oregon was more effective at suppressing the plant (James et al. 1982, McEvoy & Rudd 1993, McEvoy et al. 1993). Long-term caging studies are perhaps less desirable with plants, as caging might affect plant health by reducing light and increasing humidity.

In contrast to the above design, in which cages are used to exclude the herbivore being evaluated, an alternate design uses inclusion cages to compare caged patches of plants lacking the herbivore to plants in other cages into which the herbivore has been introduced. This approach was taken by Center et al. (2007) to evaluate the impact of the melaleuca pysllid (*Boreioglycaspis melaleucae* Moore) on survival of seedling melaleuca trees. A constant density of 15 nymphs per plant increased seedling mortality during the test from 5 to 60%. A limitation to this method is that the researcher artificially sets the insect density and so care must be taken to ensure that the densities tested are within the naturally occurring range.

Insecticide check method for use in weed control projects

Insecticides can also be used to demonstrate the effect of herbivorous arthropods in the control of weed populations (Lonsdale et al. 1995, Adair & Holtkamp 1999, Dhileepan 2001, Goolsby et al. 2004a). Insecticides are applied to some plots (from which the herbivore of interest is thus excluded chemically) and these plots are compared to similar unsprayed plots in which the herbivore is naturally present. Balciunas and Burrows (1993) assessed the effect of native insects by growing 60 saplings of *M. quinquenervia*, a target for biological weed control in Florida, in pots in the plant's natural habitat in Australia. Insecticides were used to protect half of the saplings. Treated saplings showed greater height and biomass within 6 months. Most damage was caused by insects that exhibited only low levels of herbivory, but which collectively significantly and rapidly reduced plant growth. Goolsby et al. (2004a) used miticides to exclude the eriophyid *F. perrepae*, which he studied as a possible biological control agent for the invasive Old World climbing fern, *Lygodium microphyllum* (Cav.) R. Br., in Florida, USA. Dhileepan (2001) used insecticide exclusion experiments at two sites to assess the effects of biological control insects on density of *P. hysterophorus*. Plant density was 90% lower in plots not treated with insecticide than in insecticide-treated plots at one site, but no difference was observed at the second site. When using pesticides to exclude herbivores from weed populations, it must be determined both that the pesticides actually exclude the target insects and that the pesticides have no direct effects on the plant itself. This approach can be used both in the native range (to determine whether local species are suppressing the plant's density) and in the recipient country after an agent's introduction, to see whether it is suppressing the pest. A drawback to this approach in the native range is that all herbivorous insects affecting the plant, not just a particular potential agent of interest, are likely to be present in one plot and excluded from the other.

Performance and weed population growth modeling

A rather simple modeling approach, known as matrix modeling, may assist in determining which plant life-history stages should be targeted by biological control and also in estimating the probable impact of a particular agent (Shea & Kelly 1998). Such models can help determine the population growth rates of the plants, sensitivities, elasticities (the relative contribution of a particular stage to the population growth rate), and transient dynamics of stage-specific demographic events, and the effects of biological control agents on the plant's population growth. These models analyze the shifts between life stages (e.g. in a size-structured model from seed bank to small plants, from small to large plants, or from large plants back to seed bank, etc.), which allows determination of critical transitions wherein the population growth of the plant is most likely to be sensitive to externalities (see Shea & Kelly 1998 and Caswell 1989 for more in-depth information).

Failure-time analysis originated from industrial reliability testing (Fox 1993), where is it important to know such things as the number of times that a particular mechanical part can be used before it fails. In ecological studies, it involves repeated observations of a uniquely identified individual to determine when an event of interest (e.g. death, flowering, migration, etc.) has occurred. This analytical approach can be quite useful for the analysis of survival curves or life-table data (Fox 1993). Because these data are not normally distributed, they are not suited for more typical analyses (Fox 1993). Two types of regression models are used to analyze these data, which make different assumptions about the effect of covariates. Accelerated failure-time models assume that covariates accelerate the lives of individuals by interacting with the treatments to produce earlier failure times. Proportional hazards models assume that covariates make the individuals more susceptible to the treatments. In the first case, treatments change the timing of the failure; in the second case treatments change the chance of failure.

Comparative demographic matrix models of plant populations between several areas, such as the native and invaded areas, or one invaded area with another, can be used to identify how plant populations in distinct areas might vary as to which stages were limiting to population growth. This can then be used to provide insight into stages most likely to be effective targets for biological control agents. Use of this approach, for example, showed that growth of populations of the invasive thistle *Carduus nutans* L. in New Zealand were driven by rapid transitions of early life stages, while in Australia, fecundity of the plant was of less importance than survival of the rosettes (Shea et al. 2005). Such differences may mean that agents directed at one plant stage (such as seed feeders) may be successful in one area, but fail in another.

SEPARATING EFFECTS OF A COMPLEX OF NATURAL ENEMIES

Some of the evaluation methods discussed above, such as the insecticide check method, measure the impact of the whole natural enemy complex. When information is desired on the impact of one specific group or species, additional experiments or sampling are usually required. Methods to divide total mortality into parts assignable to specific species or guilds of natural enemies differ but can be based on various types of physical evidence left following their attack on the pest, or on the use of selective-exclusion devices that allow attack by subsets of the complex.

Parasitoids of arthropods

Separating the contributions of members of a guild of parasitoids is relatively straightforward. Samples of the pest from field populations can be reared and adult parasitoids identified to species. For systems in which parasitoids enter diapause requiring lengthy rearing, immature stages can be identified to species based on DNA markers derived from known sources (reviewed in Greenstone 2006). If some members of the parasitoid guild cause additional host deaths through host feeding, those deaths will need to be counted and added to the rate of parasitism to appreciate the full impact of the species.

Pathogens of arthropods

Division of total mortality from diseases into portions resulting from attack by specific pathogens of arthropods is very similar to the example of parasitoids in that samples of the pests can be collected, some of which will be infected at the time of collection. These organisms can be reared and pathogens responsible for each dead host can be obtained for identification. Techniques for detecting infections of specific pathogens in hosts in early stages of disease include electrophoresis, antigen–antibody methods (e.g. ELISA and related techniques), and DNA detection methods (Keating et al. 1989, McGuire & Henry 1989, Hegedus & Khachatourians 1993, Shamin et al. 1994). These techniques offer advantages of speed and directly measure the underlying marginal attack rates for the pathogens.

Predaceous arthropods

Predation often leaves no physical evidence, unless the prey is located in a structure (plant parts, galls, leafmines) that is durable and retains evidence of the predator's attack (Sturm et al. 1990). Consequently, in most cases indirect methods must be used. Two general approaches have been developed. One approach (the **top-down method**) consists of measuring total losses from predation suffered by the pest population, and then by a variety of methods assigning portions of total losses to specific predators or groups of predators. The other approach (the **bottom-up method**) starts with observations on the numbers of various types of predators in a system and uses information on the foraging abilities and feeding capacity of specific predator species to estimate the impact a given predator species is having on the pest (O'Neil & Stimac 1988b).

Top-down method

The first step in this approach is to measure the total losses due to predation. For example, predator exclusion can be used to create prey populations with and without exposure to the predator complex present. Differences in survivorship of the prey between these two subpopulations provides a measure of the total mortality from predation. Chiverton (1986), for example, used barriers to exclude ground-dwelling predators attacking cereal aphids in the UK. Field exposures of natural or artificially established prey cohorts is another way to estimate the impact of predation (e.g. Hazzard et al. 1991).

The second step in this process is to break up the effect into species-specific components. For example, by varying the types of cages or barriers used, or dimensions of mesh in wire or nets, exclusion can sometimes be limited to specific predator groups, allowing their effects to be quantified separately. Campbell and Torgersen (1983) were able to use combinations of bird netting and sticky barriers to separately quantify the effects of predation by birds versus ants on larvae and pupae of western spruce budworm, *Choristoneura occidentalis* Freeman.

Bottom-up method

The bottom-up method seeks to estimate the relative importance of each predator species present in the pest's habitat (Whitcomb 1981). The first step is to develop a list of predator species present (Bechinski & Pedigo 1981). Such lists may be quite long, running to tens or hundreds of species. No single sample method or time of sampling will catch all species of predators in the habitat. Furthermore, numbers caught may be partially due to how the biology of a given species interacts with the sampling method and will not solely be a reflection of a species' density in the habitat. Therefore, it is important to use various sampling approaches in the early phase of the predator inventory and to consider the results of several methods in rating the abundance of any single species relative to the other members of the predator guild.

The second step is to obtain information on which predators are actually consuming the target prey in the field. Methods to do so are direct observation and detection in predators of some label indicative of the prey species. Direct observation is simple but time-consuming, and consists of stationing an observer near a patch of prey and waiting for predation to occur (Kiritani et al. 1972, Godfrey et al. 1989). Some important predators may feed at night and so observations must be made then as well as in the day.

One class of labels used to detect predation includes those that can be introduced (usually in rearing diet) into laboratory-reared prey, which are then exposed to predators in the field. Markers include some fat-soluble dyes such as Calco Oil Red (Elvin et al. 1983) and distinctive elements such as rubidium (Cohen & Jackson 1989). In each case, these materials when fed to prey reared in the laboratory will transfer in quantities sufficient for later detection in predators that eat the marked prey. Dyes are detected during dissections of predators, by their color. Detection of rubidium requires atomic absorption spectrophotometry. In the case of rubidium, this marker may also be applied as a spray to plants in the field. The material is taken up by plants, then by herbivores feeding on the plants, and finally by predators feeding on the self-marked herbivores. This allows field studies of predator movement, for example, to be conducted at a field scale (e.g. Prasifka et al. 2004). Predators or prey may also be marked by feeding or spraying them with rabbit immunoglobulin G (IgG), which is a readily available material and one that can later be detected via standard ELISA methods using anti-rabbit IgG antibodies (Hagler & Miller 2002).

Another class of labels is the tissue of the prey species itself. There are two general approaches to detecting prey tissue. One approach is to detect prey proteins, using antibodies raised against specific prey proteins. The other general approach is to detect segments of the prey's DNA (see Harwood & Obyrcki 2005 for a comparison of these approaches as applied to study of aphid predators). Under the first approach, antibodies against prey antigens (proteins) are used to determine if a given predator has recently ingested proteins from the prey. Many techniques have been developed for this type of analysis (Sunderland 1988). Test features of importance are ease of use (speed and cost), sensitivity (minimum detectable quantity of antigen), and specificity of the reaction (freedom from false-positive results). The last issue is especially important. If antisera are prepared against blends of prey proteins, the probability is very great that other potential prey will cross-react to the antiserum, misleading the investigation. Use of monoclonal antibody technology provides a means to prepare antisera to a single antigen determinant of a particular protein of the prey species (Hagler et al. 1994, Greenstone 1996). Once available, such an antiserum must be tested for cross-reactivity against protein mixes from the other potential prey in the habitat to estimate the potential rate of false positives.

Tests using antisera as marks may give qualitative or quantitative results, depending on the test itself. Development of quantitative assays (to score the number of prey eaten by a sampled predator) is complicated by many factors, including meal size, time since ingestion, temperature, species differences, and sensitivity of the test. Some approaches to quantification are discussed by Hagler and Naranjo (1997) and Chen et al. (2000).

Regardless of what type of markers are detected in predator guts, such detection initially only establishes a list of those predators that are actually eating a particular prey. With additional data such as predator density and consumptive capacity, it is feasible to make tentative estimates of the species' importance by estimating the predator population's daily consumption of prey per unit area within the crop or habitat. Formulae exist to calculate minimum and maximum daily rates of predation per unit area of habitat per predator species. The minimum rate formula (Dempster 1967) calculates the rate as predation rate per day = predator density × proportion of predators giving a positive reaction to prey antigen/the number of days antigen remains detectable. The maximum rate formula (Rothschild 1966) calculates the rate as predation rate per day = predator density × proportion of predators giving a positive reaction to prey antigen × the mean number of prey eaten per day in the laboratory by the predator when prey are abundant. Leathwick and Winterbourn (1984) used these formulae to calculate the effects of a series of predators on lucerne aphids (*Acyrthosiphum* spp.) in New Zealand. Sopp (1987) modeled the effects of time and temperature on antigen disappearance, to predict the number of prey ingested from the amount of antigen detected in the ELISA test. Wratten (1987) provides an overview of the principles of the evaluation of predation using these methods.

Plant pathogens

The occurrence of plant pathogens can be measured by collecting samples of diseased tissue and culturing the pathogen for identification. For this process, as with identification of arthropod pathogens, species identifications may be based on rearing and culturing. For well known species, dissections and microscopic examination may suffice, or in less obvious cases, use may be made of the electrophoretic, antigen–antibody, or DNA detection methods mentioned above. Effects of specific plant pathogens can best be assessed by experiments comparing relevant aspsects of plant health under conditions of exposure and non-exposure to the pathogen, under defined environmental conditions.

Herbivorous arthropods

Partitioning the total impact of herbivory affecting a plant into impacts associated with each herbivore species begins with field surveys to develop a comprehensive list of species in the herbivore complex (Sheppard et al. 1991). Such data must be based on a suitably wide variety of sampling techniques and must be quantified in terms of numbers per unit area so that comparisons can be made between organisms affecting various parts of the plant. These data, when combined with estimates of consumption rates, allow some comparisons to be made between herbivores, although caution is advised. Per capita consumption is a very poor indicator of the potential impact by an herbivorous species on the plant because plant tissues (e.g. meristematic compared with foliar or young compared with old foliage, etc.) vary greatly in importance. Single-species experiments in which effects on plant growth, survival, competitiveness, and reproduction are compared between plots having and plots lacking the herbivore of interest should be performed and then comparisons made between herbivore species on this basis. Development of paired plant life tables for plots having and lacking specific herbivores (such as recently introduced weed control agents), potentially through the use of cages able to exclude some but not other herbivores (James et al. 1982, McEvoy & Rudd 1993, McEvoy et al. 1993) is an effective way to quantify impacts of specific herbivores on their host plants.

ECONOMIC ASSESSMENT OF BIOLOGICAL CONTROL

Biological control projects must also be assessed in terms of their economic consequences. Augmentative projects are economically successful if pests are controlled at a competitive price and if sales of natural enemies provide adequate profits to producers to sustain the production of the natural enemy. Conservation methods of biological control can readily be evaluated economically by comparing production costs and crop yields under a conservation management system and some other approach. Unfortunately, the general economics of conservation biological control have never been assessed.

Many classical biological control projects, in contrast, have been evaluated economically, and summaries exist that define the average profitability of such work for a number of countries or sets of projects (Andres 1977, Harris 1979, Ervin et al. 1983, Norgaard 1988, Voegele 1989, Tisdell 1990, Bangsund et al. 1999, Hill & Greathead 2000, van den Berg et al. 2000, Bokonon-Ganta et al. 2002, Nordblom et al. 2002, de Groote et al. 2003, McConnachie et al. 2003, van Wilgen et al. 2004). For projects directed against arthropod pests of crops, the impact of the project on crop yield and on the crop's profitability to the farmer need separate evaluation.

A popular approach to estimating outcomes is to calculate benefit/cost ratios. For biological control projects such ratios have exceeded 100:1. Projects conducted in Australia, discussed by Tisdell (1990), averaged 10.6:1, compared to 2.5:1 for chemical control projects. Estimates for weed control projects in South Africa reached as high as 4333:1 for the golden wattle (*Acacia pycnantha* Bentham) project. Benefit/cost ratios increase over time for successful projects because: (1) in the case of crop pests, additional years of crops are protected or (2) for pests of rangeland or other wild areas, if the project checks the spread of the pest, then areas either are not infested that would otherwise have become infested or if the pest is already widespread, formerly infested areas over time come under control (as the natural enemy spreads) (Nordblom et al. 2002).

Benefit/cost ratios, however, do not reveal the absolute magnitude of the benefits achieved. Kinds of benefits are also quite variable. In the six projects discussed by van Wilgen et al. (2004) the economic value of water (otherwise lost to invasive plants infesting watersheds) accounted for 70% of the benefits. In contrast, suppression of leafy spurge (*E. esula*) in the northern plains of North America returned benefits in the form of increased forage expressed as "animal unit months." Control of red waterfern (*Azolla filiculoides* Lamarck) in South Africa was expressed in terms of reduced costs associated with providing drinking water to animals (McConnachie et al. 2003). Economic analyses for projects in which the agent has not yet reached all infested areas can show how many additional releases would be economically justified (Nordblom et al. 2002).

Most difficult to estimate are benefits to wildlife or even vaguer, the value of natural communities returned to more pristine natural vegetation. Assigning economic values to natural areas freed of weeds by biological control projects is more difficult than calculating the value of crop yields, but studies quantifying ecosystem services may provide a framework for some cases.

Part 8

CONSERVING BIOLOGICAL CONTROL AGENTS IN CROPS

Chapter 21

PROTECTING NATURAL ENEMIES FROM PESTICIDES

PROBLEMS WITH PESTICIDES

Pesticides are routinely used as the tool of first choice to control crop pests in the USA, Europe, Japan, and much of the developing world. So biological control agents in crops generally must coexist with pesticides or perish. How survival of natural enemies can be enhanced, despite pesticide use, is the focus of this chapter. The modern pesticide industry is rooted in World War II, when chemists developed products to kill insects, like mosquitoes and lice, to protect troops. The first and best of these was DDT, whose use saved thousands of lives by suppressing incipient typhus (a louse-borne disease) outbreaks among soldiers and crowded groups of displaced civilians (Cushing 1957). Immediately after the war, corporations recognized that chemical pest control was feasible and potentially profitable. The emerging pesticide industry focused on the discovery, mass production, and marketing of chemicals to kill insects, weeds, and protect against crop diseases. The new products were very popular and pesticide use increased rapidly in the late 1940s and 1950s. Products like DDT, 2-4-D (an herbicide), and captan (a fungicide) revolutionized farming by giving farmers highly effective tools to protect their crops and profits spawned a new corporate business organized for the explicit purpose of creating pesticides for use by farmers and homeowners. Initially, only the benefits of pesticide were seen: immediate suppression of pests to unheard-of levels. Improved ways to use pesticides – better ways to formulate them, better machines to apply them – were rapidly developed as scores of new products were rushed to market. A generation of farmers grew up for whom use of synthetic pesticides represented the norm for pest control.

Within a few years, however, problems with these synthetic pesticides were recognized by some foresighted ecologists (Carson 1962). One of the earliest pesticide failures was the development of populations that through selection and evolution were able to tolerate formerly lethal doses of poison (**pesticide resistance**). This problem was recognized quickly but was dealt with by finding effective replacement pesticides.

Another problem that soon emerged was the quick return of sprayed pest populations to damaging levels (**pest resurgence**). Also, new groups such as spider mites, which previously were considered nothing more than minor problems, became major pests requiring treatment on a regular basis (**secondary pest outbreak**). The ecological basis of these two problems took longer to understand. By the 1960s and 1970s, however, these problems had been shown by research to be due to the destruction of natural enemies in crop fields (DeBach 1974). For example in Japan, application of synthetic pyrethroids to peach orchards made problems with the eriophyid *Aculus fockeui* (Napela et Trouessard) worse by destroying the predatory mite *Amblyseius eharai* Amitai et Swirski (Kondo & Hiramatsu 1999).

As the volume of pesticides used (especially insecticides) increased, so did the contact between wildlife and pesticides and their residues. Harm to wildlife and contamination from pesticides developed as a major environmental crisis in the 1960s. Chlorinated hydrocarbons like DDT adversely affected the reproductive systems of raptors and wading birds, causing birds to lay eggs with abnormally thin shells. Eagle, falcon, osprey, and heron populations declined and even disappeared from large areas because residues of stable, fat-soluble pesticides circulating in the environment polluted their food chains (Graham 1970).

The problem of environmental pesticide residues was solved by banning most chlorinated pesticides in the 1970s and replacing them with organophosphates and carbamates. These new compounds, however, were actually more toxic to vertebrates and frequently caused illness in farm workers. Government programs in the USA were developed during the 1970s and 1980s to make these compounds safer by training farmers and farm workers in their use (the US-EPA pesticide applicator training program). In the 1990s, many organophosphate and carbamate pesticides uses were ended in the USA to improve food safety. In response, pesticide companies developed a wide array of new pesticides that were safer to both people and wildlife. Many of these compounds also were at least partially selective, allowing increased survival of natural enemies in crops. This timetable of events transpired in the USA, but many countries later followed these same steps.

In this chapter, we discuss first how pesticide use can be counterproductive by promotion of pest outbreaks, destruction of natural enemies, and development of pesticide resistance. We review ways in which pesticides can harm natural enemies and how such impacts might be lessened by finding and using **physiologically selective pesticides**. We show how, in some cases, even non-selective compounds may be made selective by modifying their formulations, or their time, place, or manner of application (**ecological selectivity**). We end by describing how the use of genetically modified crops that express toxins from the bacterium *Bacillus thuringiensis* Berliner has made some crops dramatically more favorable to natural enemies.

SUPER PESTS AND MISSING NATURAL ENEMIES

Resistance to pesticides

Pesticide resistance develops because those individuals in a population that are most tolerant of exposure to the chemical are those that survive and reproduce, leaving their genes better represented in the next generation. Mechanisms of pesticide resistance include enhanced detoxification and reduced cuticular penetration into the insect, among others. Pest populations may develop over several generations that can no longer easily be killed by one or more pesticides. Many insects, weeds, and plant pathogens became resistant to pesticides after 1945, some to many pesticides (Brent 1987, Georghiou & Legunes-Tejeda 1991) (Figure 21.1).

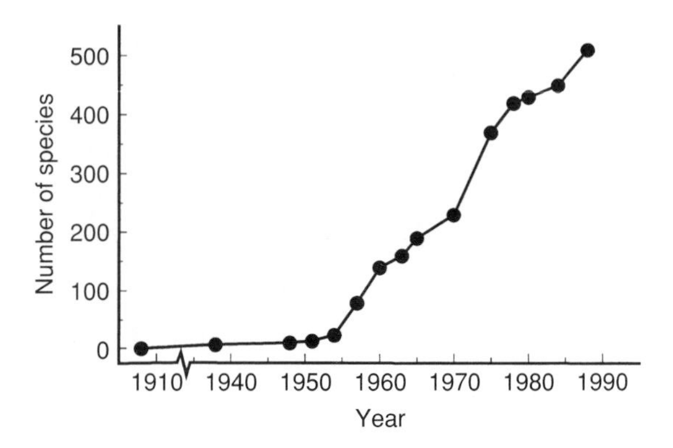

Figure 21.1 Cumulative number of cases of resistance to pesticides in arthropods. After Georghiou and Legunes-Tejeda (1991), reprinted from Van Driesche and Bellows (1996), with permission from Kluwer.

Key features affecting the development of pesticide resistance include the proportion of the population that is exposed to the pesticide and the intensity of any loss of fitness (in the absence of pesticides) that comes along with pesticide resistance. The proportion of a pest's breeding population that is exposed to a pesticide differs for many reasons, including its occurrence on hosts outside the crop and the level of treatment applied to the crop. The apple maggot [*Rhagoletis pomonella* (Walsh)], for example, did not develop resistance to commonly used orchard cover sprays such as azinphosmethyl (despite 40 years of use) because the breeding population of this fly occurs in untreated areas on wild hosts. In contrast, the Colorado potato beetle [*Leptinotarsa decemlineata* (Say)] repeatedly and rapidly developed resistance in the eastern USA, in part because it had virtually no wild hosts and tended to overwinter close to potato fields (e.g. French et al. 1992). Many pesticide products have similar chemisty so a pest resistant to one member of a class (like the carbamates) is likely to be resistant to others (termed **cross-resistance**).

Farmers may respond when pests develop resistance by increasing dosage, changing or alternating pesticides, or combining several pesticides. They may abandon chemical control altogether in favor of management systems based on biological control, including conservation of natural enemies, if resistance prevents control of the pest. Alternatively, if it is the natural enemies that develop pesticide resistance, it may be possible to conserve them in the crop even with continued pesticide use.

Hormoligosis

Another possible influence of chemicals on pest populations is **hormoligosis**. This term applies when sublethal levels of pesticides or elevated levels of crop fertilization induce higher pest reproductive rates, shorter developmental times, or a shift to earlier reproduction. Fecundity of citrus thrips, *Scirtothrips citri* (Moulton), increased significantly when thrips were reared on leaves with 21-day-old dicofol residues and 32-, 41-, and 64-day-old residues of malathion (Morse & Zareh 1991) (Figure 21.2). Lowery and Sears (1986) found that treatment of green peach aphid [*Myzus persicae* (Sulzer)] adults with sublethal doses of azinphosmethyl increased their fecundity 20–30%. For some sucking arthropods, such as spider mites, increased nitrogen levels in foliage from high levels of fertilization can cause higher survival rates, more rapid growth, and increased fecundity (van de Vrie & Boersma 1970, Hamai & Huffaker 1978, Wermelinger et al. 1985).

Pest resurgence

A quick return of pests to damaging levels sometimes follows the routine use of broad-spectrum pesticides.

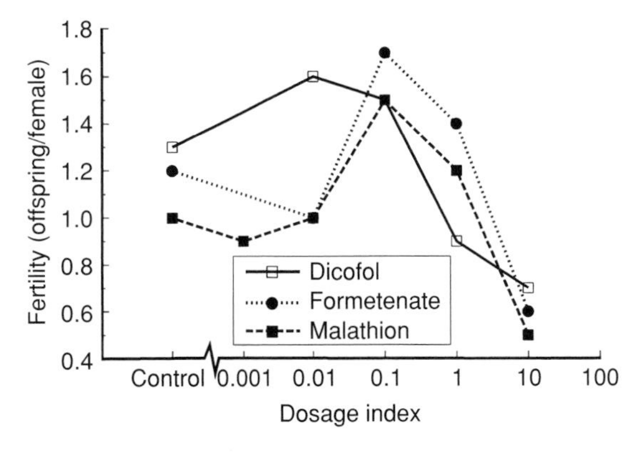

Figure 21.2 Fertility of citrus thrips, *Scirtothrips citri* (Moulton), as affected by dosage of three pesticides. At low doses, fertility was not different from controls; at high doses, fertility was reduced, but at intermediate doses, fertility was significantly higher than for controls. After Morse and Zareh (1991), reprinted from Van Driesche and Bellows (1996), with permission from Kluwer.

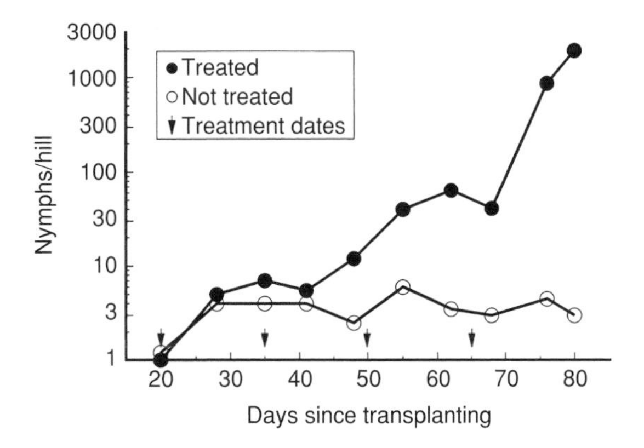

Figure 21.3 Resurgence of rice brown planthopper, *Nilaparvata lugens* (Ståhl), in rice fields treated with insecticides compared with fields not receiving treatments. After Heinrichs et al. (1982), reprinted from Van Driesche and Bellows (1996), with permission from Kluwer.

This phenomenon is termed pest resurgence and occurs because natural enemies are often more sensitive to pesticides than the pests they attack (Croft 1990). This may be due to lower levels of detoxification enzymes or a higher body-surface-to-mass ratio (leading to greater relative absorption of residues per unit of body weight). If the parasitoids and predators that normally attack a pest are destroyed, those pests that survive the pesticide application will live longer and have more offspring as a consequence. This allows the pest's numbers to quickly rebound to high levels. Pest resurgence has been observed in diverse crops and across many kinds of pests (Gerson & Cohen 1989, Buschman & DePew 1990, Talhouk 1991, Holt et al. 1992). One of the more widespread and dramatic examples has been that of *Nilaparvata lugens* (Stål), a planthopper found in rice crops in Asia (Figure 21.3) (Heinrichs et al. 1982).

Secondary pest outbreak

Another phenomenon associated with the use of broad-spectrum insecticides and miticides is outbreaks of insects and mites that are not normally pests. Outbreaks happen because the pesticides destroy the natural enemies that suppress these species to low densities (Figure 21.4). This phenomenon is called secondary pest outbreak. Spider mites, scales, and leafminers are examples of such secondary or pesticide-created pests (Luck

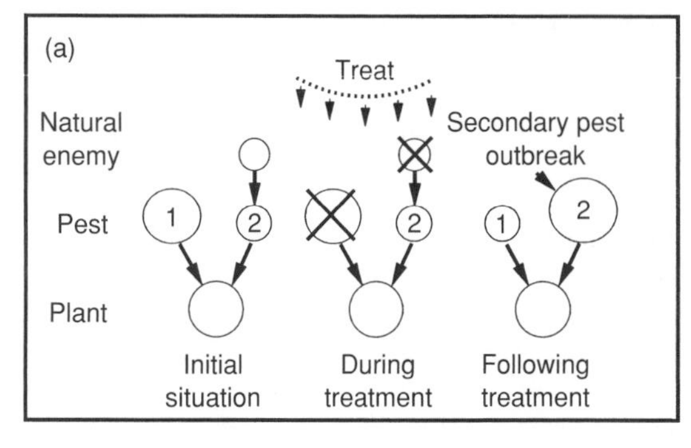

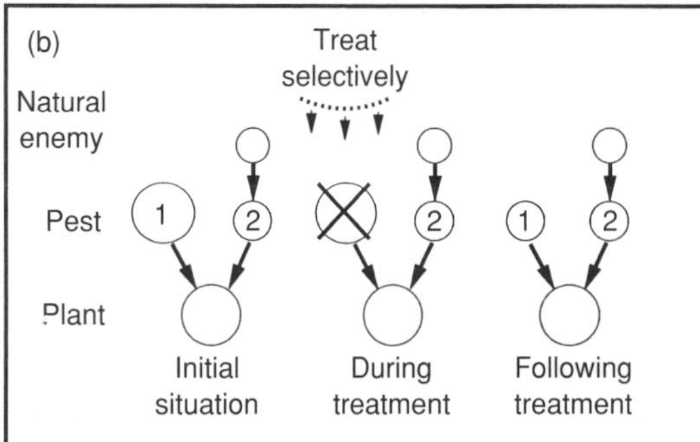

Figure 21.4 Conceptual diagram of a secondary pest outbreak. In (a), a general pesticide affects both the target pest, species 1, and also the natural enemy limiting species 2. Following treatment, species 2 undergoes population growth to pest levels in the absence of its natural enemies. In (b), a selective pesticide affects only the target pest, resulting in a situation where natural control of species 2 is not disturbed and there is no secondary pest outbreak. Reprinted from Van Driesche and Bellows (1996), with permission from Kluwer.

& Dahlsten 1975, Van Driesche & Taub 1983, DeBach & Rosen 1991). Secondary pest outbreaks differ from pest resurgence in that the causative pesticide applications do not target the secondary pest, but rather some other primary pest in the crop. From the grower's perspective, new species of pests arise in the crop that were not previously important. The crop IPM system then has to be enlarged to include controls for the new pests. Treatment of ornamental plants for caterpillars or aphids, for example, is often followed by eruptions of spider mites.

DEAD WILDLIFE AND PESTICIDE RESIDUES IN FOOD

Although problems caused by pesticides are not the focus of this book, it is important to recall their severity because they historically were, and still remain, one of the principal reasons why societies moved toward alternative pest controls such as biological control of crop pests.

Wildlife poisonings

From 1945 to about 1980, pesticide poisonings of wildlife in the USA and many other countries were common enough to be an important environmental issue that resulted in changes to pesticide laws (e.g. Hardy et al. 1986). Some poisonings occurred near the site of application and included: (1) bird kills from the ingestion of mercury or dieldrin-treated seed (Hardy et al. 1986), (2) fish kills from drift and run off into rivers or ponds adjacent to treated areas (Trotter et al. 1991), and (3) secondary kill from use of materials like sodium fluoroacetate used to kill pest vertebrates. Other wildlife deaths occurred in a more diffuse way, such as depression of acetyl cholinesterase enzyme levels (needed for normal nerve function) in songbirds in forests treated with carbamate or organophosphate pesticides (Mineau 1991). Birds with depressed enzyme levels did not die *en masse*, but succumbed, scattered over the landscape, from their inability to feed properly due to a loss of nerve function.

Other wildlife population declines were caused by widespread pesticide residues that interfered with reproductive physiology or produced sexual or developmental abnormalities. Egg-shell thinning in raptors and some other groups of birds, for example, caused some species to disappear over large areas (Burger et al. 1995). Tributyltin used in antifouling paints for boats caused feminization in marine mollusks (Horiguchi et al. 2004). Limb deformities and population declines of many amphibians occurred in the 1990s in many areas for uncertain reasons that are believed to include the effects of pesticide residues (Ankley et al. 1998).

Residues in food and the environment

Small quantities of some pesticides routinely move from treated areas into soil, water, birds, and wild animals. These residues may accumulate if they persist. The former widespread occurrence of DDT in human fat, wild fish, and other sources (Reimold & Shealy 1976, Jaga & Dharmani 2003), and the occurrence of triazine herbicides (like atrazine) in the rivers and groundwater of corn-producing areas (Pfeuffer & Rand 2004), are examples. The consequences of environmental pesticide residues varies from none to serious.

Residues of pesticides in food and drinking water are not prohibited by laws that govern pesticide use, but contamination must not exceed legal limits. Social debate over whether any levels of pesticides are acceptable in food has been one factor contributing to the increased popularity of organic food, which is produced without the use of most pesticides.

A current debate concerning pesticide residues is whether levels of parts per billion (now possible to detect and track due to improvements in analytical chemistry) of some compounds are innocuous as traditionally believed. Some pesticides resemble vertebrate hormones so the fear is that these residues could interfere with hormone function, leading to developmental and sexual abnormalities (Bustos-Obregon 2001, Palanza & vom Saal 2002).

Applicator illness

In the 1970s, to clean up environmental residues of chlorinated hydrocarbons, another group of insecticides (organophosphates and carbamates) were substituted for materials like DDT. However, many of these new nerve toxins were actually more acutely toxic to people than the chlorinated hydrocarbons they replaced. Use of these materials, especially by poorly trained applicators, resulted in many human poisonings (Graham 1970, Metcalf 1980, Dempster 1987, Newton 1988). Accidental poisonings and contamination were especially likely when farmers did not understand the toxicity of the materials they were using, when they could not read product instructions, or did not have or use the necessary protective equipment. Legislation enacted in the USA during the 1970s that required pesticide applicator safety training significantly reduced the frequency of pesticide accidents and misuse. However, applicator training and protective equipment are not consistently available in all countries. A farmer who cannot read is not able to benefit from safety information on a product's label. A poor farmer cannot purchase protective devices like respirators or special clothing for pesticide application. Although personal stories are just that, the senior author's memory of seeing barefooted, barehanded Andean potato farmers spooning Temik concentrate by hand into holes prepared for planting of potato seed pieces remains a vivid personal reminder that pesticides pose real risks to people in many parts of the world. Pesticide problems have clearly decreased in the USA over the last 40 years but this is not the case worldwide. Rather, pesticides are now in common use in many countries where they were unknown 40 years ago.

CASES WHEN PESTICIDES ARE THE BEST TOOL

Despite the deficiencies of pesticides mentioned above, there are circumstances in which pesticides are clearly superior to other forms of pest control, as for example, to control the vectors of human or animal diseases such as malaria, Lyme disease, bubonic plague, and typhus. Biological control may play a role in management of some of these problems (as in mosquito larval control), but often these programs are based on pesticides. Other critical uses for pesticides include eradication of invasive pests when they are detected early. Pesticides have been used, for example, to eradicate Mediterranean fruit fly [*Ceratitis capitata* (Wiedemann)] and the marine alga *Caleurpa taxifolia* (Vahl) C. Agardh from California, USA, and play a role in the containment/eradication of Asian long-horned beetle [*Anoplophora glabripennis* (Motschulsky)] in New York and Chicago, USA, killing beetles before they escape into natural forests. The high impact and rapid effect of pesticides, together with their ability to be precisely targeted, make them the right tool for eradication efforts.

In crops, some classes of established pests are also best managed with pesticides. Among these are pests with extremely low damage levels (e.g. direct fruit pests), pests that vector crop diseases, and pests for which biological control options are ineffective. Finally, pesticides commonly provide back up for biological and cultural controls in crops, which at times may fail or be insufficient.

HOW PESTICIDES AFFECT NATURAL ENEMIES

Insecticides can reduce the effectiveness of the natural enemies of arthropods by causing mortality or by influencing their movement, foraging, or reproductive rate (Jepson 1989, Waage 1989, Croft 1990).

Direct mortality

Many pesticides are directly toxic to important natural enemies of pest arthropods (Bartlett 1963, 1964b,

1966, Bellows & Morse 1993, Bellows et al. 1985, 1992a, Morse & Bellows 1986, Morse et al. 1987). Some pesticides may be toxic to species not suggested by the product's category. A bird repellent may be insecticidal. A fungicide may kill arthropods (e.g. sulfur is damaging to phytoseiid mites) or affect their reproduction or movement (dithiocarbamate fungicides that reduce phytoseiid reproduction rates). Herbicides may kill beneficial nematodes applied for insect control (e.g. Forschler et al. 1990). Therefore it is important to assume, until data to the contrary become available, that any pesticide, of whatever type, might affect a natural enemy (Hassan 1989). Even materials often thought of as non-toxic, such as soaps or oils, which may be safe to humans, may be injurious to natural enemies or cause environmental harm. Oils, for example, when applied to scale species, are likely to reduce emergence of scale parasitoids as well as cause mortality to scales (Meyer & Nalepa 1991).

The degree of mortality caused by a pesticide to a natural enemy population will depend on both physiological and ecological factors. Physiological selectivity implies differential intrinsic toxicity of the compound to the pest and the natural enemy. Chemicals vary greatly in their inherent toxicity to given species (Figure 21.5) (Jones et al. 1983, Smith & Papacek 1991). Some materials have been discovered that are effective against pests but relatively harmless to arthropod natural enemies (see www.koppert.com and consult the Side Effects List for examples).

Ecological selectivity results from how a pesticide is formulated and applied. Any factor that reduces contact of the natural enemy with the pesticide can confer ecological selectivity. Materials, for example, that have short residuals after application, or that act only as stomach poisons, can be ecologically selective. Non-selective materials can be used selectively if they are applied in spatial patterns or at times that limit contact with natural enemies.

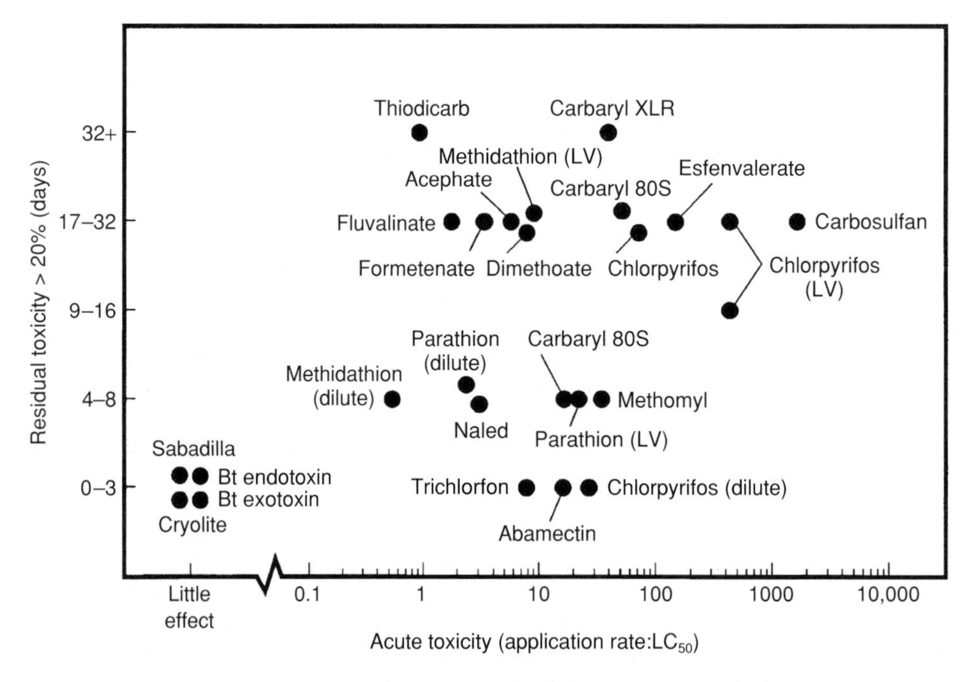

Figure 21.5 Toxicity to *Aphytis melinus* (DeBach) of pesticides used in citrus agriculture. The horizontal axis is immediate (acute) toxicity on leaves bearing freshly deposited residues, and the vertical axis is length of residual toxicity. Note the wide range of both acute toxicities and length of residual action. After Bellows and Morse (1993), reprinted from Van Driesche and Bellows (1996), with permission from Kluwer.

Non-lethal harm

Besides suffering increased mortality, natural enemies may become less effective after pesticide use if sublethal doses of pesticides shorten their longevity, decrease developmental rates, reduce foraging efficiency, are repellent, or lower reproduction. Some kinds of indirect effect can be detected from laboratory assays (Croft 1990, Van Driesche et al. 2006).

Reduced fecundity

Some pesticides that do not kill pests nevertheless lower their reproduction (e.g. Van Driesche et al. 2006a). Hislop and Prokopy (1981) found that the fungicide benomyl caused complete sterility to females of the predatory mite *Neoseiulus fallacis* (Garman) and predicted that benomyl use in apple orchards would cause mite outbreaks, which it did. The fungicides thiophanate-methyl and carbendazim inhibit oviposition by *Phytoseiulus persimilis* Athias-Henriot (Dong & Niu 1988). Several insect growth regulators reduce fecundity of coccinellids or sterilize their eggs (Hattingh & Tate 1995, 1996).

Repellency

Some materials that are not directly toxic to certain natural enemies may make treated surfaces or hosts repellent, causing natural enemies to leave. The herbicides diquat and paraquat, for example, made treated soils in vineyards repellent to the predaceous mite *Typhlodromus pyri* (Scheuten) (Boller et al. 1984). Hoddle et al. (2001b) found that among several insect growth regulators, dried residues of those formulated with petroleum distillates were repellant to the whitefly parasitoid *Eretmocerus eremicus* Rose and Zolnerowich, while materials formulated as dry wettable powders were not.

Accumulation of sublethal doses

In addition to the above, natural enemies may also be harmed by accumulating small amounts of pesticide until a lethal threshold is reached. For example, to find the survivors of a pesticide application, natural enemies may have to search more foliage, increasing their exposure to pesticide residues. Accumulation may also occur if predators feed on prey that have ingested sublethal quantities of pesticides. For example *Rodolia cardinalis* (Mulsant) in citrus may be harmed as it feeds on large numbers of cottony cushion scales (*Icerya purchasi* Maskell), each of which may contain a small amount of pesticide (Grafton-Cardwell & Gu 2003).

SEEKING SOLUTIONS: PHYSIOLOGICAL SELECTIVITY

One approach to limiting the harm to key natural enemies from pesticides is to use only materials that, while able to kill the pest, are relatively harmless to the natural enemies of concern. If feasible, substitution of a microbial pesticide (see Chapter 24) for a chemical pesticide can enhance crops or other situations as habitat for predators or parasitoids. For example, when applications of the entomopathogenic fungus *Beauveria bassiana* (Saccardo) Petch replaced pyrethrin treatments for fly control in poultry houses, numbers of adult house fly (*Musca domestica* L.) decreased, and numbers of fly larvae were reduced by half. This reduction seemed to result from larger populations of predatory histerid beetles, which increased by 43–66% (depending on life stage) after *B. bassiana* replaced pyrethrin (Kaufman et al. 2005).

Alternatively, a search can be made for the most compatible conventional chemical pesticide, starting with a focus on materials like stomach poisons (rather than contact poisons) or insect growth regulators. Beyond such groups, more conventional pesticides do vary in their toxicity to natural enemies. However, screening pesticides available in a particular crop against local natural enemies, although valuable, is not commonly done because: (1) the small number of pesticides registered for a specific crop may preclude success, (2) data generated lack broad application because the suite of registered pesticides will vary by crop and country, and (3) the natural enemies of interest will vary by crop and country and toxicity of a pesticide to one natural enemy does not predict its toxicity to another (Bellows & Morse 1993). Even populations of the same natural enemy species collected from different locations may differ in their susceptibility to a pesticide (Rosenheim & Hoy 1986, Rathman et al. 1990, Havron et al. 1991). Only local screeing of a crop's major pesticide–natural enemy combinations will define which materials might be used safely. Such screening was done, for example, to find materials to kill brown planthopper, but not harm spiders, on rice in the Philippines (Thang et al. 1987). Data on effects of pesticides on natural enemies are not required as part of pesticide registration, except by the European Union.

The most common measure used to express susceptibility to a pesticide is the LC_{50}, which is the concentration of solution applied to a treated surface that kills half of the test organisms during a period of time (usually 24 or 48 h). The ratio of the LC_{50} values of the natural enemy and the pest, or that of the natural enemy to the recommended application rate for a pesticide, is a useful comparative measure of the selectivity of a pesticide (Morse & Bellows 1986, Bellows & Morse 1993). Some pesticides are highly toxic to natural enemies only soon after application, whereas others, of varying toxicity, may be extremely persistent. Tests with residues of various ages are needed to define how long any particular natural enemy will be at risk from a pesticide application (Bellows et al. 1985, Morse et al. 1987). Assessment of natural enemy performance (ability to encounter and subdue prey successfully or, for parasitoids, to locate and oviposit in hosts) is a better indicator of the total effect of pesticide residues than is mortality because it also incorporates the sublethal effects of pesticides on natural enemies.

Methods for assessing compatibility of pesticides with natural enemies range from laboratory tests, through semi-field tests, to field studies (see Vogt 1994). Laboratory methods include treatment of natural enemies through ingestion of pesticide or pesticide-treated materials, topical application, and placement of natural enemies on surfaces with pesticide residues. Test results are sensitive to the precise conditions selected for the assay (insect age, sex, and rearing history, temperature, relative humidity, and ventilation of the test environment, and the formulation, purity, and dosage of the test material) (Croft 1990). Standard methodologies have been developed (Hassan 1977, 1980, 1985, 1989, Morse & Bellows 1986, Hassan et al. 1987, Vogt 1994). Basic to all tests is the comparison of the pest and natural enemy organism under the same conditions to determine whether differences in susceptibility exist.

Methods include the slide-dip technique in which small organisms such as mites are fixed to tape on a glass slide and then dipped in a pesticide solution. For other species, pesticide residues can be presented in vials or Petri dishes on glass, sand, or leaves. If presented on foliage, plant materials may be sprayed in the laboratory or field, and then presented to insects in cages of varying size, either when the spray solution dries or after aging under field or standardized laboratory conditions. Field tests involve assessing impacts on natural enemy populations when whole fields or plots are treated with pesticide. Large plots (replicated over time) are often necessary in field tests because mobile natural enemies can move among small plots resulting in poor separation of treatment effects (Brown 1989, Smart et al. 1989).

Natural enemies in general are sensitive to pesticides, but some groups, such as lacewings (*Chrysoperla* spp., Chrysopidae, Neuroptera) are uncharacteristically tolerant of pesticides (e.g. Grafton-Cardwell & Hoy 1985). Also, mites and other natural enemies may through selection develop pesticide resistance in crop fields that have a long history of pesticide application.

PESTICIDE-RESISTANT NATURAL ENEMIES

Populations of pesticide-resistant natural enemies have sometimes developed through natural selection in regularly sprayed crops like apples. Resistant populations may also be created artificially in the laboratory. Pesticide-resistant populations have been found (or selected) for several predatory mites: *Metaseiulus occidentalis* (Nesbitt) (Croft 1976, Hoy et al. 1983), *P. persimilis* (Fournier et al. 1988), *T. pyri* and *Amblyseius andersoni* (Chant) (Penman et al. 1979, Genini & Baillod 1987), and *Neoseiulus fallacis* (Garman) (Whalon et al. 1982). Pesticide-resistant parasitoids include an aphid parasitoid (*Trioxys pallidus* Haliday) (Hoy & Cave 1989), a leaf miner parasitoid [*Diglyphus begini* (Ashmead) (Rathman et al. 1990)], and some scale parasitoids [*Aphytis holoxanthus* DeBach (Havron et al. 1991) and *Aphytis melinus* DeBach (Rosenheim & Hoy 1986)].

Some pesticide-resistant natural enemies have been moved to new locations for establishment where regular pesticide applications are required. Initial establishment of resistant strains can be fostered by prior destruction through pesticide application of any existing susceptible population of the same species (Hoy et al. 1990). Generally, multi-year persistence of the resistant strain is necessary for effective use in outdoor crops. This has been achieved in some commercial fields or orchards where pesticide applications are made (Hoy 1982, Hoy et al. 1983, Caccia et al. 1985). Regular pesticide application is necessary to prevent cross-breeding of the resistant strain with the susceptible wild types, and prevent the resistant population's displacement (Downing & Moilliet 1972). Trials in the UK with an organophosphate-resistant strain of *T. pyri* showed survival of the predator in orchards treated with organophosphate insecticides at levels sufficient to control *Panonychus ulmi* (Koch) and *Aculus schlechtendali* (Nalepa). In a pyrethroid-treated orchard (a pesticide this strain was not resistant to) the resistant *T. pyri* was scarce and did not suppress pest mites (Solomon et al. 1993).

There is less evolutionary pressure against resistant strains used in greenhouse augmentative control programs because typically there is no susceptible population present with which the released agents would cross-breed, and the resistant agents can be reapplied as needed. Resistant strains of *P. persimilis* in greenhouse crops have been found to persist for the life of the crop (Fournier et al. 1988). The nematode *Heterorhabditis bacteriophora* Poinar has been selected for resistance to common nematicides such as avermectin, allowing them to used when crops must be treated for control of plant parasitic nematodes (Glazer et al. 1997).

Natural enemies used in augmentative control programs can also be modified for other attributes intended to improve performance (see Chapter 25). For example, some nematode strains have been selected for increased tolerance to heat or desiccation (Shapiro-Ilan et al. 2005).

ECOLOGICAL SELECTIVITY: USING NON-SELECTIVE PESTICIDES WITH SKILL

If no pesticide is available for a particular crop that is compatible with key natural enemies, it may still be possible to find ways to make the available pesticides somewhat selective by changing the ways they are formulated or applied. Ecological selectivity is achieved by reducing the contact between the pesticide and key natural enemies (Hull & Beers 1985).

Reduced dosages

Lowering the dosage may make a pesticide less damaging to natural enemies (Poehling 1989). However, this may provide too little control of the pest because natural enemies are often more sensitive to pesticides than are herbivores.

Selective formulations and materials

Formulation affects exposure. Granular formulations applied to the soil, for example, generally do not contact natural enemies foraging on foliage (unless applied materials have fumigation activity, see below) and hence many natural enemies are unaffected by granular formulations (Heimbach & Abel 1991). Granular materials, however, are designed for the purpose of producing pesticide residues in the topsoil and thus contact of the pesticide with soil-dwelling natural enemies like carabids may be extensive. Some materials such as chlorpyrifos, when applied as granules in citrus groves for ant control, have fumigation activity and kill natural enemies in the tree's foliage through volatilization.

Systemic pesticides move internally in treated plants and leave no external residue. Such materials do not harm natural enemies, provided they do not also consume plant sap (Bellows et al. 1988). Stomach poisons kill only if ingested and are less likely to harm natural enemies than pesticides that kill by contact (Bartlett 1966). Stomach poisons include the toxins of *B. thuringiensis*, some plant alkaloids (Bellows et al. 1985, Bellows & Morse 1993), and mineral compounds such as cryolite (Bellows & Morse 1993).

Limiting the areas treated

Reduced coverage or selective placement of pesticides can protect natural enemies. Treatment of alternate rows in apple orchards instead of entire blocks controls mobile orchard pests, yet allows greater survival of coccinellids, such as the mite predator *Stethorus punctum* (LeConte) (Hull et al. 1983). DeBach (1958) successfully controlled purple scale, *Lepidosaphes beckii* (Newman), in citrus by applying oil to every third row on a 6-month cycle. This approach provided satisfactory control of this species without destroying the natural enemies of other citrus pests. Velu and Kumaraswami (1990) found treatment of alternate rows in cotton to provide effective pest control and, for some of the chemicals tested, to enhance parasitism levels of key pests. In contrast, Carter (1987) found that strip spraying of cereals in the UK did not provide satisfactory control of aphids when strips were 12 m wide because natural enemies did not colonize the sprayed strips quickly enough to suppress aphid resurgence.

Limiting applications in time

In principle, carefully timed applications of non-persistent pesticides might spare natural enemies. Whereas some adults might be killed, these would be replaced through emergence from cocoons, mummies, or other protective stages after toxic residues have dissipated. Persistence of pesticides varies greatly. Materials such as diazinon or azinphosmethyl leave toxic residues on foliage for several weeks. Other materials, such as pyrethrin, degrade in hours or days. Timing can be manipulated by either: (1) reducing application frequency so crop foliage is not always toxic to natural enemies or (2) timing applications specifically to avoid periods when natural enemies are in vulnerable life stages. Gage and Haynes (1975), for example, used temperature-driven models of insect development to time pesticide applications against adult cereal leaf beetles, *Oulema melanopus* (L.), treating after emergence of beetles and before that of the parasitoid *Tetrastichus julis* (Walker). This system conserved this important parasitoid, whereas the previous approach of directing pesticide applications at the first generation of cereal leaf beetle larvae (the stage attacked by the parasitoid) did not.

System redesign

The most complete way to reduce harm from pesticides is to eliminate use of broad-spectrum materials by substituting alternative pest control methods such as traps, mating disruption with pheromones, or cultural controls. Replacement of broad-spectrum cover sprays for control of codling moth, *Cydia pomonella*, in pear (*Pyrus communis* L.) orchards in Oregon, USA, with mating disruption through the use of pheromones raised the densities of the predaceous hemipteran *Deraeocorus brevis piceatus* Knight and the lacewing *Chrysoperla carnea* (Stephens) and lowered densities of the key secondary pests, pear psylla, *Psylla pyricola* Förster, by 84%. The proportion of fruit contaminated by psyllid honeydew dropped from 9.7 to 1.5% (Westigard & Moffitt 1984).

TRANSGENIC BT CROPS: THE ULTIMATE ECOLOGICALLY SELECTIVE PESTICIDE

Transgenic crops, expressing *B. thuringiensis* (Bt) toxins, are the ultimate selective pesticide. Their use has dramatically increased conservation of natural enemies in key crops (cotton, corn, soybeans). Practical use of Bt crops became possible based on DNA transfer and expression technology developed in the 1990s. The discovery of suitable promoter genes that stimulated high levels of expression of transferred genes was a key development. This allowed crops to be created that produced high enough levels of Bt toxins in target tissues to control key pests (Shelton et al. 2002).

Many Bt toxins exist and these vary with respect to the exact species of pests for which they are lethal. For example, cotton containing the Cry1Ac protein (Bollgard®), which has been grown in the USA since 1996, provides control of the key cotton lepidopteran pests [*Heliothis virescens* (F.), *Pectinophora gossypiella* (Saunders), and *Helicoverpa zea* (Boddie)] (Moar et al. 2003). Given the previous eradication (by areawide pesticide treatments and attract-and-kill traps) of the boll weevil (*Anthonomis grandis grandis* Boheman), this meant that Bt cotton had no uncontrolled boll pests. Bt cotton is producible with significantly fewer pesticide applications than conventional cotton. In the USA, Williams (1999) estimated that (across six states, comparing 1995 as the pre-Bt year with 1998) the number of insecticide applications dropped from an average of 4.8 to 1.9, a reduction of 60%. Similarly in China, the use of Bt cotton is estimated to have reduced pesticide use in the crop by 60–80% (Xia et al. 1999) In Bt sweetcorn, pesticide reductions of 75–100% are possible (Dively & Rose 2003). This reduction in insecticide dramatically improved the crop habitat for natural enemies.

Grower adoption of Bt crops varies by region, with strongest adoption in the USA, Canada, China, and Argentina. In US cotton, Bt varieties comprise up to 80% of production by region. The global area of Bt crops increased more than 25-fold between 1996 and 2000, reaching some 44.2 million ha, which represented a very rapid rate of growth in area (James 2002).

Bt crops are thus reducing pesticide use on a scale that outstrips all other IPM efforts to shift crop production away from pesticides. Although studies tracking Bt residues in insects have demonstrated that Bt toxins can be acquired by non-target natural enemies, such as predators that sometimes feed on plants, or predators that eat intoxicated pests, such residues are rare and small (Torres et al. 2006). Field studies of Bt crops compared to the same crop managed with conventional pesticides show that natural enemies of all groups either increase or remain the same in Bt crops (Dively & Rose 2003, Moar et al. 2003, Naranjo & Ellsworth 2003, Head et al. 2005, Naranjo 2005, Naranjo et al. 2005). The only natural enemy taxa that decrease are those specialized to attack the target pest (Venditti & Steffey 2003), which is purely a side effect of the intended control of the host. Also, studies have shown that Bt toxins do not persist or accumulate in soil where Bt crops are planted during successive years (Dubelman et al. 2005).

Bt crops are an extremely positive development promoting the conservation of natural enemies in crops. Should resistance develop, alternative Bt genes or combinations of genes appear to offer methods to reimpose control. Prevention of development of resistance is being attempted by the maintenance of non-Bt blocks of the crop nearby as sources of susceptible pests to overflood and mate with any incipient resistant individuals from Bt blocks. An ecological by-product of the reduction of conventional pesticide use in Bt crops has been, however, that some secondary pests, such as cotton fleahopper [*Pseudatomoscelis seriatus* (Reuter)], have increased in importance. However, these species are of minor importance compared to key pests of the crop.

Chapter 22

ENHANCING CROPS AS NATURAL ENEMY ENVIRONMENTS

Farmers manage their crop fields for efficient production, usually with little consideration of the needs of the species that attack the crop's pests. Crop fields may, therefore, become unfavorable environments for some natural enemies. **Conservation biological control** is the attempt to improve this situation, making crop fields more hospitable for parasitoids and predators (Barbosa 1998, Pickett & Bugg 1998, Gurr et al. 2004) whenever this can be done without loss of productivity. Conservation biological control practitioners seek ways to alter cropping systems that restore the features needed by natural enemies. For successful use of this approach several things must be true: (1) the lack of some key attributes in the crop environment must contribute substantially to the pest problem, (2) restoration of the missing attributes must be possible without compromising production, and (3) the cost of restoring the attributes must not be higher than other forms of pest control available to growers.

In this chapter, we frame the discussion of conservation biological control around five potential inadequacies that crops might pose for natural enemies: (1) the crop species or variety may be a poor substrate for the natural enemy because it physically or chemically impedes natural enemy foraging or inhibits normal development, (2) the physical environment in the crop field may be too harsh, (3) key sources of natural enemy nutrition may be lacking, (4) opportunities for natural enemy reproduction may be constrained by the absence of hosts or prey, or (5) diversity, connectivity, or refuges needed to supply natural enemies to colonize newly planted crops may be inadequate. Potential solutions to each of these problems are discussed. We also discuss some additional farming practices, such as destruction of crop residues, that influence natural enemy populations.

PROBLEM 1: UNFAVORABLE CROP VARIETIES

It is on the crop plants that many natural enemies spend a significant part of their lifetime and where they must find their hosts or prey. Crop species and varieties differ in many ways that can matter to natural enemies, including the exact nature of the plant's physical surface (as foraging space or refuge), the chemistry of the plant's tissues in terms of toxins that may reach the natural enemies through the pest herbivore, and the presence or absence of nectaries or other sources of nutrition useful to natural enemies (see Ode 2006 for a review). Additionally, the chemistry of the plant, as it interacts with pest herbivory, may be significant in determining whether infested plants emit odors able to attract foraging natural enemies from a distance.

Plant breeders might avoid making plants less favorable if they understand how crop features influence natural enemies, especially when the loss of positive features is merely an accident and not the intent of the breeding program. More actively, it might sometimes be possible to make plants more favorable to key natural enemies by enhancing plant production of attractive volatiles or production of nectar or other resources needed by natural enemies (Bottrell & Barbosa 1998, Vinson 1999, Cortesero et al. 2000). Creation of varieties favorable to natural enemies, and their subsequent use by farmers, will then improve the crop as habitat for natural enemies. Below we approach this goal by developing an understanding of what crop features are detrimental to natural enemies, thus showing by contrast what features a favorable crop plant should possess.

Unfavorable crop surfaces

The crop plant's surface is the arena on which the natural enemy must forage. Various surface features might affect natural enemies, including the kinds and density of trichomes (and associated chemical exudates) (Simmons & Gurr 2005) and the presence of waxy blooms. High trichome density on cucumber, for example, lowered whitefly parasitism by *Encarsia formosa* Gahan by reducing the walking speed of foraging females on leaves (Hua et al. 1987). Similarly, high densities of two kinds of trichomes on tomatoes increases the entanglement rate of lacewing larvae (Simmons & Gurr 2004) and parasitoids (Figure 22.1). The interaction between natural enemies and leaf features such as trichomes, however, varies with the kind of natural enemies and their needs. The phytoseiid mites *Typhlodromus pyri* Scheuten and *Amblyseius aberrans* (Oudemans) released into Italian vineyards with several grape varieties become more abundant on varieties with hairy leaf undersurfaces than on those with glabrous surfaces, presumably because the leaf hairs created a layer of air with higher humidity (Duso 1992).

Even for natural enemy–pest combinations in which trichomes (particularly those with sticky exudates) are harmful, this effect may be less severe in the field than in laboratory tests (Obrycki & Tauber 1984), for several reasons. First, some natural enemies will be larger-bodied species that will be less affected by sticky trichomes. Second, the exudates on trichomes may be removed or rendered less effective in the field by rain or dust. Thus one should proceed with caution in extrapolation of laboratory studies of crop feature effects, relying on field studies to verify practical importance.

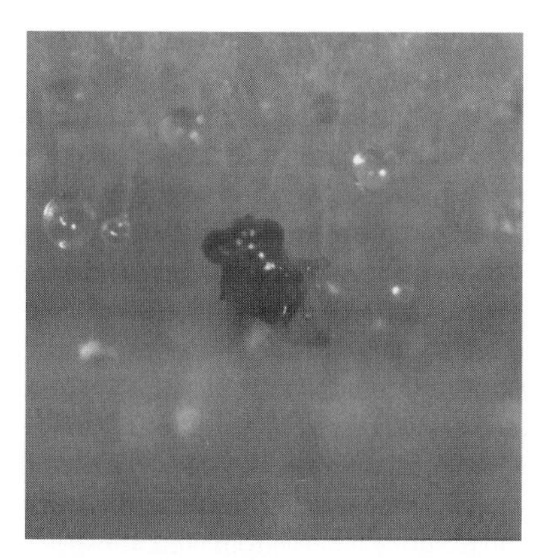

Figure 22.1 Leaf surface architecture strongly affects small parasitoids. Here the parasitoid *Encarsia luteola* Howard has become entrapped and killed by the sticky exudates (see droplets) on trichomes. Photograph courtesy of David Headrick.

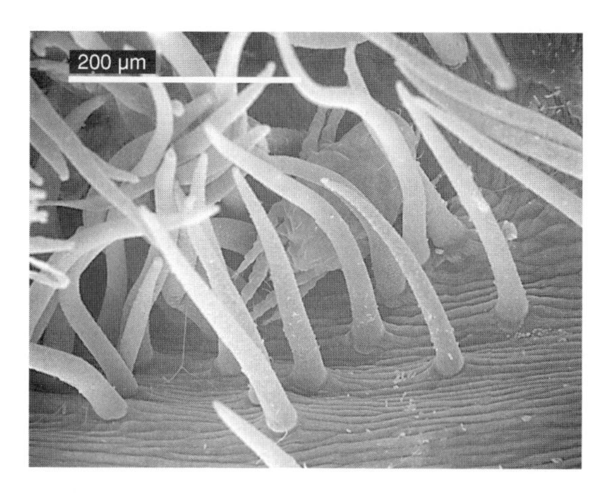

Figure 22.2 Densities of phytoseiid mites are higher on plants whose leaf architecture include domatia. These are pits or pockets, often enclosed by trichomes, that provide protection and higher relative humidity. Here we see a phytoseiid mite in a domatia. Photograph courtesy of Andrew Norton.

Another surface feature, the waxy bloom typically found on cabbage plants, has been shown to reduce the ability of some predators such as larvae of the lacewing *Chrysoperla plorabunda* (Fitch) to move effectively, compared to "glossy" mutants lacking the typical waxy bloom. This decrease in predator mobility reduced the rate of predation on diamondback moth [*Plutella xylostella* (L.)] neonates on normal compared to glossy mutant cabbage (Eigenbrode et al. 1999). These sort of effects also operate across plant species. Searching is often least effective on plants whose leaves have dense upright or hook-shaped hairs (Shah 1982), or glabrous, slippery surfaces, which caused some ladybird beetles to fall more frequently (Grevstad & Klepetka 1992).

In addition to crop surface features that harm natural enemies, other features provide refuges that may increase the local abundance of natural enemies. Acarodomatia are tufts of non-glandular trichomes or pit-like structures located at junctions of major veins on leaves of grape and other plants (Figure 22.2). Phytoseiid mites enter and remain in such areas, laying more eggs on leaves with natural acarodomatia than on leaves with artificially sealed acarodomatia. Plants with more or larger acarodomatia have higher numbers of phytoseiid mites per unit area (Walter & O'Dowd 1992, English-Loeb et al. 2002). More predators on leaves with domatia can translate into lower densities of herbivores (Agrawal et al. 2000), but this is not always the case (Agrawal 1997).

Toxic tissues

The chemistry of a plant's tissues can dramatically affect the suitability of herbivores as hosts for parasitoids. An extreme example of this is California red scale, *Aonidiella aurantii* (Maskell), on sago palm (*Cycas revoluta* Thunb.) compared with citrus. Both plants are suitable for the scale, but scale on sago palm is completely unsuitable for the parasitoid *Habrolepis rouxi* Compere, which experiences 100% mortality of its immature stages, compared with 3–17% when reared in the same scale on citrus (Smith 1957). In tomatoes, the secondary plant compound α-tomatine inhibits the entomopathogenic fungus *Nomuraea rileyi* (Farlow) Sampson (Gallardo et al. 1990). The same compound also reduces the rate of adult emergence of the ichneumonid parasitoid *Hyposoter exiguae* (Viereck), which attacks larvae of *Heliothis zea* (Boddie). The intensity of this effect is determined by the ratio of α-tomatine to total sterols, which have a protective effect. This ratio varies by five-fold among tomato cultivars (Campbell & Duffey 1981), suggesting that such varieties would differ significantly in suitability for this parasitoid.

Lack of attractive volatiles

Parasitoids commonly locate hosts from a distance by detecting and moving towards the blend of volatiles emanating from plants damaged by feeding of specific herbivores (see Chapter 3). For example, *Cotesia marginiventris* (Cresson), a braconid parasitoid of various pest Lepidoptera, finds its hosts by flying towards odors of the host–plant complex associated with recent larval feeding (Turlings et al. 1991). Pests on the crop would become undiscoverable to their parasitoids on new cultivars if these no longer emitted their characteristic attractive compounds. Whereas opportunities to improve the chemical attractiveness of crops may be few, these relationships have to be considered in plant breeding so that new varieties do not lose their abilities to attract parasitoids.

Nectarless crops

Parasitoids feed on plant nectar from either floral or extrafloral glands. Crop varieties with nectaries retain parasitoids more effectively than varieties without nectaries (Stapel et al. 1997). Thus, breeding varieties without nectaries, as is sometimes suggested as a means to suppress pests (which may also feed on nectar), makes crops less suitable as habitats for parasitoids.

SOLUTION 1: BREEDING NATURAL ENEMY-FRIENDLY CROPS

All of the issues mentioned above are driven by the attributes of the crop variety planted by the growers. The grower can modify outcomes only by switching among existing varieties. Varieties, however, may vary in other ways likely to matter to consumers and growers such as yield, flavor, time to maturity, disease resistance, or tolerance to adverse soil conditions. The choices available to growers are the traditional regional varieties, plus new material imported or developed by plant breeding programs.

Development of new varieties by breeders should strive to include "natural enemy compatibility" into the mix of features examined when shaping a new variety. One source of such information is the use of data from screening programs in which natural enemies are allowed to forage for sentinel hosts placed at equal density on various cultivars. The effects of various trichome conditions in *Lycopersicon* (tomato) species on the lacewing *Mallada signata* [Schneider] (a species produced commercially in Australia) were examined by Simmons and Gurr (2004). Similarly, such information was used to guide efforts to increase parasitism of whiteflies on greenhouse cucumbers by developing a variety with a moderate trichome density that would produce a parasitoid walking speed optimal for parasitism (Hua et al. 1987).

Attempts to make super-hairy or nectarless crop varieties, intended to be unfavorable to pests, may be counterproductive if they strongly suppress the level of natural control in the crop. Breeding programs need to balance features favorable to natural enemies with attributes useful for pest resistance in view of the most likely pest complex in the production area. In cotton, for example, it was found that glabrous varieties were more suitable for *Trichogramma* wasps and lacewing larvae than were varieties with intermediate or high trichome densities. Also glabrous cotton was more resistant to *H. zea*. However, high-density varieties were more resistant to pest Hemiptera [*Lygus* sp., *Pseudatomoscelis seriatus* (Reuter)]. So, in areas where pest Hemiptera were not important, glabrous cotton would be the best variety, but in areas with bug pests, use of varieties with intermediate levels of trichomes would be a better choice (Treacy et al. 1986, 1987).

Alternatively, it may be feasible to artificially treat crops with compounds [such as methyl salicylate, methyl jasmonate, or (Z)-3-hexenyl acetate] that are attractive to natural enemies (James & Grasswitz 2005) or with compounds such as jasmonic acid that induce plants to produce such volatiles (Lou et al. 2005).

PROBLEM 2: CROP FIELDS PHYSICALLY DAMAGING TO NATURAL ENEMIES

Hot, dry soils

In many regions, soils become hot and dry following cultivation and exposure to the sun. This makes the soil surface unfavorable for soil-dwelling groups such as carabid and staphylinid beetles. This commonly occurs when crops that do not produce closed canopies able to shade the soil (such as some vegetable crops) are grown in hot climates. Also, hot conditions may develop temporarily after whole fields are mowed, such as when alfalfa or other forage crops are harvested.

Mechanical disturbance from soil tillage

Tillage for weed control also is physically disruptive to species that enter spaces within the soil for shelter or to pass some stage in their development. Carabids, for example, often enter earthworm tunnels to seek damper, cooler soil, but such tunnels are destroyed by cultivation. Plowing of soil in Swedish rape crops reduced emergence of parasitoids of rape pollen beetles, *Meligethes* sp., by 50 and 100% in spring and winter crops (Nilsson 1985). Direct-drilling of winter wheat crops that follow rape (*Brassica napus* L.), rather than plowing, can conserve pollen beetle parasitoids. In Ontario, Canada, an outbreak of cereal leaf beetle, *Oulema melanopus* (L.), occurred when crops were tilled immediately after harvest instead of the more common practice of using cereals as a companion crop with alfalfa, without tillage (Ellis et al. 1988). Tillage killed 95% of the key cereal leaf beetle parasitoid *Tetrastichus julis* (Walker), which was absent in the outbreak area, but parasitized 74–90% of the pest's larvae in other parts of Ontario.

Also, soil is a major reservoir for viral and fungal pathogens, which may be affected by tillage. Hepialid pasture pests (*Wiseana* spp.) in New Zealand cause greater damage in recently tilled pastures because cultivation buried the nucleopolyhedrosis virus of these pests, lowering disease rates. As reseeded pastures age, viral levels build back up, increasing pest mortality (Longworth & Kalmakoff 1977).

SOLUTION 2: COVER CROPS, MULCHING, NO-TILL FARMING, STRIP HARVESTING

Methods for natural enemy conservation must be developed in response to cultivation practices in a given area, the options that exist for natural enemy conservation, and which are acceptable to growers. Discovery of such solutions requires local research. However, problems of physically unfavorable crop fields can generally be reduced by minimizing soil exposure to sun by employing cover crops and no-till farming (in which direct seeding through crop residues along with herbicide applications take the place of tillage).

Cover crops and weeds in crops

Soil under cover crops or weeds is often cooler and moister, with more free water compared to areas of bare, tilled soil. Ground covers are more likely to enhance generalist predators than specialist parasitoids. Where ground covers successfully reduce pest levels, enhanced natural enemy populations and reduced pest colonization or retention may both contribute to suppression. Carabid and staphylinid beetles, coccinellids, syrphids, and other predators may also feed on non-pest arthropods living on ground-cover vegetation, or the moister soil below. See Chapter 4 for more on effects of such alternate prey.

In the UK, cabbage intersown with clover (*Trifolium* sp.) supported larger, more effective populations of carabids and staphylinids (O'Donnell & Croaker 1975, Ryan et al. 1980), resulting in lower populations of cabbage root maggot (*Delia brassicae* Weidemann). In citrus groves in dry areas of China, ground covers of tropical ageratum, *Ageratum conyzoides* L., lowered the temperature and raised the relative humidity, making the habitat more favorable for the predaceous mite *Amblyseius eharai* Amstai and Swirski (Zhang & Olkowski 1989).

On the negative side, ground covers may compete with crops for moisture or nutrients, increasing irrigation costs or reducing crop yield. Pests may also benefit from the inclusion of additional plant species in the crop field, as ground covers or intercrops. Ground-cover species and sowing densities must be locally tested to determine their value in improving the habitat for natural enemies compared with the pests.

Weeds at subeconomic levels can be manipulated to produce "volunteer ground covers" (Altieri & Whitcomb 1979). In sugarcane in Louisiana, USA, subcompetitive stands of broad-leaf weeds enhanced predators, especially the red imported fire ant, *Solenopsis invicta* (Buren), a major predator of the sugarcane borer [*Diatraea saccharalis* (Fabricius)] (Ali & Reagan 1985). Weeds were killed by crop competition as canopy closure occurred, and yields were enhanced 19% as compared to weed-free plots. In contrast, while weeds in maize in New Zealand raised parasitism by *Apanteles ruficrus* (Haliday) of the pest armyworm *Mythimna separata* (Walker), weeds also caused a 10-fold increase in pest density, reducing yield by 30% (Hill & Allan 1986). As with cover crops, effects of weeds are likely to be variable among locations. The weed flora is likely to vary between years (in contrast to sown ground covers) so year-to-year results also may vary. The development of genetically modified crops that are tolerant to some herbicides (e.g. Roundup Ready crops tolerant of glyphosate) may make farmers more willing to permit some early-season weed growth in crops, because later weed suppression would be feasible.

Mulching

Mulching soils beneath crops (as in orchards) can both improve the soil as a physical environment (cooler, moister) and stimulate higher diversity and density of alternative prey organisms, which might translate into higher densities of generalist predators (e.g. Mathews et al. 2004).

Direct seeding and no-till

Elimination or reduction of tillage has become a widely used agricultural practice, especially in dry areas. The driving force for adoption has not, however, been pest control, but rather conservation of soil moisture and reduction of wind erosion of soil and of fuel costs. The method has been widely adopted in dry farming regions such as the western USA. Crops are seeded by direct drilling into unbroken soil with specialized machinery. Weed competition in the seed bed is reduced by topical application of herbicides. Use of Roundup Ready crops that tolerate glyphosate facilitates no-till by allowing use of this broad-spectrum herbicide after crop and weed emergence. While not intended as a tool to better conserve natural enemies, no-till agriculture creates a more favorable environment by enhancing soil moisture, reducing soil surface temperature, and preserving soil structure. Actual impacts of Roundup Ready plants and subsequent Roundup (glyphosate) applications on natural enemies have to be assessed in specific cases (e.g. Jackson & Pitre 2004).

Strip harvesting

Strip harvesting rather than cutting whole fields at the same time can help preserve natural enemies in forage crops by preserving both the physical environment and a host or prey supply for natural enemies. Strip harvest of alfalfa, for example, helps retain populations of parasitoids of aphids, alfalfa weevil [*Hypera postica* (Gyllenhal)], and *Lygus* spp. (van den Bosch et al. 1967). Nentwig (1988) found that when German hay meadows were strip harvested, predaceous and parasitic arthropods, especially spiders, became more abundant and herbivores decreased.

PROBLEM 3: INADEQUATE NUTRITIONAL SOURCES

Many natural enemies affecting crop pests require carbohydrates and protein for growth, basic metabolism, and reproduction. If the crop does not provide these materials, natural enemies will search outside the crop for nectar, pollen, or hosts on which to feed. Such emigrants may not find their way back into the crop, reducing pest control. Similarly, small natural enemies such as *Trichogramma* sp. wasps may die more quickly if crops fail to provide resources such as nectar. Crops that provide nectar, pollen, or insects on which to feed will retain larger populations of better nourished, more fecund natural enemies. The diversity of plant species in natural communities increases these requisites and prolongs their availability. Growing crops in synchronized monocultures may concentrate flowering into single brief periods of superabundance or eliminate flowering altogether in some crops. The adequacy of crops as habitats for parasitoids will depend on the need of the particular species, the size of the crop field relative to surrounding non-crop areas, and the vegetational composition of adjacent areas. Large monocultures of crops lacking nectaries, lacking flowers, providing no pollen, and supporting few alternative prey or hosts, offer few resources for natural enemies. While the needs of mechanized, cash-crop agriculture require simplification, some retention of critical plant diversity as sources of natural enemy nutrition may be possible in well-studied systems, and is common in less intensive farming systems (Gurr et al. 2004).

Figure 22.3 Plantings near or within crops of species that produce nectar useful to parasitoids and predators are a basic strategy of conservation biological control. Here, plantings of alyssum [*Lobularia maritima* (L.) Desv.] are intended to increase densities of hoverflies (Syrphidae) to increase aphid control in a lettuce crop. Photograph courtesy of Charles Pickett, CDFA.

SOLUTION 3: ADDING NUTRITION TO CROP ENVIRONMENTS

When nutritional resources are lacking in a crop, plants may be planted inside the crop (as cover crops) or adjacent to it (in strips) as sources of nectar or pollen. Alternatively, in some crops, foods (sugar, protein hydrolysates) may be applied to the crop directly. Nectar provides sugars for fuel, while pollen and protein hydrolysates provide amino acids for reproduction.

Flowers for nectar

In nature, parasitoids and some predators obtain carbohydrates from flowers and extrafloral nectaries (Rogers 1985). Flowers also provide pollen and may occur on wild plants outside the crop, weeds in the crop, or on the crop itself. An active area of investigation is the use of planted strips of flowers as nectar sources for natural enemies (Figures 22.3 and 22.4) (Pfiffner & Wyss 2004). It is becoming clear that parasitoids are often sugar-starved in nature and do feed on flowers

Figure 22.4 Cover crops between rows lower soil temperature, raise relative humidity, and provide nectar and/or pollen, which may increase densities of generalist natural enemies. Here, phacelia (*Phacelia tanacetifolia* Bentham) is planted between rows of a vineyard in New Zealand. Photograph courtesy of Jean-Luc Dufour.

planted near crop fields. Provision of floral resources can increase natural enemy numbers (e.g. Nicholls et al. 2000, Ellis et al. 2005, Rebek et al. 2005). Flowers of different plant species, however, vary in their value to natural enemies and their tendency to produce such effects.

Less information is available on whether provision of floral resources can help control pests (Wratten et al. 2002). Planting of strips of coriander (*Coriandrum sativum* L.: Umbelliferae) and fava bean (*Vicia faba* L.: Fabaceae) around potato fields increased parasitism of potato tuberworm [*Phthorimaea operculella* (Zeller)] from 38% (16 m from flowers) to 52% (adjacent to flowers). However, flowers were counterproductive because they also were resources for the pest and the net result was an increase, not a decrease, in pest damage adjacent to the flower strips (Baggen & Gurr 1998). To resolve this problem, flower species that benefit natural enemies but are not beneficial to pests must be found. Further work in this system revealed that whereas buckwheat (*Fagopyrum esculentum* Benth.) was a non-selective resource, benefiting both the pest and parasitoid, borage (*Borago officinalis* L.) was selectively beneficial to only the parasitoid (Baggen et al. 2000). Underplanting apple trees with floral resources increased parasitism rates of light-brown apple moth [*Epiphyas postvittana* (Walker)] and reduced its damage (Irvin et al. 2006).

Pollen-shedding plants

In addition to nectar, plants in or around crops may be sources of pollen, which supplies the protein often needed by natural enemies for egg maturation. Ground covers or adjacent non-crop plants that produce abundant pollen can increase phytoseiid densities in vine and tree crops (e.g. Girolami et al. 2000, Villanueva & Childers 2004). Control of the citrus red mite [*Panonychus citri* (McGregor)] in citrus groves in China is enhanced by ground covers of tropical ageratum (*A. conyzoides*) (Zhang & Olkowski 1989). Ageratum pollen and psocids on the plant provide food for predatory mites. In Queensland, Australia, Rhodes grass (*Chloris gayana* Kunth) ground covers in citrus enhanced *Amblyseius victoriensis* (Wormersley), a predator of a pest rust mite (*Tegolophus australis* Keifer) (Smith & Papacek 1991). Also, windbreaks of *Eucalyptus torelliana* F. Mueller around citrus blocks served as reservoirs of this predatory mite (Smith & Papacek 1991). These trees support few prey mites, but supply abundant pollen, on which predatory mites feed.

In non-orchard crops, pollen-bearing plants can be planted in strips adjacent to the crop, although this obviously consumes some land. Planting of *Phacelia tanacetifolia* Betham (Hydrophyllaceae) adjacent to wheat crops to enhance syrphid flies and lower aphids in the crop gave mixed results (Hickman & Wratten 1996). In greenhouse vegetable crops, biological control of pest thrips through the use of predatory mites was achieved earlier and with greater ease on peppers (*Capsicum sativum* L.), a plant that sheds pollen, than on cucumber (*Cucumis sativus* L.), which does not (De Klerk & Ramakers 1986).

Application of food sprays

Artificial application of food sprays is another means to provide sugar, which arrests foraging natural enemies (e.g. Evans & Swallow 1993, Mensah & Madden 1994) and protein (as hydrolysate, yeast, or pollen), which attracts natural enemies (Hagen et al. 1970). Reproduction of lacewings (*Chrysopa* spp.) in cotton was enhanced by the application of hydrolyzed proteins, mixed with water and sugar (Hagen et al. 1970). In contrast, applications of mixtures of sucrose and yeast failed to increase predator numbers in apples (Hagley & Simpson 1981). In Tasmanian forest plots, provision of sugar in weather-protected stations enhanced cantharid predators, which in turn increased mortality to eggs of the pest leaf beetle *Chrysophtharta bimaculata* (Olivier) (Mensah & Madden 1994). Pollen applications increase developmental rates of some phytoseiids (McMurtry & Scriven 1964) and enhance the proportion reaching the adult stage (Osakabe 1988). Greater numbers of the predator *Amblyseius hibisci* (Chant) on citrus were correlated with increased concentrations of cattail (*Typha latifolia* L.) pollen from natural sources (Kennett et al. 1979), suggesting that artificial applications might be beneficial.

PROBLEM 4: INADEQUATE REPRODUCTION OPPORTUNITIES

Natural enemies in crops may at times lack hosts or prey, causing them to leave crop fields. Retention of natural enemies in crops field can be increased if a more consistent supply of hosts or prey is available, either in the crop or on nearby vegetation. Some species of parasitoids may require alternate hosts during periods when the target pest is not present or for overwintering (Pfannenstiel & Unruh 2003).

SOLUTION 4: CREATING OPPORTUNITIES FOR CONTACT WITH ALTERNATIVE HOSTS OR PREY

In Californian vineyards, the grape leafhopper, *Erythroneura elegantula* Osborn, is attacked by the egg parasite *Anagrus epos* Girault. This parasitoid occurs in adequate numbers only in vineyards near riparian stands of wild blackberry (*Rubus* spp.). These plants support another leafhopper, *Dikrella californica* (Lawson), that is an overwintering host for the parasitoid (Doutt & Nakata 1973). Enhancement of the grape leafhopper parasitoid in vineyards has been achieved by planting rows of French prunes (*Prunus* sp.) adjacent to vineyards. These trees support a third leafhopper, *Edwardsiana prunicola* (Edwards), that also serves as an overwintering host (Wilson et al. 1989). The prune leafhopper system responds better to agricultural manipulation than the blackberry leafhopper whose populations do not develop well on blackberries away from riparian habitats (Pickett et al. 1990). In apple orchards in Washington state, USA, rates of parasitism of leafrollers by the eulophid *Colpoclypeus florus* Walker are influenced by the proximity of orchards to wild rose patches, which support other species of leafrollers that act as hosts in summer and fall (Pfannenstiel & Unruh 2003). In Belgium, planting of rowan trees (*Sorbus aucuparia* L.) adjacent to apple orchards allowed development of the non-pest aphid *Dysaphis sorbi* Kaltenbach, which in turn was attacked by the parasitoid *Ephedrus persicae* Froggatt, a parasitoid capable of attacking the rosy apple aphid, *Dysaphis plantaginea* Passerini (Bribosia et al. 2005).

Predators often require more kinds of prey than the pest alone. Alternative prey can sometimes be found in the vegetation adjacent to the crop or in the leaf litter and organic matter beneath the crop. For example, weeds and wild-flowering herbs between cereal fields in the UK increased prey of the most abundant carabid (*Poecilus cupreus* L.), thereby enhancing the carabid's reproduction (Zangger et al. 1994). For ground-dwelling predators, alternative prey may be increased by balancing chemical fertilizer use with animal or green manures to increase the detrivores associated with organic matter. Adequate levels of organic matter in soil are needed to maintain prey for carabid beetles (Purvis & Curry 1984, Hance & Gregoire-Wibo 1987) and laelapid predaceous mites (*Androlaelaps* and *Stratiolaelaps* species) that feed on eggs of *Diabrotica* spp., which are pests of maize (Chiang 1970).

PROBLEM 5: INADEQUATE SOURCES OF NATURAL ENEMY COLONISTS

New fields of annual crops need to be colonized by natural enemies, while perennial crops typically do not. As annual crop fields become larger and landscape vegetational diversity decreases, sources from which natural enemies can immigrate may disappear or be located too far from the crop fields. Crop permanency, vegetational diversity and refuges, relay plantings of the crop, crop interplanting, and landscape crop mosaics all affect the source–sink dynamics that both natural enemies and pests exhibit when colonizing new crop fields.

Crop impermanency

Perennial crops such as coconuts, apples, and citrus persist in the same physical location for many years. This stability allows local development of perennial natural enemy populations that can persist without a colonization phase. This habitat stability may promote biological control because it eliminates the time lag often seen in annual row crops, in which natural enemies arrive too long after pest populations to maintain or suppress pests to acceptable levels during short cropping cycles.

In contrast, annual crops must be colonized after planting. If fields are too isolated from sources of key natural enemies, then colonizing natural enemies are likely to arrive after colonization by pests, remain numerically inferior to the pest, and exert less control. The phytoseiid mite *Amblyseius scyphus* Shuster and Pritchard is the key natural enemy of Banks grass mite [*Oligonychus pratensis* (Banks)] in sorghum [*Sorghum bicolor* (L.) Moench] in west Texas, USA. But this predator overwinters inside the straw so it is eliminated after each crop cycle when the crop residues are burned or plowed under. New sorghum fields must then be colonized by phytoseiids dispersing from those small areas of non-crop grasses where they overwinter. Control is typically inadequate because the pest mite's dispersal period is in better synchrony with the availability of young sorghum plants. As a consequence, dynamics on sorghum and on uncultivated grasses are very different (Gilstrap et al. 1979, Gilstrap 1988).

Lack of off-season refuges

Periods may occur when crop fields are unsuitable for natural enemies, such as winter in high latitudes, dry seasons in some tropical areas, or periods when the crop is not grown. Effective conservation requires that the needs of natural enemies be considered for the entire year, including such periods. Some natural enemies pass these seasons in fields on crop residues and management of residues may be important in promoting local survival of the natural enemies. In other cases, important natural enemies pass unfavorable seasons outside crop patches. It is important to know where and under what conditions this occurs to ensure that favorable sites exist for such species near crop fields. Research on the overwintering habitat requirements of carabid and staphylinid predators of cereal aphids in the UK (Thomas et al. 1992, Dennis et al. 1994) and of coccinellids in Belgium (Hemptinne 1988) illustrate the kind of studies needed to define the ecological needs of particular species.

SOLUTION 5: CROP-FIELD CONNECTIVITY, VEGETATION DIVERSITY, AND REFUGES

Relay plantings of the crop

If a large part of the local landscape is used to grow a particular crop, such as rice, corn, or sugarcane, it is likely both that the distance between crops fields will be small and, in the tropics, there will be a continuous presence of the crop throughout the year (Mogi & Miyagi 1990). If fields are planted at different dates, all stages of the crop may overlap on a local scale. This pattern creates a landscape where the crop is a fairly stable habitat, with high connectivity between patches. Such crops will be continuously available for the natural enemies adapted to them and newly planted fields will be quickly colonized from more mature nearby fields.

Opportunities also exist to promote earlier colonization of new plantings in annual crops. Vorley and Wratten (1987) showed that biological control of aphids could be improved if some cereal fields were sown earlier in the preceding fall so that they acquired and retained overwintering parasitized aphids. Parasitoids from these overwintering aphids emerged early and colonized adjacent fields of later-sown cereals as the associated aphid populations began to develop the following spring. By area, only 4% of cereals needed to be sown early to serve as early-season sources of colonizing parasitoids for other fields. Permanent, ungrazed grasslands were also effective sources of early-season parasitoids of cereal aphids.

Similar possibilities exist for other crops. Men et al. (2004) found that aphid parasitoids developing in wheat crops moved into cotton and controlled aphids there. In rape crops, management choices of farmers can strongly affect the number of parasitoids that successfully locate and colonized new crop plantings. Hokkanen et al. (1988) noted that spring parasitoid colonization of new rape fields in Finland was enhanced by locating them as closely as possible to fields sown to rape the previous year.

Landscape crop mosaics

Some natural enemies occur in several crops, feeding on several hosts or prey. In such cases, natural enemies may be enhanced in one crop by planting it near or subsequent to another crop which acts as a source of the natural enemy. Gilstrap (1988), for example, noted that in Texas, USA, the Banks grass mite (*O. pratensis*) occurs on sorghum, wheat, and grass and that an effective phytoseiid mite moves among these crops feeding on the pest. Xu and Wu (1987) report that the movement of a coccinellid from rape crops to bamboo could be encouraged by planting rape near bamboo. When the rape was harvested, the resident coccinellids would move to the bamboo in search of other prey. Corbett et al. (1991) reported that alfalfa planted next to cotton served as a reservoir for the predatory mite *Metaseiulus occidentalis* (Nesbitt), which (if inoculated into the alfalfa early in the season) increased in number in the alfalfa and migrated into adjacent cotton areas. Mark–recapture studies demonstrated that predators such as *Orius* spp. (Hemiptera: Anthocoridae) and the coccinellid *Hippodamia convergens* Guérin-Méneville frequently move between sorghum and cotton fields in

Texas, especially as sorghum grain matures. This link provides opportunities to use strategic placement of sorghum fields to enhance predator numbers in cotton (Prasifka et al. 1999).

Increased within-crop diversity through intercropping

Whereas the goal of the single-crop sequencing strategy is to promote earlier discovery and colonization of new crop patches by both pests and natural enemies (to achieve a better ratio of the two), crop-diversification strategies seek, among other effects, to delay or diminish the number of pests colonizing the crops, or reduce their retention in the crops. Intercropping diversifies agricultural fields by growing two or more crops in the same field at the same time. Crops may be either completely mixed or may be segregated into separate rows, which are alternated in some pattern (Marcovitch 1935, Andow 1991a).

Two beneficial effects theoretically result from intercropping: reduced pest discovery and retention in the crop (the **resource concentration hypothesis**), and enhanced natural enemy numbers and action (the **natural enemies hypothesis**) (Root 1973). Andow (1986, 1988), in reviewing studies of intercropping, found that herbivore densities decreased in 56% of cases, increased in 16%, and remained unaffected in 28% of cases. Determining the reasons for observed effects (the relative importance of reduced pest colonization and retention compared with increased mortality from natural enemies) is difficult and both mechanisms may operate together. Russell (1989) reviewed the effects of intercropping on natural enemy action and found higher levels of mortality from natural enemies in nine of 13 cases, lowered levels in two, and no effect in two cases. Sheehan (1986) suggested that intercropping may be more beneficial to generalist species of natural enemies than specialists, which may perform better in pure cultures of the crop attacked by their host or prey species. No general characteristics exist that can be used to construct pest-suppressive crop mixtures. Rather, each potential combination of crops must be evaluated in the local environment to determine whether it is of value in light of the specific crops, their pests, and natural enemies.

Further, the economic value of pest reduction from vegetation diversification may, in any specific case, be potentially offset by competition among the crop species and by reduction in mechanization of the farming system. In intercrops reviewed by Andow (1991a, 1991b) where herbivores were reduced, yields were not improved for cole crops, yields improved in most bean intercrops, but results were mixed in alfalfa.

Refuges in or near crops

If crop fields are unable to provide for the needs of natural enemies, even with the above sorts of efforts at diversification, refuges can be created in or near the crop fields. Grass-sown raised earth dykes (**beetle banks**) (Figure 22.5) in English cereal fields provided overwintering sites for predators of cereal aphids, enhancing their numbers in adjacent crop areas the following year (Thomas 1990, Thomas et al. 1991, MacLeod et al. 2004). Windbreaks of *E. torelliana* around peach [*Prunus persica* (L.) Batsch] orchards in southern New South Wales, Australia, provided overwintering refuges and enhanced colonization of orchards by predatory mites in the spring (James 1989). In general, studies should be routinely conducted of the overwintering, dry season, or other off-season needs of key natural enemies.

Figure 22.5 In English cereal fields, outbreaks of aphids are an important problem. To conserve carabid beetles, which are important aphid predators, raised earthen banks are created within fields and planted to perennial grasses. These strips, called beetle banks, are not plowed and provide permanent habitat for carabids. Photograph courtesy of John Holland, The Game Conservancy Trust.

OTHER PRACTICES THAT CAN AFFECT NATURAL ENEMIES

In the preceding sections, the crop as an environment for the natural enemy was considered from the point of view of how basic needs of the natural enemies may be affected. This discussion, however, is incomplete in that other farming practices can affect natural enemies, often in several ways simultaneously. In the following sections we discuss some of these practices and how they might influence natural enemy populations.

Irrigation

Irrigation raises humidity in the crop, and this may be important in making the environment more favorable for some kinds of natural enemies. For example, it may be possible to promote epidemics of insect fungal pathogens by manipulation of irrigation or greenhouse watering patterns. Efficacy of *Verticillium lecanii* (Zimmerman) Viegas [reclassified now as *Lecanicillium muscarium* (Petch) Zare and W. Gams] applications in greenhouses for aphid or whitefly control can be enhanced by manipulating the crop foliage density, watering, and night-time temperatures so as to maintain the high humidity needed for germination of pathogen spores (Hall 1985). Epizootics of the entomopathogenic fungi *Erynia neoaphidis* Remaudière and Hennebert and *Erynia radicans* (Brefeld) occurred in pea aphids, *Acyrthosiphon pisum* (Harris), on ground covers in pecan (*Carya illinoensis* Koch) orchards in Georgia, USA, that employed overhead, but not drip, irrigation (Pickering et al. 1989). Significant potential exists to manipulate crop relative humidity and wetting periods (through crop spacing and irrigation practices) to enhance arthropod disease levels (Harper 1987). Flooding is used also in some crops to control pests. Flooding was evaluated by Whistlecraft and Lepard (1989) as a means to control the onion pest *Delia antiqua* (Meigen), but was found to be damaging to the key parasitoid *Aleochara bilineata* (Gyllenhal).

Harvest or pruning methods and timing

Block and strip harvest

In alfalfa, strip harvesting can be used to conserve natural enemies in the crop and enhance biological control. Hossain et al. (2001) found higher predation on sentinel eggs of *Helicoverpa* spp. in unharvested compared with harvested strips of alfalfa. Predation on sentinel eggs placed in harvested strips declined with distance from the unharvested strips, suggesting that predators were moving from the unharvested areas into the cut areas (Hossain et al. 2002). In Sweden, willow is grown in a coppice system to produce biomass for energy production. The crop is affected by several defoliating chrysomelid beetles, which in turn are suppressed by predators, especially the mirid *Orthotylus marginalis* (Reuter). The crop is harvested by cutting the tops off the plants in winter every 3–5 years. Regrowth foliage is highly attractive to the herbivorous beetles and this, coupled with a lowered predator density after harvest, leads to pest outbreaks. It has been suggested that adjacent blocks of willow be harvested asynchronously to conserve predators and reduce outbreaks (Björkman et al. 2004).

Plant-flush pest synchronization

A synchronized flush of new growth may follow after some woody plant crops are pruned. Young foliage is often higher-quality food for insects, especially sucking species. A surge in pest population growth rate may follow that exceeds the ability of parasitoids to numerically respond quickly. High pest densities may result from this imbalance. To prevent these events, growers can use alternate-row pruning (which staggers growth of succulent new foliage, which is attractive as oviposition sites for pests such as whiteflies). This approach prolongs the induction of increased pest populations, allowing more time for parasitoids to respond. Biological control of whiteflies in lemon orchards was improved in coastal California by use of alternate-row pruning (Rose & DeBach 1992).

Crop-residue destruction

In many crops, residues left after harvest are disposed of by burning or tillage. In some instances, these practices may be done to gain an explicit benefit but in other cases, crop-residue destruction has no definite function other than being the traditional method to clear the soil surface for the next planting or to facilitate harvest. In some instances, crop-residue management can affect key natural enemies. In Indian sugarcane, several parasitoids [*Epiricania melanoleuca* Fletcher, *Ooencyrtus*

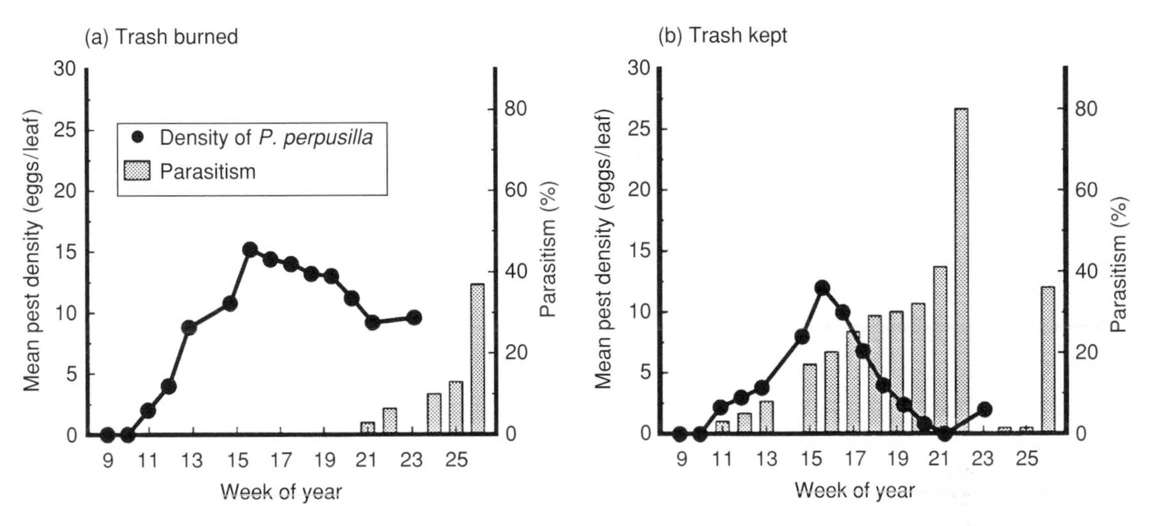

Figure 22.6 Management of post-harvest crop residues can affect natural enemy conservation. Population density of *Pyrilla perpusilla* (Walker) and egg parasitism of it by *Parachrysocharis javensis* (Girault) in sugarcane fields where the trash was either burned (a) or left in the field (b) following harvest. After Mohyuddin (1991), reprinted from Van Driesche and Bellows (1996) with permission from Kluwer.

papilionis Ashmead, *Parachrysocharis javensis* (Girault)] of the sugarcane leafhopper *Pyrilla perpusilla* Walker are eliminated when crop residues are burned. Studies show that if crop residues are left unburned and spread back on the field after burning, parasitoids can be conserved at levels able to control the pest (Joshi & Sharma 1989, Mohyuddin 1991) (Figure 22.6).

Control of species that harm natural enemies

Ants or other predators reduce the efficacy of natural enemies in some crops. If ants are controlled, natural enemies may increase (e.g. James et al. 1999) and in some cases be able to suppress the pest. For example, Argentine ant [*Linepithema humile* (Mayr)], the big-headed ant [*Pheidole megacephala* (Fabricius)], and *Lasius niger* L. interfere with the action of natural enemies by physically attacking and removing immature stages of some predators (such as larvae of coccinellids) and interfering with the host-searching and oviposition activities of some parasitoids. In some cases, ants are present because they collect honeydew from colonies of insects such as soft scales (Coccidae), mealybugs, whiteflies, and aphids. Even pests that do not produce honeydew, such as armored scales (Diaspidae) and some mites, can be affected by ants. Ants may drive off parasitoids of scale insects or attack predators such as the larvae of the mite-feeding coccinellid *Stethorus picipes* Casey, which feeds on citrus red mite, *P. citri* (Haney et al. 1987).

The suppressive influence of ants on natural enemy effectiveness has been demonstrated for various scales (DeBach et al. 1951, 1976, Steyn 1958, Samways et al. 1982, Bach 1991), as well as aphids and mealybugs (Banks & Macaulay 1967, DeBach & Huffaker 1971, Cudjoe et al. 1993). Restoration of effective biological control in such cases depends on control of the ant species involved, often through the application of pesticides to ant nests or tree trunks, or the application of sticky barriers to tree trunks. Musgrove and Carman (1965), Markin (1970a, 1970b), and Kobbe et al. (1991) provide information on the biology and control of the Argentine ant, one of the species most frequently interfering with natural enemies. Samways (1990) describes a method of sticky-banding trees to control pest-tending ants that is not phytotoxic to tree bark.

Provision of artificial shelters for natural enemies

Artificial nests made of polyethylene bags have been used to manipulate ant (*Dolichoderus thoracicus* Smith) populations in cocao (*Theobroma cacao* L.) plantations in Malaysia (Heirbaut and van Damme 1992).

Figure 22.7 Nest of a weaver ant [*Oecophylla smaragdina* (F.)]. Photograph courtesy of Grace Kim.

Planting of favorable host trees near orchards encourages nest formation by weaver ants [*Oecophylla smaragdina* (F.)] in Asia (Figure 22.7).

Empty cans placed in fruit trees have been used to augment earwig (Dermaptera) numbers in fruit trees (Schonbeck 1988), and straw bundles have been used to enhance numbers of spiders in new plantings of rice (Shepard et al. 1989). Boxes have been used to provide overwintering sites for adults of *Chrysoperla carnea* (Stephens) (Sengonca & Frings 1989). Overwintering of *M. occidentalis* in apple orchards in China (introduced from California) occurred only after overwintering sites were created around tree trunks. These sites consisted of either waste cotton held against tree trunks by plastic sheets, or piles of leaf and grass litter piled at the base of trees (Deng et al. 1988). Populations of insectivorous forest birds have been enhanced through the provision of nesting boxes (Bruns 1960). Barn owl (*Tyto alba* L.) densities in Malaysian oil palm plantations have been increased by providing nesting boxes, thus enhancing rat control (Mohd 1990).

CONCLUSIONS

Many theoretical predictions suggest that it should be possible to improve crop environments for natural enemies and enhance their pest control potential. However, whether or not any particular change to the crop or its manner of production will be useful must be determined by local experimentation. Economics of these programs of natural enemy conservation also depend on local circumstances, and such economic issues often determine whether or not a particular practice will be adopted by growers. At present, this approach to use of natural enemies is being studied to determine the extent of its potential practical application.

Part 9

BIOPESTICIDES

Chapter 23

MICROBIAL PESTICIDES: ISSUES AND CONCEPTS

HISTORY OF MICROBIAL INSECTICIDES

Understanding of insect diseases began in the nineteenth century (Kirby & Spence 1815), not in relation to insect pest control, but to control diseases of commercial species such as the silkworm, *Bombyx mori* (L.). Agostino Bassi was the first to demonstrate experimentally the infectious nature of insect disease in his 1835 study of the white muscardine disease of silkworms, caused by the fungus *Beauveria bassiana* (Balsamo) Vuillemin. More work was done on other silkworm diseases in 1865–70 by Louis Pasteur in France. The first suggestion to use insect pathogens as microbial insecticides was made in 1836 by Bassi, who proposed that putrefied cadavers of diseased insects could be mixed with water and sprayed on foliage to kill insects. The first field trials of this concept were conducted in 1884 by Elie Metchnikoff, who mass-produced conidia of *Metarhizium anisopliae* (Metchnikoff) Sorokin and applied them in field tests against larvae of the sugar-beet curculio, *Cleonus punctiventris* (Germar), causing 55–80% mortality.

The most successful microbial insecticides ultimately proved to be products based on toxins of the bacterium *Bacillus thuringiensis* Berliner (Bt), a species first discovered in Japan by Ishiwata (1901). The history of Bt, from this early discovery to use of transgenic crops expressing Bt toxins, is summarized by Federici (2005). Briefly, the bacterium was named in Germany by Berliner (1915) after its rediscovery there as pathogen of flour moths, *Anagasta kuehniella* (Zeller). French researchers, as an outgrowth of studies on diseases of silkworm larvae, developed the first Bt-based bioinsecticide (Sporeine) during the 1930s (Jacobs 1951). In the 1950s, investigations on Bt were started in California, USA, by E.A. Steinhaus. A critical early discovery was that the pathogen possessed crystalline parasporal bodies that were toxic to some insects (Hannay 1953). Isolation of many new, but subtly different, forms of the pathogen by various workers occurred quickly and caused confusion until de Barjac and Bonnefoi (1962, 1968) developed a classification system based on flagellar antigens. At about the same time Dulmage (1981) with Burges set up international standards for bioassays of new isolates and their comparison to a standard strain. From about 1965 to 1981, commercial development of Bt products was carried out by two principal companies (Abbot Labs and Sandoz Corporation), which developed products such as Dipel and Thuricide. During this same period, new subspecies of *B. thuringiensis* were discovered with activity against pests other than Lepidoptera larvae. The most important new subspecies were *B. thuringiensis israelensis* (Goldberg & Margalit 1977) with activity against larvae of Nematoceran flies (e.g. mosquitoes and blackflies; van Essen & Hembree 1980) and *B. thuringiensis morrisoni* strain *tenebrionis* (Krieg et al. 1983) with activity against some scarab and chrysomelid larvae, including Colorado potato beetle [*Leptinotarsa decemlineata* (Say)] (see Entwistle et al. 1993 and Glare & O'Callaghan 2000 for reviews).

Bacillus thuringiensis kills its hosts by producing toxins that selectively bind with receptor sites on the midgut microvilli. Insect death is caused by intoxication, which may be accompanied by invasion of the hemocael by vegetative bacterial cells (Schnepf et al. 1998). The most commonly used form of Bt is the HD1 isolate of *B. thuringiensis kurstaki*, which produces four major endotoxins, designated as Cry1Aa, Cry1Ab, Cry1Ac, and Cry 2Aa. Another important subspecies,

B. thuringiensis israelensis, produces Cry4Aa, Cry4Ba, Cry11Aa, and the toxin Cyt1Aa, a cytolytic toxin unrelated to the Cry proteins (Federici 2007).

After 1981, newly developed molecular tools were applied to this pathogen to create genetically modified *B. thuringiensis* strains, and ultimately, Bt crops. The discovery that the genes for Bt toxins were located on plasmids, not the *B. thuringiensis* chromosome, allowed easier cloning of Bt toxin genes (Schnepf & Whitely 1981). A classification scheme for Bt toxins was developed (Hofte & Whitely 1989), grouping them as either cry (crystal) or cyt (cytolytic) toxins. Studies followed of the natural variation in Bt toxins, their mode of action, specificity, and coding genes. Based on these advances, molecular technology was used to improve Bt as a biopesticide, first by creating strains that combined toxins from two or more separate natural sources. This was followed by insertion of the more useful toxins into crop plants (Fischhoff et al. 1987, Perlak et al. 1990, Koziel et al. 1993), an activity in which the Monsanto Company played the dominant role. A critical technical hurdle in creating Bt crops was to increase Bt expression in plants to levels toxic to the target pests, which was accomplished by altering the genes to optimize expression (Perlak et al. 1991). The safety of Bt crops to non-target organisms has been amply demonstrated (Shelton et al. 2002, O'Callaghan et al. 2005) and Bt crops are widely used in the USA and many other countries. By 2005, over 50% of the cotton acreage and 40% of the corn in the USA were Bt varieties (Federici 2005). A consequence of such adoption has been the economic failure of companies seeking to promote the biopesticide use of Bt formulations on these same crops, although its use continues on other crops and for targets such as mosquitoes.

Success of *B. thuringiensis* products stimulated commercial efforts with other pathogens, including fungi and viruses. A directory of microbial pesticides currently registered in the Organization for Economic Cooperation and Development (OECD) countries (most of Europe, the USA, Canada, New Zealand, Australia, Japan, Korea, Mexico, and Turkey) is available online (see Kabaluk & Gazdik 2004). For each product this list provides the name of the microbe, the target pests, the countries in which it is registered, the manufacturer, and a web link for further information on the product. Biopesticides, however, currently account for about 1% of the pesticide market. Of this, Bt products account for 80%.

WHAT MAKES A PATHOGEN A LIKELY BIOPESTICIDE?

Ease and cost of rearing

To have any chance at being commercially successful as a microbial insecticide, a pathogen must be easy to mass produce at low cost. The most important factor affecting rearing cost is whether or not living hosts are required. *Bacillus thuringiensis*, the most successful mass-produced entomopathogen, can be grown on fermentation media (a non-living mix of nutritional substances). In contrast, *Paenibacillus* (formerly *Bacillus*) *popilliae* (Dutky), a pathogen of Japanese beetle (*Popillia japonica* Newman) that attracted attention because of the importance of this beetle in the USA, requires living hosts for spore production. This dramatically increased production costs and, along with its high specificity, prevented this pathogen from becoming a major commercial success. Other aspects of production, such as the ability of an agent to grow in liquid media, or development of simple methods for local production by farmers in rural parts of developing countries, can also affect cost. Cost of production is a function of labor costs and the technology, which can change. Rearing media using cheaper ingredients, such as locally produced cereals, can reduce production costs (Hoti & Balaraman 1990), but locally produced products may lack the high and consistent quality that farmers demand.

Degree of host specificity and pathogenicity

Pathogens that make effective microbial pesticides are species with a reasonable level of specificity and high activity against one or more critical pests of a major commodity, which ensures an adequate-sized market. Research on microbial pesticides was begun in pursuit of pest control products that would be more compatible with natural enemies than chemical pesticides. High specificity was valued because it ensured pathogens would affect only the target pest and thus be easy to integrate into pest management systems. If host specificity is too high, the market may be too small to support commercial production, except when the target pest is one of great importance on a crop grown on extensive acreage. Most insect baculoviruses, for example, have host ranges limited to a few species.

Baculoviruses with broader host ranges do exist, such as the *Autographa californica* nucleopolyhedrovirus, which attacks at least 43 species in 11 families of insects (Payne 1986). However, this particular virus is weakly infectious except in a few noctuid moth caterpillars.

In principle, genetic engineering can be used to broaden the host spectrum of pathogens. For example, strains of *B. thuringiensis* specific for certain types of host (subspecies *kurstaki* for Lepidoptera, subspecies *israelensis* for Diptera, subspecies *tenebrionis* for Coleoptera) can be manipulated so that the host ranges of several strains are combined (Crickmore et al. 1990, Gelernter 1992) in a single organism. Although this has been done, no modified product has yet been a dramatic success.

Pathotypes exist within most pathogen species and these vary in the amount of material needed for control. Since pathogens are relatively expensive, use of more virulent pathotypes reduces costs by lowering the quantity that must be applied. Production costs for *B. thuringiensis* are comparable to those for modern chemical pesticides such as imidacloprid and spinosad.

Suitability of the pathogen for the intended site of application

Physical conditions at the application site can affect the efficacy of insect pathogens. In general some of these limitations are characteristic of whole groups: nematodes desiccate easily; fungi need humid conditions for conidial germination; viruses are degraded in a few days by ultraviolet light. To be suitable for an intended use, a pathogen must be tolerant of conditions commonly found at application sites. Nematodes, for example, are most suitable for use in moist habitats such as soil and inside plant tissues for control of leafminers or borers. Variation also exists among species within pathogen groups that can affect suitability for particular application sites. Black vine weevil [*Otiorhynchus sulcatus* (Fabricius)] is an important nursery pest in the USA and Europe and in some production areas the soil temperatures are rather low. However, the nematode species that were first commercialized were not highly effective in cool soil. However, *Heterorhabditis marelatus* (Liu & Berry 1996), a species discovered subsequently, is more effective at low soil temperatures (Berry et al. 1997).

OVERVIEW OF OPTIONS FOR REARING PATHOGENS

Pathogens may be reared either in intact living hosts (*in vivo*) or fermentation media (*in vitro*); however, rearing *in vivo* is rarely practical commercially. Viruses can also be reared in live insect cell cultures. From the earliest days, pathologists have recognized that dependency on living hosts limits large-scale production. Some groups of pathogens, however, are difficult or impossible to rear outside of living hosts. These include all the baculoviruses, many Entomophthoraceae fungi, some bacteria, and some nematodes. Pathogens that must be reared in living hosts require more labor for their production because this process is difficult to automate and lacks economy of scale.

Rearing in live hosts

The process of rearing in live hosts (apart from cell lines) requires: (1) the mass-rearing of an insect host, (2) steps to infect the host and produce the pathogen, and (3) methods to harvest and process the pathogen. Step one begins with choosing a convenient host in which to propagate the pathogen. Ideally this should be the target pest, but may not be if that species is difficult to rear and the pathogen can be grown in another species that is more convenient. (If, however, the pathogen is produced in an alternative host, there is a risk of adaptation to that host and loss of infectivity in the pest.) Production of the rearing host on living plants entails higher costs and presence of other organisms; therefore, whenever possible host insects are reared on artificial diet.

Step two, inoculation of the host and pathogen growth, begins by treating the host with the infective stage of the pathogen, often by simply contaminating the host's food with the pathogen. The goal is to obtain the greatest yield per host, which can be affected by the dose applied and host age. If too high a pathogen concentration is applied or used too early, hosts may die young, giving lower yield.

The final step, harvesting and purification of the pathogen, must be inexpensive and retain pathogen viability. Depending on the pathogen, host cadavers can be vacuum-aspirated, dried, and ground up (for viruses), rinsed (to harvest spores of fungus), or (for nematodes) placed in a moist arena to trap emerging nematodes as they exit the host and crawl into water. The harvested pathogens then must be stabilized in a medium and at a temperature favorable for their survival.

Rearing in fermentation media or cell lines

For pathogens that do not require live organisms, production may occur in fermentation media or insect cell cultures. Fermentation media are used for some bacteria and fungi. Such media consist of carbohydrates (like rice or grain wastes), protein, vitamins, minerals, salts, and antibiotics. Exact blends depend on the pathogen being reared and the cost and local availability of materials. For bacteria, fermentation media are liquid, which allows them to be manipulated with tanks and pumps, giving an economy of scale in production. Many fungi do not produce conidia when submerged. Therefore, rearing fungi requires a two-step system in which mycelium is grown in liquid culture and then placed on solid media for conidia production. Alternatively, fungi might be produced using as the infective unit structures that do grow in liquid (mycelial fragments, blastospores, resting spores, chlamydospores). This latter approach usually requires different formulation methods to stabilize the infective pathogen stage so that it retains its viability. For viruses, insect cell cultures are a liquid medium that provides live cells for attack and reproduction, but this system is not practical for production of viruses as biopesticides. Details of production systems for types of pathogens are discussed in Chapter 24.

AGENT QUALITY: FINDING IT, KEEPING IT, IMPROVING IT

Start cultures with high-quality agents

Discovery of new microbial agents may be the result of chance, laboratory screening, or field surveys. Chance discoveries of useful new agents have included the finding of *B. thuringiensis* subspecies *israelensis*, a strain pathogenic to mosquitoes, and the nematode *Steinernema riobrave* Cabanillas et al., a species effective against pupae of *Heliothis zea* (Boddie) (Cabanillas & Raulston 1994). Screening programs may also be used to find pathogens effective against a specific pest by examining the activity of existing laboratory collections of isolates of a pathogen for activity against the target pest. Kawakami (1987), for example, screened 61 isolates of *Beauveria brongniartii* (Saccardo) Petch for pathogenicity against the mulberry pest *Psacothea hilaris* (Pascoe). Field surveys, however, are the basic source of new pathogen isolates. New isolates effective against a specific target pest may be encountered by collecting large numbers of the target pest in the field, searching for dead or moribund specimens, and examining them by microbial culturing techniques. Koch's postulates (isolate, infect, re-isolate) must then be followed to confirm pathogenicity. New generalist pathogens can be encountered with less-specific field surveys. Wax moth larvae, for example, can be placed in soil as baits to explore for new nematodes (e.g. Deseo et al. 1988, Hara et al. 1991). This approach can be used to find nematodes or fungi pre-adapted to particular soil conditions (hot, cold, dry, wet, etc).

Retaining agent quality

A mass-rearing culture of a pathogen may over time become contaminated with other microbes, become less productive (in terms of pathogen production per unit of medium), or lose its virulence to the target pest. In commercial pathogen production, periodic testing is required to detect contamination, especially by human pathogens (Jenkins & Grzywacz 2000). Changes in yield can be monitored by counting the number of pathogens produced per host or unit of medium. Virulence can be measured with bioassays against the pest, with comparison to a standard strain or the original isolate of the pathogen.

Microbial agents can lose infectivity after being reared on artificial media for many generations. Repeated rearing of the fungus *Nomuraea rileyi* (Farlow) by conidial transfer led to a loss of virulence to *Anticarsia gemmatalis* Hübner larvae in 16 generations. Loss of virulence, however, was only associated with propagation of conidia, as no loss of virulence in this species was seen in up to 80 passages based on mycelia transfers (Morrow et al. 1989). Attenuation following prolonged artificial propagation has been observed in at least seven other species of fungi (Hajek et al. 1990b). Similarly, baculovirus produced in alternative hosts may lose infectivity in the original host, as occurred with silkworm (*B. mori*) virus when reared for 18 generations in Asiatic rice borer, *Chilo suppressalis* (Walker) (Aizawa 1987).

Infectivity lost by prolonged rearing in fermentation media can be restored by periodically restarting the culture with pathogens from living hosts or an infective isolate held in long-term storage. This approach is used to maintain the infectivity of the horntail nematode, *Deladenus (Beddingia) siricidicola* (Bedding), which if reared continuously on its fungal host loses the ability to infect insects. Such loss of infectivity led to a major breakdown of a control program against the horntail *Sirex noctilio* (Fabricius) in Australian forestry during the 1980s (Haugen 1990). This situation was resolved by recollecting a virulent nematode strain from nature and using it in mass production. To prevent a re-occurrence of attenuation, production of nematodes used to infect new pine plantations is done using material that is periodically renewed from a frozen culture of the infective nematode strain (Bedding 1993).

Another interesting case in retention of quality in a microbe being produced as a pesticide is that of *Serratia entomophila* Grimmont, Jackson, Ageron, and Noonan. This pathogen has been produced in New Zealand since 1990 for the control of the native grub *Costelytra zealandica* (White), a pasture pest (Jackson 1994). This pathogen suffered two problems when mass produced. First was a tendency for cultures to be overtaken by non-virulent strains. A process of certifying starter cultures was developed to ensure that only virulent cells are used in commercial fermenters. This process relied on visually detecting the specific plasmid in which genes for virulence are localized. This is further confirmed by quality-control assays, verifying that grubs have been inoculated with the pathogenic strain (Pearson & Jackson 1995). A second issue for rearing this species was contamination of the fermenters with viruses that attack bacteria (i.e. phages), which can cause production to collapse. This problem was solved by locating a mutant strain that could not be attacked by the phage but which still caused disease in the target (Grkovic et al. 1995).

Genetic improvement of pathogens

Nematodes and microbes can potentially be improved in a variety of characteristics, such as infectivity rate to a given host, host range, lethality, and pesticide resistance. Improvements are also possible for characteristics affecting production, such as yield of spores or rate of growth under production conditions. Gaugler et al. (1989) used laboratory selection to enhance host-finding of *Steinernema carpocapsae* (Weiser) by 20–27-fold. Entomopathogenic fungi have been genetically modified for fungicide resistance (Goettel et al. 1990). Baculoviruses have been modified to increase the speed of kill by inserting genes for venom (Bonning & Hammock 1996, Cory 2000).

MEASURING THE EFFICACY OF MICROBIAL PESTICIDES

Efficacy is a crucial issue for biopesticides. Trials to measure efficacy are similar to testing a chemical pesticide. One applies the products when and where desired and measures either percentages of pests killed or changes in numbers of pests before and after application compared to an untreated control. Some fungal and nematode pathogens are able to recycle (reproduce for additional generations after application) at application sites. Evaluations can have various objectives, including: (1) comparison of species or strains to identify the best agent for a particular pest, (2) comparison of different formulations or application methods, (3) measurement of sensitivity to variation in environmental factors, or (4) measurement of pathogen persistence after application.

Comparisons among agents and formulations

Frequently, several pathogens may be available to control the same pest. Should growers use *Steinernema feltiae* (Filipjev) or *S. carpocapsae* to suppress fungus gnats in greenhouse flower crops? Should a forester use *B. thuringiensis* or the baculovirus Gypchek® to control gypsy moth larvae? Answers to such questions come from field trials, such as those run by Capinera et al. (1988) and Wright et al. (1988) to identify the best nematode species for their particular pests of interest. Such trials typically compare aspects such as variation in the dosage applied and the formulation used. Wright et al. (1988), for example, in their tests of nematode species, considered rates of nematodes spanning an eight-fold range. Capinera et al. (1988) compared three methods of delivery of nematodes for cutworm control: calcium alginate capsules, wheat-bran baits, and aqueous suspensions.

Use in the field also requires some knowledge of how often the pathogen must be applied and how best to time applications. Tatchell and Payne (1984), for example, found that because *Pieris rapae* (L.) larvae varied in age in cole crop fields multiple applications of virus gave better control than one application. In Kenya, moth catches in pheromone traps were used to time applications of *B. thuringiensis* to control neonate larvae of *Spodoptera exempta* (Walker) (Broza et al. 1991). Integration of pathogens with pesticides may be explored as a means to lower pesticide usage. For example, trials with low rates of imidacloprid and nematodes for scarab beetle larval control showed that the combination was more effective than either alone (Koppenhöfer & Kaya 1998).

Effects of environmental factors

In the field, biopesticide efficacy will be affected by factors that change coverage, pathogen survival, or infectivity. Thatch, for example, reduces movement of nematodes applied as water applications on to turf (Georgis 1990), reducing the number of nematodes that reach grubs in the root zone. Dense canopies or hairy leaves can reduce deposition rates of products on leaves, reducing effectiveness. Survival of many kinds of microbial agents is reduced by ultraviolet light or excessive dryness. In a field trial in the UK, more than two-thirds of the granulovirus applied on cabbage against *P. rapae* was deactivated in a single day (Tatchell & Payne 1984). The degree to which pathogens that do contact hosts succeed in infecting them will depend on the agent applied, the formulation, and physical conditions at the time of application. Many fungi, for example, must have high humidities for a critical period after spores land on the host for conidia to germinate and penetrate the integument (Connick et al. 1990). Since weather is a local matter, field trials must be run where the pest is to be controlled.

Persistence of agent impact due to agent reproduction

Most microbial insecticides degrade steadily after application, but some are capable of reproducing under field conditions. For example, Allard et al. (1990) found that infection by the fungus *M. anisopliae* of the sugarcane froghopper, *Aeneolamia varia* Fabricius var. *saccharina*, remained higher in treated plots than in control plots for up to 6 months after a single application. In sugarcane in Australia, a single application of the same fungus provided commercial levels of control of the pest *Antitrogus* sp. for more than 30 months (Samuels et al. 1990). *Beauveria brongnartii* applied to soil in Switzerland to control the cockchafer *Melolontha melolontha* L. persists in soils for several years if grubs are present (Kessler et al. 2004). Jackson and Wouts (1987) found that the degree of control of the grass grub *C. zealandica* provided by applications of the nematode *Heterorhabditis* sp. in New Zealand increased from 9 to 56% over an 18-month period, indicating an increase in nematodes at the site over time through reproduction. An economic analysis in Tasmania of control of the pasture pest *Adoryphorus couloni* (Burmeister) showed that a single treatment with *M. anisopliae* persisted for 5–10 years, which made its use economical compared to the cost of renovation of insect-damaged pasture or use of chemical control (Rath et al. 1990).

DEGREE OF MARKET PENETRATION AND FUTURE OUTLOOK

Many factors affect the market potential of pathogens as microbial insecticides over and above the degree to which they control their target hosts. The potential profitability of a possible product and the extent of public subsidies both influence how much research effort is devoted to a pathogen's development as a biopesticide. The potential for sales is influenced by competing options at the time, specifically if other effective options are available for the same task. In addition, legal factors affect the economics of developing biopesticides, especially costs for product registration and extent of patent protection available. The influence of such forces on product development is illustrated by Huber (1990), who recounts the twists and turns between the 1963 discovery of a granulovirus of the codling moth, *Cydia pomonella* (L.), in Mexico, and the marketing of it decades later in Germany as Granupom®. In some cases, local production of microbial pesticides can help increase their use by reducing costs and need for foreign currency (Bhumiratana 1990). Developing a local program to rear the *Anticarsia gemmatalis* virus in Brazil increased soybeans treated with this virus from 2000 ha in 1982–3 to over 1,000,000 ha in 1989–90 (Moscardi 1990); however, this program received extensive government subsidies.

Table 23.1 Pathogens registered as insecticides

Species of microbe	Pests controlled
Bacteria	
Paenibacillus popilliae	Japanese beetle larvae
Bacillus thuringiensis subsp. *kurstaki*	Lepidopteran larvae
B. thuringiensis subsp. *israelensis*	Dipteran larvae
B. thuringiensis subsp. *tenebrionis*	Coleopteran larvae
B. thuringiensis subsp. *aizawai*	Lepidopteran larvae
Fungi	
Beauveria bassiana	Whiteflies, aphids, and other pests
Beauveria brongnartii	Cockchafer beetle grubs
Lecanicillium muscarium (Petch) Zare and W. Gams (formerly given as *Verticillium lecanii*)	Aphids and thrips
Lagenidium giganteum	Mosquito larvae
Metarhizium anisopliae strain ESF1	Cockroach and fly control
Paecilomyces fumosoroseus	Whiteflies
Viruses	
Granulovirus	Leafroller
Granulovirus	Codling moth
Granulovirus	Indian meal moth
NPV from *Autographica californica*	Caterpillars
NPV from *Anagrapha falcifera*	Caterpillars
NPV from Douglas-fir tussock moth	Douglas-fir tussock moth larvae
NPV from *Spodoptera exigua*	Caterpillars

NPV, nucleopolyhedrovirus. Data from Kabaluk and Gazdik (2004).

Kinds and numbers of registered products

In 2004, some 117 products, representing some 20 pathogens (species or strains) were registered in one or more countries of the OECD (a consortium of some 40 countries) (Table 23.1). Registered products contained two bacteria (*P. popilliae* and *B. thuringiensis*, including four subspecies: *B. thuringiensis azawi*, *B. thuringiensis israelensis*, *B. thuringiensis kurstaki*, and *B. thuringiensis tenebrionis*), six fungi [*B. bassiana*, *B. brongniartii*, *Lecanicillium muscarium* (Petch) Zare and W. Gams (formerly given as *Verticillium lecanii*), *Lagenidium giganteum* Couch, *M. anisopliae*, and *Paecilomyces fumosoroseus* (Wize) Brown and Smith], and seven baculoviruses (three granuloviruses, and four nucleopolyhedroviruses). However, just one agent (*B. thuringiensis kurstaki*) accounted for 57 of the 117 products (Kabaluk & Gazdik 2004).

Size of the market

In the absence of government subsidies, the biggest factor influencing development of a pathogen as a microbial pesticides is its potential for sales. For highly specific agents, commercial development is only likely for pathogens that kill key pests of crops grown in large areas such as cotton, maize, and soybeans (Huber 1986), or are widespread forest pests. Microbial pesticides are unlikely to exist for pests of small-area specialty crops unless the pathogen is already produced for another, larger market. The use of *B. thuringiensis* subspecies *israelensis* for control of flies in mushroom houses and sewage plants, for example, is feasible only because this agent is already being produced for mosquito control. Products for public-sector uses, such as for the control of defoliators of public forests, may be feasible if public funds are used to support the development, registration, and production of the product (Morris 1980). This approach has been suggested by Canadian foresters, who propose that governmental agencies produce several baculoviruses of key forest pests and make them available at cost to regional forest managers when outbreaks occur.

Competition with pesticides

Microbial products must compete with existing chemical pesticides for market share. Opportunities to do so may exist when a chemical's use is prohibited by government; chemicals fail due to resistance; a microbial pesticide is highly effective and cheaper than existing chemical pesticides; or pesticide-caused problems, such as secondary pest outbreaks, become severe in a crop.

To promote use of biopesticides, the variability of control by microbial pesticides should be minimized by research on factors that affect efficacy, adjusting either the formulation or the directions for use as needed. Second, extension agents must educate growers to understand that neither extremely high levels of kill nor rapid kill are truly necessary for effective pest control in most crops. Educational efforts should stress that microbial pesticides often cause a rapid cessation of pest feeding and a long-term reduction in pest reproduction rates. Sustained moderate levels of mortality from microbial pesticides combine well with conservation of predators and parasitoids, leaving some pests to serve as their hosts or prey. However, adoption of biopesticides may be inhibited in crops with multiple pest species because the microbial insecticide may kill only some species. In such cases, it is typically cheaper and easier for growers to use a chemical pesticide if it is able to control the entire pest complex.

Legal factors

Costs of registering pest control products with governments and the availability of patent protection strongly affect the feasibility of developing microbial pesticides, especially for smaller markets. The relative success of nematodes as bioinsecticides is due in part to the lack of need for product registration with this group of organisms (Hominick & Reid 1990) in most countries. Patent protection is available for newly commercialized viruses and bacteria, but most of the species in production are in fact not patented. Patent protection is not available for fungi or nematodes. Patents may be obtained for technology used in rearing, formulating, or applying such organisms, or novel use patterns.

Chapter 24

USE OF ARTHROPOD PATHOGENS AS PESTICIDES

In this chapter we review bacteria, fungi, and nematodes from the perspective of their current or potential use as biopesticides. Only some species in each group actually have such potential, while others may be important in natural or classical biological control.

BACTERIA AS INSECTICIDES

Bacterial biology

Bacteria are unicellular organisms that have rigid cell walls. They may be rods, spheres (cocci), spirals, or have no fixed shape. The species causing disease in arthropods are discussed by Tanada and Kaya (1993). Most pathogenic bacteria enter arthropod hosts when contaminated food is ingested. Such bacteria multiply in the gut, producing enzymes (such as lecithinase and proteinases) and toxins, which damage midgut cells and facilitate invasion of the hemocoel. The exact course of events following infection varies by type of bacterium. In general, after bacteria invade the hemocoel, they multiply, killing the host by either septicemia, toxins, or both. In many cases, before dying, hosts lose their appetite and cease feeding. Diseased hosts may discharge watery feces or vomit. Insects killed by bacteria often darken and become soft. Tissues may become viscous and have a putrid odor. *Photorhabdus* and *Xenorhabdus* species, bacteria associated with nematodes that attack insects, cause hosts to turn red or other characteristic colors and lack putrid odors. For these groups, host cadavers remain intact, dry out, and harden. Some bacteria are transmitted from parent to offspring in or on the eggs, as for example, *Serratia marcescens* Bizio in the brown locust, *Locustana pardalina* (Walker) (Prinsloo 1960).

The hemocoel, for many types of bacteria, is the characteristic site for infection in arthropods. Several mechanisms exist that permit bacteria to reach the hemocoel. Some species in the genus *Bacillus* produce crystalline toxic proteins that help the bacteria penetrate the midgut epithelial cells. Penetration begins with binding of these toxins to receptors on the insect's midgut cells, followed by formation of gated, cation-selective channels. These processes lead to the destruction of the transmembrane electrical potential, with subsequent osmotic lysis and death of the midgut cells (Aronson & Shai 2001). Modes of action of *Bacillus thuringiensis* Berliner endotoxins in insects are reviewed by Gill et al. (1992), Aronson and Shai (2001), and Butko (2003).

Many groups of bacteria, however, lack such toxins and normally exist as saprophytes in the insect gut or other habitats. When the host is stressed, however, these bacteria (e.g. *Proteus*, *Serratia*, *Pseudomonas* spp.) multiply more extensively in the gut and are more likely to enter the hemocoel. Some specialized pathogenic bacteria in the genera *Xenorhabdus* and *Photorhabdus* are symbionts living inside insect pathogenic nematodes. These bacteria gain entrance to the insect's hemocoel through the physical penetration of the insect by their nematode host (see section below on nematodes).

The bacterial species of greatest interest as a microbial insecticide is *Bacillus thuringiensis*, which produces toxins that paralyze and then kill the invaded host (Honée & Visser 1993). Natural epizootics of this species occur in granaries and it is believed that *B. thuringiensis* evolved in association with grain-feeding insects. Applications of this pathogen in other contexts, however, do not start self-perpetuating epizootics due to low spore production and ineffective horizontal transmission. Therefore, non-target risk from Bt applications is limited to those individuals in susceptible taxa that actually contact and ingest the applied material (e.g. Wagner et al. 1996, Rastall et al. 2003). Other species, such as *Paenibacillus* (formerly *Bacillus*) *popilliae* (Dutky), are more effective in horizontal transmission and can maintain disease cycles in arthropod populations for years under favorable conditions.

Mass rearing of bacteria

Paenibacillus popilliae is of interest because it attacks the Japanese beetle, an important turf and ornamental pest. However, it does not produce spores when grown on artificial media (Stahly & Klein 1992). Therefore, vegetative cells of this pathogen grown on artificial media, or spores collected from infected wild larvae, must be injected into the hemocoel of a live grub to produce spores (Dulmage & Rhodes 1971). This makes the product expensive, inhibiting large-scale commercial use. In contrast, *B. thuringiensis* can readily be reared in liquid artificial media (Figure 24.1) containing fish meal, molasses, corn-steep liquor solids, or cottonseed flour. Bacterial spores and associated toxins can be recovered by filtration, centrifugation, or precipitation. Production is typically carried out in 40,000–120,000-l fermenters, allowing rapid production of large quantities (Federici 2007).

Figure 24.1 The ability to produce *Bacillus thuringiensis* Berliner in liquid media is a key to its commercial success. Here, we see a small-scale liquid fermenter. Commerical production occurs in tanks of up to 120,000 l. Photograph courtesy of D. Cooper, reprinted from Van Driesche and Bellows (1996) with permission from Kluwer.

Production of *B. thuringiensis* subspecies *israelensis* that infects mosquito larvae, while possible on fermentation media, is relatively expensive. Newer media (Poopathi & Kumar 2003, Prabakaran & Balaraman 2006) have been developed that greatly reduce the cost of production, which should make the use of *B. thuringiensis israelensis* for mosquito control in developing nations affordable if consistent product quality can be achieved.

Formulation of bacterial insecticides

Most bacterial bioinsecticides contain *B. thuringiensis*. Formulations of *B. thuringiensis* must be ingested to be effective, and most products are directed against larval stages. Most *B. thuringiensis* products contain both live spores and toxins. Spores are relatively stable and are marketed as both wettable powders and liquids. Most *B. thuringiensis* products are formulated to be applied as water sprays to foliage. Some formulations use starch granules to encapsulate spores and other additives such as stickers, ultraviolet light protectants, or feeding stimulants. Formulations of *B. thuringiensis* subspecies *israelensis*, for control of mosquitoes and blackflies, are applied as liquids to aquatic habitats (Mulla et al. 1990) or as briquettes, which can be tossed into mosquito breeding areas. Genes of *B. thuringiensis* have also been introduced into major crops such as cotton and corn, causing toxins to be produced in plant foliage and protecting plants from foliage-feeding pests (see Chapters 21 and 22).

Storage of bacterial insecticides

Bacillus thuringiensis spores and toxins are stable at room temperature and do not require refrigeration (Glare & O'Callaghan 2000), giving this material storage properties as good as chemical pesticides.

Environmental limitations of bacteria

Bacillus thuringiensis products are not sensitive to dryness, although ultraviolet light can inactivate spores. For most *B. thuringiensis* products efficacy decreases a few days after application. They are stomach poisons and only kill susceptible caterpillars that actually ingest *B. thuringiensis* spores or toxin by consuming treated foliage or the mosquito larvae that ingest spores or toxin attached to filterable food items in the water.

Level of efficacy and adoption of bacterial insecticides

Many pest control scientists assumed during the 1980s and early 1990s that genetic engineering of *B. thuringiensis* strains would soon lead to a wide array of products capable of controlling numerous types of pests, thereby replacing pesticides for many uses. This did not happen, in large part because these *B. thuringiensis* products targeted pests that were ultimately controlled by Bt crop plants. This pre-empted the major markets and led to economic failure of companies producing otherwise effective *B. thuringiensis* products. As a consequence, *B. thuringiensis* products (apart from Bt plants) have remained a tiny part of the insecticide market (<1%), used mostly for integrated crop protection in orchards (Figure 24.2), organic crops, and niche markets where conventional pesticides are not desired.

Bt plants, however, are major pest control products, used on nearly half of all US corn and cotton. Other large-scale uses of Bt products include spraying pest Lepidoptera in forests by governmental agencies. In Canada, *B. thuringiensis* has replaced chemical pesticides for control of spruce budworm, *Choristoneura fumiferana* (Clemens), as a means to reduce harm to forest birds. Another major use of *B. thuringiensis israelensis* has been as an important component in the immensely successful public health campaign, mainly in West Africa, against the human disease called river blindness. This disease is caused by a filarial worm vectored by blackflies. Applications of *B. thuringiensis israelensis* to blackfly breeding sites (mainly rivers), as a replacement for chemical pesticides (after resistence developed), were part of a program that broke the transmission cycle of this pathogen, improving health for millions of people (Kurtak et al. 1989, Guillet et al. 1990, Agoua et al. 1991, Boatin & Richards 2006).

Figure 24.2 Application of *Bacillus thuringiensis* Berliner to an almond orchard to control navel orangeworm [*Amyelois transitella* (Walker)]. Photograph courtesy of P.V. Vail, reprinted from Van Driesche and Bellows (1996) with permission from Kluwer.

FUNGI AS BIOPESTICIDES

Fungal biology

Organisms that show characteristics of fungi are phylogenetically diverse and currently are classified into two kingdoms, the Straminipila (formerly Chromista) and the Eumycota (true fungi). The Straminipila includes insect pathogens in the group known as the Oomycota (e.g. *Lagenidium*), whereas the Eumycota includes insect pathogens from the Zygomycota (e.g. *Entomophthora, Entomophaga, Neozygites*), Ascomycota (e.g. *Cordyceps*) and Deuteromycota (e.g. *Beauveria, Metarhizium, Lecanicillium*).

Morphologically, fungi may occur as single cells (such as yeasts) or branched filaments (hyphae) that form mats (mycelia). Fungi may reproduce sexually, asexually, or both. Sexual reproduction involves some sort of fusion between two structures such as gametes or hyphae. The conidial spore is the most commonly used infective stage of fungal microbial pesticides. Other stages – mycelial fragments and blastospores – have been investigated but without significant applications. Commercial mycopesticides are based primarily on conidia of deuteromycotans. Host entry is usually through the integument. Most fungi do not invade hosts through the gut even if conidia are ingested. Host ranges of fungi vary from narrow to broad, but some species with broad host ranges may consist of a series of more specific pathotypes.

Fungal infections begin after conidia or other infective stages randomly make contact with a susceptible host via movement by wind, rain, or animals, or, in the case of biopesticides, by direct application to the target. Following contact, adhesion and germination of the conidia on the host's cuticle must occur. The physical and chemical properties of the insect's cuticle affect this process, influencing the host range of the fungus. Adhesion of conidia is often promoted by mucilaginous materials. The conidium, after it is deposited on the insect's cuticle and under appropriate humidity, produces a germ tube that breaches the host's integument. The penetration hypha (germination tube) exerts physical pressure on an area partly degraded by the previous release of cuticle-digesting enzymes. Fully hardened cuticle presents a greater barrier to fungal penetration than does new cuticle, making insects more susceptible after a molt.

There is great variation in fungal infections, but the following description is typical for zygomycotans and deuteromycotans. The fungus reproduces quickly after entering an insect's body cavity and kills the host. Fungi may grow as hyphae, yeast-like bodies, and wall-less protoplasts. Protoplasts help overwhelm host defenses because they are not recognized by the immune system. Yeast-like bodies produce toxins that help suppress immune reactions. After the host dies, fungi grow as saprophytes in the cadaver, forming an extensive mycelium. Conidiophores emerge from the cadaver under appropriate conditions of humidity and temperature and produce conidia, which is the stage typically harvested for mycopesticides. Temperatures of 20–30°C are most favorable for fungal infections. Conditions of high humidity (above 90%), but without free water, are often required for conidial germination and for conidial production.

Mass rearing of fungi

Mycopesticides are made from species that will grow on non-living media. Most species must be produced on solid media, with the fungus growing as a surface mat and producing conidia on aerial hyphae. Natural substances such as rice or bran are suitable rearing media. Conidia are harvested by washing fungal mats with distilled water. Effective control of target pests with fungi typically requires 10^5–10^6 conidia per cm^2 of leaf surface or, per cm^3 of soil. Production of this conidia quantity consumes 10–15 kg of rearing substrate per hectare (Federici 2007), making treatment of large areas of field crops expensive (Feng et al. 1994). Use is likely to be most practical on high-value crops such as organic wine [as in California, USA, for control of glassy-winged sharpshooter, *Homalodisca coagulata* (Say)], where the production value per unit of crop plant is very high (Federici 2007).

Production of fungi on solid media lacks a satisfactory economy of scale or potential for automation. Only a few species [such as *Beauveria bassiana* (Balsamo) Vuillemin and *Hirsutella thompsonii* Fisher] will sporulate in submerged culture (Dulmage & Rhodes 1971, van Winkelhoff & McCoy 1984). This problem can be partially resolved by a two-step culturing process in which submerged cultures are first used to produce a large quantity of mycelium, which is then placed on solid media to grow conidia (McCoy et al. 1988).

An alternative method for the commercial production of entomopathogenic fungi involves basing products on mycelial fragments or blastospores, which can readily be produced in liquid culture. This approach has been explored with *H. thompsonii*, and a patented process has been developed in which mycelia can be produced in submerged culture, dried, and stored under refrigeration until applied (McCoy et al. 1975, McCabe & Soper 1985). For *Paecilomyces fumosoroseus* (Wize) Brown and Smith, new media have been developed for the production of blastospores. This production system has a series of favorable characteristics, including short fermentation times and high yields of stable blastospores that remain viable and infective after drying (Jackson et al. 2003).

Formulation of fungi

Bateman (2004) discusses technological factors affecting development of mycopesticides. Fungal conidia need contact with the host integument to initiate infection. Stickers that promote conidial adhesion to the target are, therefore, likely to be important components of many fungal biopesticides. Wetting agents are commonly used in pesticides to help spread the product over the body of the pest by reducing electrostatic interactions that cause clumping. However, wetting agents may reduce attachment and viability of fungal conidia and must be checked for compatibility (Connick et al. 1990). Nutrients, such as powdered milk and dried egg protein may be added to mycopesticides to promote hyphal growth after conidia germinate. Nutritional supplements increase infection in some cases (Curtis et al. 2003), but in others they impede infection by stimulating saprophytic growth of the fungus.

Vegetable or mineral oils may be added to formulations to conserve water in the conidia so as to promote better germination. Bateman et al. (1993) found that formulating *Metarhizium flavoviride* Gams and Rozsypal in cottonseed oil reduced the LD_{50} of the pathogen to the desert locust, *Schistocerca gregaria* Forskal, by over 99%. Performance of oil formulations compared to water formulations was especially enhanced in arid environments (relative humidities less than 35%). Field trials under arid conditions in Niger produced satisfactory results (Bateman 1992). Formulation of fungal spores in oils also provides partial protection against degradation by ultraviolet light (Moore et al. 1993). Formulation of *Metarhizium anisopliae* (Metchnikoff) Sorokin with vegetable oils at one-twentieth of their insecticidal rates greatly enhanced control of the whitefly *Trialeurodes vaporariorum* (Westwood) by improving conidial adherence and distribution on target insects and protecting conidial viability. Adding oil increased insect mortality in a laboratory assay from 25–30% to 94–8% (Malsam et al. 2002).

Granular formulations of vegetative cells of entomopathogenic fungi such a *M. anisopliae* have also been developed (Storey et al. 1990) and appear promising for use against cutworms and other insects that feed at the soil surface. Non-granular formulations must be used for products intended to deliver fungal conidia to insects feeding on foliage. Some fungal species' conidia germinate rapidly and prematurely in water, so liquid formulations are not usable. In such cases, dust or wettable powder formulations may be used.

Storage of fungi

Storage properties of fungi used for insect control vary depending on the species and infective pathogen stage. Conidia of species such as *B. bassiana* are stable and may be stored at room temperature. Formulation of conidia in oil or kerosene improves product shelf life (Bateman 1992, Bateman et al. 1993). Blastospores of *Lecanicillium muscarium* (Petch) Zare and W. Gams (formerly given as *Verticillium lecanii*) (Vertalec® and Mycotal®) must be stored dry under refrigeration and are viable for several months (Bartlett & Jaronski 1988). The water mold mosquito pathogen *Lagenidium giganteum* Couch produces oospores that can be harvested and stored in dry form for many months, producing infective zoospores when re-wetted (Latgé et al. 1986). However, the production of zoospores by the oospores is erratic and inconsistent, making it difficult to use this water mold as a mycopesticide.

Environmental limitations on use of fungal pesticides

The principal limitation on efficacy of fungal pathogens is not host range, as many species are polyphagous, but rather failure of applied conidia to germinate and induce a high level of infection in hosts. In part this is an issue of coverage (enough conidia must land on and stick to each host's cuticle). But, more fundamentally, it is an issue of unfavorable conditions for germination on the host. Although exact requirements for conidial germination vary among species and strains of entomopathogenic fungi, many species require high humidity (>80%) for relatively long periods (12–24 h). Consequently, fungal pathogens work less well in areas that do not routinely have very high humidity.

Level of efficacy and adoption of fungal pesticides

Fungal microbial pesticides have a poor record of use by growers due to cost per hectare and variation in product efficacy due to poor infection under dry conditions. Only a few species of fungi have been registered as pesticides, despite research on many species. Six fungi – *B. bassiana*, *B. brongniartii*, *M. anisopliae*, *L. muscarium*, *P. fumosoroseus*, and *L. giganteum* – have been

Figure 24.3 Application of the fungus *Beauveria bassiana* (Balsamo) Vuillemin to coffee for control of coffee berry borer [*Hypothenemus hampei* (Ferrari)] in Colombia. Photograph courtesy of A. Bustillo.

registered for use in one or more of the OECD countries as mycoinsecticides (see Table 23.1).

Mycoinsecticides are more likely to succeed if developed as niche products to solve specific pest problems such as *B. brongniartii* against the European cockchafer *Melolontha melolontha* L. (Kessler et al. 2004) or for control of coffee berry borer [*Hypothenemus hampei* (Ferrari)] in organic coffee production (Figure 24.3) (Neves & Hirose 2005) rather than as broad-spectrum, general-purpose pesticides competing directly against established pesticides for market share. Use on low-value field crops seems especially unlikely due to required application rates of 4^9–4^{13} conidia per hectare (Federici 1999). The current capacity of commercial production systems is not adequate to treat areas as large as 20,000 hectares per week or higher, which might be needed for pests of field crops.

Another potentially viable business model for development of fungal insecticides is for work to be done by public agencies or with public funds. The development of Green Muscle by CABI BioScience researchers, with funds from governmental donor agencies from developed countries, was a project to find a fungal product to control migratory locusts in Africa and other areas. The goal was to address a major, transnational agricultural pest and replace harmful pesticides with an environmentally benign material. The large area affected by locusts stimulated donor nations to provide enough aid to support needed research and development on a range of topics, including the initial screening of fungal species and isolates, work on formulations for preservation of viability during storage and after application, and field tests of efficacy against a variety of locust species in areas with a range of climates. The fungus *M. anisopliae* var. *acridum* (formerly *M. flavoviride*) was found to be an effective species in field trials (Magalhães et al. 2000, Zhang et al. 2000, Kassa et al. 2004), and to have good storage properties when formulated in mixtures of vegetable and mineral oils (Bateman 1992, Bateman et al. 1993). Donor aid for this work has reached its end and any further development or use of these improved products now depends on national governments.

Potential of plant pathogenic fungi as bioherbicides

The above discussion concerns fungi as insect pathogens. Potentially, fungi might also be used as bioherbicides (see Charudattan 2001). Most efforts have been based on the use of species native to the area of intended use. Methods and issues for the production of plant pathogenic fungi are essentially the same as for entomopathogenic fungi (Boyette et al. 1991, Stowell 1991). Commercial success of such products, however, has so far not been achieved, in large part because the products require special storage, may not be easy for growers to purchase and use, may have limited markets, or are not competitive with chemical herbicides.

Whereas up to eight mycoherbicides have gained product registration (Charudattan 2001), few, if any, have succeeded commercially. The mycoherbicide DeVine® was formerly marketed for control of strangler vine (*Morrenia odorata* Lindle) in citrus in Florida, USA. This product contained chlamydospores of the fungus *Phytophthora palmivora* (Butler) Butler formulated as a liquid concentrate. The material had to be held under refrigeration until applied and had a shelf life of only about 6 weeks (Boyette et al. 1991). Commercial use was initially possible because the product was marketed in a small region to a specific set of users (Kenney 1986). A second fungal plant pathogen, *Colletotrichum gloeosporioides* (Penig) and Saccardo in Penzig f. spp. *clidemiae*, was marketed in the USA as Collego® for

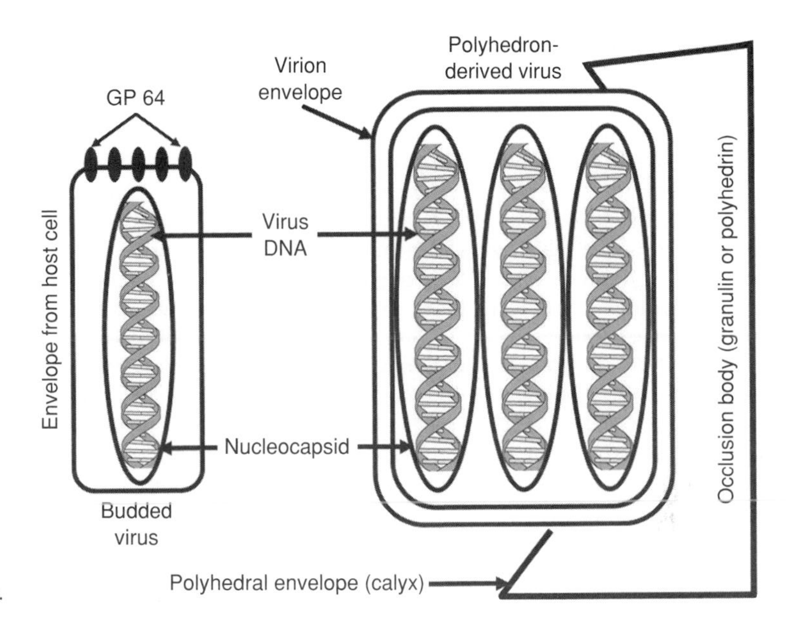

Figure 24.4 Diagram of a baculovirus.

control of northern jointvetch [*Aeschynomene virginica* (L.)] in rice and soybeans (Trujillo et al. 1986, Templeton 1992). However, eventually the manufacturers of both products ceased production for business reasons.

Two other fungi registered for use as mycoherbicides are still available. Smolder®, containing the fungus *Alternaria destruens* Simmons (Simmons 1998), is marketed for control of parasitic dodder (*Cuscuta gronovii* Willd. ex J.A. Schultes) in cranberries (Bewick et al. 1987, Hopen et al. 1997). Also, *Chondrostereum purpureum* (Pers. ex Fr.) Pouzar is sold as BioChon® for control of stump sprouting of broad-leaved trees (de Jong 2000, Conlin 2002, Becker et al. 2005). *Fusarium oxysporum* Schl. ("Foxy") is being developed for control of parasitic *Striga* plants in grain crops (Elzein et al. 2004).

VIRUSES AS INSECTICIDES

Virus biology

All viruses replicate inside host cells using the host's protein-synthesizing metabolism and materials (Matthews 1991). All viruses used as microbial insecticides are baculoviruses. Groups placed in the baculoviruses include the nucleopolyhedroviruses (NPVs) and the granulosviruses (GVs). These DNA viruses are all obligate intracellular pathogens and attack only arthropods (Figure 24.4). Baculoviruses consist of a genome within a **capsid** (protective protein coat), which collectively are called a **nucleocapsid**. The nucleocapsid becomes a mature virus particle (called a **virion**) after being coated with a lipid bilayer envelope. Viruses may occur singly with enveloped nucleocapsids or each nucleocapsid may contain multiple viruses. The virions are further embedded in a protective protein matrix to form larger masses termed **occlusion bodies**, which are called **polyhedra** for NPVs and **granules** for GVs. Occlusion bodies are usually 2–4 μm in diameter and are visible with a compound microscope.

Baculoviruses enter hosts when larvae eat contaminated food. The high pH of the insect midgut dissolves the protein occlusion bodies of NPVs, liberating virions. Virion envelopes fuse with the cell membranes of gut microvilli, and nucleocapsids enter host cells. The nucleocapsid infects the nucleus of the midgut cell, which is the primary site of infection where viral replication occurs and virion progeny are produced. These virions acquire an envelope and enter the hemocoel. These viral progeny are not occluded in the midgut cells in caterpillars (Lepidoptera), but are occluded in sawflies (Hymenoptera).

In the hemocoel, infection is caused by a non-occluded form of virus (referred to as a budded virus). Caterpillars (Lepidoptera) and sawfly larvae (Hymenoptera) are the usual hosts of NPV baculoviruses. After initial midgut infection, baculoviruses create secondary infections in many other tissues (fat body, hypodermis, trachea, blood cells) in most hosts, and the virions produced at these secondary sites are occluded by the matrix protein. NPVs in sawflies, in contrast, infect only the midgut tissue and the viral progeny from this tissue are occluded. Sawfly larvae can, therefore, pass occluded virions in their feces, enhancing transmission to other sawfly larvae. In contrast caterpillars are infectious only after they die and disintegrating cadavers release occluded virions.

Infected host larvae continue feeding, but at lower rates, up until a few days before they die. Hosts typically die 5–21 days after infection, depending on the host species. Some species of infected larvae move upward on the plant before dying, a behavior that facilitates the horizontal transmission of viruses through food contamination. Dead hosts typically become flaccid and the integument ruptures, liberating occlusion bodies containing virions, which fall down to contaminate lower foliage. Consumption of this contaminated foliage by new hosts completes the transmission cycle. Epizootics may follow if hosts are abundant.

Transmission of *Oryctes* virus (an unclassified, non-occluded DNA virus) in rhinoceros beetle, *Oryctes rhinoceros* (L.), is unusual in that adults vector the pathogen to larvae of the next generation. Transmission occurs when larvae contact the feces of sick adult beetles. Infected adult beetles can live as long as 30 days and females spread virus in their feces when they visit communal oviposition sites in rotting coconut logs, where larvae from earlier ovipositions of other females are found. Use of this virus is based on inoculation of local areas, with control persisting for extended periods (Jackson et al. 2005).

Mass rearing of insect viruses

Viruses, being obligate pathogens, can only develop in live hosts, either intact animals such as caterpillars or live cell cultures. See Ignoffo (1973) and Bell (1991) for descriptions of NPV mass rearing (Figure 24.5). For mass producton, host larvae are reared in cups of artificial diet and infected by spraying virus on the diet 1 week after host eggs are added to diet cups. At the end of the second week most larvae are dead. Cadavers can then be collected, homogenized, strained through cheese cloth, and the virus particles harvested via centrifugation. Optimal viral inoculation rates can be determined by comparing yields from a series of different viral doses per cup. Low doses may not infect all larvae. High doses kill larvae while they are still small, reducing viral yield per larva. The cost of rearing baculovirus has been calculated as US$0.02 (1991 dollars) per host, 80% of which was for labor. In Brazil, laboratory production was replaced with outdoor virus farming, in which natural host outbreaks are located, infected, and virus-infected insects later harvested.

Insect cell cultures can be used to rear insect viruses (Granados et al. 1987, King et al. 1988, Lynn et al. 1990, Lenz et al. 1991), but cost of production is higher than for rearing *in vivo* and is not a practical means of rearing these viruses for pest control. Rather, cell lines are primarily used by the pharmaceutical industry to rear genetically modified NPVs for the production of materials for medical use.

Formulation of viruses

Simple filtrates of crushed virus-killed cadavers mixed with water, if stored under refrigeration or frozen, usually perform as well or better than more complicated formulations. However, such a simple approach is not useful for production of a product for commercial use, which must be stored for up to 6 months and possess physical characteristics that permit application using various types of machinery. Commercial formulation of baculovirus products seeks to produce material with stable physical properties (no caking or clogging) suitable for application with conventional pesticide application machinery. In addition, commercial product formulations often include materials with special functions such as spreaders, ultraviolet light protectants, and food items intended to stimulate consumption by the pest (Young & Yearian 1986).

Several methods have been used to formulate commercial baculovirus products. The first of these is freeze-drying of the virus. Clumping may be prevented by first mixing the host cadavers with lactose. A second approach is mixing attapulgite clay with the virus in a water suspension, which is then sprayed and allowed to dry. This process yields a stable wettable powder in which the virus is microencapsulated by a coating of clay. A third approach is to microencapsulate virus occlusion bodies with materials such as methyl cellulose or gelatin (Young & Yearian 1986).

Materials that act as ultraviolet light protectants for viruses include a variety of dyes, especially Congo Red (Shapiro & Robertson 1990), starch encapsulation (Ignoffo et al. 1991), and optical brighteners (Shapiro &

(a)

(b)

(c)

Figure 24.5 Commercial production of baculovirus entails first mass rearing a suitable living host [a: here spruce budworm, *Choristonerua fumiferana* (Clemens)]. This is followed by (b) collection of the virus-bearing cadavers, and (c) grinding and lyophilizing the cadavers to produce a stabilized virus preparation. Photograph courtesy of J.C. Cunningham, reprinted from Van Driesche and Bellows (1996) with permission from Kluwer.

Robertson 1992). Adding such optical brighteners as Leucophor BS® and Phorwite AR® reduced the LC_{50} concentration for the virus of *Lymantria dispar* (L.) 400–1800-fold, depending on the material. In practice, these additives have not been found to be cost-effective under field conditions.

Another approach to making the most of expensive viral products is to develop methods of application other than broadcast foliar treatments. Ignoffo et al. (1980) found that if cabbage seedlings were dipped at planting in a suspension of *Trichoplusia ni* virus, pathogen activity remained high for up to 84 days. This approach reduced the quantity of virus needed for treatment and minimized labor and machinery costs.

Storage of viruses

In general, occlusion bodies of most NPVs are stable when frozen or refrigerated and can remain viable for years.

Environmental limitations of viruses

Baculoviruses degrade when exposed to light and air. This degradation is slowed by the protein coat of NPVs, but degradation still limits how long an application remains effective. Ultraviolet light is the principal cause of viral degradation. Materials like optical brighteners that absorb ultraviolet light could, after approval for inclusion in microbial pesticide products, be added to protect baculovirus.

Level of efficacy and adoption of viral insecticides

At least seven baculoviruses are currently registered for use in OECD countries (Kabaluk and Gazdik 2004) (see Table 23.1) and dozens more have been the object of research leading toward their use as microbial pesticides (Moscardi 1999).

The first viral pesticide for insect control was Elcar® (*Helicoverpa/Heliothis* NPV), which by 1975 was registered in the USA for use against *Helicoverpa zea* (Broddie). However, this product failed commercially because of the introduction of a new class of insecticides, the synthetic pyrethroids, which gave rapid mortality and killed a wide array of insect pests. In contrast, the high host specificity of Elcar and its slow rate of kill were viewed as product defects, so the product was discontinued in 1982. A new formulation of this virus was reintroduced in 1996 as GemStar® and this product is used to control both *H. zea* and the related species *Helicoverpa armigera* (Hübner) (Moscardi 1999). This virus is also produced in China, where it is applied to several hundred thousand hectares of cotton and other crops annually to control these same pests. NPVs from various species of *Spodoptera* caterpillars are used in many countries to control armyworms in maize, rice, wheat, and vegetable crops. In Europe and the USA, the product SPOD-X® is available to control *Spodoptera exigua* (Hübner) in greenhouse floral crops. The NPV of *Heliothis virescens* (F.) is registered in several countries but not widely used, except in Australia, where its use is required as a component of resistance management for Bt cotton. Elsewhere its use is declining because of the availability of Bt cotton (Federici 2007).

A granulovirus of codling moth is marketed for use on apples and is applied to about 60,000 ha annually, mostly in Europe (Moscardi 1999). This virus, however, does not provide commercial levels of fruit protection when pest pressure is high (Arthurs et al. 2005).

Viruses of various forest defoliators (caterpillars and sawflies) have also been mass produced, with subsidies from governmental agencies. Among these are TM BioControl-1®, which is the NPV of the Douglas fir tussock moth [*Orygia pseudotsugata* (McDunnough)] and the virus affecting the gypsy moth, branded as GypChek by the USDA. This virus was mass produced by the US Forest Service and enough virus was produced for up to 200,000 ha (Martignoni 1999). This material can be stored indefinitely in a frozen state and stockpiled material can be held until needed. GypChek has been produced and used by state and federal forestry agencies for control of gypsy moth outbreaks. Production costs have been funded with public money and this is not a commercial product.

The baculovirus that has been used most extensively, also with government subsidy, has been that of the soybean caterpillar *Anticarsia gemmatalis* Hübner in Brazil. A product containing this virus was developed with support from the Brazilian government and various universities and is used annually on several million hectares of soybeans. Production of the virus is based on infecting naturally occurring populations of the pest in farmers' fields and then harvesting the caterpillar cadavers. Using this approach, up to 35 tons of cadavers have been harvested in individual years for formulation into viral pesticide (Moscardi 1999).

Strategies that might make baculovirus products more competitive economically include: (1) mixing baculoviruses with low dosages of insecticides (down to one-sixth of the labeled rate) and (2) adding materials that enhance viral activity (e.g. boric acid, chitinase, neem extract) or protect virus from environmental degradation (optical brighteners) (Moscardi 1999).

NEMATODES FOR INSECT CONTROL

Nematode biology

Nematode infections usually occur in the hemocoel, but some groups such as the Phaenopsitylenchidae (e.g. *Deladenus*) and Iotonchiidae (e.g. *Paraiotonchium*) may invade the sexual organs, causing debilitation, infertility, castration, or death. Obligate parasitic nematodes of these sorts are relatively host-specific, being associated with one or a small group of hosts. Other nematodes, such as the steinernematids and heterorhabditids, however, have broad host ranges.

Nematodes that parasitize insects are translucent, and usually elongate and cylindrical. The body is covered with an elastic cuticle, but is not segmented. Nematodes are multicellular animals that possess well-developed excretory, nervous, digestive, muscular, and reproductive systems. They do not have circulatory or respiratory systems. The digestive system consists of a mouth, buccal cavity, intestine, rectum, and anus. Nematode taxonomy is based largely on sexual characters of adults; consequently, immature stages are difficult to identify without molecular techniques.

Nematodes are diverse and are found in nearly all habitats. Nematode interactions with insects range from phoresy to parasitism. Some nematodes, such as *Deladenus* (*Beddingia*) *siricidicola* (Bedding), have complex life histories with both parasitic and free-living cycles. However, the commercially reared insect-parasitic nematodes (*Steinernema* and *Heterorhabditis* spp.) have simple life cycles.

For the commercially reared nematode families (Steinernematidae and Heterorhabditidae), the **infective juvenile stage** (IJ), or **dauer stage**, is the only free-living life stage (the only one that occurs outside the insect host). The dauer stage is the third juvenile stage and is the stage that infects new hosts. This is the life stage found in commercial nematode products. This stage seeks a host and enters it via natural openings or through thin sections of cuticle. Within a few hours of host penetration, infective juveniles release symbiotic bacteria, then molt to the fourth stage, and later to the adult. In the genus *Steinernema*, adults mate and females produce eggs. The eggs hatch, develop through to adults, which again produce eggs. These eggs usually develop into infective juveniles. There are usually three generations inside a single host. In the genus *Heterorhabditis*, infective juveniles develop into hermaphrodites that produce eggs. The next generation has three sexes: males, true females, and hermaphrodites. The rest of the life cycle is the same as for *Steinernema*.

Host finding by nematodes may be an active process in which nematodes move toward and recognize hosts using cues such as bacterial gradients, host fecal components, or carbon dioxide (Grewal et al. 1993), or compounds released from plant roots in response to root herbivory (Rasmann et al. 2005). Nematode species vary in their host-searching strategies, some being ambush predators and others, active hunters (Kaya et al. 1993). For steinernematid and heterorhabditid nematodes, host penetration is an active process in which juveniles directly enter the mouth, anus, or spiracles or use proteases to penetrate the integument. Nematode infection is in the hemocoel. Infection produces relatively few external signs prior to death. Internal effects of infection, however, may be profound. Sterility is induced by several groups of nematodes, including *D. siricidicola*, the species used to suppress wood wasps in Australia.

Of nine families of nematodes parasitic on insects, only the Steinernematidae and Heterorhabditidae can be reared cheaply enough for commercial use. These families can be reared easily if provided with their symbiotic bacteria and a non-living medium. Steinernematidae and Heterorhabditidae kill their hosts in 2 to 3 days, a much shorter time than for other groups of nematodes. This occurs because these nematodes have symbiotic bacteria in their guts (*Xenorhabdus* spp., *Photorhabdis* spp.) that kill hosts by septicemia (Burnell & Stock 2000). Infective juvenile nematodes reach the hemocoel by penetrating the midgut wall or the host integument. *Xenorhabdus* spp. or *Photorhabdis* spp. bacteria, released into the host hemocoel by nematode defecation, then kill the host. Nematodes feed on the symbiotic bacteria and mature to reproductive adults. After several generations, infective juvenile nematodes exit the decomposing host cadaver. Further details on the biology of specific groups of nematodes are given in Gaugler and Kaya (1990), Kaya (1993), and Tanada and Kaya (1993).

Mass rearing of entomopathogenic nematodes

All nematodes can be reared in living hosts. For example, heterorhabditid and steinernematid nematodes, the groups of greatest commercial interest, may be reared in larvae of the greater wax moth, *Galleria mellonella* (L.). Methods for rearing the insect hosts, initiating nematode infection, harvesting, and storing the juvenile nematodes of these families have been described (Dutky et al. 1964, Woodring & Kaya 1988, Lindegren et al. 1993). Nematodes are harvested by allowing them to swim away from the host cadaver into a collection device. This system is relatively expensive with costs of about US$1.00 (1990 dollars) per million infective juveniles.

For commercial production of heterorhabditid and steinernematid nematodes, non-living media can be used in large-scale, automated systems. Glaser et al. (1940) were the first to attempt large-scale rearing of such nematodes in non-living media. Such media must: (1) use sterile ingredients to avoid unwanted bacterial contamination, (2) retain the nematode's specific symbiotic bacterium (*Xenorhabdus* spp., *Photorhabdus* spp.), and (3) provide all necessary nutrients for growth (Lunau et al. 1993).

Historically there were three challenges to development of large-scale, efficient nematode rearing: (1) identifying inexpensive nutrients, (2) identifying culture conditions that promoted high yields, and (3) using liquid rather than solid culture media (Friedman 1990). Effective media are now known, the composition of which are trade secrets of the producers. To support rearing in liquid media in large tanks it was necessary to mechanically add oxygen, taking into account susceptibility of nematodes to damage from shearing caused by stirring or bubbling. Methods to do so effectively have now been developed and commercial producers routinely use 10,000-l or larger fermenters for nematode production (Ehlers et al. 1998).

Figure 24.6 Nematodes for the citrus root weevil (*Diaprepes abbreviatus* L.) can be applied through the irrigation system, using microjets at the base of trees, which place nematodes directly over the root zone. Photograph courtesy of Steve LaPointe, USDA-ARS.

Formulation and application of nematodes

Nematodes have been formulated in many different ways, including being combined with alginate, clay, activated charcoal, gel-forming polyacrylamides, vermiculite, peat, evaporetardants, or ultraviolet protectants, being placed on sponges or in baits, and being stored in anhydrobiotic form (Georgis 1990). See Shapiro-Ilan et al. (2006) for a review of application technology and constraints imposed by the environmental limitations of nematodes. Formulations are intended to prolong nematode survival during storage, enhance ease of handling, or improve performance after application. Development of a flowable concentrate formulation, for example, eliminated the need to dissolve a carrier matrix and suspend nematodes prior to application. In general nematodes are effective only when applied to soil or when they enter plant tissues (as against borer or leafminer targets). Borers in stems of cane berries, for example, can be targets for nematodes because nematodes applied as a spray enter tunnels in canes where pest larvae feed (Miller & Bedding 1982). Nematodes may be directed against insects that attack the roots of such plants as cabbage by applying nematodes to seedlings prior to planting. They are then immediately in position to protect the plants. In turf, penetration of nematodes through the thatch into the plant root zone is critical for effective control. Nematode movement downward can be enhanced on small areas such as golf courses by irrigating after the application is made (Shetlar et al. 1988). Irrigation may not be possible at a larger scale, such as pastures, because of the large quantities of water needed. Berg et al. (1987), however, describe a mechanical device that uses a drill to introduce nematodes into the root zone, reducing the water needed from 20,000 to 1520 l/ha. In citrus, nematodes may be applied through irrigation water directed at the roots of the trees (Figure 24.6). Attempts to develop formulations that would allow nematodes to be applied against free-feeding foliar pests have generally not been successful, except in the humid tropics.

Storage of nematodes

Heterorhabditid and steinernematid nematodes survive well for a number of months if refrigerated and stored in thin, moist, well aerated layers. With some exceptions, steinernematids survive best when stored at 5–10°C and heterorhabditids at 10–15°C (Georgis 1990). Chen and Glazer (2005) report that hyperosmotic solutions (to partially dehydrate and immobilize nematodes, preventing movement that would use up energy) coupled with encapsulation in alginate granules (to conserve the water remaining in the nematodes) produced nematodes that survived well for up to 6 months when stored at room temperature and 100% relative humidity. Nematodes formulated in this manner had 96–100% survival for 6 months at 23°C, compared to only 10–15% for nematodes stored in water alone or alginate granules without treatment. The infection rate from nematodes formulated in this manner and stored for 6 months was 23%, comparable to fresh nematodes and much greater than the 2% infectivity of nematodes formulated just with alginate granules and stored for the same period.

Environmental limitations of nematodes

The principal limitation on the use of nematodes is their requirement for water as a medium in which to move toward hosts and their sensitivity to dryness and ultraviolet light, limiting their use to soil and other moist habitats. These basic features of their biology are drawbacks that seem unlikely to be overcome by technology.

Level of efficacy and adoption of nematodes for insect control

Nematodes as biopesticides have achieved a stable, if small, niche in pest control. They often work well against soil pests and they fit into the philosophy of organic farming. See Georgis et al. (2006) for a pest-by-pest review of the efficacy of augmentative use of nematodes. The US-EPA does not require nematode products to be registered as pesticides, which lowers the cost of bringing new products to market. Also, discoveries continue to be made of new nematode species that are able to attack important new pests, or do so under soil conditions that were unfavorable for earlier-commercialized species. These new species and strains make it possible to expand the market for nematodes. *Steinernema scarabaei* Stock and Koppenhofer, for example, is a newly discovered species that appears to be effective against more scarab grub pests of turf than previously available species (Koppenhöfer & Fuzy 2003). Similarly, *Steinernema riobrave* Cabanillas, Poiner, and Raulston, discovered in the mid-1990s, functions well in hot soils and has been found to provide improved control in Florida of the citrus root weevil, *Diaprepes abbreviatus* (L.) (Bullock et al. 1999). In humid climates such as Indonesia, foliar sprays of nematodes can be alternated with *B. thuringiensis* applications for control of diamondback moth [*Plutella xylostella* (L.)] to forestall development of Bt resistance (Schroer et al. 2005).

SAFETY OF BIOPESTICIDES

Most microbes and nematodes used as biological control agents occur naturally in many environments, often in large quantities during epizootics. Yet despite such potential human exposure, the medical literature does not record cases of these agents infecting people.

In many countries, including the USA, those of the European Union, Russia, and Japan, commercial microbial pesticides must be registered as pesticide products with the appropriate governmental agency. Registration requires their safety to be demonstrated to the regulatory agency before being marketed. Registration requirements generate information that a microbial product, as actually manufactured and offered for sale, is safe for use as recommended on the label. The information required for registration of microbial products differs from the information required for the registration of chemical pesticides. At a minimum, data are needed to: (1) identify the pathogen, (2) define the methods used to produce it, (3) demonstrate that the commercial product is free from contamination by other, potentially dangerous, microbes, and (4) demonstrate that the pathogen is not infectious in humans or domestic animals.

In addition, studies on the fate of the pathogen in the environment or of its effect on non-target organisms may be needed (see Betz et al. 1990). See, for example, the assessment of the effect of *B. thuringiensis* var. *israelenis* on non-target aquatic organisms (Merritt et al. 1989, Welton & Ladle 1993). Countries with commercial production of silkworms or other arthropods may require that preparations of *B. thuringiensis* not contain live spores, but only pathogen-derived toxins (Aizawa 1990). Testing procedures for microbial agents have been developed for estimating risks to plants (Campbell & Sands 1992), fish and crustaceans (Spacie 1992), birds (Kerwin 1992), mammals (Siegel & Shadduck 1992), and non-target insects and acari (Fisher & Briggs 1992).

Local systems for pathogen production may be developed in countries that do not require governmental registration of microbial pesticide products (Antía-Londoño et al. 1992). Pathogen production at the village or farm level, or by national in-country producers, should be monitored by government health agencies to ensure that systems, as operated, produce high-quality preparations of the intended pathogen, free of other microbial agents.

Requirements for registration of microbial pesticides have been summarized for the USA (Environmental Protection Agency 1983, Betz et al. 1990), Europe (Quinlan 1990), and Japan (Aizawa 1990). Although each country's requirements differ somewhat and change over time, the broad theme is to treat microbial pesticides under the same laws as pertain to chemical pesticides and to vary the data requirements to allow for differences between chemicals and infectious agents.

Safety of bacteria

The safety of Bt toxins to many organisms is based on a series of requirements to achieve a toxic effect. First, these are stomach poisons and are not toxic to any organisms unless ingested (in contrast to most insecticides). Second, the activation of Bt crystals requires an alkaline gut (pH above 8), as is found in caterpillars but not vertebrates. Following activation, midgut insect proteases must cleave the toxin and the toxin must then bind to glycoprotein receptors on midgut microvillar membranes. The requirement for this series of events renders these toxins harmless to most organisms.

The β-endotoxin produced by some strains of *B. thuringiensis* is toxic to mice and chickens, but strains used for pest control lack the ability to produce this toxin (Podgwaite 1986). Strains in commercial use do not infect humans or other vertebrates. Laboratory tests on *Bacillus sphaericus* and *B. thuringiensis* var. *israelenis* (Shadduck et al. 1980, Siegel & Shadduck 1990a) and *Clostridium bifermentans* Weinberg and Séguin serovar *malaysia* (Thiery et al. 1992) indicated that these bacteria do not cause any pathogenic effects in vertebrates. The literature on *B. sphaericus* and *B. thuringiensis* (Siegel & Shadduck 1990b, 1990c) indicates that these microbes are safe for use as pest control agents in circumstances involving human exposure.

Most *B. thuringiensis* strains will kill non-target insects that are closely related to the target pest. For example *B. thuringiensis kurstaki* is capable of killing many species of Lepidoptera. Miller (1990) assessed the effect of *B. thuringiensis* var. *kurstaki* applications on non-target forest Lepidoptera in Oregon; some species found in control areas were absent from treated areas, but the degree of impact was lower than that from chemical pesticide applications. Applications of *B. thuringiensis* to deciduous forests in the Appalachian Mountains of the eastern USA reduced densities of some non-target caterpillars (Wagner et al. 1996, Rastall et al. 2003). Caterpillars such as silkworms are susceptible to some, but not all, strains of *B. thuringiensis*. Non-target insects not closely related to the target typically are unaffected. For example, neither *B. sphaericus* nor *B. thuringiensis* affects honey bees under field conditions (Vandenberg 1990).

Bacillus thuringiensis var. *israelensis*, when applied to aquatic systems, kills larvae of flies in the families Chironomidae, Dixidae, and Certopogonidae. Densities of these groups may be moderately to severely reduced (Flexner et al. 1986). Merritt et al. (1989) evaluated the non-target consequences of the application of *B. thuringiensis israelensis* to rivers in Michigan, USA, for control of blackfly larvae and found no detectable effects on: (1) numbers of dead aquatic non-target insects drifting downstream, (2) numbers of bottom-dwelling insects in samples, (3) growth or mortality of caged mayfly larvae, or (4) mortality or feeding of various fish, especially rock bass. Collectively, these data suggest little impact of *B. thuringiensis israelensis* applications on streams, apart from effects on blackflies. A comprehensive world review of the non-target effects of *B. thuringiensis israelensis* suggests minor potential for effects on aquatic foodwebs (Boisvert & Boisvert 2000; see also Glare & O'Callaghan 2000). Many studies have demonstrated that the effects of *B. thuringiensis* on non-target organisms in or near crops is negligible, especially in comparison to the use of conventional pesticides (Sears et al. 2001, O'Callaghan et al. 2005), greatly improving crops as habitats for natural enemies.

Safety of fungi

Of the various fungi that have been developed for commercial use as pest control agents, most have shown no infectivity to humans or other vertebrates (Podgwaite 1986). No harm was observed in mice fed or exposed to *Nomuraea rileyi* (Farlow) (Ignoffo et al. 1979), rats, rabbits, and guinea pigs fed or exposed to *H. thompsonii* (McCoy & Heimpel 1980), or in mice injected with *L. muscarium* (formerly *lecanii*) (Podgwaite 1986) or *L. giganteum* (Kerwin et al. 1990). However, *B. bassiana* has been reported to cause allergies in humans (York 1958) and is an opportunistic pathogen in humans and other mammals (Burges 1981b). Also, two species of

Conidiobolus in the Entomophthorales have been reported to be pathogenic in humans (Wolf 1988).

The potential toxicity of chemicals secreted by fungi, especially during production in nutrient rich culture media, constitutes a risk separate from that of direct infections. An array of potential secondary metabolites have been recognized from species of *Beauveria*, *Metarhizium* and other groups, including destruxins, efrapeptins, oosporein, beauvericin, and beauveriolides (Strasser et al. 2000). The risks from the secondary metabolites associated with particular fungi are difficult to generalize and should be assessed on a case-by-case basis. Strasser et al. (2000) provide an overview of these classes of metabolites and their properties.

Mortality of non-target invertebrates from external contact with spores of bioinsecticidal fungi is typically less than 10% (Flexner et al. 1986). Higher mortality can occur if fungal spores are ingested. Larvae of *Cryptolaemus montrouzieri* Mulsant suffered 50% mortality when fed spores of *B. bassiana*. Adult ladybird beetles, however, were not affected (Flexner et al. 1986). Honeybee workers experienced 29% mortality when fed spores of *H. thompsonii* (Cantwell & Lehnert 1979). Both *B. bassiana* and *M. anisopliae* infect silkworms [*Bombyx mori* (L.)] and have been associated with honeybee kills following field applications (Podgwaite 1986). Granular mycelial formulations of fungi appear relatively safe to non-target organisms.

Safety of viruses

Baculoviruses pose no health risks to vertebrates. Several NPVs have been tested extensively using over 24 mammalian, avian, or fish species, and none was able to infect vertebrates (Burges et al. 1980, Podgwaite 1986). Granuloviruses have been tested less extensively, but available data suggest that these only infect Lepidoptera.

Risks of baculoviruses to non-target insects also appear to be low to nil. Most baculoviruses have narrow host ranges, typically infecting only species in one or a few related genera, usually in one family. Consequently, more distantly related invertebrates (other orders, or other families) are not at risk from virus applications (Podgwaite 1986). A few baculoviruses with wider host ranges, however, have been found, such as the *Autographa californica* NPV, which infects at least 43 species of Lepidoptera.

Safety of nematodes

Nematodes are considered safe to humans and other vertebrates by most governments and are consequently exempted from pesticide product registration laws. Rats exposed by mouth or injection to *Steinernema carpocapsae* (Weiser) showed no signs of pathogenicity, toxicity, or infection (Gaugler & Boush 1979).

Nematodes in the Steinernematidae and Heterorhabditidae have broad physiological host ranges within the insects. However, risks to non-target species from nematode applications are believed to be low (Akhurst 1990, Jansson 1993), in part because nematodes have limited motility and are restricted to specific environments due to intolerance of dryness and other unfavorable physical conditions (Georgis et al. 1991). *Steinernema carpocapsae* has been shown to have no effect on intact earthworms (*Aporrectodea* sp.) (Capinera et al. 1982). Georgis et al. (1991) did not observe any harm to non-target soil arthropods in golf-course turf, maize or cabbage fields, or cranberry (*Vaccinium macrocarpon* Aiton) bogs from applications of steinernematid or heterorhabditid nematodes. However, entomopathogenic nematode applications have been shown to reduce plant-parasitic nematode populations in laboratory, greenhouse, and field trials.

Genetically modified pathogens

Genetic engineering can be used to alter microbial pathogens for biological control. Past projects have altered the virulence or host ranges of some baculoviruses (Betz 1986, Wood & Granados 1991) and the bacterium *B. thuringiensis* (Gelernter 1992). More rapid cessation of feeding by baculovirus-infected hosts has been achieved by incorporating scorpion toxin genes into the *Autographa californica* NPV that code for production of an insect-specific neurotoxin (Stewart et al. 1991).

In principle, viral agents with overly broad invertebrate host ranges might put native moths or butterflies at risk. Williamson (1991), for example, estimated that 5–10% of Britain's Lepidoptera would be susceptible to a strain of the *Autographa californica* virus that has been modified to expand its host range. He recommended further genetic modifications, such as removal of the polyhedral gene, to render the virus incapable of sustained persistence in the wild. Field trials of modified

Autographa californica virus indicated that such a system of removing genes for production of polyhedron protein production makes this virus non-persistent (Possee et al. 1990). Efficacy under field conditions is reduced because such non-occluded viruses are rapidly inactivated. Co-occlusion (in which modified viruses and wild-type virus are used to simultaneously infect hosts to produce virus of both strains in shared occlusion bodies) has been proposed as a strategy for formulating such non-occluded viruses to permit their effective use (Wood et al. 1994). Wood and Granados (1991) give an overview of the potential uses of genetically modified baculoviruses. Modified viruses would likely pose more or less of a threat if modifications imposed on them increase or decrease their intrinsic fitness. Assessments of some viruses that have been modified for improved pest control indicate that modified viruses are less fit than their corresponding wild types, reducing any risks they might pose (Cory 2000).

Genetically modified NPVs, however, have not been commercialized and it seems unlikely that they will be. Bt crops have removed most market incentives to do so and government approval for registration of viruses possessing genes for such things as scorpion venom seems unlikely.

Part 10

AUGMENTATIVE BIOLOGICAL CONTROL

Chapter 25

BIOLOGICAL CONTROL IN GREENHOUSES

HISTORICAL BEGINNINGS

Greenhouses were among the first environments in which the idea of artificially releasing natural enemies was proposed. Kirby and Spence (1815) advocated rearing ladybird beetles for aphid control. Actual use began in 1926, when Speyer (1927) in England began rearing *Encarsia formosa* Gahan (Figure 25.1) for control of greenhouse whitefly [*Trialeurodes vaporariorum* (Westwood)] (Figure 25.2) in tomatoes. Speyer learned of this parasitoid from a grower who had noticed black (parasitized) whitefly pupae on his plants. This parasitoid was used by tomato growers for 20 years, until new insecticides caused growers to lose interest (Hussey 1985) and adopt pesticides for nearly all pest control.

Figure 25.1 The aphelinid *Encarsia formosa* Gahan, is commonly used for control of the greenhouse whitefly [*Trialeurodes vaporariorum* (Westwood)]. Photograph courtesy of Jack Kelly Clark, University of California IPM Photo Library.

Figure 25.2 The greenhouse whitefly [*Trialeurodes vaporariorum* (Westwood)]. Photograph courtesy of Les Shipp.

By the late 1950s, another major greenhouse pest, the two-spotted spider mite (*Tetranychus urticae* Koch), became uncontrollable due to pesticide resistance (Bravenboer 1960). About the same time, a German orchid grower found the predatory mite *Phytoseiulus persimilis* Athias-Henriot (Figure 25.3) in orchids from Chile and noticed it feeding on spider mites. Rearing this predator started an insectary industry for European greenhouses (Bravenboer & Dosse 1962). By the 1960s, pesticides were also failing to control greenhouse whitefly, and this stimulated the rediscovery of earlier work with *E. formosa*.

Greenhouse biological control rebirth in the 1970s provided a solution to these problems with pesticide-resistant mites and whiteflies. Initially insectaries were small operations run by growers to provide sources of *P. persimilis* and *E. formosa* for their own use but any

Figure 25.3 The phytoseiid *Phytoseiulus persimilis* Athias-Henriot is the most commonly used predator for control of two-spotted spider mites (*Tetranychus urticae* Koch). Photograph courtesy of Jack Kelly Clark, University of California IPM Photo Library.

excess was sold. A Dutch grower, J. Koppert, started a business that grew to become the world's largest insectary. Better product availability, coupled with an advisory service for growers supported by this company, led to more effective and wider use of biological control in European greenhouses, which launched the insectary industry as it currently exists.

The insectary industry aims to produce large numbers of natural enemies for release where they are absent or too scarce to provide effective pest control. Two release approaches were developed. **Inoculative releases** are meant only to seed the crop with the natural enemy, with control being provided later after the natural enemies reproduce for several generations. If natural enemies are not expected to reproduce and control is expected from releases of large numbers of the agent, the approach is called **inundative or mass release**.

Adoption of biological control in greenhouses is significant, but far from widespread. Estimates of area covered by greenhouses vary depending where one places the cut-off point on the spectrum from large, permanent glasshouses to year-round, heated plastic houses to unheated seasonal plastic tunnels. A conservative estimate is about 400,000 ha of greenhouses worldwide (van Lenteren 2000a), but China may have up to 2,000,000 ha (nearly all unheated, seasonal plastic tunnels) (Zheng et al. 2005). The portion of this area on which biological control is used is small, either 5% (excluding China) or 0.1% (China included). Biological control is mainly used in vegetable crops – 30,000 ha – mostly in north temperate areas (11,000 ha being in China) (van Lenteren & Woets 1988, van Lenteren 2000a, 2000b, Zheng et al. 2005). In addition, biological control is used on 1000 ha of ornamental crops (van Lenteren 2000a, 2000b) and a small amount of warm-region vegetable crops. The number of natural enemies reared commercially has increased from one in 1968 to more than 100 by 2006.

WHEN ARE GREENHOUSES FAVORABLE FOR BIOLOGICAL CONTROL?

Biological control for use in greenhouses was originally developed in northern European vegetable crops in greenhouses with relatively sophisticated construction. Efforts to directly apply the approaches developed there to flower crops, hot climates, or low-technology greenhouses have not worked well. Biological control is likely to be more successful in: (1) long-term rather than short-term crops, (2) vegetables rather than ornamentals, (3) crops having few pests other than the one targeted for biological control, (4) crops in which the target pest does not attack the part of the plant that is sold, (5) crops in which the targeted pest does not transmit plant diseases, and (6) well-screened greenhouses in regions with cold winters.

Long-term crops

Biological control was pioneered in long-term crops such as 4–8-month-long tomato or cucumber crops (Figure 25.4). Such crops permit approaches based on inoculative releases of a small number of natural enemies at the start of the crop. Long-duration crops allow natural enemy populations to build for several generations until, through a numerical advantage based on faster reproduction, they eventually suppress the pest. In contrast in many flower crops, the short growing period (4–6 weeks) only allows for one or two

Figure 25.4 View of a modern vegetable production greenhouse filled with young cucumber plants. Photograph courtesy of Les Shipp.

Figure 25.5 A floriculture greenhouse filled with poinsettia plants, one of relatively few flower crops grown as monocultures. Photograph courtesy of Peter Krause, Texas A&M University.

generations of the natural enemy, which is insufficient for significant population increase of most natural enemies. Consequently, releases in short-term crops have to be massive and frequent because little can be expected from reproduction. This increases the cost and may render biological control unaffordable.

Vegetable crops

The major greenhouse vegetable crops (tomatoes, peppers, and cucumbers), in addition to being long-duration crops, are affected principally by **indirect pests** of foliage, not pests of the fruit. Therefore the threshold level of pests that can be tolerated without economic losses is quite high. In tomatoes, for example, in northern Europe, whiteflies cause economic losses only at densities high enough to restrict photosynthesis, which is hundreds or more whiteflies per leaf. In contrast, in flower crops, the foliage is usually part of the product, and therefore whitefly densities that would be of little concern in a vegetable greenhouse become unacceptable for flower production.

Size of the pest complex

Crops that harbor large pest complexes make for more difficult settings for biological control. Poinsettia (Figure 25.5), in contrast, is favorable because it has only one important pest (whiteflies). Biological control programs for crops with many pests may fail because of the required effort and higher cost, or there may be no effective natural enemies for some of the pest species. In that case, the biological control program may be abandoned if an incompatible pesticide has to be used.

Direct and indirect pests

Indirect pests, which do not attack the saleable part of the plant, are better targets for biological control because more of them can be safely tolerated. Pests on rose foliage, for example, are of little consequence if confined to the lower branches, which are not included with cut roses. Spider mites occur low on rose plants grown in the bent-cane system, making the use of predator mites feasible.

Non-plant disease vectors

Species that do not transmit plant pathogens are better targets for biological control than those that do, because the tolerable threshold level of a vector may be too low to attain using biological control. Plant varieties tolerant of the pathogen may be combined with

Table 25.1 Parasitoids commonly used for control of greenhouse pests

Parasitoid	Host
Aphidius colemani Viereck, Braconidae	Aphids
Aphidius ervi Haliday, Braconidae	Aphids
Dacnusa sibirica Telenga, Braconidae	Leafminers
Diglyphus isaea (Walker), Eulophidae	Leafminers
Encarsia formosa Gahan, Aphelinidae	Greenhouse whitefly
Eretmocerus eremicus Rose and Zolnerowich, Aphelinidae	Silverleaf and greenhouse whiteflies
Eretmocerus mundus Mercet, Aphelinidae	Silverleaf and tobacco whiteflies

biological control to manage vectors. The Q strain of sweetpotato whitefly [*Bemisia tabaci* (Gennadius)] transmits tomato yellow leafcurl virus (TYLCV) in Spanish tomato greenhouses. The aphelinid *Eretmocerus mundus* Mercet can provide effective control of this strain when combined with virus-tolerant cultivars and insect screening (Stansly et al. 2004).

Cold- and warm-climate greenhouses

Greenhouses in cold climates (northern Europe, Canada, etc.) are more favorable for biological control than those in warm climates (in southern Europe, Japan, etc.). In cold climates, winter temperatures eliminate outside populations of whiteflies, aphids, thrips, etc., cutting off potential invasions. In contrast, in warm climates, greenhouses are open to maximize ventilation and are often surrounded by crops or natural vegetation that harbor pests populations. Pest levels in warm-climate greenhouses, therefore, reflect not just events in the crop, but also influxes of outdoor pests at unforeseen moments.

Greenhouse structure and design

Better-built greenhouses can assist biological control. Greenhouses with insect screening can reduce pest invasions, facilitating biological control in warm climates. Computer regulation of temperature and humidity can help avoid damaging conditions, such as low humidities, which in cucumbers can damage essential predator mite populations that otherwise can control thrips (Shipp et al. 1996). Hoop house or tunnel greenhouses, in contrast, may experience frequent episodes of overheating or excessively high or low humidity.

NATURAL ENEMIES AVAILABLE FROM THE INSECTARY INDUSTRY

For names of species of natural enemies commonly sold for use in greenhouses see Tables 25.1–25.4 (see below) and Hunter (1997), or consult websites of major insectaries. Some natural enemy suppliers rear and distribute, whereas others are just distributors. To sell a new species of natural enemy, producers must invent an inexpensive mass-rearing method and there must be a potentially profitable market for the species. Markets for producers are being fragmented because of restrictions placed on international sales, due to governmental concerns over potential non-target impacts from imported species. Vendors wishing to sell products across national borders must demonstrate that shipments are correctly identified, consistent in their content, and free from all types of contaminants. Also, an evaluation must be made as to whether the species has the potential to permanently establish outdoors where sold and whether it would matter. This is forcing commercialization of local species that are duplicative of existing products. Some countries, especially in Europe (e.g. Switzerland and Austria; Bigler 1997; Blümel & Womastek 1997) have for several years required registration of parasitoid and predator products, reflecting an increasing trend as more countries adopt similar requirements. Registration for predators or parasitoids was not required as of 2006 in the USA. In the following sections we discuss the most commonly used natural enemies and nematodes.

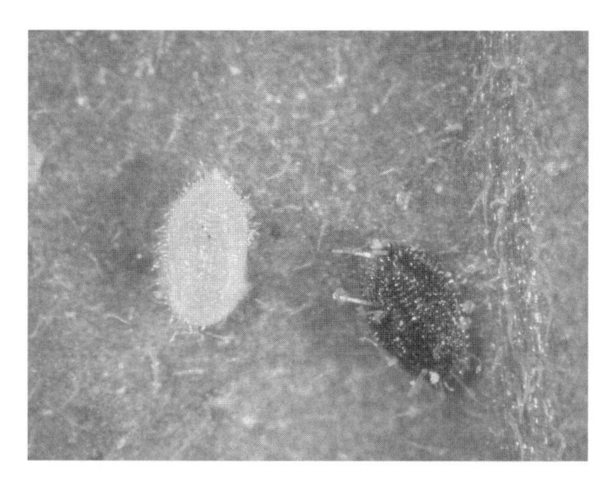

Figure 25.6 Pupae of the greenhouse whitefly [*Trialeurodes vaporariorum* (Westwood)] turn black (in contrast to a normal cream color) when parasitized by *Encarsia formosa* Gahan, facilitating monitoring of parasitism in the greenhouse. Photograph courtesy of G. Zilahi-Balogh.

Figure 25.7 Whitefly parasitoids, here *Eretmocerus eremicus* Rose and Zolnerowich, are often sold as parasitized whitefly pupae glued to release cards that can be hung in the crop canopy. Photograph courtesy of G. Zilahi-Balogh.

Parasitoids

Parasitoids are sold primarily for control of aphids, leafminers, and whiteflies (Table 25.1), and are more efficient than predators. Parasitoids exist in nature that provide control of additional pests, including various scales or mealybugs, but the market for these is too small to allow for commercial production.

Aphidius colemani

This parasitoid attacks green peach aphid [*Myzus persicae* (Sulzer)] and cotton aphid (*Aphis gossypii* Glover), but not foxglove aphid [*Aulacorthum solani* (Kaltenbach)], which is also a frequent greenhouse pest. It is sold as parasitized aphid mummies. Open rearing units (called banker plants) can be used to establish parasitoids in advance of aphid infestations; they consist of cereal plants infested with a grass-feeding aphid suitable as a host for *A. colemani*. These are placed in the greenhouse when the crop is started and parasitoids are released to initiate a population. Use of banker plants can reduce cost and improve control, but needs careful monitoring.

Encarsia formosa

This parasitoid is widely used for control of greenhouse whitefly (Hoddle et al. 1998a). All wasps are females and have black and yellow bodies. Most eggs are laid in older nymphs and one generation takes 20 days at 23°C. Wasps are reared on the target pest on tobacco plants. Parasitized *T. vaporariorum* nymphs turn black (Figure 25.6). This species and *Eretmocerus eremicus* are sold as pupae glued to release cards that can be hung in the crop (Figure 25.7).

Eretmocerus eremicus

This lemon-colored parasitoid (Figure 25.8) was commercialized to combat *Bemisia argentifolii* Bellows and Perring on poinsettia. There are both males and females (50:50 sex ratio). Females lay eggs beneath nymphs and young larvae burrow into hosts. Females preferentially attack second and early third instars, and one generation takes 17–20 days. In commercial poinsettia, most host suppression is due to host-feeding, not parasitization.

Eretmocerus mundus

This parasitoid is native to the Mediterranean region and in Spanish tomato greenhouses is more effective against the Q strain of *B. tabaci* than *E. eremicus* (Stansly et al. 2004). It parasitizes all whitefly instars but prefers second instars (Jones & Greenberg 1998).

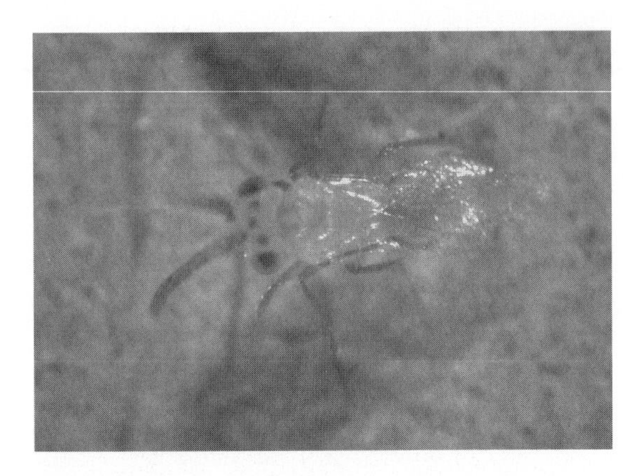

Figure 25.8 The whitefly parasitoid *Eretmocerus eremicus* Rose and Zolnerowich is used primarily against *Bemisia tabaci* (Gennadius) (strain B or Q). Photograph courtesy of Les Shipp.

Dacnusa sibirica

Dacnusa sibirica Telenga is an internal parasitoid of larvae of *Liriomyza bryoniae* (Kaltenbach), *Liriomyza huidobrensis* (Blanchard), *Liriomyza trifolii* (Burgess), and *Phytomyza syngenesiae* (Hardy). Females lay more eggs under cool conditions, making this species well suited to cool winter crops (Minkenberg & van Lenteren 1986). Adults parasitize first and second instars, which continue to feed. Parasitoids develop in older larvae and emerge from pupae. This species does not host-feed.

Diglyphus isaea

This leafminer parasitoid attacks *L. bryoniae*, *L. huidobrensis*, *L. trifolii*, and *P. syngenesiae* (Minkenberg & van Lenteren 1986, Johnson & Hara 1987, Heinz & Parrella 1990). Unlike *D. sibirica* it an avid host-feeder. It is an external parasitoid that prefers older larvae (second or third instars) and pupates in the leafmine. It is well adapted to warm temperatures and is used in areas not cool enough for *D. sibirica*.

Predatory mites

Spider mites (Tetranychidae), broad mites [*Polyphagotarsonemus latus* (Banks), Tarsonematidae] and cyclamen mites [*Phytonemus pallidus* (Banks) Tarsonemidae] are important greenhouse pests. Biological mite control is based on use of predatory mites (Table 25.2).

Phytoseiulus persimilis

This is the species (Figure 25.3) most widely used for control of spider mites. It does not feed on broad mites. It frequently consumes all available prey and dies out, requiring periodic additional releases. Biweekly applications are used preventatively in ornamental crops in Florida, USA. When used curatively, applications should be concentrated near the densest mite infestations. Low relative humidity (<50%) and high temperature (>32°C) are unfavorable. Strains may be purchased that are resistant to some pesticides.

Neoseiulus (=Amblyseius) cucumeris

This Type IV predator mite (Figure 25.9) (McMurtry & Croft 1997) can feed on both eggs of spider mites and pollen, allowing it to increase on plants shedding pollen even in the absence of prey. It is used extensively against western flower thrips [*Frankliniella occidentalis* (Pergande)], but only kills young larvae. It also feeds on cyclamen and broad mites. A sachet formulation

Table 25.2 Predatory mites commonly used in greenhouses

Predator*	Prey
Galendromus (=Metaseiulus=Typhlodromus) occidentalis (Nesbitt)	Spider mites
Hypoaspis aculeifer Canestrini and *Hypoaspis miles* (Berlese)	Fungus gnats and western flower thrips
Neoseiulus (=Amblyseius) californicus (McGregor)	Spider mites
Neoseiulus (=Amblyseius) cucumeris (Oudemans)	Thrips, cyclamen, mites, and broad mites
Phytoseiulus persimilis Athias-Henriot	Spider mites

*Phytoseiidae, except *Hypoaspis* species, which are Laelapidae.

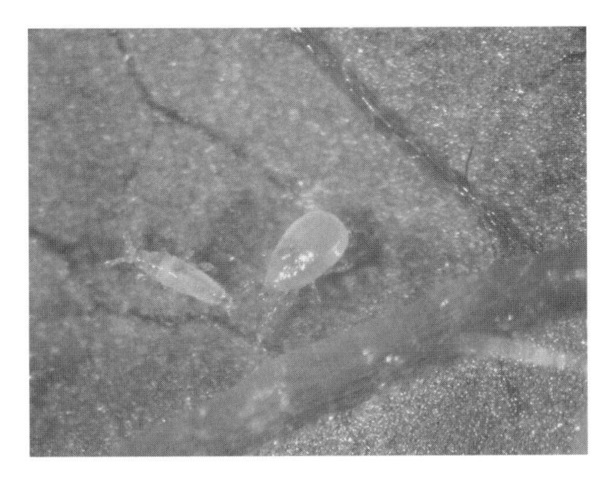

Figure 25.9 The phytoseiid *Neoseiulus* (=*Amblyseius*) *cucumeris* (Oudemans) is the predator most commonly used in greenhouses for control of thrips. Photograph courtesy of Les Shipp.

Figure 25.10 The phytoseiid *Neoseiulus* (=*Amblyseius*) *cucumeris* (Oudemans) can be released by sprinkling bran containing mites, or, as shown here, by placing open rearing units called sachets in the greenhouse. Such sachets contain a non-pest prey mite and food source (grain) and continue to produce mites that leave the sachet for up to 6 weeks. Photograph courtesy of Andrew Chow.

(Figure 25.10), containing predators and grain mites, produces predators for 6 weeks. It is most effective on long-term crops like peppers that produce pollen, but is also used in cucumbers, eggplants, melons, and ornamental crops. It provides only partial control of western flower thrips in spring flower crops in the northeastern USA, even at three or four times the recommended rate (Van Driesche et al. 2006b).

Predaceous insects

Some predaceous insects are produced commercially for control of aphids, mealybugs, mites, whiteflies, or thrips (Table 25.3). Some are effective against their target prey, but others are not.

Aphidoletes aphidimyza (predatory midge)

This fly's larvae are aphid predators, used in vegetable and ornamental crops. Adults are weak fliers, crepuscular,

Table 25.3 Predatory insects commonly used in greenhouses

Predator	Prey
Aphidoletes aphidimyza (Rondani), Cecidomyiidae (gall midge)	Aphids
Chrysoperla (=*Chrysopa*) *carnea* (Stephens), Chrysopidae (common green lacewing)	General predator
Cryptolaemus montrouzieri Mulsant, Coccinellidae (mealybug destroyer)	Various scales and mealybugs
Feltiella acarisuga (Vallot), Cecidomyiidae (gall midge)	Mites
Harmonia axyridis (Pallas), Coccinellidae (ladybeetle)	Aphids
Macrolophus caliginosus Wagner, Miridae	Whiteflies
Orius spp. Anthocoridae (minute pirate bugs)	General predator

and eat nectar and honeydew. Mating occurs on spider webs and eggs are laid near aphids. Larvae eat three to 50 aphids daily and pupate in soil. Greenhouses with plastic or concrete floors are unfavorable because the larvae cannot reach soil. Midges are sold as pupae, which are sprinkled on moist substrates. Non-diapausing strains should be used when days are short and cool.

Chrysoperla spp. (green lacewings)

Green lacewing larvae eat aphids, mealybugs, thrips, and whiteflies. Adults are light-green, have transparent wings with strong veins, and feed on honeydew, nectar, and pollen (Hagen 1964). Lacewings rarely reproduce in greenhouses and are mass released for control. Eggs can be produced cheaply, and equipment has been developed for their mechanical application. Larvae are more efficacious but difficult to rear due to cannibalism.

Cryptolaemus montrouzieri

This beetle can control citrus mealybug [*Planococcus citri* (Risso)], which lays eggs in ovisacs, but is ineffective against species such as longtailed mealybug [*Pseudococcus longispinus* (Targioni-Tozzetti)] that produce live nymphs, because the predator lays its eggs in ovisacs. Larvae and adults feed on all mealybug stages, but dense prey populations are required to sustain a population.

Feltiella acarisuga

This fly's larvae feed on all stages of the two-spotted spider mite, *T. urticae*. Adults are excellent searchers, and in Florida this species is often first to find mite infestations on outdoor plants. The optimal relative humidity for this species is around 90%.

Harmonia axyridis

This ladybird overwinters in homes, where it is a minor pest. It feeds on aphids on trees and shrubs (Lamana & Miller 1998). Insectaries sell larvae, which are black with orange-yellow spots. It tolerates low temperatures and can be used in unheated greenhouses.

Macrolophus caliginosus

This whitefly predator is widely used in European tomato crops. Bugs can feed on crop foliage, which allows them to establish and increase even when whiteflies are scarce. Plant feeding may cause minor damage. This species is not approved for use in the USA because of risk to plants.

Orius spp.

These anthocorids feed on thrips, mites, aphids, whiteflies, lepidopteran eggs, pollen, and plant sap, but are released mainly against thrips. *Orius* multiply and are successful thrips predators in crops such as pepper that produce abundant pollen. Conversely, they are ineffective in crops where pollen is limited, when short photoperiods induce diapause, or if the cropping cycle is too short.

Entomopathogenic nematodes

Species of Steinernematidae and Heterorhabditidae can be mass reared and are marketed for use outdoors and in greenhouses (Table 25.4). Species vary as to what pests they attack and what soil temperature or wetness is suitable.

Steinernema carpocapsae

One of the first nematodes to be commercialized, this nematode attacks various insects that occur in suitably moist habitats. Soil pests in greenhouses, such as fungus gnats (*Bradysia* spp., Sciaridae), are the principal target. *Steinernema carpocapsae* (Weiser) has a longer shelf life than *Heterorhabditis* spp. and is effective between 22 and 28°C.

Steinernema feltiae

Pot drenches of this nematode are used to control fungus gnats. It is most effective in moist soil between 15 and 20°C. It has a relatively short shelf life. Foliar applications are used for western flower thrips (Buitenhuis & Shipp 2005).

Heterorhabditis bacteriophora

This species is used to control larvae of black vine weevil, *Otiorhynchus sulcatus* (F.), in potted *Taxus* and *Rhododendron*. Control declines below 20°C. It has a short shelf life, and the infective stage does not persist in soil.

Table 25.4 Some commercially available nematodes and their target pests

Nematode species	Target pests
Steinernema carpocapsae (Weiser) (S)	Caterpillars, beetle larvae, some flies, and other soil-dwelling insects
Steinernema feltiae (=*bibionis*) (Filipjev) (S)	Various soil-dwelling insects, including fungus gnats
Heterorhabditis bacteriophora Poinar (H)	Manure flies, caterpillars, weevil larvae, and other soil-dwelling insects
Heterorhabditis megidis Poinar, Jackson, & Klein (H)	Various soil-dwelling insects

H, Heterorhabditidae; S, Steinernematidae.

Heterorhabditis megidis

This species is also used against black vine weevil larvae but is effective at lower soil temperatures (12°C) and remains active up to 4 weeks. Good control of black vine weevil in Ireland was achieved in bagged strawberries in unheated plastic tunnels with soil temperatures of 11–13°C (Lola-Luz et al. 2005).

GROWERS' COMMITMENT TO CHANGE

Growers have adopted biological control: (1) to protect pollinators, (2) because of control failures from insecticide resistance, (3) to protect worker health and avoid post-application re-entry restrictions associated with pesticides, and (4) to sell produce to the organic market.

Greenhouse tomatoes were pollinated by hand at great expense until the 1980s when methods were developed to rear bumblebees, which are excellent tomato pollinators. Bumblebee pollination reduced costs and improved fruit quality and yield. Because bumblebees are sensitive to pesticides, tomato growers using them had to replace pesticides with biological control for whiteflies and other tomato pests.

Pesticide resistance caused some growers to adopt biological control. The predator *P. persimilis*, the first mass-reared natural enemy, was commercialized specifically because of a need to control miticide-resistant two-spotted spider mites.

Applicator or worker poisonings increased when organophosphate pesticides replaced organochlorines during the 1970s because of the higher acute mammalian toxicity of these new pesticides. Pesticide laws were passed in many countries to reduce these problems, requiring applicator training, use of protective equipment, and imposing mandatory re-entry times on treated areas. These measures made pesticides less convenient and called attention to their risks. Some growers adopted biological control to avoid such difficulties.

Consumer desires to avoid pesticide residues in food caused growers to emphasize their use of integrated pest management (IPM), low-residue, or organic farming techniques. Labeling crops as organic or low-residue produce was formally tied to use of biological control in some countries, which encouraged more growers to use natural enemies for control because of higher prices for organic products.

REQUIREMENTS FOR SUCCESS: EFFICACY AND LOW COST

For augmentative biological control to be preferred by growers over pesticides, natural enemies must control the target pests consistently when used as directed and must be priced competitively with other pest control options, in order for biological control to make economic sense.

Is the natural enemy effective?

To be effective, parasitoids or predators must locate and attack the pest under typical greenhouse crop conditions. Each species of natural enemy is unique in how well it works for any given pest/crop/environment combination. Small differences in host preferences, rates of increase, or climatic tolerances, can make one natural enemy highly effective while another is a failure. Even strains of a single species can vary in important features such as critical photoperiod for diapause induction (Havelka & Zemek 1988), parasitism rate (Pak and van Heiningen 1985, Antolin 1989), or pesticide resistance (Rosenheim & Hoy 1986, Inoue et al. 1987).

Laboratory tests can identify potentially effective agents for new problems (Hassan 1994). Steps in such screening (van Lenteren & Woets 1988) are: (1) eliminating species with obvious defects for the intended use, (2) confirming the agent can develop to the adult stage on the target pest, (3) confirming that the agent will attack the pest on the crop, (4) verifying (for agents used in inoculative programs) that the agent's rate of population increase is greater than that of the pest, and (5) verifying that the agent is safe to other beneficial organisms used in the crop.

After preliminary laboratory studies, greenhouse trials are needed to determine that the agent can locate and attack the target pest efficiently under greenhouse conditions (e.g. Hoddle et al. 1998b). For example, the best species of parasitoid to control *B. tabaci* has been the subject of extensive research. While *E. formosa* does attack *B. tabaci*, it does not do so efficiently. Consequently *E. eremicus* was brought into commercial production, based on research that showed it to be more effective (Hoddle et al. 1997a, 1997b, Hoddle & Van Driesche 1999, Van Driesche et al. 1999, Van Driesche & Lyon 2003). *Eretmocerus eremicus* was developed for use on poinsettia in northern climates. Even with *T. vaporariorum*, research has shown that *E. ere*micus is more effect than *E. formosa* during winter months in temperate climates (Zilahi-Balogh et al. 2006). In a different context, tomatoes in a warm climate (Spain), *E. mundus* was been found to be more effective (Stansly et al. 2004) and has come into commercial production for that market.

Can the natural enemy be reared with no loss of quality?

Quality of a mass-reared agent may decline over time (van Lenteren 2003). Potential deterioration of behaviors required for foraging or attack on pests can be prevented by monitoring production with standardized tests. Tests can assess overall performance of the agent or focus on specific component attributes such as walking speed or parasitism rates. The quality of *Trichogramma* species reared for use against *Ostrinia nubilalis* (Hübner), for example, can be monitored by releasing wasps into greenhouses where host eggs have been placed on maize plants. The test measures the ability of wasps to fly to maize, find eggs, oviposit, and develop successfully (Bigler 1994). Measures of success (numbers of wasps reaching the plant in a fixed time, numbers of egg masses discovered, percentage of eggs attacked, percentage of wasps emerging from parasitized eggs) can be compared to the performance of the original colony. Quality-control tests have been developed for most major natural enemies used in greenhouses, and these tests should be used regularly by major producers (Nicoli et al. 1994).

Major producers recognize the need to rear agents under conditions that preserve essential traits, but trade-offs exist between conditions that favor rearing efficiency and those that produce agents of the highest quality (Boller 1972). When managing a mass-rearing colony, several factors must be considered: (1) genetics, (2) nutrition, (3) prevention of contamination, and (4) opportunities for exposure to host kairomones. Also, in some cases, specific cultures can be genetically improved for use as natural enemies.

Genetics

The same genetic processes that affect colonies reared in support of classical biological control (see Chapter 19) affect the quality of mass-reared populations: founder effects, drift, and selection (Mackauer 1972, Roush 1990b). Founder effects and drift are caused by starting colonies with too few individuals or by population bottlenecks caused by crashes in the rearing colony, which normally are not concerns in mass-rearing colonies. Selection for survival under laboratory conditions, with concurrent reduction in fitness for the wild environment, is the major potential problem in mass-rearing facilities. Mass-reared natural enemies often experience high host densities, unnatural foods, prey, or hosts, artificial light, and the absence of normal host cues. Under such conditions, parasitoids may be selected for reduced flight because hosts are easy to find by walking or may come to prefer the kairomones of an artificial rearing host rather than those of the target pest.

Nutrition

The hosts or foods used in natural enemy colonies can influence the size, vigor, fecundity, sex ratio, and host-recognition abilities of the agents produced. Some agents can be reared on the natural host, but some predators, such as generalist phytoseiids, may need other foods for a balanced diet, such as pollens or an alternative prey species (James 1993). For other natural enemies, rearing on the target pest is not practical and a more easily reared species is substituted. However, natural enemies reared on an alternate host may lose their ability to find, recognize, or attack the target pest (e.g. Matadha et al. 2005). Dicke et al. (1989) found that the mite *Amblyseius potentillae* (Garman), when reared on bean pollen (*Vicia faba* L.), preyed less on apple rust mite, *Aculus schlechtendali* (Nalepa), compared to a colony reared on two-spotted spider mite. In contrast, the predatory bug *Geocoris punctipes* (Say), reared for 6 years on artificial diet, showed no change in host preferences compared to wild individuals (Hagler & Cohen 1991).

Preventing contamination

Mass-rearing cultures are vulnerable to contamination. Pathogens, once present, spread well in such colonies because extensive contact among individuals and their waste products promotes pathogen transfer (Bjørnson & Schütte 2003). Microsporidia are transmitted both horizontally and vertically and reduce fertility and longevity without causing immediate death (Kluge & Caldwell 1992). Infected colonies are difficult to clean up. Hyperparasitoids may invade cultures of parasitoids and predators (Gilkeson et al. 1993). Cross-contamination between cultures of two or more similar species can also be a problem. *Neoseiulus cucumeris* (Oudemans) and *Amblyseius mckenziei* Schuster and Pritchard, for example, were difficult to rear in the same facility because of cross-contamination.

Contact with host kairomones

Host kairomones used for prey recognition influence natural enemy behaviors (Vet & Dicke 1992) (see Chapter 3). Agents reared on artificial diets or alternate hosts may lack contact with the target pest's kairomones (Noldus 1989), reducing field performance. When important cues are understood, pre-release conditioning of natural enemies may be feasible. Natural enemies can be conditioned, if shipped as adults, by providing opportunities to contact the pest, or the kairomone as an isolated chemical. Contact with hosts may not be feasible, however, if agents are sold as immature stages rather than adults.

Genetic improvement

Mass-reared natural enemies can be subjected to selection pressures for improvement. Nematodes have been selected for enhanced movement and host-finding (Gaugler et al. 1989); predatory flies and mites, for reduced rates of diapause (Gilkeson & Hill 1986); and various natural enemies, for resistance to pesticides (Roush & Hoy 1981, Hoy & Cave 1988). Agents used in greenhouses are not subject to continual natural selection after release because they are periodically released on to new crops. Genetic improvements that have been successful are the production of non-diapausing lines of phytoseiids and *Aphidoletes aphidimyza* (Rondani) for use in northern greenhouses during winter and development of pesticide-resistant strains of predatory mites for use in crops requiring pesticides for other pests.

Is the agent's price competitive with other options?

Growers are more willing to adopt natural enemies if they are priced competitively with pesticides. Price competitiveness depends on rearing cost, cost of control with pesticides, and the value of the crop. Direct comparisons of costs between natural enemies and pesticides are complex because allowances have to be made for differences in labor and convenience (set-up and clean-up time, cost of special protective equipment and worker training for pesticides) as well as benefits from the natural enemies' carryover to future crops. These features may not be obvious to growers, however, especially those without personal experience with biological control. In greenhouse tomatoes in western Europe, biological control was less expensive than chemical control for all major pests (whiteflies, spider mites, thrips, and leafminers) (van Lenteren 1989). Grower adoption, however, may be very sensitive to relative costs and adoption may be delayed until the difference between biological and chemical controls is small (Van Driesche et al. 2002c, Van Driesche & Lyon 2003). After growers switch to biological control, they often comment how the quality of their plants improves, with better fruit yield and quality or more vibrant color in flower crops.

METHODS FOR MASS REARING PARASITOIDS AND PREDATORS

Mass-rearing systems have to be efficient or the product will be expensive and growers will not use it. Three rearing methods exist: (1) rearing on the target host and plant, (2) rearing on alternative hosts or non-living foods, and (3) rearing in artificial hosts.

Rearing on the target host

Natural systems rear the agent on the target pest on its normal food plant. The mite *P. persimilis*, for example, can be economically reared on the two-spotted spider mite, reared on bean plants (Fournier et al. 1985; see Gilkeson 1992 for a review of mass-rearing methods for phytoseiids). Other phytoseiids can also be reared efficiently using natural prey (Friese et al. 1987). The parasitoid *E. formosa* is reared commercially on its natural host (*T. vaporariorum*) on tobacco (Popov et al. 1987). Natural rearing systems are feasible for some parasitoids and predators of *Liriomyza* leafminers, thrips, scales, aphids, and mealybugs. However, for many species, high labor costs make natural rearing systems impractical. This is especially so when necessary plants or herbivores are slow-growing or expensive to produce, the host is cannibalistic or susceptible to disease when crowded, or the natural enemy itself is cannibalistic.

Rearing on alternate hosts or non-living foods

Costs may be reduced by finding less-expensive substitutes at either the plant or herbivore trophic level in a rearing system. For example, winter squash may be used to rear some diaspidid scales instead of their woody hosts (Rose 1990). Alternative hosts are commonly used for parasitoids. *Trichogramma* spp., for example, are widely employed in augmentative programs (see Chapter 26), including some in greenhouses. These parasitoids could not be reared inexpensively on their target pests, but are cheap to rear in eggs of stored product moths [such as *Anagasta kuehniella* (Zeller) and *Sitotroga cerealella* (Olivier)] or silkworm moths (Laing & Eden 1990). In some cases, a parasitoid's host can be reared on artificial diet, reducing the cost. Also, rearing on alternate hosts may be useful because it eliminates potential contamination of the product with life stages of the target pest, which may pose an invasion threat in some countries.

For predators, non-prey foods may be used. The lygaeid bug *G. punctipes*, for example, has been reared successfully on liver and ground beef at costs as low as US$0.63 per 1000 insects (Cohen 1985). The phytoseiid *Amblyseius teke* Pritchard and Baker has been reared on a diet of honey, egg yolk, Wesson's salt, and water (Ochieng et al. 1987). Alternating live prey with pollen is an effective method of rearing *Neoseiulus fallacis* (Garman) at a lower cost than using only live prey (Zhang & Li 1989). Castañé et al. (2006) found that cysts of brine shrimp (*Artemia* sp.) are an excellent and inexpensive prey for mass rearing *Macrolophus caliginosus* Wagner.

Rearing in artificial hosts

Successful rearing of parasitoids in artificial hosts containing only non-living media confined in an artificial membrane has long been a basic research goal in parasitoid physiology. Attempts have been made with species of *Trichogramma* (Trichogrammatidae), *Brachymeria* (Chalcididae), *Catolaccus* (Pteromalidae), *Eucelatoria* (Tachinidae), and *Trichogramma* (Trichogrammidae), among others (Hoffman et al. 1975, Nettles et al. 1980, Thompson 1981, Guerra & Martinez 1994, Nordlund et al. 1997, Dahlan & Gordh 1998, Dindo et al. 2001). The process consists of creating an artificial host (liquid diet inside a cell of some sort), inducing oviposition by placing kairomones on the artificial host, and obtaining development of offspring through to adult emergence, with emerging adults exhibiting normal mating and fecundity. The first two steps have been worked out for several parasitoid/host systems. Rearing media may contain insect-derived ingredients or be fully defined diets with no insect components. *Bracon mellitor* Say and *Catolaccus grandis* Burks have been reared *in vitro* on diets containing only defined biochemicals, minerals, and chicken egg yolk (Guerra et al. 1993). Rearing success, however, is often significantly improved by inclusion of insect extracts (e.g. Dindo et al. 2001). The quality of parasitoids reared in artificial hosts must be assessed in field trials (Liu et al. 1985, Dai et al. 1988). It is assumed that artificial rearing systems will have lowered production costs due to greater mechanization. Practical results are still in the future.

PRACTICAL USE OF NATURAL ENEMIES

To use augmentative biological control, growers need to know how to: (1) choose and order the correct natural enemy, (2) receive and handle shipments, (3) assess the product's quality, (4) release the agent correctly, and (5) monitor for impact on the pest.

Ordering natural enemies

Catalogs and websites of natural enemy vendors list the agents recommended for each target pest (e.g. see www.koppert.com). On the Koppert website, for example, one can click on the spider mite picture and see a list of products sold for their control. Five agents are listed [*Dicyphus hesperus* Knight, *M. caliginosus*, *Neoseiulus* (=*Amblyseius*) *californicus* (McGregor), *Feltiella acarisuga* (Vallot) and *P. persimilis*]. One can then click on any agent, such as *P. persimilis*, and see the package sizes (number of animals per container, in this case 2000 mites per bottle, packed in wood chips) and recommended application rates and frequencies. For additional information contact local agricultural extension services. Lists of distributors serving a given region may be available (e.g. Hunter 1997, for the USA).

Shipping

Natural enemies must be shipped to the user quickly (2–4 days) and should not be exposed to hot or excessively dry conditions in transit. Express postal service and private carriers such as UPS, DHL, and Federal Express are typically used. Tracking slip numbers help to locate missing packages and prevent delays. Shipping boxes are designed to avoid crushing and overheating, often being made of Styrofoam. In summer, cold packs may be included. Moist sponges may be added to packages to reduce risk of desiccation. For some species, adding honey or other food to the shipping containers allows natural enemies to feed immediately upon emergence.

Storage

Natural enemies should be released immediately, but if not, should be stored in a cool place. Species vary, but in general, agents should be stored at about 5°C and used within 2–3 days for best results. The midge *A. aphidimyza* can be stored at 1°C for up to 2 months with less than 10% mortality, but requires pre-storage conditioning of 10 days at 5°C (Gilkeson 1990). *Neoseiulus cucumeris* can be stored for 10 weeks at 9°C with 63% survival (Gillespie & Ramey 1988). Storage of *P. persimilis* is improved by the addition of food, even at low temperatures, but bran or vermiculite reduces survival by promoting mold (Morewood 1992). Diapausing individuals survive longer than non-diapausing ones. Diapausing adults of *Chrysoperla carnea* (Stephens) survived at 5°C for 31 weeks (Tauber et al. 1993). Although growers should not store products for such periods, longer storage allows insectaries to stockpile production, reducing cost.

Assessing quality and application rate

Growers should inspect the contents of packages upon receipt to verify they have received the proper species, that they are alive, and that the number present matches the order. Simple methods have been devised to assess shipments. For example, for predator mites shipped in bran, take out part of the contents (say, 2% by weight or volume) and place it in a pile on white paper. Use a head-mounted magnifier (like Optivisor®) to count mites as they crawl out of the material. Finish by using a small brush to check the pile for other live mites. Then multiply by 50 and compare that number with the stated contents.

For *E. formosa*, the number of pupae received is rarely different from that stated because dosing of release cards is mechanized. For this species, the key observation is percent emergence. Place one card (with 50 or 100 pupae) in a glass jar with a tight lid and hold in dim light for 1 week. Then count the number of wasps dead in the jar and calculate percent emergence. Obviously this information is available only after the release is made. Using such methods, one can calculate the number of natural enemies present per container or card and then adjust the amount ordered up or down as needed to release the desired number.

Releasing natural enemies

To be effective, growers must release the right number of natural enemies, in the right manner. Using the methods discussed above, it is possible to accurately calculate the number being released and make any necessary adjustments. Release rate, frequency, and timing should follow either the recommendations of the manufacturer or those of public institutions or extension services.

How a release is made matters because it is possible to kill natural enemies by placing them where they quickly become wet or overheated. Also, their efficacy can vary depending on how well they are dispersed. Heinz (1998) found that for *A. colemani*, control was best if release points in a large chrysanthemum greenhouse were 3.25 m or less apart. Similarly, cards of *E. formosa* pupae or sachets of *N. cucumeris* must be properly dispersed for good control. With non-flying agents (like mites), hanging baskets must be treated individually. Natural enemies formulated in bulk carrier material may be scattered by hand, with a granular-pesticide dispenser (Ables 1979, Fournier et al. 1985), or with modified leaf blowers (Van Driesche et al. 2002b). See Mahr (2000) for a review of options for mechanical application of natural enemies. Some natural enemies, such as *P. persimilis*, are applied by hand to pest hot spots. *Orius* releases are most effective if made mid-day when conditions are hottest and driest (Zhang & Shipp 1998).

Banker plants are plants infested with a non-pest, alternative host for a natural enemy, which acts like an open rearing unit in the greenhouse. The banker plant system has been used for aphid parasitoids such as *A. colemani* and for leafminer parasitoids (Bennison 1992, van Lenteren 1995, Jacobson & Croft 1998, Schoen 2000). The intent is to allow the natural enemy to increase in number before the pest colonizes the crop, improving control and reducing cost.

Release rates: how are they determined?

One of the weaknesses of augmentative biological control is that efficacy and release rates are often not justified by strong experimental data. Recommendations should be based on field trials under conditions of actual use. Such data are sometimes available for key pests of major crops grown as monocultures (like many vegetables) in traditional production areas (like The Netherlands, Spain, and Canada). However, for minor crops or greenhouses in areas where local research has not been done, recommendations are educated guesses based on work done elsewhere. In The Netherlands, sale of biological control agents is allowed only after proof of field efficacy. Producers must submit data from trials under practical conditions (such as commercial greenhouses) that demonstrate the agent is effective. Most countries have no such requirement.

Monitoring pest levels during the biocontrol program

To be successful, growers should monitor to determine results of releases. Shortly after the first release, growers should check crops for signs of natural enemy reproduction, such as mummies for aphid parasitoids, or blackened whitefly pupae for *E. formosa*, or increasing numbers of predator mites on leaves. Later, growers should track pest densities over time. Yellow sticky cards can be used to monitor whiteflies or fungus gnats. For non-flying pests, the pest scout must turn leaves or take other samples and count the pests. Extension services in many locations provide guidance on what to count, how to make the counts, and how to interpret them. The general goal is to see whether pest numbers are changing and whether the current density is below the damage threshold. In some areas, commercial scouting services are available.

PROGRAMS WITH DIFFERENT BIOLOGICAL CONTROL STRATEGIES

Here, we discuss five pest control programs that illustrate the major approaches used in greenhouses: (1) a preventative program (for fungus gnats), (2) an inoculative program (*E. formosa* control of whitefly on vegetables), (3) a mass release program (*E. eremicus* control of whitefly on poinsettia), (4) an integrated program (supplementing a partially effective natural enemy with a compatible pesticide), and (5) a banker plant program (*A. colemani* for aphids in flower crops).

1. Preventative pest control: fungus gnat control

Some pests, such as fungus gnats (*Bradysia* spp.), are nearly always present and their suppression is built into the crop's management. Fungus gnats are often not damaging, but may become so. Control in flower crops is based on pesticide drenches of the root zone, which sometimes can disrupt biological control of other pests. Soil drenches with *B. thuringiensis israelensis* or nematodes, or the release of *Hypoaspis* mites, may be substituted to avoid this disruption. Applications should start early and be repeated several times. Effectiveness can be monitored using potato plugs placed in pots for larvae or yellow sticky traps for adults.

Bacillus thuringiensis israelensis can be applied with a conventional sprayer or through the irrigation system at the rate of 0.58–1.16 billion International Toxic Units per 100 l of solution for light infestations and at higher rates for heavy infestations. Applications of *B. thuringiensis israelensis* can cause up to 92% larval mortality of fungus gnats (Osborne et al. 1985), but have little effect on shore flies [*Scatella stagnalis* (Fallén)], a common additional soil-dwelling greenhouse pest.

Steinernema carpocapsae and *S. feltiae* can reduce fungus-gnat densities (Nedstam & Burman 1990, Lindquist & Piatkowski 1993) when applied at fewer than 0.4 billion nematodes per hectare (Georgis 1990). Oetting and Latimer (1991) found that *S. carpocapsae* survived a wide range of potting media, plant growth regulators, pH levels, fertilizers, and salts. Nematodes can be applied with a pesticide sprayer, provided the tank is first rinsed with water. Nematodes can withstand pressures up to 300 psi (21.09 kg/cm^2) and can be delivered with any spray nozzles producing droplets 50 μm or larger in diameter. Sprayers that generate temperatures over 32°C harm nematodes. Other methods of delivery include application through irrigation systems and spreading of granular formulations directly on the potting media.

Hypoaspis miles (Berlese) is shipped in mixtures of sphagnum moss, vermiculite, and grain mites (as food for the predators). Predators are applied by sprinkling product on to the soil or media soon after planting and before fungus gnats have infested the media (Chambers et al. 1993).

2. Inoculative release: *E. formosa* for whitefly control in vegetables

In long-term crops with high tolerance of the pest, biological control can be achieved in some cases merely by seeding the newly planted crop with the necessary natural enemy and allowing it to increase in number over time. The classic example of such seasonal inoculative release is *E. formosa* for control of greenhouse whitefly (*T. vaporariorum*) in tomato and sweet pepper crops in temperate climates (Woets and van Lenteren 1976, van Lenteren et al. 1977, van Lenteren & Woets 1988, van Lenteren 1995). Releases of *E. formosa* are started at planting in anticipation of a whitefly population and continue at the rate of one parasitoid pupa per plant per week until parasitized nymphs are seen, at which time release rates are reduced, based on the level of parasitism observed. Seasonal inoculative releases have lower costs because fewer agents are purchased. Most pest control is achieved by natural enemies reared at no cost during the crop.

For seasonal inoculative programs to work, there must be enough time for several generations of the parasitoid during the cropping cycle, usually 4 months or more. Second, the crop must be able to tolerate some build up of the pest, which is likely to happen while the natural enemy is increasing. Whiteflies in tomato, for example, may increase up to a 1000-fold before *E. formosa* exerts control (Foster & Kelly 1978). This is acceptable with *T. vaporariorum* in northern Europe, but could not be tolerated with the Q strain of *B. tabaci* in Spain because it transmits an important viral disease. A disease-tolerant cultivar must be grown to make seasonal inoculative releases effective in Spanish tomatoes. Also, the crop must not require perfect control by a prescribed date (as with flower crops produced for specific holidays) because there is considerable variation in the time required for the parasitoid to control the pest.

3. Mass release: *E. eremicus* for whitefly control in poinsettia

Mass release is a strategy used in shorter-term crops with a low tolerance for the pest, conditions that are not suitable for inoculative programs. *Eretmocerus eremicus* is used in this manner to control the B strain of *B. tabaci* (=*B. argentifolii*) on poinsettia in the northeastern USA (Hoddle & Van Driesche 1999, Van Driesche et al. 1999, 2002c, Van Driesche & Lyon 2003). With the mass-release approach, pests are controlled by the parasitoids actually released and reproduction is not required or anticipated. In poinsettia, whitefly populations must be maintained at low levels throughout the crop cycle. This is achieved by making up to 14 weekly releases of 0.5 females per plant. Released parasitoids act both as predators and parasitoids, and it is mainly their host feeding that keeps the population below the required threshold (Van Driesche et al. 1999, 2002c, Van Driesche & Lyon 2003). To be cost-competitive with pesticides, ultra-low rates of parasitoid release are combined with a compatible mid-crop use of pesticides (insect growth regulators) (see example 4, below).

4. Integrated control: whitefly parasitoids + insect growth regulators on poinsettia

Integrated control programs (chemicals plus natural enemies) are used when natural enemies alone are not fully effective or effective rates are too high to be cost-competitive. This deficiency is resolved by supplementing the natural enemy release with use of a compatible pesticide. The use of *E. eremicus* in poinsettia, as discussed above, is such a case. For complete control based on parasitoids alone, a release rate of three females per plant per week is required (Hoddle & Van Driesche 1999). However, at this release rate, the biological control program is not competitively priced with pesticides. If, however, the density of the whitefly population is reduced at mid-crop by a double application of an insect growth regulator, 0.5 females per plant per week is effective (Van Driesche et al. 2001, 2002c, Van Driesche & Lyon 2003), a cost-competitive rate. The insect growth regulators used do not affect adult parasitoids (Hoddle et al. 2001b), the stage providing pest control.

5. Banker plant strategy (*A. colemani* for aphids)

In general, natural enemies are most effective when pests are scarce, as releases then give the highest natural enemy/pest ratio. Banker plants are tools to pre-establish a natural enemy in advance of the pest's invasion of the crop. Banker plants are infested with a non-pest species that is a host for the natural enemy. The use of banker plants seeded with *A. colemani* can control several major pest aphids in flower crops. Banker plants consist of pots of rye grass infested with a grain aphid (that feeds only on monocots) on which mummies of *A. colemani* are placed at the beginning of the crop. This allows a breeding colony of the parasitoid to develop in advance of pest aphid invasions. Parasitoids from grain aphids also forage on the crop, killing newly present pest aphids, preventing population growth. For this system to work well, the bank plants must be well maintained, the pest aphid must be susceptible to *A. colemani*, and the greenhouse must not experience high temperatures (>32°C).

INTEGRATION OF MULTIPLE BIOCONTROL AGENTS FOR SEVERAL PESTS

Biological control of two or more pests may be required in some crops. Each pest may require several natural enemies. Incompatibility among biological control agents may occur (intraguild predation) in crops with large pest complexes, or growers may lose interest if biological control becomes too costly or complicated.

Risk of intraguild predation

Some biological control agents occasionally attack and eat other agents (Rosenheim et al. 1995). Predators may eat other predators or consume parasitized hosts. When laboratory tests pit one predator against another, intraguild predation is common. For example, the predatory bug *Orius tristicolor* (White) will consume the predaceous mite *P. persimilis* (Cloutier & Johnson 1993). Both of these agents could be employed in the same greenhouse crop so such an interaction could happen. The importance of such friction between natural enemies will be species-specific and vary with the crop. Some combinations can reduce pest control. The effectiveness of *E. formosa* in Italian greenhouses was reduced by the introduction of the *E. pergandiella* Howard, which is a facultative hyperparasitoid of *E. formosa* (Gabarra et al. 1999).

Grower fatigue

When multiple pests must be controlled by several biological control programs all running simultaneously, growers may tire of the difficulties involved. Problems may arise, for example, if natural enemies are available for only some pests. First, chemicals used for species without effective natural enemies may destroy the biological control agents released for other pests, or make biological control unnecessary by controlling the entire pest complex. Second, the time and cost of using natural enemies increase sharply when many species are required. Third, invasions of new pests create a control crisis until biological control options for them are worked out.

SAFETY OF NATURAL ENEMY RELEASES IN GREENHOUSES

The safety of new biological control agents should be evaluated relative to the agent's potential to be a nuisance pest itself or have adverse effects on people or crops, or, if established outdoors, to harm non-target invertebrates.

Potential to cause nuisance problems

Natural enemies should not bite, sting, contaminate food, or enter houses. The coccinellid *Harmonia axyridis* (Pallas) is an Asian coccinellid now used in greenhouses that has become a domestic nuisance pest in the USA and Europe. This occurs because this species can establish permanent outdoor populations, which then enter homes in large numbers in the fall of the year to overwinter (Bathon 2003). Its use in new countries is not recommended.

Effects on humans

Apart from allergies, there are no known risks to human health from parasitoids or predators used in greenhouses. These agents are a distinct improvement over many pesticides. However, workers responsible for mass rearing or release of natural enemies may be exposed to high levels of insect parts, which become airborne and be inhaled, or may contact the skin (Cipolla et al. 1997). Such is the case, for example, with the mite *N. cucumeris*, released in large quantities against thrips (Groenewould et al. 2002). Workers should avoid inhalation of dust from natural enemy products and protect skin of forearms from direct contact. The safety of nematodes used as biopesticides is quite high (see Chapter 24).

Effects on crop plants

Risk of damage to the crop plant from released parasitoids is nil. Risk from releasing predators is usually low but some predaceous Hemiptera (e.g. *Macrolophus* and *Dicyphus*) (Lucas & Alomar 2002, Shipp & Wang 2006) and mites do feed on plants when prey are scarce. Whether this occurs frequently enough to be of any importance must be evaluated on a case-by-case basis. Some biological control products also pose some risk of spreading the rearing host from the mass culture. If this species is a pest and is present in the product even at very low levels, it may be moved to new regions by natural enemy purchases. The rearing of the parasitoid *E. mundus* on the Q strain of the whitefly *B. tabaci* in Spain is such a case. If even a small number of live whitefly pupae are included in the product, it could spread the whitefly, which would have large economic consequences due to pesticide resistance in this strain and its ability to transmit plant diseases not transmitted by the B strain.

Risks to native non-target species

Releases of some natural enemies may be incompatible with operations such as farming of silk moths, birdwing butterflies, or other Lepidoptera. *Trichogramma* releases or applications of biopesticides in the vicinity of such activities may harm the farmed insects. More importantly, augmentative biological control agents may establish in the environment, which might permanently affect some native species (Frank & McCoy 1994, van Lenteren et al. 2003). The European mantid, *Mantis religiosa* L., established in the USA following its sale, as did the predaceous mite *P. persimilis* in California (McMurtry et al. 1978) and Australia, and the braconid parasitoid *A. colemani* in Germany (Adisu et al. 2002). The European bumblebee, *Bombus terrestris* (L.), used as a pollinator in greenhouse tomatoes, has established outdoors on Hokkaido, Japan (Inari et al. 2005). Outdoor establishment of these natural enemies may affect native insects by feeding on or parasitizing them, or competing for scarce resources. Native lacewings of islands such as Hawaii, USA (Tauber et al. 1992), for example, might be reduced in density if highly competitive exotic lacewings became established.

Chapter 26

AUGMENTATIVE RELEASE OF NATURAL ENEMIES IN OUTDOOR CROPS

The principles governing greenhouse and outdoor releases of reared natural enemies are the same, but outdoor environments are typically more complex and less under the control of the manager. Outdoor releases of natural enemies have been employed in corn, cotton, soybeans, sugarcane, citrus, apple, other orchard crops, vegetables, strawberries, ornamental foliage plants, forests, and animal-rearing facilities. The method is applied annually on large areas (up to 32 million ha; Li 1994). The natural enemies applied to the largest area have been *Trichogramma* egg parasitoids. Other natural enemies applied to much smaller areas include predatory mites (phytoseiids), parasitoids of filth flies, and a variety of generalist predators (such as species of ladybird beetles and green lacewings).

Development of new augmentative biological controls depends both on effective mass-rearing methods and a scientific understanding of the agent's biology and ecology, especially the agent's dispersal, host seeking, and fecundity. This knowledge allows a preliminary estimate to be made of a release rate and pattern that might be successful, which must then be assessed in the crop.

Four questions are important to determine whether an augmentative release of a natural enemy is successful. (1) Did it actually suppress the pest density significantly and prevent damage? (2) Was it cost-effective relative to potential pest damage? (3) Did ecological factors or release pattern affect efficacy or cost? (4) Is use competitive with other available control options, such as pesticides or transgenic plants?

Biological efficacy and the ecological factors affecting efficacy and cost can be assessed in field trials. In the case histories that follow, we consider how successful various release programs have been. In theory, even natural enemies that are only partially effective can be made fully effective if the release rate is increased sufficiently. However, higher release rates mean higher costs, which quickly become unaffordable.

The economic cost of pest control using a natural enemy is determined by its efficacy (how high a release rate is required and how much labor to make the release), by efficiency of the rearing method (how cheaply the agent can be reared), and by governmental policies concerning public support for natural enemy rearing facilities or direct financial subsidies made as payments to farmers using the natural enemies. Government policies on product registration or natural enemy importation (for non-native species) can also affect the cost and availability of natural enemy products.

The governments of some countries (most noticeably the former USSR, China, Mexico, Brazil, and India) altered the economics of natural enemy use on outdoor crops by using public funds to build rearing facilities for agent production. Natural enemies were typically given to farmers or sold at low prices, which encouraged their use. In some areas with concentrated production of sugarcane, coffee, or citrus, industry associations have built natural enemy rearing facilities that supply natural enemies to association members at better prices. Governmental or grower-association support for natural enemy rearing facilities typically lowers the natural enemy's cost, increasing its competitiveness with pesticides or other pest control methods. In such cases, the biological control program may be abandoned if support is withdrawn. In some parts of the European Union (e.g. Germany), farmers using some natural enemies (e.g. *Trichogramma* sp.) receive a per hectare payment that represents the difference between the cost of the natural enemy and pesticides for the same job. This promotes the use of the biological control agent by eliminating the price disadvantage with other pest control methods.

In assessing what works one must carefully separate such intertwining issues as the inherent cost to rear the agent, agent efficacy (including how many are needed per hectare to control the pest), and rearing subsidies. The use of a natural enemy is context-dependent. Often a natural enemy is employed because pests have become pesticide-resistant or pesticide use does not fit the grower's practices or target market (as in organic farming). In these circumstances, a natural enemy release may be both biologically and economically effective. But, should a new pesticide be registered for control of the target pest, growers may abandon use of the natural enemy if the new pesticide is less expensive and compatible with the market objectives. Also, since labor costs are a large part of natural enemy rearing, augmentative biological control will either require access to inexpensive labor or be organized for high productivity.

To clarify how the above-mentioned forces interact, we present the history of several natural enemy release programs that have achieved various levels of success, discussing how biological issues, economic factors, and governmental policies have led to the success or failure of the program.

TRICHOGRAMMA WASPS FOR MOTH CONTROL

Overview of approach

Trichogramma wasps (Hymenoptera: Trichogrammatidae) (Figure 26.1) have long been found parasitizing the eggs of important pest moths. Although several hundred species have been described (Pinto & Stouthamer 1994), mass rearing for pest control has concentrated on five species – *Trichogramma evanescens* Westwood, *Trichogramma dendrolimi* Matsumura, *Trichogramma pretiosum* Riley, *Trichogramma brassicae* (=*maidis*) Bezdenko, and *Trichogramma nubilale* Ertle and Davis (Smith 1996) – with some use of another 10 species. Nearly all usage targets eggs of pest moths

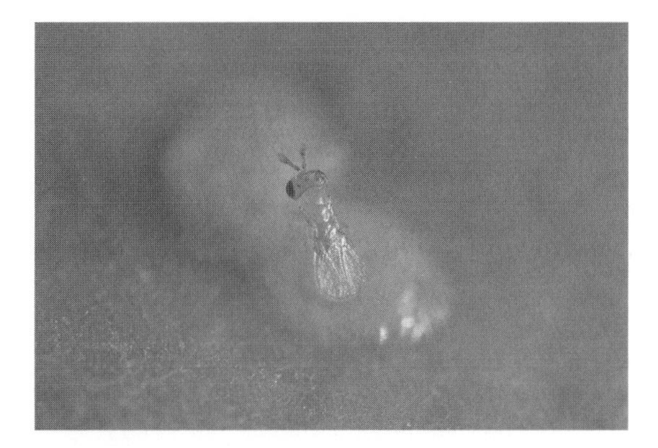

Figure 26.1 The egg parasitoid *Trichogramma platneri* Nagarkatii is being studied as a tool for suppressing codling moth [*Cydia pomonella* (L.)] in California. Photograph courtesy of Jack Kelly Clark, University of California IPM Photo Library.

whose larvae feed on a variety of crops. The most frequently targeted pests have been borers in maize (*Ostrinia* species) or sugarcane (*Chilo* or *Diatraea* spp.), or *Helicoverpa*/*Heliothis* species whose larvae tunnel in cotton bolls, fruits, or vegetables.

The general approach is to release large numbers of *Trichogramma* (as parasitized host eggs) when the pest is laying eggs. The intent is to cause high levels of egg parasitism (>80%) to reduce later larval damage. Since the target pests are dispersed over large amounts of space and foliage, high numbers of *Trichogramma* must be released (50,000–300,000 per hectare) if a high percentage of the pests' eggs are to be found before the wasps die, which usually happens in 3–7 days. Several releases at regular intervals are often necessary. To support releases of this magnitude, mass-rearing procedures are required that can produce many millions of wasps per week. Such large-scale rearing is typically not economically feasible using the target pests of interest because those species would have to be reared on plants or artificial diet. The commercial use of *Trichogramma* became possible because methods were developed to rear these parasitoids in alternate hosts: either eggs of: (1) moths whose larvae feed on inexpensive grain or (2) silkworm moths available as byproducts of the silk industry (see Greenberg et al. 1998 for a review of these methods).

In much of the world, *Trichogramma* are reared on eggs of grain moths such as *Sitotroga cerealella* (Olivier), *Ephestia kuehneilla* Zeller, or *Corcyra cephalonica*

(Stainton). Of these, *S. cerealella* seems to be the poorest rearing host, producing the smallest wasps. This rearing approach was developed by Flanders (1930) in California, USA, and further improved for large-scale industrial production in the USSR (1940–70) (Meyer 1941, Telenga & Schepetilnikova 1949, Lebedev 1970). Moths are reared on grain (wheat or rice) (Figure 26.2a) and their eggs harvested. This process yields millions of eggs at a very low cost because the rearing drums require little space, are handled with machinery, and are held dry and without light. Moth eggs are separated from grain by shaking the drums, which have screened bottoms, and collecting the eggs as they fall. Eggs are then placed in trays in cabinets containing gravid female wasps, where they are parasitized. Parasitized eggs (Figure 26.2b) are collected and formulated for sale either as eggs stuck on cards, as loose eggs to be broadcast, or as eggs in release boxes (Figure 26.3) that provide protection from weather and predators. Wasps emerge in the field from these parasitized eggs.

The size of the rearing host affects parasitoid size (and hence fecundity) so a large rearing host – if available – is beneficial. The large eggs of silk moths, especially the Chinese oak silkworm, *Antheraea pernyi* (Guérin-Méneville), are a cheap by-product of the silk industry in China, and can be used to rear *Trichogramma* species. Wasps reared in such large eggs are believed to have better adult longevity and fecundity.

Government subsidies have been important in securing farmer acceptance of *Trichogramma* releases as an approach to pest control. Rearing facilities in China were frequently village-level cottage industries on government-organized collective farms. *Trichogramma* rearing was frequently included as part of the expected work on such farms. In the former USSR, specialized public factories were built to produce large quantities of *Trichogramma* for use on cooperative farms. In other countries, such as Mexico, state or federal natural enemy rearing facilities were built to provide farmers with natural enemies at a lower cost than would have been possible otherwise. In sugarcane-producing areas, natural enemy rearing facilities have been built by grower associations to supply association members with natural enemies. In Europe, *Trichogramma* wasps for use against borers in maize are reared by private insectaries. In some countries, support for mass rearing has declined due to changes in government (Russia) or social conditions (China).

In North America, prices for *Trichogramma* produced by private commercial firms are generally about US$200 (1996 rates) per million (Smith 1996). Wasps produced by government-subsidized insectaries cost less. *Trichogramma* release programs that depend on such subsidies are sustainable only with continued governmental support. In principle, research on

(a)

(b)

Figures 26.2 To mass rear *Trichogramma* wasps, a suitable source of host eggs is required. (a) Special cabinets for semi-automated rearing of *Ephestia kuehneilla* Zeller is one approach to obtaining host eggs; (b) eggs are then placed in sting cabinets for parasitization of the eggs, which later appear black. Photographs courtesy of Mario Waldburger, Agroscope Reckenholz-Tänikon Research Station, Switzerland.

Figure 26.3 Whereas parasitized eggs bearing *Trichogramma* pupae may simply be sprinkled in the crop, several release devices, as in this photograph, have been invented that may be hung on plants or scattered on the soil. Photograph courtesy of Mario Waldburger, Agroscope Reckenholz-Tänikon Research Station, Switzerland.

many aspects of the use of *Trichogramma* spp. such as rearing efficiency, or procedures to make lower doses effective in the field, could reduce cost (Parra & Zucchi 2004).

Use of *Trichogramma* in particular crops

Use in sugarcane

Releases of *Trichogramma* spp. in this crop are targeted at lepidopteran borers (especially *Diatraea* and *Chilo* spp.) that are pests in tropical areas. In some cases, release programs have clearly been ineffective. *Trichogramma fasciatum* (Perkins), for example, was released in Barbados from 1930 to 1958 for suppression of *Diatraea saccharalis* (Fabricius), but when releases were stopped in 1958, there was no subsequent increase in pest levels, suggesting that the previous releases had no pest control value (Alam et al. 1971). In contrast, other efforts seem credible and scientifically documented. In the Indian Punjab, for example, releases of *Trichogramma chilonis* Ishii at 50,000 per hectare at 10-day intervals provides about 50% reduction in damage from the borer *Chilo auricilius* Dudgeon, reducing losses from 13% in the untreated control to 6%. Seasonally, 11–12 releases were needed from July to October (Brar et al. 1996, Shenhmar et al. 2003). In other trials, with higher pest pressure, similar releases reduced infestation rates to 13%, from 38% in untreated controls (Shenhmar & Brar 1996). In Uttar Pradesh (India), similar releases of this parasitoid increased yield by 5–21 tons/ha (Singhal et al. 2001). In Pakistan, 5 years of trials on an aggregate of 50,400 ha of sugarcane found that releases of *T. chilonis* reduced borer infestations to 3–5% compared to 9–33% in untreated controls (Ashraf et al. 1999). In Brazil, releases of *Trichogramma galloi* Zucchi (at 200,000 per hectare) for control of the sugarcane borer, *D. saccharalis*, have been assessed, either alone or in combination with releases of the larval parasitoid *Cotesia flavipes* (Cameron). Combining three weekly releases of the egg parasitoid *T. galloi* with one release of the larval parasitoid *C. flavipes* (at 6000 per hectare) reduced borer infestation by 60% compared to an untreated control. Use of *Trichogramma* alone reduced the infestation rate by only 33% (Botelho et al. 1999).

Various trials with *Trichogramma* releases in sugarcane have sought to enhance control by manipulating the release method. Releases in which parasitized eggs were protected (from predators or adverse physical conditions) within a capsule led to a higher effective wasp release rate compared to parasitized eggs glued to cards and hung on plants (Rajendran & Hanifa 1998, Shenhmar et al. 1998, Pinto et al. 2003).

To assess such release programs as described above, it is important to have information on the cost of the natural enemies and the crop's value to see whether increased crop yield exceeds the cost of the releases. Any negative effects associated with insecticide use (like secondary pest outbreak) that do not occur with biological control should also be considered. Ultimately, it is important to determine whether the biological control program returns higher profits to the grower than other options. For example, Tanwar et al. (2003) found during trials in Uttar Pradesh that a chemically intensive program increased crop yield to higher levels than intensive releases of *T. chilonis* and *C. flavipes*. However, a cost analysis was not provided, so it is not clear whether growers benefited from higher profits.

Ultimately, the efficacy of augmentative biological control programs is determined by local pest complexes and farming practices. Profitability of natural enemy releases is further determined by local pricing of the natural enemy, the crop, local tolerance for various levels of crop quality, market objectives, and competing pest control options such as pesticides or *Bacillus thuringiensis* varieties. Labor is a large part of natural enemy costs, so systems that might be feasible in developing countries, may be uneconomical in developed countries. Such influences are not only local but changeable. A formerly successful program may be replaced by a new pesticide, new pest-resistant variety, or other new management practice or a formerly uncompetitive natural enemy species may become desirable when pesticide resistance, a new pest invasion, lowered natural enemy rearing prices, or increased demand for organic produce change the circumstances of crop production.

Use in cotton

In cotton, the targets for *Trichogramma* releases have been various bollworms [such as *Helicoverpa* spp., *Pectinophora gossypiella* (Saunders), and *Earias insulana* (Boisduval)]. Studies have been conducted in India, Egypt, Australia, and the USA, among others. In India, *T. chilonis* is applied weekly, eight times per season, at 150,000 parasitized eggs per hectare per release as part of a larger integrated pest management (IPM) package for the crop (since the pest complex includes among other things, various sucking pests not susceptible to *Trichogramma*). Use of *T. chilonis* reportedly reduced bollworm damage by 70% and increased yield by 45% over insecticides alone (Brar et al. 2002). In Egypt, releases of *Trichogrammatoidea bactrae* Nagaraja, applied in alternating sequence with an insect growth regulator, gave up to 64% control of the spiny bollworm (*E. insulana*), although results were highly variable by year (Mesbah et al. 2003). *T. evanescens* in the same system only lowered damage to 19–26% (even when combined with *B. thuringiensis* applications), compared to 29–38% in untreated controls (Mansour 2004). In the USA, releases of *Trichogramma exiguum* Pinto and Platner in North Carolina significantly increased parasitism (from 25% in controls to 67% in release fields). However, this did not significantly reduce larval numbers or damage (Suh et al. 2000a) because higher egg mortality was offset by a subsequent decrease in larval mortality (Suh et al. 2000b). This illustrates that evaluations of biological efficacy of egg parasitoids must include assessment of larval density and damage and not just be based on rates of egg parasitism achieved.

A further complication in this crop is the present availability of Bt cotton, which is widely adopted in some countries. Bt cotton provides better control of major cotton pests and is therefore likely to replace use of *Trichogramma* where it is available.

Use in maize

Trichogramma releases are used in field maize and sometimes sweetcorn to control stalk-boring larvae [e.g. *Ostrinia nubilalis* (Hübner) in Europe and North America; *Ostrinia furnacalis* Guenée in Asia] or species of *Helicoverpa* that tunnel in the cobs. Use of *Trichogramma* on this crop is widespread and somewhat better studied than in other systems. Here we contrast use in western/central Europe, the USA, and China to illustrate how local circumstances – both biological and political – can affect the use of the same pest control approach.

Europe In Germany, *T. brassicae* is reared and released annually on about 11,000 ha of field maize (Zimmermann 2004) for control of *O. nubilalis*. This use pattern has been refined over the last 25 years and improvements have been made in various aspects, including release methods (Albert et al. 2001). The recommended rate is two releases of 150,000 individuals per hectare per growing season for areas with two pest generations per year or three releases of 50,000 each for areas with only one pest generation. These rates reduce damage from the borer by 65–75% (Hassan & Zhang 1999). Use, however, occurs on less than 1% of German maize. Furthermore, use is subsidized by the government, which pays farmers the cost differential between using *Trichogramma* and using pesticides. Currently, *Trichogramma* releases are approximately twice as expensive as the use of pesticides. Government subsidies are approximately €50 per hectare (Degenhardt et al. 2003). Currently Bt maize is prohibited in Germany, but if allowed it would likely displace use of *Trichogramma* due to improved profitability (Degenhardt et al. 2003). Use of *T. brassicae* occurs on a larger acreage in France, but the treated area still constitutes a very small percentage of the crop. Efficacy is similar to Germany, but cost is more competitive, being only about 10% greater than the cost of control with pesticides. Grower use is not dependent on government subsidies.

The USA There has been little use of *Trichogramma* in maize by growers in the USA. Research studies, however, have identified *Trichogramma ostriniae* Pang and Chen as a species that may be effective against corn borer (Wang et al. 1999, Kuhar et al. 2002). Trials in New York found that a single inoculative release of 75,000 per hectare of this species in sweetcorn reduced the percentage of cobs infested with *O. nubilalis* from approximately 13% in untreated controls to 6% (the threshold used by growers is 5%) (Wright et al. 2002). This augmentative release strategy is somewhat unique in that one release made when the corn plant is small results in parasitoid establishment and reproduction, which provides season-long pest suppression.

These results suggest some potential for use of this parasitoid in sweetcorn. *Trichogramma ostriniae* became commercially available in the USA in 2005 and some time will be required to determine the level of grower adoption. Organic growers are most likely to adopt this approach. In field maize, Bt corn has already been widely adopted in the USA and is likely to preclude significant use of *Trichogramma* releases by conventional growers.

China In contrast to Europe and the USA, in China use of *Trichogramma* spp. against pests of maize reportedly occurs over large areas, especially in northern China where over 4 million ha are treated (Wang et al. 2005). The principal parasitoids are *T. dendrolimi* and *T. chilonis*, both of which can be mass reared on eggs of the oak silkworm. Interest also exists in mass producing *T. ostriniae* on *S. cereallela*. Both *T. dendrolimi* and *T. chilonis* have also been reared in artificial host eggs, and this method is being used for commercial production against *O. nubilalis* and *H. armigera* (Feng et al. 1999, Wang 2001). Control of corn borer in northern China (where there is only one pest generation per year) is based on two seasonal releases of 150,000–300,000 *T. dendrolimi* per hectare. Releases are timed by monitoring pupation of the pest and achieve 60–85% parasitism (Piao & Yan 1996). Control in more southern areas, where there are multiple generations per year, requires additional releases. Use of *Trichogramma* releases in China is a basic part of the IPM system for managing pests of maize (Wang et al. 2003).

Conclusions from the maize case history The sharp contrast between the use of *Trichogramma* in China and the USA, with Europe being somewhat intermediate, illustrates that financial support by government for *Trichogramma* production or releases can strongly affect usage. Rearing opportunities and cost remain a key issue for augmentative uses of parasitoids. The incidental availability of a favorable cheap rearing host (oak silkworm) and a labor force available for the inexpensive production of parasitoids have greatly stimulated use of the method in China. Mechanized methods based on artificial eggs also seem to offer potential, but it is yet not possible to evaluate the efficacy of this approach. To date, use of artificial rearing hosts has not yet come to dominate production, even in China. Interest in the method in the USA for use in field corn has been pre-empted by the earlier development and deployment of Bt corn. However, grower use of *Trichogramma* in sweetcorn remains possible. Producers most likely to adopt this approach are organic growers and conventional growers whose area is too small to justify the purchase of high-clearance sprayers needed to apply pesticides to the crop.

Use in processing tomatoes

Trichogramma releases have also been used for control of *Helicoverpa zea* (Broddie) in processing tomatoes. In Mexico, some 4000 ha of tomatoes are treated with *T. pretiosum* (at rates of 100,000/ha per week for up to 9 weeks per season), in combination with mating disruption and *B. thuringiensis* for *Keiferia lycopersicella* (Walsingham) and *Spodoptera exigua* (Hübner), respectively (Trumble & Alvarado-Rodriquez 1998). This practice has become well established in several Mexican states (Sinaloa, Baja California), producing tomatoes with an acceptable level of damage (<3%) at costs well below conventional pesticide spray programs. Parasitoids are produced cheaply either in state-run facilities or in facilities provided by tomato processors. The significance of the *Trichogramma* releases separate from the rest of the IPM package has not been established. Also, how much the economical viability of the use of this parasitoid depends on subsidies from the government or processor-run insectaries is unclear.

Other natural enemies (*T. chilonis* and *T. pretiosum*, and *Trichogramma brasiliense* Ashmead) have been used to control *H. armigera* on tomato in various parts of India. Release rates in the 50,000–100,000/ha range produced parasitism levels of about 40%, with damage levels similar to pesticide-treated fields and enhanced fruit yields (Praveen & Dhandapani 2003, Kumar et al. 2004).

Use in apple and walnut against codling moth

The codling moth [*Cydia pomonella* (L.)] has been investigated as a potential target for *Trichogramma platneri* Nagarkatii in apple, pear, and walnut orchards in the USA and Canada. Unlike the pests discussed previously, this insect is a direct pest of a very high-value product, with a very low damage threshold (≈1% infestation for apples). Releases of approximately 200,000 wasps per hectare reduces damage 60% relative to controls. This may suffice when pest pressure is low (Mills et al. 2000), but significant damage may result when pest pressure is high (Cossentine & Jensen 2000). Currently, this system is not competitive with pesticides, which can reduce damage by 80–100%. Also, the cost (US$300 per hectare in 1998) exceeds that of chemical control. Further research might improve efficacy and greater use might reduce cost (Mills 1998). Aerial application methods to apply the parasitoid in orchards have been studied (Figure 26.4). In this system, an alternative control system – mating disruption – has been developed that might either compete or be integrated with parasitoid releases.

Use in forests

In Canada, outbreaks of native defoliators such as the spruce budworm [*Choristoneura fumiferana* (Clemens)] occur periodically; however, public policy has been enacted that restricts chemical pesticide application by air to public forests. Such prohibition creates a market for microbial pesticides or augmentative parasitoid releases. Aerial application of *Trichogramma minutum* Riley was investigated during the 1980s and early 1990s for possible use against this pest in eastern Canada (Smith et al. 1990, Smith 1996). Smith et al. (2001) showed that budworm larval numbers can be reduced by 70% during the year of release, reducing defoliation by 50%. Later work showed that parasitoid releases enhance parasitism by native tachinids that are believed to regulate spruce budworm densities (Bourchier & Smith 1998). However, the release rates necessary to achieve this level of biological control were quite high, exceeding 10 million wasps per hectare and the resulting cost was estimated at the time to be around CAN$200 per hectare, which exceeded the cost of control with aerial applications of *B. thuringiensis* (S.M. Smith, personal communication). Consequently, this approach has not been adopted, although the situation was confounded by the concurrent collapse of spruce budworm populations in eastern Canada (1995 to present), obviating the need for control by either method.

(a)

(b)

(c)

Figures 26.4 Devices are under development for the aerial application of *Trichogramma* (as parasitized host eggs) in orchards: (a) release device on wing, (b) plane, and (c) close up of release device. Photographs courtesy of Nick Mills.

USE OF PREDATORY PHYTOSEIID MITES

Reasons for need: natural enemy destruction and miticide resistance

Spider mites became pests in many crops after 1950, due to greatly increased use of pesticides that killed spider mite predators. Spider mite problems later intensified due to development of miticide resistance, especially on crops such as apple and strawberry. The increased pest status of spider mites led to research on the role of phytoseiids and other predators in the natural regulation of spider mites. One branch of this work sought ways to restore natural control by modifying pesticide use in ways that would allow phytoseiids and other spider mite predators to increase. But another response was the commercial rearing of various phytoseiids for release in crops such as strawberries, outdoor floral crops, and some high-price vegetables like eggplant. A variation on the above responses in a few crops was the inoculative release of pesticide-resistant strains of phytoseiids in orchards. In Chapters 21 and 22, we discuss conservation biological control efforts aimed at restoring healthy field populations of mite natural enemies. Here we discuss inoculative and mass releases of insectary-reared phytoseiids.

Commonly used phytoseiids

Some species of phytoseiids can be reared cheaply. *Phytoseiulus persimilis* Athias-Henriot can be reared commercially on spider mites on bean leaves. *Neoseiulus cucumeris* (Oudeman) can be mass reared on grain mites or pollen. The number of species of phytoseiids that have been reared and considered for use in augmentative biological control has increased over time. Species such as *P. persimilis*, *Galendromus occidentalis* (Nesbitt), *Neoseiulus californicus* (McGregor), and *Neoseiulus fallacis* (Garman) have been investigated in many different countries and crops, while other species have been studied only in particular regions.

Phytoseiulus persimilis

This was the first phytoseiid to be commercially mass reared and is widely used in greenhouse vegetables (see Chapter 25). It feeds only on spiders mites. It is produced on two-spotted spider mites (*Tetranychus urticae* Koch), which are allowed to develop on young bean plants sown in greenhouses with dirt floors. Predator mites are introduced after large spider mite populations have developed and are later harvested by picking infested leaves. Use of this species is most successful if relative humidity is about 50% and the temperature is below 32°C (Osborne et al. 2004). This predator can detect spider mite colonies from a distance; it quickly locates them, and consumes most of the colony. It is better suited for suppression of small, high-density patches of mites than for prevention of population increase over large areas. Outdoor use of this species includes application to foliage plants grown in shade houses in Florida, USA (Osborne 1987, Osborne et al. 1998). It has been used in Australia since 1984 to prevent outbreaks of *T. urticae* and *Tetranychus ludeni* Zacher in field strawberries (Waite 2001). This predator has also been found to be effective for control of two-spotted spider mites in hops in the UK, when used at the rate of 10 mites per plant, in combination with the ovicidal acaricide clofentizine (Lilley & Campbell 1999).

In Australia, strong support from extension personnel increased grower acceptance of this predator in strawberries. Releases were based on the system of release called pest-in-first, in which some plants are first inoculated with spider mites. Release of prey increases the reliability and efficacy of biological control by promoting predator establishment. In Florida, *P. persimilis* has been used in strawberries since 1999 because of spider mite resistance to abamectin, the major miticide (Price et al. 2002a). However, while adoption of biological mite control in Florida initially increased from 15 to 30% of strawberry producers, pesticide companies responded by developing new miticides [e.g. Savey (hexythiazox) 50 WP] and pesticides recaptured lost market share (Price et al. 2002b). This illustrates the dynamic and sometimes unsustainable nature of augmentative biological control. Since mite releases must compete against all alternatives, their use may increase or decrease suddenly.

Neoseiulus californicus

This mite is more tolerant of low humidity than *P. persimilis* (Osborne et al. 1998) and can survive longer without food. This makes it possible to use this species preventatively when pest mites are still scarce. It can be reared on spider mites (Henrickson 1980) or on pollen (Dindo 1995). It is very mobile and has been used on strawberries, ornamental foliage crops, and fruit trees (Castagnoli & Simoni 2003). Releases of 2000 mites per tree provided control of persea mite (*Oligonychus perseae* Tuttle, Baker and Abbatiello) in Californian avocado equivalent to application of a horticultural oil (Hoddle et al. 2000). However, the cost of mite applications was 10-fold higher than aerially applied horticultural oil, so mites are not used by growers.

Galendromus occidentalis

This species has been tested for control of spider mites in several crops, including apple (Croft & MacRae 1992), hops (Strong & Croft 1995), avocado (Hoddle et al. 1999), and cotton (Colfer et al. 2004), but only with minimal success.

Neoseiulus cucumeris

This mite can be reared cheaply on grain mites and is extensively used in greenhouse vegetable crops in cold climates for control of thrips (Shipp & Ramakers 2004). It has a broad diet and readily feeds and reproduces on pollen. Outdoor use is limited, but it has been found to partially control the tarsonemid mite *Phytonemus pallidus* (Banks) in outdoor strawberries in Finland (Petrova et al. 2002). Tarsonemid mites are not controlled by species such as *P. persimilis* that are commonly released for control of spider mites (Fitzgerald & Easterbrook 2003).

Various pesticide-resistant predatory mites

In North America, researchers have attempted to increase populations of *N. fallacis* and *Typhlodromus pyri* Scheuten in apple or peach orchards through inoculative releases of pesticide-resistant strains. Releases of 500–2000 *N. fallacis* per tree (Prokopy & Christie 1992, Lester et al. 1999) failed to lower spider mite densities, but studies in Canadian apple orchards (Hardman et al. 2000) suggest that releases of *T. pyri* are more effective. This may be due to the overwintering habits of these mites; *T. pyri* overwinters on orchard trees, but *N. fallacis* overwinters in other habitats, making its winter survival needs more complex. In Japan, a pyrethroid-resistant strain of *Neoseiulus womersleyi* (Schicha) has been tested in tea plantations for control of *Tetranychus kanzawai* Kishida when concurrent applications of pyrethroids are required for leafhoppers and thrips (Mochizuki 2002).

Lessons from use of predatory mites

The use of phytoseiids has been characterized by several features. First, there are many phytoseiids and each country is likely to have locally available species that might be commercialized. Indeed, there is political pressure to do so to prevent field establishment of exotic phytoseiids, although no harm has been demonstrated from instances in which such establishment has occurred. Different phytoseiids are likely to emerge as the most promising species in different locations, given variations in local climate and other factors. Local research is often required to sort out the existing options.

Phytoseiids do vary in their degree of diet specialization. Some species are strict spider mite feeders, while others feed on several types of prey and may eat large amounts of pollen. A scheme classifying phytoseiids into groups based on diet has been developed (McMurtry & Croft 1997). Extreme spider mite specialists like *P. persimilis* are best if curative control is required, but may not persist if prey densities are low. In contrast, species able to feed on a broader range of foods may be better adapted to develop and persist early in the crop season when prey are scarce, providing better long-term control.

Second, details of phytoseiid biology have been shown to be crucially important. The superior control provided by *T. pyri* compared with *N. fallacis* in apple in the northeastern USA, for example, is attributable to the overwintering biology of these two species. Third, resistance to acaricides has been a key force in driving interest in outdoor use of predatory mites. Growers, however, readily abandon biological control when new acaricides are developed to which mites are not resistant. This introduces instability into the market for phytoseiids, making their production more costly and reducing availability. Fourth, many crops have pests other than just spider mites. These may include mites (in strawberry, groups such as tarsonematids) that are not controlled by commercially produced phytoseiids, or species from other pest groups (in tea, leafhoppers and thrips). In such cases, a more complex IPM program may be required. Pesticide-resistant phytoseiids provide a potential means to solve this problem.

CONTROL OF FILTH FLIES

Flies that breed in manure in or around animal production facilities have been targeted for control through the conservation or augmentation of larval or pupal parasitoids or predators (Rutz 1986, Petersen & Greene 1989, Rutz & Patterson 1990). The main pest flies have been the house fly (*Musca domestica* L.) in various situations (poultry, dairy, and cattle feedlot) and the stable fly [*Stomoxys calcitrans* (L.)] in cattle feedlots.

Various pteromalid wasps have been found associated with these flies in animal production areas (e.g. Rutz & Axtell 1980, 1981). Potential biological control agents have included species mainly in the genera *Muscidifurax*, *Spalangia*, *Pachycrepoideus*, and *Nasonia*, such as *Muscidifurax raptor* Girault and Sanders, *Muscidifurax zaraptor* Kogan, *Muscidifurax raptorellus* Kogan, *Spalangia cameroni* Perkins, *Spalangia endius* Walker, and *Nasonia vitripennis* (Walker). These parasitoids generally are native where studied, but one, *M. raptorellus*, is a gregarious, highly promising species that appears to have invaded the USA (where it was assessed as a biological control agent) from South America (Antolin et al. 1996).

Field trials for control of manure-breeding flies have been run in poultry houses, cattle feedlots, and dairies. Rutz and Axtell (1979) reported that release of 40,000 *M. raptor* per week in poultry houses was partially effective, lowering fly numbers in narrow-caged layer houses, but not in high-rise caged-layer houses. In poultry facilities, manure varies in wetness and depth of deposition, and both of these factors affect levels of parasitism (Geden 1999, 2002). Because of this important variability, releases of combinations of parasitoids with complementary niche characteristics appear to improve efficacy (Geden & Hogsette 2006).

Early trials in cattle feedlots tested various species of parasitoids (Stage & Petersen 1981), including *S. endius* (Petersen et al. 1983), which failed to increase parasitism rates. This parasitoid appears to perform well only in warmer climates. In contrast, releases of *M. zaraptor* elevated parasitism from 2 to 38% at the highest release rate tested (37,000 per week for 15 weeks) (Petersen et al. 1995). The greatest impact (96% parasitism) was observed with a single release of 200,000 *M. raptorellus* (Petersen & Currey 1996). Other release rates and intervals between releases varied in their impact, but all showed that parasitism could be elevated to the 40–80% range and held there for several weeks. In dairy facilities in New York, releases of 10,000–12,000 *M. raptor*, released as parasitized host pupae, reduced fly levels by 50% (Geden et al. 1992).

Efforts have been made to identify pesticides potentially compatible with parasitoid releases (Scott et al. 1988, 1991). Parasitoids of manure-breeding flies continue to be sold by commercial insectaries. There are about a dozen common species of fly parasitoids in natural systems, but the most effective are *S. cameroni* and any of the three common species of *Muscidifurax*. *Nasonia vitripennis* is sometimes included in commercial shipments because it is a common contaminant in mass-rearing colonies, but it is generally regarded as ineffective (Patterson et al. 1981, Rutz & Patterson 1990). Although no data are available on the percentage of livestock and poultry producers using parasitoids, the market for commercially produced parasitoids has remained steady during the last 15 years in the USA and Europe, perhaps because the most common target pest (*M. domestica*) readily develops resistance to pesticides.

OTHER EXAMPLES OF SPECIALIZED AGENTS

In addition to the groups discussed above, various other specialized parasitoids and predators have been considered for use in outdoor augmentative biological control. Most of these species have been developed by university or government researchers for use against specific pests.

Scales and mealybugs in citrus

Scales

Citrus-producing areas around the world have been invaded repeatedly by exotic pests. Most of these – especially scales, whiteflies, and leafminers – have been successfully suppressed with classical biological control (Bennett et al. 1976). This has been the dominant method of biological control applied in the crop in southern California. However, several key natural enemies that provide permanent biological control in southern California do not persist in the San Joaquin Valley of California due to climate. In response, growers in this region have come to rely on pesticides. Augmentative biological control has therefore been suggested as an alternative approach in San Joaquin Valley citrus (Luck et al. 1996).

In lemons, release of 50,000–200,000 adults of the aphelinid *Aphytis melinus* DeBach successfully suppressed California red scale, *Aonidiella aurantii* (Maskell)

(Diaspididae). Releases were economically competitive with pesticides and, unlike pesticides, did not harm natural enemies of other citrus pests, which avoided secondary pest outbreaks (Moreno & Luck 1992, Luck et al. 1996). Parasitoid releases resulted in fruit of equal or better quality than the traditional broad-spectrum chemical program with a 40% reduction in pest control costs (Luck et al. 1996). However, since the invasion of California by the glassy-winged sharpshooter [*Homalodisca coagulata* (Say)], pesticides have been applied to suppress this pest in citrus (its major breeding area, even though it is not damaging to citrus), to protect grape production from Pierce's disease, which is vectored by this leafhopper. These pesticide applications make use of augmentative biological control in citrus with sharpshooter populations difficult, even though biological control of this pest is improving.

Other pests in California citrus for which augmentative biological control has been pursued include two soft scales (Coccidae): the citricola scale, *Coccus pseudomagnoliarum* (Kuwana), and the black scale, *Saissetia oleae* (Olivier). In each case several species of *Metaphycus* parasitoids (Encyrtidae) [*Metaphycus helvolus* (Compere) and *Metaphycus* nr. *favus* (Howard)] were assessed (Bernal et al. 1999, Schweizer et al. 2002, 2003a, 2003b), comparing early-, mid- and late-season releases. No release system, however, provided highly effective control of either target pest. In part, the level of control was mediated by the effect release timing had on the size of the scale available for oviposition, because smaller hosts tended to yield a disproportionate percentage of male parasitoids in the next generation, reducing efficacy. The same phenomenon affects mass-rearing colonies (Weppler et al. 2003). These studies highlight the relatively small role augmentative biological control has in citrus as compared to classical biological control, but success against the red scale illustrates that in some cases, pesticide-disruption problems can be eliminated by switching to augmentative parasitoid releases, where they are economical and biologically effective.

Mealybugs

In general, mealybugs are amenable to permanent suppression by classical biological control and many successful cases exist (e.g. Clausen 1978). However, mealybugs may fail to come under classical biological control in some areas, particularly the less tropical parts of their ranges, because of high winter mortality of key natural enemies. In California, the citrus mealybug [*Planococcus citri* (Risso)] and the citrophilus mealybug [*Pseudococcus calceolariae* (Maskell)] were both uncontrolled invaders in coastal citrus districts during the early decades of the twentieth century. Although parasitoids existed that attacked these pests, they were insufficient. Similarly, the effectiveness of the coccinellid *Cryptolaemus montrouzeri* Mulsant, a significant predator of these mealybugs, was limited by winter. Smith and Armitage (1926) developed a method for mass rearing this predator by rearing mealybugs on sprouted potatoes. Inoculation in spring with 10 beetles per tree provided effective control in areas where control was inadequate. This led a large number of insectaries to produce this species for use in California citrus (Bennett et al. 1976). This practice continued from the 1930s through the 1960s at a large scale (with releases of up 42 million beetles annually), but the volume of use declined greatly when one of the mealybugs, *P. calceolariae*, came under effective classical biological control through additional parasitoid introductions (Kennett et al. 1999). Currently, just a small number of insectaries continue to produce *C. montrouzeri*, which is released to control localized outbreaks of *P. citri* as they occur.

Egg parasitoids of plant bugs in strawberry

Damage in California to strawberries by a western mirid, *Lygus hesperus* Knight, has some features that suggested the bug might potentially be a viable target for augmentative biological control. The crop has a very high value per hectare, with significant *Lygus* damage (fruits with deformities caused by plant bug feeding are not marketable). As such, even relatively expensive pest controls could be economically feasible if more effective than conventional pesticide applications. Both pesticide applications and parasitoid releases are made less effective by significant immigration into the crop of *Lygus* bugs from other crops or natural vegetation. This increases pest pressure and requires a nearly continuous-acting form of pest protection. Even with multiple pesticide applications, *Lygus* density in trials was reduced only by 45% compared to controls (Udayagiri et al. 2000a).

The natural enemy seen as potentially useful was the mymarid egg parasitoid, *Anaphes inole* (Girault). On some plants this wasps parasitizes up to 100% of plant bug eggs. In strawberry, weekly releases of 14,800 parasitoids per hectare produced 50% parasitism of sentinel eggs, but 4920 per hectare caused very little parasitism (6–7%). Parasitism was high only for a few days even in the plots with higher release rates and then dropped to background levels. Plant bug numbers were reduced by 43% and fruit injury by 22% (Norton & Welter 1996). Some improvements later allowed weekly releases of 6000 wasps per hectare to perform better, causing 65% parasitism (Udayagiri et al. 2000a). Increasing release frequency to twice per week was more effective, but the increase was marginal and not proportionate to the release rate. In part, the lowered efficacy of this parasitoid on strawberry occurred because *Lygus* eggs laid in the fruit (especially the receptacle) were partially protected from parasitoid attack (Udayagiri et al. 2000a). Based on these findings, two further efforts were made. One focused on improving rearing methods to reduce cost (Smith & Nordlund 2000) and the other on finding pesticides that might be compatible with parasitoid releases (Udayagiri et al. 2000b). Neither of these efforts produced any significant increase in the feasibility of this system. None of the currently available pesticides for *Lygus* suppression is compatible with *A. inole*.

This program illustrates how crop features can alter achievable parasitism rates (here due to protection of eggs laid in fruits), how the between-crop movement of pests can structure the nature of the pest challenge, and how lack of compatible pesticides for use on the crop may limit integrated control.

Stink bugs in Brazilian soybeans

In Brazil, a group of stinkbugs [mainly *Nezara viridula* (L.), *Piezodorus guildinii* (Westwood), and *Euschistus heros* (Fabricius)] attack soybeans and reduce seed production. The scelionid egg parasitoid *Trissolcus basalis* (Wollaston) has been reared in the laboratory and used to test the level of control achievable from its release. Release of 15,000 wasps per hectare into a young soybean crop (used as a trap crop to the main field) resulted in a 54% reduction in bug density in the trap crop and a 58% reduction in the main crop (Corrêa-Ferreira & Moscardi 1996). Releases in trap crop nurseries delayed invasion of the main field and lowered resulting bug populations and led to higher seed quality. Mass rearing of the parasitoid is done on *N. viridula* eggs and further studies have suggested that this parasitoid may be useful in soybean IPM in some parts of Brazil (Corrêa-Ferreira et al. 2000). The economics of mass rearing compared with the value of the pest control achieved have not been reported.

Boll weevils in Texas, Mexican, and Brazilian cotton

The boll weevil (*Anthonomus grandis* Boheman) is a key pest of cotton in the USA and Mexico, and – since its invasion in 1983 – in Brazil. It is attacked in its native range (southern Mexico and northern Central America) by the pteromalid wasp *Catolaccus grandis* (Burks) (Figure 26.5), which is a parasitoid of older larvae in cotton squares and bolls. Trials in south Texas, USA (Summy et al. 1995, 1997), Mexico (Vargas-Camplis et al. 2000), and Brazil (Ramalho et al. 2000) have clearly demonstrated that releases of 700–2000 females per hectare per week (for ≈8 weeks) can cause high levels of mortality (70–90%). Parasitoids have higher population growth rates in the field than the pest and have good searching ability. These features resulted in suppression of boll weevil

Figure 26.5 The pteromalid parasitoid *Catolaccus grandis* (Burks) can be reared in its natural host the boll weevil (*Anthonomus grandis* Boheman), itself reared on artificial diets in artificial cells as shown here. While effective, the economics of this system have not been competitive with continuing eradication of the boll weevil in the USA, although investigations continue in Brazil. Photograph courtesy of Randy Coleman.

infestations in bolls to levels below the economic injury level (Summy et al. 1995).

A limiting factor in the ability to employ this parasitoid commercially has been the relatively high cost of rearing. An *in vitro* rearing method for producing host larvae in diet-filled cells has been developed, but parasitoids reared on such hosts have lower quality than those reared on normal boll weevil larvae (Morales-Ramos et al. 1998). Rearing *in vivo* in the bean bruchid *Callosobruchus maculates* F. as an alternative host is feasible, but has been unsatisfactory because, after several generations, parasitoids lose their preference for boll weevil larvae in favor of the bean weevil (Rojas et al. 1999). In Brazil, another weevil [*Euscepes postfaciatus* (Fairmaire)] has been employed successfully as an alternative rearing host (Ramalho et al. 2000).

This parasitoid was not adopted for use in Texas because of an eradication program against the weevil. Use of the parasitoid may, however, prove economically and socially feasible in Brazil.

GENERALIST PREDATORS SOLD FOR NON-SPECIFIC PROBLEMS

A variety of predators and a few parasitoids are sold not as solutions for specific problems but rather as pest control products for the general public or as potential solutions for classes of pests, such as "aphids," in a range of crops. The use of several species of green lacewings (*Chrysopa*) and ladybird beetles illustrates this approach, which has little value and limited scientific justification.

Ladybird beetles

Several species of coccinellids (ladybird beetles) are widely sold as general-purpose predators of aphids. Two of the more widely marketed species are *Hippodamia convergens* Guerin and *Harmonia axyridis* (Pallas). Homeowners purchase *H. convergens* for control aphids in gardens (Lind 1998). Some experimental assessments exist of this species. It controlled *Aphis spiraecola* Patch on ornamental firethorn (*Pyracantha coccinea* (L.) Roem var. *lalandei*) in Maryland, USA, but had no effect on woolly aphids in the genus *Eriosoma* (Raupp et al. 1994). *Hippodamia convergens* is part of a pecan IPM program in New Mexico, USA, to suppress pecan aphids (LaRock & Ellington 1996). However, rapid dispersal away from release sites can make this species ineffective (e.g. in California outdoor chrysanthemums; Dreistadt & Flint 1996). Releases on roses of hundreds to thousands of beetles per plant were necessary to control the rose aphid in a trial in California (Flint & Dreistadt 2005). This trial showed that to obtain effective control with this pest, releases on the order of 2300 beetles/m^2 were required, in dramatic contrast to the rate recommended by insectaries of 11–22 beetles/m^2. This illustrates that many minor uses of insectary-reared insects, as recommended by the producing companies, are not supported by adequate research and probably do not work.

The coccinellid *H. axyridis* has been used in both greenhouses and outdoors for general control of aphids, but its use is now discouraged because it establishes outdoor populations that enter homes and may displace native ladybirds. Nevertheless, it is still sold and has been studied for release in several crops, including melons in Italy (Orlandini & Martellucci 1997), faba beans in Egypt (El-Arnaouty et al. 2000), and red currants in The Netherlands (Balkhoven & van Zuidam 2002). In addition to the environmental problems caused by this species, the pest control it provides is very expensive; for example €569 per km^2 of currants (Balkhoven & van Zuidam 2002).

Green lacewings

Green lacewing larvae readily feed on aphids and other soft-bodied pests in many crops (McEwen et al. 2001). Numerous efforts have been made to assess the potential of various *Chrysoperla* and *Chrysopa* species, including use of *Chrysoperla carnea* (Stephens) for azalea lace bugs [*Stephanitis pyrioides* (Scott)] in nurseries (Shrewsbury & Smith-Fiola 2000) and against *Scirtothrips perseae* Tuttle, Baker and Abbatiello in avocado (Hoddle & Robinson 2004); use of *Chrysoperla rufilabris* (Bermeister) for control of longtailed mealybug [*Pseudococcus longispinus* (Targioni Tozzetti)] in interior landscape plantings (Goolsby et al. 2000b) and in cotton against cotton aphid (*Aphis gossypii* Glover) (Knutson & Tedders 2002); and use of *Chrysoperla plorabunda* (Fitch) for control of brown citrus aphid [*Toxoptera citricida* (Kirkaldy)] (Michaud 2001).

Of these above examples, use of *C. carnea* in avocado in California against thrips failed, in part because

Figure 26.6 Mechanical application of green lacewing eggs in Californian avocado orchards. Photograph courtesy of Mark Hoddle.

mechanically applied eggs (Figure 26.6) or larvae fell to the ground and failed to find hosts before larvae starved, which occurred in 1–2 days (Hoddle & Robinson 2004).

Release of *C. rufilabris* against aphids on cotton in Texas failed; even at 160,000 eggs per hectare there was no discernable effect on aphid density (Knutson & Tedders 2002). Release of 116–275 larvae of *C. plorabunda* against brown citrus aphid in Florida failed to produce differences in the rate of aphid colony maturation between release and control trees (Michaud 2001). Release of *C. rufilabris* eggs (when supplied with grain moth eggs as initial food) in interior landscapes in pothos ivy plantings in Texas buildings, however, did suppress longtailed mealybug populations for 4 weeks (Goolsby et al. 2000b). Also, release of 10 *C. carnea* larvae per plant in nurseries in Maryland provided acceptable control of azalea lacebugs (Shrewsbury & Smith-Fiola 2000).

These varied results suggest that in field settings lacewing larvae are relatively ineffective, in part due to the physical and biotic complexity of the environment into which they are placed. Obstacles include high food needs, poor contact with the target host, cannibalism, and naturally occurring lacewings or other predators. In contrast, in simpler settings such as interiorscapes and outdoor plant nurseries greater impacts have been demonstrated following lacewing releases.

Part 11

OTHER TARGETS AND NEW DIRECTIONS

Chapter 27

VERTEBRATE PESTS

Vertebrates have been extremely successful colonizers in many areas, through accidental or deliberate introduction. Vertebrates have been relocated for food, to be hunted, to assist with hunting, or to help control pests (Long 2003), and rodents have been moved extensively on ships. Many vertebrates have become major pests (Vitousek et al. 1996). While the number of vertebrate pests is small, their impact on agriculture and conservation is high. Poisoning, shooting, or trapping is possible but expensive and temporary (Hone 1994, Williams & Moore 1995). A few species (rabbits, cats) have been subject to classical biological control programs, which has been controversial because of concerns that the released pathogens may put humans or wildlife at risk and because of an aversion to causing suffering to warm-blooded animals. In general, there are four potential ways to achieve biological control of vertebrates: (1) use of predacious vertebrates, (2) release of parasites into vertebrate populations lacking them, (3) introduction of novel pathogens, and (4) immunocontraception mediated by an infectious host-specific vector (Hoddle 1999).

PREDATORS AS VERTEBRATE CONTROL AGENTS

Generalist vertebrate predators were introduced by private individuals as biological control agents many times in the nineteenth and early twentieth centuries. These predators usually failed to control their targets and frequently had disastrous impacts on non-target wildlife, especially on islands (Case 1996). For example, the small Indian mongoose, released in Hawaii, USA, to suppress rats in sugarcane, had little effect on rats (Cagne 1988) but must now be poisoned to prevent its attack on native birds (Loope et al. 1988).

Under some circumstances, introduced predators can regulate target vertebrates if prey densities are first suppressed by other factors. After poisoning programs in New Zealand in the 1950s and 1960s substantially reduced rabbit densities, ferrets and cats maintained rabbits at low levels (Newsome 1990). Similarly in Australia, red foxes and cats can maintain rabbit populations at low densities once prolonged hot summers have caused rabbit populations to crash because of food shortage (Newsome et al. 1989, Newsome 1990). The suppressive action of predators on rabbits in Australia has been demonstrated through experiments in which foxes and cats were removed by night shooting. Predator removal resulted in a rapid increase in rabbit population growth (Newsome et al. 1989, Sinclair 1996). Regulation of prey by predators if prey densities fall within specific low-density bounds has been termed the predator pit (May 1977). For the rabbit/fox system in Australia, a predator pit operates at densities of 8–15 rabbits per km of linear transect. Below these densities foxes utilize alternate food sources (e.g. native animals) and above this critical density rabbit populations escape regulation by predators (Newsome 1990).

Efficacy of native or previously introduced predators can be enhanced through habitat modification. Adding nesting boxes for barn owls (*Tyto alba* L. var. *javanica*) reduced crop damage from rats in Malaysian oil palm plantations (Wahid et al. 1996), in combination with rodenticide campaigns. In *Pinus radiata* Don plantations in Chile, barn owl efficiency was enhanced by clearing strips 4 m wide between trees to favor owl flight and constructing resting perches for surveillance (Muñoz & Murúa 1990).

PARASITES AS VERTEBRATE CONTROL AGENTS

The potential of parasites, such as helminths, lice, ticks, and fleas, to regulate vertebrate host populations was proposed in 1911 (Lack 1954) and demonstrated theoretically with Lotka–Voltera models (Anderson & May 1978, May & Anderson 1978, May 1980). In the laboratory, introduction of the nematode *Heligmosomoides polygyrus* Dujardin, under ideal conditions for transmission, reduced mouse densities by 94% in comparison to controls. Reduction of nematode transmission rates and elimination of parasites with helminthicides allowed infected mouse populations to increase (Scott 1987). However, the population densities and intensities of infection in this study were higher than for wild mice. In Australia, epidemiology studies of parasites in wild mice found that parasites did not regulate mice populations (Singleton et al. 2005).

Parasite regulation of other vertebrates has been observed under field conditions (Scott & Dobson 1989). Population cycles in red grouse [*Lagopus lagopus scoticus* (Latham)] in Scottish heathlands are controlled by the parasitic helminth *Trichostrongylus tenuis* (Cobbold) (Dobson & Hudson 1994). The regulatory effect of *T. tenuis* has been demonstrated by reducing parasite infestations with helminthicides in experimental birds. Treated grouse showed increased overwintering survival, clutch sizes, and hatching rates when compared with untreated birds (Dobson & Hudson 1994).

In cereal growing regions of southeastern Australia, the house mouse is an introduced pest that erupts (Figure 27.1) every 7–9 years (Singleton & McCallum 1990, McCallum 1993) causing more than AUS$50 million in losses (Beckman 1988, Singleton 1989). Outbreaks are driven by seed availability, which is affected by rainfall. Mouse populations crash when food is exhausted (Singleton 1989). Saunders & Giles (1977) suggested that drought removes the regulating effect of natural enemies (when mice are too rare to attract predation) and disease, and later when rains increase, low predation and increased natality permit mouse numbers to rebound.

Figure 27.1 Plagues of mice erupt periodically in Australia. Photograph courtesy of Grant Singleton.

The potential for biological control of house mice in cereal-producing areas with the nematode *Capillaria hepatica* (Bancroft) has been investigated (Singleton et al. 1995). The nematode has a direct life cycle that requires host death for transmission. Female nematodes deposit eggs in the host's liver but they do not embryonate. Eggs are liberated from the liver when mice die or via necrophagy by mice or arthropods. Nematode eggs then embryonate in habitats such as mouse burrows. Infective embryonated eggs are consumed when mice preen contaminated body areas (Singleton et al. 1991, 1995). Nematode infection lowers mouse natality and weening success (Singleton & Spratt 1986, Spratt & Singleton 1986, McCallum & Singleton 1989, Singleton & McCallum 1990). However, experiments in enclosures and on a large scale with increasing populations of free-ranging mice have failed to demonstrate long-term regulation of mouse population growth from releases of *C. hepatica* eggs. Transmission of *C. hepatica* in treated populations is not density-dependent and is influenced by soil temperatures, aridity, and the requirement of host death for egg release which reduces the efficacy of this agent (Barker et al. 1991, Singleton & Chambers 1996). Furthermore, low rat (*Rattus norvegicus* and *Rattus rattus*) numbers in cereal-growing regions of Australia may contribute to the non-persistence of *C. hepatica* because rats are a major reservoir of the nematode (Singleton et al. 1991). Infection rates in rats in urban areas range from 40 to 80% (Childs et al. 1988, Singleton et al. 1991).

Island populations of introduced vertebrates often have low parasite loads compared to their source populations (Dobson & May 1986), because either island populations were started with uninfected animals by chance, or islands lacked necessary intermediate hosts. Invasive sparrows and starlings in North America have fewer than half as many parasites as in Europe. Introduced rats, goats, and cats on oceanic islands also exhibit simplified parasite faunas (Dobson 1988). Fewer parasites, coupled with presumed low genetic diversity and reduced selection pressures for resistance to parasites, may make these island vertebrate populations vulnerable to introduced host-specific parasites. Host-specific parasites also may have the potential to reduce reproduction and longevity of pest reptiles

(Dobson 1988) and amphibian species (Freeland 1985). The potential of hemogregarine parasites (vector-transmitted blood protozoa) has been investigated, for example, for control of the brown tree snake, a pest on Guam (see Chapter 7) (Telford 1999).

PATHOGENS AS VERTEBRATE CONTROL AGENTS

Vertebrate pathogens – viruses, bacteria, and protozoans – often exhibit epizootic (i.e. boom or bust) population cycles (Anderson 1979, McCallum 1994). Their potential to regulate vertebrate densities by reducing the longevity and fecundity has been demonstrated with models and perturbation experiments (Smith 1994). Models suggest that pathogens of intermediate virulence would be the most effective biological control agents (Anderson 1982) because of persistent transmission. The most contagious pathogens are those spread by water, air, or vectors, or are associated with high-density host populations. Pathogens with lower transmission rates usually are spread by host-to-host contact or are associated with low-density host populations (Ebert & Herre 1996).

Two pathogens of rabbits (myxoma virus and rabbit hemorrhagic virus) and one of cats (feline parvo virus) are the only agents used in successful biological control programs against vertebrate pests. Other viruses, particularly sexually transmitted pathogens, may have potential for effective use.

Figure 27.2 A rabbit infected with myxoma virus. Photograph courtesy of Invasive Animals Cooperative Research Centre, Landcare, New Zealand.

Figure 27.3 Plagues of rabbits were common in Australia before the introduction of two viral pathogens. Photograph courtesy of CSIRO.

Myxomatosis and biological control of rabbits

The myxoma virus (*Leporipoxvirus*, Poxviridae) was first recognized in 1896 when European rabbits in Uruguay died of a disease that caused myxoma-like tumors on their head and ears (Figure 27.2) (Fenner & Marshall 1957, Fenner & Ratcliffe 1965, Fenner 1994). The virus's native host in South America is the forest rabbit [*Sylvilagus brasiliensis* (L.)], but in this host, the virus causes only benign fibromas. Mosquitoes vector the disease among forest rabbits in South America.

Myxoma virus has been released in Australia, Europe, Chile, and Argentina to kill European rabbits, a noxious pest in these regions (Figure 27.3). In Australia, before the establishment of myxoma virus, rabbits caused annual losses of AUS$600 million (Robinson et al. 1997, Bomford & Hart 2004). Losses included damage to crops, reduction of sheep forage (Vere et al. 2004), and destruction of native plants, including the endangerment of at least 17 plant species (Bomford & Hart 2004). Native animals are also affected when rabbits compete with indigenous herbivores for food and by sustaining exotic predator populations that feed on native animals (Gibb & Williams 1994, Myers et al. 1994, Robinson et al. 1997).

Following preliminary investigation in the UK and on an island quarantine station in Australia, the myxoma virus was established on mainland Australia in 1950 (Fenner 1994) and within 2 years was present over most of the rabbit's range (Fenner & Ratcliffe 1965). The virus initially reduced the estimated 600 million rabbits in Australia by 75–95%. Locally, efficacy was dependent on climate, rabbit population susceptibility, and presence of vectors. The myxoma virus–rabbit system in Australia proved to be very dynamic and within a few years of the initial panzootic, the myxoma virus declined in virulence compared to the original strain, which had killed >99% of laboratory rabbits in about 11 days. Concurrently, genetic resistance by rabbits also increased (Fenner & Marshall 1957, Fenner & Ratcliffe 1965). Rabbit populations eventually stabilized at around 300 million (50% control).

In Australia, mosquitoes were the dominant vector of the myxoma virus, but in Europe, the rabbit flea, *Spilopsylus cuniculi* (Dale), proved to be an important vector. This flea was introduced into Australia in 1968 and it increased the geographic distribution of the disease. However, this flea could not persist in dry areas (<200 mm rainfall), and the xeric-adapted Spanish rabbit flea, *Xenopsylla cunicularis* Smit, was introduced in 1993 (Fenner & Ross 1994). The European rabbit flea was also introduced into the subantarctic Kerguelen Islands in 1987. Island rabbits with antibodies to myxoma virus increased from 34% (pre-1987) to 85% in 1998, suggesting that *S. cuniculi* increased exposure to the virus (Chekchak et al. 2000).

In New Zealand, attempts to establish the myxoma virus (1951–3) failed because of inclement weather and lack of arthropod vectors. Further attempts at establishment were not undertaken because poisoning programs reduced rabbits adequately, and the New Zealand public was not in favor of using myxoma virus on humanitarian grounds (Gibb & Williams 1994).

Rabbit hemorrhagic disease and biological control of rabbits

Emergence of a new virus

In 1984, a second highly contagious viral disease, rabbit hemorrhagic disease (RHD; also known as rabbit calicivirus disease), was observed in Angora rabbits shipped from Germany to China (Liu et al. 1984). The RHD virus belongs to the Caliciviridae (Ohlinger et al. 1990, Parra & Prieto 1990). Mortality rates are higher in rabbits over 8 weeks of age; younger rabbits often survive and may develop antibodies to RHD virus (Nagesha et al. 1995). Studies on rabbit sera collected in 1961 from Czechoslovakia and Austria suggest that the RHD virus probably evolved from a non-pathogenic European strain (Nowotny et al. 1997). RNA sequencing suggests that avirulent RHD strains may have been present for centuries before becoming virulent (Moss et al. 2002). In 1986, RHD appeared in Italy and killed 38 million rabbits. It spread rapidly throughout Europe (Chasey 1994), most likely from movement of live rabbits and rabbit products. Outbreaks of RHD followed in Mexico (Gregg et al. 1991) and Réunion Island, most likely spread by shipments of frozen rabbits from China (Chasey 1994).

Biological control programs with RHD

European rabbits appear to be the only animals susceptible to infection by RHD virus, and vaccines have been developed to protect domestic rabbits (Boga et al. 1997). Cottontail rabbits (*Sylvilagus* spp.), black-tailed jack rabbits (*Lepus californicus* Gray), volcano rabbits (*Romerolagus diazi* Ferrari-Pérez) (Gregg et al. 1991), and hares (Gould et al. 1997) are not affected by RHD virus. The limited host range of RHD virus makes it an obvious candidate to kill European rabbits in New Zealand and Australia. A joint biological control program between these two countries using RHD virus was initiated in 1989 and a strain of virus from the Czech Republic was imported into Australian quarantine facilities in 1991 (Robinson & Westbury 1996) and tested on domestic livestock (horses, cattle, sheep, deer, goats, pigs, cats, dogs, and fowl), noxious exotic vertebrates (foxes, hares, ferrets, rats, and mice), native mammals (eight species), birds (five species), and reptiles (one species). There was no evidence for viral replication, clinical symptoms, or lesions in any species tested (Gould et al. 1997). Artificial inoculation of RHD virus into North Island brown kiwis (*Apteryx australis mantelli* Bartlett) and lesser short-tailed bats (*Mystacina tuberculata* Gray), species of concern in New Zealand, also failed to produce disease (Buddle et al. 1997).

The host-specificity of RHD virus to the European rabbit, rapidity of action, and the capacity for infection from contact with infected rabbits, feed, feces, or from a contaminated environment (O'Brien 1991) prompted further evaluation of this biological control agent. Field studies were initiated at a quarantine station on Wardang Island, off the south coast of Australia in 1995 (Rudzki 1995, Robinson & Westbury 1996). RHD breached the island's quarantine and appeared on mainland Australia within a year, probably spread by bush flies carried by on onshore winds (Lawson 1995, McColl et al. 2002). Attempts at containment failed (Seife 1996) and within 2 months 5 million rabbits died in South Australia. Mortality was 80–95% in dry areas (Anderson 1995), compared to 65% elsewhere (Anon 1997b). Mortality varied by region from 50 to 90%.

Vectors included flies, mosquitoes, and rabbit fleas (McColl et al. 2002). Approximately 70% of rabbits that survive RHD outbreaks in higher-rainfall areas developed antibodies and a demographic shift towards younger rabbits was observed. In temperate habitats, rabbit numbers rebounded to pre-RHD levels within two breeding seasons (Bruce et al. 2004). Attacks of generalist predators such as foxes on native wildlife did not increase when rabbit populations declined. However, effects of increased attacks could have been masked due to concurrent drought in the study area (Saunders et al. 2004).

Economic benefits of RHD in Australia

The cost/benefit ratios for RHD to Australian agriculture were 2.9:1 and 32:1 for 25 and 50% reductions in rabbit numbers, respectively (Vere et al. 2004). Use of the rabbit poison 1080 (sodium fluoroacetate) declined by 83% in New South Wales (saving $1.2 million per year) and by 24–73% in South Australia (saving $0.56 million per year) (Saunders et al. 2002). Conversely, rabbit farmers (an industry worth $1.66 million in Australia) have been burdened with vaccine costs of $3–15 per rabbit (Saunders et al. 2002). However, the general opinion is that RHD has greatly benefited Australian agriculture (Saunders et al. 2002) and has had major conservation benefits in Australia's arid zones.

RHD in New Zealand

RHD virus was smuggled illegally into New Zealand by farmers in 1997 and disseminated with contaminated baits (Parkes et al. 2002, Forrester et al. 2003). The virus was quickly spread over large areas, making containment impossible. Accepting the situation, the New Zealand government sanctioned releases of the Czech V351 strain of RHD virus into new areas (Forrester et al. 2003). In New Zealand, RHD has reduced rabbit densities in some areas by 50–90%, while having no impact in other locations (Parkes et al. 2002). Rabbit populations that suffered substantial RHD mortality were further lowered by predation (Reddiex et al. 2002). Consequently, rabbit grazing subsequently declined by 77% in parts of South Island. Reduced rabbit numbers correlated with declines in ferrets and feral cats and increases in other herbivores such as hares and possums. Predation on eggs of native birds by exotic predators increased in some areas after rabbit populations declined (Norbury et al. 2002). Population models suggest that over the long term RHD will reduce rabbit densities in New Zealand by about 75% (Barlow et al. 2002).

Biological control of feral cats

Cats on oceanic islands are a major threat to ocean birds. On Marion Island in the Indian Ocean six cats abandoned in 1949 (Howell 1984) increased to 3000 by 1977 and were increasing by 23% per year (van Rensburg et al. 1987). These cats killed 450,000 seabirds yearly and were probably responsible for the local extinction of the common diving petrel, *Pelecanoides urinatrix* (Gmelin) (Bloomer & Bester 1992). On the Kerguelen Islands, five cats increased to 20,000 and killed 3 million sea birds per year (Courchamp & Sugihara 1999). Such island cat populations have few pathogens and most individuals are immunologically naïve to cat-specific disease agents (Courchamp & Sugihara 1999). Surveys on Marion Island detected feline herpes and corona viruses, but not the highly contagious feline parvo virus (Howell 1984). In 1977, 93 feral cats collected from the island were inoculated with parvo virus and returned to the population (Howell 1984). Disease lowered cat numbers by 82% in 5 years by reducing fecundity and increasing juvenile mortality (van Rensburg et al. 1987). Hunting and trapping then became feasible (Bloomer & Bester 1992) and were incorporated into an eradication program (Courchamp & Sugihara 1999).

Other cat pathogens also have potential for use as biological control agents, such as feline immunodeficiency virus (FIV) and feline leukemia virus (FeLV). These pathogens may be even more efficient than parvo virus because they persist for longer in the host before causing death, providing more time for transmission. Also, these viruses are transmitted through behaviors that favor continued transmission even at very low population densities. Models even suggest that feline leukemia could eradicate immune-naïve cat populations on islands (Courchamp & Sugihara 1999).

Sexually transmitted diseases

Sexually transmitted diseases are often host-specific, require physical contact for transmission, and can reduce survival, conception rates, and numbers of offspring born or successfully weaned (Smith & Dobson 1992). Host population density does not affect persistence or rate of spread as the requirement for physical contact for transmission enhances the ability of parasites and pathogens to persist in low-density populations or in solitary species such as predators. This property, together with long infectious periods, and vertical transmission (infective propagules are passed from mother to offspring), greatly enhances the ability of sexually transmitted diseases to persist in low-density host populations (Smith & Dobson 1992). These desirable attributes mean that sexually transmitted diseases may have potential for biological control of vertebrate pests.

NEW AVENUES FOR BIOLOGICAL CONTROL OF VERTEBRATES

Concept of immunocontraception

Many vertebrates cannot be suppressed through biological control because they lack effective, host-specific natural enemies, or control agents pose unacceptable risk for non-target impacts, or there is strong societal pressure not to cause suffering to animals, especially with debilitating disease agents. Consequently, new avenues for vertebrate pest management are being pursued and the most intriguing control strategy is immunocontraception based on the use of genetically modified host-specific pathogens to deliver sterilizing antigens to target pests. The aim of immunocontraception research is to develop a sterilizing vaccine and a self-delivery mechanism. Lower fertility in target populations is achieved with a vaccine that expresses the pest species' own egg or sperm proteins, inducing an immune response. The induced antibodies block fertilization by interfering with sperm mobility or binding sites on egg surfaces (Ylönen 2001). Exotic pests being considered as targets for immunocontraception include brushtail possums, cats, grey squirrels, foxes, mice, rabbits, ferrets (Figure 27.4), and stoats (Barlow 2000, Parkes & Murphy 2004, Hardy et al. 2006).

Mode of action

In vertebrates, proteins associated with male and female gametes are potentially foreign antigens if introduced into the body outside of the reproductive tract. Exposure to male reproductive antigens during copulation does not stimulate females to develop antibodies, but subcutaneous or intramuscular inoculation of sperm

Figure 27.4 Ferrets are among the invasive European vertebrates in New Zealand under consideration for biological control with immunocontraception. Photograph courtesy of Invasive Animals Cooperative Research Centre, Landcare, New Zealand.

into females causes high antibody titers, inducing permanent or temporary infertility (Robinson & Holland 1995). Once an immune response occurs, antibodies bind to sperm during mating and cause agglutination or immobilization of sperm. Antibodies may also prevent fertilization of the egg (Shulman 1995).

Antibodies may also be raised in females against proteins from the female zona pellucida, which is a protective layer around the oocyte (Barber & Fayrer-Hosken 2000). Non-reproductive tract inoculation of females with zona preparations leads to infertility (Millar et al. 1989). Whereas zona glycoproteins can be different between taxonomic classes (Kalaydjiev et al. 2000), they tend to be similar among species in the same class. For example, non-specific pig zona preparations cause infertility in humans, primates, dogs, rabbits, horses, and deer (Robinson & Holland 1995). Research on immunocontraception research seeks host-specific zona glycoproteins that would not cause sterility in non-target species, but low variability among zona glycoproteins may make it difficult to find the desired level of specificity (Millar et al. 1989).

Applications of immunocontraception

Immunocontraception, via baits or injections, has been used for wildlife population control, such as free-ranging horses (*Equus caballus* L.) (Kirkpatrick et al. 1992, 1997) and elephants (*Loxodonta africana* and *Elephas maximus*; Fayrer-Hosken et al. 2000). Free-ranging feral mares inoculated by dart gun with porcine zona pellucida showed depressed urinary estrogen concentrations and failure to ovulate. Immunocontraception was reversible after four consecutive years of treatment but prolonged treatment (5–7 years) with zona preparations caused irreversible sterility (Kirkpatrick et al. 1992, 1997). Similar results have been achieved with porcine zona pellucida inoculations in white-tail deer [*Odocoileus virginianus* (Zimmerman)] (Kirkpatrick et al. 1997, Kirkpatrick & Frank 2005).

Delivery of sterilizing antigens

Contraceptive antigens can be delivered to target animals in several ways, including: (1) mechanical delivery with darts or injections, (2) ingested baits, and, potentially, (3) infections by self-spreading, genetically modified pathogens (Tyndale-Biscoe 1994a, 1994b, Polkinghorne et al. 2005, Hardy et al. 2006). Injection provides a strong immunocontraceptive with no risk to non-target organisms, but it is very expensive. To control the estimated 300,000 wild horses in Australia with dart-delivered porcine zona pellucida would cost AUS$20 per horse per year, compared to 50 cents for permanent control with a bullet (Tyndale-Biscoe 1991). Lethal control provides immediate reductions in pest numbers and their damage, and is directly observable. Population control via immunocontraception, in contrast, is delayed, and a high proportion of the population must be sterilized to produce an effect, and environmental or economic damage caused by sterilized animals continues until an appreciable population decline occurs.

Baits have the advantage of not requiring individual contact with each treated animal. Foods favored by the target species are formulated with microencapsulated antigens. Antigens must escape early digestion and reach the lower gastrointestinal tract intact, where they stimulate a response in the mucosal immune system. This in turn induces mucosal immunity in the reproductive tract of females and causes sterilization (Bradley et al. 1997). This approach was pursued for fox control in Australia for more than 10 years but an effective bait specific to foxes that is stable and easy to manufacture has not been developed. However, use of baits to disseminate rabies vaccines to foxes in Europe demonstrates the potential for this approach (Bradley et al. 1997). Non-target impact is a concern because most antigens in current use may not be sufficiently species-specific. For herbivores, a variation on baiting is a proposal to genetically modify plants to express the desired immunocontraceptive antigens and sow plants or distribute them in the range of the target pest. Transgenic plants such as carrots or maize could be harvested and placed in fenced watering points allowing access for pests while excluding people and livestock (Smith et al. 1997). Carrots are currently used to deliver toxins to kill brushtail possums in New Zealand, and poisoned carrots are regularly distributed over 90% of possum range either aerially or in bait stations. Transgenic carrots expressing antigens may be used in a similar manner. It has been estimated that sterilizing antigens in transgenic carrots could provide possum control if 50% of the population were sterilized (Polkinghorne et al. 2005). However, adoption of sterilizing baits seems unlikely as poison delivery is a much more effective and rapid form of control that is already in widespread use and currently is publicly acceptable.

Host-specific pathogens genetically modified to express pest-specific antigens are potentially the solution to the problems discussed above to disseminating immunocontraceptive materials (Tyndale-Biscoe 1994a, 1994b, Barlow 2000). To be effective, the pathogen must carry foreign DNA coding for the target pest's gametic antigens, as well as promoters to express the foreign genes and cytokines to enhance effectiveness (Tyndale-Biscoe 1994a). Engineered pathogens should not interfere with sexual behavior or social organization, as this might lead to increased breeding by non-sterilized individuals of lower social status (Caughley et al. 1992, Tyndale-Bisoce 1994a, Robinson & Holland 1995).

Potential pathogens for antigen delivery

The myxoma virus, murine cytomegalovirus, mousepox virus, ectromelia virus, vaccinia virus, and canine herpes virus have been investigated as gamete antigen-delivery agents for rabbits, mice, and foxes in Australia (McCallum 1996, Shellam 1994, Tyndale-Bisoce 1994a, Jackson et al. 2001, Gu et al. 2004, Hardy et al. 2006). The ability of novel recombinant myxoma virus strains to compete and spread in field situations has been demonstrated by monitoring the spread of a strain containing identifiable gene deletions (Robinson et al. 1997). Recombinant myxoma viruses expressing zona pellucida antigens have been developed and a sterilizing effect has been demonstrated in the laboratory (Gu et al. 2004).

In the absence of arthropod vectors, sexually transmitted diseases are superior to non-sexually transmitted ones for antigen dispersal because multiple matings with sterilized females increases the competitiveness of the engineered agent with non-sterilizing strains. The potential impact of immunocontraception is further enhanced if the sterilizing agent causes limited host mortality and there is little naturally occurring immunity to the sexually transmitted disease (Barlow 1997). Sexually transmitted herpes-type viruses are proposed as vectors to spread sterilizing antigens in brushtail possums in New Zealand (Barlow 1994, 1997). The borna disease virus, which causes wobbly possum disease in New Zealand, may also be suitable to genetic engineering and use against this pest (Atkinson 1997, Bertschinger et al. 2000).

Ethics and risks of using immunocontraception

Vertebrate pathogens engineered to cause immunocontraception offer the possibility of pest control without killing or causing suffering and would reduce the use of vertebrate-killing toxins and their associated non-target impacts. This would be particularly helpful for the control of pest vertebrates in suburbs, parks, or other areas where lethal controls may no longer be legal or safe (Kirkpatrick et al. 1997, Williams 1997).

The method, however, entails several potential risks. First, viruses might mutate after release and infect non-target species (Anderson 1997), particularly if they exchange genetic material with untransformed wild types (Angulo & Cooke 2002). Under such conditions it may be impossible to contain and eradicate a mutant virus from an infected animal population (Tyndale-Biscoe 1995). Second, sterilizing viruses might disperse to areas where the target species is not a pest (Tyndale-Biscoe 1994a, Henzell & Murphy 2002). For example, viruses engineered to sterilize invasive marsupials in New Zealand might reach Australia and infect endangered wildlife (McCallum 1996, Rodger 1997).

Third, resistance to the infectious agent may develop through natural selection, threatening the long-term viability of this technique (Magiafoglou et al. 2003). In theory, use of multiple agents that act in different ways (e.g. using agents that cause sterilization, alter levels of reproductive hormones, or affect lactation) could make the development of resistance less likely (Jolly 1993, Tyndale-Biscoe 1994a, Cowan 1996, Cowan & Tyndale-Biscoe 1997, Magiafoglou et al. 2003).

Fourth, in many countries, the general public is not comfortable with the use of genetic engineering, particularly the manipulation of viruses infective in vertebrates. Such fears could easily delay or prevent field trials and widespread application (Lovett 1997). Regulatory legislation, such as the Gene Technology Act of 2000 in Australia will constrain all field testing of sterilizing micro-organisms until all non-target risks have been assessed (Hardy et al. 2006).

Fifth, objectives of different research programs using recombinant pathogens but targeting the same animals may conflict. Rabbits have become rare in parts of Europe because of the myxoma and RHD viruses. Lower rabbit numbers have adversely affected recreational hunting and endangered predatory species such as imperial eagles (*Aquila adalberti*) and the Iberian lynx (*Lynx pardinus*), which rely primarily on rabbits for food (Angulo & Cooke 2002). European researchers are engineering the myxoma virus to vaccinate wild European rabbits against myxomatosis and RHD to conserve European rabbits and their predators. Concurrently, research in Australia is attempting to engineer the myxoma virus to sterilize European rabbits and control population growth. The research goals of these two myxoma research programs are diametrically opposed. International guidelines on the deployment of genetically engineered pathogens for sterilizing vertebrates may need to be developed to prevent conflicts over release and spread beyond political boundaries (Angulo & Cooke 2002, Parkes & Murphy 2004).

CONCLUSIONS

Biological control of vertebrates is limited by several factors. First, they have few highly specific natural enemies. The most effective are pathogens, which have been used successfully against rabbits and small island populations of cats. Second, strong public concerns exist over: (1) potential non-target impacts on native wildlife, (2) suffering of targeted mammals, and (3) the concept of sterilization with genetically modified pathogens.

Nevertheless, real opportunities do exist to use vertebrate biological control to solve important social, agricultural, and conservation problems. Many vertebrate pests such as feral goats, pigs, horses, rabbits, mice, foxes, dogs, and cats have been well studied by humans and much veterinarian information is available on their diseases and vaccines are available for many of them. On islands, biological control programs could be initiated by simply reassociating disease-causing parasites or pathogens with isolated populations (Dobson & May 1986). Use of genetically engineered natural enemies is a special case of vertebrate biological control, but is a promising additional tool. Research with agents that cause immunocontraception will likely increase with further advances in molecular biology. Application of this approach will ultimately depend on both technical and social factors and the utility of immunoncontraception has yet to be demonstrated conclusively.

Chapter 28

EXPANDING THE BIOLOGICAL CONTROL HORIZON: NEW PURPOSES AND NEW TARGETS

Invasive species pose an increasing threat to a diversity of aquatic and terrestrial environments. Threatened habitats are not only those that support agricultural crops, recreational pursuits, and human habitation (i.e. urban areas) but also areas of critical conservation importance. Species threatening conservation of nature include not only the groups to which biological control has been applied in the protection of agriculture and forestry (insects, mites, and weeds) but also a diverse set of additional groups, including crustaceans, platyhelminths, mollusks, and vertebrates (fish, birds, amphibians, reptiles, and mammals). The threat of invasive land planaria (Figure 28.1), for example, is both novel and serious, especially the New Zealand species that is reducing earthworm populations in the British Isles (Cannon et al. 1999).

Figure 28.1 Invasive land planaria, such as this Floridian invader *Bipalium kewense* Moseley, are a novel group of invaders. Photograph courtesy of P.M. Choate.

Invasion biology is now a mainstream and widely recognized branch of applied ecology. Parties interested in invasive species and their management include ecologists, biological control scientists, conservationists, political officials, agricultural producers, and the public. Invasive species problems and their management are regularly discussed in the media, particularly in Australia, New Zealand, and North America.

The traditional targets of insect classical biological control have been pests of agriculture and forestry. Weed biological control projects, however, have a long history of projects in wild lands and water bodies. Some new arthropod projects are now focused on pests of conservation areas or pests with both economic and conservation significance, such as the imported fire ant [*Solenopsis invicta* (Burden)] in the southern USA. Emerging projects are also assessing the feasibility of controlling invasive crustacea such as the European green crab, *Carcinus maenas* (L.). However, applying classical biological control to marine organisms would be ground breaking in several areas, including assessment of host specificity of entirely new kinds of natural enemies, development of methods to evaluate threats to marine non-target organisms, and solving the complicated issues affecting the measurement of efficacy in the open recruitment systems characteristic of the populations of many marine species.

Other non-traditional targets for classical biological control are potentially very diverse, including snails, slugs, frogs, snakes, planaria, and other groups. Use of biological control for such unfamiliar targets is potentially controversial and some older projects have drawn strong criticism from some prominent ecologists. The potential risks and limitations of using natural enemies for invasive pests in these categories need careful consideration and pertinent controversial issues will be outlined in this chapter.

TARGETING WEEDS AND ARTHROPOD PESTS OF NATURAL AREAS

Biological control is often the best and sometimes the only feasible technology for controlling invasive species in wildlands (Headrick & Goeden 2001). Suppression of weeds in natural areas is currently the dominant application of biological control in support of conservation. Wildland weed biological control grew out of projects directed against weeds of rangelands and agriculture (McFadyen 1998). In the Florida Everglades, USA, biological control is currently being used against melaleuca [*Melaleuca quinquenervia* (Cavanilles)], an invasive tree that alters water-table levels and displaces native plants and wildlife (Center et al. 1997b, Goolsby et al. 2000a). Similar programs are being conducted in New Zealand's Tongariro National Park, a World Heritage Area, where an exotic European heather, *Calluna vulgaris* (L.), is being targeted with heather beetles, *Lochmaea suturalis* (Thomson), which feed exclusively on this weed (Syrett et al. 2000). Other weeds of conservation importance in the USA that are current targets of biological control programs are purple loosestrife, *Lythrum salicaria* L. (Blossey et al. 2001b), Brazilian peppertree, *Schinus terebinthfolius* Raddi (Medal et al. 1999), and salt cedar, *Tamarix* spp. (Milbrath & De-Loach 2006).

Several biological control introductions have also been directed against a variety of invasive arthropods threatening to native plants and animals, as described below.

- Introduced scale insects, *Carulaspis minima* (Targioni-Tozzetti) and *Insulaspis pallida* (Maskell), caused extreme declines of the endemic Bermuda cedar, *Juniperus bermudiana* L., and natural enemies were used in a control program for this pest (Cock 1985).
- On the island of St. Helena, an ensign scale, *Orthezia insignis* Browne, threatened the survival of an endemic gumwood tree, *Commidendrum robustum* (Roxb.) DC, until it was brought under successful biological control by the introduction from Africa of the coccinellid *Hyperaspis pantherina* Fürsch (Fowler 2004).
- In the eastern USA, an Asian hemlock-feeding adelgid, *Adelges tsugae* Annand, is killing large numbers of native eastern hemlock trees, *Tsuga canadensis* (L.) over an extensive area. A biological control program using predatory coccinellids (*Scymnus* spp.) and derodontids (*Laricobius* spp.) is underway (Lu & Montgomery 2001).
- An exotic Mexican weevil, *Metamasius callizona* (Chevrolat), is attacking and killing threatened species of bromeliads in Florida. This pest was introduced through bromeliad importations. Biological control with a newly discovered tachinid fly, *Admontia* sp., may be the only feasible solution in natural areas (Frank & Thomas 1994, Frank 1999, Salas & Frank 2001, Frank & Cave 2005).
- In New Zealand, nectar-feeding birds are being outcompeted for beech scale honeydew in South Island forests by highly aggressive introduced yellow jackets (*Vespula vulgaris* L.). A specialized ichneumonid parasitoid [*Sphecophaga vesparum vesparum* (Curtis)] that attacks yellow jacket brood has been established to reduce wasp densities in forests (Barlow et al. 1996).
- The vedalia beetle, *Rodolia cardinalis* Mulsant, has been released in the Galápagos National Park to protect native plants there that are threatened by cottony cushion scale. Rigorous safety testing (Causton et al. 2004) demonstrated such introduction posed no risk to the native species.
- In the northeastern USA, releases of the eulophid parasitoid *Tetrastichus setifer* Thomson against the lily leafbeetle, *Lilioceris lilii* Scopoli (Coleoptera: Chrysomelidae), will most likely provide significant protection for rare native lilies, which are vulnerable to attack by this exotic European pest (Tewksbury et al. 2005).

TARGETING "NON-TRADITIONAL" INVASIVE PESTS

Natural enemy regulation of animal and plant populations is not unique to terrestrial weeds, insects, and mites. Ecological studies in many systems provide evidence for such regulation. Therefore it is reasonable to consider extending biological control theory and technology to more taxa of organisms, such as marine species and freshwater or terrestrial snails. The following sections discuss emerging biological control programs for several non-traditional pest groups.

Marine pests

Many introduced marine species are ecologically and economically important pests. There are few management options, however, when such species are widely established. Currently the principles of classical biological control as derived from work in terrestrial systems are being considered for application to introduced marine pests (Lafferty & Kuris 1996). These efforts are in various stages of planning or implementation and include: (1) viral or microbial control of harmful algal blooms, (2) predatory control of the ctenophore *Mnemiopsis leidyi* (Agassiz) in the Black Sea, (3) parasitic castration by ciliates of the predatory starfish *Asterias amurensis* Lütken in Australia, (4) the use of sacoglossan sea slugs to control the alga *Caulerpa taxifolia* (Vahl) C. Agardth in the Mediterranean, and (5) parasitic castration of the European green crab, *C. maenas*, in California, USA, and Australia (Secord 2003).

Implementation of these projects is proceeding cautiously because some attributes of marine systems differ importantly from those of the terrestrial and freshwater systems in which classical biological control concepts and models were developed. Some important unique features of marine systems (Secord 2003) are: (1) hyper-dispersive larval and adult stages of some species, (2) dependence of parasites on intermediate hosts, (3) higher uncertainty about community structure and species interactions, (4) unique biomechanics of the saltwater environment, and (5) the large size and openness of marine ecosystems.

Figure 28.2 European green crabs, *Carcinus maenas* (L.), grow to much larger sizes in Californian waters (left), compared with crabs of sizes typical in Europe (right). Photograph courtesy of Jeff Goddard.

Figure 28.3 The parasitic castrating barnacle, *Sacculina carcini* (Thompson), is a potential biological control agent for European green crab, *Carcinus maenas* (L.). The barnacle appears as a sponge-like growth on the rear, underside of this crab. Photograph courtesy of Todd C. Huspeni.

Invasive crabs

The European green crab, *C. maenas*, is a very successful marine invader, having established large populations on the coasts of North America, South Africa, and parts of Australia. It is a food competitor of shorebirds and has harmed commercial clam rearing and crabbing (Cohen et al. 1995, Grosholz & Ruiz 1996, Grosholz et al. 2000). Studies show that a significant reason for its success has been escape from its natural enemies, notably parasites. One visual consequence of this lowered parasitism is the increased size of *C. maenas* in invaded areas (Figure 28.2), which permits higher reproduction rates than in its European home range (Torchin et al. 2001). The most likely candidate for successful biological control of *C. maenas* is the parasitic castrator *Sacculina carcini* (Thompson) (Rhizocephala: Sacculinidae) (Figure 28.3) (Lafferty & Kuris 1996), which is specific to portunid crabs and one species of the closely related Pirimelidae (Høeg & Lutzen 1985). *Sacculina carcini* has severe effects on its host's growth, morphology, physiology, and behavior. Additionally, this barnacle prevents reproduction by male and female *C. maenas* crabs, and induces feminization of male crabs (Thresher et al. 2000).

Feasibility of biological control of green crab is being investigated in the USA and Australia. Laboratory host-specificity tests are being used to measure effects of *S. carcini* on native crabs and then predict real risks under field conditions (Thresher et al. 2000, Goddard et al. 2005). Work in California suggests that some native crab species may be at risk of attack if exposed to high densities of infective *S. carcini* stages. However, these attacked native crabs are not suitable hosts and the parasite failed to reproduce in them (Goddard et al. 2005).

Before *C. maenas* can be released, better data are needed to: (1) quantify the risk to non-target crabs and (2) assess the likely population level impacts of the barnacle on green crab. To assess risks to non-target species, larger-scale laboratory tests with Californian species of crabs are needed to determine the consequences of exposure to large numbers of infective juvenile barnacles from heavily infested green crabs. Laboratory experiments and field studies have determined that the inability of larval *S. carcini* barnacles to locate and settle on a non-target crab is the primary determinant of *S. carcini* host specificity (A. Kuris, personal communication). To better assess potential efficacy, studies are needed on the effects of local transplantation of the barnacle into green crab populations in Europe lacking the parasite.

Finally, other natural enemies may need to be assessed. For example, the parasitic castrator *Portunion maenadis* Giard (Isopoda: Entoniscidae), parasitoid-like flatworms, and obligate nemertean egg predators may all have potential for providing some control of *C. maenas* while posing little risk to non-target crustaceans (Goddard et al. 2005).

Killer alga

Caulerpa taxifolia is a marine alga native to several tropical areas of the world (Meinesz 1999, pp. 209–18; see also the Nova website, www.pbs.org/wgbh/nova/algae/), and an Australian strain of this species has established invasive populations in the Mediterranean Sea and along the east coast of Australia (outside of its native Australian range). Incipient populations have also been reported in Japan and California, but these populations failed to establish either because water temperatures were too cold (Komatsu et al. 2003) or eradication programs were conducted that tentatively appear to have been successful (Anderson 2005).

In the Mediterranean, a cold-water-tolerant strain of *C. taxifolia* "escaped" from a marine research institute in the early 1980s, and via vegetative growth and fragmentation this toxic alga now covers thousands of hectares of sea bottom with dense algal meadows, to the detriment of native flora and fauna (Secord 2003). Such vast infestations are not amenable to chemical or physical control, and natural enemies, in particular sacoglossan sea slugs, have been investigated as possible control agents (Thibaut & Meinesz 2000, Thibaut et al. 2001). A major drawback with sea slugs from tropical Atlantic habitats is their apparent intolerance of cold water temperatures typical of the Mediterranean in winter (Thibaut & Meinesz 2000, Thibaut et al. 2001) and modeling suggests that high densities of sea slugs at cool water temperatures would be required to provide biological control of *C. taxifolia* (Coquillard et al. 2000).

Invasive or disease-vectoring snails

Medically important freshwater snails

Certain aquatic snails are intermediate hosts for trematode worms. Human schistosomiasis, caused by parasitic blood flukes, affects about 200 million people worldwide, inflicting considerable morbidity and some mortality. *Schistosoma mansoni* infected over 1 million people in Puerto Rico as recently as the 1960s (Wright 1973). Interest in biological control possibilities began in the 1950s (Michelson 1957). Laboratory studies revealed that the large snail *Marisa cornusrietis* was an effective predator of egg masses, juveniles, and sometimes adults of the schistosome intermediate host snail, *Biomphalaria glabrata* Say. *Marisa cornuarietis* also reduced the availability of food and oviposition sites for *B. glabrata*. The control campaign in Puerto Rico was carefully planned and was monitored for 15 years (Ferguson 1978). Where environmentally possible, pesticides were used to temporarily reduce or eliminate local *B. glabrata* populations. *Marisa cornuarietis* was then introduced into the habitats to prevent return of the pest snails. At some sites, environmental alterations (ditching, concrete slopes) and environmental management (aquatic weed reduction) were also employed to reduce habitat suitability for aquatic snails. Other developments in Puerto Rico over the same time span also contributed to the reduction to near elimination of schistosomiasis on the island. Urbanization and economic development raised living standards and greatly improved water sanitation. Although it is not possible to fully separate the effects of biological control snails from these other public health improvements, it seems certain that only *M. cornuarietis* would have so significantly reduced the presence of *B. glabrata* in aquatic systems that it is now considered rare (Giboda et al. 1997).

In Kenya, the Louisiana crayfish, *Procambarus clarkii* (Girard), has been manipulated to suppress the snail *Bulinus africanus* (Krauss), the intermediate host of *Schistosoma haematobium*, the causal agent of urinary schistosomiasis (Mkoji et al. 1999). Laboratory and pond studies showed that crayfish were voracious predators of the snail. A pilot study demonstrated that addition of crayfish to village ponds caused *B. africanus* populations to decline precipitously, reducing infections in local children by 60–80% compared to the children in an untreated village. Louisiana crayfish were introduced to East Africa for aquaculture purposes and have spread throughout Kenya and, via the Nile, into Egypt. These introductions were largely unregulated and the crayfish has likely caused environmental damage (Lodge et al. 2005). However, given that this invasive species is already widely distributed in the region and given that most urinary schistosomiasis is transmitted in small village impoundments, of little ecological value, that lack crayfish, introducing crayfish to such ponds might reduce human disease with little increase in ecological harm.

Terrestrial snails

Attempts at biological control of herbivorous land snails with predatory snails have resulted in disastrous impacts on non-target snails including the extinction of several geographically localized species of tree snails. The best-documented such case is the attempted control of the giant African land snail, *Achatina fulica* (Bowdich), in tropical countries with *Gonaxis quadrilateralis* (Preston) from East Africa and *Euglandia rosea* (Férrusac) from Florida, USA. These predators have failed to control the target pest (Christensen 1984, Gerlach 2001), have caused the extinction of numerous native snail species (Clarke et al. 1984, Coote & Loéve 2003), and in turn have become unwanted exotic invaders (Civeyrel & Simberloff 1996, Cowie 2001).

In contrast, the suppression in California of the European brown garden snail, *Helix aspersa* Müller (Helicidae), by the self-introduced facultative carnivorous snail *Rumina decollata* (L.) (Fisher & Orth 1985) is widely believed to be a case of successful biological snail control. However, some authorities dispute this interpretation because pest suppression has not been adequately quantified and outcomes were inconsistent and may have been attributable to causes other than *R. decollata* (Cowie 2001).

Insect parasitoids, rather than predatory snails, may be a better option for control of pest snails. In south and western Australia, four introduced Mediterranean helicid snails have become serious agricultural pests (Coupland & Baker 1995), damaging or contaminating crops and interfering with cattle grazing (Coupland & Baker 1995).

A biological control program against these snails has investigated dipteran parasitoids in the Sciomyzidae and Sarcophagidae from Europe that have potential for use in Australia (Coupland & Baker 1994, 1995, Coupland et al. 1994). The most promising species is the sciomyzid *Pherbellia cinerella* (Fallén), which prefers pasture habitats. The climate of this fly's home range is similar to that of areas in Australia where control is needed. *Pherbellia cinerella* attacks and kills endemic Australian snails in no-choice feeding tests. However, its strong preference for open pastures may reduce its impact in the non-pasture habitats of endemic Australian snails (see the CSIRO European Laboratory website, www.csiro-europe.org/snails.html). The sarcophagid *Sarcophaga penicillata* (Villeneuve) parasitizes aestivating helicid snails (Coupland & Baker 1994). In France, its attack rates are low (4%), but the fly is subjected to heavy hyperparasitism (79%), which may indicate potential for greater impact in Australia once hyperparasitoids are eliminated. Safety testing with 38 species of Australian snails indicated minimal risk, and *S. penicillata* was released in South Australia in 2000. Impact evaluations are on-going (Baker 2000).

CONCLUSIONS

Biological control of non-traditional pests is an emerging area, the benefits and difficulties of which have yet to be appreciated fully. This will change as the success of current projects is evaluated and new projects are undertaken. Currently there are no precedent-setting examples of use of biological control against such groups as non-mammalian vertebrates (e.g. pest amphibians, reptiles, etc.), freshwater and marine crustaceans and mollusks, or platyhelminths. Biologists studying these invasive organisms may be unfamiliar with the concept of biological control and its potential benefits, or wary of introducing another unwanted invasive species (Van Driesche 1994). The concepts and technologies developed for weed and insect biological control projects provide a starting point from which further development may allow successful application to non-traditional targets such as marine species.

Chapter 29

FUTURE DIRECTIONS

Biological control, in each of its four methods of application, will continue to grow. How this happens will vary among countries due to differences in technology, economies, and cultural values. Expanded use of biological control is, however, not guaranteed, nor are technological solutions inevitable for the issues faced by some forms of biological control. Below are our thoughts on the possible future of each of the major approaches to biological control.

CLASSICAL BIOLOGICAL CONTROL

The need for this type of work is already large due to a backlog of high-impact invasive species that could be controlled with this technology but have not been. Also, new invaders continue to establish and spread. Therefore, it seems relatively certain that use of this approach will continue to expand. As invasive species invade new regions, countries with no previous history of classical biological control may initiate projects. Thus the pool of countries that have historically conducted classical biological control will expand to include more countries, including island nations and countries in tropical or developing areas. Larger countries with sufficient capital and skilled expertise are likely to develop an indigenous capacity for classical biological control, whereas smaller, less-developed nations may contract work to technical groups such as CABI BioScience.

Concerns over risks of imported natural enemies are likely to continue to grow, leading to increased legal oversight of the importation and release process. In some countries, legal oversight may raise significant, even prohibitive barriers to natural enemy importations. For example, in the USA, such concern is currently slowing the application of biological control in some areas, particularly Hawaii. In developing countries, concern over risks to non-target species may be judged less significant if invasive species affect critical food production or other essential resources.

For the foreseeable future, weed biological control is likely to be implemented against a wider range of weeds than arthropod pests because the process of evaluating risks for weed biological control agents is better understood and laboratories and infrastructure for such projects are better developed, at least in countries like Australia, Canada, New Zealand, and the USA.

Major constraints to the expansion of use of classical biological control include: (1) interagency confusion and lack of regulatory guidance, which is a prominent issue for arthropod biological control in the USA, (2) administrative barriers to protect and profit from biodiversity, which have been raised by some countries reluctant to freely export natural enemies as needed, and (3) poor understanding on the part of the public of the invasive species problem in a way that allows reasoned comparison of risks and benefits of proposed biological control projects.

CONSERVATION BIOLOGICAL CONTROL

Conservation biological control covers two rather different activities: protecting natural enemies from pesticides and enhancing crops as natural enemy habitats. The former activity is tied to the integrated pest management (IPM) movement. To the degree that there continues to be public interest in lowering use of pesticides, as distinct from abolition of pesticide use (organic farming), then research on how to integrate pesticides and natural enemies in crops will continue in universities and government laboratories. If the public swings toward the belief either: (1) that pesticides are so bad that all farming should be organic farming or (2) that pesticides are no longer much of a problem (since newer products have tended to be less toxic to people and safer to the environment), then the drive to do the necessary research on pesticide–natural enemy interactions may falter. The most noticeable advance in this area in recent years has been the development and widespread use of Bt crops, which has greatly reduced use of pesticides in crops.

In contrast to protecting natural enemies from pesticides, enhancing crops as natural enemy habitats requires more from plant breeders and farmers. Either plant breeders need to create new crop varieties that include more natural enemy-favorable features, or farmers need to spend time, money, land, water, and labor to enhance the crop environment for biological control agents. So far in industrialized countries, farmers have only been willing to engage in crop habitat enhancement when government subsidies pay them to do so. To date, there are virtually no well documented cases in which improved cropping habitats for natural enemies consistently and economically achieve adequate control of specific target pests. Among researchers and their graduate students, this topic is currently a popular research area. The push–pull system for using trap crops in Africa to control corn borers is an example of an effective non-pesticidal pest control system that is at least partly based on conservation biological control. However, other robust examples are lacking.

Constraints on adoption of conservation biological control are likely to include: (1) loss of concern by the public about risks of pesticides, (2) failure of research to find modifications that can be cheaply imposed on crops that make them so much better for natural enemies that tangible pest control results, (3) the reluctance of many farmers to divert attention to managing natural enemies, or (4) an unwillingness by growers to spend money to enhance natural enemy habitat.

AUGMENTATION BIOLOGICAL CONTROL

Augmentative biological control in greenhouse vegetable crops in cold climates is already well developed. Its use will certainly continue. A challenge to maintaining currently effective programs will be adapting to new pest invasions. While future success cannot be guaranteed, in the past the industry has adapted successfully to the invasions of both a new whitefly [*Bemisia tabaci* (Gennadius)] and an important thrips [*Frankliniella occidentalis* (Pergande)]. Development of effective biological control programs for greenhouse vegetable crops in warm production areas is underway and likely to be successful, with adequate levels of research, which seem forthcoming. Development of biological control programs for use in flower production, however, seems much less likely. At best, use will be limited to particularly favorable species (e.g. roses and poinsettias), with very limited use in most short-term crops with rapid turnover.

Constraints likely to affect greenhouse augmentative biological control will be: (1) disruptions to existing programs by invasions of new pests, likely to come from greenhouses established in new parts of the world that have not previously been common sources of plants in international trade (African or Asian nations, for example); and (2) legal barriers to international marketing of effective natural enemies because of risks of permanent establishment of such exotic species out of doors, with consequent potential for non-target impacts.

Outdoor use of augmentative biological control is not likely to expand, largely because mass rearing of natural enemies will prove too costly and not sufficiently effective. Indeed, total usage is more likely to go down, rather than up, given that some considerable amount of current use of *Trichogramma* wasps (the major natural enemy used outdoors augmentatively) receives government subsidies in many countries in one manner or another and that these subsidies are more likely to decline than increase. Constraints on outdoor augmentative use of biological control include: (1) high cost of rearing many natural enemies relative to value of crop protected or competiting pest control options, (2) insufficient research on release rates, timings, and application technologies, and (3) low efficacy. Technical solutions to some of these constraints are possible, but if the immediate past (1970–2005) is taken as a guide, it is not likely that enough new efficicous agents and application strategies will be developed to result in a greater adoption of outdoor augmentative biological control.

BIOPESTICIDES

Biopesticides have remained niche products rather than replacing pesticides as once predicted. This is not likely to change. However, some pathogen groups have been more successful than others. Of arthropod pathogens, bacteria and nematodes have been used the most. Creation of Bt plants has been the single biggest change in agriculture in recent time and certainly the biggest application of entomopathogens (at least their insect-killing products) to plant protection. Use of *Bacillus thuringiensis* Berlinger and *Bacillus sphaericus* Neide to control of mosquitoes and blackflies is likely to continue and expand. Nematodes, while aimed at relatively small markets, have increased steadily, and the number of species produced commercially has grown. New nematode products target new pests or previously difficult abiotic environments. Nematode use is likely to be stable and increase modestly.

In contrast, fungal biopesticides have generally failed to become common pest control products, mostly because of high dosage requirements, high variability in efficacy due to sensitivity to environmental conditions, and production difficulties or high cost. In theory, products might become more reliable through technical improvements to formulations and better strain selection. However, a little product unreliability goes a long way to foster grower rejection. Rearing of fungi is not as easy as for bacteria because spore formation usually does not occur in submerged liquid culture. Consequently, either a two-step (liquid/solid) rearing system must be used, or the inoculum used must shift from spores to mycelial fragments. In developing nations with low labor costs, labor-intensive production of fungal entomopathogens may be cost effective. However, inadequate focus on quality control at such rearing facilities can result in low product quality, fluctuating efficacy, contamination, or inaccurate species identification, any of which may reduce product demand.

Commercial use of viruses remains minimal and will likely remain so because of the limitations of: (1) narrow host specificity, (2) high *in vivo* rearing cost, (3) sensitivity to ultraviolet light and dryness, and (4) loss of fitness in transgenic insect viruses. Government-subsidized production of various baculoviruses has potential for use to control forest pests on public land. The high cost of rearing viruses, due to the inability to rear them outside of living hosts, is likely a permanent obstacle to their commercial use.

CONCLUSIONS

Biological control, especially classical or inoculative efforts, will be needed in the future even more than now as problems with an ever greater diversity of invasive species continue to grow at an alarming pace. Its practice, however, has grown legally more complicated and safety expectations and costs have increased greatly. Demands for lengthy, complicated host specificity testing will increase, so many feasible programs are likely to pass out of reach of the single reasearcher's laboratory and may in the future only be possible in specialized laboratories with cooperative teams available to cover the many aspects of the work. This book is dedicated to today's practitioners and the students they train who will be inspired to do the work.

REFERENCES

Abera Abera-Kanyamuhungu, A.M., Hasyim, A., Gold, C.S., and Van Driesche, G.G. (2006) Field surveys in Indonesia for natural enemies of the banana weevil, *Cosmopolites sordidus* (Germar). *Biological Control* **37**: 16–24.

Ables, J.R. (1979) Methods for the field release of insect parasites and predators. *Transactions of the American Society Agricultural Engineers* **22**: 59–62.

Abraham, Y.J., Moore, D., and Godwin, G. (1990) Rearing and aspects of biology of *Cephalonomia stephanoderis* and *Prorops nasuta* (Hymenoptera: Bethylidae), parasitoids of the coffee berry borer, *Hypothenemus hampei* (Coleoptera: Scolytidae). *Bulletin of Entomological Research* **80**: 121–8.

Adair, R.J. and Scott, J.K. (1993) Biology and host specificity of *Ageniosa electoralis* (Coleoptera: Chrysomelidae), a prospective biological control agent for *Chrysanthemoides monilifera* (Asteraceae). *Biological Control* **3**: 191–8.

Adair, R.J. and Scott, J.K. (1997) Distribution, life history, and host specificity of *Chrysolina picturata* and *Chrysolina* sp. B (Coleoptera: Chrysomelidae), two biological control agents for *Chrysanthemoides monilifera* (Compositae). *Bulletin of Entomological Research* **87**: 331–41.

Adair, R.J. and Holtkamp, R.H. (1999) Development of a pesticide exclusion technique for assessing the impact of biological control agents for *Chrysanthemoides monilifera*. *Biocontrol Science and Technology* **9**: 383–90.

Adams, B.J., Fodor, A., Koppenhöfer, H.S., Stackebrandt, E., Stock, S.P., and Klein, M.G. (2006) Biodiversity and systematics of nematode-bacterium entomopathogens. *Biological Control* **37**: 32–49.

Adisu, B., Starý, P., Freier, B., and Büttner, C. (2002) *Aphidius colemani* Vier. (Hymenoptera: Braconidae, Aphidiinae) detected in cereal fields in Germany. *Anzeiger für Schädlingskunde* **75**: 89–94.

Adlung, K. (1966) A critical evaluation of the European research on use of red wood ants (*Formica rufa* group) for the protection of forests against harmful insects. *Zeitschrift für Angewandte Entomologie* **57**: 167–89.

Agoua, H., Quillevere, D., Back, C. et al. (1991) Evaluation of means of control against black flies in the setting of the OCP programme (Onchocerciasis Control Programme). *Annales de la Societe Belge de Medecine Tropicale* **71** (supplement 1): 49–63.

Agrawal, A.A. (1997) Do leaf domatia mediate a plant-mite mutualism? An experimental test of the effects on predators and herbivores. *Ecological Entomology* **22**: 371–6.

Agrawal, A.A., Karban, R., and Colfer, R.G. (2000) How leaf domatia and induced plant resistance affect herbivores, natural enemies, and plant performance. *Oikos* **89**: 70–80.

Agustí, N., Unruh, T.R., and Welter, S.C. (2003) Detecting *Cacopsylla pyricola* (Hemiptera: Psyllidae) in predator guts using COI mitochondrial markers. *Bulletin of Entomological Research* **93**: 179–85.

Aizawa, K. (1987) Strain improvement of insect pathogens. In: Maramorosch, K. (ed.), *Biotechnology in Invertebrate Pathology and Cell Culture*, pp. 3–11. Academic Press, San Diego, CA.

Aizawa, K. (1990) Registration requirements and safety considerations for microbial pest control agents in Japan. In: Laird, M., Lace, L.A., and Davidson, E.W. (eds.), *Safety of Microbial Insecticides*, pp. 31–39. CRC Press, Boca Raton, FL.

Akhurst, R.J. (1990) Safety to nontarget invertebrates of nematodes of economically important pests. In: Laird, M., Lace, L.A., and Davidson, E.W. (eds.), *Safety of Microbial Insecticides*, pp. 233–40. CRC Press, Boca Raton, FL.

Alam, M.M., Bennett, F.D., and Carl, K.P. (1971) Biological control of *Diatraea saccharalis* (F.) in Barbados by *Apanteles flavipes* Cam. and *Lixophaga diatraeae* T.T. *Entomophaga* **16**: 151–8.

Albert, R., Dannemann, K.R., and Hassan, S.A. (2001) Twenty five years of biological control of the corn stem borer: looking back and forward to successful use of *Trichogramma* parasites in Germany. *Mais* **29(33)**: 106–9 (in German).

Aldrich, J.R., Kochansky, J.P., and Abrams, C.B. (1984) Attractant for a beneficial insect and its parasitoids: pheromone of the predatory spined soldier bug, *Podisus maculiventris* (Hemiptera: Pentatomidae). *Environmental Entomology* **13**: 1031–6.

Ali, A.D. and Reagan, T.E. (1985) Vegetation manipulation impact on predator and prey populations in Louisiana sugarcane ecosystems. *Journal of Economic Entomology* **78**: 1409–14.

AliNiazee, M.T. and Croft, B.A. (1999) Biological control in deciduous fruit crops. In: Bellows, Jr., T.S. and Fisher, T.W.

(eds.), *Handbook of Biological Control*, pp. 743–59. Academic Press, San Diego, CA.

Allard, G.B., Chase, C.A., Heale, J.B., Isaac, J.E., and Prior, C. (1990) Field evaluation of *Metarhizium anisopliae* (Deuteromycotina: Hyphomycetes) as a mycoinsecticide for control of sugarcane froghopper, *Aeneolamia varia saccharina* (Hemiptera: Cercopidae). *Journal of Invertebrate Pathology* **55**: 41–6.

Allee, W.C. (1931) *Animal Aggregations: a Study in General Sociology*. University of Chicago Press, Chicago, IL.

Allee, W.C., Emerson, A.E., Park, O., Park, T., and Schmidt, K.P. (1949) *Principles of Animal Ecology*. W.B. Saunders Co., Philadelphia, PA.

Alleyne, M. and Wiedenmann, R.N. (2001) Encapsulation and hemocyte numbers in three lepidopteran stemborers parasitized by *Cotesia flavipes*-complex endoparasitoids. *Entomologia Experimentalis et Applicata* **100**: 279–93.

Altieri, M.A. and Whitcomb, W.H. (1979) The potential use of weeds in the manipulation of beneficial insects. *Hortscience* **14**: 12–18.

Alvarez, J.M. and Hoy, M.A. (2002) Evaluation of the ribosomal ITS2 DNA sequences in separating closely related populations of the parasitoid *Ageniaspis* (Hymenoptera: Encyrtidae). *Annals of the Entomological Society of America* **95**: 250–6.

Alyokin, A. and Sewell, G. (2004) Changes in a lady beetle community following the establishment of three alien species. *Biological Invasions* **6**: 463–71.

Anable, M.E., McClaran, M.P., and Ruyle, G.B. (1992) Spread of introduced Lehmann lovegrass *Eragrostis lehmanniana* Nees. in southern Arizona, USA. *Biological Conservation* **61**: 181–8.

Andersen, M.C., Ewald, M., and Northcott, J. (2005) Risk analysis and management decisions for weed biological control agents: Ecological theory and modeling results. *Biological Control* **35**: 330–7.

Anderson, I. (1995) Runaway rabbit virus kills millions. *New Scientist* **152**: 4.

Anderson, I. (1997) Alarm greets contraceptive virus. *New Scientist* **154**: 4.

Anderson, J.L. and Whitlatch, R. (2003) Testimony before the 108th US Congress, Fisheries, Wildlife, and Oceans Subcommittee on Chesapeake Bay Oyster Introduction Efforts. United States Congressional Record, Washington, D.C.

Anderson, L.W.J. (2005) California's reaction to *Caulerpa taxifolia*: a model for invasive species rapid response. *Biological Invasions* **7**: 1003–16.

Anderson, R.M. (1979) Parasite pathogenicity and the depression of host equilibria. *Nature* **279**: 150–2.

Anderson, R.M. (1982) Theoretical basis for the use of pathogens as biological control agents. *Parasitology* **84**: 3–33.

Anderson, R.M. and May, R.M. (1978) Regulation and stability of host-parasite population interactions I. Regulatory processes. *Journal of Animal Ecology* **47**: 219–47.

Anderson, R.M. and May, R.M. (1980) Infectious diseases and population cycles of forest insects. *Science* **210**: 658–61.

Anderson, R.M. and May, R.M. (1981) The population dynamics of microparasites and their invertebrate hosts. *Philosophical Transactions of the Royal Society of London, Series B* **291**: 451–524.

Andow, D.A. (1986) Plant diversification and insect population control in agroecosystems. In: Pimentel, D. (ed.), *Some Aspects of Integrated Pest Management*, pp. 277–86. Cornell University Press, Ithaca, NY.

Andow, D.A. (1988) Management of weeds for insect manipulation in agroecosystems. In: Altieri, M.A. and Liebman, M. (eds.), *Weed Management in Agroecosystems: Ecological Approaches*, pp. 265–301. CRC Press, Boca Raton, FL.

Andow, D.A. (1991a) Yield loss to arthropods in vegetationally diverse agroecosystems. *Environmental Entomology* **20**: 1228–35.

Andow, D.A. (1991b) Control of arthropods using crop diversity. In: Pimentel, D.P. (ed.), *CRC Handbook of Pest Management in Agriculture*, 2nd edn., vol. 1, pp. 257–84. CRC Press, Boca Raton, FL.

Andres, L.A. (1977) The economics of biological control of weeds. *Aquatic Biology* **3**: 111–13.

Andres, L.A., Davis, C.J., Harris, P., and Wapshere, A.J. (1976) Biological control of weeds. In: Huffaker, C.B. and Messenger, P.S. (eds.), *Theory and Practice of Biological Control*, pp. 481–97. Academic Press, New York.

Andrewartha, H.G. and Birch, L.C. (1954) *The Distribution and Abundance of Animals*. University of Chicago Press, Chicago, IL.

Angalet, G.W., Tropp, J.M., and Eggert, A.N. (1979) *Coccinella septempunctata* L. in the continental United States: recolonization and notes on its ecology. *Environmental Entomology* **8**: 896–901.

Angulo, E. and Cooke, B. (2002) First synthesize new viruses then regulate their release? The case of the wild rabbit. *Molecular Ecology* **11**: 2703–9.

Ankley, G.T., Tietge, J.E., DeFoe, D.L. et al. (1998) Effects of ultraviolet light and methoprene on survival and development of *Rana pipiens*. *Environmental Toxicology and Chemistry* **17**: 2530–42.

Anon (1988) The fruit flies: one more victory. *Citrograph* **73(5)**: 85.

Anon (1992) *Expert Consultation on Guidelines for Introduction of Biological Control Agents*. FAO Rome, Italy, September 17–19, 1991.

Anon (1997a) Code of conduct for the import and release of exotic biological control agents. *Biocontrol News and Information* **18(4)**: 119N–24N.

Anon (1997b) Rabbit virus vectors named. *Science* **278**: 229.

Anon (2000) *NAPPO Regional Standards for Phytosanitary Measures (RSPM) #12. Guidelines for Petition for Release of Exotic Entomophagous Agents for the Biological Control of Pests*. The Secretariat of the North American Plant Protection Organization, Ottawa.

Anon (2001) *NAPPO Regional Standards for Phytosanitary Measures (RSPM) #7. Guidelines for Petition for Release of Exotic Phytophagous Agents for the Biological Control of Weeds.* The Secretariat of the North American Plant Protection Organization, Ottawa.

Anon (2004) Invasive species: how to identify emerald ash borer. *Journal of Forestry* **102**: 4–5.

Antía-Londoño, O.P., Posada-florez, F., Busillo-Pardey, A.E., and González-Garciá, M.T. (1992) Produccion en finca del hongo *Beauveria bassiana* para el control de la broca del café. No. 182. Pub. by Cenicafe, Chinchiná, Caldas, Colombia, Octubre, 1992.

Antolin, M.F. (1989) Genetic considerations in the study of attack behavior of parasitoids, with reference to *Muscidifurax raptor* (Hymenoptera: Pteromalidae). *Florida Entomologist* **72**: 15–32.

Antolin, M.F., Guertin, D.S., and Petersen, J.J. (1996) The origin of gregarious *Muscidifurax* (Hymenoptera: Pteromalidae) in North America: an analysis using molecular markers. *Biological Control* **6**: 766–82.

Arnett, A.E. and Louda, S.M. (2002) Re-test of *Rhinocyllus conicus* host specificity, and the prediction of ecological risk in biological control. *Biological Conservation* **106**: 251–7.

Arnett, R.H. (1968) *The Beetles of the United States (a Manual for Identification).* The American Entomological Institute, Ann Arbor, MI.

Arnett, Jr., R.H. (1985) *American Insects.* Van Nostrand Reinhold Co., New York.

Aronson, A.I. and Shai, Y. (2001) Why *Bacillus thuringiensis* insecticidal toxins are so effective: unique features of their mode of action. *FEMS Microbiology Letters* **195(1)**: 1–8.

Arthington, A.H. and Lloyd, L.N. (1989) Introduced poeciliids in Australia and New Zealand. In: Meffe, G.K. and Snelson, Jr., F.F. (eds.), *Ecology and Evolution of Livebearing Fishes (Poecilidae)*, pp. 333–48. Prentice Hall, Englewood Cliffs, NJ.

Arthurs, S.P., Lacey, L.A., and Fritts, Jr., R. (2005) Optimizing use of codling moth granulovirus: effects of application rate and spraying frequency on control of codling moth larvae in Pacific Northwest apple orchards. *Journal of Economic Entomology* **98**: 1459–68.

Ash, G.J., Cother, E.J., and Tarleton, J. (2004) Variation in lanceleaved waterplantain (*Alisma lanceolatum*) in southeastern Australia. *Weed Science* **52**: 413–17.

Ashfaq, M., Braun, L., Hegedus, D., and Erlandson, M. (2004) Estimating parasitism levels in *Lygus* spp. (Hemiptera: Miridae) field populations using standard and molecular techniques. *Biocontrol Science and Technology* **14**: 731–5.

Ashraf, M., Fatima, B., Hussain, T., and Ahmad, N. (1999) Biological control: an essential component of IPM programme for sugarcane borers. In: *Symposium on Biological Control in the Tropics*, pp. 38–42, MARDI Training Centre, Serdang, Selangor, Malaysia, March 18–19, 1999.

Askew, R.R. (1971) *Parasitic Insects.* American Elsevier Pub. Co., New York.

Askew, R.R. and Shaw, M.R. (1986) Parasitoid communities: their size, structure, and development. In: Waage, J. and Greathead, D. (eds.), *Insect Parasitoids*, pp. 225–64. Academic Press, London.

Atkinson, I.A.E. (1985) The spread of commensal species of *Rattus* to oceanic islands and their effects on island avifaunas. In: Moors, P.J. (ed.), *Conservation of Island Birds: Case Studies for the Management of Threatened Island Species*, pp. 35–81. Proceedings of a symposium held at the XVIII ICBP World Conference in Cambridge, 1982. ICBP Technical Publication No. 3.

Atkinson, K. (1997) New Zealand grapples with possum virus. *Search* **28**: 260.

Atlegrim, O. (1989) Exclusion of birds from bilberry stands: impact on insect larval density and damage to the bilberry. *Oecologia* **79**: 136–9.

Auer, C. (1968) Erst Ergebnisse einfacher stochastischer modelluntersuchungen uber die ursachen der populationsbewegung des grauen larchenwicklers *Zeiraphera diniana*, Gn. (=*Z. griseana* Hb.) im Oberengadin, 1949/66. *Zeitschrift für Angewandte Entomologie* **62**: 202–35.

Augustinos, A.A., Stratikopoulos, E.E., Zacharopoulou, A., and Mathiopoulos, K.D. (2002) Polymorphic microsatellite markers in the olive fly, *Bactrocera oleae*. *Molecular Ecology Notes* **2**: 278–80.

Aukema, B.H., Dahlsten, D.L., and Raffa, K.F. (2000) Improved population monitoring of bark beetles and predators by incorporating disparate behavioral responses to semiochemicals. *Environmental Entomology* **29**: 618–29.

Austin, A.D., Johnson, N.F., and Dowton, M. (2005) Systematics, evolution, and biology of scelionid and platygastrid wasps. *Annual Review of Entomology* **50**: 553–82.

Avilla, J., Albajes, R., Alomar, O., Castañe, C., and Gabarra, R. (2004) Biological control of whiteflies on vegetable crops. In: Heinz, K.M., Van Driesche, G.G., and Parrella, M.P. (eds.), *Biocontrol in Protected Culture*, pp. 171–84. Ball Publishing, Batavia, IL.

Avise, J.C. (2000) *Phylogeography: the History and Formation of Species.* Harvard University Press, Cambridge, MA.

Avise, J.C. (2004) *Molecular Markers, Natural History, and Evolution.* Sinauer Associates, Sunderland, MA.

Axtell, R.C. (1981) Use of predators and parasites in filth fly IPM programs in poultry housing. In: *Status of Biological Control of Filth Flies*, pp. 26–43. Proceedings of a Workshop, February 4–5, 1981, Gainesville, Florida. USDA, Washington, D.C.

Baars, J.-R. and Neser, S. (1999) Past and present initiatives on the biological control of *Lantana camara* (Verbenaceae) in South Africa. In: Olckers, T. and Hill, M.P. (eds.), *Biological Control of Weeds in South Africa (1990–1998)*, pp. 21–33. African Entomology Memoir No. 1. Entomological Society of Southern Africa, Hatfield.

Babcock, C.S. and Heraty, J.M. (2000) Molecular markers distinguishing *Encarsia formosa* and *Encarsia luteola*

(Hymenoptera: Aphelinidae). *Annals of the Entomological Society of America* **93**: 738–44.

Babendreier, D., Bigler, F., and Kuhlmann, U. (2005) Methods used to assess non-target effects of invertebrate biological control agents of arthropod pests. *BioControl* **50**: 821–70.

Bach, C.E. (1991) Direct and indirect interactions between ants (*Pheidole megacephala*), scales (*Coccus viridis*) and plants (*Pluchea indica*). *Oecologia* **87**: 233–9.

Baggen, L.R. and Gurr, G.M. (1998) The influence of food on *Copidosoma koehleri* (Hymenoptera: Encyrtidae), and the use of flowering plants as a habitat management tool to enhance biological control of potato moth, *Phthorimaea operculella* (Lepidoptera: Gelechiidae). *Biological Control* **11**: 9–17.

Baggen, L.R., Gurr, G.M., and Meats, A. (2000) Field observations on selective food plants in habitat manipulation for biological control of potato moth by *Copidosoma koehleri* Blanchard (Hymenoptera: Encrytidae). In: Anon, *Hymenoptera: Evolution, Biodiversity, and Biological Control*, pp. 388–95. CSIRO Publishing, Collingwood, Victoria.

Bai, B. and Mackauer, M. (1991) Recognition of heterospecific parasitism: competition between aphidiid (*Aphidius ervi*) and aphelinid (*Aphelinus asychis*) parasitoids of aphids (Hymenoptera: Aphidiidae, Aphelinidae). *Journal of Insect Behavior* **4**: 333–45.

Bais, H.P., Vepachedu, R., Gilroy, S., Callaway, R.M., and Vivanco, J.M. (2003) Allelopathy and exotic plant invasion: from molecules and genes to species interactions. *Science* **301**: 1377–80.

Baker, D.A., Loxdale, H.D., and Edwards, O.R. (2003) Genetic variation and founder effects in the parasitoid wasp, *Diaeretiella rapae* (M'intosh) (Hymenoptera: Braconidae: Aphidiidae), affecting its potential as a biological control agent. *Molecular Ecology* **12**: 3303–11.

Baker, G. (2000) Release of fly spells disaster for snails. *Farming Ahead* **105**: 49.

Baker, R.H.A. (2002) Predicting the limits to the potential distribution of alien crop pests. In: Hallman, G.J. and Schwalbe, C.P. (eds.) *Invasive Arthropods in Agriculture. Problems and Solutions*, pp. 207–41. Science Publishers, Enfield, NH.

Baker, R., Cannon, R., Bartlett, P., and Barker, I. (2005) Novel strategies for assessing and managing the risks posed by invasive alien species to global crop production and biodiversity. *Annals of Applied Biology* **146**: 177–91.

Balch, R.E. and Bird, F.T. (1944) A disease of the European spruce sawfly, *Gilpinia hercyniae* (Htg.) and its place in natural control. *Science in Agriculture* **25**: 65–80.

Balciunas, J.K. and Burrows, D.W. (1993) The rapid suppression of the growth of *Melaleuca quinquenervia* saplings in Australia by insects. *Journal of Aquatic Plant Management* **31**: 265–70.

Balciunas, J.K. and Smith, L. (2006) Prerelease efficacy assessment, in quarantine, of a tephritid gall fly considered as a biological control agent for Cape-ivy (*Delairea odorata*). *Biological Control* **39**: 516–24.

Balciunas, J.K., Burrows, D.W., and Purcell, M.F. (1994a) Insects to control melaleuca. I: Status of research in Australia. *Aquatics* **16**: 10–13.

Balciunas, J.K., Burrows, D.W., and Purcell, M.F. (1994b) Field and laboratory host ranges of the Australian weevil, *Oxyops vitiosa* (Coleoptera: Curculionidae), a potential biological control agent for the paperbark tree, *Melaleuca quinquenervia*. *Biological Control* **4**: 351–60.

Balciunas, J.K., Burrows, D.W., and Purcell, M.F. (1996) Comparison of the physiological and realized host-ranges of a biological control agent from Australia for the control of the aquatic weed *Hydrilla verticillata*. *Biological Control* **7**: 148–58.

Bale, J.S. and Walters, K.F.A. (2001) Overwintering biology as a guide to the establishment potential of non-native arthropods *in the U.K. In: Atkinson, D. and Thorndyke, M. (eds.), Animal Developmental Ecology*, pp. 343–54. BIOS Science Publishers, Oxford.

Baliraine, F.N., Bonizzoni, M., Guglielmino, C.R. et al. (2004) Population genetics of the potentially invasive African fruit fly species, *Ceratitis rosa* and *Ceratitis fasciventris* (Diptera: Tephritidae). *Molecular Ecology* **13**: 683–95.

Balirwa, J.S., Chapman, C.A., Chapman, L.J. et al. (2003) Biodiversity and fishery sustainability in the Lake Victoria basin: an unexpected marriage? *BioScience* **53**: 703–15.

Balkhoven, J. and van Zuidam, K. (2002) Biological control of hennep aphid is possible, but expensive. *Fruitteelt (Den Haag)* **92(12)**: 10–11.

Baltensweiler, W. and Fischlin, A. (1988) The larch budmoth in the Alps. In: Berryman, A.A. (ed.), *Dynamics of Forest Insect Populations: Patterns, Causes, Implications*, pp. 331–51. Plenum Press, New York.

Bangsund, D.A., Leistritz, F.L., and Leitch, J.A. (1999) Assessing economic impacts of biological control of weeds: the case of leafy spurge in the northern Great Plains of the United States. *Journal of Environmental Management* **56**: 35–43.

Banks, C.J. and Macaulay, E.D.M. (1967) Effects of *Aphis fabae* Scop. and its attendant ants and insect predators on yields of field beans (*Vica faba* L.). *Annals of Applied Biology* **60**: 445–53.

Barber, M.R. and Fayrer-Hosken, R.A. (2000) Possible mechanisms of mammalian immunocontraception. *Journal of Reproductive Immunology* **46**: 103–24.

Barbosa, P. (1998) *Conservation Biological Control*. Academic Press, San Diego, CA.

Barker, S.C., Singleton, G.R., and Spratt, D.M. (1991) Can the nematode *Capillaria hepatica* regulate abundance in wild house mice? Results of enclosure experiments in southeastern Australia. *Parasitology* **103**: 439–49.

Barlow, N.D. (1994) Predicting the effect of a novel vertebrate biocontrol agent: a model for viral-vectored immunocontraception of New Zealand possums. *Journal of Applied Ecology* **31**: 454–62.

Barlow, N.D. (1997) Modeling immunocontraception in disseminating systems. *Reproduction, Fertility and Development* **9**: 51–60.

Barlow, N.D. (2000) The ecological challenge of immunocontraception: editor's introduction. *Journal of Applied Ecology* **37**: 897–902.

Barlow, N.D., Moller, H., and Beggs, J.R. (1996) A model for the effect of *Sphecophaga vesparum vesparum* as a biological control agent of the common wasp in New Zealand. *Journal of Applied Ecology* **33**: 31–44.

Barlow, N.D., Barron, M.C., and Parkes, J. (2002) Rabbit haemorrhagic disease in New Zealand: field test of a disease-host model. *Wildlife Research* **29**: 649–53.

Barlow, N.D., Kean, J.M., and Goldson, S.L. (2003) Biological control lessons: modeling successes and failures in New Zealand. In: Van Driesche, R.G. (ed.), *Proceedings of the First International Symposium on Biological Control of Arthropods*, pp. 105–7, January 14–18, 2002, Honolulu, Hawaii. FHTET-03-05. USDA Forest Service, Morgantown, WV.

Barnes, H.F. (1929) Gall midges as enemies of aphids. *Bulletin of Entomological Research* **20**: 433–42.

Barratt, B.I.P. (2004) *Microctonus* parasitoids and New Zealand weevils: comparing laboratory estimates of host ranges to realized host ranges. In: Van Driesche, R.G. and Reardon, R. (eds.), *Assessing Host Ranges for Parasitoids and Predators Used for Classical Biological Control: a Guide to Best Practice*, pp. 103–20. FHTET-2004-03. USDA Forest Service, Morgantown, WV.

Barratt, B.I.P., Evans, A.A., Ferguson, C.M., Barker, G.M., McNeill, M.R., and Phillips, C.B. (1997) Laboratory non-target host range of the introduced parasitoids *Microctonus aethiopoides* and *M. hyperodae* (Hymenoptera: Braconidae) compared with field parasitism in New Zealand. *Environmental Entomology* **26**: 694–702.

Barratt, B.I.P., Blossey, B., and Hokkanen, H.M.T. (2006) Post-release evaluation of non-target effects of biological control agents. In: Bigler, F., Babendreir, D., and Kuhlmann, U. (eds.), *Environmental Impact of Invertebrates for Biological Control of Arthropods*, pp. 166–86. CABI Publishing, Wallingford.

Barron, M.C., Barlow, N.D., and Wratten, S.D. (2003) Non-target parasitism of the endemic New Zealand red admiral butterfly (*Bassaris gonerilla*) by the introduced biological control agent *Pteromalus puparum*. *Biological Control* **27**: 329–35.

Bartlett, B.R. (1963) The contact toxicity of some pesticide residues to hymenopterous parasites and coccinellid predators. *Journal of Economic Entomology* **56**: 694–8.

Bartlett, B.R. (1964a) Patterns in the host-feeding habit of adult parasitic Hymenoptera. *Annals of the Entomological Society of America* **57**: 344–50.

Bartlett, B.R. (1964b) The toxicity of some pesticide residues to adult *Amblyseuis hibisci*, with a compilation of the effects of pesticides upon phyotseiid mites. *Journal of Economic Entomology* **57**: 559–63.

Bartlett, B.R. (1966) Toxicity and acceptance of some pesticides fed to parasitic Hymenoptera and predatory coccinellids. *Journal of Economic Entomology* **59**: 1142–9.

Bartlett, B.R. (1978) Margarodidae. In: Clausen, C.P. (ed.), *Introduced Parasites and Predators of Arthropod Pests and Weed: a World Review*, pp. 132–6. USDA Agriculture Handbook No. 480. USDA, Washington, D.C.

Bartlett, B.R. and van den Bosch, R. (1964) Foreign exploration for beneficial organisms. In: DeBach, P. and Schlinger, E.I. (eds.), *Biological Control of Insect Pests and Weeds*, pp. 283–304. E.I. Reinhold Publishing Corporation, New York.

Bartlett, M.C. and Jaronski, S.T. (1988) Mass production of entomogenous fungi for biological control of insects. In: Burge, M.N. (ed.), *Fungi in Biological Control Systems*, pp. 61–85. Manchester University Press, Manchester.

Bartoli, P. and Boudouresque, C.F. (1997) Transmission failure of parasites (Digenea) in sites colonized by the recently introduced invasive alga *Caulerpa taxifolia*. *Marine Ecology, Progress Series* **154**: 253–60.

Barton Browne, L. and Withers, T.M. (2002) Time-dependent changes in the host-acceptance threshold of insects: implications for host specificity testing of candidate biological control agents. *Biocontrol Science and Technology* **12**: 677–93.

Baruch, Z., Pattison, R.R., and Goldstein, G. (2000) Responses to light and water availability of four invasive Melastomataceae in the Hawaiian islands. *International Journal of Plant Sciences* **161**: 107–18.

Bateman, R.P. (1992) Controlled droplet application of mycoinsecticides: an environmentally friendly way to control locusts. *Antenna* **16(1)**: 6–13.

Bateman, R. (2004) Constraints and enabling technologies for mycopesticide development. *Outlooks on Pest Management* **15**: 64–9.

Bateman, R. and Chapple, A. (2001) The spray application of mycopesticide formulations. In: Butt, T.M., Jackson, C., and Magan, N. (eds.), *Fungi as Biocontrol Agents: Progress, Problems, and Potential*, pp. 289–309. CABI Bioscience, Silwood Park, Ascot.

Bateman, R.P., Carey, M., Moore, D., and Prior, C. (1993) The enhanced infectivity of *Metarhizium flavoviride* in oil formulations to desert locusts at low humidities. *Annals of Applied Biology* **122**: 145–52.

Bathon, H. (2003) Invasive natural enemy species, a problem for biological plant protection. *DgaaE Nachrichten* **17(1)**: 8.

Batra, S.W.T., Schroeder, D., Boldt, P.E., and Mendl, W. (1986) Insects associated with purple loosestrife (*Lythrum salicaria* L.) in Europe. *Proceedings of the Entomological Society of Washington* **88**: 748–59.

Baumann, L. and Baumann, P. (1989) Expression in *Bacillus subtilis* of the 51- and 42-kilodalton mosquitocidal toxin genes of *Bacillus sphaericus*. *Applied Environmental Microbiology* **55**: 252–3.

Baumann, L., Broadwell, A.H., and Baumann, P. (1988) Sequence analysis of the mosquitocidal toxin genes encoding 51.4- and 41.9-kilodalton proteins from *Bacillus sphaericus* 2362 and 2297. *Journal of Bacteriology* **170**: 2045–50.

Baumann, P., Baumann, L., Bowditch, R.D., and Broadwell, A.H. (1987) Cloning of the gene for the larvicidal toxin of *Bacillus sphaericus* 2362: evidence for a family of related sequences. *Journal of Bacteriology* **169**: 4061–67.

Baumann, P., Clark, M.A., Baumann, L., and Broadwell, A.H. (1991) *Bacillus sphaericus* as a mosquito pathogen: properties of the organism and its toxins. *Microbiology Reviews* **55**: 425–36.

Bax, N. (1999) Eradicating a dreissenid from Australia. *Dreissenia* **10**: 1–5.

Bay, E.C., Berg, C.O., Chapman, H.C., and Legner, E.F. (1976) Biological control of medical and veterinary pests. In: Huffaker, C.B. and Messenger, P.S. (eds.), *Theory and Practice of Biological Control*, pp. 457–79. Academic Press, New York.

Beard, R.L. (1940) Parasite castration of *Anasa tristis* DeG. by *Trichopoda pennipes* Fab., and its effect on reproduction. *Journal of Economic Entomology* **33**: 269–72.

Becerra, J.X. (1997) Insects on plants: macroevolutionary chemical trends in host use. *Science* **276**: 253–6.

Bechinski, E.J. and Pedigo, L.P. (1981) Ecology of predaceous arthropods in Iowa soybean agroecosystems. *Environmental Entomology* **10**: 771–8.

Beckage, N.E. (1985) Endocrine interactions between endoparasitic insects and their hosts. *Annual Review of Entomology* **30**: 371–413.

Beckage, N.E. and Gelman, D.B. (2004) Wasp parasitoid disruption of host development: implications for new biologically based strategies for insect control. *Annual Review of Entomology* **49**: 299–330.

Becker, E., Shamoun, S.F., and Hintz, W.E. (2005) Efficacy and environmental fate of *Chondrostereum purpureum* used as a biological control for red alder (*Alnus rubra*). *Biological Control* **33**: 269–77.

Beckman, R. (1988) Mice on the farm. *Rural Research* **138**: 23–7.

Bedding, R.A. (1984) Nematode parasites of Hymenoptera. In: Nickle, W.R. (ed.), *Plant and Insect Nematodes*, pp. 755–95. Marcel Dekker, New York.

Bedding, R.A. (1993) Biological control of *Sirex noctilio* using the nematode *Deladenus siricidola*. In: Bedding, R., Akhurst, R., and Kaya, H. (eds.), *Nematodes and the Biological Control of Insect Pests*, pp. 11–20. CSIRO Publishing, Melbourne.

Beddington, J.R. (1975) Mutual interference between parasites or predators and its effect on searching efficiency. *Journal of Animal Ecology* **44**: 331–40.

Beddington, J.R., Free, C.A., and Lawton, J.H. (1978) Characteristics of successful natural enemies in models of biological control of insects. *Nature* **273**: 513–19.

Bedford, G.O. (1986) Biological control of the rhinoceros beetle (*Oryctes rhinoceros*) in the South Pacific by baculovirus. *Agriculture, Ecosystems and the Environment* **15**: 141–7.

Beebee, T.J.C. and Rowe, G. (2004) *An Introduction to Molecular Ecology*, Oxford University Press, Oxford.

Beegle, C.C. and Yamamoto, T. (1992) Invitation paper (C. P. Alexander Fund): history of *Bacillus thuringiensis* Berliner research and development. *The Canadian Entomologist* **124**: 587–616.

Beirne, B.P. (1975) Biological control attempts by introductions against pest insects in the field in Canada. *The Canadian Entomologist* **107**: 225–36.

Beirne, B.P. (1984) Biological control of the European fruit lecanium, *Lecanium tiliae* (Homoptera: Coccidae), in British Columbia. *Journal of the Entomological Society of British Columbia* **81**: 28.

Beirne, B.P. (1985) Avoidable obstacles to colonization in classical biological control of insects. *Canadian Journal of Zoology* **63**: 743–7.

Bell, H.A., Kirkbride-Smith, A.E., Marris, G.C., and Edwards, J.P. (2004) Teratocytes of the solitary endoparasitoid *Meteorus gyrator* (Hymenoptera: Braconidae): morphology, numbers and possible functions. *Physiological Entomology* **29**: 335–43.

Bell, M.R. (1991) *In vivo* production of a nuclear polyhedrosis virus utilizing tobacco budworm and a multicellular larval rearing container. *Journal of Entomological Science* **26**: 69–75.

Bell, W.J. (1990) Searching behavior patterns in insects. *Annual Review of Entomology* **35**: 447–67.

Bellotti, A.C., Mesa, N., Serrano, M., Guerrero, J.M., and Herrera, C.J. (1987) Taxonomic inventory and survey activity for natural enemies of cassava green mites in the Americas. *Insect Science and its Application* **8**: 845–9.

Bellows, Jr., T.S. (1993) Introduction of natural enemies for suppression of arthropod pests. In: Lumsden, R. and Vaughn, J. (eds.), *Pest Management: Biologically Based Technologies*, pp. 82–89. American Chemical Society, Washington, D.C.

Bellows, Jr., T.S. and Birley, M.H. (1981) Estimating developmental and mortality rates and stage recruitment from insect stage frequency data. *Researches on Population Ecology* **23**: 232–44.

Bellows, Jr., T.S. and Legner, E.F. (1993) Foreign exploration. In: Van Driesche, R.G. and Bellows, Jr., T.S. (eds.), *Steps in Classical Arthropod Biological Control*, pp. 25–41. Thomas Say Publications in Entomology. Entomological Society of America, Lanham, MD.

Bellows, Jr., T.S. and Morse, J.G. (1993) Toxicity of pesticides used in citrus to *Aphytis melinus* DeBach (Hymenoptera: Aphelinidae) and *Rhizobius lophanthae* (Blaid.) (Coleoptera: Coccinellidae). *The Canadian Entomologist* **125**: 987–94.

Bellows, Jr., T.S. and Van Driesche, R.G. (1999) Life table construction and analysis for evaluating biological control agents. In: Bellows, Jr., T.S. and Fisher, T.W. (eds.), *Handbook of Biological Control*, pp. 199–223. Academic Press, San Diego, CA.

Bellows, Jr., T.S., Ortiz, M., Owens, J.C., and Huddleston, E.W. (1982) A model for analyzing insect stage-frequency data when mortality varies with time. *Researches on Population Ecology* **24**: 142–56.

Bellows, Jr., T.S., Morse, J.G., Hadjidemetriou, D.G., and Iwata, Y. (1985) Residual toxicity of four insecticides used for control of citrus thrips (Thysanoptera: Thripidae) on three beneficial species in a citrus agroecosystem. *Journal of Economic Entomology* **78**: 681–6.

Bellows, Jr., T.S., Morse, J.G., Gaston, L.K., and Bailey, J.B. (1988) The fate of two systemic insecticides and their impact on two phytophagous and a beneficial arthropod in a citrus agroecosystem. *Journal of Economic Entomology* **81**: 899–904.

Bellows, Jr., T.S., Van Driesche, R.G., and Elkinton, J.S. (1989) Extensions to Southwood and Jepson's graphical method of estimating numbers entering a stage for calculating mortality due to parasitism. *Researches on Population Ecology* **31**: 169–84.

Bellows, Jr., T.S., Paine, T.D., Gould, J.R. et al. (1992a) Biological control of ash whitefly: a success in progress. *California Agriculture* **46(1)**: 24, 27–8.

Bellows, Jr., T.S., Van Driesche, R.G., and Elkinton, J.S. (1992b) Life-table construction and analysis in the evaluation of natural enemies. *Annual Review of Entomology* **37**: 587–614.

Bennett, F.D. (1971) Current status of biological control of the small moth borers of sugarcane, *Diatraea* spp. (Lep.: Pyralidae). *Entomophaga* **16**: 111–24.

Bennett, F.D. and Zwolfer, H. (1968) Exploration for natural enemies of the water hyacinth in northern South America and Trinidad. *Hyacinth Control Journal* **7**: 44–52.

Bennett, F.D., Cochereau, P., Rosen, D., and Wood, B.J. (1976) Biological control of pests of tropical fruits and nuts. In: Huf-faker, C.B. and Messenger, P.S. (eds.), *Theory and Practice of Biological Control*, pp. 359–95. Academic Press, New York.

Bennison, J.A. (1992) Biological control of aphids on cucumbers, use of open rearing systems or "banker plants" to aid establishment of *Aphidius matricariae* and *Aphidoletes aphidimyza*. *Mededelingen van de Faculteit Landouwwetenschappen, Universiteit Gent* **57**: 457–66.

Benson, J., Van Driesche, R.G., Pasquale, A., and Elkinton, J. (2003) Introduced braconid parasitoids and range reduction of a native butterfly in New England. *Biological Control* **28**: 197–213.

Benz, G. (1974) Negative Ruckkoppelung durch Raum- und Nahrungskonkurrenz sowie zyklische Veranderung der Nahrungsgrundlage als Regelprinzip in der Populationsdynamik des Grauen Larchenwicklers, *Zeiraphera diniana* (Guenée) (Lep. Torticidae). *Zeitschrift für Angewandte Entomologie* **76**: 196–228.

Benz, G. (1987) Environment. In: Fuxa, J.R. and Tanada, Y. (eds.), *Epizootiology of Insect Diseases*, pp. 177–214. John Wiley and Sons, New York.

Ben-Ze'ev, I., Kenneth, R.G., and Bitton, S. (1981) The Entomophthorales of Israel and their arthropod hosts. *Phytoparasitica* **9**: 43–50.

Berdegue, M., Trumble, J.T., Hare, J.D., and Redak, R.A. (1996) Is it enemy-free space? The evidence for terrestrial insects and freshwater arthropods. *Ecological Entomology* **21**: 203–17.

Berg, G.N., Williams, P., Bedding, R.A., and Akhurst, R.J. (1987) A commercial method of application of entomopathogenic nematodes to pasture for controlling subterranean insect pests. *Plant Protection Quarterly* **2(4)**: 174–7.

Berliner, E. (1915) Ueber die schlaffsucht der *Ephestia kuhniella* und *Bac. thuringiensis* n. sp. *Zeitschrift für Angewandte Entomologie* **2**: 21–56.

Bernal, J.S., Luck, R.F., and Morse, J.G. (1999) Augmentative release trials with *Metaphycus* spp. (Hymenoptera: Encyrtidae) against citricola scale (Homoptera: Coccidae) in California's San Joaquin Valley. *Journal of Economic Entomology* **92**: 1099–1107.

Bernays, E.A. and Chapman, R.F. (1994) *Host-plant Selection by Phytophagous Insects*. Chapman and Hall, New York.

Berry, R.E., Liu, J., and Groth, E. (1997) Efficacy and persistence of *Heterorhabditis marelatus* (Rhabditida: Heterorhabditidae) against root weevils (Coleoptera: Curculionidae) in strawberry. *Environmental Entomology* **26**: 465–70.

Bertschinger, H., Cowan, P., Kay, D., Liu, I., and Parkes, J. (2000) Mammal biocontrol: the hunt continues. *Biocontrol News and Information* **21**: 89N–93N.

Betz, F.S. (1986) Registration of baculoviruses as pesticides. In: Granados, R.R. and Federici, B.A. (eds.), *The Biology of Baculoviruses: Volume II. Practical Application for Insect Control*, pp. 203–22. CRC Press, Boca Raton, FL.

Betz, F.S., Forsyth, S.F., and Stewart, W.E. (1990) Registration requirements and safety considerations for microbial pest control agents in North America. In: Laird, M., Lace, L.A., and Davidson, E.W. (eds.), *Safety of Microbial Insecticides*, pp. 3–10. CRC Press, Boca Raton, FL.

Bewick, T.A., Binning, L.K., Stevenson, W.R., and Stewart, J. (1987) A mycoherbicide for control of swamp dodder (*Cuscuta gronovii* Willd.) Cusctaceae. In: Anon, *Proceedings of the 4th International Symposium on Parasitic Flowering Plants*, pp. 93–104. Marburg, Germany.

Bhumiratana, A. (1990) Local production of *Bacillus sphaericus*. In: de Barjac, H. and Sutherland, D.J. (eds.), *Bacterial Control of Mosquitoes and Black Flies: Biochemistry, Genetics, and Application of Bacillus thuringiensis israelensis and Bacillus sphaericus*, pp. 272–83. Rutgers University Press, New Brunswick, NJ.

Bickel, D.J. and Hernandez, M.C. (2004) Neotropical *Thrypticus* (Diptera: Dolichopodidae) reared from water hyacinth, *Eichhornia crassipes*, and other Pontederiaceae. *Annals of the Entomological Society of America* **97**: 437–49.

Bigler, F. (1994) Quality control in *Trichogramma* production. In: Wajnberg, É. and Hassan, S.A. (eds.), *Biological Control with Egg Parasitoids*, pp. 93–144. CABI Publishing, Wallingford.

Bigler, F. (1997) Use and registration of macroorganisms for biological control crop protection. *Bulletin OEPP* **27(1)**: 95–102.

Bigler, F. and Kölliker-Ott, E. (2006) Balancing environmental risks and benefits: a basic approach. In: Bigler, F., Babendreir, D., and Kuhlmann, U. (eds.), *Environmental Impact of Invertebrates for Biological Control of Arthropods*, pp. 273–86. CABI Publishing, Wallingford.

Bigler, F., Bale, J.S., Cock, M.J.W. et al. (2005) Guidelines on information requirements for import and release of invertebrate biological control agents in European countries. *Biocontrol News and Information* **26(4)**: 115N–23N.

Bigler, F., Babendreir, D., and Kuhlmann, U. (eds.) (2006) *Environmental Impact of Invertebrates for Biological Control of Arthropods*. CABI Publishing, Wallingford.

Biron, D.G., Landry, B.S., Nénon, J.P., Coderre, D., and Boivin, G. (2000) Geographic origin of an introduced pest species, *Delia radicum* (Diptera: Anthomyiidae), determined by RAPD analysis and egg micromorphology. *Bulletin of Entomological Research* **90**: 23–32.

Bishop, L. and Riechert, S.E. (1990) Spider colonization of agroecosystems: mode and source. *Environmental Entomology* **19**: 1738–45.

Björkman, C., Bommarco, R., Eklund, K., and Höglund, S. (2004) Harvesting disrupts biological control of herbivores in a short-rotation coppice system. *Ecological Applications* **14**: 1624–33.

Bjørnson, S. and Schütte, C. (2003) Pathogens of mass-produced natural enemies and pollinators. In: van Lenteren, J.C. (2003) *Quality Control and Production of Biological Control Agents: Theory and Testing Procedures*, pp. 133–65. CABI Publishing, Wallingford.

Blissard, G.W. and Rohrmann, G.F. (1990) Baculovirus diversity and molecular biology. *Annual Review of Entomology* **35**: 127–55.

Bloomer, J.P. and Bester, M.N. (1992) Control of feral cats on sub-Antarctic Marion Island, Indian Ocean. *Biological Conservation* **60**: 211–19.

Blossey, B. (2002) Purple loosestrife. In: Van Driesche, R.G., Blossey, B., Hoddle, M., Lyon, S., and Reardon, R. (eds.), *Biological Control of Invasive Plants in the Eastern United States*, pp. 149–57. FHTET-2002-04. USDA Forest Service, Morgantown, WV.

Blossey, B. and Notzold, R. (1995) Evolution of increased competitive ability in invasive nonindigenous plants: a hypothesis. *Journal of Ecology* **83**: 887–9.

Blossey, B. and Kamil, J. (1996) What determines the increased competitive ability of invasive non-indigenous plants? In: Moran, V.C. and Hoffmann, J.H. (eds.), *Proceedings of the 9th International Symposium on Biological Control of Weeds. Stellenbosch, South Africa*, pp. 3–9, January 19–26, 1996, Stellenbosch, South Africa. University of Cape Town, Rondebosch.

Blossey, B. and Schat, M. (1997) Performance of *Galerucella calmariensis* (Coleoptera: Chrysomelidae) on different North American populations of purple loosestrife. *Environmental Entomology* **26**: 439–45.

Blossey, B., Malecki, R.A., Schroeder, D., and Skinner, L. (1996) A biological control programme using insects against purple loosestrife, *Lythrum salicaria*, in North America. In: Moran, V.C. and Hoffmann, J.H. (eds.), *Proceedings of the IX International Symposium on Biological Control of Weeds*, pp. 351–5, January 19–26, 1996, Stellenbosch, South Africa. University of Cape Town, Rondebosch.

Blossey, B., Casagrande, R., Tewksbury, L., Landis, D.A., Wiedenmann, R.N., and Ellis, D.R. (2001a) Nontarget feeding of leaf-beetles introduced to control purple loosestrife (*Lythrum salicaria* L.). *Natural Areas Journal* **21**: 368–77.

Blossey, B., Skinner, L.C., and Taylor, J. (2001b) Impact and management of purple loosestrife (*Lythrum salicaria*) in North America. *Biodiversity and Conservation* **10**: 1787–1807.

Blumberg, D. (1997) Parasitoid encapsulation as a defense mechanism in the Coccoidea (Homoptera) and its importance in biological control. *Biological Control* **8**: 225–36.

Blumberg, D. and Luck, R.F. (1990) Differences in the rates of superparasitism between two strains *of Comperiella bifasciata* (Howard) (Hymenoptrea: Encrytidae) parasitizing California red scale (Homoptera: Diapididae): an adaptation to circumvent encapsulation? *Annals of the Entomological Society of America* **83**: 591–7.

Blümel, S. and Womastek, R. (1997) Authorization requirements for organisms as plant protection products in Austria. *Bulletin OEPP* **27(1)**: 127–31.

Blumenthal, D. (2005) Interrelated causes of plant invasion. *Science* **310**: 243–4.

Boatin, B.A. and Richards, Jr, F.O. (2006) Control of human parasitic diseases. *Advances in Parasitology* **61**: 349–94.

Boettner, G.H., Elkinton, J.S., and Boettner, C. (2000) Effects of a biological control introduction on three nontarget native species of saturniid moths. *Conservation Biology* **14**: 1798–1806.

Boga, J.A., Alonso, J.M.M., Casais, R., and Parra, F. (1997) A single dose immunization with rabbit haemorrhagic disease virus major capsid protein produced in *Saccharomyces cerevisiae* induces protection. *Journal of General Virology* **78**: 2315–18.

Bohonak, A.J., Davies, N., Villablanca, F.X., and Roderick, G.K. (2001) Invasion genetics of New World medflies: testing alternative colonization scenarios. *Biological Invasions* **3**: 103–11.

Boisvert, M. and Boisvert, J. (2000) Effects of *Bacillus thuringiensis* var. *israelensis* on target and nontarget organisms: a review of laboratory and field experiments. *Biocontrol Science and Technology* **10**: 517–61.

Bokonon-Ganta, A.H. and Neuenschwander, P. (1995) Impact of the biological control agent *Gyranusoidea tebygi* Noyes (Hymenoptera: Encrytidae) on the mango mealybug, *Rastrococcus invadens* Williams (Homoptera: Pseudococcidae), in Benin. *Biocontrol Science and Technology* **5**: 95–107.

Bokonon-Ganta, A.H., de Groote, H., and Neuenschwander, P. (2002) Socio-economic impact of biological control of mango mealybug in Benin. *Agriculture, Ecosystems and Environment* **93**: 367–78.

Boldt, P.E. and Drea, J.J. (1980) Packaging and shipping beneficial insects for biological control. *FAO Plant Protection Bulletin* **28(2)**: 64–71.

Boller, E. (1972) Behavioral aspects of mass-rearing of insects. *Entomophaga* **17**: 9–25.

Boller, E.F., Janser, E., and Potter, C. (1984) Testing of the side-effects of herbicides used in viticulture on the common spider mite *Tetranychus urticae* and the predaceous mite *Typhlodromus pyri* under laboratory and semi-field conditions. *Zeitschrift für Pflanzenkrankheiten und Pflanzenschutz* **91**: 561–8.

Bomford, M. and Hart, Q. (2004) Non-indigenous vertebrates in Australia. In: Pimentel, D. (ed.), *Biological Invasions: Economic and Environmental Costs of Alien Plant, Animal, and Microbe Species*, pp. 25–44. CRC Press, Boca Raton, FL.

Bon, M.C., Hurard, C., Gaskin, J., and Risterucci, A.M. (2005) Polymorphic microsatellite markers in polyploid *Lepidium draba* L. ssp *draba* (Brassicaceae) and cross-species amplification in closely related taxa. *Molecular Ecology Notes* **5**: 68–70.

Bonning, B.C. and Hammock, B.D. (1996) Development of recombinant baculoviruses for insect control. *Annual Review of Entomology* **41**: 191–210.

Borgemeister, C., Djossou, F., Adda, C. et al. (1997) Establishment, spread, and impact of *Teretriosoma nigrescens* (Coleoptera: Histeridae), an exotic predator of the larger grain borer (Coleoptera: Bostrichidae) in southwestern Benin. *Environmental Entomology* **26**: 1405–15.

Borghuis, A., Pinto, J.D., Platner, G.R., and Stouthamer, R. (2004) Partial cytochrome oxidase II sequences distinguish the sibling species *Trichogramma minutum* Riley and *Trichogramma platneri* Nagarkatti. *Biological Control* **30**: 90–4.

Bossdorf, O., Auge, H., Lafuma, L., Rogers, W.E., Siemann, E., and Prati, D. (2005) Phenotypic and genetic differentiation between native and introduced plant populations. *Oecologia* **144**: 1–11.

Botelho, P.S.M., Parra, J.R.P., das Chagas Neto, J.F., and Oliveira, C.P.B. (1999) Association of the egg parasitoid *Trichogramma galloi* Zucchi (Hymenoptera: Trichogrammatidae) with the larval parasitoid *Cotesia flavipes* (Cam.) (Hymenoptera: Braconidae) to control the sugarcane borer *Diatraea saccharalis* (Fabr.) (Lepidoptera: Crambidae). *Anais da Sociedade Entomologica do Brasil* **28**: 491–6.

Bottrell, D.G. and Barbosa, P. (1998) Manipulating natural enemies by plant variety selection and modification: a realistic strategy? *Annual Review of Entomology* **43**: 347–67.

Boudouresque, C.F. and Verlaque, M. (2002) Biological pollution in the Mediterranean Sea: invasive versus introduced macrophytes. *Marine Pollution Bulletin* **44**: 32–8.

Bourchier, R.S. and Smith, S.M. (1998) Interactions between large-scale inundative releases of *Trichogramma minutum* (Hymenoptera: Trichogrammatidae) and naturally occurring spruce budworm (Lepidoptera: Tortricidae) parasitoids. *Environmental Entomology* **27**: 1273–9.

Box, G.E.P. and Jenkins, G.M. (1976) *Time Series Analysis: Forecasting and Control*. Holden Day, Oakland, CA.

Boyette, C.D., Quimby, Jr., P.C., Connick, Jr., W.J., Daigle, D.J., and Fulgham, F.E. (1991) Progress in the production, formulation, and application of mycoherbicides. In: TeBeest, D.O. (ed.), *Microbial Control of Weeds*, pp. 209–22. Chapman and Hall, New York.

Boykin, L.S. and Campbell, M.V. (1982) Rate of population increase of the two-spotted spider mite (Acari: Tetranychidae) on peanut leaves treated with pesticides. *Journal of Economic Entomology* **75**: 966–71.

Bradley, M.P., Hinds, L.A., and Bird, P.H. (1997) A bait delivered immunocontraceptive vaccine for the European fox (*Vulpes vulpes*) by the year 2002? *Reproduction, Fertility and Development* **9**: 111–16.

Brady, B.L. (1981) Fungi as parasites of insects and mites. *Biocontrol News and Information* **2**: 281–96.

Brar, K.S., Shenhmar, M., Bakhetia, D.R.C. et al. (1996) Bioefficacy of *Trichogramma chilonis* Ishii (Hymenoptera: Trichogrammatidae) for the control of *Chilo auricilius* Dudgeon on sugarcane in Punjab. *Plant Protection Bulletin (Faridabad)* **48(1/4)**: 9–10.

Brar, K.S., Sekhon, B.S., Singh, J., Shenhmar, M., and Joginder Singh (2002) Biocontrol-based management of cotton bollworms in the Punjab. *Journal of Biological Control* **16**: 121–4.

Braun, A.R., Guerrero, J.M., Bellotti, A.C., and Wilson, L.T. (1987a) Relative toxicity of permethrin to *Mononychellus progresivus* Doreste and *Tetranychus urticae* Koch (Acari: Tetranychidae) and their predators *Amblyseius limonicus* Garman and McGregor (Acari: Phytoseiidae) and *Oligota minuta* Cameron (Coleoptera: Staphylinidae): bioassays and field validation. *Environmental Entomology* **16**: 545–50.

Braun, A.R., Guerrero, J.M., Bellotti, A.C., and Wilson, L.T. (1987b) Evaluation of possible nonlethal side effects of permethrin used in predator exclusion experiments to evaluate *Amblyseius limonicus* (Acari: Phytoseiidae) in biological control of cassava mites (Acari: Tetranychidae). *Environmental Entomology* **16**: 1012–18.

Braun, A.R., Bellotti, A.C., Guerrero, J.M., and Wilson, L.T. (1989) Effect of predator exclusion on cassava infested with tetranychid mites (Acari: Tetranychidae). *Environmental Entomology* **18**: 711–14.

Bravenboer, L. (1960) De chemische en biologische bestrijding van de Spintmijt *Tetranychus urticae* Koch. *Publikatie Proefstn Groenten-en Fruiteelt onder Glas te Naaldwijk* **75**: 85.

Bravenboer, L. and Dosse, G. (1962) *Phytoseiulus riegeli* Dosse als predator einiger shcadmilben aus der *Tetranychus urticae* gruppe. *Entomologia Experimentalis et Applicata* **5**: 291–304.

Brede, E.G. and Beebee, T.J.C. (2005) Polymerase chain reaction primers for microsatellite loci in the semi-aquatic grasshopper, *Cornops aquaticum*. *Molecular Ecology Notes* **5**: 914–16.

Brent, K.J. (1987) Fungicide resistance in crops – its practical significance and management. In: Brent, K.J. and Atkin, R.K. (eds.), *Rational Pesiticide Use, Proceedings of the 9th Long Ashton Symposium*, pp. 137–51. Cambridge University Press, Cambridge.

Brewer, R.H. (1971) The influence of the parasite *Comperiella bifasciata* How. on populations of two species of armoured scales, *Aonidiella aurantii* (Maskell) and *A. citrina* (Coq.), in South Australia. *Australian Journal of Zoology* **19**: 53–63.

Bribosia, E.D. Bylemans, Mignon, M., and Van Impe, G. (2005) In-field production of parasitoids of *Dysaphis plantaginea* by using the rowan aphid *Dysaphis sorbi* as substitute host. *BioControl* **50**: 601–10.

Briese, D.T. (1989a) Natural enemies of carduine thistles in New South Wales. *Journal of the Australian Entomological Society* **28**: 125–6.

Briese, D.T. (1989b) Host-specificity and virus-vector potential of *Aphis chloris* (Hemiptera: Aphididae), a biological control agents for St. John's wort in Australia. *Entomophaga* **34**: 247–64.

Briese, D.T. (1996) Potential impact of the stem-boring weevil *Lixus cardui* on the growth and reproductive capacity of *Onopordum* thistles. *Biocontrol Science and Technology* **6**: 251–61.

Briese, D.T. (1999) Open field host-specificity tests: is "natural" good enough for risk assessment? In: Withers, T.M., Barton Browne, L., and Stanley, J. (eds.), *Host Specificity Testing in Australasia: Towards Improved Assays for Biological Control*, pp. 44–59. Queensland Department of Natural Resources, Coorparoo, DC, Queensland.

Briese, D.T. (2003) The centrifugal phylogenetic method used to select plants for host-specificity testing of weed biological control agents: can and should it be modernized? In: Spafford Jacob, H. and Briese, D.T. (eds.), *Improving the Selection, Testing, and Evaluation of Weed Biological Control Agents*, pp. 23–33. CRC for Australian Weed Management, Glen Osmond.

Briese, D.T. (2005) Translating host-specificity test results into the real world: the need to harmonize the yin and yang of current test procedures. *Biological Control* **35**: 208–14.

Briese, D.T. (2006a) Can an *a priori* strategy be developed for biological control? The case of *Onopordum* spp. thistles in Australia. *Australian Journal of Entomology* **45**: 306–22.

Briese, D.T. (2006b) Host specificity testing of weed biological control agents: initial attempts to modernize the centrifugal phylogenetic method. In: Hoddle, M.S., and Johnson, M. (eds.), *The Fifth California Conference on Biological Control*, pp. 32–9, July 25–27, 2006, Riverside, CA. University of California, Riverside, CA.

Briese, D.T. and Milner, R.J. (1986) Effect of the microsporidian *Pleistophora schubergi* on *Anaitis efformata* (Lepidoptera: Geometridae) and its elimination from a laboratory colony. *Journal of Invertebrate Pathology* **48**: 107–16.

Briese, D.T. and Cullen, J.M. (2001) The use and usefulness of mites in biological control of weeds. In: Halliday, R.B., Walter, D.E., Proctor, H.C., Norton, R.A., and Colloff, M.J. (eds.), *Acarology: Proceedings of the 10th International Congress*, pp. 453–63. CSIRO Publishing, Melbourne.

Briese, D.T. and Walker, A. (2002) A new perspective on the selection of test plants for evaluating the host-specificity of weed biological agents: the case of *Deuterocampta quadrijuga*, a potential insect control agent of *Heliotropium amplexicaule*. *Biological Control* **25**: 273–87.

Briese, D.T., Sheppard, A.W., Zwölfer, H., and Boldt, P.E. (1994) Structure of the phytophagous insect fauna of *Onopordum* thistles in the northern Mediterranean basin. *Biological Journal of the Linnean Society* **53**: 231–53.

Briese, D.T., Sheppard, A.W., and Reifenberg, J.M. (1995) Open-field host-specificity testing for potential biological control agents of *Onopordum* thistles. *Biological Control* **5**: 158–66.

Briese, D.T., Pettit, W.J., and Walker, A.D. (1996) Multiplying cages: a strategy for the rapid redistribution of agents with slow rates of increase. In: Moran, V.C. and Hoffmann, J.H. (eds.), *Proceedings of the 9th International Symposium on Biological Control of Weeds*, pp. 243–7, January 19–26, 1996, Stellenbosch, South Africa. University of Cape Town, Rondebosch.

Briese, D.T., Thomann, T., and Vitou, J. (2002a) Impact of the rosette crown weevil *Trichosirocalus briesei* on the growth and reproduction of *Onopordum* thistles. *Journal of Applied Ecology* **39**: 688–98.

Briese, D.T., Pettit, W.J., Swirepik, A., and Walker, A. (2002b) A strategy for the biological control of *Onopordum* spp. thistles in south-eastern Australia. *Biocontrol Science and Technology* **12**: 121–36.

Briese, D.T., Zapater, M., Andorno, A., and Perez-Camargo, G. (2002c) A two-phase open-field test to evaluate the host-specificity of candidate biological control agents for *Heliotropium amplexicaule*. *Biological Control* **25**: 259–72.

Briese, D.T., Pettit, W.J., and Walker, A. (2004) Evaluation of the biological control agent, *Lixus cardui*, on *Onopordum* thistles: experimental studies on agent demography and impact. *Biological Control* **31**: 165–71.

Briggs, C.J. and Godfray, H.C.J. (1995) The dynamics of insect-pathogen interactions in stage structured populations. *American Naturalist* **145**: 855–87.

Briggs, C.J. and Godfray, H.C.J. (1996) The dynamics of insect-pathogen interactions in seasonal environments. *Theoretical Population Biology* **50**: 149–77.

Britton, K.O., Orr, D., and Jianghua Sun (2002) Kudzu. In: Van Driesche, R.G., Blossey, B., Hoddle, M., Lyon, S., and Reardon, R. (eds.), *Biological Control of Invasive Plants in the Eastern United States*, pp. 325–30. FHTET-2002-04. USDA Forest Service, Morgantown, WV.

Brower, A.V.Z. and Desalle, R. (1994) Practical and theoretical considerations for choice of a DNA sequence region in insect molecular systematics, with a short review of published studies using nuclear gene regions. *Annals of the Entomological Society of America* **87**: 702–16.

Brower, J.H. and Press, J.W. (1988) Interactions between the egg parasite *Trichogramma pretiosum* (Hymenoptera: Trichogrammatidae) and a predator, *Xylocoris flavipes* (Hemiptera: Anthocoridae) of the almond moth *Cadra cautella* (Lepidoptera: Pyralidae). *Journal of Entomological Science* **23**: 342–9.

Brown, C.J., Blossey, B., Maerz, J.C., and Joule, S.J. (2006) Invasive plant and experimental venue affect tadpole performance. *Biological Invasions* **8**: 327–38.

Brown, D.W. and Goyer, R.A. (1982) Effects of a predator complex on lepidopterous defoliators of soybean. *Environmental Entomology* **11**: 385–9.

Brown, K.C. (1989) The design of experiments to assess the effects of pesticides on beneficial arthropods in orchards: replication versus plot size. In: Jepson, P.C. (ed.), *Pesticides and Non-Target Invertebrates*, pp. 71–80. Intercept, Wimborne.

Brown, M.D., Watson, T.M., Carter, J., Purdie, D.M., and Kay, B.H. (2004) Toxicity of VectoLex (*Bacillus sphaericus*) products to selected Australian mosquito and nontarget species. *Journal of Economic Entomology* **97**: 51–8.

Brown, M.W. and Miller, S.S. (1998) Coccinellidae (Coleoptera) in apple orchards of eastern West Virginia and the impact of invasion by *Harmonia axyridis*. *Entomological News* **109(2)**: 143–51.

Browne, L.B. and Withers, T.M. (2002) Time-dependent changes in the host-acceptance threshold of insects: implications for host specificity testing of candidate biological control agents. *Biocontrol Science and Technology* **12**: 677–93.

Broza, M., Brownbridge, M., and Sneh, B. (1991) Monitoring secondary outbreaks of the African armyworm in Kenya using pheromone traps for timing *Bacillus thuringiensis* application. *Crop Protection* **10**: 229–33.

Bruce, J.S., Twigg, L.E., and Gray, G.S. (2004) The epidemiology of rabbit haemorrhagic disease and its impact on rabbit populations in south-western Australia. *Wildlife Research* **31**: 31–49.

Bruns, H. (1960) The economic importance of birds in forests. *Bird Study* **7**: 193–208.

Bruzzese, E. (1995) Present status of biological control of European blackberry (*Rubus fruticosa* aggregate) in Australia. In: Delfosse, E.S. and Scott, R.R. (eds.) *Biological Control of Weeds: Proceedings of the VIII International Symposium on Biological Control of Weeds*, pp. 297–99, February 2–7, 1992, Canterbury, New Zealand. DSIR/CSIRO, Melbourne.

Buckingham, G.R. (1996) Biological control of alligatorweed, *Alternanthera philoxeroides*, the world's first aquatic weed success story. *Castanea* **61**: 231–43.

Buckingham, G.R. (1998) Proposed field release of the Australian sawfly *Lophyrotoma zonalis* Rohwer (Hymenoptera: Pergidae) for control of the Australian melaleuca or paperbark tree, *Melaleuca quinquenervia* (Cav.) S.T. Blake (Myrtales: Myrtaceae). Petition to the Interagency Technical Advisory Group on the Biological Control of Weeds (unpublished document).

Buckingham, G.R. (2001) Quarantine host range studies with *Lophyrotoma zonalis*, an Australian sawfly of interst for biological control of melaleuca, *Melaleuca quinquenervia*, in Florida. *BioControl* **46**: 363–86.

Buckingham, G.R. and Passoa, S. (1985) Flight muscle and egg development in waterhyacinth weevils. In: Delfosse, E.S. (ed.), *Proceedings of the VIth Internation Symposium on Biological Control of Weeds*, pp. 497–510, August 19–25, 1984, Vancouver, Canada. Agriculture Canada, Ottawa.

Buckingham, G.R., Okrah, E.A., and Thomas, M.C. (1989) Laboratory host range tests with *Hydrellia pakistanae* (Diptera: Ephydridae), an agent for biological control of *Hydrilla verticillata* (Hydrocharitaceae). *Environmental Entomology* **18**: 164–71.

Buddle, B.M., de Lisle, G.W., McColl, K., Collins, B.J., Morrissy, C., and Westbury, H.A. (1997) Response of the North Island brown kiwi, *Apteryx australis mantelli* and the lesser short-tailed bat, *Mystacina tuberculata* to a measured dose of rabbit haemorrhagic disease virus. *New Zealand Veterinary Journal* **45**: 109–13.

Buitenhuis, R. and Shipp, J.L. (2005) Efficacy of entomopathogenic nematode *Steinernema feltiae* (Rhabditida: Steinernematidae) as influenced by *Frankliniella occidentalis* (Thysanoptera: Thripidae) developmental stage and host plant stage. *Journal of Economic Entomology* **98**: 1480–5.

Bullock, D.J., North, S.G., Dulloo, M.E., and Thorsen, M. (2002) The impact of rabbit and goal eradication on the ecology of Round Island, Mauritius. In: Veitch, C.R. and Cout, M.N. (eds.), *Turning the Tide: the Eradication of Invasive Species*, pp. 60–70. IUCN SSC Invasive Species Specialist Group. IUCN, Gland; www.hear.org/articles/turningthetide/turningthetide.pdf.

Bullock, R.C., Pelosi, R.R., and Keller, E.E. (1999) Management of citrus root weevils (Coleoptera: Curculionidae) on Florida citrus with soil-applied entomopathogenic nematodes (Nematoda: Rhabditida). *Florida Entomologist* **82**: 1–7.

Burbutis, P.P., Erwin, N., and Ertle, L.R. (1981) Reintroduction and establishment of *Lydella thompsoni* and notes on other parasites of the European corn borer in Delaware. *Environmental Entomology* **10**: 779–81.

Burge, M.N. (ed.) (1988) *Fungi in Biological Control Systems*. Manchester University Press, Manchester.

Burger, J., Viscido, K., and Gochfeld, M. (1995) Eggshell thickness in marine birds in the New York bight – 1970s to 1990s. *Archives of Environmental Contamination and Toxicology* **29**: 187–91.

Burger, J.M.S., Huang, Y., Hemerik, L., van Lenteren, J.C., and Vet, L.E.M. (2006) Flexible use of patch-leaving mechanisms in a parasitoid wasp. *Journal of Insect Behavior* **19**: 155–70.

Burges, H.D. (ed.) (1981a) *Microbial Control of Pests and Plant Diseases*. Academic Press, New York.

Burges, H.D. (1981b) Safety, safety testing and quality control of microbial pesticides. In: Burges, H.D. (ed.), *Microbial Control of Pests and Plant Diseases*, pp. 738–69. Academic Press, London.

Burges, H.D., Croizier, G., and Huber, J. (1980) A review of safety tests on baculoviruses. *Entomophaga* **25**: 329–40.

Burks, R.A. and Pinto, J.D. (2002) Reproductive and electrophoretic comparisons of *Trichogramma californicum* Nagaraja and Nagarkatti with the *T. minutum* complex (Hymenoptera: Trichogrammatidae). *Proceedings of the Entomological Society of Washington* **104**: 33–40.

Burnell, A.M. and Stock, S.P. (2000) *Heterorhabditis, Steinernema* and their bacterial symbionts – lethal pathogens of insects. *Nematology* **2**: 31–42.

Burrows, D.W. and Balciunas, J.K. (1997) Biology, distribution and host range of the sawfly *Lophyrotoma zonalis* (Hym.: Pergidae), a potential biological control agent for the paperbark tree, *Melaleuca quinquenervia*. *Entomophaga* **42**: 299–313.

Buschman, L.L. and DePew, L.J. (1990) Outbreaks of Banks grass mite (Acari: Tetranychidae) in grain sorghum following insecticide application. *Journal of Economic Entomology* **83**: 1570–4.

Bustillo, A.E. and Drooz, A.T. (1977) Cooperative establishment of a Virginia (USA) strain of *Telenomus alsophilae* on *Oxydia trychiata* in Colombia. *Journal of Economic Entomology* **70**: 767–70.

Bustos-Obregon, E. (2001) Adverse effects of exposure to agropesticides on male reproduction. *Acta Pathologica, Microbiologica et Immunologica* **109**: S233–42.

Butko, P. (2003) Cytolytic toxin Cyt1A and its mechanism of membrane damage: data and hypotheses. *Applied and Environmental Microbiology* **69**: 2415–22.

Byrne, F.J. and Toscano, N.C. (2006) Detection of *Gonatocerus ashmeadi* (Hymenoptera : Mymaridae) parasitism of *Homalodisca coagulata* (Homoptera : Cicadellidae) eggs by polyacrylamide gel electrophoresis of esterases. *Biological Control* **36**: 197–202.

Byrne, M.J., Currin, S., and Hill, M.P. (2002) The influence of climate on the establishment and success of the biocontrol agent *Gratiana spadicea*, released on *Solanum sisymbriifolium* in South Africa. *Biological Control* **24**: 128–34.

Cabanillas, H.E. and Raulston, J.R. (1994) Evaluation of the spatial pattern of *Steinernema riobravis* in corn plots. *Journal of Nematology* **26**: 25–61.

Caccia, R., Baillod, M., Guignard, E., and Kreiter, S. (1985) Introduction d'une souche de *Amblyseius andersoni* Chant (Acari: Phytoseiidae) resistant a l'azinphos, dans la lutte contre les acariens phytophages en viticulture. *Revue Suisse Viticulture, Arboriculture, et Horticulture* **17**: 285–90.

Cade, W. (1975) Acoustically orienting parasitoids: fly phonotaxis to cricket song. *Science* **190**: 1312–13.

Cagne, W.C. (1988) Conservation priorities in Hawaiian natural systems. *BioScience* **38**: 264–71.

Cain, S.A. (1943) Criteria for the indication of the center of origin in plant geographical studies. *Torreya* **43**: 132–54.

Calder, A.A. and Sands, D.P.A. (1985) A new Brazilian *Cyrtobagous* Hustache (Coleoptera: Curculionidae) introduced into Australia to control salvinia. *Journal of the Australian Entomological Society* **24**: 57–64.

Callcott, A.M.A. and Collins, H.L. (1996) Invasion and range expansion of red imported fire ant (Hymnoptera: Formicidae) in North America from 1918–1995. *Florida Entomologist* **79**: 240–51.

Caltagirone, L.E. and Doutt, R.L. (1989) The history of the vedalia beetle importation to California and its impact on the development of biological control. *Annual Review of Entomology* **34**: 1–16.

Cameron, P.J., Hill, R.L., Bain, J., and Thomas, W.P. (1989) *A Review of Biological Control of Invertebrate Pests and Weeds in New Zealand 1847–1987*. Commonwealth Institute of Biological Control, Technical Communication no. 10. Commonwealth Agricultural Bureaux, Farnham Royal, Slough.

Campbell, A., Frazer, B.D., Gilbert, N., Gutierrez, A.P., and Mackauer, M. (1974) Temperature requirements of some aphids and their parasites. *Journal of Applied Ecology* **11**: 431–8.

Campbell, B.C. and Duffey, S.S. (1981) Alleviation of alphatomatine-induced toxicity to the parasitoid *Hyposoter exiguae* by phytosterols in the diet of the host, *Heliothis zea*. *Journal of Chemical Ecology* **7**: 927–46.

Campbell, C.L. and Sands, D.C. (1992) Testing the effects of the microbial agents on plants. In: Levin, M.A., Seidler, R.J., and Rogul, M. (eds.), *Microbial Ecology: Principles, Methods, and Applications*, pp. 689–705. McGraw-Hill, New York.

Campbell, R.W. (1975) *The Gypsy Moth and its Natural Enemies*. United States Department of Agriculture Information Bulletin no. 381. USDA, Washington, D.C.

Campbell, R.W. and Torgersen, T.R. (1983) Compensatory mortality in defoliator population dynamics. *Environmental Entomology* **12**: 630–2.

Canard, M. and Volkovich, T.A. (2001) Outlines of lacewing development. In: McEwen, P., New, T.R., and Whittington, A.E. (eds.), *Lacewings in the Crop Environment*, pp. 130–53. Cambridge University Press, Cambridge.

Cannon, R.J.C., Baker, R.H.A., Taylor, M.C., and Moore, J.P. (1999) A review of the status of the New Zealand flatworm in the UK. *Annals of Applied Biology* **135**: 597–614.

Cantwell, G.E. and Lehnert, T. (1979) Lack of effect of certain microbial insecticides on the honeybee. *Journal of Invertebrate Pathology* **33**: 381–2.

Capinera, J.L., Blue, S.L., and Wheeler, G.S. (1982) Survival of earthworms exposed to *Neoaplectana carpocapsae* nematodes. *Journal of Invertebrate Pathology* **39**: 419–21.

Capinera, J.L., Pelissier, D., Menout, G.S., and Epsky, N.D. (1988) Control of black cutworm, *Agrotis ipsilon* (Lepidoptera: Noctuidae), with entomogenous nematodes (Nematoda:

Steinernematidae, Heterorhabditidae). *Journal of Invertebrate Pathology* **52**: 427–35.

Cardé, R. and Lee, H.-P. (1989) Effect of experience on the responses of the parasitoid *Brachymeria intermedia* (Hymenoptera: Chalcididae) to its host, *Lymantria dispar* (Lepidoptera: Lymantriidae) and to kairomone. *Annals of the Entomological Society of America* **82**: 653–7.

Carey, J.R. (1989) The multiple decrement life table: a unifying framework for cause-of-death analysis in ecology. *Oecologia* **78**: 131–7.

Carey, J.R. (1992) The Mediterranean fruit fly in California: taking stock. *California Agriculture* **46(1)**: 12–17.

Carey, J.R. (1993) *Applied Demography for Biologists, with Special Emphasis on Insects.* Oxford University, Press, New York.

Carpenter, D. and Cappuccino, N. (2005) Herbivory, time since introduction and the invasiveness of exotic plants. *Journal of Ecology* **93**: 315–21.

Carruthers, R.I. and Soper, R.S. (1987) Fungal diseases. In: Fuxa, J.R. and Tanada, Y. (eds.), *Epizootiology of Insect Diseases*, pp. 357–416. John Wiley and Sons, New York.

Carruthers, R.I. and Hural, K. (1990) Fungi as naturally occurring entomopathogens. In: Baker, R.R. and Dunn, P.E. (eds.), *New Directions in Biological Control, Alternatives for Suppressing Agricultural Pests and Diseases*, pp. 115–38. Alan R. Liss, New York.

Carson, R. (1962) *Silent Spring*. Houghton Mifflin, New York.

Carter, M.C.A., Robertson, J.L., Haack, R.A., Lawrence, R.K., and Hayes, J.L. (1996) Genetic relatedness of North American populations of *Tomicus piniperda* (Coleoptera: Scolytidae). *Journal of Economic Entomology* **89**: 1345–53.

Carter, N. (1987) Management of cereal aphid (Hemiptera: Aphididae) populations and their natural enemies in winter wheat by alternate strip spraying with a selective insecticide. *Bulletin of Entomological Research* **77**: 677–82.

Casas, J. (1989) Foraging behaviour of a leafminer parasitoid in the field. *Ecological Entomology* **14**: 257–65.

Casas, J., Swarbrick, S., and Murdoch, W.W. (2004) Parasitoid behaviour: predicting field from laboratory. *Ecological Entomology* **29**: 657–65.

Case, T.J. (1996) Global patterns in the establishment and distribution of exotic birds. *Biological Conservation* **78**: 69–96.

Castagnoli, M. and Simoni, S. (2003) *Neoseiulus californicus* (McGregor) (Acari: Phytoseiidae): survey of biological and behavioral traits of a versatile predator. *Redia* **86**: 153–64.

Castañé, C., Quero, R., and Riudavets, J. (2006) The brine shrimp *Artemia* sp. as alternative prey for rearing the predatory bug *Macrolophus caliginosus*. *Biological Control* **38**: 405–12.

Caswell, H. (1989) *Matrix Populaton Models: Construction, Analysis, and Interpretation.* Sinauer Associates, Sunderland, MA.

Caudell, J.N., Whittier, J., Conover, M.R., Fall, M.W., and Jackson, W.B. (2002) The effects of haemogregarine-like parasites on brown tree snakes (*Boiga irregularis*) and slatey-grey snakes (*Stegonotus cucullatus*) in Queensland, Australia. *International Biodeterioration and Biodegradation* **49**: 113–19.

Caughley, G., Pech, R., and Grice, D. (1992) Effect of fertility control on a population's productivity. *Wildlife Research* **19**: 623–7.

Causton, C.E. (2004) Predicting the field prey range of an introduced predator, *Rodolia cardinalis* Mulsant, in the Galápagos. In: Van Driesche, R.G. and Reardon, R. (eds.), *Assessing Host Ranges for Parasitoids and Predators Used for Classical Biological Control: a Guide to Best Practice*, pp. 195–223. FHTET-2004-03. USDA Forest Service, Morgantown, WV.

Causton, C.E., Lincango, M.P., and Poulson, T.G.A. (2004) Feeding range studies of *Rodolia cardinalis* (Mulsant), candidate biological control agent of *Icerya purchasi* Maskell in the Galápagos Islands. *Biological Control* **29**: 315–25.

Causton, C.E., Sevilla, C.R., and Porter, S.D. (2005) Eradication of the little fire ant, *Wasmannia auropunctata* (Hymenoptera: Formicidae) from Marchena Island, Galápagos: on the edge of success? *Florida Entomologist* **88**: 159–68.

Center, T.D. (1981) Biological control and its effect on production and survival of waterhyacinth leaves. In: Delfosse, E.S. (ed.), *Proceedings of the Vth International Symposium on Biological Control of Weeds*, pp. 393–410, July 22–29, 1980, Brisbane, Australia. CSIRO Publishing, Melbourne.

Center, T.D. (1985) Leaf life tables: A viable method for assessing sublethal effects of herbivory on waterhyacinth shoots. In: Delfosse, E.S. (ed.), *Proceedings of the VIth International Symposium on Biological Control of Weeds*, pp. 511–24, August 19–25, 1984, Vancouver, Canada. Agriculture Canada, Ottawa.

Center, T.D. and Van, T.K. (1989) Alteration of water hyacinth (*Eichhornia crassipes* [Mart.] Solms) leaf dynamics and phytochemistry by insect damage and plant density. *Aquatic Botany* **35**: 181–95.

Center, T.D., Grodowitz, M.J., Cofrancesco, A.F., Jubinsky, G., Snoddy, E., and Freedman, J.E. (1997a) Establishment of *Hydrellia pakistanae* (Diptera: Ephydridae) for the biological control of the submersed aquatic plant *Hydrilla verticillata* (Hydrocharitaceae) in the southeastern United States. *Biological Control* **8**: 65–73.

Center, T.D., Frank, J.H., and Dray, F.A. (1997b) Biological control. In: Simberloff, D., Schmitz, D.C., and Brown, T.C. (eds.), *Strangers in Paradise: Impact and Management of Nonindigenous Species in Florida*, pp. 245–63. Island Press, Washington, D.C.

Center, T.D., Van, T.K., Rayachhetry, M.B. et al. (2000) Field colonization of the melaleuca snout beetle (*Oxyops vitiosa*) in south Florida. *Biological Control* **19**: 112–23.

Center, T.D., Hill, M.P., Cordo, H., and Julien, M.H. (2002) Waterhyacinth. In: Van Driesche, R.G., Blossey, B., Hoddle, M., Lyon, S., and Reardon, R. (eds.), *Biological Control of Invasive Plants in the Eastern United States*, pp. 41–64. FHTET-2002-04. USDA Forest Service, Morgantown, WV.

Center, T.D., Van, T.K., Dray, F.A. et al. (2005) Herbivory alters competitive interactions between two invasive aquatic plants. *Biological Control* **33**: 173–185.

Center, T.D., Pratt, P.D., Tipping, P.W. et al. (2006) Field colonization, population growth, and dispersal of *Boreioglycaspis melaleucae* Moore, a biological control agent of the invasive tree *Melaleuca quinquenervia* (Cav.) Blake. *Biological Control* **39**: 363–74.

Center, T.D., Pratt, P.D., Tipping, P.W. et al. (2007) Initial impacts and field validation of host range for *Boreioglycaspis melaleucae* Moore (Hemiptera: Psyllidae), a biological control agent of the invasive tree *Melaleuca quinquenervia* (Cav.) Blake (Myrtales: Myrtaceae: Leptospermoideae). *Environmental Entomology* **36**: 569–76.

Chambers, R.J., Wright, E.M., and Lind, R.J. (1993) Biological control of glasshouse sciarid larvae (*Bradysia* spp.) with the predatory mite, *Hypoaspis miles* on cyclamen and poinsettia. *Biocontrol Science and Technology* **3**: 285–93.

Chandler, L.D., Gilstrap, F.E., and Browning, H.W. (1988) Evaluation of the within-field mortality of *Liriomyza trifolii* (Diptera: Agromyzidae) on bell pepper. *Journal of Economic Entomology* **81**: 1089–96.

Chang, G.C. (1996) Comparison of single versus multiple species of generalist predators for biological control. *Environmental Entomology* **25**: 207–13.

Chapuis, J.L., Boussès, P., and Barnaud, G. (1994) Alien mammals, impact, and management in the French subantarctic islands. *Biological Conservation* **67**: 97–104.

Charles, J.-F., Nielsen-LeRoux, C., and Delécluse, A. (1996) *Bacillus sphaericus* toxins: molecular biology and mode of action. *Annual Review of Entomology* **41**: 451–72.

Charudattan, R. (2001) Biological control of weeds by means of plant pathogens: significance for integrated weed management in modern agro-ecology. *BioControl* **46**: 229–60.

Charudattan, R., Perkins, B.D., and Littell, R.C. (1978) Effects of fungi and bacteria on the decline of arthropod-damaged waterhyacinth (*Eichhornia crassipes*) in Florida. *Weed Science* **26**: 101–7.

Chasey, D. (1994) Possible origin of rabbit haemorrhagic disease in the United Kingdom. *The Veterinary Record* **135**: 496–9.

Chekchak, T., Chapuis, J.L., Pisanu, B., and Bousses, P. (2000) Introduction of the rabbit flea, *Spilopsylus cuniculi* (Dale), to a subantarctic island (Kerguelen Archipelago) and its assessment as a vector of myxomatosis. *Wildlife Research* **27**: 91–101.

Chelliah, S., Fabellar, L.T., and Heinrichs, E.A. (1980) Effects of sub-lethal doses of three insecticides on the reproductive rate of the brown planthopper, *Nilaparvata lugens*, on rice. *Environmental Entomology* **9**: 778–80.

Chen, S. and Glazer, I. (2005) A novel method for long-term storage of the entomopathogenic nematode *Steinernema feltiae* at room temperature. *Biological Control* **32**: 104–10.

Chen, Y.K., Giles, K.L., Payton, M.E., and Greenstone, M.H. (2000) Identifying key cereal aphid predators by molecular gut analysis. *Molecular Ecology* **9**: 1887–98.

Chen, Y., Giles, K.L., and Greenstone, M.H. (2002) Molecular evidence for a species complex in the genus *Aphelinus* (Hymenoptera: Aphelinidae). *Annals of the Entomological Society of America* **95**: 29–34.

Cherwonogrodzky, J.W. (1980) Microbial agents as insecticides. *Residue Reviews* **76**: 73–96.

Chesson, J. (1989) The effect of alternative prey on the functional response of *Notonecta hoffmani*. *Ecology* **70**: 1227–35.

Chiang, H.C. (1970) Effects of manure applications and mite predation on corn rootworm populations in Minnesota. *Journal of Economic Entomology* **63**: 934–6.

Childs, J.E., Glass, G.E., and Korch, Jr., G.W. (1988) The comparative epizootiology of *Capillaria hepatica* (Nematoda) in urban rodents from different habitats of Baltimore, Maryland. *Canadian Journal of Zoology* **66**: 2769–75.

Chiverton, P.A. (1986) Predatory density manipulation and its effects on populations of *Rhopalosiphum padi* (Hom.: Aphididae) in spring barley. *Annals of Applied Biology* **109**: 49–60.

Chiverton, P.A. (1987) Effects of exclusion barriers and inclusion trenches on polyphagous and aphid specific predators in spring barley. *Journal of Applied Entomology* **103**: 193–203.

Choh, Y., Shimoda, T., Ozawa, R., Dicke, M., and Takabayashi, J. (2004) Exposure of lima bean leaves to volatiles from herbivore-induced conspecific plants results in emission of carnivore attractants: active or passive process? *Journal of Chemical Ecology* **30**: 1305–17.

Christensen, C.C. (1984) Are *Euglandina* and *Gonaxis* effective agents for biological control of the giant African snail in Hawaii? *American Malacological Bulletin* **2**: 98–9.

Cilliers, C.J., Zeller, D., and Strydom, G. (1996) Short- and long-term control of water lettuce (*Pistia stratiotes*) on seasonal water bodies and on a river system in the Kruger National Park. *Hydrobiologia* **340**: 173–9.

Cilliers, C.J., Hill, M.P., Ogwang, J.A., and Ajuonu, O. (2003) Aquatic weeds in Africa and their control. In: Neuenschwander, P., Borgemeister, C., and Langewald, J. (eds.), *Biological Control in IPM Systems in Africa*, pp. 161–78. CABI Publishing, Wallingford.

Ciociola, Jr., A.I., Zucchi, R.A., and Stouthamer, R. (2001) Molecular key to seven Brazilian species of *Trichogramma* (Hymenoptera: Trichogrammatidae) using sequences of the ITS2 region and restriction analysis. *Neotropical Entomology* **30**: 259–62.

Cipolla, C., Lugo, G., Sassi, C. et al. (1997) Hypersensitivity and allergic disease in a group of workers employed in breeding insects for biological pest control. *Medicina del Lavoro* **88(3)**: 220–5.

Civeyrel, L. and Simberloff, D. (1996) A tale of two snails: is the cure worse than the disease? *Biodiversity and Conservation* **5(10)**: 1231–52.

Clark, S.E., Van Driesche, R.G., Sturdevant, N., Elkinton, J., and Buonaccorsi, J.P. (2001) Effects of site characteristics and release history on establishment of *Agapeta zoegana* (Lepidoptera: Cochylidae) and *Cyphocleonus achates*

(Coleoptera: Curculionidae), root-feeding herbivores of spotted knapweed, *Centaurea maculosa*. *Biological Control* **22**: 122–30.

Clarke, A.R., Armstrong, K.E., Carmichael, A.E. et al. (2005) Invasive phytophagous pests arising through a recent tropical evolutionary radiation: the *Bactrocera dorsalis* complex of fruit flies. *Annual Review of Entomology* **50**: 293–319.

Clarke, B., Murray, J., and Johnson, M.S. (1984) The extinction of endemic species by a program of biological control. *Pacific Science* **38(2)**: 97–104.

Clarke, R.D. and Grant, P.R. (1968) An experimental study of the role of spiders as predators in a forest litter community. Part I. *Ecology* **49**: 1152–4.

Clausen, C.P. (1962) *Entomophagous Insects*. McGraw-Hill Co., New York.

Clausen, C.P. (ed.) (1978) *Introduced Parasitoids and Predators of Arthropod Pests and Weeds: a World Review*. Agriculture Handbook No. 480. USDA, Washington, D.C.

Clement, S.L. and Sobhian, R. (1991) Host-use patterns of capitulum-feeding insects of yellow starthistle: results from a garden plot in Greece. *Environmental Entomology* **20**: 724–30.

Clement, M. and Cristofaro, S.L. (1995) Open-field tests in host-specificity determination of insects for biological control of weeds. *Biocontrol Science and Technology* **5**: 395–406.

Clifford, K.T., Gross, L., Johnson, K., Marin, N., and Shaheen, K.J. (2003) Slime-trail tracking in the predatory snail, *Euglandia rosea*. *Behavioral Neuroscience* **117**: 1086–95.

Cloutier, C. and Johnson, S.G. (1993) Predation by *Orius tristicolor* (Hemiptera: Anthocoridae) on *Phytoseiulus persimilis* (Acarina: Phytoseiidae): testing for compatibility between biocontrol agents. *Environmental Entomology* **22**: 477–82.

Cock, M.J.W. (ed.) (1985) *A Review of Biological Control of Pests in the Commonwealth Caribbean and Bermuda up to 1982*. Commonwealth Institute of Biological Control, Technical Communication no. 9. Commonwealth Agricultural Bureaux, Farnham Royal, Slough.

Cock, M.J.W. (2003) Risks of non-target impact versus stakeholder benefits in classical biological control of arthropods: selected case studies from developing countries. In: Van Driesche, R.G. (ed.), *Proceedings of the First International Symposium on Biological Control of Arthropods*, pp. 25–33, January 14–18, 2002, Honolulu, Hawaii. FHTET-03-05. USDA Forest Service, Morgantown, WV.

Coetzee, J., Byrne, M., and Hill, M. (2003) Failure of *Eccritotarsus catarinensis*, a biological control agent of waterhyacinth, to persist on pickerelweed, a non-target host in South Africa, after forced establishment. *Biological Control* **28**: 229–36.

Cohen, A.C. (1985) Simple methods for rearing the insect predator *Geocoris punctipes* (Heteroptera: Lygaeidae) on a meat diet. *Journal of Economic Entomology* **78**: 1173–5.

Cohen, A.C. and Jackson, C.G. (1989) Using rubidium to mark a predator, *Geocoris punctipes*. (Hemiptera: Lygaeidae). *Journal of Entomological Science* **24**: 57–61.

Cohen, A.N., Carlton, J.C., and Fountain, M.C. (1995) Introduction, dispersal and potential impacts of the green crab *Carcinus maenas* in San Francisco Bay, California. *Marine Biology* **122**: 225–37.

Colborn, T., Dumanoski, D., and Petersen Meyers, J. (1997) *Our Stolen Future. Are We Threatening our Fertility, Intelligence and Survival?* Plume, Penguin Group, New York.

Colfer, R.G. and Rosenheim, J.A. (2001) Predation on immature parasitoids and its impact on aphid suppression. *Oecologia* **126**: 292–304.

Colfer, R.G., Rosenheim, J.A., Godfrey, L.D., and Hsu, C.L. (2003) Interactions between the augmentatively released predaceous mite *Galendromus occidentalis* (Acari: Phytoseiidae) and naturally occurring generalist predators. *Environmental Entomology* **32**: 840–52.

Colfer, R.G., Rosenheim, J.A., Godfrey, L.D., and Hsu, C.L. (2004) Evaluation of large-scale releases of western predatory mite for spider mite control in cotton. *Biological Control* **30**: 1–10.

Coll, M. and Bottrell, D.G. (1991) Microhabitat and resource selection of the European corn borer (Lepidoptera: Pyralidae) and its natural enemies in Maryland field corn. *Environmental Entomology* **20**: 526–33.

Coll, M. and Bottrell, D.G. (1992) Mortality of European corn borer larvae by natural enemies in different corn microhabitats. *Biological Control* **2**: 95–103.

Colunga-Garcia, M. and Gage, S. (1998) Arrival, establishment, and habitat use of the multicolored Asian lady beetle (Coleoptera: Coccinellidae) in a Michigan landscape. *Environmental Entomology* **27**: 1574–80.

Colvin, B.A., Fall, M.W., Fitgerald, L.A., and Loope, L.L. (2005) *Review of Brown Treesnake Problems and Control Programs: Report of Observations and Recommendations*. Prepared at the request of the U.S. Department of Interior, Office of Insular Affairs, for the Brown Treesnake Control Committee, March 2005. www.invasivespeciesinfo.gov/animals/bts.shtml.

Conlin, T. (2002) *An Evaluation of Two Chondrostereum purpureum Carrier Formulations Used for the Control of Sitka Alder in a 41-year-old Not Satisfactorily Restocked Stand (ESSFwc1), Kootenay Forest District, Nelson Forest Region*. Extension Note – British Columbia Ministry of Forests, vol. 61. British Columbia Ministry of Forests, Victoria.

Connick, Jr., W.J., Lewis, J.A., and Quimby, Jr., P.C. (1990) Formulation of biocontrol agents for use in plant pathology. In: Baker, R.R. and Dunn, P.E. (eds), *New Directions in Biological Control*, pp. 345–72. UCLA Symposium, Alan Liss Pub., New York.

Cook, A. (1989) The basis of food choice by the carnivorous snail *Euglandina rosea*. *Monograph British Crop Protection Council* **41**: 367–72.

Coombs, E.M. and Wilson, L.M. (2004) *Phrydiuchus tau*. In: Coombs, E.M., Clark, J.K., Piper, G.L., and Cofrancesco, Jr., A.F. (eds), *Biological Control of Invasive Plants in the United States*, pp. 264–7. Oregon State University Press, Corvallis, OR.

Coombs, E.M., Clark, J.K., Piper, G.L., and Cofrancesco, Jr., A.F. (eds.) (2004) *Biological Control of Invasive Plants in the United States*. Oregon State University Press, Corvallis, OR.

Coombs, M. (2004) Overwintering survival, starvation resistance, and post-diapause reproductive performance of *Nezara viridula* (L.) (Hemiptera: Pentatomidae) and its parasitoid *Trichopoda giacomellii* Blanchard (Diptera: Tachinidae). *Biological Control* **30**: 141–8.

Coombs, M. and Sands, D.P.A. (2000) Establishment in Australia of *Trichopoda giacomellii* (Blanchard) (Diptera: Tachinidae), a biological control agents for *Nezara viridula* (L.) (Hemiptera: Pentatomidae). *Australian Journal of Entomology* **39**: 219–22.

Coop, L.B. and Berry, R.E. (1986) Reduction in variegated cutworm (Lepidoptera: Noctuidae) injury in peppermint by larval parasitoids. *Journal of Economic Entomology* **79**: 1244–8.

Coote, T. and Loève, E. (2003) From 61 to five: endemic tree snails of the Society Islands fall prey to an ill-judged biological control programme. *Oryx* **37(1)**: 91–6.

Coppel, H.C. and Mertins, J.W. (1977) *Biological Insect Pest Suppression*. Springer-Verlag, New York.

Coquillard, P., Thibaut, T., Hill, D.R.C., Gueugnot, J., Mazel, C., and Coquillard, Y. (2000) Simulation of the mollusc Ascoglossa *Elysia subornata* population dynamics: application to the potential biocontrol of *Caulerpa taxifolia* growth in the Mediterranean Sea. *Ecological Modelling* **135**: 1–16.

Corbett, A., Leigh, T.F., and Wilson, L.T. (1991) Interplanting alfalfa as a source of *Metaseiulus occidentalis* (Acari: Phytoseiidae) for managing spider mites in cotton. *Biological Control* **1**: 188–96.

Cordo, H.A. and DeLoach, C.J. (1976) Biology of the waterhyacinth mite in Argentina. *Weed Science* **24**: 245–9.

Corn, J.G., Story, J.M., and White, L.J. (2006) Impacts of the biological control agent *Cyphocleonus achates* on spotted knapweed, *Centaurea maculosa*, in experimental plots. *Biological Control* **37**: 75–81.

Cornell, H.W. and Hawkins, B.A. (1995) Survival patterns and mortality sources of herbivorous insects: some demographic trends. *American Naturalist* **145**: 563–593.

Cornell, H.V. and Hawkins, B.A. (2003) Herbivore responses to plant secondary compounds: a test of phytochemical coevolution theory. *American Naturalist* **161**: 507–22.

Corrêa-Ferreira, B.S. and Moscardi, F. (1996) Biological control of stink bugs by inoculative releases of *Trissolcus basalis*. *Entomologia Experimentalis et Applicata* **79**: 1–7.

Corrêa-Ferreira, B.S., Domit, L.A., Morales, L., and Guimarães, R.C. (2000) Integrated soybean pest management in micro basins in Brazil. *Integrated Pest Management Reviews* **5(2)**: 75–80.

Corrigan,, J.E., Mackenzie, D.L., and Simser, L. (1998) Field observations of non-target feeding by *Galerucella calmariensis* (Coleoptera: Chrysomelidae), an introduced biological control agent of purple loosestrife, *Lythrum salicaria* (Lythraceae). *Proceedings of the Entomological Society of Ontario* **129**: 99–106.

Cortesero, A.M., Stapel, J.O., and Lewis, W.J. (2000) Understanding and manipulating plant attributes to enhance biological control. *Biological Control* **17**: 35–49.

Cory, J.S. (2000) Assessing the risks of releasing genetically modified virus insecticides: progress to date. *Crop Protection* **19**: 779–85.

Cossentine, J.E. and Jensen, L.B.M. (2000) Releases of *Trichogramma platneri* (Hymenoptera: Trichogrammatidae) in apple orchards under a sterile codling moth release program. *Biological Control* **18**: 179–86.

Costello, S.L., Pratt, P.D., Rayamajhi, M.B., and Center, T.D. (2003) Arthropods associated with above-ground portions of the invasive tree *Melaleuca quinquenervia* in south Florida, USA. *Florida Entomologist* **86**: 300–22.

Coulson, J.R. and Soper, R.S. (1989) Protocols for the introduction of biological control agents in the U.S. In: Kahn, R.P. (ed.), *Plant Protection and Quarantine, vol. III. Special Topics*, pp. 1–35. CRC Press, Boca Raton, FL.

Coulson, J.R., Klaasen, W., Cook, R.J. et al. (1982) Notes on biological control of pests in China, 1979. In: Anon, *Biological Control of Pests in China*. USDA, Washington, D.C.

Coulson, J.R., Soper, R.S., and Williams, D.W. (eds.) 1991. *Biological Control Quarantine Needs and Procedures, Appendix III. Proposed ARS Guidelines for Introduction and Release of Exotic Organisms for Biological Control*. Proceedings of a Workshop. USDA, Agricultural Research Service 99. USDA, Washington, D.C.

Coupland, J. and Baker, G. (1994) Host distribution, larviposition behavior and generation time of *Sarcophaga penicillata* (Diptera: Sarcophagidae), a parasitoid of conical snails. *Bulletin of Entomological Research* **84**: 185–9.

Coupland, J. and Baker, G. (1995) The potential of several species of terrestrial Sciomyzidae as biological control agents of pest helicid snails in Australia. *Crop Protection* **14**: 573–6.

Coupland, J., Espiau, A., and Baker, G. (1994) Seasonality, longevity, host choice, and infection efficiency of *Salticella fasciata* (Diptera: Sciomyzidae), a candidate for the biological control of pest helicid snails. *Biological Control* **4**: 32–7.

Courchamp, F. and Sugihara, G. (1999) Modeling the biological control of an alien predator to protect island species from extinction. *Ecological Applications* **9**: 112–23.

Courtenay, Jr., W.R. (1997) Nonindigenous fishses. In: Simberloff, D., Schmitz, D.C., and Brown, T.C. (eds.), *Strangers in Paradise*, pp. 109–22. Island Press, Washington, D.C.

Courtenay, Jr., W.R. and Meffe, G.K. (1989) Small fishes in strange places: a review of introduced poeciliids. In: Meffe, G.K. and Snelson, Jr., F.F. (eds.), *Ecology and Evolution of Livebearing Fishes (Poecilidae)*, pp. 319–31. Prentice Hall, Englewood Cliffs, NJ.

Coutts, A.D.M., Moore, K.M., and Hewitt, C.L. (2003) Ships' sea chests: an overlooked transfer mechanism for non-indigenous marine species? *Marine Pollution Bulletin* **46**: 1510–13.

Cowan, P.E. (1996) Possum biocontrol: prospects for fertility control. *Reproduction, Fertility and Development* **8**: 655–60.

Cowan, P.E. and Tyndale-Biscoe, C.H. (1997) Australian and New Zealand mammal species considered to be pests or problems. *Reproduction, Fertility and Development* **9**: 27–36.

Cowie, R.H. (2001) Can snails ever be effective and safe biocontrol agents? *International Journal of Pest Management* **47**: 23–40.

Cowley, J.M. (1983) Lifecycle of *Apion ulicis* (Coleoptera: Apionidae), and gorse seed attack around Auckland, New Zealand. *New Zealand Journal of Zoology* **10**: 83–6.

Cowling, R. and Richardson, D. (1995) *Fynbos: South Africa's Unique Floral Kingdom*. Institute for Plant Conservation, Fernwood Press, Vlaeberg.

Craig, T.P., Price, P.W., and Itami, J.K. (1986) Resource regulation by a stem-galling sawfly on the arroyo willow. *Ecology* **67**: 419–25.

Crawford, H.S. and Jennings, D.T. (1989) Predation by birds on spruce budworm *Choristoneura fumiferana*: functional, numerical and total responses. *Ecology* **70**: 152–63.

Crawley, M.J. (1983) *Herbivory. The Dynamics of Animal-Plant Interactions*. University of California Press, Berkeley, CA.

Crawley, M.J. (1989) Insect herbivores and plant population dynamics. *Annual Review of Entomology* **34**: 531–64.

Crickmore, N., Nicholls, C., Earp, D.J., Hodgman, T.C., and Ellar, D.J. (1990) The construction of *Bacillus thuringiensis* strains expressing novel entomocidal delta endotoxin combinations. *Biochemical Journal* **270**: 133–6.

Cristoforo, M., Sale, F., Campobasso, G., Knutson, L., and Sbordoni, V. (1998) Biology and host preference of *Nephopteryx divisella* (Lepidoptera: Pyralidae): candidate agent for biological control of leafy spurge complex in North America. *Environmental Entomology* **27**: 731–5.

Croft, B.A. (1976) Establishing insecticide-resistant phytoseiid mite predators in deciduous tree fruit orchards. *Entomophaga* **21**: 383–99.

Croft, B.A. (1990) *Arthropod Biological Control Agents and Pesticides*. John Wiley and Sons, New York.

Croft, B.A. and Barnes, M.M. (1971) Comparative studies on four strains of *Typhlodromus occidentalis*. III. Evaluations of release of insecticide-resistant strains into an apple orchard ecosystem. *Journal of Economic Entomology* **64**: 845–50.

Croft, B.A. and MacRae, I.V. (1992) Persistence of *Typhlodromus pyri* and *Metaseiulus occidentalis* (Acari: Phytoseiidae) on apple after inoculative release and competition with *Zetzellia mali* (Acari: Stigmaeidae). *Environmental Entomology* **21**: 1168–77.

Cronk, Q.C.B. and Fuller, J.L. (1995) *Plant Invaders*. Chapman and Hall, London.

Crooks, J.A. (2002) Characterizing ecosystem-level consequences of biological invasions: the role of ecosystem engineers. *Oikos* **97**: 153–66.

Crooks, J.A. (2005) Lag times and exotic species: the ecology and management of biological invasions in slow-motion. *Ecoscience* **12**: 316–29.

Cross, A.E. and Noyes, J.S. (1995) Dossier on *Anagyrus kamali* Moursi, biological control agent for the pink mealybug, *Maconellicoccus hirsutus*, in Grenada. International Institute of Biological Control, Silwood Park, Acot (unpublished document).

Crowe, M.L. and Bourchier, R.S. (2006) Interspecific interactions between the gall-fly *Urophora affinis* Frfld. (Diptera: Tephritidae) and the weevil *Larinus minutus* Gyll. (Coleoptera: Curculionidae), two biological control agents released against spotted knapweed, *Centaurea stobe* L. ssp. *micranthos*. *Biocontrol Science and Technology* **16(3/4)**: 417–30.

Cruttwell-McFadyen, R.E. (1998) Biological control of weeds. *Annual Review of Entomology* **43**: 369–93.

CSIRO (1970) *The Insects of Australia*. Melbourne University Press, Carlton, Victoria.

Cudjoe, A.R., Neuenschwander, P., and Copland, M.J.W. (1993) Interference by ants in biological control of the cassava mealybug *Phenacoccus manihoti* (Hemiptera: Pseudococcidae) in Ghana. *Bulletin of Entomological Research* **83**: 15–22.

Cullen, J.M. (1995) Predicting effectiveness: fact and fantasy. In: Delfosse, E.A. and Scott, R.R. (eds.), *Proceedings of the Eight International Symposium on Biological Control of Weeds*, pp. 103–9, February 2–7, 1992, Lincoln University, Canterbury, New Zealand. DSIR/CSIRO, Melbourne.

Cullen, J.M. and Delfosse, E.S. (1985) *Echium plantagineum*: catalyst for conflict and change in Australia. In: Delfosse, E.S. (ed.), *Proceedings of the VIth International Symposium on Biological Control of Weeds*, pp. 249–92, August 19–25, 1984, Vancouver, British Columbia, Canada. Agriculture Canada, Ottawa.

Culliney, T.W., Beardsley, Jr., J.W., and Drea, J.J. (1988) Population regulation of the Eurasian pine adelgid (Homoptera: Adelgidae) in Hawaii. *Journal of Economic Entomology* **81**: 142–7.

Culver, J.J. (1919) *A Study of Compsilura concinnata, an Imported Tachinid Parasitoid of the Gipsy Moth and the Browntail Moth*. Bulletin No. 766, USDA, Washington, D.C.

Curtis, J.E., Price, T.V., and Ridland, P.M. (2003) Initial development of a spray formulation which promotes germination and growth of the fungal entomopathogen *Verticillium lecanii* (Zimmerman) Viegas (Deuteromycotina: Hyphomycetes) on capsicum leaves (*Capsicum annuum grossum* Sendt. Var California Wonder) and infection of *Myzus persicae* Sulzer (Homoptera: Aphididae). *Biocontrol Science and Technology* **13**: 35–46.

Cushing, E.C. (1957) *History of Entomology in World War II*. Smithsonian Institute Publication no. 4294. Smithsonian Institute, Washington, D.C.

Daane, K.M., Malakar-Kuenen, R., Guillén, M., Bentley, W.J., Bianchi, M., and Gonzalez, D. (2003) Abiotic and biotic pest refuges hamper biological control of mealybugs in California vineyards. In: Van Driesche, R.G. (ed.), *Proceedings of the First International Symposium on Biological Control of Arthropods*, pp. 389–98, January 14–18, 2002,

Honolulu, Hawaii. FHTET-03-05. USDA Forest Service, Morgantown, WV.

Dadswell, L.P., Abbott, W.D., and McKenzie, R.A. (1985) The occurrence, cost and control of sawfly larval (*Lophyrotoma interrupta*) poisoning of cattle in Queensland 1972–1981. *Australian Veterinary Journal* **62**: 94–7.

Dahlan, A.N. and Gordh, G. (1998) Development of *Trichogramma australicum* Girault (Hymenoptera: Trichogrammatidae) in eggs of *Helicoverpa armigera* (Hübner) (Lepidoptera: Noctuidae) and in artificial diet. *Australian Journal of Entomology* **37**: 254–64.

Dahlsten, D.L. and Mills, N.J. (1999) Biological control of forest insects. In: Bellows, T.S. and Fisher, T.W. (eds.), *Handbook of Biological Control*, pp. 761–88. Academic Press, San Diego, CA.

Dahlsten, D.L., Daane, K.M., Paine, T.D. et al. (2005) Imported parasitic wasp helps control red gum lerp psyllid. *California Agriculture* **59**: 229–34.

Dai, K.J., Zhang, L.W., Ma, Z.J. et al. (1988) Research and utilization of artificial egg for propagation of parasitoid *Trichogramma*. *Colloques de l'INRA* **43**: 311–18.

Dajoz, R. (2002) *The Coleoptera. Carabids and Tenebrionids: Ecology and Biology*. Editions Tec and Doc, Paris.

Danforth, B.N., Lin, C.-P., and Fang, J. (2005) How do insect nuclear ribosomal genes compare to protein-coding genes in phylogenetic utility and nucleotide substitution patterns? *Systematic Entomology* **30**: 549–62.

D'Antonio, C.M. and Vitousek, P.M. (1992) Biological invasions by exotic grasses, the grass/fire cycle, and global change. *Annual Review of Ecology and Systematics* **23**: 63–87.

Darlington, Jr., P.J. (1957) *Zoogeography: the Geographical Distribution of Animals*. John Wiley and Sons, New York.

Davies, A.P., Lange, C.L., and O'Neill, S.L. (2006) A rapid single-step multiplex method for discriminating between *Trichogramma* (Hymenoptera: Trichogrammatidae) species in Australia. *Journal of Economic Entomology* **99**: 2142–5.

Davies, D.H. and Siva-Jothy, M.T. (1991) Encapsulation in insects: polydnaviruses and encapsulation-promoting factors. In: Gupta, A.P. (ed.), *Immunology of Insects and other Arthropods*, pp. 119–32. CRC Press, Boca Raton, FL.

Davies, K.A. and Giblin-Davis, R.M. (2004) The biology and associations of *Fergusobia* (Nematoda) from the *Melaleuca leucadendra*-complex in eastern Australia. *Invertebrate Systematics* **18**: 291–319.

Davis, C.J. (1964) The introduction, propagation, liberation, and establishment of parasites to control *Nezara viridula* variety *smaragdula* (Fabricius) in Hawaii (Heteroptera: Pentatomidae). *Proceedings of the Hawaiian Entomological Society* **18**: 369–75.

Davis, D.E., Myers, K., and Hoy, J.B. (1976) Biological control among vertebrates. In: Huffaker, C.B. and Messenger, P.S. (eds.), *Theory and Practice of Biological Control*, pp. 501–19. Academic Press, New York.

Davis, M.A., Grime, J.P., and Thompson, K. (2000) Fluctuating resources in plant communities: a general theory of invisibility. *Journal of Ecology* **88**: 528–34.

Day, M.D. (1999) Continuation trials: their use in assessing the host range of a potential biological control agent. In: Withers, T.M., Barton Browne, L., and Stanley, J. (eds.), *Host Specificity Testing in Australasia: Towards Improved Assays for Biological Control*, pp. 11–19. Queensland Department of Natural Resources, Coorparoo, DC, Queensland.

Day, M.D. and Neser, S. (2000) Factors influencing the biological control of *Lantana camara* in Australia and South Africa. In: Spencer, N.R. (ed.), *Proceedings of the Xth International Symposium on Biological Control of Weeds*, pp. 897–908, July 4–14, 1999, Montana State University, Bozeman, Montana. Montana State University Press, Bozeman, MT.

Day, M.D., Broughton, S., and Hannan-Jones, M.A. (2003) Current distribution and status of *Lantana camara* and its biological control agents in Australia, with recommendations for further biocontrol introductions into other countries. *Biocontrol News and Information* **24(3)**: 63N–76N.

Day, W.H. (1996) Evaluation of biological control of the tarnished plant bug (Hemiptera: Miridae) in alfalfa by the introduced parasite *Peristenus digoneutis* (Hymenoptera: Braconidae). *Environmental Entomology* **25**: 512–18.

Day, W.H. (2005) Changes in abundance of native and introduced parasites (Hymenoptera: Braconidae), and of the target and non-target plant bug species (Hemiptera: Miridae), during two classical biological control programs in alfalfa. *Biological Control* **33**: 368–74.

Day, W.H., Prokrym, D.R., Ellis, D.R., and Chianese, R.J. (1994) The known distribution of the predator *Propylea quatuordecimpunctata* (Coleoptera: Coccinellidae) in the United States, and thoughts on the origin of this species and five other exotic lady beetles in eastern North America. *Entomological News* **105(4)**: 244–56.

Day, W.H., Tropp, J.M., Eaton, A.T., Romig, R.F., Van Driesche, R.G., and Chianese, R.J. (1998) Geographic distributions of *Peristenus conradi* and *P. digoneutis* (Hymenoptera: Braconidae), parasites of the alfalfa plant bug and the tarnished plant bug (Hemiptera: Miridae) in the northeastern United States. *Journal of the New York Enotmological Society* **106**: 69–75.

Day, W.H., Eaton, A.T., Romig, R.F., Tilmon, K.J., Mayer, M., and Dorsey, T. (2003) *Peristenus digoneutis* (Hymenoptera: Braconidae), a parasite of *Lygus lineolaris* (Hemiptera: Miridae) in northeastern United States alfalfa, and the need for research. *Entomological News* **114 (2)**: 105–11.

Deans, A.R. (2005) Annotated catalog of the world's ensign wasp species (Hymenoptera: Evaniidae). *Contributions of the American Entomological Institute* **34(1)**. American Entomological Institute, Ann Arbor, MI.

DeBach, P. (1958) Application of ecological information to control citrus pests in California. *Proceedings of the Xth International Congress of Entomology* **3**: 187–94.

DeBach, P. (ed.) (1964a) *Biological Control of Insect Pests and Weeds*. Reinhold Publishing Corporation, New York.

DeBach, P. (1964b) The scope of biological control. In: DeBach, P. and Schlinger, E.I. (eds.), *Biological Control of Insect Pests and Weeds*, pp. 3–20. Reinhold Publishing Corporation, New York.

DeBach, P. (1974) *Biological Control by Natural Enemies*. Cambridge University Press, London.

DeBach, P. and Huffaker, C.B. (1971) Experimental techniques for evaluation of the effectiveness of natural enemies. In: Huffaker, C.B. (ed.), *Biological Control*, pp. 113–40. Plenum Press, New York.

DeBach, P. and Rosen, D. (1991) *Biological Control by Natural Enemies*, 2nd edn, pp. 140–8. Cambridge University Press, Cambridge.

DeBach, P. and Sundby, R.A. (1963) Competitive displacement between ecological homologues. *Hilgardia* **34**: 105–66.

DeBach, P., Fleschner, C.A., and Dietrick, E.J. (1951) A biological check method for evaluating the effectiveness of entomophagous insects in the field. *Journal of Economic Entomology* **44**: 763–6.

DeBach, P., Rosen, D., and Kennett, C.E. (1971) Biological control of coccids by introduced natural enemies. In: Huffaker, C.B. (ed.), *Biological Control*, pp. 165–94. Academic Press, New York.

DeBach, P., Huffaker, C.B., and MacPhee, A.W. (1976) Evaluation of the impact of natural enemies. In: Huffaker, C.B. and Messenger, P.S. (eds.), *Theory and Practice of Biological Control*, pp. 255–85. Academic Press, New York.

de Barjac, H. (1978) Un nouveau candidat a la lutte biologique contre les moustiques: *Bacillus thuringiensis* var. *israelensis*. *Entomophaga* **23**: 309–19.

de Barjac, H. and Bonnefoi, A. (1962) Essai de classification biochemique et serologique de 24 sourches de *Bacillus* du type *B. thuringiensis*. *Entomophaga* **7**: 5–31.

de Barjac, H. and Bonnefoi, A. (1968) A classification of strains of *Bacillus thuringiensis* with a key to their differentiation. *Journal of Invertebrate Pathology* **11**: 335–47.

De Barro, P.J. and Driver, F. (1997) Use of RAPD PCR to distinguish the B biotype from other biotypes of *Bemisia tabaci* (Gennadius) (Hemiptera: Aleyrodidae). *Australian Journal of Entomology* **36**: 149–52.

de Boer, J.G. and Dicke, M. (2005) Information use by the predatory mite *Phytoseiulus persimilis* (Acari: Phytoseiidae), a specialized natural enemy of herbivorous spider mites. *Applied Entomology and Zoology* **40**: 1–12.

Debolt, J.W. (1991) Behavioral avoidance of encapsulation by *Leiophron uniformis* (Hymenoptera: Braconidae)., a parasitoid of *Lygus* spp. (Hemiptera: Miridae): relationship between host age, encapsulating ability, and host acceptance. *Annals of the Entomological Society of America* **84**: 444–6.

De Bruijn, S.L. and Bork, E.W. (2006) Biological control of Canada thistle in temperate pastures using high density rotational cattle grazing. *Biological Control* **36**: 305–15.

Dech, J.P. and Nosko, P. (2002) Population establishment, dispersal, and impact of *Galerucella pusilla* and *G. calmariensis*, introduced to control purple loosestrife in central Ontario. *Biological Control* **23**: 228–36.

De Clerck-Floate, R.A., Wikeem, B., and Bourchier, R.S. (2005) Early establishment and dispersal of the weevil, *Mogulones cruciger* (Coleoptera: Curculionidae) for biological control of houndstongue (*Cynoglossum officinale*) in British Columbia, Canada. *Biocontrol Science and Technology* **15**: 173–90.

De Clercq, P., Mohaghegh, J., and Tirry, L. (2000) Effect of host plant on the functional response of the predator *Podisus nigrispinus* (Heteroptera: Pentatomidae). *Biological Control* **18**: 65–70.

Degenhardt, H., Horstmann, F., and Mulleder, N. (2003) Bt-maize in Germany: experiences with cultivation from 1998 to 2002. *Mais* **31(2)**: 75–7.

de Groote, H., Ajuonu, O., Attignon, S., Djessou, R., and Neuenschwander, P. (2003) Economic impact of biological control of water hyacynth in southern Benin. *Ecological Economics* **45**: 105–17.

de Hoog, G.S. (1972) The genera *Beauveria, Isaria, Tritirachium*, and *Acrodontium* gen. Nov. *Studies in Mycology* **1**: 1–41.

de Jong, M.D. (2000) The BioChon story: deployment of *Chrondrostereum purpureum* to suppress stump sprouting in hardwoods. *Mycologist* **14(2)**: 58–62.

De Klerk, M.L. and Ramakers, P.M.J. (1986) Monitoring population densities of the phytoseiid predator *Amblyseius cucumeris* and its prey after large-scale introduction to control *Thrips tabaci* on sweet pepper. *Mededelingen Faculteit Landbouwwetenschappen, Rijksunniversiteit Gent* **51(3a)**: 1045–8.

de Leon, J.H. and Jones, W.A. (2005) Genetic differentiation among geographic populations of *Gonatocerus ashmeadi*, the predominant egg parasitoid of the glassy-winged sharpshooter, *Homalodisca coagulata*. *Journal of Insect Science (Tucson)* **5**: 1–9.

de Leon, J.H., Jones, W.A., and Morgan, D.J.W. (2004) Molecular distinction between populations of *Gonatocerus morrilli*, egg parasitoids of the glassy-winged sharpshooter from Texas and California: do cryptic species exist? *Journal of Insect Science (Tucson)* **4**: 7.

Delfosse, E.S. (1985) *Echium plantagineum* in Australia: effects of a major conflict of interest. In: Delfosse, E.S. (ed.), *Proceedings of the VIth International Symposium on Biological Control of Weeds*, pp. 293–9, August 19–25, 1984, University of British Columbia, Vancouver. Agriculture Canada, Ottawa.

Delfosse, E.S. (1990) Biological control and the cane toad syndrome. *Australian Natural History* **23(6)**: 480–9.

Delfosse, E.S. (2005) Risk and ethics in biological control. *Biological Control* **35**: 319–29.

DeLoach, C.J. (1976) *Neochetina bruchi*, a biological control agent of waterhyacinth: host specificity in Argentina. *Annals of the Entomological Society of America* **69**: 635–42.

DeLoach, C.J. (1978) Considerations in introducing foreign biotic agents to control native weeds of rangelands. In: Freeman, T.E. (ed). *Proceedings of the IVth International Symposium on Biological Control of Weeds*, pp. 39–50, August 30–September 2, 1976, Gainesville, Florida. Institute of Food and Agricultural Sciences, University of Florida, Gainesville, FL.

DeLoach, C.J. (1980) Prognosis for biological control of weeds of southwestern U.S. rangelands. In: Delfosse, E.S. (ed.), *Proceedings of the Vth International Symposium on Biological Control of Weeds*, pp. 179–199. CSIRO Publishing, Brisbane.

DeLoach, C.J. (1985) Conflicts of interest over beneficial and undesirable aspects of mesquite (*Prosopis* spp.) in the United States, as related to biological control. In: Delfosse, E.S. (ed.), *Proceedings of the VIth International Symposium on Biological Control of Weeds*, pp. 301–4. Canada Agriculture, Vancouver, British Columbia,.

DeLoach, C.J. and Carruthers, R.I. (2004) Saltcedar. In: Coombs, E.M., Clark, J.K., Piper, G.L., and Cofrancesco, Jr., A.F. (eds.), *Biological Control of Invasive Plants in the United States*, pp. 311–16. Oregon State University Press, Corvallis, OR.

DeLoach, C.J., Gerling, D., Fornasari, L. et al. (1996) Biological control programme against saltcedar (*Tamarix* spp.) in the United States of America: progress an problems. In: Moran, V.C. and Hoffmann, J.H. (eds.), *Proceedings of the 9th International Symposium on Biological Control of Weeds*, pp. 253–60, 19–26 January 1996, Stellenbosch, South Africa. University of Cape Town, Rondebosch.

DeLoach, C.J., Lewis, P.A., Herr, J.C., Carruthers, R.I., Tracy, J.L., and Johnson, J. (2003) Host specificity of the leaf beetle *Diorhabda elongata deserticola* (Coleoptera: Chrysomelidae) from Asia, a biological control agent for saltcedars (*Tamarix*: Tamaricaceae) in the western United States. *Biological Control* **27**: 117–47.

DeLoach, C.J., Carruthers, R.I., Dudley, T.L. et al. (2004) First results for control of saltcedar (*Tamarix* spp.) in the open field in the western United States. In: Cullen, J.M., Briese, D.T., Kriticos, D.J., Lonsdale, W.M., Morin, L., and Scott, J.K. (eds.), *Proceedings of the XI International Symposium on Biological Control of Weeds*, pp. 505–13. CSIRO Entomology, Canberra.

Dempster, J.P. (1956) The estimation of the numbers of individuals entering each stage during the development of one generation of an insect population. *Journal of Animal Ecology* **25**: 1–5.

Dempster, J.P. (1967) The control of *Pieris rapae* with DDT, I. The natural mortality of the young stages of *Pieris*. *Journal of Applied Ecology* **4**: 485–500.

Dempster, J.P. (1987) Effects of pesticides on wildlife and priorities in future studies. In: Brent, K.J. and Atkin, R.K. (eds.), *Rational Pesiticide Use, Proceedings of the 9th Long Ashton Symposium*, pp. 17–25. Cambridge University Press, Cambridge.

De Nardo, E.A.B. and Hopper, K.R. (2004) Using the literature to evaluate parasitoid host ranges: a case sudy of *Macrocentrus grandii* (Hymenoptera: Braconidae) introduced into North America to control *Ostrinia nubilalis* (Lepidoptera: Crambidae). *Biological Control* **31**: 280–95.

Den Boer, P.J. (ed.) (1971) *Disperal and Dispersal Power of Carabid Beetles*. Miscellaneous Paper no. 8. Agricultural University of Wageningen, Wageningen.

Den Boer, P.J., Theile, H.U., and Weber, F. (eds.) (1979) *On the Evolution of Behaviour in Carabid Beetles*. Miscellaneous Paper no. 18. Agricultural University of Wageningen, Wageningen.

Deng, X., Zheng, Z.Q., Zhang, N.X., and Jia, X.F. (1988) Methods of increasing the winter-survival of *Metaseiulus occidentalis* (Acari: Phytoseiidae) in northwest China. *Chinese Journal of Biological Control* **4**: 97–101.

Dennill, G.B. (1985) The effect of the gall wasp *Trichilogaster acaciaelongifoliae* (Hymenoptera: Pteromalidae) on reproductive potential and vegetative growth of the weed *Acacia longifolia*. *Agriculture, Ecosystems and Environment* **14**: 53–61.

Dennill, G.B. and Donnelly, D. (1991) Biological control of *Acacia longifolia* and related weed species (Fabaceae) in South Africa. *Agriculture, Ecosystems and Environment* **37**: 115–35.

Dennill, G.B., Donnelly, D., Stewart, K., and Impson, F.A.C. (1999) Insect agents used for the biological control of Australian *Acacia* species and *Paraserianthes lophantha* (Willd.) Nielsen (Fabaceae) in South Africa. In: Olckers, T. and Hill, M.P. (eds.), *Biological Control of Weeds in South Africa*, pp. 45–54. African Entomology Memoir No. 1. Entomological Society of Southern Africa.

Dennis, B. and Taper, M.L. (1994) Density dependence in time series observations of natural populations: estimation and testing. *Ecological Monographs* **64**: 205–24.

Dennis, P., Thomas, M.B., and Sotherton, N.W. (1994) Structural features of field boundaries which influence the overwintering densities of beneficial arthropod predators. *Journal of Applied Ecology* **31**: 361–70.

Denno, R.F., Lewis, D., and Gratton, C. (2005) Spatial variation in the relative strength of top-down and bottom-up forces: causes and consequences for phytophagous insect populations. *Annals Zoologici Fennici* **42**: 295–311.

Denoth, M., Frid, L., and Myers, J.H. (2002) Multiple agents in biological control: improving the odds? *Biological Control* **24**: 20–30.

Desender, K., Baert, L., Maelfait, J.-P., and Verdyck, P. (1999) Conservation on Volcán Alcedo (Galápagos): terrestrial invertebrates and the impact of introduced feral goats. *Biological Conservation* **87**: 303–10.

Deseo, K.V., Fantoni, P., and Lazzari, G.L. (1988) Presenza di nematodi entomopatogeni (*Steinernema* spp., *Heterorhabditis* spp.) nei terreni agricoli in Italia. *Atti Giornate Fitopatologia* **2**: 269–80.

Dexter, R.R. (1932) The food habits of the imported toad, *Bufo marinus*, in the sugar cane sections of Puerto Rico.

Proceedings of the 4th Congress of the International Society of Sugar Cane Technologists, San Juan, Bulletin **74**: 2–6.

Dhileepan, K. (2001) Effectiveness of introduced biocontrol insects on the weed *Parthenium hysterophorus* (Asteraceae) in Australia. *Bulletin of Entomological Research* **91**: 167–76.

Dhileepan, K., Teviño, M., Donnelly, G.P., and Raghu, S. (2005) Risk to non-target plants from *Charidotis auroguttata* (Chrysomelidae: Coleoptera), a potential biocontrol agent for cat's claw creeper, *Macfadyena unguis-cati* (Bignoniaceae), in Australia. *Biological Control* **32**: 450–60.

Dicke, M. (1988) Microbial allelochemicals affecting the behavior of insects, mites, nematodes, and protozoa in different trophic levels. In: Barbosa, P. and Letourneau, D.K. (eds.), *Novel Aspects of Insect-Plant Interactions*, pp. 125–63. John Wiley and Sons, New York.

Dicke, M. and Groenveld, A. (1986) Hierarchical structure in kairomone preference of the predatory mite *Amblyseius potentillae*: dietary component of indispensable diapause induction aspects of prey location behavior. *Ecological Entomology* **11**: 131–8.

Dicke, M., van Lenteren, J.C., Boskamp, G.J.F., and van Dongen-Van Leeuwen, E. (1984) Chemical stimuli in host-habitat location by *Leptopilina heterotoma* (Thompson) (Hymenoptera: Eucolidae), a parasite of *Drosophila*. *Journal of Chemical Ecology* **10**: 695–712.

Dicke, M., van Lenteren, J.C., Boskamp, G.J.F., and van Voorst, R. (1985) Intensification and prolongation of host searching in *Leptopilina heterotoma* (Thompson) (Hymenoptera: Eucolidae) through a kairomone produced by *Drosophila melanogaster*. *Journal of Chemical Ecology* **11**: 125–36.

Dicke, M., Sabelis, M.W., and Groenveld, A. (1986) Vitamin A deficiency modifies response of the predatory mite *Amblyseius potentillae* to volatile kairomone of two spotted spider mite, *Tetranychus urticae*. *Journal of Chemical Ecology* **12**: 1389–96.

Dicke, M., de Jong, M., Alers, M.P.T., Stelder, F.C.T., Wunderink, R., and Post, J. (1989) Quality control of mass-reared arthropods: nutritional effects on performance of predatory mites. *Journal of Applied Entomology* **108**: 462–75.

Diehl, J. and McEvoy, P.B. (1990) Impact of cinnabar moth (*Tyria jacobaeae*) on *Senecio trianularis*, a non-target native plant in Oregon. In: Delfosse, E.S. (ed.), *Proceedings of the VIIth International Symposium on Biological Control of Weeds*, pp. 119–26. Ministero dell'Agricoltura e delle Foreste and CSIRO, Rome.

Dindo, M.L. (1995) Arthropod predator and parasitoid rearing. *Informatore Fiotpatologico* **45(7/8)**: 18–23.

Dindo, M.L., Farneti, R., and Baronio, P. (2001) Rearing of the pupal parasitoid *Brachymeria intermedia* on veal homogenate-based artificial diets: evaluation of factors affecting effectiveness. *Entomologia Experimentalis et Applicata* **100**: 53–61.

Dively, G.P. and Rose, R. (2003) Effects of Bt transgenic and conventional insecticide control on the non-target natural enemy community in sweet corn. In: Van Driesche, R.G. (ed.), *Proceedings of the First International Symposium on Biological Control of Arthropods*, pp. 265–74, January 14–18, 2002, Honolulu, Hawaii. FHTET-03-05. USDA Forest Service, Morgantown, WV.

Dixon, A.F.G. (2000) *Insect Predator-Prey Dynamics: Ladybird Beetles and Biological Control*. Cambridge University Press, Cambridge.

Doane, C.C. (1976) Ecology of pathogens of the gypsy moth. In: Anderson, J.F. and Kaya, H.K. (eds.), *Perspectives in Forest Entomology*, pp. 285–93. Academic Press: New York.

Dobson, A.P. (1988) Restoring island ecosystems: the potential of parasites to control introduced mammals. *Conservation Biology* **2**: 31–9.

Dobson, A.P. and May, R.M. (1986) Patterns of invasion by pathogens and parasites. In: Mooney, H.A. and Drake, J.A. (eds.), *Ecology of Biological Invasions of North America and Hawaii*, pp. 58–76. Springer-Verlag, New York.

Dobson, A.P. and Hudson, P.J. (1994) Population biology of *Trichostrongylus tenuis* in the red grouse *Lagopus lagopus scoticus*. In: Scott, M.E. and Smith, G. (eds.), *Parasitic and Infectious Diseases, Epidemiology and Ecology*, pp. 310–19. Academic Press, San Diego, CA.

Dodd, A.P. (1940) *TheBiological Campaign against Prickly Pear*. Commonwealth Prickly Pear Board, Brisbane.

Dodd, S.L. and Stewart, A. (2003) RAPD-PCR and UP-PCR techniques distinguish a *Pithomyces chartarum* isolate with biocontrol capabilities against *Botrytis cinerea* on grape (*Vitis vinifera*). *New Zealand Journal of Crop and Horticultural Science* **31**: 55–64.

Dodd, S.L., Hill, R.A., and Stewart, A. (2004) A duplex-PCR bioassay to detect a *Trichoderma virens* biocontrol isolate in non-sterile soil. *Soil Biology and Biochemistry* **36**: 1955–65.

Dong, H.F. and Niu, L.P. (1988) Effect of four fungicides on the establishment and reproduction of *Phytoseiulus persimilis* (Acari: Phytoseiidae). *Chinese Journal of Biological Control* **4**: 1–5.

Donnelly, B.A. and Phillips, T.W. (2001) Functional response of *Xylocoris flavipes* (Hemiptera: Anthocoridae): effects of prey species and habitat. *Environmental Entomology* **30**: 617–24.

Doutt, R.L. (1959) The biology of parasitic Hymenoptera. *Annual Review of Entomology* **3**: 161–82.

Doutt, R.L. (1964) Biological characteristics of entomophagous adults. In: DeBach, P. and Schlinger, E. (eds.), *Biological Control of Insect Pests and Weeds*, pp. 145–67. Reinhold Publishing Corporation, New York.

Doutt, R.L. and Nakata, J. (1973) The *Rubus* leafhopper and its egg parasitoid: an endemic biotic system useful in grape-pest management. *Environmental Entomology* **3**: 381–6.

Doutt, R.L., Annecke, D.P., and Tremblay, E. (1976) Biology and host relationships of parasitoids. In: Huffaker, C.B. and Messenger, P.S. (eds.), *Theory and Practice of Biological Control*, pp. 143–68. Academic Press, New York.

Dowell, R.V., Fitzpatrick, G.E., and Reinert, J.A. (1979) Biological control of citrus blackfly in southern Florida. *Environmental Entomology* **8**: 595–7.

Downie, D.A. (2002) Locating the sources of an invasive pest, grape phylloxera, using a mitochondrial DNA gene genealogy. *Molecular Ecology* **11**: 2013–26.

Downing, R.S. and Moilliet, T.K. (1972) Replacement of *Typhlodromus occidentalis* by *T. caudiglans* and *T. pyri* (Acari: Phytoseiidae) after cessation of sprays on apple trees. *The Canadian Entomologist* **104**: 937–40.

Doyle, R.D., Grodowitz, M., Smart, R.M., and Owens, C. (2002) Impact of herbivory by *Hydrellia pakistanae* (Diptera: Ephydridae) on growth and photosynthetic potential of *Hydrilla verticillata*. *Biological Control* **24**: 221–9.

Dray, Jr., F.A. and Center, T.D. (1992) Biological control of *Pistia stratiotes* L. (waterlettuce) using *Neohydronomus affinis* Hustache (Coleoptera: Curculionidae). U.S. Army Corps of Engineers, Aquatic Plant Control Program Technical Report A-92-1. US Army Waterways Experiment Station, Vicksburg, MS.

Dreistadt, S.H. and Dahlsten, D.L. (1989) Gypsy moth eradication in Pacific coast states: history and eradication. *Bulletin of the Entomological Society of America* **35(2)**: 13–19.

Dreistadt, S.H. and Flint, M.L. (1996) Melon aphid (Homoptera: Aphididae) control by inundative convergent lady beetle (Coleoptera: Coccinellidae) release on chrysanthemum. *Environmental Entomology* **25**: 688–97.

Driessen, G. and Hemerik, L. (1992) The time and egg budget of *Leptopilina calvipes*, a parasitoid of larval *Drosophila*. *Ecological Entomology* **17**: 12–27.

Drooz, A.T., Bustillo, A.E., Fedde, G.F., and Fedde, V.H. (1997) North American egg parasite successfully controls a different host in South America. *Science* **197**: 390–1.

Duan, J.J., Ahmad, M., Joshi, K., and Messing, R.H. (1997) Evaluation of the impact of the fruit fly parasitoid *Diachasmimorpha longicaudata* (Hymenoptera: Braconidae) on a nontarget tephritid, *Eutreta xanthochaeta* (Diptera: Tephritidae). *Biological Control* **8**: 58–64.

Duan, J.J., Purcell, M.F., and Messing, R.H. (1998) Association of the opine parasitoid *Diachasmimorpha tryoni* (Hymenoptera: Braconidae) with the lantana gall fly (Diptera: Tephritidae) on Kauai. *Environmental Entomology* **27**: 419–26.

Dubelman, S., Ayden, B.R., Bader, B.M., Brown, C.R., Jiang, C., and Vlachos, D. (2005) Cry1Ab protein does not persist in soil after 3 years of sustained Bt corn use. *Environmental Entomology* **34**: 915–21.

Dudley, T.L. and DeLoach, C.J. (2004) Saltcedar (*Tamarix* spp.), endangered species, and biological control of weed control – can they mix? *Weed Technology* **18** (supplement): 1542–51.

Dudley, T.L. and Kazmer, D.J. (2005) Field assessment of the risk posed by *Diorhabda elongata*, a biocontrol agent for control of saltcedar (*Tamarix* spp.), to a nontarget plant, *Frankenia salina*. *Biological Control* **35**: 265–75.

Dulmage, H.T. (1981) Insecticidal activity of isolates of *Bacillus thuringiensis* and their potential for pest control. In: Burges, H.D. (ed.), *Microbial Control of Pests and Diseases*, pp. 193–222. Academic Press, New York.

Dulmage, H.T. and Rhodes, R.A. (1971) Production of pathogens in artificial media. In: Burges, H.D. and Hussey, N.W. (eds.), *Microbial Control of Insects and Mites*, pp. 507–40. Academic Press, New York.

Duso, C. (1992) Role of *Amblyseius aberrans* (Oud.), *Typhlodromus pyri* Scheuten and *Amblyseius andersoni* (Chant) (Acari: Phytoseiidae) in vineyards. *Journal of Applied Entomology* **114**: 455–62.

Dussourd, D.E. (1993) Foraging with finesse: caterpillar adaptations for circumventing plant defenses. In: Stamp, N.E. and Casey, T.M. (eds.), *Caterpillars: Ecological and Evolutionary Contraints on Foraging*, pp. 92–131. Chapman and Hall, New York.

Dutky, S.R., Thompson, J.V., and Cantwell, G.E. (1964) A technique for mass propagation of the DD-136 nematode. *Journal of Insect Pathology* **6**: 417–22.

Dwyer, G. (1991) The roles of density, stage, and patchiness in the transmission of an insect virus. *Ecology* **72**: 559–74.

Dwyer, G. and Elkinton, J.S. (1995) Host dispersal and the spatial spread of insect pathogens. *Ecology* **76**: 1262–75.

Dwyer, G., Elkinton, J.S., and Hajek, A.E. (1998) Spatial scale and the spread of a fungal pathogen of gypsy moth. *American Naturalist* **152**: 485–94.

Dwyer, G., Dushoff, J., and Yee, S.H. (2004) The combined effects of pathogens and predators on insect outbreaks. *Nature* **430**: 341–5.

Dysart, R.J., Maltby, H.L., and Brunson, M.H. (1973) Larval parasites of *Oulema melanipus* in Europe and their colonization in the United States. *Entomophaga* **18**: 133–67.

Eberhardt, L.L. (1970) Correlation, regression, and density-dependence. *Ecology* **51**: 306–10.

Ebert, D. and Herre, E.A. (1996) The evolution of parasitic diseases. *Parasitology Today* **12**: 96–101.

Echendu, T.N.C. and Hanna, R. (2000) Evaluation of the impact of *Typhlodromalus aripo* (De Leon), a predator of the cassava green mite (*Mononychellus tanajoa* [Bondar]) on cassava. In: Dicke, M.C., Ajayi, O., Okunade, S.O., Okoronkwo, N.O., and Abba, A.A. (eds.), *Proceedings of ESN 30th Annual Conference*, pp. 75–81, October 4–7, 1999, Kano, Nigeria. Entomological Society of Nigeria, Zaria.

Edwards, O.R. and Hoy, M.A. (1995) Monitoring laboratory and field biotypes of the walnut aphid parasite, *Trioxys pallidus*, in population cages using RAPD-PCR. *Biocontrol Science and Technology* **5**: 313–27.

Ehler, L.E. (1990) Introduction strategies in biological control of insects. In: Mackauer, M., Ehler, L.E., and Roland, J. (eds.), *Critical Issues in Biological Control*, pp. 111–34. Intercept, Andover.

Ehler, L.E. (1995) Biological control of obscure scale (Homoptera: Diaspididae) in California: an experimental approach. *Environmental Entomology* **24**: 779–95.

Ehler, L.E. and Miller, J.C. (1978) Biological control in temporary agroecosystems. *Entomophaga* **23**: 207–12.

Ehlers, R.-U., Lunau, S., Drasomil-Osterfeld, K., and Osterfeld, K.H. (1998) Liquid culture of the entomopathogenic nematode-bacterium complex *Heterorhabditis megidis/ Photorhabdus luminescens*. *Biocontrol* **43**: 77–86.

Eigenbrode, S.D., Kabalo, N.N., and Stoner, K.A. (1999) Predation, behavior, and attachment by *Chrysoperla plorabunda* larvae on *Brassica oleracea* with different surface waxblooms. *Entomologia Experimentalis et Applicata* **90**: 225–35.

Eikenbary, R.D. and Rogers, C.E. (1974) Importance of alternative hosts in establishment of introduced parasites. *Proceedings of the Tall Timbers Conference on Ecological Animal Control by Habitat Management* **5**: 119–33.

El-Arnaouty, S.A., Beyssat-Arnaouty, V., Ferran, A., and Galal, H. (2000) Introduction and release of the coccinellid *Harmonia axyridis* Pallas for controlling *Aphis craccivora* Koch on faba beans in Egypt. *Egyptian Journal of Biological Pest Control* **10**: 129–36.

Elkinton, J.S. (2000) Detecting stability and causes of change in population density. In: Boitani, L. and Fuller, T. (eds.), *Research Techniques in Ethology and Animal Ecology: Uses and Misuses*, pp. 191–200. Columbia University Press, New York.

Elkinton, J.S. (2003) Population ecology. In: Cardé, R.T. and V. Resh (eds.), *Encyclopedia of Insects*, pp. 933–44. Academic Press, San Diego, CA.

Elkinton, J.S., Buonaccorsi, J.P., Bellows, Jr., T.S., and Van Driesche, R.G. (1992) Marginal attack rate, *k*-values, and density dependence in the analysis of contemporaneous mortality factors. *Researches on Population Ecology* **34**: 29–44.

Elkinton, J.S., Healy, W.H., Buonaccorsi, J.P. et al. (1996) Interactions between gypsy moths, white-footed mice and acorns. *Ecology* **77**: 2332–42.

Elliot, N.C., Kieckhefer, R.W., and Kauffman, W.C. (1991) Estimating adult coccinellid populations in wheat fields by removal, sweepnet, and visual count sampling. *The Canadian Entomologist* **123**: 13–22.

Ellis, C.R., Kormos, B., and Guppy, J.C. (1988) Absence of parasitism in an outbreak of the cereal leaf beetle, *Oulema melanopus* (Coleoptera: Chrysomelidae), in the central tobacco growing area of Ontario. *Proceedings of the Entomological Society of Ontario* **119**: 43–6.

Ellis, J.A., Walter, A.D., Tooker, J.F. et al. (2005) Conservation biological control in urban landscapes: manipulating parasitoids of bagworm (Lepidoptera: Psychidae) with flowering forbs. *Biological Control* **34**: 99–107.

Elsey, K.D. (1974) Influence of plant host on searching speed of two predators. *Entomophaga* **19**: 3–6.

Elton, C.S. (1958) *The Ecology of Invasions by Animals and Plants*. Chapman and Hall, New York.

Elvin, M.K., Stimac, J.L., and Whitcomb, W.H. (1983) Estimating rates of arthropod predation on velvetbean caterpillar larvae in soybeans. *Florida Entomologist* **66**: 319–30.

Elzein, A., Kroschel, J., and Müller-Stöver, D. (2004) Effects of inoculum type and propagule concentration on shelf life of pest formulations containing *Fusarium oxysporum* Foxy 2, a potential mycoherbicide agent for *Striga* spp. *Biological Control* **30**: 203–11.

Elzen, G.W., Williams, H.J., and Vinson, S.B. (1986) Wind tunnel flight responses by the hymenopterous parasitoid *Campoletis sonorensis* to cotton cultivars and lines. *Entomologia Experimentalis et Applicata* **42**: 285–9.

Embree, D.G. (1960) Observations on the spread of *Cyzenis albicans* (Fall.) (Tachinidae: Diptera), an introduced parasite of the winter moth, *Operophtera brumata* (L.) (Geometridae: Lepidoptera), in Nova Scotia. *The Canadian Entomologist* **92**: 862–4.

Embree, D.G. (1965) The population dynamics of the winter moth in Nova Scotia, 1954–1962. *Memoirs of the Entomological Society of Canada* **46**: 1–57.

Embree, D.G. (1966) The role of introduced parasites in the control of the winter moth in Nova Scotia. *The Canadian Entomologist* **98**: 1159–68.

Embree, D.G. (1971) The biological control of the winter moth in eastern Canada by introduced parasites. In: Huffaker, C.B. (ed.), *Biological Control*, pp. 217–26. Plenum Press, New York.

Engeman, R.M. and Vince, D.S. (2001) Objectives and integrated approaches for the control of brown tree snakes. *Integrated Pest Management Reviews* **6**: 59–76.

Engeman, R.M., Vince, D.S., Nelson, G., and Muña, E. (2000) Brown tree snakes effectively removed from a large plot of land on Guam by perimeter trapping. *International Biodeterioration and Biodegradation* **45**: 139–42.

English-Loeb, G., Norton, A.P., and Walker, M.A. (2002) Behavioral and population consequences of acarodomatia in grapes on phytoseiid mites (Mesostigmata) and implications for plant breeding. *Entomologia Experimentalis et Applicata* **104**: 307–19.

Enkerli, J., Widmer, F., and Keller, S. (2004) Long-term persistence of *Beauveria brongniartii* strains applied as biocontrol agents against European cockchafer larvae in Switzerland. *Biological Control* **29**: 115–23.

Entwistle, P.F. (1983) Control of insects by virus diseases. *Biocontrol News and Information* **43(3)**: 203–25.

Entwistle, P.F., Cory, J.S., Bailey, M.J., and Higgs, S. (eds.) (1993) *Bacillus thuringiensis, an Environmental Pesticide: Theory and Practice*. John Wiley and Sons, New York.

Environmental Protection Agency (1983) *Title 40, Protection of Environment*, Chapter 1, *Environmental Protection Agency*, Subchapter E, *Pesticide Programs (OPP-30063A)*, Part 158, *Data Requirements for Pesticide Registration*. Environmental Protection Agency, Washington, D.C.

Erbilgin, N., Dahlsten, D.L., and Chen, P.Y. (2004) Intraguild interactions between generalist predators and an introduced parasitoid of *Glycaspis brimblecombei* (Homoptera: Psylloidae). *Biological Control* **31**: 329–37.

Erlandson, M., Braun, L., Baldwin, D., Soroka, J., Ashfaq, M., and Hegedus, D. (2003) Molecular markers for *Peristenus* spp. (Hymenoptera: Bracondiae), parasitoids associated with *Lygus* spp. (Hemiptera: Miridae). *The Canadian Entomologist* **135**: 71–83.

Ervin, R.T., Moffitt, L.J., and Meyerdirk, D.E. (1983) Comstock mealybug (Homoptera: Pseudococcidae): cost analysis of a biological control program in California. *Journal of Economic Entomology* **76**: 605–9.

Erwin, T.L., Ball, G.E., Whitehead, D.R., and Halpern, A.L. (1979) *Carabid Beetles: their Evolution, Natural History, and Classification.* Proceedings of the First International Symposium of Carabidology. Smithsonian Institution, Washington, D.C., USDA, August 21–23, 1976. Dr. W. Junk, N.V. Publishers, The Hague.

Etzel, L.K., Levinson, S.O., and Andres, L.A. (1981) Elimination of *Nosema* in *Galeruca rufa*, a potential biological control agent for field bindweed. *Environmental Entomology* **10**: 143–6.

Eubanks, M.D. and Denno, R.F. (2000) Host plants mediate omnivore-herbivore interactions and influence prey suppression. *Ecology* **81**: 936–47.

Evans, E.W. and Swallow, J.G. (1993) Numerical responses of natural enemies to artificial honeydew in Utah alfalfa. *Environmental Entomology* **22**: 1392–1401.

Evans, K.J., Morin, L., Bruzzese, E., and Roush, R.T. (2004) Overcoming limits on rust epidemics in Australian infestations of European blackberry. In: Cullen, J.M., Briese, D.T., Kriticos, D.J., Lonsdale, W.M., Morin, L., and Scott, J.K. (eds.), *Proceedings of the XIth International Symposium on Biological Control of Weeds*, pp. 514–19, April 27–May 2, 2003, Canberra, Australia. CSIRO Entomology, Canberra.

Evans, R.A. (2004) *Hemlock Ecosystems and Hemlock Woolly Adelgid at the Delaware Water Gap National Recreation Area.* www.fs.fed.us/na/morgantown/fhp/hwa/pub/HemRpt03_USFS_website.pdf.

Evans, R.A., Johnson, E., Shreiner, J. et al. (1996) Potential impacts of hemlock woolly adelgid (*Adelges tsugae*) on eastern hemlock (*Tsuga canadensis*) ecosystems. In: Salom, S.M., Tigner, T.C., and Reardon, R.C. (eds.), *Proceedings of the First Hemlock Wooly Adelgid Review*, pp. 42–57, Charlottsville, Virginia, 1995. FHTET-96-10. USDA Forest Service, Morgantown, WV.

Everest, J.W., Miller, J.H., Ball, D.M., and Patterson, M.G. (1991) *Kudzu in Alabama.* Alabama Cooperative Extension Service Circular ANR-65. Auburn University, Auburn, AL.

Facon, B., Pointier, J.P., Glaubrecht, M., Poux, C., Jarne, P., and David, P. (2003) A molecular phylogeography approach to biological invasions of the New World by parthenogenetic Thiarid snails. *Molecular Ecology* **12**: 3027–39.

Faeth, S.H. and Simberloff, D. (1981) Population regulation of a leaf-mining insect, *Cameraria* sp. nov., at increased field densities. *Ecology* **62**: 620–4.

Fayrer-Hosken, R.A., Grobler, D., Van Altena, J.J., Bertschinger, H.J., and Kirkpatrick, J.F. (2000) Immunocontraception of African elephants. *Nature* **407**: 149.

Federici, B.A. (1991) Viewing polydnaviruses as gene vectors of endoparasitic Hymenoptera. *Redia* **74**: 387–92.

Federici, B.A. (1999) A perspective on pathogens as biological control agents for insect pests. In: Bellows, Jr., T.S. and Fisher, T.W. (eds.), *Handbook of Biological Control*, pp. 517–48. Academic Press, San Diego, CA.

Federici, B.A. (2005) Insecticidal bacteria: an overwhelming success for invertebrate pathology. *Journal of Invertebrate Pathology* **89**: 30–8.

Federici, B.A. (2007) Bacteria as biological control agents for insects: economics, engineering, and environmental safety. In: Gressel, J. and Vurro, M. (eds.), *Novel Biotechnologies for Biocontrol Enhancement and Management*, pp. 25–51. Springer, Dordrecht.

Feener, Jr., D.H. and Brown, B.V. (1992) Reduced foraging of *Solenopsis geminata* (Hymenoptera: Formicidae) in the presence of parasitic *Pseudacton* spp. (Diptera: Phoridae). *Annals of the Entomological Society of America* **85**: 80–4.

Feener, Jr., D.H. and Brown, B.V. (1997) Diptera as parasitoids. *Annual Review of Entomology* **42**: 73–97.

Fellows, D.P. and Newton, W.E. (1999) Prescribed fire effects on biological control of leafy spurge. *Journal of Range Management* **52**: 489–93.

Feng, J.G., Tao, X., Zhang, A.-S., Yu, Y., Zhang, C.-W., and Cui, Y.-Y. (1999) Studies on using *Trichogramma* spp. reared on artificial host egg to control pests. *Chinese Journal of Biological Control* **15**: 97–9.

Feng, M.G., Poprawski, T.J., and Khachatourians, G.G. (1994) Production, formulation, and application of the entomopathogenic fungus *Beauveria bassiana* for insect control. *Biocontrol Science and Technology* **4**: 3–34.

Fenner, F. (1994) Myxomatosis. In: Scott, M.E. and Smith, G. (eds.), *Parasitic and Infectious Diseases, Epidemiology and Ecology*, pp. 337–46. Academic Press, San Diego, CA.

Fenner, F. and Marshall, I.D. (1957) A comparison of the virulence for European rabbits (*Oryctolagus cuniculus*) of strains of myxoma virus recovered in the field in Australia, Europe, and America. *Journal of Hygiene* **55**: 149–91.

Fenner, F. and Ratcliffe, F.N. (1965) *Myxomatosis.* Cambridge University Press Cambridge.

Fenner, F. and Ross, J. (1994) *Myxomatosis.* In: Thompson, H.V. and King, C.M. (eds.), *The European Rabbit, the History and Biology of a Successful Colonizer*, pp. 205–39. Oxford University Press, Oxford.

Ferguson, F.F. (1978) *The Role of Biological Agents in the Control of Schistosome-bearing Snails.* United States Department of Health, Eucation and Welfare, Atlanta, GA.

Ferreras, P. and MacDonald, D.W. (1999) The impact of American mink, *Mustela vison*, on water birds in the upper Thames. *Journal of Applied Ecology* **36**: 701–8.

Ferron, P. (1978) Biological control of insect pests by entomogenous fungi. *Annual Review of Entomology* **23**: 409–42.

Fiaboe, M.K., Chabi-Olaye, A., Gounou, S., Smith, H., Borgemeister, C., and Schulthess, F. (2003) *Sesamia calamistis* calling behavior and its role in host finding of egg parasitoids *Telenomus busseolae*, *Telenomus isis*, and *Lathromeris ovicida*. *Journal of Chemical Ecology* **29**: 921–9.

Field, R.P. and Darby, S.M. (1991) Host specificity of the parasitoid *Sphecophaga vesparum* (Curtis) (Hymenoptera: Ichneumonidae), a potential biological control agent of the social wasps *Vespula germanica* (Fabricius) and *V. vulgaris* (L.) (Hymenoptera: Vespidae) in Australia. *New Zealand Journal of Zoology* **18**: 193–7.

Fillman, D.A. and Sterling, W.L. (1983) Killing power of the red imported fire ant (Hymen.: Formicidae): a key predator of the boll weevil (Col.: Curculionidae). *Entomophaga* **28**: 339–44.

Fischhoff, D.A., Bowdish, K.S., Perlak, F.J. et al. (1987) Insect tolerant tomato plants. *Bio/Technology* **5**: 807–13.

Fisher, S.W. and Briggs, J.D. (1992) Testing of microbial pest control agents in nontarget insects and acari. In: Levin, M.A., Seidler, R.J., and Rogul, M. (eds.), *Microbial Ecology: Principles, Methods, and Applications*, pp. 761–77. McGraw-Hill, New York.

Fisher, T.W. and Orth, R.E. (1985) *Biological Control of Snails*. Occasional Papers, no. 1. Department of Entomology, University of California, Riverside, CA.

Fitzgerald, J. and Easterbrook, M. (2003) Phytoseiids for control of spider mites, *Tetranychus urticae*, and tarsonemid mite, *Phytonemus pallidus*, on strawberry in UK. *Bulletin OILB/SROP* **26(2)**: 107–11.

Flaherty, D.L. and Huffaker, C.B. (1970) Biological control of Pacific mites and Willamette mites in the San Joaquin Valley vineyard. Part I. Role of *Metaseiulus occidentalis*. Part II. Influence of dispersion patterns of *Metaseiulus occidentalis*. *Hilgardia* **40**: 267–330.

Flanagan, G.J., Hills, L.A., and Wilson, C.G. (2000) The successful control of spinyhead sida, *Sida acuta* [Malvaceae], by *Calligrapha pantherina* (Col: Chrysomelidae) in Australia's Northern Territory. In: Spencer, N.R. (ed.), *Proceedings of the Xth International Symposium on Biological Control of Weeds*, pp. 35–41, July 4–14, 1999. Montana State University, Bozeman, MT.

Flanders, S.E. (1930) Mass production of egg parasites of the genus *Trichogramma*. *Hilgardia* **4**: 464–501.

Flanders, S.E. (1960) The status of San Jose scale parasitization (including biological notes). *Journal of Economic Entomology* **53**: 757–9.

Fleming, J.G.W. and Summers, M.D. (1991) Polydnavirus DNA is integrated in the DNA of its parasitoid wasp host. *Proceedings of the National Academy of Sciences USA* **88**: 9770–4.

Fleschner, C.A. (1954) Biological control of avocado pests. *California Avocado Society Yearbook* **38**: 125–9.

Fleschner, C.A., Hall, J.C., and Ricker, D.W. (1955) Natural balance of mite pests in an avocado grove. *California Avocado Society Yearbook* **39**: 155–62.

Flexner, J.L., Lighthart, B., and Croft, B.A. (1986) The effects of microbial pesticides on non-target, beneficial arthropods. *Agriculture, Ecosystems, and Environment* **16**: 203–54.

Flint, M.L. and Dreistadt, S.H. (2005) Interactions among convergent lady beetle (*Hippodamia convergens*) releases, aphid populations, and rose cultivar. *Biological Control* **34**: 38–46.

Florentine, S.K., Raman, A., and Dhileepan, K. (2005) Effects of gall induction by *Epiblema strenuana* on gas exchange, nutrients, and energetics in *Parthenium hysterophorus*. *BioControl* **50**: 787–801.

Foelix, R.F. (1982) *Biology of Spiders*. Harvard University Press, Cambridge, MA.

Follett, P.A., Johnson, M.T., and Jones, V.P. (2000) Parasitoid drift in Hawaiian pentatomids. In: Follett, P.A. and Duan, J.J. (eds.), *Nontarget Effects of Biological Control*, pp. 77–109. Kluwer Academic Publishers, Boston, MA.

Forno, I.W. and Bourne, A.S. (1984) Studies in South America of arthropods on the *Salvinia auriculata* complex of floating fern and their effects on *S. molesta*. *Bulletin of Entomological Research* **74**: 609–21.

Forrester, N.L., Boag, B., Moss, S.R. et al. (2003) Long-term survival of New Zealand rabbit haemorrhagic disease virus RNA in wild rabbits, revealed by RT-PCR and phylogenetic analysis. *Journal of General Virology* **84**: 3079–86.

Forschler, B.T., All, J.N., and Gardner, W.A. (1990) *Steinernema feltiae* activity and infectivity in response to herbicide exposure in aqueous and soil environments. *Journal of Invertebrate Pathology* **55**: 375–9.

Foster, G.N. and Kelly, A. (1978) Initial density of glasshouse whitefly (*Trialeurodes vaporariorum* [Westwood]), Hemiptera) in relation to the success of suppression by *Encarsia formosa* Gahan (Hymenoptera) on glasshouse tomatoes. *Horticultural Research* **8**: 55–62.

Fournier, D., Millot, P., and Pralavorio, M. (1985) Rearing and mass production of the predatory mite *Phytoseiulus persimilis*. *Entomologia Experimentalis et Applicata* **38**: 97–100.

Fournier, D., Pralavorio, M., Coulon, J., and Berge, J.B. (1988) Fitness comparison in *Phytoseiulus persimilis* strains resistant and susceptible to methidathion. *Experimental and Applied Acarology* **5**: 55–64.

Fowler, M.C. and Robson, T.O. (1978) The effects of the food preferences and stocking rates of grass carp (*Ctenopharyngodon idella* Val.) on mixed plant communities. *Aquatic Botany* **5**: 261–76.

Fowler, S.V. (2004) Biological control of an exotic scale, *Orthezia insignis* Browne (Homoptera: Ortheziidae), saves the endemic gumwood tree, *Commidendrum robustum* (Roxb.) DC. (Asteraceae) on the island of St. Helena. *Biological Control* **29**: 367–74.

Fowler, S.V., Syrett, P., and Hill, R.L. (2000) Success and safety in the biological control of environmental weeds in New Zealand. *Austral Ecology* **25**: 553–62.

Fox, G.A. (1993) Failure-time analysis: emergence, flowering, survivorship, and other waiting times. In: Scheiner, S.M.

and Gurevitch, J. (eds.), *Design and Analysis of Ecological Experiments*, pp. 253–89. Chapman and Hall, New York.

Fox, T.B., Landis, D.A., Cardoso, F.f., and Difonzo, C.D. (2004) Predators suppress *Aphis glycines* Matsumura population growth in soybean. *Environmental Entomology* **33**: 608–18.

Frank, J.H. (1999) Bromeliad feeding weevils. *Selbyana* **20**: 40–8.

Frank, J.H. and McCoy, E.D. (1994) Commercial importation into Florida of invertebrate animals as biological control agents. *Florida Entomologist* **71**: 1–20.

Frank, J.H. and Thomas, M.C. (1994) *Metamasius callizona* (Chevrolat) (Coleoptera: Curculionidae) an immigrant pest, destroys bromeliads in Florida. *Canadian Entomologist* **126**: 673–82.

Frank, J.H. and Cave, R. (2005) *Metamasius callizona* is destroying Florida's native bromeliads. In: Hoddle, M.S. (compiler), *Proceedings of the Second International Symposium on Biological Control of Arthropods, Davos, Switzerland*, pp. 91–101. USDA-FS Forest Health Technology Team, Morgantown, WV.

Franks, S.J., Kral, A.M., and Pratt, P.D. (2006) Herbivory by introduced insects reduces growth and survival of *Melaleuca quinquenervia* seedlings. *Environmental Entomology* **35**: 366–72.

Frazer, B.D., Gilbert, N., Nealis, V., and Raworth, D.A. (1981) Control of aphid density by a complex of predators. *The Canadian Entomologist* **113**: 1035–41.

Freckleton, R.P., Watkinson, A.R., Green, R.E., and Sutherland, W.J. (2006) Census error and the detection of density dependence. *Journal of Animal Ecology* **75**: 837–51.

Freeland, J.R. (2005) *Molecular Ecology*. John Wiley and Sons, Chichester.

Freeland, W.J. (1985) The need to control cane toads. *Search* **16**: 211–15.

French, N.M., Heim, D.C., and Kennedy, G.G. (1992) Insecticide resistance patterns among Colorado potato beetle, *Leptinotarsa decemlineata* (Say) (Coleoptera: Chrysomelidae), populations in North Carolina. *Pesticide Science* **36**: 95–100.

Frick, K.E. (1974) Biological control of weeds: introduction, history, theoretical and practical applications. In: Maxwell, F.G. and Harris, F.A. (eds.), *Proceedings of the Summer Institute on Biological Control of Plants, Insects and Diseases*, pp. 204–23. University of Mississippi, Jackson, MS.

Frick, K.E. and Quimby, Jr., P.C. (1977) Biocontrol of purple nutsedge by *Bactra verutana* Zeller in a greenhouse. *Weed Science* **25**: 13–17.

Frick, K.E. and Chandler, J.M. (1978) Augmenting the moth (*Bactra verutana*) in field plots for early-season suppression of purple nutsedge (*Cyperus rotundus*). *Weed Science* **26**: 703–10.

Friedman, M.J. (1990) Commercial production and development. In: Gaugler, R. and Kaya, H. (eds.), *Entomopathogenic Nematodes in Biological Control*, pp. 153–72. CRC Press, Boca Raton, FL.

Friese, D.D., Megevand, B., and Yaninek, J.S. (1987) Culture maintenance and mass production of exotic phytoseiids. *Insect Science and its Application* **8**: 875–8.

Fritts, T.H. and Rodda, G.H. (1998) The role of introduced species in the degradation of island ecosystems: a case history of Guam. *Annual Review of Ecology and Systematics* **29**: 113–40.

Fritts, T.H., Fall, M.W., and Jackson, W.B. (2002) Economic costs of electrical system instability and power outages caused by snakes on the island of Guam. *International Biodeterioration and Biodegradation* **49**: 93–100.

Froud, K.J. and Stevens, P.S. (1997) Life table comparison between the parasitoid *Thripobius semiluteus* and its host greenhouse thrips. In: O'Callaghan, M. (ed.), *Proceedings of the Fiftieth New Zealand Plant Protection Confernce*, pp. 232–5, August 18–21, 1997, Canterbury, New Zealand. New Zealand Plant Protection Society, Auckland.

Fry, J.M. (compiler) (1989) *Natural Enemy Databank, 1987. A Catalogue of Natural Enemies of Arthropods Derived from Records in CIBC Natural Enemy Databank*. CABI Publishing, Wallingford.

Fuxa, J.R. (1990) New directions for insect control with baculoviruses. In: Baker, R.R. and Dunn, P.E. (eds.), *New Directions in Biological Control: Alternatives for Suppressing Agricultural Pests and Diseases*, pp. 97–113. Alan R. Liss, New York.

Fuxa, J.R. and Tanada, Y. (eds.) (1987) *Epizootiology of Insect Diseases*. John Wiley and Sons, New York.

Gabarra, R., Arno, J., Alomar, O., and Albajes, R. (1999) Naturally occurring populations of *Encarsia pergandiella* (Hymenoptera: Aphelinidae) in tomato greenhouses. *Bulletin OILB/SROP* **22(1)**: 85–8.

Gage, S.H. and Haynes, D.L. (1975) Emergence under natural and manipulated conditions of *Tetrastichus julis*, an introduced larval parasite of the cereal leaf beetle, with reference to regional population management. *Environmental Entomology* **4**: 425–35.

Gallardo, F., Boethel, D.J., Fuxa, J.R., and Richter, A. (1990) Susceptibility of *Heliothis zea* (Boddie) larvae to *Nomuraea rileyi* (Farlow) Samson: Effects of alpha-tomatine at the third trophic level. *Journal of Chemical Ecology* **16**: 1751–9.

Gaponyuk, I.L. and Asriev, E.A. (1986) *Metaseiulus occidentalis* in vineyards. *Zashchita Rastenii* **8**: 22–3.

Gaskin, J.F. (2003) Molecular systematics and the control of invasive plants: a case study of *Tamarix* (Tamaricaceae). *Annals of the Missouri Botanical Garden* **90(1)**: 109–18.

Gasperi, G., Bonizzoni, M., Gomulski, L.M. et al. (2002) Genetic differentiation, geneflow and the origin of infestations of the Medfly, *Ceratitis capitata*. *Genetica* **116**: 125–35.

Gassmann, A. and Kok, L.-T. (2002) Must thistle (nodding thistle). In: Van Driesche, R.G., Blossey, B., Hoddle, M., Lyon, S., and Reardon, R. (eds.), *Biological Control of Invasive Plants in the Eastern United States*, pp. 229–45. FHTET-2002-04. USDA Forest Service, Morgantown, WV.

Gaugler, R. and Boush, G.M. (1979) Nonsusceptibility of rats to the entomogenous nematode *Neoaplectana carpocapsae*. *Environmental Entomology* **8**: 658–60.

Gaugler, R. and Kaya, H.K. (eds.) (1990) *Entomopathogenic Nematodes in Biological Control*. CRC Press, Boca Raton, FL.

Gaugler, R., Campbell, J.F., and McGuire, T.R. (1989) Selection for host-finding in *Steinernema feltiae*. *Journal of Invertebrate Pathology* **54**: 363–72.

Gauld, I. and Bolton, B. (eds.) (1988) *The Hymenoptera*. Oxford University Press, Oxford.

Gautam, R.D. (2003) Classical biological control of pink hibiscus mealybug, *Maconellicoccus hirsutus* Green in the Caribbean. *Plant Protection Bulletin (Faridabad)* **55(1/2)**: 1–8.

Geden, C.J. (1999) Host location by house fly (Diptera: Muscidae) parasitoids in poultry manure at different moisture levels and host densities. *Environmental Entomology* **28**: 755–60.

Geden, C.J. (2002) Effect of habitat depth on host location by five species of parasitoids (Hymenoptera: Pteromalidae, Chalcididae) of house flies (Diptera: Muscidae) in three types of substrate. *Environmental Entomology* **31**: 411–17.

Geden, C.J. and Hogsette, J.A. (2006) Suppression of house flies (Diptera: Muscidae) in Florida poultry houses by sustained releases of *Muscidifurax raptorellus* and *Spalangia cameroni* (Hymenoptera: Pteromalidae). *Environmental Entomology* **35**: 75–82.

Geden, C.J., Rutz, D.A., Miller, R.W., and Steinkraus, D.C. (1992) Suppression of house flies (Diptera: Muscidae) on New York and Maryland dairies using releases of *Muscidifurax raptor* (Hymenoptera: Pteromalidae) in an integrated management program. *Environmental Entomology* **21**: 1419–26.

Gelernter, W.D. (1992) Application of biotechnology for improvement of *Bacillus thuringiensis* based on products and their use for control of lepidopteran pests in the Caribbean. *Florida Entomologist* **75**: 484–93.

Genini, M. and Baillod, M. (1987) Introduction de souches resitantes de *Typhlodromus pyri* (Scheuten) et *Amblyseius andersoni* Chant (Acari: Phytoseiidae) en vergers de pommiers. *Revue Suisse Viticulture, Arboriculture, et Horticulture* **19**: 115–23.

Genton, B.J., Kotanen, P.M., Cheptou, P.-O., Adolphe, C., and Shykoff, J.A. (2005) Enemy release but no evolutionary loss of defence in a plant invasion: an inter-continental reciprocal transplant experiment. *Oecologia* **146**: 404–14.

Georghiou, G. and Legunes-Tejeda, A. (1991) *The Occurrence of Resistance to Pesticides in Arthropods*. Food and Agriculture Organization of the United Nations, Rome, Italy.

Georgis, R. (1990) Formulation and application technology. In: Gaugler, R. and Kaya, H. (eds.), *Entomopathogenic Nematodes in Biological Control*, pp. 173–91. CRC Press, Boca Raton, FL.

Georgis, R., Kaya, H.K., and Gaugler, R. (1991) Effect of steinernematid and heterorhabditid nematodes (Rhabditida: Steinernematidae and Heterorhabditidae) on nontarget arthropods. *Environmental Entomology* **20**: 815–22.

Georgis, R., Koppenhöfer, A.M., Lacey, L.A. et al. (2006) Successes and failures in the use of parasitic nematodes for pest control. *Biological Control* **38**: 103–23.

Gerlach, J. (2001) Predator, prey, and pathogen interactions in introduced snail populations. *Animal Conservation* **4**: 203–9.

Gerling, D., Roitberg, B.D., and Mackauer, M. (1988) Behavioral defense mechanisms of the pea aphid. In: *Parasitoid Insects, European Workshop*, September 7–10, 1987, Lyon, France. *Colloques de l'INRA* **48**: 55–6.

Gerson, U. (1992) Perspectives of non-phytoseiid predators for the biological control of plant pests. *Experimental and Applied Acarology* **14**: 383–91.

Gerson, U. and Cohen, E. (1989) Resurgences of spider mites (Acari: Tetranychidae) induced by synthetic pyrethroids. *Experimental and Applied Acarology* **6**: 29–46.

Gerson, U. and Smiley, R.L. (1990) *Acarine Biocontrol Agents, an Illustrated Key and Manual*. Chapman and Hall, New York.

Gibb, J.A. (1962) Tinbergen's hypothesis of the role of specific searching images. *Ibis* **104**: 106–11.

Gibb, J.A. and Williams, J.M. (1994) The rabbit in New Zealand. In: Thompson, H.V. and King, C.M. (eds.), *The European Rabbit, the History and Biology of a Successful Colonizer*, pp. 158–204. Oxford University Press, Oxford.

Giblin-Davis, R.M., Scheffer, S.J., Davies, K.A. et al. (2003) Coevolution between *Fergusobia* and *Fergusonina* mutualists. *Nematology Monographs and Perspectives* **2**: 407–17.

Giboda, M., Malek, E.A., and Correa, R. (1997) Human schistosomiasis in Puerto Rico: reduced prevalence rate and absence of *Biomphalaria glabrata*. *American Journal of Tropical Medicine and Hygiene* **57**: 564–8.

Gibson, A.P., Huber, J.T., and Woolley, J.B. (eds.) (1997) *Annotated Keys to the Genera of Nearctic Chalcidoidea (Hymenoptera)*. NRC Research Press, Ottawa.

Gilbert, L.E. and Morrison, L.W. (1997) Patterns of host specificity in *Pseudacteon* parasitoid flies (Diptera: Phoridae) that attack *Solenopsis* fire ants (Hymenoptera: Formicidae). *Environmental Entomology* **26**: 1149–54.

Gilkeson, L.A. (1990) Cold storage of the predatory mite *Aphidoletes aphidimyza* (Diptera: Cecidomyiidae). *Journal of Economic Entomology* **83**: 965–70.

Gilkeson, L.A. (1991) State of the art: Biological control in greenhouses. In: McClay, A.S. (ed.), *Proceedings of the Workshop on Biological Control of Pests in Canada*, pp. 3–8, October 11–12, 1990, Calgary, Alberta. AECV91-P1. Alberta Environmental Centre, Vegreville, Alberta.

Gilkeson, L.A. (1992) Mass rearing of phytoseiid mites for testing and commercial applications. In: Anderson, T.E. and Leppla, N.C. (eds.), *Advances in Insect Rearing for Research and Pest Management*, pp. 489–506. Westview Press, Boulder, CO.

Gilkeson, L.A. and Hill, S.B. (1986) Genetic selection for and evaluation of nondiapause lines of the predatory midge *Aphidoletes aphidimyza* (Rondani) (Diptera: Cecidomyiidae). *The Canadian Entomologist* **118**: 867–79.

Gilkeson, L.A., McLean, J.P., and Dessart, P. (1993) *Aphanogmus fulmeki* Ashmead (Hymenoptera: Ceraphronidae), a parasitoid of *Aphidoletes aphidimyza* (Rondani) (Diptera: Cecidomyiidae). *The Canadian Entomologist* **125**: 161–2.

Gill, S.S., Cowles, E.A., and Pietrantonio, P.V. (1992) The mode of action of *Bacillus thuringiensis* endotoxins. *Annual Review of Entomology* **37**: 615–36.

Gillespie, A.T. (1988) Use of fungi to control pests of agricultural importance. In: Burge, M.N. (ed.) (1988) *Fungi in Biological Control Systems*, pp. 37–60. Manchester University Press, Manchester.

Gillespie, D.R. and Ramey, C.A. (1988) Life history and cold storage of *Amblyseius cucumeris* (Acarina: Phytoseiidae). *Journal of the Entomological Society of British Columbia* **85**: 71–6.

Gillespie, J.J., Munro, J.B., Heraty, J.M., Yoder, M.J., Owen, A.K., and Carmichael, A.E. (2005) A secondary structural model of the 28S rRNA expansion segments D2 and D3 for chalcidoid wasps (Hymenoptera : Chalcidoidea). *Molecular Biology and Evolution* **22**: 1593–1608.

Gillock, H.H. and Hain, F.P. (2001/2002) A historical overview of North American gypsy moth controls, chemical and biological, with emphasis on the pathogenic fungus, *Entomophaga maimaiga*. *Reviews in Toxicology* **4**: 105–28.

Gilreath, M.E. and Smith, Jr., J.W. (1988) Natural enemies of *Dactylopius confusus* (Homoptera: Dactylopiidae): exclusion and subsequent impact on *Opuntia* (Cactaceae). *Environmental Entomology* **17**: 730–8.

Gilstrap, F.E. (1988) Sorghum-corn-Johnsongrass and Banks grass mite: a model for biological control in field crops. In: Harris, M.K. and Rogers, C.E. (eds.), *The Entomology of Indigenous and Naturalized Systems in Agriculture*, pp. 141–58. Westview Press, Boulder, CO.

Gilstrap, F.E., Summy, K.R., and Friese, D.D. (1979) The temporal phenology of *Amblyseius scyphus*, a natural predator of Banks grass mite in west Texas. *The Southwestern Entomologist* **4**: 27–34.

Girolami, V., Borella, E., Di Bernardo, A., and Malagnini, V. (2000) Positive influence on phytoseiid mites of allowing the grassy interrow to flower. *Informatore Agrario* **56**: 71–3.

Glare, T.R. and O'Callaghan, M. (2000) *Bacillus thuringiensis: Biology, Ecology, and Safety*. John Wiley and Sons, Chichester.

Glaser, R.W., McCoy, E.E., and Girth, H.B. (1940) The biology and economic importance of a nematode parasitic in insects. *Journal of Parasitology* **26**: 479–95.

Glazer, I., Salame, L., and Segal, D. (1997) Genetic enhancement of nematicide resistance in entomopathogenic nematodes. *Biocontrol Science and Technology* **7**: 499–512.

Gnanvossou, D., Hanna, R., Yaninek, J.S., and Toko, M. (2005) Comparison of life history traits of three neotropical phytoseiid mites maintained on plant-based diets. *Biological Control* **35**: 32–9.

Goddard, J.H.R., Torchin, M.E., Kuris, A.M., and Lafferty, K.D. (2005) Host specificity of *Sacculina carcini*, a potential biological control agent of the introduced European green crab *Carcinus maenas* in California. *Biological Invasions* **7**: 895–912.

Godfray, H.C.J. (1994) *Parasitoids: Behavioural and Evolutionary Ecology*. Princeton University Press, Princeton, NJ.

Godfray, H.C.J. and Hassell, M.P. (1988) The population biology of insect parasitoids. *Science Progress* **72**: 531–48.

Godfray, H.C.J. and Waage, J.K. (1991) Predictive modeling in biological control: the mango mealybug (*Rastrococcus invadens*) and its parasitoids. *Journal of Applied Ecology* **28**: 434–53.

Godfray, H.C. and Pacala, S.W. (1992) Aggregation and the population dynamics of parasitoids and predators. *American Naturalist* **140**: 30–40.

Godfrey, K.E., Whitcomb, W.H., and Stimac, J.L. (1989) Arthropod predators of velvetbean caterpillar, *Anticarsia gemmatalis* Hübner (Lepidoptera: Noctuidae), eggs and larvae. *Environmental Entomology* **18**: 118–23.

Goeden, R.D. (1971) The phytophagous insect fauna of milk thistle in southern California. *Journal of Economic Entomology* **64**: 1101–4.

Goeden, R.D. (1974) Comparative survey of the phytophagous insect faunas of Italian thistle, *Carduus pycnocephalus*, in southern California and southern Europe relative to biological weed control. *Environmental Entomology* **3**: 464–74.

Goeden, R.D. (1978) Part II. Biological control of weeds. In: Clausen, C.P. (ed.) (1978) *Introduced Parasitoids and Predators of Arthropod Pests and Weeds: a World Review*, pp. 357–414. Agricultural Handbook no. 480. USDA, Washington, D.C.

Goeden, R.D. (1983) Critique and revision of Harris' scoring system for selection of insect agents in biological control of weeds. *Protection Ecology* **5**: 287–301.

Goeden, R.D. and Ricker, D.W. (1968) The phytophagous insect fauna of Russian thistle (*Salsola kali* var. *tenuifolia*) in southern California. *Annals of the Entomological Society of America* **61**: 67–72.

Goeden, R.D. and Louda, S.M. (1976) Biotic interference with insects imported for weed control. *Annual Review of Entomology* **21**: 325–42.

Goeden, R.D. and Kok, L.T. (1986) Comments on a proposed "new" approach for selecting agents for the biological control of weeds. *The Canadian Entomologist* **118**: 51–8.

Goettel, M.S., Leger, R.J.S., Bhairi, S. et al. (1990) Pathogenicity and growth of *Metarhizium anisopliae* stably transformed to benomyl resistance. *Current Genetics* **17**: 129–32.

Goh, K.S., Berberet, R.C., Young, L.J., and Conway, K.E. (1989) Mortality of *Hypera postica* (Coleoptera: Curculionidae) in Oklahoma caused by *Erynia phytonomi* (Zygomycetes: Entomophthorales). *Environmental Entomology* **18**: 964–9.

Goldberg, L.F. and Margalit, J. (1977) A bacterial spore demonstrating rapid larvicidal activity against *Anopheles sergentii*, *Uranotaenia unguiculata*, *Culex univittatus*, *Aedes aegypti*, and *Culex pipiens*. *Mosquito News* **37**: 317–24.

Goldschmidt, T. (1996) *Darwin's Dreampond: Drama in Lake Victoria*. MIT Press, Cambridge, MA.

Goldson, S.L., McNeill, M.R., Phillips, C.B., and Proffitt, J.R. (1992) Host specificity testing and suitability of the parasitoid *Microctonus hyperodae* (Hym.: Braconidae, Euphorinae) as a biological control agent of *Listronotus bonariensis* (Col.: Curculionidae) in New Zealand. *Entomophaga* **37**: 483–98.

Goldson, S.L., Proffitt, J.R., and Barlow, N.D. (1993) *Sitona discoideus* (Gyllenhal) and its parasitoid *Microctonus aethiopoides* Loan: a case study in successful biological control. In: Corey, S.A., Dall, D.J., and Milne, W.M. (eds.), *Pest Control and Sustainable Agriculture*, pp. 236–9. CSIRO Publishing, Melbourne.

Goldson, S.L., McNeill, M.R., Proffitt, J.R., and Barrat, B.I.P. (2005) Host specificity testing and suitability of a European biotype of the braconid parasitoid *Microctonus aethiopoides* as a biological control agent against *Sitona lepidus* (Coleoptera: Curculionidae) in New Zealand. *Biocontrol Science and Technology* **115**: 791–813.

Gollasch, S. (2002) The importance of ship hull fouling as a vector of species introductions into the North Sea. *Biofouling* **18**: 105–21.

González, D. and Gilstrap, F.E. (1992) Foreign exploration: assessing and prioritizing natural enemies and consequences of pre-introduction studies. In: Kauffman, W.C. and Nechols, J.E. (eds.), *Selection Criteria and Ecological Consequences of Importing Natural Enemies*, pp. 53–70. Proceedings Thomas Say Publications in Entomology. Entomological Society of America, Lanham, MD.

González-Hernandez, H., Johnson, M.W., and Reimer, N.J. (1999) Impact of *Pheidole megacephala* (F.) (Hymenoptera: Formicidae) on the biological control of *Dysmicoccus brevipes* (Cockerell) (Homoptera: Pseudococcidae). *Biological Control* **15**: 145–52.

Goodsell, J.A. and Kats, L.B. (1999) Effect of introduced mosquitofish on Pacific treefrogs and the role of alternative prey. *Conservation Biology* **13**: 921–4.

Goolsby, J.A. (2004) Potential distribution of the invasive old world climbing fern, *Lygodium microphyllum*, in North and South America. *Natural Areas Journal* **24**: 351–3.

Goolsby, J.A. (2007) Epilogue. In: Gould, J., Goolsby, J.J., and Hoelmer, K. (eds.), *Classical Biological Control of Bemisia tabaci (Biotype B): A Review of the Interagency Research and Implementation Program in the United States, 1992–2001*. USDA-ARS, New Orleans, LA (in press).

Goolsby, J.A., Ciomperlik, M.A., Legaspi, Jr., B.C., Legaspi, J.C., and Wendel, L.E. (1998) Laboratory and field evaluation of exotic parasitoids of *Bemisia tabaci* (Gennadius) (Biotype 'B') (Homoptera: Aleyrodidae) in the lower Rio Grande Valley of Texas. *Biological Control* **12**: 127–35.

Goolsby, J.A., Ciomperlik, M.A., Kirk, A.A. et al. (1999) Predictive and empirical evaluation for parasitoids of *Bemisia tabaci* (Biotype 'B'), based on morphological and molecular systematics. In: Austin, A.D. and Dowton, M. (eds.), *Hymenoptera: Evolution, Biodiversity, and Biological Control*, pp. 347–58. CSIRO Publishing, Collingwood, Victoria.

Goolsby, J.A., Makinson, J., and Purcell, M. (2000a) Seasonal phenology of the gall-making fly *Fergusonina* sp. (Diptera: Fergusoninidae) and its implications for biological control of *Melaleuca quinquenervia*. *Australian Journal of Entomology* **39**: 336–43.

Goolsby, J.A., Rose, M., Morrison, R.K., and Woolley, J.B. (2000b) Augmentative biological control of longtailed mealybug by *Chrysoperla rufilabris* (Burmeister) in the interior plantscape. *Southwestern Entomologist* **25**: 15–19.

Goolsby, J.A., Zonneveld, R., and Bourne, A. (2004a) Prerelease assessment of impact on biomass production of an invasive weed, *Lygodium microphyllum* (Lygodiaceae: Pteridophyta), by a potential biological control agent, *Floracarus perrepae* (Acariformes: Eriophyidae). *Environmental Entomology* **33**: 997–1002.

Goolsby, J.A., Makinson, J.R., Hartley, D.M., Zonneveld, R., and Wright, A.D. (2004b) Pre-release evaluation and host-range testing of *Floracarus perrepae* (Eriophyidae) genotypes for biological control of Old World climbing fern. In: Cullen, J.M., Briese, D.T., Kriticos, D.J., Lonsdale, W.M., Morin, L., and Scott, J.K. (eds.), *Proceedings of the XIth International Symposium on Biological Control of Weeds*, pp. 113–16, April 27–May 2, 2003, Canberra, Australia. CSIRO Entomology, Canberra.

Goolsby, J.A., DeBarro, P.J., Kirk, A.A. et al. (2005a) Post-release evaluation of biological control of *Bemisia tabaci* biotype "b" in the USA and the development of predictive tools to guide introductions for other countries. *Biological Control* **32**: 70–7.

Goolsby, J.A., Alexander Jesudasan, R.W., Jourdan, H., Muthuraj, B., Bourne, A.S., and Pemberton, R.W. (2005b) Continental comparisons of the interaction between climate and the herbivorous mite, *Floracarus perrepae* (Acari: Eriophyidae). *Florida Entomologist* **88**: 129–34.

Goolsby, J.A., Zonneveld, R., Makinson, J.R., and Pemberton, R.W. (2005c) Host-range and cold temperature tolerance of *Floracarus perrepae* Knihinicki and Boczek (Acari: Eriophyidae), a potential biological-control agent of *Lygodium microphyllum* (Pteridophyta: Lygodiaceae). *Australian Journal of Entomology* **44**: 321–30.

Goolsby, J.A., van Klinken, R.D., and Palmer, W.A. (2006a) Maximising the contribution of native-range studies towards the identification and prioritization of weed biocontrol agents. *Australian Journal of Entomology* **45**: 276–85.

Goolsby, J.A., DeBarro, P.J., Makinson, J.R., Pemberton, R.W., Hartley, D.M., and Frohlich, D.R. (2006b) Matching the origin of an invasive weed for selection of a herbivore haplotype for a biological control programme. *Molecular Ecology* **15**: 287–97.

Gordh, G. (1977) Biosystematics of natural enemies. In: Ridgeway, R.L. and Vinson, S.B. (eds.), *Biological Control by Augmentation of Natural Enemies*, pp. 125–48. Plenum Press, New York.

Gordon, D.R. (1998) Effects of invasive, non-indigenous plant species on ecosystem processes: lessons from Florida. *Ecological Applications* **8**: 975–89.

Gould, A.R., Kattenbelt, J.A., Lenghaus, C. et al. (1997) The complete nucleotide sequence of rabbit haemorrhagic disease virus (Czech strain C V351): use of the polymerase chain reaction to detect replication in Australian vertebrates and analysis of viral population sequence variation. *Virus Research* **47**: 7–17.

Gould, J.R., Elkinton, J.S., and Wallner, W.E. (1990) Density-dependent suppression of experimentally created gypsy moth, *Lymantria dispar* (Lepidoptera: Lymantriidae), populations by natural enemies. *Journal of Animal Ecology* **59**: 213–33.

Gould, J.R., Bellows, Jr., T.S., and Paine, T.D. (1992a) Population dynamics of *Siphoninus phillyreae* in California in the presence and absence of a parasitoid, *Encarsia partenopea*. *Ecological Entomology* **17**: 127–34.

Gould, J.R., Bellows, Jr., T.S., and Paine, T.D. (1992b) Evaluation of biological control of *Siphoninus phillyreae* (Haliday) by the parasitoid *Encarsia partenopea* (Walker), using life table analysis. *Biological Control* **2**: 257–65.

Goulet, H. and Huber, J.T. (eds.) (1993) *Hymenoptera of the World: An Identification Guide to the Families*. Canada Communications Group, Ottawa.

Gozlan, S., Millot, P., Rousset, A., and Fournier, D. (1997) Test of the RAPD-PCR method to evaluate the efficacy of augmentative biological control with *Orius* (Het., Anthocoridae). *Entomophaga* **42**: 593–604.

Grafton-Cardwell, E.E. and Hoy, M.A. (1985) Intraspecific variability in response to pesticides in the common green lacewing, *Chrysoperla carnea* (Stephens) (Neuroptera: Chrysopidae). *Hilgardia* **53(6)**: 1–32.

Grafton-Cardwell, E.E. and Ouyang, Y. (1995) Manipulation of the predacious mite, *Euseius tularensis* (Acari: Phytoseiidae), with pruning for citrus thrips control. In: Parker, B.L., Skinner, M., and Lewis, T. (eds.), *Thrips Biology and Management: Proceedings of the 1993 International Conference on Thysanoptera*, pp. 251–4. Plenum Publishing Co., London.

Grafton-Cardwell, E.E. and Ouyang, Y. (1996) Influence of citrus leaf nutrition on survivorship, sex ratio, and reproduction of *Euseius tularensis* (Acari: Phytoseiidae). *Environmental Entomology* **25**: 1020–5.

Grafton-Cardwell, E.E. and Gu, P. (2003) Conserving vedalia beetle, *Rodolia cardinalis* (Mulsant) (Coleoptera: Coccinellidae), in citrus: a continuing challenge as new insecticides gain registration. *Journal of Economic Entomology* **96**: 1388–98.

Grafton-Cardwell, E.E., Ouyang, Y., and Bugg, R.L. (1999) Leguminous cover crops to enhance population development of *Euseius tularensis* (Acari: Phytoseiidae) in citrus. *Biological Control* **16**: 73–80.

Graham, Jr., F. (1970) *Since Silent Spring*. Houghton Mifflin Co., Boston, MA.

Granados, R.R. and Federici, B.A. (eds.) (1986) *The Biology of Baculoviruses: Volume I. Biological Properties and Molecular Biology and Volume II. Practical Application for Insect Control*. CRC Press, Boca Raton, FL.

Granados, R.R., Dwyer, K.G., and Derksen, A.C.G. (1987) Production of viral agents in invertebrate cell cultures. In: Maramorosch, K. (ed.), *Biotechnology in Invertebrate Pathology and Cell Culture*, pp. 167–81. Academic Press, San Diego, CA.

Grandgirard, J., Petit, J.N., Hoddle, M.S., Roderick, G.K., and Davies, N. (2006) Successful biocontrol of (Hemiptera: Cicadellidae) in French Polynesia. In: Hoddle, M.S. and Johnson, M. (eds.), *The Fifth California Conference on Biological Control*, pp. 145–7, July 25–7, 2006, Riverside, California. University of California, Riverside, CA.

Grapputo, A., Boman, S., Lindstrom, L., Lyytinen, A., and Mappes, J. (2005) The voyage of an invasive species across continents: genetic diversity of North American and European Colorado potato beetle populations. *Molecular Ecology* **14**: 4207–19.

Gravena, S. and Sterling, W.L. (1983) Natural predation on the cotton leafworm (Lepidoptera: Noctuidae). *Journal of Economic Entomology* **76**: 779–84.

Gray, A. (2004) Genetically modified crops: broader environmental issues. *Journal of Commercial Biotechnology* **10(3)**: 234–40.

Greany, P.D. and Oatman, E.R. (1972) Demonstration of host discrimination in the parasite *Orgilus lepidus* (Hymenoptera: Bracondiae). *Annals of the Entomological Society of America* **65**: 375–6.

Greany, P.D., Tumlinson, J.H., Chambers, D.L., and Bousch, G.M. (1977) Chemically-mediated host finding by *Biosteres (Opius) longicaudatus*, a parasitoid of tephritid fruit fly larvae. *Journal of Chemical Ecology* **3**: 189–95.

Greathead, D.J. (1968) Biological control of lantana – a review and discussion of recent developments in East Africa. *Pest Articles and News Summaries, Section C* **14**: 167–75.

Greathead, D.J. (1986a) Parasitoids in classical biological control. In: Waage, J. and Greathead, D. (eds.), *Insect Parasitoids*, pp. 289–318. Academic Press, London.

Greathead. D.G. (1986b) Opportunities for biological control of insect pests in tropical Africa. *Revue d'Zoologie Africaine* **100**: 85–96.

Greathead, D.J. and Greathead, A.H. (1992) Biological control of insect pests by parasitoids and predators: the BIOCAT database. *Biocontrol News and Information* **13(4)**: 61N–8N.

Green, D.S. (1984) A proposed origin of the coffee leaf-miner, *Leucoptera coffeella* (Guérin-Méneville) (Lepidoptera: Lyonetiidae). *Bulletin of the Entomological Society of America* Spring issue: 30–1.

Greenberg, S.M., Morrison, R.K., Nordlund, D.A., and King, E.G. (1998) A review of the scientific literature and methods for production of factitious hosts for use in mass rearing of *Trichogramma* spp. (Hymenoptera: Trichogrammatidae) in the former Soviet Union, the United States, Western Europe, and China. *Journal of Entomological Science* **33**: 15–32.

Greenstone, M.H. (1996) Serological analysis of arthropod predation: past, present, and future. In: Symondson, W.O.C.

and Liddell, J.E. (eds.), *The Ecology of Agricultural Pests*, pp. 265–300. Chapman and Hall, London.

Greenstone, M.H. (1999) Spider predation: How and why we study it. *Journal of Arachnology* **27**: 333–42.

Greenstone, M.H. (2006) Molecular methods for assessing insect parasitism. *Bulletin of Entomological Research* **96**: 1–13.

Greenstone, M.H., Rowley, D.L., Heimbach, U., Lundgren, J.G., Pfannenstiel, R.S., and Rehner, S.A. (2005) Barcoding generalist predators by polymerase chain reaction: carabids and spiders. *Molecular Ecology* **14**: 3247–66.

Gregg, D.A., House, C., Myer, R., and Berninger, M. (1991) Viral haemorrhagic disease of rabbits in Mexico: epidemiology and viral characterization. *Revue Scientifique et Technique Office International des Epizooties* **10**: 435–51.

Grenier, S. (1988) Applied biological control with tachinid flies (Diptera: Tachinidae): A review. *Anzeiger für Schädlingskunde, Pflanzenschutz, Umweltschutz* **61**: 49–56.

Grevstad, F.S. (1999a) Factors influencing the chance of population establishment: Implications for release strategies in biocontrol. *Ecological Applications* **9**: 1439–47.

Grevstad, F.S. (1999b) Experimental invasions using biological control introductions: the influence of release size on the chance of population establishment. *Biological Invasions* **1**: 313–23.

Grevstad, F.S. and Klepetka, B.W. (1992) The influence of plant architecture on the foraging efficiencies of a suite of ladybird beetles feeding on aphids. *Oecologia* **92**: 399–404.

Grevstad, F.S. and Herzig, A.L. (1997) Quantifying the effects of distance and conspecifics on colonization: experiments and models using the loosestrife leaf beetle, *Galerucella calmariensis*. *Oecologia* **110**: 60–8.

Grewal, P.S., Gaugler, R., and Selvan, S. (1993) Host recognition by entomopathogenic nematodes: behavioral response to contact with host feces. *Journal of Chemical Ecology* **19**: 1219–31.

Grewal, P.S., Ehlers, R.U., and Shapiro-Ilan, D.I. (eds.) (2005) *Nematodes as Biocontrol Agents*. CABI Publishing, Wallingford.

Grewal, P.S., Bornstein-Forst, S., Burnell, A.M., Glazer, I., and Jagdale, G.B. (2006) Physiological, genetic, and molecular mechanisms of chemoreception, thermobiosis, and anhydrobiosis in entomopathogenic nematodes. *Biological Control* **38**: 54–65.

Grissell, E.E. and Schauff, M.E. (1990) *A Handbook of the Families of Nearctic Chalcidoidea (Hymenoptera)*. Entomological Society of America, Washington, D.C.

Grkovic, S., O'Callaghan, M., and Mahanty, H.K. (1995) Characterization of *Serratia entomophila* bacteriophages and the phage-resistant mutant strain BC4B. *Applied and Environmental Microbiology* **61**: 4160–6.

Groenewould, G.C.M., de Graaf in 't Veld, C., van Oorschot-van Nes, A.J. et al. (2002) Prevalence of sensitization to the predatory mite *Amblyseius cucumeris* as a new occupational allergen in horticulture. *Allergy* **57**: 614–19.

Grosholz, E.D. and Ruiz, G.M. (1996) Predicting the impact of introduced marine species: lessons from the multiple invasions of the European green crab, *Carcinus maenas*. *Biological Conservation* **78**: 59–66.

Grosholz, E.D., Ruiz, G.M., Dean, C.A., Shirley, K.A., Maron, J.L., and Connors, P.G. (2000) The implications of a nonindiginous marine predator in a California bay. *Ecology* **81**: 1206–24.

Gross, P. (1991) Influence of target pest feeding niche on success rates in classical biological control. *Environmental Entomology* **20**: 1217–27.

Gross, P. (1993) Insect behavioral and morphological defenses against parasitoids. *Annual Review of Entomology* **28**: 251–73.

Grossman, A.H., Breemen, M., van Holtz, A. et al. (2005) Searching behavior of an omnivorous predator for novel and native host plants of its herbivores: a study on arthropod colonization of eucalyptus in Brazil. *Entomologia Experimentalis et Applicata* **116**: 135–42.

Grout, T.G. and Richards, G.I. (1990) The influence of windbreak species on citrus thrips (Thysanoptera: Thripidae) populations and their damage to South African citrus orchards. *Journal of the Entomological Society of South Africa* **53**: 151–7.

Grout, T.G. and Richards, G.I. (1991a) The dietary effect of windbreak pollens on longevity and fecundity of a predaceous mite *Euseius addoensis addoensis* (Acari: Phytoseiidae) found in citrus orchards in South Africa. *Bulletin of Entomological Research* **82**: 317–20.

Grout, T.G. and Richards, G.I. (1991b) Value of pheromone traps for predicting infestations of red scale, *Aonidiella aurantii* (Maskell) (Hom., Diaspididae), limited by natural enemy activity and insecticides used to control citrus thrips, *Scirtothrips aurantii* Faure (Thys., Thripidae). *Journal of Applied Entomology* **111**: 20–7.

Gruner, D.S. (2005) Biotic resistance to an invasive spider conferred by generalist insectivorous birds on Hawaii Island. *Biological Invasions* **7**: 541–6.

Gu, W., Holland, M., Janssens, P., Seamark, R., and Kerr, P. (2004) Immune response in rabbit ovaries following infection of a recombinant myxoma virus expressing rabbit zona pellucida protein B. *Virology* **318**: 516–23.

Guerra, A.A. and Martinez, S. (1994) An *in vitro* rearing system for the propagation of the ectoparasitoid *Catolaccus grandis*. *Entomologia Experimentalis et Applicata* **72**: 11–16.

Guerra, A.A., Robacker, K.M., and Martinez, S. (1993) *In vitro* rearing of *Bracon mellitor* and *Catolaccus grandis* with artificial diets devoid of insect components. *Entomologia Experimentalis et Applicata* **68**: 303–7.

Guerra, G.P. and Kosztarab, M. (1992) *Biosystematics of the Family Dactylopiidae (Homoptera: Coccineae) with Emphasis on the Life Cycle of Dactylopius coccus Costa*. Studies on the Morphology and Systematics of Scale Insects no. 16. Virginia Agricultural Experiment Station, Bulletin no. 92–1. Virginia Polytechnic Institute and State University, Blacksburg, VA.

Guillet, P., Kurtak, D.C., Phillippon, B., and Meyer, R. (1990) Use of *Bacillus thuringiensis israelensis* for onchocerciasis control in West Africa. In: de Barjac, H. and Sutherland, D.J. (eds.), *Bacterial Control of Mosquitoes and Black Flies: Biochemistry, Genetics, and Applications of Bacillus thuringiensis israelensis and Bacillus sphaericus*, pp. 187–201. Rutgers University Press, New Brunswick, NJ.

Gurr, G.M., Wratten, S.D., and Altieri, M.A. (2004) *Ecological Engineering for Pest Management, Advances in Habitat Manipulation for Arthropods*. Cornell University Press, Ithaca, NY.

Gutierrez, A., Neuenschwander, P., Schulthess, F. et al. (1988) Analysis of biological control of cassava pests in Africa. II. Cassava mealybug *Phenacoccus manihoti*. *The Journal of Applied Ecology* **25**: 921–40.

Gwiazdowski, R.A., Van Driesche, R.G., Desnoyers, A. et al. (2006) Possible geographic origin of beech scale, *Cryptococcus fagisuga* (Hemiptera: Eriococcidae), an invasive pest in North America. *Biological Control* **39**: 9–18.

Gyenge, J.E., Trumper, E.V., and Edelstein, J.D. (1997) Design of sampling planes of predatory arthropods in alfalfa with fixed precision levels. *CEIBA* **38**: 23–8.

Hadfield, M.G. and Mountain, B.S. (1981) A field study of a vanishing species, *Achatinella mustelina* (Gastropoda, Pulmonata), in the Waianae Mountains of Oahu. *Pacific Science* **34**: 345–58.

Hadfield, M.G., Miller, S.E., and Carwile, A.H. (1993) The decimation of endemic Hawai'an tree snails by alien predators. *American Zoologist* **33**: 610–22.

Haenfling, B., Carvalho, G.R., and Brandl, R. (2002) mt-DNA sequences and possible invasion pathways of the Chinese mitten crab. *Marine Ecology Progress Series* **238**: 307–10.

Hagen, K.S. (1964) Nutrition of entomophagous insects and their hosts. In: DeBach, P. (ed.), *Biological Control of Insect Pests and Weeds*, pp. 356–80. Chapman and Hall, London.

Hagen, K.S. and van den Bosch, R. (1968) Impact of pathogens, parasites, and predators on aphids. *Annual Review of Entomology* **13**: 325–84.

Hagen, K.S., Sawall, Jr., E.F., and Tassen, R.L. (1970) The use of food sprays to increase effectiveness of entomophagous insects. *Proceedings of the Tall Timbers Conference on Ecological Animal Control by Habitat Management* **2**: 59–81.

Hagen, K.S., Bombosch, S., and McMurtry, J.A. (1976) The biology and impact of predators. In: Huffaker, C.B. and Messenger, P.S. (eds.), *Theory and Practice of Biological Control*, pp. 93–142. Academic Press, New York.

Hagen, K.S., Mills, N.J., Gordh, G., and McMurtry, J.A. (1999) Terrestrial arthropod predators of insect and mite pests. In: Bellows, Jr., T.S. and Fisher, T.W. (eds.), *Handbook of Biological Control*, pp. 383–503. Academic Press, San Diego, CA.

Hagimori, T., Abe, Y., Date, S., and Miura, K. (2006) The first finding of a *Rickettsia* bacterium associated with parthenogenesis induction among insects. *Current Microbiology* **52**: 97–101.

Hagler, J.R. and Cohen, A.C. (1991) Prey selection by *in vitro*- and field-reared *Geocoris punctipes*. *Entomologia Experimentalis et Applicata* **59**: 201–5.

Hagler, J.R. and Naranjo, S.E. (1997) Measuring the sensitivity of an indirect predator gut content ELISA: detectability of prey remains in relation to predator species, temperature, time, and meal size. *Biological Control* **9**: 112–19.

Hagler, J.R. and Miller, E. (2002) An alternative to conventional insect marking procedures: detection of a protein mark on pink bollworm by ELISA. *Entomologia Experimentalis et Applicata* **103**: 1–9.

Hagler, J.R., Naranjo, S.E., Bradley-Dunlop, D., Enriquez, F.J., and Henneberry, T.J. (1994) A monoclonal antibody to pink bollworm (Lepidoptera: Gelechiidae) egg antigen: a tool for predator gut analysis. *Annals of the Entomological Society of America* **87**: 85–90.

Hagley, E.A.C. and Simpson, C.M. (1981) Effect of food sprays on numbers of predators in an apple orchard. *The Canadian Entomologist* **113**: 75–7.

Hågvar, E.B. (1991) Ecological problems in the establishment of introduced predators and parasites for biological control. *Acta Entomologica Bohemoslovaca* **88**: 1–11.

Hågvar, E.B. and Hofsvang, T. (1989) Effect of honeydew and hosts on plant colonization by the aphid parasitoid *Ephedrus cerasicola*. *Entomophaga* **34**: 495–501.

Hajek, A.E., Humber, R.A., Elkinton, J.S., May, B., Walsh, S.R.A., and Silver, J.C. (1990a) Allozyme and RFLP analyses confirm *Entomophaga maimaiga* responsible for 1989 epizootics in North American gypsy moth populations. *Proceedings of the National Academy of Sciences USA* **87**: 6979–82.

Hajek, A.E., Humber, R.A., and Griggs, M.H. (1990b) Decline in virulence of *Entomophaga maimaiga* (Zygomycetes: Entomophthorales) with repeated *in vitro* subculture. *Journal of Invertebrate Pathology*. **56**: 91–7.

Hajek, A.E., Butler, L., Walsh, S.R.A. et al. (1996) Host range of the gypsy moth (Lepidoptera: Lymantriidae) pathogen *Entomophaga maimaiga* (Zygomycetes: Entomophthorales) in the field versus laboratory. *Environmental Entomology* **25**: 709–21.

Hale, L.D. (1970) Biology of *Icerya purchasi* and its natural enemies in Hawaii. *Proceedings of the Hawaiian Entomological Society* **20**: 533–50.

Hall, R.A. (1985) Whitefly control by fungi. In: Hussey, N.W. and Scopes, N. (eds.), *Biological Pest Control, the Glasshouse Experience*, pp. 116–24. Cornell University, Ithaca, NY.

Hall, R.A. and Papierok, B. (1982) Fungi as biological control agents of arthropods of agricultural and medical importance. *Parasitology* **84**: 205–40.

Hall, R.W. and Ehler, L.E. (1979) Rate of establishment of natural enemies in classical biological control. *Bulletin of the Entomological Society of America* **25**: 280–2.

Hall, R.W., Ehler, L.E., and Bisabri-Ershadi, B. (1980) Rate of success in classical biological control of arthropods. *Bulletin of the Entomological Society of America* **26**: 111–14.

Hamai, J. and Huffaker, C.B. (1978) Potential of predation by *Metaseiulus occidentalis* in compensating for increased, nutritionally induced, power of increase of *Tetranychus urticae*. *Entomophaga* **23**: 225–37.

Hance, Th. and Gregoire-Wibo, C. (1987) Effect of agricultural practices on carabid populations. *Acta Phytopathologica et Entomologica Hungarica* **22**: 147–60.

Haney, P.B., Luck, R.F., and Moreno, D.S. (1987) Increases in densities of the citrus red mite, *Panonychus citri* (Acarina: Tetranychidae), in association with the Argentine ant, *Iridomyrmex humilis* (Hymenoptera: Formicidae), in southern California citrus. *Entomophaga* **32**: 49–57.

Hanks, L.M., Gould, J.R., Paine, T.D., and Millar, J.G. (1995) Biology and host relations of *Avetianella longoi*, an egg parasitoid of the eucalyptus longhorned borer. *Annals of the Entomological Society of America* **88**: 666–71.

Hanks, L.M., Paine, T.D., and Millar, J.G. (1996) Tiny wasp helps protect eucalypts from eucalyptus longhorned borer. *California Agriculture* **50**: 14–16.

Hanks, L.M., Paine, T.D., Millar, J.G., Campbell, C.D., and Schuch, U.K. (1999) Water relations of host trees and resistance to the phloem-boring beetle *Phoracantha semipunctata* F. (Coleoptera: Cerambycidae). *Oecologia* **119**: 400–7.

Hanks, L.M., Millar, J.G., Paine, T.D., and Campbell, C.D. (2000) Classical biological control of the Australian weevil *Gonipterus scutellatus* Gyll. (Coleoptera: Curculionidae) in California. *Environmental Entomology* **29**: 369–75.

Hannay, C.L. (1953) Crystalline inclusions in aerobic spore-forming bacteria. *Nature* **172**: 1004–6.

Hansen, R.W., Spencer, N.R., Fornasari, L., Quimby, Jr., P.C., Pemberton, R.W., and Nowierski, R.M. (2004) Leafy spurge. In: Coombs, E.M., Clark, J.K., Piper, G.L., and Cofrancesco, Jr., A.F. (eds.), *Biological Control of Invasive Plants in the United States*, pp. 233–62. Oregon State University Press, Corvallis, OR.

Hanski, I. (1989) Metapopulation dynamics: does it help to have more of the same? *Trends in Ecology and Evolution* **4**: 113–14.

Hanson, P.E. and Gauld, I.D. (eds.) (1995) *The Hymenoptera of Costa Rica*. Oxford University Press, Oxford.

Hara, A.H., Gaugler, R., Kaya, H.K., and LeBeck, L.M. (1991) Natural populations of entomopathogenic nematodes (Rhabditida: Heterorhabditidae, Steinernematidae) from the Hawaiian Islands. *Environmental Entomology* **20**: 211–16.

Hardman, J.M., Moreau, D.L., Snyder, M., Gaul, S.O., and Bent, E.D. (2000) Performance of a pyrethroid-resistant strain of the predator mite *Typhlodromus pyri* (Acari: Phytoseiidae) under different insecticide regimes. *Journal of Economic Entomology* **93**: 590–604.

Hardy, A.R., Fletcher, M.R., and Stanley, P.I. (1986) Pesticides and wildlife: twenty years of vertebrate wildlife incident investigations by MAFF. *State Veterinary Journal* **40(117)**: 182–92.

Hardy, C.M., Hinds, L.A., Kerr, P.J. et al. (2006) Biological control of vertebrate pests using virally vectored immunocontraception. *Journal of Reproductive Immunology* **71**: 102–11.

Harley, K.L.S., Forno, I.W., Kassulke, R.C., and Sands, D.P.A. (1984) Biological control of water lettuce. *Journal of Aquatic Plant Management* **22**: 101–2.

Harmon, J.P. and Andow, D.A. (2004) Indirect effects between shared prey: predictions for biological control. *BioControl* **49**: 605–26.

Harper, J.D. (1987) Applied epizootiology: microbial control of insects. In: Fuxa, J.R. and Tanada, Y. (eds.), *Epizootiology of Insect Diseases*, pp. 473–96. John Wiley and Sons, New York.

Harper, J.L. (1977) *Population Biology of Plants*. Academic Press, New York.

Harper, J.L. (1981) The concept of population in modular organisms. In: May, R.M. (ed.), *Theoretical Ecology, Principals and Applications*, pp. 53–77. Sinauer Associates, Sunderland, MA.

Harris, P. (1973) The selection of effective agents for the biological control of weeds. *The Canadian Entomologist* **105**: 1495–1503.

Harris, P. (1977) Biological control of weeds: from art to science. In: Freeman, T.E. (ed.), *Proceedings of the IVth International Symposium on Biological Control of Weeds*, pp. 85–6, August 30–September 2, 1976, Gainesville, Florida. Institute of Food and Agricultural Sciences, University of Florida, Gainesville, FL.

Harris, P. (1979) Cost of biological control of weeds by insects in Canada. *Weed Science* **27**: 242–50.

Harris, P. (1980a) Establishment of *Urophora affinis* Frfld. and *U. quadrifasciata* (Meig.) (Diptera: Tephritidae) in Canada for the biological control of diffuse and spotted knapweed. *Zeitschrift fur Angewandte Entomologie* **89**: 504–14.

Harris, P. (1980b) Effects of *Urophora affinis* Frfld. and *U. quadrifasciata* (Meig.) (Diptera: Tephritidae) on *Centaurea diffusa* Lam. and *C. maculosa* Lam. (Compositae). *Zeitschrift fur Angewandte Entomologie* **90**: 190–201.

Harris, P. (1981) Stress as a strategy in the biological control of weeds. In: Papavizas, G.C. (ed.), *Beltsville Symposia in Agricultural Research. 5. Biological Control in Crop Protection*, pp. 333–40. Allanheld, Osmun, Totowa, NJ.

Harris, P. (1984) *Carduus nutans* L., nodding thistle and *C. acanthoides*, plumeless thistle (Compositae). In: Kelleher, J.S. and Hulme, M.A. (eds.), *Biological Control Programmes Against Insects and Weeds in Canada 1969–1980*, pp. 159–69. Commonwealth Agricultural Bureaux, London.

Harris, P. (1985) Biocontrol of weeds: bureaucrats, botanists, beekeepers and other bottlenecks. In: Delfosse, E.S. (ed.), *Proceedings of the VIth International Symposium on Biological Control of Weeds*, pp. 3–12, August 19–25, 1984, Vancouver, Canada. Agriculture Canada, Ottawa.

Harris, P. (1988) Environmental impact of weed-control insects. *BioScience* **38**: 542–8.

Harris, P. (1989) Practical considerations in a classical biocontrol of weeds program. In: *Proceedings of the International*

Symposium on Biological Control Implementation, pp. 23–31, April 4–6, 1989, McAllen, Texas. North American Plant Protection Organization Bulletin no. 6. North American Plant Protection Organization, Ottawa.

Harris, P. (1990) Environmental impact of introduced biological control agents. In: Mackauer, M., Ehler, L.E., and Roland, J. (eds.), *Critical Issues in Biological Control*, pp. 289–300. Intercept, Andover.

Harris, P. (1996) Effectiveness of gall inducers in weed biological control. *The Canadian Entomologist* **128**: 1021–55.

Harris, P. and Myers, J.H. (1984) *Centaurea diffusa* Lam. and *C. maculosa* Lam. *s. lat.*, diffuse and spotted knapweed (Compositae). In: Kelleher, J.S. and Hulme, M.A. (eds.), *Biological Control Programmes against Insects and Weeds in Canada 1969–1980*, pp. 127–37. Commonwealth Agricultural Bureaux, London.

Harris, V.E. and Todd, J.W. (1980) Male-mediated aggregation of male, female and 5th instar southern green stink bugs and concomitant attraction of a tachinid parasite, *Trichopoda pennipes*. *Entomologia Experimentalis et Applicata* **27**: 117–26.

Harrison, S. (1997) Persistent, localized outbreaks in the western tussock moth *Orgyia vetusta*: the roles of resource quality, predation and poor dispersal. *Ecological Entomology* **22**: 158–66.

Hart, A.J., Bale, J.S., Tullet, A.G., Worland, M.R., and Walters, F.K.A. (2002) The effects of temperature on the establishment potential of the predatory mite *Amblyseius californicus* McGregor (Acari: Phytoseiidae) in the UK. *Journal of Insect Physiology* **48**: 593–9.

Harvey, C.T. and Eubanks, M.D. (2005) Intraguild predation of parasitoids by *Solenopsis invicta*: a non-disruptive interaction. *Entomologia Experimentalis et Applicata* **114**: 127–35.

Harwood, J.D. and Obrycki, J.J. (2005) Quantifying aphid predation rates of generalist predators in the field. *European Journal of Entomology* **102**: 335–50.

Harwood, J.D., Sunderland, K.D., and Symondson, W.O.C. (2004) Prey selection by linyphiid spiders: molecular tracking of the effects of alternative prey on rates of aphid consumption in the field. *Molecular Ecology* **13**: 3549–60.

Hasan, S. (1981) A new strain of the rust fungus *Puccinia chondrillina* for biological control of skeleton weed in Australia. *Annals of Applied Biology* **99**: 119–24.

Hasan, S. and Wapshere, A.J. (1973) The biology of *Puccinia chrondrilla*, a potential biological control agent of skeleton weed. *Annals of Applied Biology* **74**: 325–32.

Hassan, S.A. (1977) Standardized techniques for testing side-effects of pesticides on beneficial arthropods in the laboratory. *Zeitschrift für Pflanzenkrankheiten und Pflanzenschutz* **84**: 158–63.

Hassan, S.A. (1980) A standard laboratory method to test the duration of harmful effects of pesticides on egg parasites of the genus *Trichogramma* (Hymenoptera: Trichogrammatidae). *Zeitschrift für Pflanzenkrankheiten und Pflanzenschutz* **89**: 282–9.

Hassan, S.A. (1985) Standard methods to test the side-effects of pesticides on natural enemies of insects and mites developed by the IOBC/WPRS Working Group "Pesticides and Beneficial Organisms." *Bulletin of OEPP/EPPO* **15**: 214–55.

Hassan, S.A. (1989) Testing methodology and the concept of the IOBC/WPRS working group. In: Jepson, P.C. (ed.), *Pesticides and Non-Target Invertebrates*, pp. 1–18. Intercept, Wimborne.

Hassan, S.A. (1994) Strategies to select *Trichogramma* species for use in biological control. In: Wajnberg, É. and Hassan, S.A. (eds.), *Biological Control with Egg Parasitoids*, pp. 55–71. CABI Publishing, Wallingford.

Hassan, S.A. and Zhang, W.Q. (2001) Variability in quality of *Trichogramma brassicae* (Hymenoptera: Trichogrammatidae) from commercial suppliers in Germany. *Biological Control* **22**: 115–21.

Hassan, S.A., Albert, R., Bigler, F. et al. (1987) Results of the third joint pesticide testing programme by the IOBC/WPRS Working Group "Pesticides and Beneficial Organisms." *Journal of Applied Entomology* **103**: 92–107.

Hassell, M.P. (1971) Mutual interference between searching insect parasites. *Journal of Animal Ecology* **40**: 473–86.

Hassell, M.P. (1980) Foraging strategies, population models and biological control: a case study. *Journal of Animal Ecology* **49**: 603–28.

Hassell, M.P. (2000) *The Spatial and Temporal Dynamics of Host-Parasitoid Interactions*. Oxford University Press, Oxford.

Hassell, M.P. and Huffaker, C.B. (1969) The appraisal of delayed and direct density-dependence. *The Canadian Entomologist* **101**: 353–61.

Hassell, M.P. and May, R.M. (1973) Stability in insect host-parasite models. *Journal of Animal Ecology* **42**: 693–726.

Hassell, M.P. and May, R.M. (1974) Aggregation of predators and insect parasites and its effect on stability. *Journal of Animal Ecology* **43**: 567–94.

Hassell, M.P. and Comins, H.N. (1977) Sigmoid functional responses and population stability. *Theoretical Population Biology* **14**: 62–7.

Hatherly, I.S., Hart, A.J., Tullett, A.G., and Bale, J.S. (2005) Use of thermal data as a screen for the establishment potential of non-native biological control agents in the U.K. *BioControl* **50**: 687–98.

Hattingh, V. and Samways, M.J. (1991) Determination of the most effective method for field establishment of biocontrol agents of the genus *Chilocorus* (Coleoptera: Coccinellidae). *Bulletin of Entomological Research* **81**: 169–74.

Hattingh, V. and Tate, B.A. (1995) Effects of field-weathered residues of insect growth regulators on some Coccinellidae (Coleoptera) of economic importance as biological control agents. *Bulletin of Entomological Research* **85**: 489–93.

Hattingh, V. and Tate, B.A. (1996) The effects of insect growth regulator use on IPM in Southern African citrus. *Proceedings of the International Society of Citriculture* **1**: 523–5.

Haugen, D.A. (1990) Control procedures for *Sirex noctilio* in the Green Triangle: review from detection to severe outbreak (1977–1987). *Australian Forestry* **53**: 24–32.

Havelka, J. and Zemek, R. (1988) Intraspecific variability of aphidophagous gall midge *Aphidoletes aphidimyza* (Rondani) (Dipt.: Cecidomyiidae) and its importance for biological control of aphids. 1. Ecological and morphological characteristics of populations. *Journal of Applied Entomology* **105**: 280–8.

Havill, N.P., Montgomery, M.E., Yu, G., Shiyake, S., and Caccone, A. (2006) Mitochondrial DNA from hemlock woolly adelgid (Hemiptera: Adelgidae) suggests cryptic speciation and pinpoints the source of the introduction to eastern North America. *Annals of the Entomological Society of America* **99**: 195–203.

Havron, A., Rosen, D., Prag, H., and Rossler, Y. (1991) Selection for pesticide resistance in *Aphytis holoxanthus*, a parasite of the Florida red scale. *Entomologia Experimentalis et Applicata* **61**: 221–8.

Hawkins, B.A. and Gross, P. (1992) Species richness and population limitation in insect parasitoid-host systems. *American Naturalist* **139**: 417–23.

Hawkins, B.A., Cornell, H.V., and Hochberg, M.E. (1997) Predators, parasitoids, and pathogens as mortality agents in phytophagous insect populations. *Ecology* **78**: 2145–52.

Hawkins, B.A., Mills, N.J., Jervis, M.A., and Price, P.W. (1999) Is biological control a natural phenomenon? *Oikos* **86**: 493–506.

Haye, T., Goulet, H., Mason, P.G., and Kuhlmann, U. (2005) Does fundamental host range match ecological host range? *Biological Control* **35**: 55–67.

Haye, T., Kuhlmann, U., Goulet, H., and Mason, P.G. (2006) Controlling Lygus plant bugs (Heteroptera: Miridae) with European *Peristenus relictus* (Hymenoptera: Braconidae) in Canada – risky or not? *Bulletin of Entomological Research* **96**: 187–96.

Hayes, K.R. (1998) Ecological risk assessment for ballast water introductions: A suggested approach. *ICES Journal of Marine Science* **55**: 201–12.

Hays, D.B. and Vinson, S.B. (1971) Acceptance of *Heliothis virescens* (F.) as a host by the parasite *Cardiochiles nigriceps* Viereck (Hymenoptera: Braconidae). *Animal Behavior* **19**: 344–52.

Hazzard, R.V. and Ferro, D.N. (1991) Feeding response of *Coleomegilla maculata* (Coleoptera: Coccinellidae) to eggs of Colorado potato beetle (Coleoptera: Chrysomelidae) and green peach aphids (Homoptera: Aphidae). *Environmental Entomology* **20**: 644–51.

Hazzard, R.V., Ferro, D.N., Van Driesche, R.G., and Tuttle, A.F. (1991) Mortality of eggs of Colorado potato beetle (Coleoptera: Chrysomelidae) from predation by *Coleomegilla maculata* (Coleoptera: Coccinellidae). *Environmental Entomology* **20**: 841–8.

Head, G., Moar, W., Eubanks, M. et al. (2005) A multiyear, large-scale comparison of arthropod populations on commercially managed Bt and non-Bt cotton fields. *Environmental Entomology* **34**: 1257–66.

Headrick, D.H. and Goeden, R.D. (2001) Biological control as a tool for ecosystem management. *Biological Control* **21**: 249–57.

Heads, P.A. and Lawton, J.H. (1983) Studies on the natural enemy complex of the holly leaf miner: the effects of scale on the detection of aggregative responses and the implications for biological control. *Oikos* **40**: 267–76.

Heard, T.A. (2000) Concepts in insect host-plant selection behavior and their application to host specificity testing. In: Van Driesche, R., Heard, T., McClay, A., and Reardon, R. (eds.), *Proceedings of a Session: Host Specificity Testing of Exotic Arthropod Biological Control Agents – The Biological Basis for Improvement in Safety*, pp. 1–10. FHTET-99-1. USDA Forest Service, Morgantown, WV.

Heard, T.A. and Pettit, W. (2005) Review and analysis of the surveys for natural enemies of *Mimosa pigra*: what does it tell us about surveys for broadly distributed hosts? *Biological Control* **34**: 247–54.

Heard, T.A., O'Brian, C.W., Forno, I.W., and Burcher, J.A. (1998) *Chalcodermus persimilis* O'Brien n. sp. (Coleoptera: Curculionidae): description, biology, host range, and suitability for biological control of *Mimosa pigra* L. (Mimosaceae). *Transactions of the American Entomological Society* **124**: 1–11.

Heard, T.A., Paynter, Q., Chan, R., and Mira, M. (2005) *Malacorhinus irregularis* for biological control of *Mimosa pigra*: host specificity, life cycle and establishment in Australia. *Biological Control* **32**: 252–62.

Hedley, J. (2004) The International Pant Protection Convention and alien species. In: Miller, M.L. and Rabian, R.N. (eds.), *Harmful Invasive Species: Legal Responses*, pp. 185–201. Environmental Law Institute, Washington, D.C.

Hegedus, D.D. and Khachatourians, G.G. (1993) Construction of cloned DNA probes for the specific detection of the entomopathogenic fungus *Beauveria bassiana* in grasshoppers. *Journal of Invertebrate Pathology* **62**: 233–40.

Heidari, M. and Copland, M.J.W. (1993) Honeydew: a food resource or arrestant for the mealybug predator *Cryptolaemus montrouzieri*? *Entomophaga* **38**: 63–8.

Heidger, C. and Nentwig, W. (1989) Augmentation of beneficial arthropods by strip-management. 3. Artificial introduction of a spider species which preys on wheat pest insects. *Entomophaga* **34**: 511–22.

Heimbach, U. and Abel, C. (1991) Side effects of soil insecticides in different formulations on some beneficial arthopods. *Verhandlung der Gesellschaft für Ökologie* **19**: 163–70.

Heimpel, G.E., Rosenheim, A.J., and Mangel, M. (1996) Egg limitation, host quality, and dynamic behaviour by a parasitoid in the field. *Ecology* **77**: 2410–20.

Heinrichs, E.A., Aquino, G.B., Chelliah, S., Valencia, S.L., and Reissig, W.H. (1982) Resurgence of *Nilaparvata lugens* (Stål) populations as influenced by method and timing of insecticide applications in lowland rice. *Environmental Entomology* **11**: 78–84.

Heinz, K.H. and Parrella, M. (1990) Biological control of insect pests on greenhouse marigolds. *Environmental Entomology* **19**: 825–35.

Heinz, K.M. (1998) Dispersal and dispersion of aphids (Homoptera: Aphididae) and selected natural enemies in spatially subdivided greenhouse experiments. *Environmental Entomology* **27**: 1029–38.

Heirbaut, M. and van Damme, P. (1992) The use of artificial nests to establish colonies of the black cocoa ant (*Dolichoderus thoracicus* Smith) used for biological control of *Helopeltis theobromae* Mill. in Malaysia. *Mededelingen van de Faculteit Landbouwwetenschappen* **57**: 533–42.

Hemptinne, J.-L. (1988) Ecological requirements for hibernating *Propylea quatuordecimpunctata* (l.) and *Coccinella septempunctata* (Col.: Coccinellidae). *Entomophaga* **33**: 238–45.

Henaut, Y., Alauzet, C., and Lambin, M. (2002) Effects of starvation on the search path characteristics of *Orius majusculus* (Reuter) (Het. Anthocoridae). *Journal of Applied Entomology* **126**: 501–3.

Henderson, L. (2001) *Alien Weeds and Invasive Plants.* Plant Protection Research Institute Handbook no. 12. Agricultural Research Council, Pretoria.

Hendrickson, Jr., R.M. (1980) Continuous production of predaceous mites in the greenhouse. *Journal of the New York Entomological Society* **88**: 252–6.

Hendrickson, Jr., R.M., Barth, S.E., and Ertle, L.R. (1987) Control of relative humidity during shipment of parasitic insects. *Journal of Economic Entomology* **80**: 537–9.

Henneman, M.L. and Memmott, J. (2004) Infiltration of a Hawaiian community by introduced biological control agents. *Science* **293**: 1314–16.

Henzell, R. and Murphy, E. (2002) Rabbits and possums in the GMO potboiler. *Biocontrol News and Information* **23**: 89N–96N.

Hérard, F., Keller, M.A., Lewis, W.J., and Tumlinson, J.H. (1988) Beneficial arthropod behavior mediated by airborne semiochemicals. IV. Influence of host diet on host-orientated flight chamber responses of *Microplitis demolitor* Wilkinson. *Journal of Chemical Ecology* **14**: 1597–1606.

Heraty, J. (2004) Molecular systematics, chalcidoidea and biological control. In: Ehler, L., Sforza, R., and Mateille, T. (eds.), *Genetics, Evolution and Biological Control*, pp. 39–71. CABI Publishing, Wallingford.

Herms, D.A., Stone, A.K., and Chatfield, J.A. (2004) *Emerald Ash Borer: the Beginning of the End of Ash in North America?* Special Circular no. 193, pp. 62–71, Ohio Agricultural Research and Development Center, Wooster, OH.

Herren, H.R. and Neuenschwander, P. (1991) Biological control of cassava pests in Africa. *Annual Review of Entomology* **36**: 257–84.

Herren, H.R., Bird, T.J., and Nadel, D.J. (1987) Technology for automated aerial release of natural enemies of the cassava mealybug and cassava green mite. *Insect Science and its Application* **8**: 883–5.

Herrnstadt, C., Gaertner, F., Gelernter, W., and Edwards, D.L. (1987) *Bacillus thuringiensis* isolate with activity against Coleoptera. In: Maramorosch, K. (ed.), *Biotechnology in Invertebrate Pathology and Cell Culture*, pp. 101–13. Academic Press, New York.

Hickman, J.M. and Wratten, S.D. (1996) Use of *Phacelia tanacetifolia* strips to enhance biological control of aphids by hoverfly larvae in cereal fields. *Journal of Economic Entomology* **89**: 832–40.

Higashiura, Y. (1989) Survival of eggs in the gypsy moth, *Lymantria dispar*. I. Predation by birds. *Journal of Animal Ecology* **58**: 403–12.

Hight, S.D. (1990) Available feeding niches in populations of *Lythrum salicaria* (purple loosestrife) in the northeastern United States. In: Delfosse, E.S. (ed.), *Proceedings of the VII International Symposium on Biological Control of Weeds*, pp. 269–78, March 6–11, 1988, Rome, Italy. Ist. Sper. Patol. Veg.

Hill, D.S. (1975) *Agricultural Insect Pests of the Tropics and their Control.* Cambridge University Press, Cambridge.

Hill, G. and Greathead, D. (2000) Economic evaluation in classical biological control. In: Perrings C., Williamson, M., and Dalmazzone, S. (eds.), *The Economics of Biological Invasions*, pp. 208–23. Edward Elgar, Cheltenham.

Hill, M.G. and Allan, D.J. (1986) The effects of weeds on armyworm in maize. *Proceedings of the 39th New Zealand Weed and Pest Control Conference*, pp. 260–3, New Zealand Weed and Pest Control Society, Palmerston North.

Hill, M.G., Nang'ayo, F.L.O., and Wright, D.J. (2003) Biological control of the larger grain borer, *Prostephanus truncatus* (Coleoptera: Bostrichidae), in Kenya using a predatory beetle, *Teretrius nigrescens* (Coleoptera: Histeridae). *Bulletin of Entomological Research* **93**: 299–306.

Hill, M.P. (1997) The potential for the biological control of the floating aquatic fern, *Azolla filiculoides* Lamarck (red waterfer/rooiwatervaring) in South Africa. WRC Report no. KV 100/97. Water Research Commission, Pretoria.

Hill, M.P. (1998) Life history and laboratory host range of *Stenopelmus rufinasus* Gyllenhal (Coleoptera: Curculionidae) a natural enemy for *Azolla filiculoides* Lamarck (Azollaceae) in South Africa. *BioControl* **43**: 215–24.

Hill, M.P. (1999) Biological control of red water fern, *Azolla filiculoides* Lamarck (Pteridophylla: Azollaceae) in South Africa. In: Olckers, T. and Hill, M.P. (eds.), *Biological Control of Weeds in South Africa (1990–1998)*, pp. 119–24. African Entomology Memoir no. 1. Entomological Society of South Africa, Johannesburg.

Hill, M.P. and Hulley, P.E. (1995) Host-range extension by native parasitoids to weed biocontrol agents introduced to South Africa. *Biological Control* **5**: 297–302.

Hill, M.P. and Cilliers, C.J. (1999) *Azolla filiculoides* Lamarck (Pteridophyta: Azollaceae), its status in South Africa and control. *Hydrobiologia* **415**: 203–6.

Hill, R.L. and Stone, C. (1985) Spider mites as control agents for weeds. In: Helle, W. and Sabelis, M.W. (eds.), *Spider*

Mites: their Biology, Natural Enemies and Control, vol. 1B, pp. 443–8. Elsevier, Amsterdam.

Hill, R.L., Grindell, J.M., Winks, C.J., Sheat, J.J., and Hayes, L.M. (1991) Establishment of gorse spider mite as a control agent for gorse. In: Anon. *Proceedings of the 44th New Zealand Weed and Pest Control Conference*, pp. 31–4. New Zealand Weed and Pest Control Society, Pamerston North.

Hillis, D.M. and Dixon, M.T. (1991) Ribosomal DNA: molecular evolution and phylogenetic inference. *The Quarterly Review of Biology* **66**: 411–53.

Hinks, C.F. (1971) Observations of larval behaviour and avoidance of encapsulation of *Perilampus hyalinus* (Hymenoptera: Perilampidae) parasitic in *Neodriprion lecontei* (Hymenoptera: Diprionidae). *The Canadian Entomologist* **103**: 182–7.

Hislop, R.G. and Prokopy, R.J. (1981) Integrated management of phytophagous mites in Massachusetts (USA) apple orchards. 2. Influences of pesticides on the mite predator *Amblyseius fallacis* under laboratory and field conditions. *Protection Ecology* **3**: 157–72.

Hochberg, M.E. and Ives, A.R. (eds.) (2000) *Parasitoid Population Biology*. Princeton University Press, Princeton, NJ.

Hoddle, M.S. (1999) Biological control of vertebrate pests. In: Bellows, Jr., T.S. and Fisher, T.W. (eds.), *Handbook of Biological Control*, pp. 955–74. Academic Press, San Diego, CA.

Hoddle, M.S. (2003) Predation behaviors of *Franklinothrips orizabensis* (Thysanoptera: Aeolothripidae) towards *Scirtothrips perseae* and *Heliothrips haemorrhoidalis* (Thysanoptera: Thripidae). *Biological Control* **27**: 323–8.

Hoddle, M.S. (2004) The potential adventive geographic range of glassy-winged sharpshooter, *Homalodisca coagulata*, and the grape pathogen *Xylella fastidiosa*: implications for California and other grape growing regions of the world. *Crop Protection* **23**: 691–9.

Hoddle, M.S. (2006) Historical review of control programs for *Levuana iridescens* (Lepidoptera: Zygaenidae) in Fiji and examination of possible extinction of this moth by *Bessa remota* (Diptera: Tachinidae). *Pacific Science* **60**: 439–53.

Hoddle, M.S. and Robinson, L. (2004) Evaluation of factors influencing augmentative releases of *Chrysoperla carnea* for control of *Scirtothrips perseae* in California avocado orchards. *Biological Control* **31**: 268–75.

Hoddle, M.S. and Van Driesche, R.G. (1999) Evaluation of *Eretmocerus eremicus* and *Encarsia formosa* Beltsville strain in commercial greenhouses for biological control of *Bemisia argentifolii* on colored poinsettia plants. *Florida Entomologist* **82**: 556–69.

Hoddle, M., Van Driesche, R.G., and Sanderson, J. (1997a) Biological control of *Bemisia argentifolii* (Homoptera: Aleyrodidae) on poinsettia with inundative releases of *Eretmocerus eremicus* (Hymenoptera: Aphelinidae): do release rates and plant growth affect parasitism? *Bulletin of Entomological Research* **88**: 47–58.

Hoddle, M., Van Driesche, R.G., and Sanderson, J. (1997b) Biological control of *Bemisia argentifolii* (Homoptera: Aleyrodidae) on poinsettia with inundative releases of *Encarsia formosa* (Hymenoptera: Aphelinidae): are higher release rates necessarily better? *Biological Control* **10**: 166–79.

Hoddle, M., Van Driesche, R.G., and Sanderson, J. (1998a) Biology and utilization of the whitefly parasitoid *Encarsia formosa*. *Annual Review of Entomology* **43**: 645–9.

Hoddle, M.S., Van Driesche, R.G., Elkinton, J.S., and Sanderson, J.P. (1998b) Discovery and utilization of *Bemisia argentifolii* patches by *Eretmocerus eremicus* and *Encarsia formosa* (Beltsville strain) in greenhouses. *Entomologia Experimentalis et Applicata* **87**: 15–28.

Hoddle, M.S., Aponte, O., Kerguelen, V., and Heraty, J. (1999) Biological control of *Oligonychus perseae* (Acari: Tetranychidae) on avocado: I. Evaluating release timings, recovery, and efficacy of six commercially available phytoseiids. *International Journal of Acarology* **25**: 211–19.

Hoddle, M.S., Robinson, L., and Virzi, J. (2000) Biological control of *Oligonychus perseae* (Acari: Tetranychidae) on avocado: III. Evaluating the efficacy of varying release rates and release frequency of *Neoseiulus californicus* (Acari: Phytoseiidae). *International Journal of Acarology* **26**: 203–14.

Hoddle, M.S., Jones, J., Oishi, K., Morgan, D., and L. Robinson. (2001a) Evaluation of diets for the development and reproduction of *Franklinothrips orizabensis* (Thysanoptera: Aeolothripidae). *Bulletin of Entomological Research* **91**: 273–80.

Hoddle, M.S., Van Driesche, R.G., Lyon, S.M., and Sanderson, J.P. (2001b) Compatibility of insect growth regulators with *Eretmocerus eremicus* (Hymenoptera: Aphelinidae) for whitefly (Homoptera: Alyerodidae) control on poinsettia: I. Laboratory Assays. *Biological Control* **20**: 122–31.

Hoddle, M.S., Morse, J.G., Phillips, P.A., Faber, B.A., and Jetter, K.M. (2002a) Avocado thrips: a new challenge for growers. *California Agriculture* **56**: 103–7.

Hoddle, M.S., Nakahara, S., and Phillips, P.A. (2002b) Foreign exploration for *Scirtothrips perseae* (Thysanoptera: Thripidae) and associated natural enemies on avocado (*Persea americana* Miller). *Biological Control* **24**: 251–65.

Hoddle, M.S., Oevering, P., Phillips, P.A., and Faber, B.A. (2004) Evaluation of augmentative releases of *Franklinothrips orizabensis* for control of *Scirtothrips perseae* in California avocado orchards. *Biological Control* **30**: 456–65.

Hodek, I. (1970) Coccinellids and modern pest management. *BioScience* **20**: 543–52.

Hodek, I. (1973) *Biology of the Coccinellidae*. Dr. W. Junk, N.V. Publishers, The Hague.

Hodek, I. (ed.) (1986) *Ecology of Aphidophaga*. Proceedings of the 2nd symposium held at Zvíkovské Podhradií, September 2–8, 1984. Dr. W. Junk Publishers, Dordrecht.

Hodek, I. and Honìk, A. (1996) *Ecology of Coccinellidae*. Kluwer Academic Publishers, Dordrecht.

Hodkinson, I.D. (1974) The biology of the Psylloidea (Homoptera): a review. *Bulletin of Entomological Research* **64**: 325–39.

Hodkinson, I.D. (1999) Biocontrol of eucalyptus psyllid *Ctenarytaina eucalypti* by the Australian parasitoid *Psyllaephagus pilosus*: a review of current programmes and their success. *Biocontrol News and Information* **20(40)**: 129N–34N.

Høeg, J. and Lutzen, J. (1985) *Crustacea: Rhizocephala. Marine Invertebrates of Scandinavia, no. 6*. Norwegian University Press, Oslo.

Hoelmer, K.A. and Kirk, A.A. (2005) Selecting arthropod biological control agents against arthropod pests: Can the science be improved to decrease the risk of releasing ineffective agents? *Biological Control* **34**: 255–64.

Hoelzel, A.R. (ed.) (1998) *Molecular Genetic Analysis of Populations: a Practical Approach*. Oxford University Press, Oxford.

Hoffman, J.D., Ignoffo, C.M., and Dickerson, W.A. (1975) *In vitro* rearing of the endoparasitic wasp *Trichogramma pretiosum*. *Annals of the Entomological Society of America* **68**: 335–6.

Hoffmann, J.H. (1990) Interactions between three weevils species in the biocontrol of *Sesbania punicea* (Fabaceae): the role of simulation models in evaluation. *Agriculture, Ecosystems and Environment* **32**: 77–87.

Hoffmann, J.H. (1996) Biological control of weeds: the way forward, a South African perspective. In: Stirton, C.H. (ed.), *Weeds in a Changing World International Symposium*, pp. 77–89, November 20, 1995, Brighton, England. British Crop Protection Council Monograph no. 64. British Crop Protection Council, Farnham.

Hoffmann, J.H. and Moran, V.C. (1992) Oviposition patterns and the supplementary role of a seed-feeding weevil, *Rhyssomatus marginatus* (Coleoptera: Curculionidae), in the biological control of a perennial leguminous weed, *Sesbania punicea*. *Bulletin of Entomological Research* **82**: 343–7.

Hoffmann, J.H. and Moran, V.C. (1995) Localized failure of a weed biological control agent attributed to insecticide drift. *Agriculture, Ecosystems and Environment* **52**: 197–203.

Hoffmann, J.H. and Moran, V.C. (1998) The population dynamics of an introduced tree, *Sesbania punicea*, in South Africa, in response to long-term damage caused by different combinations of three species of biological control agents. *Oecologia* **114**: 343–8.

Hoffmann, M.P., Wilson, L.T., Zalom, F.G., and Hilton, R.J. (1991) Dynamic sequential sampling plan for *Helicoverpa zea* (Lepidoptera: Noctuidae) eggs in processing tomatoes: parasitism and temporal patterns. *Environmental Entomology* **20**: 1005–12.

Hoffmann, J.H., Moran, V.C., and Zeller, D.A. (1998a) Long-term population studies and the development of an integrated management programme for control of *Opuntia stricta* in Kruger National Park, South Africa. *Journal of Applied Ecology* **35**: 156–60.

Hoffmann, J.H., Moran, V.C., and Zeller, D.A. (1998b) Evaluation of *Cactoblastis cactorum* (Lepidoptera: Phycitidae) as a biological control agent of *Opuntia stricta* (Cactaceae) in the Kruger National Park, South Africa. *Biological Control* **12**: 20–4.

Hofte, H. and Whiteley, H.R. (1989) Insecticidal crystal proteins of *Bacillus thuringiensis*. *Microbiological Review* **53**: 242–55.

Hokkanen, H.M.T. and Pimentel, D. (1984) New approach for selecting biological control agents. *The Canadian Entomologist* **116**: 1109–21.

Hokkanen, H.M.T. and Pimentel, D. (1989) New associations in biological control: theory and practice. *The Canadian Entomologist* **121**: 829–40.

Hokkanen, H., Husberg, G.B., and Söderblom, M. (1988) Natural enemy conservation for the integrated control of the rape blossom beetle *Meligethes aeneus* F. *Annales Agricultura Fenniae* **27**: 281–93.

Hölldobler, B. and Wilson, E.O. (1990) *The Ants*. The Belknap Press of Harvard University Press, Cambridge, MA.

Holling, C.S. (1959) Some characteristics of simple types of predation and parasitism. *The Canadian Entomologist* **91**: 385–98.

Holling, C.S. (1965) The functional response of predators to prey density and its role in mimicry and population regulation. *Memoirs of the Entomological Society of Canada* **45**: 3–60.

Holst, N. and Meikle, W.G. (2003) *Teretrius nigrescens* against larger grain borer, *Prostephanus truncatus*, in African maize stores: biological control at work? *Journal of Applied Ecology* **40**: 307–19.

Holt, J., Wareing, D.R., and Norton, G.A. (1992) Strategies of insecticide use to avoid resurgence of *Nilaparvata lugens* (Homoptera; Delphacidae) in tropical rice: a simulation analysis. *Journal of Economic Entomology* **85**: 1979–89.

Holt, R.D. (1977) Predation, apparent competition, and the structure of prey communities. *Theoretical Population Biology* **12**: 197–229.

Holt, R.D. and Hochberg, M.E. (2001) Indirect interactions, community modules, and biological control: a theoretical perspective. In: Wajnberg, É., Scott, J.K., and Quimby, P.C. (eds.), *Evaluating Indirect Effects of Biological Control*, pp. 13–38. CABI Publishing, Wallingford.

Hominick, W.M. and Reid, A.P. (1990) Perspectives on entomopathogenic nematology. In: Gaugler, R. and Kaya, H. (eds.), *Entomopathogenic Nematodes in Biological Control*, pp. 327–45. CRC Press, Boca Raton, FL.

Hone, J. (1994) *Analysis of Vertebrate Pest Control*. Cambridge University Press, Cambridge.

Honée, G. and Visser, B. (1993) The mode of action of *Bacillus thuringiensis* crystal proteins. *Entomologia Experimentalis et Applicata* **69**: 145–55.

Hood, W.G. and Naiman, R.J. (2000) Vulnerability of riparian zones to invasion by exotic vascular plants. *Plant Ecology* **148**: 105–14.

Hoogendoorn, M. and Heimpel, G.E. (2001) PCR-based gut content analysis of insect predators: using ribosomal ITS-1 fragments from prey to estimate predation frequency. *Molecular Ecology* **10**: 2059–67.

Hopen, H.J., Caruso, F.L., and Bewick, T.A. (1997) Control of dodder in cranberry, *Vaccinium macrocarpon*, with a pathogen-based bioherbicide. In: *Proceedings of the Sixth International Symposium on Vaccinium culture*, pp. 427–8. Orno, ME.

Hopper, K.R. and Roush, R.T. (1993) Mate finding, dispersal, number released, and the success of biological control introductions. *Ecological Entomology* **18**: 321–31.

Hopper, K.R., Roush, R.T., and Powell, W. (1993) Management of genetics of biological-control introductions. *Annual Review of Entomology* **38**: 27–51.

Horiguchi, T., Li, Z., Uno, S. et al. (2004) Contamination of organotin compounds and imposex in mollusks from Vancouver, Canada. *Marine Environmental Research* **57**: 75–88.

Hossain, Z., Gurr, G.M., and Wratten, S.D. (2001) Habitat manipulation in lucerne (*Medicago sativa* L.): strip harvesting to enhance biological control of insect pests. *International Journal of Pest Management* **47**: 81–7.

Hossain, Z., Gurr, G.M., Wratten, S.D., and Raman, A. (2002) Habitat manipulation in lucerne, *Medicago sativa*: arthropod population dynamics in harvested and "refuge" crop strips. *Journal of Applied Ecology* **39**: 445–54.

Hoti, S.L. and Balaraman, K. (1990) Utility of cheap carbon and nitrogen sources for the production of a mosquito-pathogenic fungus, *Lagenidium*. *Indian Journal of Medical Research, Section A, Infectious Diseases* **91**: 67–9.

Howard, F.W., Pemberton, R.W., Hamon, A. et al. (2002) Lobate lac scale, *Paratachardina lobata lobata* (Hemiptera: Stenorrhycha: Coccoidea: Kerridae). Featured Creatures, University of Florida, http://creatures.ifas.ufl.edu/orn/scales/lobate_lac.htm.

Howard, L.O. and Fiske, W.F. (1911) The importation into the United States of the parasites of the gipsy-moth and the brown-tail moth. USDA, Bureau of Entomology Bulletin no. 91. USDA, Washington, D.C.

Howard, R.W. and Flinn, P.W. (1990) Larval trails of *Cryptolestes ferrugineus* (Coleoptera: Cucujidae) as kairomonal host-finding cues for the parasitoid *Cephalonomia waterstoni* (Hymenoptera: Bethylidae). *Annals of the Entomological Society of America* **83**: 239–44.

Howarth, F.G. (1983) Classical biocontrol: panacea or Pandora's box. *Proceedings of the Hawaiian Entomological Society* **24(2/3)**: 239–44.

Howarth, F.G. (1991) Environmental impacts of classical biological control. *Annual Review of Entomology* **36**: 485–509.

Howell, P.G. (1984) An evaluation of the biological control of the feral cat *Felis catus* (Linnaeus, 1758). *Acta Zoologica Fennica* **172**: 111–13.

Hoy, M.A. (1982) Aerial dispersal and field efficacy of a genetically improved strain of the spider mite predator *Metaseiulus occidentalis*. *Entomologia Experimentalis et Applicata* **32**: 205–12.

Hoy, M.A. (1990) Pesticide resistance in arthropod natural enemies: variability and selection responses. In: Roush, R.T. and Tabashnik, B.E. (eds.), *Pesticide Resistance in Arthropods*, pp. 203–36. Chapman and Hall, New York.

Hoy, M.A. (1994) *Insect Molecular Genetics*. Academic Press, San Diego, CA.

Hoy, M.A. and Cave, F.E. (1988) Guthion-resistant strain of walnut aphid parasite. *California Agriculture* **42(4)**: 4–5.

Hoy, M.A. and Cave, F.E. (1989) Toxicity of pesticides used on walnuts to a wild and azinphosmethyl-resistant strain of *Trioxys pallidus* (Hymenoptera: Aphididae). *Journal of Economic Entomology* **82**: 1585–92.

Hoy, M.A., Westigard, P.H., and Hoyt, S.C. (1983) Release and evaluation of laboratory-selected, pyrethroid-resistant strains of the predaceous mite *Typhlodromus occidentalis* (Acarina: Phytoseiidae) into southern Oregon pear orchards and Washington apple orchards. *Journal of Economic Entomology* **76**: 383–8.

Hoy, M.A., Cave, F.E., Beede, R.H. et al. (1990) Release, dispersal, and recovery of a laboratory-selected strain of the walnut aphid parasite *Trioxys pallidus* (Hymenoptera: Aphidiidae) resistant to azinphosmethyl. *Journal of Economic Entomology* **83**: 89–96.

Hoyt, S.C. and Caltagirone, L.E. (1971) The developing programs of integrated control of pests of apples in Washington and peaches in California. In: Huffaker, C.B. (ed.), *Biological Control*, pp. 395–421. Plenum Press, New York.

Hua, L.Z., Lammes, F., van Lenteren, J.C., Huisman, P.W.T., van Vianen, A., and de Ponti, O.M.B. (1987) The parasite–host relationship between *Encarsia formosa* Gahan (Hymenoptera, Aphelinidae) and *Trialeurodes vaporariorum* (Westwood) (Homoptera: Aleyrodidae). XXV. Influence of leaf structure on the searching activity of *Encarsia formosa*. *Journal of Applied Entomology* **104**: 297–304.

Huber, J. (1986) Use of baculoviruses in pest management programs. In: Granados, R.R. and Federici, B.A. (eds.), *The Biology of Baculoviruses: Volume II. Practical Application for Insect Control*, pp. 181–202. CRC Press, Boca Raton, FL.

Huber, J. (1990) History of CPGV as a biological control agent – its long way to a commercial viral pesticide. In: Pinnock, D.E. (ed.), *Vth International Colloquium on Invertebrate Pathology and Microbial Control*, pp. 424–7, August 20–24, 1990, Adelaide, Australia. Department of Entomology, University of Adelaide, Glen Osmond, South Australia.

Hufbauer, R.A. and Roderick, G.K. (2005) Microevolution in biological control: mechanisms, patterns and processes. *Biological Control* **35**: 227–39.

Hufbauer, R.A., Bogdanowicz, S.M., and Harrison, R.G. (2004) The population genetics of a biological control introduction: mitochondrial DNA and microsatellite variation in native and introduced populations of *Aphidius ervi*, a parasitoid wasp. *Molecular Ecology* **13**: 337–48.

Huffaker, C.B. and Kennett, C.E. (1956) Experimental studies on predation: (1) Predation and cyclamen mite populations on strawberries in California. *Hilgardia* **26**: 191–222.

Huffaker, C.B. and Messenger, P.S. (1964) The concept and significance of natural control. In: DeBach, P. (ed.), *Biological Control of Insect Pests and Weed*, pp. 74–117. Chapman and Hall, London.

Huffaker, C.B. and Kennett, C.E. (1969) Some aspects of assessing efficiency of natural enemies. *The Canadian Entomologist* **101**: 425–47.

Huffaker, C.B. and Messenger, P.S. (eds.) (1976) *Theory and Practice of Biological Control.* Academic Press, New York.

Huffaker, C.B., Hamai, J., and Nowierski, R.M. (1983) Biological control of puncturevine, *Tribulus terrestris* in California after twenty years of activity of introduced weevils. *Entomophaga* **28**: 387–400.

Hughes, R.F. and Denslow, J.S. (2005) Invasion by a N_2-fixing tree alters function and structure in wet lowland forests of Hawaii. *Ecological Applications* **15**: 1615–28.

Hull, L.A. and Beers, E.H. (1985) Ecological selectivity: modifying chemical control practices to preseve natural enemies. In: Hoy, M.A. and Herzog, D.C. (eds.), *Biological Control in Agricultural IPM Systems*, pp. 103–22. Academic Press, Orlando, FL.

Hull, L.A., Hickey, K.D., and Kanour, W.W. (1983) Pesticide usage patterns and associated pest damage in commercial apple orchards of Pennsylvania. *Journal of Economic Entomology* **76**: 577–83.

Humber, R.A. (1981) An alternative view of certain taxonomic criteria used in the Entomophthorales (Zygomycetes). *Mycotaxon* **13**: 191–240.

Hunter, C.D. (1997) *Suppliers of Beneficial Organisms in North America.* California Environmental Protection Agency, Sacramento, CA.

Hunt-Joshi, T.R., Blossey, B., and Root, R.B. (2004) Root and leaf herbivory on *Lythrum salicaria*: implications for plant performance and communities. *Ecological Applications* **14**: 1574–8.

Hunt-Joshi, T.R., Root, R.B., and Blossey, B. (2005) Disruption of weed biological control by an opportunistic mirid predator. *Ecological Applications* **15**: 861–70.

Hurd, H. (1993) Reproductive disturbances induced by parasites and pathogens of insects. In: Beckage, N.E., Thompson, S.N., and Federici, B.A. (eds.), *Parasites and Pathogens of Insects, Volume I. Parasites*, pp. 87–105. Academic Press, New York.

Hurst, G.D.D. and Jiggins, F.M. (2000) Male-killing bacteria in insects: Mechanisms, incidence, and implications. *Emerging Infectious Diseases* **6**: 329–36.

Hurst, G. and Jiggins, F. (2005) Problems with mitochondrial DNA as a marker in population, phylogeographic and phylogenetic studies: the effects of inherited symbionts. *Proceedings of the Royal Society of London Series B Biological Sciences* **272**: 1525–34.

Hussey, N.W. (1985) History of biological control in protected culture. In: Hussey, W.N. and Scopes, N. (eds.), *Biological Pest Control: the Glasshouse Experience*, pp. 11–22. Cornell University Press, Ithaca, NY.

Ignoffo, C.M. (1973) Development of a viral insecticide: concept to commercialization. *Experimental Parasitology* **33**: 380–406.

Ignoffo, C.M., Garcia, C., Kapp, R.W., and Coate, W.B. (1979) An evaluation of the risks to mammals of the use of an entomopathogenic fungus, *Nomuraea rileyi*, as a microbial insecticide. *Environmental Entomology* **8**: 354–9.

Ignoffo, C.M., Garcia, C., Hostetter, D.L., and Pinnell, R.E. (1980) Transplanting: a method of introducing an insect virus into an ecosystem. *Environmental Entomology* **9**: 153–4.

Ignoffo, C.M., Shasha, B.S., and Shapiro, M. (1991) Sunlight ultraviolet protection of the *Heliothis* nuclear polyhedrosis virus through starch-encapsulation technology. *Journal of Invertebrate Pathology* **57**: 134–6.

Iline, I.I. and Phillips, C.B. (2004) Allozyme markers to help define the South American origins of *Microctonus hyperodae* (Hymenoptera: Braconidae) established in New Zealand for biological control of Argentine stem weevil. *Bulletin of Entomological Research* **94**: 229–34.

Impson, F.A.C. and Moran, V.C. (2004) Thirty years of exploration for and selection of a succession of *Melanterius* weevil species for biological control of invasive Australian acacias in South Africa: should we have done anything differently? In: Cullen, J.M.D., Briese, T., Kriticos, D.J., Lonsdale, W.M., Morin, L., and Scott, J.K. (eds.) *Proceedings of the XIth International Symposium on Biological Control of Weeds* pp. 127–34, April 27–May 2, 2003, Canberra, Australia. CSIRO Entomology, Canberra.

Impson, F.A.C., Moran, V.C., and Hoffmann, J.H. (1999) A review of the effectiveness of seed-feeding bruchid beetles in the biological control of mesquite, *Prosopis* species (Fabaceae), in South Africa. *African Entomology Memoir* **1**: 81–8.

Inari, N., Nagamitsu, T., Kenta, T., Goka, K., and Hiura, T. (2005) Spatial and temporal pattern of introduced *Bombus terrestris* abundance in Hokkaido, Japan, and its potential impact on native bumblebees. *Population Ecology* **47**: 77–82.

Inoue, K., Osakabe, M., and Ashihara, W. (1987) Identification of pesticide-resistant phytoseiid mite populations in citrus orchards, and on grapevines in glasshouses and vinyl-houses (Acarina: Phytoseiidae). *Japanese Journal of Applied Entomology and Zoology* **31**: 398–403.

Irvin, N.A., Scarratt, S.L., Wratten, S.D., Frampton, C.M., Chapman, R.B., and Tylianakis, J.M. (2006) The effects of floral understoreys on parasitism of leafrollers (Lepidoptera: Tortricidae) on apples in New Zealand. *Agricultural and Forest Entomology* **8**: 25–34.

Ishibashi, N. and Kondo, E. (1990) Behavior of infective juveniles. In: Gaugler, R. and Kaya, H. (eds.), *Entomopathogenic*

Nematodes in Biological Control, pp. 139–50. CRC Press, Boca Raton, FL.

Ishiwata, S. (1910) On a type of severe flacherie (sotto disease). *Dainihon Sanshi Kaiho* **114**: 1–5.

Ives, A.R., Kareiva, P., and Perry, R. (1993) Response of a predator to variation in prey density at three hierarchical scales: lady beetles feeding on aphids. *Ecology* **74**: 1929–38.

Ives, W.G.H. (1976) The dynamics of larch sawfly (Hymenoptera: Tenthredinidae) populations in southeastern Manitoba. *The Canadian Entomologist* **108**: 701–30.

Jackson, M.A., Cliquet, S., and Iten, L.B. (2003) Media and fermentation processes for the rapid production of high concentrations of stable blastospores of the bioinsecticide fungus *Paecilomyces fumosoroseus*. *Biocontrol Science and Technology* **13**: 23–33.

Jackson, R.E. and Pitre, H.N. (2004) Influence of Round Up Ready® soybean production systems and glyphosate applications on pest and beneficial insects in wide-row soybean. *Journal of Agricultural and Urban Entomology* **21**: 61–70.

Jackson, R.J., Ramsay, A.J., Christensen, C.D., Beaton, S., Hall, D.F., and Ramshaw, I.A. (2001) Expression of mouse interleukin-4 by a recombinant ectromelia virus suppresses cytolytic lymphocyte responses and overcomes genetic resistance to mousepox. *Journal of Virology* **75**: 1205–10.

Jackson, T.A. (1990) Commercial development *of Serratia entomophilia* as a biocontrol agent for the New Zealand grass grub. In: Pinnock, D.E. (ed.), *Vth International Colloquium on Invertebrate Pathology and Microbial Control*, p. 15, August 20–24, 1990, Adelaide, Australia. Department of Entomology, University of Adelaide, Glen Osmond, South Australia.

Jackson, T.A. (1994) Development of biopesticides – lessons from Invade®, a commercial microbial control agent for the New Zealand grass grub. In: Monsour, C.J., Reid, S., and Teakle, R.E. (eds.), *Biopesticides: Opportunities for Australian Industry*. Proceedings of the 1st Brisbane Symposium, June 9–10, 1994.

Jackson, T.A. and Wouts, W.M. (1987) Delayed action of an entomophagous nematode (*Heterorhabditis* sp. [V16]) for grass grub control. *Proceedings of the New Zealand Weed and Pest Control Conference*, pp. 33–5. New Zealand Weed and Pest Control Society, Palmerston North.

Jackson, T.A., Crawford, A.M., and Glare, T.R. (2005) *Oryctes* virus – time for a new look at a useful biocontrol agent. *Journal of Invertebrate Pathology* **89**: 91–4.

Jacobs, S.E. (1951) Bacteriological control of the flour moth, *Ephestia kuehniella* Z. *Proceedings of the Society of Applied Bacteriology* **13**: 83–91.

Jacobson, R.J. and Croft, P. (1998) Strategies for the control of *Aphis gossypii* Glover (Hom.: Aphididae) with *Aphidius colemani* Viereck (Hym.: Braconidae) in protected cucumbers. *Biocontrol Science and Technology* **8**: 377–87.

Jaffe, M. (1994) *And No Birds Sing*. Simon and Schuster, New York.

Jaga, K. and Dharmani, C. (2003) Global surveillance of DDT and DDE levels in human tissues. *International Journal of Occupational Medicine and Environmental Health* **16**: 7–20.

Jakes, K.A., Donoghue, P.J., and Whittier, J. (2003) Ultrastructure of *Hepatozoon boiga* (Mackerras, 1961) nov. comb. from brown tree snakes, *Boiga irregularis*, from northern Australia. *Parasitology Research* **3**: 225–31.

Jamal, E. and Brown, G.C. (2001) Orientation of *Hippodamia convergens* (Coleoptera: Coccinellidae) larvae to volatile chemicals associated with *Myzus nicotianae* (Homoptera: Aphidiidae). *Environmental Entomology* **30**: 1012–16.

James, C. (2002) Global review of commercialized transgenic crops: 2001. Feature: Bt. Cotton. *ISAAA Briefs* no. 26.

James, D.G. (1989) Overwintering of *Amblyseius victoriensis* (Womersley) (Acarina: Phytoseiidae) in southern New South Wales. *General Applied Entomology* **21**: 51–5.

James, D.G. (1990) Biological control of *Tetranychus urticae* Koch (Acarina: Tetranychidae) in southern New South Wales peach orchards: the role of *Amblyseius victoriensis*. *Australian Journal of Zoology* **37**: 645–55.

James, D.G. (1993) Pollen, mould mites, and fungi: improvements to mass rearing of *Typhlodromus doreenae* and *Amblyseius victoriensis*. *Experimental and Applied Acarology* **14**: 271–6.

James, D.G. (2003) Synthetic herbivore-induced plant volatiles as field attractants for beneficial insects. *Environmental Entomology* **32**: 977–82.

James, D.G. (2005) Further field evaluation of synthetic herbivore-induced plant volatiles as attractants for beneficial insects. *Journal of Chemical Ecology* **31**: 481–95.

James, D.G. (2006) Methyl salicylate is a field attractant for the goldeneyed lacewing, *Chrysopa oculata*. *Biocontrol Science and Technology* **16**: 107–10.

James, D.G. and Whitney, J. (1993) Mite populations on grapevines in southeastern Australia: implications for biological control of grapevine mites. *Experimental and Applied Acarology* **17**: 259–70.

James, D.G. and Price, T.S. (2004) Field-testing of methyl salicylate for recruitment and retention of beneficial insects in grapes and hops. *Journal of Chemical Ecology* **30**: 1613–28.

James, D.G. and Grasswitz, T.R. (2005) Synthetic herbivore-induced plant volatiles increase field captures of parasitic wasps. *BioControl* **50**: 871–80.

James, D.G., Stevens, M.M., O'Malley, K.J., and Faulder, R.J. (1999) Ant foraging reduces the abundance of beneficial and incidental arthropods in citrus canopies. *Biological Control* **14**: 121–6.

James, R.R., McEvoy, P.B., and Cox, C.S. (1982) Combining the cinnabar moth (*Tyria jacobaeae*) and the ragwort flea beetle (*Longitarsus jacobaeae*) for control of ragwort (*Senecio jacobaea*): an experimental analysis. *Journal of Applied Ecology* **29**: 589–96.

Jansson, R.K. (1993) Introduction of exotic entomopathogenic nematodes (Rhabditida: Heterorhabditidae and Steinernematidae) for biological control of insects: potential and problems. *Florida Entomologist* **76**: 82–96.

Janzen, D.H. (1979) New horizons in the biology of plant defenses. In: Rosenthal, G.A. and Janzen, D.H. (eds.), *Herbivores: their Interaction with Secondary Plant Metabolites*, pp. 331–50. Academic Press, New York.

Jarvis C.H. and Baker, R.H.A. (2001) Risk assessment for nonindigenous pests: I. Mapping the outputs of phenology models to assess the likelihood of establishment. *Diversity & Distributions* **7**: 223–35.

Jenkins, N.E. and Grzywacz, D. (2000) Quality control of fungal and viral biocontrol agents – assurance of product performance. *Biocontrol Science and Technology* **10**: 753–77.

Jepson, P.C. (ed.) (1989) *Pesticides and Non-Target Invertebrates*. Intercept, Wimborne.

Jervis, M.A. and Kidd, N.A.C. (1986) Host-feeding strategies in hymenopteran parasitoids. *Biological Reviews* **61**: 395–434.

Jervis, M. and Kidd, N. (1996) *Insect Natural Enemies: Practical Approaches to their Study and Evaluation*. Chapman and Hall, London.

Jervis, M.A. and Ferns, P.N. (2004) The timing of egg maturation in insects: ovigeny index and initial egg load as measures of fitness and resource allocation. *Oikos* **107**: 449–60.

Jeyaprakash, A. and Hoy, M.A. (2000) Long PCR improves *Wolbachia* DNA amplification: wsp sequences found in 76% of sixty-three arthropod species. *Insect Molecular Biology* **9**: 393–405.

Jhansi, L.V., Krishnaih, K., Lingaiah, T., and Pasalu, I.C. (2000) Rice leafhopper and planthopper honeydew as a source of host searching kairomone for the mirid predator, *Cyrtorhinus lividpennis* (Reuter) (Hemiptera: Miridae). *Journal of Biological Control* **14**: 7–13.

Johnson, D.M. and Stiling, P.D. (1996) Host specificity of *Cactoblastis cactorum* (Lepidoptera: Pyralidae), an exotic *Opuntia*-feeding moth, in Florida. *Environmental Entomology* **25**: 743–8.

Johnson, D. and Stiling, P.D. (1998) Distribution and dispersal of *Cactoblastis cactorum* (Lepidoptera: Pyralidae), an exotic *Opuntia*-feeding moth, in Florida. *Florida Entomologist* **81**: 12–22.

Johnson, M.T., Follett, P.A., Taylor, A.D., and Jones, V.P. (2005) Impacts of biological control and invasive species on a non-target native Hawaiian species. *Oecologia* **142**: 529–40.

Johnson, M.W. and Hara, A.H. (1987) Influence of host crop on parasitoids (Hymenoptera) of *Liriomyza* spp. (Diptera: Agromyzidae). *Environmental Entomology* **16**: 339–44.

Johnson, N. (2005) *Catalog of the World Proctotrupoidea, excluding Platygastridae*. Memoirs of the American Entomological Institute 51.

Jolivet, P. and Verma, K.K. (2002) *Biology of Leaf Beetles*. Intercept, Andover.

Jolly, S.E. (1993) Biological control of possums. *New Zealand Journal of Zoology* **20**: 335–9.

Jones, D. (1985) Endocrine interaction between host (Lepidoptera) and parasite (Cheloninae: Hymenoptera): is the host or the parasite in control? *Annals of the Entomological Society of America* **78**: 141–8.

Jones, D., Jones, G., Van Steenwyk, R.A., and Hammock, B.D. (1982) Effect of the parasite *Copidosoma truncatellum* on development of it host *Trichoplusia ni*. *Annals of the Entomological Society of America* **75**: 7–11.

Jones, D., Snyder, M., and Granett, J. (1983) Can insecticides be integrated with biological control agents of *Trichoplusia ni* in celery? *Entomologia Experimentalis et Applicata* **33**: 290–6.

Jones, H.P., Williamhenry, R., Howald, G.R., Tershy, B., and Croll, D. (2005) Predation of artificial Xanthus's murrelet (*Synthliboramphus hypoleucus scrippsi*) nests before and after black rat (*Rattus rattus*) eradication. *Environmental Conservation* **32**: 320–5.

Jones, S.A., Hodges, R.J., Birkinshaw, L.A., and Hall, D.R. (2004) Responses of *Teretrius nigrescens* towards the dust and frass if its prey, *Prostephanus truncates*. *Journal of Chemical Ecology* **30**: 1629–46.

Jones, W.A. (1988) World review of the parasitoids of the southern green stink bug, *Nezara viridula* (L.) (Heteroptera: Pentatomidae). *Annals of the Entomological Society of America* **81**: 262–73.

Jones, W.A. and Greenberg, S.M. (1998) Suitability of *Bemisia argentifolii* (Homoptera: Aleyrodidae) instars for the parasitoid *Eretmocerus mundus* (Hymenoptera: Aphelinidae). *Environmental Entomology* **27**: 1569–73.

Jongejans, E., Sheppard, A.W., and Shea, K. (2006) What controls the population dynamics of the invasive thistle *Carduus nutans* in its native range? *Journal of Applied Ecology* **43**: 877–86.

Jonsen, I.D., Bourchier, R.S., and Roland, J. (2001) The influence of matrix habitat on *Aphthona* flea beetle immigration to leafy spurge patches. *Oecologia* **127**: 287–94.

Joshi, J. and Vrieling, K. (2005) The enemy release and EIC hypothesis revisted: incorporating the fundamental difference between specialist and generalist herbivores. *Ecology Letters* **8**: 704–14.

Joshi, R.K. and Sharma, S.K. (1989) Augmentation and conservation of *Epiricania melanoleuca* Fletcher, for the population management of sugarcane leafhopper, *Pyrilla perpusilla* Walker, under arid conditions of Rajasthan. Indian Sugar **39(8)**: 625–8.

Jousson, O., Pawlowski, J., Zaninetti, L. et al. (2000) Invasive alga reaches California. *Nature* **408**: 157–8.

Julien, M.H. (1981) Control of aquatic *Alternanthera philoxeroides* in Australia; another success for *Agasicles hygrophila*. In: Delfosse, E.S. (ed.), *Proceedings of the V International Symposium on Biological Control of Weeds*, pp. 583–8,

July 22–9, 1980, Brisbane, Australia. CSIRO Publishing, Melbourne.

Julien, M.H. (1982) *Biological Control of Weeds: a World Catalogue of Agents and their Target Weeds.* Commonwealth Institute of Biological Control, Commonwealth Agricultural Bureaux, Farnham Royal, Slough.

Julien, M.H. and Griffiths, M.W. (eds.) (1998) *Biological Control of Weeds: a World Catalogue of Agents and their Target Weeds*, 4th edn. CABI Publishing, Wallingford.

Julien, M.H., Kerr, J.D., and Chan, R.R. (1984) Biological control of weeds: an evaluation. *Protection Ecology* **7**: 3–25.

Julien, M.H., Center, T.D., and Tipping, P.W. (2002) Floating fern (salvinia). In: Coombs, E.M., Clark, J.K., Piper, G.L., and Cofrancesco, Jr., A.F. (eds.), *Biological Control of Invasive Plants in the United States*, pp. 17–32. Oregon State University Press, Corvallis, OR.

Kabaluk, T. and Gazdik, K. (2004) *Directory of Microbial Pesticides for Agricultural Crops in the OECD Countries.* Agriculture and Agri-Food Canada. www.agr.gc.ca/env/pdf/cat_e.pdf.

Kainoh, Y., Tatsuki, S., Sugie, H., and Tamaki, Y. (1989) Host egg kairomones essential for egg-larval parasitoid, *Ascogaster reticulatus* Watanabe (Hymneoptera: Bracondiae). II. Identification of internal kairomone. *Journal of Chemical Ecology* **15**: 1219–29.

Kainoh, Y., Tatsuki, S., and Kusano, T. (1990) Host moth scales: a cue for host location for *Ascogaster reticulatus* Watanabe (Hymenoptera: Braconidae). *Applied Entomology and Zoology* **25**: 17–25.

Kairo, M.T.K., Pollard, G.V., Peterkin, D.D., and Lopez, V.F. (2000) Biological control of the hibiscus mealybug, *Maconellicoccus hirsutus* Green (Hemiptera: Pseudococcidae) in the Caribbean. *Integrated Pest Management Reviews* **5**: 241–54.

Kalaydjiev, S.K., Vasilevska, M., and Nakov, L. (2000) Common egg envelope antigens are limited to animal class. *Theriogenology* **53**: 1467–75.

Kamal, M. (1951) Biological control projects in Egypt, with a list of introduced parasites and predators. *Bulletin de la Société Fouad I d'Entomologie* **35**: 205–20.

Kankare, M., Van Nouhuys, S., and Hanski, I. (2005) Genetic divergence among host-specific cryptic species in *Cotesia melitaearum* aggregate (Hymenoptera: Braconidae), parasitoids of checkerspot butterflies. *Annals of the Entomological Society of America* **98**: 382–94.

Karban, R. and Myers, J.H. (1989) Induced plant responses to herbivory. *Annual Review of Ecology and Systematics* **20**: 331–48.

Kareiva, P. and Perry, R. (1989) Leaf overlap and ability of ladybird beetles to search among plants. *Ecological Entomology* **14**: 127–9.

Kareiva, P. and Sahakian, R. (1990) Tritrophic effects of a simple architectural mutation in pea plants. *Nature* **345**: 433–4.

Kassa, A., Stephan, D., Vidal, S., and Zimmermann, G. (2004) Laboratory and field evaluation of different formulations of *Metarhizium anisopliae* var. *acridum* submerged spores and aerial conidia for the control of locusts and grasshoppers. *BioControl* **49**: 63–81.

Katovich, E.J.S., Becker, R.L., and Ragsdale, D.W. (1999) Effect of *Galerucella* spp. on survival of purple loosestrife (*Lythrum salicaria*) roots and crowns. *Weed Science* **47**: 360–5.

Kaufman, P.E., Reasor, C., Rutz, D.A., Ketzis, J.K., and Arends, J.J. (2005) Evaluation of *Beauveria bassiana* applications against adult house fly, *Musca domestica*, in commercial caged-layer poultry facilities in New York state. *Biological Control* **33**: 360–7.

Kawakami, K. (1987) The use of an entomogenous fungus, *Beauveria brongniartii*, to control the yellow-spotted longicorn beetle, *Psacothea hilaris*. In: *Biological Pest Control for Field Crops.* Summaries of papers presented at the International Seminar on Biological Pest Control for Field Crops, pp. 38–9, August–September, 1986, Kyushu, Japan. Extension Bulletin no. 257. ASPAC Food and Fertilizer Technology Center for the Asian and Pacific Region, Taipei, Taiwan.

Kaya, H.K. (1985) Entomogenous nematodes for insect control in IPM systems. In: Hoy, M.A. and Herzog, D.C. (eds.), *Biological Control in Agricultural IPM Systems*, pp. 283–302. Academic Press, New York.

Kaya, H.K. (1993) Entomogenous and entomopathogenic nematodes in biological control. In: Evans, K., Trudgill, D.L., and Webster, J.M. (eds.), *Plant Parasitic Nematodes in Temperate Agriculture*, pp. 565–91. Commonwealth Agricultural Bureaux International, Cambridge University Press, Cambridge.

Kaya, H.K. and Gaugler, R. (1993) Entomopathogenic nematodes. *Annual Review of Entomology* **38**: 181–206.

Kaya, H.K., Burlando, T.M., and Thurston, G.S. (1993) Two entomopathogenic nematode species with different search strategies for insect suppression. *Environmental Entomology* **22**: 859–64.

Kazmer, D.J. (1991) Isoelectric-focusing procedures for the analysis of allozymic variation in minute arthropods. *Annals of the Entomological Society of America* **84**: 332–9.

Kazmer, D.J. and Luck, R.F. (1995) Field tests of the size-fitness hypothesis in the egg parasitoid *Trichogramma pretiosum*. *Ecology* **76**: 412–25.

Kazmer, D.J., Hopper, K.R., Coutinot, D.M., and Heckel, D.G. (1995) Suitability of random amplified polymorphic DNA for genetic markers in the aphid parasitoid, *Aphelinus asychis* Walker. *Biological Control* **5**: 503–12.

Keating, S.T., Burand, J.P., and Elkinton, J.S. (1989) DNA hybridization assay for detection of gypsy moth nuclear polyhedrosis virus in infected gypsy moth (*Lymantria dispar* L.) larvae. *Applied and Environmental Microbiology* **55**: 2749–54.

Keller, M.A. (1987) Influence of leaf surfaces on movements by the hymenopterous parasitoid *Trichogramma exiguum*. *Entomologia Experimentalis et Applicata* **43**: 55–9.

Kenis, M. and Mills, N.J. (1994) Parasitoids of European species of the genus *Pissodes* (Col.: Curculionidae) and their potential for the biological control of *Pissodes strobi* (Peck) in Canada. *Biological Control* **4**: 14–21.

Kenis, M., Tomov, R., Svatos, A. et al. (2005) The horse-chestnut leaf miner in Europe – prospects and constraints for biological control. In: Hoddle, M.S. (ed.), *2nd International Symposium on Biological Control of Arthropods*, pp. 77–90, September 12–16, 2005, Davos, Switzerland. FHTET-2005–08. USDA Forest Service, Morgantown, WV.

Kennett, C.E., Flaherty, D.L., and Hoffmann, R.W. (1979) Effect of wind-borne pollens on the population dynamics of *Amblyseius hibisci* (Acarina: Phytoseiidae). *Entomophaga* **24**: 83–98.

Kennett, C.E., McMurtry, J.A., and Beardsley, J.W. (1999) Biological control in subtropical and tropical crops. In: Bellows, T.S. and Fisher, T.W. (eds.), *Handbook of Biological Control*, pp. 713–42. Academic Press, San Diego, CA.

Kenney, F.D. (1986) DeVine – the way it was developed – an industrialist's view. *Weed Science* **34** (supplement 1): 15–16.

Kerguelen, V. and Hoddle, M.S. (1999) Biological control of *Oligonychus perseae* (Acari: Tetranychidae) on avocado: II. Evaluating the efficacy of *Galendromus helveolus* and *Neoseiulus californicus* (Acari: Phytoseiidae). *International Journal of Acarology* **25**: 221–9.

Kerwin, J.L. (1992) Testing the effects of microorganisms on birds, pp. 729–44. In: Levin, M.A., Seidler, R.J., and M. Rogul (eds.), *Microbial Ecology: Principles, Methods, and Applications*, pp. 729–44. McGraw-Hill, New York.

Kerwin, J.L., Dritz, D.A., and Wahino, R.K. (1990) Confirmation of the safety of *Lagenidium giganteum* (Oomycetes: Lagenidiales) to mammals. *Journal of Economic Entomology* **83**: 374–6.

Kessler, P., Enkerli, J., Schweizer, C., and Keller, S. (2004) Survival of *Beauveria brongniartii* in the soil after application as a biocontrol agent against the European cockchafer *Melolontha melolontha*. *BioControl* **49**: 563–581.

Kester, K.M. and Barbosa, P. (1992) Effects of postemergence experience on searching and landing responses of the insect parasitoid, *Coteisa congregata* (Say) (Hymenoptera: Braconidae), to plants. *Journal of Insect Behavior* **5**: 301–20.

Kfir, R. (1998) Origin of the diamondback moth (Lepidoptera: Plutellidae). *Annals of the Entomological Society of America* **91**: 164–7.

Khetan, S.K. (2001) *Microbial Pest Control*. Marcel Dekker, New York.

Kiesecker, J.M. (2002) Synergism between trematode infection and pesticide exposure: a link to amphibian limb deformities in nature? *Proceedings of the National Academy of Sciences USA* **99**: 9900–4.

Kidd, M.A. (2005) *Insects as Natural Enemies: a Practical Perspective*. Kluwer, Academic Publishers, Dordrecht.

Kiefer, H.H., Baker, E.W., Kono, T., Delfinado, M., and Styer, W.E. (1982) *An Illustrated Guide to Plant Abnormalities Caused by Eriophyid Mites in North America*, Agricultural Handbook no. 573. USDA, Washington, D.C.

Killgore, E.M., Sugiyama, L.S., Barreto, R.W., and Gardner, D.E. (1999) Evaluation of *Colletotrichum gloeosporioides* for biological control of *Miconia calvescens* in Hawaii. *Plant Disease* **83**: 964.

Kindlmann, P. and Dixon, A.F.G. (1999) Generation time ratios – determinants of prey abundance in insect predator-prey interactions. *Biological Control* **16**: 133–8.

King, E.G., Hopper, K.R., and Powell, J.E. (1985) Analysis of systems for biological control of crop arthropod pests in the U.S. by augmenation of predators and parasites. In: Hoy, M.A. and Herzog, D.C. (eds.), *Biological Control in Agriculutral IPM Systems*, pp. 201–27. Academic Press, Orlando, FL.

King, G.A., Daugulis, A.J., Faulkner, P., Bayly, D., and Goosen, M.F.A. (1988) Growth of baculovirus-infested insect cells in microcapsules to a high cell and virus density. *Biotechnology Letters* **10**: 683–8.

King, J.L. (1931) The present status of the established parasites of *Popillia japonica* Newman. *Journal of Economic Entomology* **24**: 453–62.

Kinnear, J.E., Sumner, N.R., and Onus, M.L. (2002) The red fox in Australia – an exotic predator turned biocontrol agent. *Biological Conservation* **108**: 335–9.

Kinzie, III, R.A. (1992) Predation by the introduced carnivorous snail *Euglandia rosea* (Ferussac) on endemic aquatic lymnaeid snails in Hawaii. *Biological Conservation* **60**: 149–55.

Kirby, W. and Spence, W. (1815) *An Introduction to Entomology*. Longman, Brown, Green and Longmans, London.

Kiritani, K. and Nakasuji, F. (1967) Estimations of the stage-specific survival rate in the insect population with overlapping stages. *Researches on Population Ecology* **9**: 143–52.

Kiritani, K., Kawahara, S., Sasaba, T., and Nakasuji, F. (1972) Quantitative evaluation of predation by spiders on the green rice leafhopper, *Nephotettix cinctipes* Uhler, by a sight-count method. *Researches on Population Ecology* **13**: 187–200.

Kirkpatrick, J.F. and Frank, K.M. (2005) Contraception in free ranging wildlife. In: Asa, C.S. and Porton, I.J. (eds.), *Wildlife Contraception: Issues, Methods, and Applications*, pp. 195–221. The John Hopkins University Press, Baltimore, MD.

Kirkpatrick, J.F., Liu, I.M.K., Turner, Jr., J.W., Naugle, R., and Keiper, R. (1992) Long-term effects of porcine zonae pellucidae immunocontraception on ovarian function in feral horses (*Equus caballus*). *Journal of Reproduction and Fertility* **94**: 437–44.

Kirkpatrick, J.F., Turner, Jr., J.W., Liu, I.K.M., Fyrer-Hosken, R., and Rutberg, A.T. (1997) Case studies in wildlife immunocontraception: wild and feral equids and white-tailed deer. *Reproduction, Fertility and Development* **9**: 105–10.

Kleinjan, C.A., Morin, L., Edwards, P.B., and Wood, A.R. (2004) Distribution, host range, and phenology of the rust fungus *Puccinia myrsiphylli* in South Africa. *Australasian Plant Pathology* **33**: 263–71.

Klingman, D.L. and Coulson, J.R. (1982) Guidelines for introducing foreign organisms into the United States for biological control of weeds. *Weed Science* **30**: 661–7.

Klomp, H. (1958) On the synchronization of the generations of the tachinid *Carcelia obesa* Zett. (=*rutilla* B.B.) and its host *Bupalus piniarus*. *Zeitshcrift für Angewandte Entomologie* **42**: 210–17.

Kloot, P.M. (1983) The role of common iceplant (*Mesembryanthemum crystallinum*) in the deterioration of medic pastures. *Australian Journal of Ecology* **8**: 301–6.

Kluge, R.L. and Caldwell, P.M. (1992) Microsporidian diseases and biological weed control agents: to release or not to release? *Biocontrol News and Information* **13(3)**: 43N–7N.

Knapp, R.A. and Matthews, K.R. (2000) Non-native fish introductions and the decline of the mountain yellow-legged frog from within protected areas. *Conservation Biology* **14**: 128–38.

Knutson, A.E. and Gilstrap, F.E. (1989) Direct evaluation of natural enemies of the southwestern corn borer (Lepidoptera: Pyralidae) in Texas corn. *Environmental Entomology* **18**: 732–9.

Knutson, A.E. and Tedders, L. (2002) Augmentation of green lacewing, *Chrysoperla rufilabris*, in cotton in Texas. *Southwestern Entomologist* **27**: 231–9.

Knutson, L., Sailer, R.I., Murphy, W.L., Carlson, R.W., and Dogger, J.R. (1990) Computerized data base on immigrant arthropods. *Annuals of the Entomological Society of America* **83**: 1–18.

Kobbe, B., Clark, J.K., and Dreistadt, S.H. (1991) *Integrated Pest Management of Citrus*, 2nd edn. University of California Press, Oakland, CA.

Koch, R.L. (2003) The multicolored Asian lady beetle, *Harmonia axyridis*: a review of its biology, uses in biological control, and non-target impacts. *Journal of Insect Science* outline publication no. **3.32**.

Komatsu, T., Ishikawa, T., Yamaguchi, N., Hori, Y., and Ohba, H. (2003) But next time? Unsuccessful establishment of the Mediterranean strain of the green seaweed *Caulerpa taxifolia* in the Sea of Japan. *Biological Invasions* **3**: 275–8.

Kondo, A. and Hiramatsu, T. (1999) Resurgence of the peach silver mite, *Aculus fockeui* (Napela et Trouessart) (Acari: Eriophyidae), induced by a synthetic pryrethroid fluvalinate. *Applied Entomology and Zoology* **34**: 531–4.

Koppenhöfer, A.M. and Kaya, H.K. (1998) Synergism of imidacloprid and an entomopathogenic nematode: a novel approach to white grub (Coleoptera: Scarabaeidae) control in turfgrass. *Journal of Economic Entomology* **91**: 618–23.

Koppenhöfer, A.M. and Fuzy, E.M. (2003) *Steinernema scarabaei* for the control of white grubs. *Biological Control* **28**: 47–59.

Koss, A.M. and Snyder, W.E. (2005) Alternative prey disrupt biocontrol by a guild of generalist predators. *Biological Control* **32**: 243–51.

Kovach, J. (2004) Impact of multicolored Asian lady beetles as a pest of fruit and people. *American Entomologist* **50(3)**: 159–61.

Koziel, M.G., Beland, G.L., Bowman, C., Carozzi, N., and Crenshaw, R. (1993) Field performance of elite transgenic maize plants expressing an insecticidal protein gene derived from *Bacillus thuringiensis*. *Bio/Technology* **11**: 195–200.

Krantz, G.W. (1978) *A Manual of Acarology*. Oregon State University, Corvallis, OR.

Kraus, F. (2003) Invasion pathways for terrestrial vertebrates. In: Ruiz, G. and Carlton, J. (eds.), *Invasive Species: Vectors and Management Strategies*, pp. 68–92. Island Press, Washington, D.C.

Kraus, F. and Cravalho, D. (2001) The risk to Hawaii from snakes. *Pacific Science* **55**: 409–17.

Krebs, C.J. (1999) *Ecological Methodology*. Addison-Welsey Publisher, New York.

Krebs, C.J. (2005) *Ecology: the Experimental Analysis of Distribution and Abundance*, 5th edn. Benjamin Cummings, San Francisco, CA.

Krebs, J.R. (1973) Behavioral aspects of predation. In: Bateson, P.P.G. and Klopfer, P.H. (eds.), *Perspectives in Ethology*, pp. 73–111. Plenum Press, New York.

Krieg, A., Huger, A.M., Langenbruch, G.A., and Schnetter, W. (1983) *Bacillus thuringiensis* var. *tenebrionis*: a new pathotype effective against larvae of Coleoptera. *Zeitschrift für Angewandte Entomologie* **96**: 500–8.

Krimsky, S. (2000) *Hormonal Chaos, the Scientific and Social Origins of the Environmental Endocrine Hypothesis*. The Johns Hopkins University Press, Baltimore, MD.

Krombein, K.V., Hurd, Jr., P.D., Smith, D.R., and Burks, B.D. (eds.) (1979) *Catalog of Hymenoptera in America North of Mexico*. Smithsonian Press, Washington, D.C.

Kuhar, T.P., Wright, M.G., Hoffmann, M.P., and Chenus, S.A. (2002) Life table studies of European corn borer (Lepidoptera: Crambidae) with and without inoculative releases of *Trichogramma ostriniae* (Hymenoptera: Trichogrammatidae). *Environmental Entomology* **31**: 482–9.

Kuhlmann, U. and Mason, P.G. (2003) Use of field host range surveys for selecting candidate non-target species for physiological host specificity testing of entomophagous biological control agents. In: Van Driesche, R.G. (ed.), *Proceedings of the First International Symposium on Biological Control of Arthropods*, pp. 370–7, January 14–18, 2002, Honolulu, Hawaii. USDA Forest Service, Morgantown, WV.

Kuhlman, U., Schaffner, U., and Mason, P.G. (2006a) Selection of non-target species for host specificity testing. In: Bigler, F., Babendreir, D., and Kuhlmann, U. (eds.), *Environmental Impact of Invertebrates for Biological Control of Arthropods*, pp. 15–37. CABI Publishing, Wallingford.

Kuhlmann, U., Mason, P.G., Hinz, H.L. et al. (2006b) Avoiding conflicts between insect and weed biological control: selecton of non-target species to assess host specificity of cabbage seedpod weevil parasitoids. *Journal of Applied Entomology* **130**: 129–41.

Kumar, P., Shenhmar, M., and Brar, K.S. (2004) Field evaluation of trichogrammatids for the control of *Helicoverpa armigera* (Hübner) on tomato. *Journal of Biological Control* **18**: 45–50.

Kuris, A. (2003) Did biological control cause extinction of the coconut moth, *Levuana iridescens*, in Fiji? *Biological Invasions* **5**: 133–41.

Kuris, A. and Culver, C.S. (1999) An introduced sabellid polychaete pest infesting cultured abalones and its potential spread to other California gastropods. *Invertebrate Biology* **118**: 391–403.

Kurtak, D., Back, C., Chalifour, A. et al. (1989) Impact of Bti on blackfly control in the *Onchocerciasis* control programme in West Africa. *Israel Journal of Entomology* **23**: 21–38.

Kuske, S., Babendreier, D., Edwards, P.J., Turlings, T.C.J., and Bigler, F. (2004) Parasitism of non-target Lepidoptera by mass released *Trichogramma brassicae* and its implication for the larval parasitoid *Lydella thompsoni*. *BioControl* **49**: 1–19.

Lack, D. (1954) *The Natural Regulation of Animal Numbers*. Clarendon Press, Oxford.

Lactin D.J., Holliday, N.J., Johnson, D.L., and Craigen, R. (1995) Improved rate model of temperature-dependent development by arthropods. *Environmental Entomology* **24**: 68–75.

Ladd, T.L. and McCabe, P.J. (1966) The status of *Tiphia vernalis* Rohwer, a parasite of the Japanese beetle, in southern New Jersey and southeastern Pennsylvania in 1963. *Journal of Economic Entomology* **59**: 480.

Lafferty, K.D. and Kuris, A.M. (1996) Biological control of marine pests. *Ecology* **77**: 1989–2000.

Laing, J.E. and Eden, G.M. (1990) Mass-production of *Trichogramma minutum* Riley on factitious host eggs. *Memoirs of the Entomological Society of Canada* **153**: 10–24.

Laing, J.E. and Hamai, J. (1976) Biological control of insect pests and weeds by imported parasites, predators, and pathogens. In: Huffaker, C.B. and Messenger, P.S. (eds.), *Theory and Practice of Biological Control*, pp. 685–743. Academic Press, New York.

Lake, P.S. and O'Dowd, D.J. (1991) Red crabs in rain forest, Christmas Island: biotic resistance to invasion by an exotic snail. *Oikos* **62**: 25–9.

Lamana, M.L. and Miller, J.C. (1998) Temperature-dependent development in an Oregon population of *Harmonia axyridis* (Coleoptera: Coccinellidae). *Environmental Entomology* **27**: 1001–5.

Lamine, K., Lambin, M., and Alauzet, C. (2005) Effect of starvation on the searching path of the predatory bug *Deraecoris lutescens*. *BioControl* **50**: 717–27.

Landis, D.A., Sebolt, D.C., Haas, M.J., and Klepinger, M. (2003) Establishment and impact of *Galerucella calmariensis* L. (Coleoptera: Chrysomelidae) on *Lythrum salicaria* L. and associated plant communities in Michigan. *Biological Control* **28**: 78–91.

Lang, A. (2003) Intraguild interference and biocontrol effects of generalist predators in a winter wheat field. *Oecologia* **134**: 144–53.

LaRock, D.R. and Ellington, J.J. (1996) An integrated pest management approach, emphasizing biological control, for pecan aphids. *Southwestern Entomologist* **21**: 153–66.

Latgé, J.P., Hall, R.A., Cabrera, R.I., and Kerwin, J.C. (1986) Liquid fermentation of entomogenous fungi, pp. 603–606. In: Samson, R.A., Vlak, J.M., and Peters, D. (eds.), *Fundamental and Applied Aspects of Invertebrate Pathology*. Foundation 4th International Colloquium on Invertebrate Pathology, Wageningen.

Lawrence, J.F. (1989) *A Catalog of Coleoptera of America North of Mexico. Family: Derodontidae*. USDA Agriculture Handbook no. 529–65. USDA, Washington, D.C.

Lawrence, L. (2006) A new green control for locusts now readily available to farmers. *Biocontrol News and Information* **27(1)**: 18N–19N.

Lawrence, P.O. and Lanzrein, B. (1993) Hormonal interactions between insect endoparasites and their host insects. In: Beckage, N.E., Thompson, S.N., and Federici, B.A. (eds.), *Parasites and Pathogens of Insects, Volume I. Parasites*, pp. 59–85. Academic Press, New York.

Lawson, M. (1995) Rabbit virus threatens ecology after leaping the fence. *Nature* **378**: 531.

Lawton, J.H. (1990) Biological control of plants: a review of generalizations, rules, and principles using insects as agents. In: Bassett, C., Whitehouse, L.J., and Zabkiewicz, J.A. (eds.), *Alternatives to Chemical Control of Weeds*, pp. 3–17. FRI Bulletin No. 155. New Zealand Ministry of Forestry, Wellington.

Leather, S.R., Walters, K.F.A., and Bale, J.S. (1993) *The Ecology of Insect Overwintering*. Cambridge University Press, Cambridge.

Leathwick, D.M. and Winterbourn, M.J. (1984) Arthropod predation on aphids in lucerne crop. *New Zealand Entomologist* **8**: 75–80.

Lebedev, G.I. (1970) Utilization des méthods biologique de lutte biologique contre les insects nuisibles et les mauveses herbes en Union Sovietique. *Annals of Zoology and Ecology of the Amin. Hors. Series* 17–23.

Legaspi, J.C. and O'Neil, R.J. (1993) Life history of *Podisus maculiventris* given low numbers of *Epilachna varivestis* as prey. *Environmental Entomology* **22**: 1192–1200.

Legaspi, J.C. and O'Neil, R.J. (1994) Developmental response of nymphs of *Podisus maculiventris* (Heteroptera: Pentatomidae) reared with low numbers of prey. *Environmental Entomology* **23**: 374–80.

Legaspi, J.C. and Legaspi, Jr., B.C. (1997) Life history trade-offs in insects with emphasis on *Podisus maculiventris* (Say) (Heteroptera: Pentatomidae). Thomas Say Publication, Entomological Society of America, Lanham, MD.

Legaspi, J.C., Legaspi, Jr., B.C., Carruthers, R.I. et al. (1996) Foreign exploration for natural enemies of *Bemisia tabaci* from southeast Asia. *Subtropical Plant Science* **48**: 43–8.

Leger, E.A. and Forister, M.L. (2005) Increased resistance to generalist herbivores in invasive populations of the California poppy (*Eschscholzia californica*). *Diversity and Distributions* **11**: 311–17.

Legner, E.F. (1986) The requirement for reassessment of interactions among dung beetles, symbovine flies and natural enemies. *Entomological Society of America, Miscellaneous Publications* **61**: 120–31.

Legner, E.F. and Gordh, G. (1992) Lower navel orangeworm (Lepidoptera: Phycitiidae) population densities following establishment of *Goniozus legneri* (Hymenoptera: Bethylidae) in California. *Journal of Economic Entomology* **85**: 2153–60.

Legner, E.F., Sjorgren, R.D., and Hall, I.M. (1974) The biological control of medically important arthropods. *Critical Reviews in Environmental Control* **4**: 85–113.

Lennartz, F.E. (1973) *Modes of Dispersal of Solenopsis invicta from Brazil into the Continental United States – a Study in Spatial Diffusion*. MS Thesis, University of Florida.

Lenz, C.J., McIntosh, A.H., Mazzacano, C., and Monderloh, U. (1991) Replication of *Heliothis zea* nuclear polyhedrosis virus in cloned cell lines. *Journal of Invertebrate Pathology* **57**: 227–33.

Leppla, N.C. and Ashley, T.R. (eds.) (1978) *Facilities for Insect Research and Production*. USDA Technical Bulletin no. 1576. USDA, Washington, D.C.

Lester, P.J., Thistlewood, H.M.A., Marshall, D.B., and Harmsen, R. (1999) Assessment of *Amblyseius fallacis* (Acari: Phytoseiidae) for biological control of tetranychid mites in an Ontario peach orchard. *Experimental and Applied Acarology* **23**: 995–1009.

Lever, C. (1994) *Naturalized Animals: the Ecology of Successfully Introduced Species*. Poyser, London.

Levin, S.A. (1969) Some demographic and genetic consequences of environmental heterogeneity for biological control. *Bulletin of the Entomological Society of America* **15**: 237–40.

Levine, J.M., Adler, P.B., and Yelenik, S.G. (2004) A meta-analysis of biotic resistance to exotic plant invasions. *Ecology Letters* **7**: 975–89.

Lewis, E.E., Campbell, J., Griffin, C., Kaya, H., and Peters, A. (2006) Behavioral ecology of entomopathogenic nematodes. *Biological Control* **38**: 66–79.

Lewis, P.A., DeLoach, C.J., Knutson, A.E., Tracy, J.L., and Robbins, T.O. (2003a) Biology of *Diorhabda elongata deserticola* (Coleoptera: Chrysomelidae), an Asian leaf beetle for biological control of saltcedars (*Tamarix* spp.) in the United States. *Biological Control* **27**: 101–16.

Lewis, P.A., DeLoach, C.J., Herr, J.C., Dudley, T.L., and Carruthers, R.I. (2003b) Assessment of risk to native *Frankenia* shrubs from an Asian leaf beetle, *Diorhabda elongata deserticola* (Coleoptera: Chrysomelidae), introduced for biological control of saltcedars (*Tamarix* spp.) in the western United States. *Biological Control* **27**: 148–66.

Lewis, W.J. and Martin, Jr., W.R. (1990) Semiochemicals for use with parasitoids: status and future. *Journal of Chemical Ecology* **16**: 3067–89.

Lewis, W.J. and Takasu, K. (1990) Use of learned odors by a parasitic wasp in accordance with host and food needs. *Nature* **348**: 635–6.

Lewis, W.J., Snow, J.W., and Jones, R.L. (1971) A pheromone trap for studying populations of *Cariochiles nigriceps*, a parasite of *Heliothis virescens*. *Journal of Economic Entomology* **64**: 1417–21.

Lewis, W.J., Jones, R.L., Gross, Jr., H.R., and Nordlund, D.A. (1976) The role of kairomones and other behavioral chemicals in host finding by parasitic insects. *Behavioral Biology* **16**: 267–89.

Lewis, W.J., Nordlund, D.A., Gueldner, R.C., Teal, P.E.A., and Tumlinson, J.H. (1982) Kairomones and their use for management of entomophagous insects. XIII. Kairomonal activity for *Trichogramma* spp. of abdominal tips, excretion, and a synthetic sex pheromone blend of *Heliothis zea* (Boddie) moths. *Journal of Chemical Ecology* **8**: 1323–31.

Lewis, W.J., Vet, L.E.M., Tumlinson, J.H., van Lenteren, J.C., and Papaj, D.R. (1990) Variations in parasitoid foraging behavior: essential element of a sound biological control theory. *Environmental Entomology* **19**: 1183–93.

Lewis, W.J., Tumlinson, J.H., and Krasnoff, S. (1991) Chemically mediated associative learning: an important function in the foraging behavior of *Microplitis croceipes* (Cresson). *Journal of Chemical Ecology* **17**: 1309–25.

Ley, R.R. and D'Antonio, C.M. (1998) Exotic grass invasion alters potential rates of N fixation in Hawaiian woodlands. *Oecologia* **113**: 179–87.

Li, B.P., Bateman, R., Li, G.Y., Meng, L., and Zheng, Y.A. (2000) Field trial on the control of grasshoppers in the mountain grassland by oil formulation of *Metarhizium flavoviride*. *Chinese Journal of Biological Control* **16**: 145–7.

Li, L.-Y. (1994) Worldwide use of *Trichogramma* for biological control of on different crops: a survey. In: Wajnberg, É. and Hassan, S.A. (eds.), *Biological Control with Egg Parasitoids*, pp. 37–51. CABI Publishing, Wallingford.

Liebhold, A.M. (1994) Use and abuse of insect and disease models in forest pest management: past, present, and future. In: Covington, W.W. and DeBano, L.F. (eds.), *Sustainable Ecological Systems: Implementing an Ecological Approach to Land Management*, pp. 204–10. USDA Forest Service Technical Report RM-247. USDA Forest Service, Morgantown, WV.

Liebhold, A.M. and Elkinton, J.S. (1989) Elevated parasitism in artificially augmented populations of *Lymantria dispar* (Lepidoptera: Lymantriidae). *Environmental Entomology* **18**: 986–95.

Liebhold, A.M. and Bascompte, J. (2003) The Allee effect, stochastic dynamics and the eradication of alien species. *Ecology Letters* **6**: 133–40.

Liebhold, A.M., Elkinton, J.S., Williams, D., and Muzika, R.M. (2000) What causes outbreaks of gypsy moth in North America? *Population Ecology* **42**: 257–66.

Liljesthrom, G. (1980) Nota sobre *Trichopoda giacomellii* (Blanchard) Guimaraes, 1971 (Diptera: Tachinidae), parasitoide de *Nezara viridula* (L.) 1758. Hem. Pentatomidae. *Revista de la Sociedad d'Entomologia d'Argentina* **44**: 433–9.

Lilley, R. and Campbell, C.A.M. (1999) Biological, chemical, and integrated control of two-spotted spider mite *Tetranychus urticae* on dwarf hops. *Biocontrol Science and Technology* **9**: 467–73.

Lind, P. (1998) Encouraging ladybugs. *Journal of Pesticide Reform* **18(3)**: 22–33.

Lindegren, J.E., Valero, K.A., and Mackey, B.E. (1993) Simple *in vivo* production and storage methods for *Steinernema carpocapsae* infective juveniles. *Journal of Nematology* **25**: 193–7.

Lindquist, R.K. and Piatkowski, J. (1993) Evaluation of entomopathogenic nematodes for control of fungus gnat larvae. *IOBC/WPRS Bulletin* **16**: 97–100.

Liu, J. and Berry, R.E. (1996) *Heterorhabditis marelatus* n. sp. (Rhabditida: Heterorhabditidae) from Oregon. *Journal of Invertebrate Pathology* **67**: 48–54.

Liu, S.J., Xue, H.P., Pu, B.Q., and Qian, N.H. (1984) A new viral disease in rabbits. *Animal Husbandry and Veterinary Medicine* **16**: 253–5.

Liu, Z.C., Sun, Y.R., Wang, Z.Y. et al. (1985) Field release of *Trichogramma confusum* reared on artificial host eggs against sugarcane borers. *Chinese Journal of Biological Control* **3**: 2–5.

Lodge, D.M., Rosenthal, S.K., Mavuti, K.M. et al. (2005) Louisiana crayfish (*Procambarus clarkii*) (Crustacea: Cambaridae) in Kenyan ponds: non-target effects of a potential biological control agent for schistosomiasis. *African Journal of Aquatic Science* **30**: 119–24.

Logan, J.A. (1994) In defense of big ugly models. *American Entomologist* **40**: 202–7.

Loke, W.H. and Ashley, T.R. (1984) Behavioral and biological responses of *Cotesia marginiventris* to kairomones of the fall armyworm, *Spodoptera frugiperda*. *Journal of Chemical Ecology* **10**: 521–9.

Lola-Luz, T., Downes, M., and Dunne, R. (2005) Control of black vine weevil larvae *Otiorhynchus sulcatus* (Fabricius) (Coleoptera: Curculionidae) in grow bags outdoors with nematodes. *Agriculture and Forest Entomology* **7**: 121–6.

Long, J.L. (2003) *Introduced Mammals of the World – their History, Distribution and Influence*. CSIRO Publishing, Collingwood, Victoria.

Longworth, J.F. and Kalmakoff, J. (1977) Insect viruses for biological control: an ecological approach. *Intervirology* **8**: 68–72.

Lonsdale, W.M. and Farrell, G.S. (1998) Testing the effects on *Mimosa pigra* of a biological control agent *Neurostrota gunniella* (Lepidoptera: Gracillaridae), plant competition and fungi under field conditions. *Biocontrol Science and Technology* **8**: 485–500.

Lonsdale, W.M., Harley, K.L.S., and Gillett, J.D. (1988) Seed bank dynamics in *Mimosa pigra*, an invasive tropical shrub. *Journal of Applied Ecology* **25**: 963–76.

Lonsdale, W.M., Farrell, G., and Wilson, C.G. (1995) Biological control of a tropical weed: a population model and experiment for *Sida acuta*. *Journal of Applied Ecology* **32**: 391–9.

Loope, L.L., Hamann, O., and Stone, C.P. (1988) Comparative conservation biology of oceanic archipelagoes. *BioScience* **38**: 272–82.

Lopez, E.R. and Van Driesche, R.G. (1989) Direct measurement of host and parasitoid recruitment for assessment of total losses due to parasitism in a continuously breeding species, the cabbage aphid *Brevicoryne brassicae* (L.) (Hemiptera: Aphididae). *Bulletin of Entomological Research* **79**: 47–59.

Lorvelec, O. and Pascal, M. (2005) French alien mammal eradication attempts and their consequences on the native fauna and flora. *Biological Invasions* **7**: 135–40.

Lotka, A.J. (1925) *Elements of Physical Biology*. Dover Publications, New York (reprinted in 1956).

Lou, Y.G., Du, M.H., Turling, T.C.J., Cheng, J.A., and Shan, W.F. (2005) Exogenous application of jasmonic acid induces volatile emissions in rice and enhances parasitism of *Nilaparvata lugens* eggs by the parasitoid *Anagrus nilaparvatae*. *Journal of Chemical Ecology* **31**: 1985–2002.

Louda, S.M. (1984) Herbivore effect on stature, fruiting, and leaf dynamics of a native crucifer. *Ecology* **65**: 1379–86.

Louda, S.M. (1998) Population growth of *Rhinocyllus conicus* (Coleoptera: Curculionidae) on two species of native thistles in prairie. *Environmental Entomology* **27**: 834–41.

Louda, S.M. and Potvin, M.A. (1995) Effect of inflorescence-feeding insects on the demography and lifetime fitness of a native plant. *Ecology* **76**: 229–45.

Louda, S.M., Kendall, D., Connor, J., and Simberloff, D. (1997) Ecological effects of an insect introduced for the biological control of weeds. *Science* **277**: 1088–90.

Louda, S.M., Pemberton, R.W., Johnson, M.T., and Follett, P.A. (2003a) Nontarget effects – the Achilles' heel of biological control? Retrospective analyses to reduce risk associated with biocontrol introductions. *Annual Review of Entomology* **48**: 365–96.

Louda, S.M., Arnett, A.E., Rand, T.A., and Russell, F.L. (2003b) Invasiveness of some biological control insects and adequacy of their ecological risk assessment and regulation. *Conservation Biology* **17**: 73–82.

Louda, S.M., Rand, T.A., Arnett, A.E., McClay, A.S., Shea, K., and McEachern, A.K. (2005) Evaluation of ecological risk to populations of a threatened plant from an invasive biocontrol agent. *Ecological Applications* **15**: 234–49.

Lovett, J. (1997) Birth control for feral pests. *Search* **28**: 209–11.

Lowery, D.T. and Sears, M.K. (1986) Stimulation of reproduction of the green peach aphid (Homoptera: Aphididae) by azinphosmethyl applied to potatoes. *Journal of Economic Entomology* **79**: 1530–3.

Lozier, J.D., Mills, N.J., and Roderick, G.K. (2006) Di- and trinucleotide repeat microsatellites for the parasitoid wasp, *Aphidius transcaspicus*. *Molecular Ecology Notes* **6**: 27–9.

Lu, W. and Montgomery, M.E. (2001) Oviposition, development, and feeding of *Scymnus* (*Neopullus*) *sinuanodulus* (Coleoptera: Coccinellidae): a predator of *Adelges tsugae*

(Homoptera: Adelgidae). *Annals of the Entomological Society of America* **94**: 64–70.

Lucas, É. and Alomar, O. (2002) Impact of *Macrolophus caliginosus* presence on damage production by *Dicyphus tamaninii* (Heteroptera: Miridae) on tomato fruits. *Journal of Economic Entomology* **95**: 1123–9.

Luck, R.F. (1981) Parasitic insects introduced as biological control agents for arthropod pests. In: Pimentel, D. (ed.), *CRC Handbook of Pest Management in Agriculture*, pp. 125–284. CRC Press, Boca Raton, FL.

Luck, R.F. and Dahlsten, D.L. (1975) Natural decline of a pine needle scale (*Chionaspis pinifoliae* [Fitch]), outbreak at South Lake Tahoe, California, following cessation of adult mosquito control with malathion. *Ecology* **56**: 893–904.

Luck, R.F. and Uygun, N. (1986) Host recognition and selection by *Aphytis* species: response to California red, oleander, and cactus scale cover extracts. *Entomologia Experimentalis et Applicata* **40**: 129–36.

Luck, R.F., Shepard, B.M., and Kenmore, P.E. (1988) Experimental methods for evaluating arthropod natural enemies. *Annual Review of Entomology* **33**: 367–91.

Luck, R.F., Forster, L.D., and Morse, J.G. (1996) An ecologically based IPM program for citrus in California's San Joaquin Valley using augmentative biological control. *Proceedings of the International Society of Citriculture* **1**: 499–503.

Luck, R.F., Shepard, B.M., and Kenmore, P.E. (1999) Evaluation of biological control with experimental methods. In: Bellows, Jr., T.S. and Fisher, T.W. (eds.), *Handbook of Biological Control*, pp. 225–42. Academic Press, San Diego, CA.

Lunau, S., Stoessel, S., Schmidt-Peisker, A.J., and Ehlers, R.-U. (1993) Establishment of monoxenic inocula for scaling up *in vitro* cultures of the entomopathogenic nematodes *Steinernema* spp. and *Heterorhabditis* spp. *Nematologica* **39**: 385–99.

Lundgren, J.G., Razzak, A.A., and Wiedenmann, R.N. (2004) Population responses and food consumption by predators *Coleomegilla maculata* and *Harmonia axyridis* (Coleoptera: Coccinellidae) during anthesis in an Illinois corn field. *Environmental Entomology* **33**: 958–63.

Lüthy, P. (1986) Insect pathogenic bacteria as pest control agents. In: Franz, J.M. (ed.), *Biological Plant and Health Protection: Biological Control of Plant Pests and of Vectors of Human and Animal Diseases*, pp. 201–16. International Symposium of the Akademie der Wissenschaften und der Literatur, November 15–17, 1984, Mainz, Germany. *Fortschritte der Zoologie* vol. 32. Gustav Fischer Verlag, Stuttgart.

Lynch, L.D. and Thomas, M.B. (2000) Nontarget effects in the biocontrol of insects with insects, nematodes and microbial agents: the evidence. *Biocontrol News and Information* **21(4)**: 117N–30N.

Lynch, L.D., Hokkanen, H.M.T., Babendreier, D. et al. (2002) Insect biological control and non-target effects: a European perspective. In: Wajnberg, É., Scott, J.K., and Quimby, P.C. (eds.). *Evaluating Indirect Ecological Effects of Biological Control*, pp. 99–125. CABI Publishing, Wallingford.

Lynn, D.E., Shapiro, M., Dougherty, E.M. et al. (1990) Gypsy moth nuclear polyhedrosis virus in cell culture: a likely commercial systems for viral pesticide production. In: Pinnock, D.E. (ed.), *Vth International Colloquium on Invertebrate Pathology and Microbial Control*, p. 12, August 20–24, 1990, Adelaide, Australia. Department of Entomology, University of Adelaide, Glen Osmond, South Australia.

MacArthur, R.H. and Pianka, E.R. (1966) On optimal use of a patchy environment. *American Naturalist* **100**: 603–9.

Mackauer, M. (1972) Genetic aspects of insect production. *Entomophaga* **17**: 27–48.

MacLeod, A., Evans, H.F., and Baker, R.H.A. (2002) An analysis of pest risk from an Asian longhorn beetle (*Anoplophora glabripennis*) to hardwood trees in the European community. *Crop Protection* **21**: 635–45.

MacLeod, A., Wratten, S.D., Sotherton, N.W., and Thomas, M.B. (2004) "Beetle banks" as refuges for beneficial arthropods in farmland: long-term changes in predator communities and habitat. *Agricultural and Forest Entomology* **6**: 147–54.

MacLeod, D.M. (1963) Entomophthorales infections. In: Steinhaus, E.A. (ed.), *Insect Pathology: An Advanced Treatise, Volume 2*, pp. 189–231. Academic Press, New York.

Macom, T.E. and Porter, S.D. (1996) Comparison of polygyne and monogyne red imported fire ants (Hymenoptera: Formicidae) population densities. *Annals of the Entomological Society of America* **89**: 535–43.

Madden, J.L. (1968) Behavioural responses of parasites to the symbiotic fungus associated with *Sirex noctilio* F. *Nature* **218**: 189–90.

Maddox, D.M. (1982) Biological control of diffuse knapweed (*Centaurea diffusa*) and spotted knapweed (*C. maculosa*). *Weed Science* **30**: 76–82.

Madeira, P.T., Hale, R.E., Center, T.D., Buckingham, G.R., Wineriter, S.A., and Purcell, M. (2001) Whether to release *Oxyops vitiosa* from a second Australian site onto Florida's melaleuca? A molecular approach. *BioControl* **46**: 511–28.

Maeto, K. and Kudo, S. (1992) A new euphorine species of *Aridelus* (Hymenoptera: Braconidae) associated with a subsocial bug, *Elasmucha putoni* (Heteroptera, Acanthosomatidae). *Japanese Journal of Entomology* **6**: 77–84.

Magalhães, B.P., Lecoq, M., de Faria, M.R., Schmidt, F.G.V., and Guerra, W.D. (2000) Field trial with the entomopathogenic fungus *Metarhizium anisopliae* var. *acridum* against bands of the grasshopper *Rhammatocerus schistocercoides* in Brazil. *Biocontrol Science and Technology* **10**: 427–41.

Magiafoglou, A., Schiffer, M., Hoffmann, A.A., and McKechnie, W. (2003) Immunocontraception for population control: will resistance evolve? *Immunology and Cell Biology* **81**: 152–9.

Mahr, S. (2000) Mechanized delivery of beneficial insects. *IPM Practitioner* **22(4)**: 1–5.

Maier, C.T. (1994) Biology and impact of parasitoids of *Phyllorycter blancardella* and *P. crataegella* (Lepidoptera: Gracillariidae) in northeastern North American apple orchards. In: Maier, C.T. (ed.), *Integrated Management of Tentiform Leafminers, Phyllonorycter spp. (Lepidoptera: Gracillariidae), in North American Apple Orchards*, pp. 6–24. Thomas Say Publications in Entomology. Entomological Society of America, Lanham, MD.

Malsam, O., Kilian, M., Oerke, E.-C., and Dehne, H.-W. (2002) Oils for increased efficacy of *Metarhizium anisopliae* to control whiteflies. *Biocontrol Science and Technology* **12**: 337–48.

Maltby, H.L., Stehr, F.W., Anderson, R.C., Moorehead, G.E., Barton, L.C., and Paschke, J.D. (1971) Establishment in the United States of *Anaphes flavipes*, an egg parasitoid of the cereal leaf beetle. *Journal of Economic Entomology* **64**: 693–7.

Manly, B.F.J. (1974) Estimation of stage-specific survival rates and other parameters for insect populations developing through several life stages. *Oecologia* **15**: 277–85.

Manly, B.F.J. (1976) Extensions to Kiritani and Nakasuji's method for analyzing insect stage-frequency data. *Researches on Population Ecology* **17**: 191–9.

Manly, B.F.J. (1977) The determination of key factors from life table data. *Oecologia* **31**: 111–17.

Manly, B.F.J. (1989) A review of methods for the analysis of stage-frequency data. In: McDonald, L.L., Manly, B.F.J., Lockwood, J., and Logan, J. (eds.), *Estimation and Analysis of Insect Populations*, pp. 3–69. Springer-Verlag, New York.

Mann, J. (1969) *Cactus-feeding Insects and Mites*. United States National Museum Bulletin 256, Smithsonian Institution Press, Washington, D.C.

Mann, J. (1970) *Cacti Naturalized in Australia and Their Control*. Department of Lands, Brisbane, Queensland.

Manrique-Saide, P., Ibañez-Bernal, S., Defín-González, H., and Tabla, V.P. (1998) *Mesocyclops longisetus* effects on survivorhsip of *Aedes aegypti* immature stages in car tiers. *Medical and Veterinary Entomology* **12**: 386–90.

Mansour, E.S. (2004) Effectiveness of *Trichogramma evanescens* Westwood, bacterial insecticide and their combination on the cotton bollworms in comparison with chemical insecticides. *Egyptian Journal of Biological Pest Control*. **14**: 339–43.

Mansour, F., Rosen, D., Shulov, A., and Plaut, H.N. (1980) Evaluation of spiders as biological control agents of *Spodoptera littoralis* larvae on apple in Israel. *Oecologia Applicata* **1**: 225–32.

Manzari, S., Polaszek, A., Belshaw, R., and Quicke, D.L.J. (2002) Morphometric and molecular analysis of the *Encarsia inaron* species-group (Hymenoptera: Aphelinidae), parasitoids of whiteflies (Hemiptera: Aleyrodidae). *Bulletin of Entomological Research* **92**: 165–75.

Maramorosch, K. and Sherman, K.E. (eds.) (1985) *Viral Insecticides for Biological Control*. Academic Press, New York.

Marcovitch, S. (1935) Experimental evidence on the value of strip farming as a method for the natural control of injurious insects with special reference to plant lice. *Journal of Economic Entomology* **28**: 62–70.

Markin, G.P. (1970a) Foraging behavior of the Argentine ant in a California citrus grove. *Journal of Economic Entomology* **63**: 740–4.

Markin, G.P. (1970b) The seasonal life cycle of the Argentine ant, *Iridomymrex humilis* (Hymenoptera: Formicidae), in southern California. *Annals of the Entomological Society of America* **63**: 1238–42.

Markkula, M., Tiittanen, K., Hamalainen, M., and Forsberg, A. (1979) The aphid midge *Aphidoletes aphidimyza* (Diptera: Cecidomyiidae) and it use in biological control of aphids. *Annales Entomologici Fenniae* **45**: 89–98.

Maron, J.L. and Vilà, M. (2001) When do herbivores affect plant invasion? Evidence for the natural enemies and biotic resistance hypotheses. *Oikos* **95**: 361–73.

Martel, A.L., Pathy, D.A., Madill, J.B., Renaud, C.B., Dean, S.L., and Kerr, S.J. (2001) Decline and regional extripation of freshwater mussels (Unionidae) in a a small river system invaded by *Dreissena polymorpha*: the Rideau River, 1993–200. *Canadian Journal of Zoology* **79**: 2181–91.

Martignoni, M.E. (1999) History of TM BioControl-1: the first registered virus-based product for control of a forest insect. *American Entomologist* **45(1)**: 30–7.

Martin, F.N. and Tooley, P.W. (2003) Phylogenetic relationships of *Phytophthora ramorum*, *P. nemorosa*, and *P. pseudosyringae*, three species recovered from areas in California with sudden oak death. *Mycological Research* **107(12)**: 1379–91.

Martin, Jr., W.R., Nordlund, D.A., and Nettles, Jr., W.C. (1990) Response of the parasitoid *Eucelatoria bryani* to selected plant material in an olfactometer. *Journal of Chemical Ecology* **16**: 499–508.

Mason, P.G. and Huber, J.T. (eds.) (2001) *Biological Control Programmes in Canada, 1981–2000*. CABI Publishing, Wallingford.

Matadha, D., Hamilton, G.C., Lashomb, J.H., and Zhang, J. (2005) Ovipositional preferences and functional response of parasitoids of euonymus scale, *Unaspis euonymi* (Comstock) and San Jose scale, *Quadraspidiotus perniciosus* (Comstock) (Homoptera: Diaspididae). *Biological Control* **32**: 337–47.

Mathews, C.R., Bottrell, D.G., and Brown, M.W. (2004) Habitat manipulation of the apple orchard floor to increase ground-dwelling predators and predation of *Cydia pomonella* (L.) (Lepidoptera: Tortricidae). *Biological Control* **30**: 265–73.

Matthews, R.E.F. (1991) *Plant Virology*, 3rd edn. Academic Press, San Diego, CA.

Mauchamp, A. (1997) Threats from alien plant species in the Galápagos Islands. *Conservation Biology* **11**: 260–3.

May, R.M. (1974) Biological populations with non-overlapping generations: stable points, stable cycles, and chaos. *Science* **186**: 645–7.

May, R.M. (1976) Simple mathematical models with very complicated dynamics. *Nature* **261**: 459–67.

May, R.M. (1977) Thresholds and breakpoints in ecosystems with a multiplicity of stable states. *Nature* **269**: 471–7.

May, R.M. (1978) Host-parasitoid systems in patchy environments: a phenomenological model. *Journal of Animal Ecology* **47**: 833–43.

May, R.M. (1980) Depression of host population abundance by direct life cycle macroparasites. *Journal of Theoretical Biology* **82**: 283–311.

May, R.M. and Anderson, R.M. (1978) Regulation and stability of host-parasite population interactions. II. Destabilizing processes. *Journal of Animal Ecology* **47**: 249–67.

Mays, W.T. and Kok, L.T. (2003) Population dynamics and dispersal of two exotic biological control agents of spotted knapweed, *Urophora affinis* and *U. quadrifasciata* (Diptera: Tephritidae) in southwestern Virginia from 1986 to 2000. *Biological Control* **27**: 43–52.

McCabe, D. and Soper, R.S. (1985) Preparation of an entomopathogenic fungal insect control agent. U.S. Patent 4,530,834.

McCaffrey, J.P., Campbell, C.L., and Andres, L.A. (1995) St. Johnswort. In: Nechols, L.A., Beardsley, J.W., Goeden, R.D., and Jackson, C.G. (eds.), *Biological Control in the Western United States: Accomplishments and Benefits of Regional Research Project W-84, 1964–1989*, pp. 281–5. Publication 3361. University of California, Oakland, CA.

McCall, P.J., Turlings, T.C.J., Lewis, W.J., and Tumlinson, J.H. (1993) Role of plant volatiles in host location by the specialist parasitoid *Microplitis croceipes* Cresson (Braconidae: Hymenoptera). *Journal of Insect Behavior* **6**: 625–39.

McCallum H.I. (1993) Evaluation of a nematode (*Capillaria hepatica* Bancroft, 1893) as a control agent for populations of house mice (*Mus musculus domesticus* Schwartz and Schwartz, 1943). *Revue Scientifique et Technique Office International des Epizooties* **12**: 83–93.

McCallum, H. (1994) Quantifying the impact of disease on threatened species. *Pacific Conservation Biology* **1**: 107–17.

McCallum, H. (1996) Immunocontraception for wildlife population control. *Trends in Ecology and Evolution* **11**: 491–3.

McCallum, H. and Singleton, G.R. (1989) Models to assess the potential of *Capillaria hepatica* to control population outbreaks of house mice. *Parasitology* **98**: 425–37.

McClay, A.S. (1995) Beyond "before-and-after": experimental design and evaluation in classical weed biological control. In: Delfosse, E.S. and Scott, R.R. (eds.), *Proceedings of the VIIIth International Symposium on Biological Control of Weeds*, pp. 203–9, February 2–7, 1992, Canterbury, New Zealand. DSIR/CSIRO, Melbourne.

McClay, A.S. and Balciunas, J.K. (2005) The role of pre-release efficacy assessment in selecting classical biological control agents for weeds–applying the Anna Karenina principle. *Biological Control* **35**: 197–207.

McClay, A.S., Crisp, M.D., Evans, H.C. et al. (2004) Centres of origin: do they exist, can we identify them, does it matter? In: Cullen, J.M., Briese, D.T., Kriticos, D.J., Lonsdale, W.M., Morin, L., and Scott, J.K. (eds.), *Proceedings of the XI International Symposium on Biological Control of Weeds*, pp. 619–20, April 27–May 2, 2003, Canberra, Australia. CSIRO Publishing, Canberra.

McClure, M.S. (1979) Self-regulation in populations of the elongate hemlock scale, *Fiorinia externa* (Homoptera: Diaspididae). *Oecologia* **39**: 25–36.

McClure, M.S. (1980) Competition between exotic species: scale insects on hemlock. *Ecology* **61**: 1391–1401.

McClure, M.S. (1987) Biology and control of hemlock woolly adelgid. *Connecticut Agriculture Experiment Station Bulletin* no. 851.

McClure, M.S. (1991) Density-dependent feedback and population cycles in *Adelges tsugae* (Homoptera: Adelgidae) on *Tsuga canadensis*. *Environmental Entomology* **20**: 258–64.

McClure, M.S. (1995) Using natural enemies from Japan to control hemlock woolly adelgid. *Frontiers of Plant Science* **47**: 5–7.

McClure, M.S. (1996) Biology of *Adelges tsugae* and its potential spread in the northeastern United States. In: Salom S.M., Tigner, T.C., and Reardon, R.C. (eds.), *The First Hemlock Woolly Adelgid Review*, pp. 16–25. FHTET-96-10. USDA Forest Service, Morgantown, WV.

McColl, K.A., Merchant, J.C., Hardy, J., Cooke, B.D., Robinson, A., and Westbury, H.A. (2002) Evidence for insect transmission of rabbit haemorrhagic disease virus. *Epidemiology and Infection* **129**: 655–63.

McConnachie, A.J., de Wit, M.P., Hill, M.P., and Byrne, M.J. (2003) Economic evaluation of the successful biological control of *Azolla filiculoides* in South Africa. *Biological Control* **28**: 25–32.

McConnachie, A.J., Hill, M.P., and Byrne, M.J. (2004) Field assessment of a frond-feeding weevil, a successful biological control agent of red waterfern, *Azolla filiculoides*, in southern Africa. *Biological Control* **29**: 326–31.

McCoy, C.W. (1981) Pest control by the fungus *Hirsutella thompsonii*. In: Burges, H.D. (ed.), *Microbial Control of Pests and Plant Diseases*, pp. 499–512. Academic Press, New York.

McCoy, C.W. and Heimpel, A.M. (1980) Safety of the potential mycoacaricide, *Hirsutella thompsonii*, to vertebrates. *Environmental Entomology* **9**: 47–9.

McCoy, C.W., Hill, A.J., and Kanavel, R.F. (1975) Large-scale production of the fungal pathogen *Hirsutella thompsonii* in submerged culture and its formulation for application in the field. *Entomophaga* **20**: 229–40.

McCoy, C.W., Samson, R.A., and Boucias, D.G. (1988) Entomogenous fungi. In: Ignoffo, C.M. (ed.), *CRC Handbook of Natural Pesticides. Microbial Insecticides, Part A. Entomogenous Protozoa and Fungi, vol.* 5, pp. 151–236. CRC Press, Boca Raton, FL.

McDermott, G.J. and Hoy, M.A. (1997) Persistence and containment of *Metaseiulus occidentalis* (Acari: Phytoseiidae) in Florida: risk assessment for possible releases of transgenic strains. *Florida Entomologist* **80**: 42–53.

McDonald, R.C. and Kok, L.T. (1992) Colonization and hyperparasitism of *Cotesia rubecula* (Hymen.: Braconidae), a newly introduced parasite of *Pieris rapae*, in Virginia. *Entomophaga* **37**: 223–8.

McEvoy, P.B. and Cox, C. (1991) Successful biological control of ragwort, *Senecio jacobaea*, by introduced insects in Oregon. *Ecological Applications* **1**: 430–42.

McEvoy, P.B. and Rudd, N.T. (1993) Effects of vegetation disturbances on insect biological control of tansy ragwort, *Senecio jacobaea*. *Ecological Applications* **3**: 682–98.

McEvoy, P.B., Rudd, N.T., Cox, C.S., and Huso, M. (1993) Disturbance, competition, and herbivory effects on ragwort *Senecio jacobaea* populations. *Ecological Monographs* **63**: 55–75.

McEwen, P., New, T.R., and Whittington, A.E. (eds.) (2001) *Lacewings in the Crop Environment*. Cambridge University Press, Cambridge.

McFadyen, R.E.C. (1991) Climate modeling and the biological control of weeds: one view. *Plant Protection Quarterly* **6**: 14–15.

McFadyen, R.E.C. (1998) Biological control of weeds. *Annual Review of Entomology* **43**: 369–93.

McFadyen, R.E.C. (2000) Successes in biological control of weeds. In: Spencer, N.R. (ed.), *Proceedings of the Xth International Symposium on Biological Control of Weeds*, pp. 3–14, July 4–14, 1999, Bozeman, Montana. Montana State University, Bozeman, MT.

McGregor, R.C. (1973) *The Emigrant Pests*. A report to Dr. Fancis Mulhern, Administration, Animal and Plant Health Inspection Service, Berkeley, California, Unpublished report on file at the Hawaii Department of Agriculture, Honolulu, HI. www.hear.org/articles/mcgregor1973.pdf.

McGregor, R.R. and Gillespie, D.R. (2005) Intraguild predation by the generalist predator *Dicyphus hesperus* on the parasitoid *Encarsia formosa*. *Biocontrol Science and Technology* **15**: 219–27.

McGuire, M.R. and Henry, J.E. (1989) Production and partial characterization of monoclonal antibodies for detection of entomopoxvirus from *Melanoplus sanguinipes*. *Entomologia Experimentalis et Applicata* **51**: 21–8.

McKillup, S.C., Allen, P.G., and Skewes, M.A. (1988) The natural decline of an introduced species following its initial increase in abundance; an explanation for *Ommatoiulus moreletii* in Australia. *Oecologia* **77**: 339–42.

McMurtry, J.A. (1992) The role of exotic natural enemies in the biological control of insect and mite pests of avocado in California. In: *Proceedings of the Second World Avocado Congress: the Shape of Things to Come*, pp. 247–52, April 21–26, 1991, Orange, California. California Avocado Society, Riverside, CA.

McMurtry, J.A. and Croft, B.A. (1997) Life styles of phytoseiid mites and their role as biological control agents. *Annual Review of Entomology* **42**: 291–321.

McMurtry, J.A. and Scriven, G.T. (1964) Studies on the feeding, reproduction, and development of *Amblyseius hibisci* (Acarina: Phytoseiidae) on various food substances. *Annals of the Entomological Society of America* **57**: 649–55.

McMurtry, J.A., Oatman, E.R., Phillips, P.H., and Wood, G.W. (1978) Establishment of *Phytoseiulus persimilis* (Acari: Phytoseiidae) in southern California. *Entomophaga* **23**: 175–9.

McNeill, M.R., Goldson, S.L., Proffitt, J.R., Phillips, C.B., and Addison, P.J. (2002) A description of the commercial rearing and distribution of *Microctonus hyperodae* (Hymenoptera: Braconidae) for biological control of *Listronotus bonariensis* (Kuschel) (Coleoptera: Curculionidae). *Biological Control* **24**: 167–75.

Meadow, R.H., Kelly, W.C., and Shelton, A.M. (1985) Evaluation of *Aphidoletes aphidimyza* (Dip.: Cecidomyiidae) for control of *Myzus persicae* (Hom.: Aphididae) in greenhouse and field experiments in the United States. *Entomophaga* **30**: 385–92.

Medal, J.C., Vitorino, M.D., Habeck, D.H., Gillmore, J.L., Pedrosa, J.H. and De Sousa, L.P. (1999) Host specificity of *Heteroperreyia hubrichi* Malaise (Hymenoptera: Pergidae), a potential biological control agent of Brazilian peppertree (*Schinus terebinthifolius* Raddi). *Biological Control* **14**: 60–5.

Medal, J.C., Cuda, J.P., and Gandolfo, D. (2004) *Gratiana boliviana*. In: Coombs, E.M., Clark, J.K., Piper, G.L., and Cofrancesco, A.F. (eds.), *Biological Control of Invasive Plants in the United States*, pp. 399–401. Oregon State University Press, Corvallis, OR.

Meinesz, A. (1999) *Killer Algae – the True Tale of Biological Invasion*. University of Chicago Press, Chicago, IL.

Meinesz, A. (2004) *Caulerpa taxifolia*: following its invasion. *Biofutur* **244**: 41–6.

Melching, J.S., Bromfield, K.R., and Kingsolver, C.H. (1983) The plant pathogen containment facility at Frederick, Maryland. *Plant Disease* **67**: 717–22.

Memmott, J., Fowler, S.V., and Hill, R.L. (1998) The effect of release size on the probability of establishment of biological control agents: gorse thrips (*Sericothrips staphylinus*) released against gorse (*Ulex europaeus*) in New Zealand. *Biocontrol Science and Technology* **8**: 103–15.

Memmott, J., Craze, P.G., Harman, H.M., Syrett, P., and Fowler, S.V. (2005) The effect of propagule size on the invasion of an alien insect. *Journal of Animal Ecology* **74**: 50–62.

Men, X.Y., Ge Feng, Yardim, E.N., and Parajulee, M.N. (2004) Evaluation of winter wheat as a potential relay crop for enhancing biological control of cotton aphids in seedling cotton. *BioControl* **49**: 701–14.

Mendel, Z., Golan, Y., and Madar, Z. (1984) Natural control of the eucalyptus borer, *Phoracantha semipunctata* (F.) (Coleoptera: Cerambycidae), by the Syrian woodpecker. *Bulletin of Entomological Research* **74**: 121–7.

Mensah, R.K. and Madden, J.J. (1994) Conservation of two predator species for biological control of *Chrysophtharta bimaculata* (Col.: Chrysomelidae) in Tasmanian forests. *Entomophaga* **39**: 71–83.

Merlin, M.D. and Juvik, J.O. (1992) Relationships among native and alien plants on Pacific islands with and without significant human disturbance and feral ungulates. In: Stone, C.P., Smith, C.W., and Tunison, J.T. (eds.), *Alien Plant Invasions in Native Ecosystems of Hawaii: Management and Research*, pp. 597–624. University of Hawaii Cooperative National Park Resources Studies Unit, Honolulu, HI.

Merritt, R.W., Walker, E.D., Wilzbach, M.A., Cummins, K.W., and Morgan, W.T. (1989) A broad evaluation of *B.t.i.* for black fly (Diptera: Simuliidae) control in a Michigan river: efficacy, carryover, and nontarget effects on invertebrates and fish. *Journal of the American Mosquito Control Association* **5**: 397–415.

Mesbah, A.H., Shoeb, M.A., and El-Heneidy, A.H. (2003) Preliminary approach towards the use of the egg parasitoid, *Trichogrammatoidea bactrae* Nagaraja against cotton bollworms in Egyptian cotton fields. *Egyptian Journal of Agricultural Research* **81**: 981–95.

Messenger, P.S. (1971) Climatic limitation to biological controls. *Proceedings of the Tall Timbers Conference on Ecological Animal Control by Habitat Management* **3**: 97–114.

Messenger, P.S., Biliotii, E., and van den Bosch, R. (1976) The importance of natural enemies in integrated control. In: Huffaker, C.B. and Messenger, P.S. (eds.), *Theory and Practice of Biological Control*, pp. 543–63. Academic Press, New York.

Messina, F.J. and Hanks, J.B. (1998) Host plant alters the shape of the functional response of an aphid predator (Coleoptera: Coccinellidae). *Environmental Entomology* **27**: 1196–1202.

Messing, R.H. and Wright, M.G. (2006) Biological control of invasive species: solution or pollution? *Frontiers in Ecology and the Environment* **4**: 132–40.

Messing, R., Roitberg, B., and Brodeur, J. (2006) Measuring and predicting indirect impacts of biological control: competition, displacement and secondary interactions, pp. 64–77. In: Bigler, F., Babendreir, D., and Kuhlmann, U. (eds.), *Environmental Impact of Invertebrates for Biological Control of Arthropods*. CABI Publishing, Wallingford.

Metcalf, R.L. (1980) Changing role of insecticides in crop protection. *Annual Review of Entomology* **25**: 219–56.

Meusnier, I., Valero, M., Destombe, C. et al. (2002) Polymerase chain reaction-single strand conformation polymorphism analysis of nuclear and chloroplast DNA provide evidence for recombination, multiple introductions and nascent speciation in the *Caulerpa taxifolia* complex. *Molecular Ecology* **11**: 2317–25.

Meyer, J.R. and Nalepa, C.A. (1991) Effect of dormant oil treatments on white peach scale (Homoptera: Diaspididae) and its overwintering parasite complex. *Journal of Entomological Science* **26**: 27–32.

Meyer, N.F. (1941) *Trichogramma*. Selhozhiuz, Leningrad.

Meyhöfer, R. and Hindayana, D. (2000) Effects of intraguild predation on aphid parasitoid survival. *Entomologia Experimentalis et Applicata* **97**: 115–22.

Michaud, J.P. (1999) Sources of mortality in colonies of the brown citrus aphid, *Toxoptera citricida* (Kirdaldy). *Biological Control* **44**: 347–67.

Michaud, J.P. (2001) Evaluation of green lacewings, *Chrysoperla plorabunda* (Fitch) (Neurop., Chrysopidae), for augmentative release against *Toxoptera citricida* (Hom., Aphididae) in citrus. *Journal of Applied Entomology* **125**: 383–8.

Michaud, J.P. (2002a) Classical biological control: a critical review of recent programs against citrus pests in Florida. *Annals of the Entomological Society of America* **94**: 531–40.

Michaud, J.P. (2002b) Invasion of the Florida citrus ecosystem by *Harmonia axyridis* (Coleoptera: Coccinellidae) and asymmetric competition with a native species, *Cycloneda sanguinea*. *Environmental Entomology* **31**: 827–35.

Michaud, J.P. (2003) Three targets of classical biological control in the Caribbean: success contribution, and failure. In: Van Driesche, R.G. (ed.), *Proceedings of the First International Symposium on Biological Control of Arthropods*, pp. 335–42, January 14–18, Honolulu, Hawaii. USDA Forest Service, Morgantown, WV.

Michaud, J.P. (2004) Natural mortality of Asian citrus psyllid (Homoptera: Psyllidae) in central Florida. *Biological Control* **29**: 260–9.

Michaud, J.P. and Browning, H.W. (1999) Seasonal abundance of the brown citrus aphid, *Toxoptera citricida* (Homoptera: Aphididae) and its natural enemies in Puerto Rico. *Florida Entomologist* **82**: 424–47.

Michelson, E.H. (1957) Studies on the biological control of schistosome-bearing snails. Predators and parasites of freshwater mollusks: a review of the literature. *Parasitology* **47**: 413–26.

Milberg, P. and Lamont, B.B. (1995) Fire enhances weed invasion of roadside vegetation in southwestern Australia. *Biological Conservation* **73**: 45–9.

Milbrath, L.R. and DeLoach, C.J. (2006) Host specificity of different populations of the leaf beetle *Diorhabda elongata* (Coleoptera: Chrysomelidae), a biological control agent of saltcedar (*Tamarix* spp.). *Biological Control* **36**: 32–48.

Millar, S.E., Chamow, S.M., Baur, A.W., Oliver, C., Robey, F., and Dean, J. (1989) Vaccination with a synthetic zona pellucida peptide produces long-term contraception in female mice. *Science* **246**: 935–8.

Miller, I.L. and Lonsdale, W.M. (1987) Early records of *Mimosa pigra* in the Northern Territory. *Plant Protection Quarterly* **2**: 140–2.

Miller, J.C. (1990) Effects of a microbial insecticide, *Bacillus thuringiensis kurstaki*, on nontarget lepidoptera in a spruce budworm-infested forest. *Journal of Research on Lepidoptera* **29**: 267–76.

Miller, L.A. and Bedding, R.A. (1982) Field testing of the insect parasitic nematode, *Neoaplectana bibionis* (Nematoda: Steinernematidae) against current borer moth, *Synanthedon tipuliformis* (Lep.: Sessiidae) in blackcurrants. *Entomophaga* **27**: 109–14.

Miller, L.K., Lingg, A.J., and Bulla, Jr., L.A. (1983) Bacterial, viral and fungal insecticides. *Science* **219**: 715–21.

Miller, M. and Aplet, G. (1993) Biological control: a little knowledge is a dangerous thing. *Rutgers Law Review* **45(2)**: 285–334.

Miller, M.L. and Aplet, G.H. (2005) Applying legal sunshine to the hidden regulation of biological control. *Biological Control* **35**: 358–65.

Mills, N.J. (1983) Possibilities for the biological control of *Choristoneura fumiferana* (Clemens) using natural enemies from Europe. *Biocontrol News and Information*, **4(2)**: 103–25.

Mills, N.J. (1993) Observations on the parasitoid complexes of budmoths (Lepidoptera: Tortricoidea) on larch in Europe. *Bulletin of Entomological Research* **83**: 103–12.

Mills, N. (1998) *Trichogramma:* the field efficacy of inundative biological control of the codling moth in Californian orchards. In: Hoddle, M.S. (ed.), *California Conference on Biological Control*, pp. 66–73, June 10–11, 1998, Berkeley, California.

Mills, N. (2005) Selecting effective parasitoids for biological control introductions: codling moth as a case history. *Biological Control* **34**: 274–82.

Mills, N.J. and Fischer, P. (1986) The entomophage complex of *Pissodes* weevils, with emphasis on the value of *P. validirostris* as a source of parasitoids for use in biological control. In: Roques, A. (ed.), *Proceedings of the 2nd Conference of the Cone and Seed Insects Working Party*, pp. 297–307, September 3–5, 1986, Briancon, France. Station de Zoologie Forestiere, Olivet.

Mills, N.J. and Schlup, J. (1989) The natural enemies of *Ips typographus* in central Europe: Impact and potential use in biological control. In: Kulhavy, D.L. and Miller, M.C. (eds.), *Potential for Biological Control of Dendroctonus and Ips Bark Beetles*, pp. 131–46. Center for Applied Studies, School of Forestry, Stephen F. Austin State University, Nacogdoches, TX.

Mills, N., Pickel, C., Mansfield, S. et al. (2000) Mass releases of *Trichogramma* wasps can reduce damage from codling moth. *California Agriculture* **54(6)**: 22–5.

Milne, W.M. and Bishop, A.L. (1987) The role of predators and parasites in the natural regulation of lucerne aphids in eastern Australia. *Journal of Applied Ecology* **24**: 893–905.

Milner, R.J., Soper, R.S., and Lutton, G.G. (1982) Field release of an Israeli strain of the fungus *Zoophthora radicans* (Brefeld) Batko for biological control of *Therioaphis trifolii* (Monnell) f. *maculata*. *Journal of the Australian Entomological Society* **21**: 113–18.

Minchin, D. and Gollasch, S. (2003) Fouling and ships' hulls: how changing circumstances and spawning events may result in the spread of exotic diseases. *Biofouling* **19**: 111–22.

Mineau, P. (1991) *Cholinesterase-Inhibiting Insecticides. Their Impact on Wildlife and the Environment*. Chemicals in Agriculture. Elsevier Science Publishers, Amsterdam.

Minkenberg, O.P.J.M. and van Lenteren, J.C. (1986) The leaf miners *Liriomyza trifolii* and *L. bryoniae* (Diptera: Agromyzidae), their parasites and host plants: a review. Agricultural University Wageningen Press, Wageningen.

Minkenberg, O.P.J.M., Tatar, M., and Rosenheim, J.A. (1992) Egg load as a major source of variability in insect foraging and oviposition behavior. *Oikos* **65**: 134–42.

Miura, T., Takahashi, R.M., and Wilder, W.H. (1984) Impact of the mosquitofish (*Gambusia affinis*) on a rice field ecosystem when used as a mosquito control agent. *Mosquito News* **44(4)**: 510–17.

Mkoji, G.M., Hofkin, B.V., Kuris, A.M. et al. (1999) Impact of the crayfish *Procambarus clarkii* on *Schistosoma haematobium* transmission in Kenya. *American Journal of Tropical Medicine and Hygiene* **61**: 751–9.

Mo, J., Trevino, M., and Palmer, W.A. (2000) Establishment and distribution of the rubber vine moth, *Euclasta whalleyi* Popescu-Gorj and Constantinescu (Lepidoptera: Pyralidae), following its release in Australia. *Australian Journal of Entomology* **39**: 344–50.

Moar, W.J., Eubanks, M., Freeman, B., Turnipseed, S., Ruberson, J., and Head, G. (2003) Effects of Bt cotton on biological control agents in the southeastern United States. In: Van Driesche, R.G. (ed.), *Proceedings of the First International Symposium on Biological Control of Arthropods*, pp. 275–7, January 14–18, 2002, Honolulu, Hawaii. USDA Forest Service, Morgantown, WV.

Mochiah, M.B., Ngi-Song, A.J., Overholt, W.A., and Stouthamer, R. (2002) *Wolbachia* infection in *Cotesia sesamiae* (Hymenoptera : Braconidae) causes cytoplasmic incompatibility: implications for biological control. *Biological Control* **25**: 74–80.

Mochizuki, M. (2002) Control of kanzawa spider mite, *Tetranychus kanzawai* Kishida (Acari: Tetranychidae) on tea by a synthetic pyrethroid resistant predatory mite, *Amblyseius womersleyi* Schicha (Acari: Phytoseiidae). *Japanese Journal of Applied Entomology and Zoology* **46**: 243–51.

Mogi, M. and Miyagi, I. (1990) Colonization of rice fields by mosquitoes (Diptera: Culicidae) and larvivorous predators in asynchronous rice cultivation in the Philippines. *Journal of Medical Entomology* **27**: 530–6.

Mohan, K.S. and Pillai, C.B. (1993) Biological control of *Oryctes rhinoceros* (L.) using an Indian isolate of *Oryctes baculovirus*. *Insect Science and its Application* **14**: 551–8.

Mohd, S. (1990) Barn owls (*Tyto alba*) for controlling rice field rats. *MAPPS Newsletter* **14(4)**: 51.

Mohyuddin, A.I. (1991) Utilization of natural enemies for the control of insect pests of sugar-cane. *Insect Science and Its Application* **12**: 19–26.

Mohyuddin, A.I., Inayatullah, C., and King, E.G. (1981) Host selection and strain occurrence in *Apanteles flavipes* (Cameron) (Hymenoptera: Braconidae) and its bearing on biological control of graminaceous stem-borers (Lepidoptera: Pyralidae). *Bulletin of Entomological Research* **71**: 575–81.

Monge, J.P. and Cortesoro, A.M. (1996) Tritrophic interactions among larval parasitoids, bruchids and Leguminosae seeds: influence of pre- and post-emergence learning on parasitoids' response to host and host-plant cures. *Entomologia Experimentalis et Applicata* **80**: 293–6.

Montgomery, B.R. and Wheeler, G.S. (2000) Antipredatory activity of the weevil *Oxyops vitiosa*: a biological control agent of *Melaleuca quinquenervia*. *Journal of Insect Behavior* **13**: 915–26.

Montgomery, M.E. and Lyon, S.M. (1996) Natural enemies of adelgids in North America: their prospect for biological control of *Adelges tsugae* (Homoptera: Adelgidae). In: Salom, S.M., Tigner, T.C., and Reardon, R.C. (eds.), *Proceedings, First Hemlock Woolly Adelgid Review*, pp. 89–102, October, 12, 1995, Charlottesville, VA. FHTET-96-10. USDA Forest Service, Morgantown, WV.

Montgomery, M.E., Yao, D., and Wang, H. (2000) Chinese Coccinellidae for biological control of the hemlock woolly adelgid: description of native habitat. In: McManus, K.A., Shields, K.S., and Souto, D.R. (eds.), *Proceedings, Symposium on Sustainable Management of Hemlock Ecosystems in Eastern North America*, pp. 97–102, June 22–24, 1999, Durham, New Hampshire. USDA Forest Service General Technical Report NE-267. USDA Forest Service, Morgantown, WV.

Mook, L.J. (1963) Birds and the spruce budworm. In: Morris, R.F. (ed.), *The Dynamics of Epidemic Spruce Budworm Populations*, pp. 268–91. Memoires of the Entomological Society of Canada.

Moore, D. and Prior, C. (1993) The potential of mycoinsecticides. *Biocontrol News and Information* **14(2)**: 33N–40N.

Moore, D., Bridge, P.D., Higgins, P.M., Bateman, R.P., and Prior, C. (1993) Ultra-violet radiation damage to *Metarhizium flavoviride* conidia and the protection given by vegetable and mineral oils and chemical sunscreens. *Annals of Applied Biology* **122**: 605–16.

Moore, N.F., King, L.A., and Possee, R.D. (1987) Mini review: viruses of insects. *Insect Science and its Application* **8**: 275–89.

Moore, S.D. (1989) Regulation of host diapause by an insect parasitoid. *Ecological Entomology* **14**: 93–8.

Moorehead, G.E. and Maltby, H.L. (1970) A container for releasing *Anaphes flavipes* from parasitized eggs of *Oulema melanopus*. *Journal of Economic Entomology* **63**: 675–6.

Moraes, M.C.B., Laumann, R., Sujii, E.R., Pires, C., and Borges, M. (2005) Induced volatiles in soybean and pigeon pea plants artificially infested with the neotropical brown stink bug, *Euschistus heros*, and their effect on the egg parasitoid, *Telenomus podisi*. *Entomologia Experimentalis et Applicata* **115**: 227–37.

Morales-Ramos, J.A., Rojas, M.G., Coleman, R.J., and King, E.G. (1998) Potential use of *in vitro*-reared *Catolaccus grandis* (Hymenoptera: Pteromalidae) for biological control of the boll weevil (Coleoptera: Curculionidae). *Journal of Economic Entomology* **91**: 101–9.

Moran, R.C. (1992) The story of the molesting salvinia. *Fiddlehead Forum* **19**: 26–8.

Moran, V.C., Gunn, B.H., and Walter, G.H. (1982) Wind dispersal and settling of first-instar crawlers of the cochineal insect *Dactylopius austrinus* (Homoptera: Coccoidea: Dactylopiidae). *Ecological Entomology* **7**: 409–19.

Morath, S.U., Pratt, P.D., Silvers, C.S., and Center, T.D. (2006) Herbivory by *Boreioglycaspis melaleucae* (Hemiptera: Psyllidae) accelerated foliar senescence and abscission in the invasive tree *Melaleuca quinquenervia*. *Environmental Entomology* **35**: 1372–8.

Moreau, S.J.M. and Guillot, S. (2005) Advances and prospects on biosynthesis, structures, and functions of venom proteins from parasitc wasps. *Insect Biochemistry and Molecular Biology* **35**: 1209–23.

Moreno, D.S. and Luck, R.F. (1992) Augmentative releases of *Aphytis melinus* (Hymenoptera: Aphelinidae) to suppress California red scale (Homoptera: Diaspididae) in southern California lemon orchards. *Journal of Economic Entomology* **85**: 1112–19.

Morewood, W.D. (1992) Cold storage of *Phytoseiulus persimilis* (Phytoseiidae). *Experimental and Applied Acarology* **13**: 231–6.

Morin, L., Willis, A.J., Armstrong, J., and Kriticos, D. (2002) Spread, epidemic development and impact of the bridal creeper rust in Australia: summary of results. In: Spafford Jacob, H., Dodd, J., and Moore, J.H. (eds.), *Thirteenth Australian Weeds Conference Papers and Proceedings*, pp. 385–8. Plant Protection Society of Western Australia, Perth.

Moritz, C. and Cicero, C. (2004) DNA Barcoding: promise and pitfalls. *PLoS Biology* **2**: e354.

Moriya, S., Inoue, K., and Mabuchi, M. (1989) The use of *Torymus sinensis* to control chestnut gall wasp, *Dryocosmus kuriphilus*, in Japan. *Food and Fertilizer Technology Center (FFTC) Technical Bulletin* **118**: 1–12.

Moriya, S., Shiga, M., and Adachi, I. (2003) Classical biological control of the chestnut gall wasp in Japan. In: Van Driesche, R.D. (ed.), *Proceedings of the First International Symposium on Biological Control of Arthropods*, pp. 25–33, January 14–18, 2002, Honolulu, Hawaii. FHTET-03-05. USDA Forest Service, Morgantown, WV.

Morris, M.J. (1987) Biology of the *Acacia* gall rust, *Uromycladium tepperianum*. *Plant Pathology* **36**: 100–6.

Morris, M.J. (1991) The use of plant pathogens for biological weed control in South Africa. *Agriculture, Ecosystems and Environment* **37**: 239–55.

Morris, M.J. (1999) The contribution of the gall-forming rust fungus *Uromycladium tepperianum* (Sacc.) AcAlp. to the biological control of *Acacia saligna* (Labill.) Wendl. (Fabaceae) in South Africa. *Africa Entomology Memoir* **1**: 125–8.

Morris, O.N. (1980) Entomopathogenic viruses: strategies for use in forest insect pest management. *The Canadian Entomologist* **112**: 573–84.

Morris, R.F. (1959) Single-factor analysis in population dynamics. *Ecology* **40**: 580–8.

Morris, R.F. (1963) The dynamics of epidemic spruce budworm populations. *Memoirs of the Entomological Society of Canada* **31**: 1–332.

Morrison, L.W. and Porter, S.D. (2005) Testing for population-level impacts of introduced *Pseudacteon tricuspis* flies, phorid parasitoids of *Solenopsis invicta* fire ants. *Biological Control* **33**: 9–19.

Morrow, B.J., Boucias, D.G., and Heath, M.A. (1989) Loss of virulence in an isolate of an entomopathogenic fungus, *Nomeraea rileyi*, after serial *in vitro* passage. *Journal of Economic Entomology* **82**: 404–7.

Morse, J.G. and Bellows, Jr., T.S. (1986) Toxicity of major citrus pesticides to *Aphytis melinus* (Hymenoptera: Aphelinidae) and *Cryptolaemus montrouzieri* (Coleoptera: Coccinellidae). *Journal of Economic Entomology* **79**: 311–14.

Morse, J.G. and Zareh, N. (1991) Pesticide-induced hormoligosis of citrus thrips (Thysanoptera: Thripidae) fecundity. *Journal of Economic Entomology* **84**: 1169–74.

Morse, J.G. and Hoddle, M.S. (2006) Invasion biology of thrips. *Annual Review of Entomology* **51**: 67–89.

Morse, J.G., Bellows, Jr., T.S., Gaston, L.K., and Iwata, Y. (1987) Residual toxicity of acaricides to three beneficial species on California citrus. *Journal of Economic Entomology* **80**: 953–60.

Moscardi, F. (1983) Utilizacão de *Baculovirus anticarsia* para o controle da lagarta da soya, *Anticarsia gemmatalis*. Empresa Brasiliera de Pesquira Agropecuaria, Comunicado Tecnico No. 23.

Moscardi, F. (1990) Development and use of soybean caterpillar baculovirus in Brazil. In: Pinnock, D.E. (ed.), *Vth International Colloquium on Invertebrate Pathology and Microbial Control*, pp. 184–7, August 20–24, 1990, Adelaide, Australia. Department of Entomology, University of Adelaide, Glen Osmond, South Australia.

Moscardi, F. (1999) Assessment of the application of baculoviruses for control of Lepidoptera. *Annual Review of Entomology* **44**: 257–89.

Moss, S.R., Turner, S.L., Trout, R.C. et al. (2002) Molecular epidemiology of rabbit haemorrhagic disease virus. *Journal of General Virology* **83**: 2461–7.

Mulatu, B., Applebaum, S.W., and Coll, M. (2004) A recently acquired host plant provides an oligophagous insect herbivore with enemy-free space. *Oikos* **107**: 231–8.

Mulla, M.S., Federici, B.A., and Darwazeh, H.A. (1982) Larvicidal efficacy of *Bacillus thuringiensis* serotype H-14 against stagnant-water mosquitoes and its effect on nontarget organisms. *Environmental Entomology* **11**: 788–95.

Mulla, M.S., Chaney, J.D., and Rodcharoen, J. (1990) Control of nuisance aquatic midges (Diptera: Chironomidae) with the microbial larvicide *Bacillus thuringiensis* var. *israelensis* in a man-made lake in southern California. *Bulletin of the Society for Vector Control* **15(2)**: 176–84.

Müller-Kögler, E. (1965) *Pilzkrankheiten bei Insekten*. Paul Parey, Berlin.

Müller-Schärer, H. and Schroeder, D. (1993) The biological control of *Centaurea* spp. in North America: do insects solve the problem? *Pesticide Science* **37**: 343–53.

Müller-Schärer, H., Lewinsohn, T.M., and Lawton, J.H. (1991) Search for weed biocontrol agents – when to move on? *Biocontrol Science and Technology* **1**: 271–80.

Mun, J., Bohonak, A.J., and Roderick, G.K. (2003) Population structure of the pumpkin fruit fly *Bactrocera depressa* (Tephritidae) in Korea and Japan: Pliocene allopatry or recent invasion? *Molecular Ecology* **12**: 2941–51.

Muñoz, A. and Murúa, R. (1990) Control of small mammals in a pine plantation (Central Chile) by modification of the habitat of predators (*Tyto alba*, Stringiforme and *Pseudalopex* sp. Canidae). *Acta Oecologia* **11**: 251–61.

Munro, V.M.W. and Henderson, I.M. (2002) Nontarget effect of entomophagous biocontrol: shared parasitism between native lepidopteran parasitoids and the biocontrol agent *Trigonospila brevifacies* (Diptera: Tachinidae) in forest habitats. *Environmental Entomology* **31**: 388–96.

Munroe, E. (1972) *The Moths of America North of Mexico. Fasc. 13.1A. Pyraloidea. Pyralidae (Part)*. E.W. Classey and R.B.D. Publications, London.

Murdoch, W.W. and Oaten, A. (1975) Predation and population stability. *Advances in Ecological Research* **9**: 1–131.

Murdoch, W.W. and Stewart-Oaten, A. (1989) Aggregation by parasitoids and predators: effects on equilibrium and stability. *American Naturalist* **134**: 288–310.

Murdoch, W.W., Chesson, J., and Chesson, P.L. (1985) Biological control in theory and practice. *American Naturalist* **125**: 344–66.

Murdoch, W.W., Nisbet, R.M., Gurney, W.S.C., and Reeve, J.D. (1987) An invulnerable age class and stability in delay-differential parasitoid-host models. *American Naturalist* **129**: 263–82.

Murdoch, W.W., Luck, R.F., Swarbrick, S.L., Walde, S., Yu, D.S., and Reeve, J.D. (1995) Regulation of an insect population under biological control. *Ecology* **76**: 206–17.

Murdoch, W.W., Briggs, C.J., and Nisbet, R.M. (1996) Competitive displacement and biological control in parasitoids: a model. *American Naturalist* **148**: 807–26.

Murdoch, W.W., Briggs, C.J., and Nisbet, R.M. (2003) Consumer-resource dynamics. In: Levin, S. (ed.), *Monographs in Population Biology*. Princeton University Press, Princeton, NJ.

Murdoch, W.W., Briggs, C.J., and Swarbrick, S. (2005) Host suppression and stability in a parasitoid-host system: experimental demonstration. *Science* **309**: 610–13.

Murphy, N.E. and Schaffelke, B. (2003) Use of amplified fragment length polymorphism (AFLP) as a new tool to explore the invasive green alga *Caulerpa taxifolia* in Australia. *Marine Ecology, Progress Series* **246**: 307–10.

Murray, J., Murray, E., Johnson, M.S., and Clarke, B. (1988) The extinction of *Partula* on Moorea. *Pacific Science* **42**: 150–3.

Musgrove, C.H. and Carman, G.E. (1965) Argentine ant control in citrus in southern California with granular formulations of certain chlorinated hydrocarbons. *Journal of Economic Entomology* **58**: 428–34.

Musser, F.R. and Shelton, A.M. (2003) Predation of *Ostrinia nubilias* (Lepidoptera: Crambidae) eggs in sweet corn by generalist predators and the impact of alternative foods. *Environmental Entomology* **32**: 1131–8.

Myers, J.H. (1985) How many insect species are necessary for successful biocontrol of weeds? In: Delfosse, E.S. (ed.), *Proceedings of the VIth International Symposium on Biological Control of Weeds*, pp. 77–82, August 19–25, 1984, Vancouver, Canada. Agriculture Canada, Ottawa.

Myers, J.H. and Bazely, D.R. (2003) *Ecology and Control of Introduced Plants*. Cambridge University Press, Cambridge.

Myers, K., Parer, I., Wood, D., and Cooke, B.D. (1994) The rabbit in Australia. In: Thompson, H.V. and King, C.M. (eds.), *The European Rabbit, the History and Biology of a Successful Colonizer*, pp. 108–57. Oxford University Press, Oxford.

Nadel, H. and van Alphen, J.J.M. (1987) The role of host-and host-plant odours in the attraction of a parasitoid, *Epidinocarsis lopezi*, to the habitat of its host, the cassava mealybug, *Phenacoccus manihoti*. *Entomologia Experimentalis et Applicata* **45**: 181–6.

Nafus, D.M. (1993) Movement of introduced biological control agents onto nontarget butterflies, *Hypolimnas* spp. (Lepidoptera: Nymphalidae). *Environmental Entomology* **22**: 265–72.

Nagesha, H.S., Wang, L.F., Hyatt, A.D., Morrissy, C.J., Lenghaus, C., and Westbury, H.A. (1995) Self assembly, antigenicity, and immunogenicity of the rabbit haemorrhagic disease virus (Cezchoslovakia strain V-351) capsid protein expressed in baculovirus. *Archives of Virology* **140**: 1095–1108.

Nappi, A.J. (1973) Parasitic encapsulation in insects. In: Maramorosch, K. and Shope, R.E. (eds.), *Invertebrate Immunity*, pp. 293–326. Academic Press, New York.

Nappi, A.J. and Vass, E. (1998) Hydrogen peroxide production in immune-reactive *Drosophila melanogaster*. *Journal of Parasitology* **84**: 1150–7.

Nappi, A.J., Frey, F., and Carton, Y. (2005) *Drosophila* serpin 27A is a likely target for immune suppression of the blood cell-mediated melanotic encapsulation response. *Journal of Insect Physiology* **51**: 197–205.

Naranjo, S.E. (2005) Long-term assessment of the effects of transgenic Bt cotton on the abundance of nontarget arthopod natural enemies. *Environmental Entomology* **34**: 1193–1210.

Naranjo, S.E. and Ellsworth, P.C. (2003) Arthropod communities and transgenic cotton in the western United States: implications for biological control. In: Van Driesche, R.G. (ed.), *Proceedings of the First International Symposium on Biological Control of Arthropods*, pp. 284–91, January 14–18, 2002, Honolulu, Hawaii. USDA Forest Service, Morgantown, WV.

Naranjo, S.E., Head, G., and Dively, G.P. (2005) Field studies assessing arthropod nontarget effects in Bt transgenic crops: introduction. *Environmental Entomology* **34**: 1178–80.

Nealis, V. (1985) Diapause and the seasonal ecology of the introduced parasite, *Cotesia* (*Apanteles*) *rubecula* (Hymenoptera: Braconidae). *The Canadian Entomologist* **117**: 333–42.

Nealis, V.G. (1986) Responses to host kairomones and foraging behavior of the insect parasite *Cotesia rubecula* (Hymenoptera: Braconidae). *Canadian Journal of Zoology* **64**: 2393–8.

Nechols, J.R. and Kikuchi, R.S. (1985) Host selection of the spherical mealybug (Homoptera: Pseudococcidae) by *Anagyrus indicus* (Hymenoptera: Encyrtidae): influence of host stage on parasitoid oviposition, development, sex ratio, and survival. *Environmental Entomology* **14**: 32–7.

Nechols, J.R., Andres, L.A., Beardsley, J.W., Goeden, R.D., and Jackson, C.G. (eds.) (1995) *Biological Control in the Western United States: Accomplishments and Benefits of a Regional Project W-84, 1964–1989*. DANR Publications, University of California, Oakland, CA.

Nedstam, B. and Burman, M. (1990) The use of nematodes against sciarids in Swedish greenhouses. *IOBC/WPRS Bulletin* **8(5)**: 147–8.

Neill, W.M. (1983) The tamarisk invasion of desert riparian areas. *Education Bulletin* 83–4. Desert Protective Council, Spring Valley, CA.

Nentwig, W. (1988) Augmentation of beneficial arthropods by strip-management. 1. Succession of predaceous arthropods and long-term change in the ratio of phytophagous and predaceous arthropods in a meadow. *Oecologia* **76**: 597–606.

Neser, S. (1985) A most promising bud-galling wasp, *Trichilogaster acaciaelongifoliae* (Pteromalidae), established against *Acacia longifolia* in South Africa. In: E.S. Delfosse (ed.), *Proceedings of the VIth International Symposium on Biological Control of Weeds*, pp. 797–803, August, 19–25 1984, Vancouver, Canada. Agriculture Canada, Ottawa.

Nettles, Jr., W.C., Wilson, C.M., and Ziser, S.W. (1980) A diet and methods for the *in vitro* rearing of the tachinid *Eucelatoria* sp. *Annals of the Entomological Society of America* **73**: 180–4.

Neuenschwander, P. (2003) Biological control of cassava and mango mealybugs in Africa. In: Neuenschwander, P., Borgemeister, C., and Langewald, J. (eds.), *Biological Control Systems in Africa*, pp. 45–59. CABI Publishing, Wallingford.

Neuenschwander, P. and Ajuonu, O. (1995) Measuring host finding capacity and arrestment of natural enemies of the cassava mealybug, *Phenococcus manihoti*, in the field. *Entomologia Experimentalis et Applicata* **77**: 47–55.

Neuenschwander, P., Schulthess, F., and Madojemu, E. (1986) Experimental evaluation of the efficiency of *Epidinocarsis lopezi*, a parasitoid introduced into Africa against the cassava mealybug, *Phenacoccus manihoti*. *Entomologia Experimentalis et Applicata* **42**: 133–8.

Neuenschwander, P., Hammond, W.N.O., Gutierrez, A.P. et al. (1989) Impact assessment of the biological control of the cassava mealybug, *Phenacoccus manihoti* Matile-Ferrero [Hemiptera: Pseudococcidae], by the introduced parasitoid *Epidinocarsis lopezi* (De Santis) [Hymenoptera: Encyrtidae]. *Bulletin of Entomological Research* **79**: 579–94.

Neuenschwander, P., Borgemeister, C., and Langewald, J. (eds.) (2003) *Biological Control in IPM Systems in Africa*. CABI Publishing, Wallingford.

Neves, P.M.O.J. and Hirose, E. (2005) *Beauveria bassiana* strains selection for biological control of the coffee berry borer, *Hypothenemus hampei* (Ferrari) (Coleoptera: Scolytidae). *Neotropical Entomology* **34**: 77–82.

New, T.R. (1992) *Insects as Predators*. New South Wales University Press, Kensington, New South Wales.

New, T.R. (2005) Biological control and invertebrate conservation. In: *Invertebrate Conservation and Agrocultural Ecosystems*, pp. 139–88. Cambridge University Press, Cambridge.

Newsome, A. (1990) The control of vertebrate pests by vertebrate predators. *Trends in Ecology and Evolution* **5**: 187–91.

Newsome, A.E., Parer, I., and Catling, P.C. (1989) Prolonged prey suppression by carnivores – predator-removal experiments. *Oecologia* **78**: 458–67.

Newton, I. (1988) Monitoring of persistent pesticide residues and their effects on bird populations. In: Harding, D.J.L. (ed.), *Britain Since "Silent Spring," an Update on the Ecological Effects of Agricultural Pesticides in the U.K.*, pp. 33–45. Proceedings of a symposium held in Cambridge, March 18, 1988.

Ngi-Song, A.J., Overholt, W.A., Smith, Jr., J.W., and Vinson, S.B. (1999) Suitability of new and old association hosts for the development of selected microgastrine parasitoids of gramineous stemborers. *Entomologia Experimentalis et Applicata* **90**: 257–66.

Nguyen, R., Brazzel, J.R., and Poucher, C. (1983) Population density of the citrus blackfly, *Aleurocanthus woglumi* Ashby (Homoptera: Aleyrodidae), and its parasites in urban Florida in 1979–1981. *Environmental Entomology* **12**: 878–84.

Nicholls, C.I., Parrella, M.P., and Altieri, M.A. (2000) Reducing the abundance of leafhoppers and thrips in a northern California organic vineyard through manipulation of full season floral diversity with summer cover crops. *Agricultural and Forest Entomology* **2**: 107–13.

Nicholson, A.J. (1957) The self-adjustment of populations to change. *Cold Springs Harbor Symposium Quantative Biology* **22**: 153–72.

Nicholson, A.J. and Bailey, V.A. (1935) The balance of animal populations. *Proceedings of the Zoological Society of London* **98**: 1935–98.

Nicoli, G., Benuzzi, M., and Leppla, N.C. (1994) *Seventh Workshop of the IOBC Global Working Group "Quality control of mass reared arthropods," Rimimi, Italy*, September 13–16, 1993. International Organization for Biological and Integrated Control of Noxious Animals and Plants, West Palaearctic Regional Section (IOBC/WPRS), General Secretariat Montfavet, France.

Nilsson, C. (1985) Impact of ploughing on emergence of pollen beetle parasitoids after hibernation. *Zeitschrift für Angewandte Entomologie* **100**: 302–8.

Ninkovic, V. and Pettersson, J. (2003) Searching behavior of the sevenspotted ladybird, *Coccinella septempunctata* – effects of plant-plant odor interaction. *Oikos* **100**: 65–70.

Ninkovich, V., Al Abassi, S., and Pettersson, J. (2001) The influence of aphid-induced plant volatiles on ladybird beetle searching behavior. *Biological Control* **21**: 191–5.

Nishida, T. and Napompeth, B. (1974) Trap for tephritid fruit fly parasites. *Entomophaga* **19**: 349–52.

Noldus, L.P.J.J. (1989) Semiochemicals, foraging behaviour and quality of entomophagous insects for biological control. *Journal of Applied Entomology* **108**: 425–51.

Noldus, L.P.J.J., Lewis, W.J., and Tumlinson, J.H. (1990) Beneficial arthropod behavior mediated by airborne semiochemicals. IX. Differential response of *Trichogramma pretiosum*, an egg parasitoid of *Heliothis zea*, to various olfactory cues. *Journal of Chemical Ecology* **16**: 3531–44.

Nomikou, M., Janssen, A., Schraag, R., and Sabelis, M.W. (2004) Vulnerability of *Bemisia tabaci* nymphs to phytoseiid predators: consequences for oviposition and influence of alternative food. *Entomologia Experimentalis et Applicata* **110**: 95–102.

Norbury, G., Heyward, R., and Parkes, J. (2002) Short-term ecological effects of rabbit haemorrhagic disease in the short-tussock grasslands of the South Island, New Zealand. *Wildlife Research* **29**: 599–604.

Nordblom, T.L., Smyth, M.J., Swirepik, A., Sheppard, A.W., and Briese, D.T. (2002) Spatial economics of biological control: investing in new release of insects for earlier limitation of Paterson's curse in Australia. *Agricultural Economics* **27**: 403–24.

Nordlund, D.A., Wu, Z.X., and Greenberg, S.M. (1997) *In vitro* rearing of *Trichogramma minutum* Riley (Hymenoptera: Trichogrammatidae) for ten generations, with quality assessment comparisons of *in vitro*- and *in vivo*-reared adults. *Biological Control* **9**: 201–7.

Norgaard, R.B. (1988) Economics of the cassava mealybug (*Phenacoccus manihoti*; Hom.: Pseudococcidae) biological control program in Africa. *Entomophaga* **33**: 3–6.

Norris, M.J. (1935) A feeding experiment on the adults of *Pieris rapae* Linnaeus (Lepid.: Rhop.). *Entomologist* **68**: 125–7.

North, S.G., Bullock, D.J., and Dulloo, M.E. (1994) Changes in vegetation and reptile populations on Round Island, Mauritius, following eradication of rabbits. *Biological Conservation* **67**: 21–8.

Norton, A.P. and Welter, S.C. (1996) Augmentation of the egg parasitoid *Anaphes iole* (Hymenoptera: Mymaridae) for *Lygus hesperus* (Heteroptera: Miridae) management in strawberries. *Environmental Entomology* **25**: 1406–14.

Nötzold, R., Blossey, B., and Newton, E. (1998) The influence of below ground herbivory and plant competition on growth and biomass allocation of purple loosestrife. *Oecologia* **113**: 82–93.

Nowierski, R.M. and Pemberton, R.W. (2002) Leaf spurge. In: Van Driesche, R.G., Blossey, B., Hoddle, M., Lyon, S., and Reardon, R. (eds.), *Biological Control of Invasive Plants in the Eastern United States*, pp. 181–207. FHTET-2002-04. USDA Forest Service, Morgantown, WV.

National Research Council (2002) Predicting invasions of nonindigenous plants and plant pests in the United States. National Academy Press, Washington, D.C.

Nowotny, N., Bascuñana, A., Ballagi-Pordany, D., Gavier-Widen, M., Uhlen, M., and Belak, S. (1997) Phylogenetic analysis of rabbit haemorrhagic disease and European brown hare syndrome viruses by comparison of sequences from the capsid protein gene. *Archives of Virology* **142**: 657–73.

Nuessly, G.S. and Goeden, R.D. (1984) Rodent predation on larvae of *Coleophora parthenica* (Lepidoptera: Coleophoridae), a moth imported for the biological control of Russian thistle. *Environmental Entomology* **13**: 502–8.

Nyffeler, M. and Benz, G. (1987) Spiders in natural pest control: a review. *Journal of Applied Entomology* **103**: 321–39.

Nyrop, J.P. (1988) Sequential classification of prey/predator ratios with application to European red mite (Acari: Tetranychidae) and *Typhlodromus pyri* (Acari: Phytoseiidae) in New York apple orchards. *Journal of Economic Entomology* **81**: 14–21.

O'Brien, P. (1991) The social and economic implications of RHD introduction. *Search* **22**: 191–3.

Obrycki, J.J. and Kring, T.J. (1998) Predaceous Coccinellidae in biological control. *Annual Review of Entomology* **43**: 295–321.

Obrycki, J.J. and Tauber, M.J. (1984) Natural enemy activity on glandular pubescent potato plants in the greenhouse: an unreliable predictor of effects in the field. *Environmental Entomology* **13**: 679–83.

O'Callaghan, M. and Gerard, E.M. (2005) Establishment of *Serratia entomophila* in soil from a granular formulation. Proceedings of a conference, Wellington, New Zealand, 9–11, August 2005. *New Zealand Plant Protection* **58**: 122–5.

O'Callaghan, M., Glare, T.R., Burgess, E.P.J., and Malone, L.A. (2005) Effects of plants genetically modified for insect resistance on nontarget organisms. *Annual Review of Entomology* **50**: 271–92.

Ochieng, R.S., Oloo, G.W., and Amboga, E.O. (1987) An artificial diet for rearing the phytoseiid mite *Amblyseius teke* Pritchard and Baker. *Experimental and Applied Acarology* **3**: 169–73.

Ode, P.J. (2006) Plant chemistry and natural enemy fitness: effects on herbivory and natural enemy interactions. *Annual Review of Entomology* **51**: 163–85.

O'Donnell, M.S. and Croaker, T.H. (1975) Potential of intra-crop diversity in the control of brassica pests, pp. 101–7. *Proceedings of the 8th British Insecticide and Fungicide Conference*, Brighton.

Oelrichs, P.B., MacLeod, J.K., Seawright, A.A. et al. (1999) Unique toxic peptides isolated from sawfly larvae in three continents. *Toxicon* **37**: 537–44.

Oetting, R.D. and Latimer, J.G. (1991) An entomogenous nematode, *Steinernema carpocapase*, is compatible with potting media environments created by horticultural practices. *Journal of Entomological Science* **26**: 390–4.

O'Hara, J.E. (1985) Oviposition strategies in the Tachinidae, a family of beneficial parasitic flies. *Agriculture and Forestry Bulletin, University of Alberta* **8**: 31–4.

Ohlinger, V.F., Haas, B., Meyers, G., and Thiel, H.J. (1990) Identification and characterization of the virus causing rabbit haemorrhagic disease. *Journal of Virology* **64**: 3331–6.

Ohmart, C.P. and Edwards, P.B. (1991) Insect herbivory on *Eucalyptus*. *Annual Review of Entomology* **36**: 637–57.

Oksanen, L., Fretwell, S.D., Arruda, J., and Niemela, P. (1981) Exploitation ecosystems in gradients of primary productivity. *American Naturalist* **118**: 240–61.

Oksanen, L., Oksanen, T., Ekerholm, P. et al. (1996) Structure and dynamics of arctic-subarctic grazing webs in relation to primary productivity. In: Polis, G.A. and Winemiller, K.O. (eds.), *Food Webs: Integration of Patterns and Dynamics*, pp. 231–42. Chapman and Hall, New York.

Okuda, M.S. and Yeargan, K.V. (1988) Intra- and interspecific host discrimination in *Telenomus podisi* and *Trissolcus euschisti* (Hymenoptera: Scelionidae). *Annals of the Entomological Society of America* **81**: 1017–20.

Olckers, T. and Hulley, P.E. (1995) Importance of preintroduction surveys in the biological control of *Solanum* weeds in South Africa. *Agriculture, Ecosystems, and the Environment* **52**: 179–85.

Olckers, T. and Hill, M.P. (eds.) (1999) *Biological Control of Weeds in South Africa (1990–1998)*. African Entomology Memoir no. 1. Entomological Society of Southern Africa.

Olckers, T. and Lotter, W.D. (2004) Possible non-target feeding by the bugweed lace bug, *Gargaphia decoris* (Tingidae), in South Africa: field evaluations support predictions of laboratory host-specificity studies. *African Entomology* **12**: 283–5.

Oliver, K.M., Russell, J.A., Moran, N.A., and Hunter, M.S. (2003) Facultative bacterial symbionts in aphids confer resistance to parasitic wasps. *Proceedings of the National Academy of Sciences USA* **100(4)**: 1803–7.

Oliver, K.M., Moran, N.A., and Hunter, M.S. (2005) Variation in resistance to parasitism in aphids is due to symbionts not host genotype. *Proceedings of the National Academy of Sciences USA* **102**: 12795–800.

O'Neil, R.J. (1984) Measurement and analysis of arthropod predation on velvetbean caterpillar (*Anticarsia gemmatalis*). Ph.D. Dissertation, University of Florida, Gainesville, FL.

O'Neil, R.J. (1988) Predation by *Podisus maculiventris* on Mexican bean beetle, *Epilachna varivestis*, in Indiana soybeans. *The Canadian Entomologist* **120**: 161–6.

O'Neil, R.J. (1997) The search strategy and functional response of *Podisus maculiventris* in potatoes. *Environmental Entomology* **26**: 1183–90.

O'Neil, R.J. and Wiedenmann, R.N. (1987) Adaptations of arthropod predators to agricultural systems. *Florida Entomologist* **70**: 40–8.

O'Neil, R.J. and Stimac, J.L. (1988a) Measurement and analysis of arthropod predation on velvetbean caterpillar, *Anticarsia gemmatalis* (Lepidoptera: Noctuidae), in soybeans. *Environmental Entomology* **17**: 821–6.

O'Neil, R.J. and Stimac, J.L. (1988b) Model of arthropod predation on velvetbean caterpillar (Lepidoptera: Noctuidae) larvae in soybean. *Environmental Entomology* **17**: 983–7.

Onzo, A., Hanna, R., and Sabelis, M.W. (2005) Biological control of cassava green mites in Africa: impact of the predatory mite *Typhlodromalus aripo*. *Entomologische Berichten* **65**: 2–7.

Opp. S.B. and Luck, R.F. (1986) Effects of host size on selected fitness components of *Aphytis melinus* and *A. lingnanenesis* (Hymenoptera: Aphelinidae). *Annals of the Entomological Society of America* **79**: 700–4.

Orlandini, G. and Martellucci, R. (1997) Melon: biological control of *Aphis gossypii*. *Colture Protette* **26(6)**: 33–6.

Ortega, Y.K., Pearson, D.E., and McKelvey, K.S. (2004) Effects of biological control agents and exotic plant invasion on deer mouse populations. *Ecological Applications* **14**: 241–53.

Osakabe, M. (1988) Relationships between food substances and developmental success in *Amblyseius sojaensis* Ehara (Acarina: Phytoseiidae). *Applied Entomology and Zoology* **23**: 45–51.

Osborne, K.J., Powles, R.J., and Rogers, P.L. (1990) *Bacillus sphaericus* as a biocontrol agent. *Australian Journal of Biotechnology* **4**: 205–11.

Osborne, L.S. (1987) Biological control of *Tetranychus urticae* Koch on ornamental foliage plants in Florida. *Bulletin SROP* **10(2)**: 144–8.

Osborne, L.S., Boucias, D.G., and Lindquist, R.K. (1985) Activity of *Bacillus thuringiensis* var. *israelensis* on *Bradysia coprophila* (Diptera: Sciaridae). *Journal of Economic Entomology* **78**: 922–5.

Osborne, L.S., Peña, J.E., Ridgway, R.L., and Klassen, W. (1998) Predaceous mites for mite management on ornamentals in protected culture. In: Ridgway, R.L., Hoffmann, M.P., Inscoe, N.N., and Glenister, C. (eds.), *Mass-reared Natural Enemies: Application, Regulation, and Needs*, pp. 116–38. Thomas Say Publications in Entomology: Proceedings. Entomological Society of America, Lanham, MD.

Osborne, L.S., Bolckmans, K., Landa, Z., and Peña, J. (2004) Kinds of natural enemies. In: Heinz, K.M., Van Driesche, R.G., and Parrella, M.P. (eds.), *Biocontrol in Protected Culture*, pp. 95–127. Ball Publishing, Batavia, IL.

Östman, Ö. (2004) The relative effects of natural enemy abundance and alternative prey abundance on aphid predation rates. *Biological Control* **30**: 281–7.

Outreman, Y., le Ralec, A., Wajnberg, É., and Pierre, J.S. (2005) Effects of within- and among-patch experiences on the patch-leaving decision rules in an insect parasitoid. *Behavioral Ecology and Sociobiology* **58**: 208–17.

Overmeer, W.P.J. (1985) Rearing and handling. In: Helle, W. and Sabelis, M.W. (eds.), *Spider Mites: their Biology, Natural Enemies, and Control*, vol. 1B, pp. 161–70. Elsevier, Amsterdam.

Pacala, S.W. and Hassell, M.P. (1991) The persistence of host-parasitoid associations in patchy environments. II. Evaluation of field data. *American Naturalist* **138**: 584–605.

Page, A.R., and Lacey, K.L. (2006) *Economic Impact Assessment of Australian Weed Biological Control*. CRC for Australian Weed Management Technical Series 10. CRC Weed Management, Adelaide.

Paine, R.W. (1994) *Recollections of a Pacific Entomologist, 1925–1966*. Australian Centre for International Agricultural Research, Canberra.

Paine, T.D. and Millar, J.G. (2002) Insect pests of eucalypts in California: implications of managing invasive species. *Bulletin of Entomological Research* **92**: 147–51.

Pak, G.A. and van Heiningen, T.G. (1985) Behavioural variations among strains of *Trichogramma* spp.: adaptability to field-temperature conditions. *Entomologia Experimentalis et Applicata* **38**: 3–13.

Palanza, P. and vom Saal, F. (2002) Effects of endocrine disrupters on behaviour and reproduction. In: Dell'Omo, G. (ed.), *Behavioural Ecotoxicology*, pp. 377–407. John Wiley and Sons, Chichester.

Palrang, A.T. and Grigarick, A.A. (1993) Flight response of the rice water weevil (Coleoptera: Curculionidae) to simulated habitat conditions. *Journal of Economic Entomology* **86**: 1376–80.

Papaj, D.R. and Vet, L.E.M. (1990) Odor learning and foraging success in the parasitoid *Leptopilina heterotoma*. *Journal of Chemical Ecology* **16**: 3137–50.

Parajulee, M.N., Phillips, T.W., and Hogg, D.B. (1994) Functional response of *Lyctocoris campestris* (F.) adults: effects of predator sex, prey species, and experimental habitat. *Biological Control* **4**: 80–7.

Paré, P.W. and Tumlinson, J.H. (1996) Plant volatile signals in response to herbivore feeding. *Florida Entomologist* **79**: 93–103.

Park, H.-Y., Bideshi, D.K., and Federici, B.A. (2003) Recombinant strain of *Bacillus thuringiensis* producing Cyt1A, Cry11B, and the *Bacillus sphaericus* binary toxin. *Applied and Environmental Microbiology* **69**: 1331–4.

Park, H.-W., Bideshi, D.K., Wirth, M.C., Johnson, J.J., Walton, W.E., and Federici, B.A. (2005) Recombant larvicidal bacterial with markedly improved efficacy against *Culex* vectors of West Nile virus. *American Journal of Tropical Medicine and Hygiene* **72**: 732–8.

Parker, J.D., Burkepile, D.E., and Hay, M.E. (2006) Opposing effects of native and exotic herbivores on plant invasions. *Science* **311**: 1459–61.

Parkes, J. and Murphy, E. (2004) Risk assessment of stoat control methods for New Zealand. *Science for Conservation* **237**: 5–38.

Parkes, J.P., Norbury, G.L., Heyward, R.P., and Sullivan, G. (2002) Epidemiology of rabbit haemorrhagic disease (RHD) in the South Island, New Zealand. *Wildlife Research* **29**: 543–55.

Parkman, J.P., Hudson, W.G., Frank, J.H., Nguyen, K.B., and Smart, Jr., G.C. (1993) Establishment and persistence of *Steinernema scapterisci* (Rhabditida: Steinernematidae) in field populations of *Scapteriscus* spp. mole crickets (Orthoptera: Gryllotalpidae). *Journal of Entomological Science* **28**: 182–90.

Parkman, J.P., Frank, J.H., Walker, T.J., and Schuster, D.J. (1996) Classical biological control of *Scapteriscus* spp. (Orthoptera: Gryllotalpidae) in Florida. *Environmental Entomology* **25**: 1415–20.

Parra, F. and Prieto, M. (1990) Purification and characterization of a calicivirus as the causative agent of lethal haemorrhagic disease in rabbits. *Journal of Virology* **68**: 4013–15.

Parra, J.R.P. and Zucchi, R.A. (2004) *Trichogramma* in Brazil: feasibility of use after twenty years of research. *Neotropical Entomology* **33**: 271–81.

Pasqualini, E. and Malavolta, C. (1985) Possibility of natural limitation of *Panonychus ulmi* (Koch) (Acarina, Tetranychidae) on apple in Emilia-Romagna. *Bolletino dell'Instituto di Entomologia "Guido Grandi" della Universita degli Studi di Bologna* **39**: 221–30.

Pasteels, J.M., Grégoire, J.-C., and Rowell-Ratheir, M. (1983) The chemical ecology of defense in arthropods. *Annual Review of Entomology* **28**: 263–89.

Pathak, J.P.N. (ed.) (1993) *Insect Immunity*. Kluwer Academic Publications, Boston, MA.

Patterson, R.S., Koehler, P.G., Morgan, R.B., and Harris, R.L. (eds.) (1981) *Status of Biological Control of Filth Flies*. USDA/SEA Publication, New Orleans, LA.

Paul, I., van Jaarsveld, A.S., Korsten, L., and Hattingh, V. (2005) The potential global geographic distribution of citrus black spot caused by *Guignardia citricarpa* (Kiley): likelihood of disease establishment in the European Union. *Crop Protection* **24**: 297–308.

Pavis, C., Huc, J.-A., Delvare, G., and Boissot, N. (2003) Diversity of the parasitoids of *Bemisia tabaci* B-biotype (Hemiptera: Aleyrodidae) in Guadeloupe Island (West Indies). *Environmental Entomology* **32**: 608–13.

Payne, C.C. (1986) Insect pathogenic viruses as pest control agents. In: Franz, J.M. (ed.), *Biological Plant and Health Protection: Biological Control of Plant Pests and of Vectors of Human and Animal Diseases*, pp. 183–200. International Symposium of the Akademie der Wissenschaften und der Literatur, November 15–17, 1984, Mainz, Germany. *Fortschritte der Zoologie* vol. 32. Gustav Fischer Verlag, Stuttgart.

Paynter, Q. (2005) Evaluating the impact of a biological control agent *Carmenta mimosa* on the woody wetland weed *Mimosa pigra* in Australia. *Journal of Applied Ecology* **42**: 1054–62.

Paynter, Q.E., Fowler, S.V., Gourlay, A.H. et al. (2004) Safety in New Zealand weed biocontrol: a nationwide survey for impacts on non-target plants. Proceedings of a conference, Hamilton, New Zealand, 10–12, August, 2004. *New Zealand Plant Protection* **57**: 102–7.

Pearl, R. (1927) The growth of populations. *Quarterly Review of Biology* **2**: 532–48.

Pearl, R. and Reed, L.J. (1920) On the rate of growth of the population of the United States since 1790 and its mathematical representation. *Proceedings of the National Academy of Sciences USA* **6**: 275–88.

Pearson, D.E. and Callaway, R.M. (2003) Indirect effects of host-specific biological control agents. *Trends in Ecology and Evolution* **18**: 456–61.

Pearson, D.E. and Callaway, R.M. (2005) Indirect nontarget effects of host-specific biological control agents: implications for biological control. *Biological Control* **35**: 288–98.

Pearson, D.E. and Callaway, R.M. (2006) Biological control agents elevate hantavirus by subsidizing deer mouse populations. *Ecology Letters* **9**: 443–50.

Pearson, D.E., McKelvey, K.S., and Ruggiero, L.F. (2000) Nontarget effects of an introduced biological control agent on deer mouse ecology. *Oecologia* **122**: 121–8.

Pearson, D.L. and Vogler, A.P. (2001) *Tiger Beetles: the Evolution, Ecology, and Diversity of the Cicindelids*. Comstock Publishing Associates, Cornell University Press, Ithaca, NY.

Pearson, J.F. and Jackson, T.A. (1995) Quality control management of the grass grub microbial control product, Invade®. *Proceedings of the Agronomy Society of New Zealand* **25**: 51–3.

Pedersen, B.S. and Mills, N.J. (2004) Single vs. multiple introduction in biological control: the roles of parasitoid efficiency, antagonsism, and niche overlap. *Journal of Applied Ecology* **41**: 973–84.

Pemberton, C.E. (1925) *The Field Rat in Hawaii and its Control*. Hawaiian Sugar Planters' Association Experiment Station Bulletin No. 17, Honolulu, HI.

Pemberton, R.W. (2000) Predictable risk to native plants in weed biological control. *Oecologia* **125**: 489–94.

Pemberton, R.W. (2002) Selection of appropriate future target weeds for biological control. In: Van Driesche, R.G., Blossey, B., Hoddle, M., Lyon, S., and Reardon, R. (eds.), *Biological Control of Invasive Plants in the Eastern United States*, pp. 375–86. FHTET-2002-04. USDA Forest Service, Morgantown, WV.

Pemberton, R.W. and Pratt, P.D. (2002) Skunk vine. In: Van Driesche, R.G., Blossey, B., Hoddle, M., Lyon, S., and Reardon, R. (eds.), *Biological Control of Invasive Plants in the*

Eastern United States, pp. 343–51. FHTET-2002-04. USDA Forest Service, Morgantown, WV.

Penman, D.R., Wearing, C.H., Collyer, E., and Thomas, W.P. (1979) The role of insecticide-resistant phytoseiids in integrated mite control in New Zealand. *Recent Advances in Acarology* **1**: 59–69.

Pennacchio, F. and Strand, M.R. (2006) Evolution of developmental strategies in parasitic Hymenoptera. *Annual Review of Entomology* **51**: 233–58.

Perdikis, D.C., Margaritopoulos, J.T., Stamatis, C. et al. (2003) Discrimination of the closely related biocontrol agents *Macrolophus melanotoma* (Hemiptera: Miridae) and *M. pygmaeus* using mitochondrial DNA analysis. *Bulletin of Entomological Research* **93**: 507–14.

Perfecto, I. (1991) Ants (Hymenoptera: Formicidae) as natural control agents of pests in irrigated maize in Nicaragua. *Journal of Economic Entomology* **84**: 65–70.

Perlak, F.J., Deaton, R.W., Armstrong, T.A. et al. (1990) Insect resistant cotton plants. *Bio/Technology* **8**: 939–43.

Perlak, F.J., Fuchs, R.L., Dean, D.A., McPherson, S.L., and Fischhoff, D.A. (1991) Modification of coding sequence enhances plant expression of insect control protein genes. *Proceedings of the National Academy of Sciences USA* **88**: 3325–8.

Peschken, D.P. and McClay, A.S. (1995) Picking the target: a revision of McClay's scoring system to determine the suitability of a weed for classical biological control, pp. 137–43. In: Delfosse, E.S. and Scott, R.R. (eds.), *Proceedings of the VIIIth International Symposium on Biological Control of Weeds*, February 2–7, 1992, Canterbury, New Zealand. CSIRO Publishing, Melbourne.

Peschken, D.P., DeClerck-Floate, R., and McClay, A.S. (1997) *Casida azurea* Fab. (Coleoptera: Chrysomelidae): host specificity and establishment in Canada as a weed biological control agent against the weed *Silene vulgaris* (Moench) Garcke. *The Canadian Entomologist* **129**: 949–58.

Petersen, J.J. and Currey, D.M. (1996) Timing of releases of gregarious *Muscidifurax raptorellus* (Hymenoptera: Pteromalidae) to control flies associated with confined beef cattle. *Journal of Agricultural Entomology* **13**: 55–63.

Petersen, J.J. and Greene, G.L. (1989) Potential for biological control of stable flies associated with confined livestock. *Miscellaneous Publications of the Entomological Society of America* **74**: 41–5.

Petersen, J.J., Meyer, J.A., Stage, D.A., and Morgan, P.B. (1983) Evaluation of sequential releases of *Spalangia endius* (Hymenoptera: Pteromalidae) for control of house flies and stable flies (Diptera: Muscidae) associated with confined livestock in eastern Nebraska. *Journal of Economic Entomology* **76**: 283–6.

Petersen, J.J., Watson, D.W., and Cawthra, J.K. (1995) Comparative effectiveness of three release rates for a pteromalid parasitoid (Hymenoptera) of house flies (Diptera) in beef cattle feedlots. *Biological Control* **5**: 561–5.

Peterson, P.G., McGregor, P.G., and Springett, B.P. (1994) Development of *Stethorus bifidus* in relation to temperature: implications for regulation of gorse spider mite populations. *Proceedings of the New Zealand Plant Protection Conference* **47**: 103–6.

Petrova, V., Cudare, Z., and Steinite, I. (2002) The efficiency of the predatory mite *Amblyseius cucumeris* (Acari: Phytoseiidae) as a control agent of the strawberry mite *Phytonemus pallidus* (Acari: Tarsonemidae) on field strawberry. In: Hietaranta, T., Linna, M.M., Palonen, P., and Parikka, P. (eds.), *Proceedings of the Fourth International Strawberry Symposium*, pp. 675–8, July 9–14, 2000, Tampere, Finland. *Acta Horticulturae* no. 567, vol. 2. International Society for Horticultural Science, Leuven.

Pettersson, B., Rippere, K.E., Yousten, A.A., and Priest, F.G. (1999) Transfer of *Bacillus lentimorbus* and *Bacillus popilliae* to the genus *Paenibacillus* with emended descriptions of *Paenibacillus lentimorbus* comb. nov. and *Paenibacillus popilliae* comb. nov. *International Journal of Systematic Bacteriology* **49**: 531–40.

Pfannenstiel, R.S. and Yeargan, K.V. (2002) Identification and diel activity patterns of predators attacking *Helicoverpa zea* (Lepidoptera: Ncotuideae) eggs in soybean and sweet corn. *Environmental Entomology* **31**: 232–41.

Pfannenstiel, R.S. and Unruh, T.R. (2003) Conservation of leafroller parasitoids through provision of alternate hosts in near-orchard habitats. In: Van Driesche, R.G. (ed.), *Proceedings of the First International Symposium on Biological Control of Arthropods*, pp. 256–62, January 14–18, 2002, Honolulu, Hawaii. FHTET-03-05. USDA Forest Service, Morgantown, WV.

Pfannenstiel, R.S., Hunt, R.E., and Yeargan, K.V. (1995) Orientation of a hemipteran predator to vibrations produced by feeding caterpillars. *Journal of Insect Behavior* **8**: 1–9.

Pfeuffer, R.J. and Rand, G.M. (2004) South Florida ambient pesticide monitoring program. *Ecotoxicology* **13**: 195–205.

Pfiffner, L. and Wyss, E. (2004) Use of sown wildflower strips to enhance natural enemies of agricultural pests. In: Gurr, G.M., Wratten, S.D., and Altieri, M.A. (eds.) *Ecological Engineering for Pest Management, Advances in Habitat Manipulation for Arthropods*, pp. 165–86. Cornell University Press, Ithaca, NY.

Phillips, C.B., Cane, R.P., Mee, J., Chapman, H.M., Hoelmer, K.A., and Coutinot, D. (2002) Intraspecific variation in the ability of *Microctonus aethiopoides* (Hymenoptera: Braconidae) to parasitise *Sitona lepidus* (Coleoptera: Curculionidae). *New Zealand Journal of Agricultural Research* **45**: 295–303.

Piao, Y.F. and Yan, S. (1996) Progress of mass production and field application of *Trichogramma dendrolimi*. In: Zhang, Z.L., Piao, Y.F., and Wu, J.W. (eds.), *Proceedings of the National Symposium on IPM in China*, pp. 1135–6. China Agricultural Scientech Press, Beijing.

Pickering, J., Dutcher, J.D., and Ekbom, B.S. (1989) An epizootic caused by *Erynia neoaphidis* and *E. radicans* (Zygomycetes: Entomophthoraceae) on *Acrythosiphon pisum* (Hom., Aphididae) on legumes under overhead irrigation. *Journal of Applied Entomology* **107**: 331–3.

Pickett, C.H. and Bugg, R.L. (1998) *Enhancing Biological Control, Habitat Management to Promote Natural Enemies of Agricultural Pests.* University of California Press, Berkeley, CA.

Pickett, C.H., Wilson, L.T., and Flaherty, D.L. (1990) The role of refuges in crop protection, with reference to plantings of French prune trees in a grape agroecosystem. In: Bostanian, N.J., Wilson, L.T., and Dennehy, T.J. (eds.), *Monitoring and Integrated Management of Arthropod Pests of Small Fruit Crops*, pp. 151–65. Intercept, Andover, UK.

Pierce, N.E., Kitching, R.L., Buckley, R.C., Taylor, M.F.J., and Benbow, K.F. (1987) The costs and benefits of cooperation between the Australian lycaenid butterfly, *Jalmenus evagoras*, and its attendant ants. *Behavioral Ecology and Sociobiology* **21**: 237–48.

Piggin, C.M. (1982) The biology of Australian weeds. 8. *Echium plantagineum* L. *Journal of the Australian Institute of Agricultural Science* **48**: 3–16.

Pilkington, L.J. and Hoddle, M.S. (2006) Reproductive and development biology of *Gonatocerus ashmeadi* (Hymenoptera: Mymaridae), an introduced egg parasitoid of *Homalodisca coagulata* (Hemiptera: Cicadellidae). *Biological Control* **37**: 266–75.

Pimentel, D. (1963) Introducing parasites and predators to control native pests. *The Canadian Entomologist* **95**: 785–92.

Pimentel D. and Wheeler, Jr., A.G. (1973) Species diversity of arthropod in the alfalfa community. *Environmental Entomology* **2**: 659–68.

Pinto, A., de S., Parra, J.R.P., de Oliveira, H.N., and Arrigoni, E.D.B. (2003) Comparison of release techniques of *Trichogramma galloi* Zucchi (Hymenoptera: Trichogrammatidae) to control *Diatraea saccharalis* (Fabricius) (Lepidoptera: Crambidae). *Neotropical Entomology* **32**: 311–18.

Pinto, J.D. and Stouthamer, R. (1994) Systematics of the Trichogrammatidae with emphasis on *Trichogramma*. In: Wajnberg, É. and Hassan, S.A. (eds.), *Biological Control with Egg Parasitoids*, pp. 1–36. CABI Publishing, Wallingford.

Pinto, J.D., Kazmer, D.J., Platner, G.R., and Sassaman, C.A. (1992) Taxonomy of the *Trichogramma minutum* complex (Hymenoptera, Trichogrammatidae) – allozymic variation and its relationship to reproductive and geographic data. *Annals of the Entomological Society of America* **85**: 413–22.

Pinto, J.D., Koopmanschap, A.B., Platner, G.R., and Stouthamer, R. (2002) The North American *Trichogramma* (Hymenoptera: Trichogrammatidae) parasitizing certain Tortricidae (Lepidoptera) on apple and pear, with ITS2 DNA characterizations and description of a new species. *Biological Control* **23**: 134–42.

Pintureau, B. (1990) Polymorphism biogeography and parasitic specificity of European *Trichogramma*. *Bulletin de la Societe Entomologique de France* **95**: 17–38.

Pintureau, B. (1993) Enzymatic analysis of the genus *Trichogramma* (Hymenoptera: Trichogrammatidae) in Europe. *Entomophaga* **38**: 411–31.

Piper, G.L., Coombs, E.M., Blossey, B., McEvoy, P.B., and Schooler, S.S. (2004) Purple loosestrife. In: Coombs, E.M., Clark, J.K., Piper, G.L., and Cofrancesco, Jr., A.F. (eds.), *Biological Control of Invasive Plants in the United States*, pp. 271–92. Oregon State University Press, Corvallis, OR.

Pizzol, J., Khoualdia, O., Ferran, A., Chavigny, P., and Vanlerberghe-Masutti, F. (2005) A single molecular marker to distinguish between strains of *Trichogramma cacoeciae*. *Biocontrol Science and Technology* **15**: 527–31.

Podgwaite, J.D. (1986) Effects of insect pathogens on the environment. In: Franz, J.M. (ed.), *Biological Plant and Health Protection: Biological Control of Plant Pests and of Vectors of Human and Animal Diseases*, pp. 279–87. International Symposium of the Akademie der Wissenschaften und der Literatur, November 15–17, 1984, Mainz, Germany. *Fortschritte der Zoologie* vol. 32. Gustav Fischer Verlag, Stuttgart.

Podoler, H. and Rogers, D. (1975) A new method for the identification of key factors from life-table data. *Journal of Animal Ecology* **44**: 85–114.

Poehling, H.-M. (1989) Selective application strategies for insecticides in agricultural crops. In: Jepson, P.C. (ed.), *Pesticides and Non-Target Invertebrates*, pp. 151–75. Intercept, Wimborne.

Poinar, G.O. (1986) Entomophagous nematodes. In: Franz, J.M. (ed.), *Biological Plant and Health Protection: Biological Control of Plant Pests and of Vectors of Human and Animal Diseases*, pp. 95–121. International Symposium of the Akademie der Wissenschaften und der Literatur, November 15–17, 1984, Mainz, Germany. *Fortschritte der Zoologie* vol. 32. Gustav Fischer Verlag, Stuttgart.

Polis, G.A. and Winemiller, K.O. (eds.) (1996) *Food Webs: Integration of Patterns and Dynamics.* Chapman and Hall, New York.

Polkinghorne, I., Hamerli, S., Cowan, P., and Duckworth, J. (2005) Plant-based immunocontraceptive control of wildlife – "potentials, limitations, and possums." *Vaccine* **23**: 1847–50.

Pollard, E., Lakhani, K.H., and Rothery, P. (1987) The detection of density-dependence from a series of annual censuses. *Ecology* **68**: 2046–55.

Poolman Simons, M.T.T., Sukerkropp, B.P., Vet, L.E.M., and de Moed, G. (1992) Comparison of learning in related generalist and specialist eucoilid parasitoids. *Entomologia Experimentalis et Applicata* **64**: 117–24.

Poopathi, S. and Kumar, K.A. (2003) Novel fermentation media for production of *Bacillus thuringiensis* subsp. *israelensis*. *Journal of Economic Entomology* **96**: 1039–44.

Poopathi, S., Kumar, K.A., Arunachalam, N., Tyagi, B.K., and Sekar, V. (2003) Control of *Culex quinquefasciatus* (Diptera: Culicidae) by *Bacillus sphaericus* and *B. thuringiensis* subsp. *israelensis*, produced on a new potato extract culture medium. *Biocontrol Science and Technology* **13**: 743–8.

Popov, N.A., Zabudskaja, I.A., and Burikson, I.G. (1987) The rearing of *Encarsia* in biolaboratories in greenhouse combines. *Zashchita Rastenii* **6**: 33.

Porter, S.D. (1998) Host-specific attraction of *Pseudacteon* flies (Diptera: Phoridae) to fire ant colonies in Brazil. *Florida Entomologist* **81**: 423–9.

Porter, S.D. (2000) Host specificity and risk assessment of releasing the decapitating fly, *Pseudacteon curvatus*, as a classical biological control agent for imported fire ants. *Biological Control* **19**: 35–47.

Porter, S.D. and Gilbert, L.E. (2004) Assessing host specificity and field release potential of fire ant decapitating flies (Phoridae: *Pseudacteon*). In: Van Driesche, R.G. and Reardon, R. (eds.), *Assessing Host Ranges for Parasitoids and Predators Used for Classical Biological Control: a Guide to Best Practice*, pp. 152–76. FHTET-2004-03. USDA Forest Service, Morgantown, WV.

Porter, S.D., Fowler, H.G., Campiolo, S., and Pesquero, M.A. (1995) Host specificity of several *Pseudacteon* (Diptera: Phoridae) parasites of fire ants (Hymenoptera: Formicidae) in South America. *Florida Entomologist* **78**: 70–5.

Porter, S.D., Williams, D.F., Patterson, R.S., and Fowler, H.G. (1997) Intercontinental differences in the abundance of *Solenopsis* fire ants (Hymenoptera: Formicidae): an escape from natural enemies? *Environmental Entomology* **26**: 373–84.

Porter, S.D., Nogueira de Sá, L.A., and Morrison, L.W. (2004) Establishment and dispersal of the fire ant decapitating fly *Pseudacteon tricuspis* in north Florida. *Biological Control* **29**: 179–88.

Possee, R.D., Allen, C.J., Entwistle, P.F., Cameron, L.R., and Bishop, D.H.L. (1990) Field trials of genetically engineered baculovirus insecticides. In: Anon, *Risk Assessment in Agricultural Biotechnology*, pp. 50–60. Proceedings of the International Conference, August, 1998, Oakland, CA. University of California, Berkeley, CA.

Potting, R.P.J., Vet, L.E.M., and Dicke, M. (1995) Host microhabitat location by stem-borer parasitoid *Cotesia flavipes*: the role of herbivore volatiles and locally and systemically induced plant volatiles. *Journal of Chemical Ecology* **21**: 525–39.

Pottinger, R.P. and LeRoux, E.J. (1971) The biology and dynamics of *Lithocolletis blancardella* (Lepidoptera: Gracillariidae) on apple in Quebec. *Memoirs of the Entomological Society of Canada* **77**: 1–437.

Powell, G.W., Sturko, A., Wikeem, B.M., and Harris, P. (1994) *Field Guide to the Biological Control of Weeds in British Columbia*. Land Management Handbook no. 27, Province of British Columbia. Ministry of Forests, Victoria.

Prabakaran, G. and Balaraman, K. (2006) Development of a cost-effective medium for the large scale production of *Bacillus thuringiensis* var *israelensis*. *Biological Control* **36**: 288–92.

Prasad, Y.K. (1989) The role of natural enemies in controlling *Icerya purchasi* in South Australia. *Entomophaga* **34**: 391–5.

Prasifka, J.R., Krauter, P.C., Heinz, K.M., Sansone, C.G., and Minzenmayer, R.R. (1999) Predator conservation in cotton: using grain sorghum as a source for insect predators. *Biological Control* **16**: 223–9.

Prasifka, J.R., Heinz, K.M., and Sansone, C.G. (2004) Timing, magnitude, rates, and putative causes of predator movement between cotton and grain sorghum fields. *Environmental Entomology* **33**: 282–90.

Pratt, P.D., Coombs, E.M., and Croft, B.A. (2003a) Predation by phytoseiid mites on *Tetranychus lintearius* (Acari: Tetranychidae), and established weed biological control agent of gorse (*Ulex europaeus*). *Biological Control* **26**: 40–7.

Pratt, P.D., Slone, D.H., Rayamajhi, M.B., Van, T.H., and Center, T.D. (2003b) Geographic distribution and dispersal rate of *Oxyops vitiosa* (Coleoptera: Curculionidae), a biological control agent of the invasive tree *Melaleuca quinquenervia* in south Florida. *Environmental Entomology* **32**: 397–406.

Pratt, P.D., Rayamajhi, M.B., Van, T.K., Center, T.D., and Tipping, P.W. (2005) Herbivory alters resource allocation and compensation in the invasive tree *Melaleuca quinquenervia*. *Ecological Entomology* **30**: 316–26.

Praveen, P.M. and Dhandapani, N. (2003) Development of biocontrol based pest management in tomato, *Lycopersicon esculentum* (Mill.). In: Anon, *Proceedings of the Symposium of Biological Control of Lepidopteran Pests*, pp. 267–70, July 17–18, 2002, Bangalore, India. Society for Biocontrol Advancement, Bangalore.

Press, J.W., Flaherty, B.R., and Arbogast, R.T. (1974) Interactions among *Plodia interpunctella*, *Bracon hebetor*, and *Xylocoris flavipes*. *Environmental Entomology* **3**: 183–4.

Price, J.F., Legard, D.E., and C.K. Chandler. (2002a) Two-spotted spider mite resistance to abamectin miticide on strawberry and strategies for resistance management. In: Hietaranta, T., Linna, M.M., Palonen, P., and Parikka, P. (eds.), *Proceedings of the Fourth International Strawberry Symposium*, pp. 683–5, July 9–14, 2000, Tampere, Finland. *Acta Horticulturae* No. 567, vol. 2. International Society for Horticultural Science, Leuven.

Price, J.F., Legard, D.E., Chandler, C.K., and McCord, E. (2002b) Changes in Florida strawberry production in response to twospotted spider mite resistance to Agri-Mek abamectin. In: Hokanson, S.C. and Jamieson, A.R. (eds.), *Strawberry Research to 2001. Proceedings of the 5th North American Strawberry Conference*, pp. 64–6. American Society for Horticultural Science, Alexandria, VA.

Price, P.W. (1970) Trail odours: recognition by insects parasitic in cocoons. *Science* **170**: 546–7.

Price, P.W. (1997) *Insect Ecology*. John Wiley and Sons, New York.

Prinsloo, G., Chen, Y., Giles, K.L., and Greenstone, M.H. (2002) Release and recovery in South Africa of the exotic aphid parasitoid *Aphelinus hordei* Kurdjumov (Hymenoptera: Aphelinidae) verified by the polymerase chain reaction. *BioControl* **47**: 127–36.

Prinsloo, H.E. (1960) Parasitiese mikro-organismes by die bruinsprinkaan *Locustana pardalina* (Walk.). *Suid-Afrikaanse Tydskrif vir Landbouuwwetenskap* **3**: 551–60.

Prokopy, R.J. and Webster, R.P. (1978) Oviposition-deterring pheromone of *Rhagoletis pomonella*, a kairomone for its parasitoid *Opius lectus*. *Journal of Chemical Ecology* **4**: 481–94.

Prokopy, R.J. and Christie, M. (1992) Studies on releases of mass-reared organophosphate resistant *Amblyseius fallacis* (Garm.) predatory mites in Massachusetts commercial orchards. *Journal of Applied Entomology* **114**: 131–7.

Pschorn-Walcher, H. (1963) Historical-biogeographical conclusions from host-parasite associations in insects. *Zeitschrift für Angewandte Entomologie* **51**: 208–14.

Pujol, M., Badosa, E., Cabrefiga, J., and Montesinos, E. (2005) Development of a strain-specific quantitative method for monitoring *Pseudomonas fluorescens* EPS62e, a novel biocontrol agent of fire blight. *FEMS Microbiology Letters* **249**: 343–52.

Purvis, G. and Curry, J.P. (1984) The influence of weeds and farmyard manure on the activity of carabidae and other ground-dwelling arthropods in a sugar beet crop. *Journal of Applied Ecology* **21**: 271–83.

Puttler, B., Parker, F.D., Pinnell, R.E., and Thewke, S.E. (1970) Introduction of *Apanteles rubecula* into the United States as a parasite of the imported cabbageworm. *Journal of Economic Entomology* **63**: 304–5.

Pyšek, P., Richardson, D.M., Rejmánek, M., Webster, G.L., Williamson, M., and Kirschner, J. (2004) Alien plants in checklists and floras: towards better communication between taxonomists and ecologists. *Taxon* **53**: 131–43.

Quezada, J.R. and DeBach, P. (1973) Bioecological and population studies of the cottony-cushion scale, *Icerya purchasi* Mask. and its natural enemies *Rodolia cardinalis* Mul. and *Cryptochaetum iceryae* Will., in southern California. *Hilgardia* **41**: 631–88.

Qin, Q.-L, Wang, F.-H., and Gong, H. (1999) Actions on teratocytes in coordinating the relationship between a parasitoid and its host – an overview. *Acta Entomologica Sinica* **42**: 431–8.

Quicke, D.L.J. (1997) *Parasitic Wasps*. Chapman and Hall, London.

Quinlan, R.J. (1990) Registration requirements and safety considerations for microbial pest control agents in the European 'economic community. In: Laird, M., Lace, L.A., and Davidson, E.W. (eds.), *Safety of Microbial Insecticides*, pp. 11–18. CRC Press, Boca Raton, FL.

Raghu, S. and Dhileepan, K. (2005) The value of simulating herbivory in selecting effective weed biological control agents. *Biological Control* **34**: 265–73.

Rajendran, B. and Hanifa, A.M. (1998) Efficacy of different techniques for the release of *Trichogramma chilonis* Ishiii, parasitising sugarcane internode borer, *Chilo sacchariphagus indicus* (Kapur). *Journal of Entomological Research* **22**: 355–9.

Ram, P., Tshernyshev, W.B., Afonina, V.M., and Greenberg, S.M. (1995) Studies on the strains of *Trichogramma evanescens* Westwood (Hymenoptera, Trichogrammatidae) collected from different hosts in Northern Maldova. *Journal of Applied Entomology* **119**: 79–82.

Ramalho, F.S., Medeiros, R.S., Lemos, W.P., Wanderley, P.A., Dias, J.M., and Zanuncia, J.C. (2000) Evaluation of *Catolaccus grandis* (Burks) (Hym., Pteromalidae) as a biological control agent against cotton boll weevil. *Journal of Applied Entomology* **124**: 359–64.

Rasmann, S., Köllner, T.G., Degenhardt, J. et al. (2005) Recruitment of entomopathogenic nematodes by insect-damaged maize roots. *Nature* **434**: 732–7.

Rastall, K., Kondo, V., Strazanac, J.S., and Buttler, L. (2003) Lethal effects of biological insecticide applications on non-target lepidopterans in two Appalachian forests. *Environmental Entomology* **32**: 1364–9.

Ratcliffe, N.A. (1993) Cellular defense responses of insects: unresolved problems, pp. 267–304. In: Beckage, N.E., Thompson, S.N., and Federici, B.A. (eds.), *Parasites and Pathogens of Insects, Volume I. Parasites*. Academic Press, New York.

Rath, A.C., Pearn, S., and Worladge, D. (1990) An economic analysis of production of *Metarhizium anisopliae* for control of the subterranean pasture pest *Adoryphorus couloni*. In: Pinnock, D.E. (ed.), *Vth International Colloquium on Invertebrate Pathology and Microbial Control*, p. 13, August 20–24, 1990, Adelaide, Australia. Department of Entomology, University of Adelaide, Glen Osmond, South Australia.

Rathman, R.J., Johnson, M.W., Rosenheim, J.A., and Tabashnik, B.E. (1990) Carbamate and pyrethroid resistance in the leafminer parasitoid *Diglyphus begini* (Hymenoptera: Eulophidae). *Journal of Economic Entomology* **83**: 2153–8.

Raupp, M.J., Hardin, M.R., Braxton, S.M., and Bull, B.B. (1994) Augmentative releases for aphid control on landscape plants. *Journal of Arboriculture* **20**: 241–9.

Rayamajhi, M.B., Purcell, M.F., Van, T.K., Center, T.D., Pratt, P.D., and Buckingham, G.R. (2002) Australian paperbark tree (*Melaleuca*). In: Van Driesche, R.G., Blossey, B., Hoddle, M., Lyon, S., and Reardon, R. (eds.), *Biological Control of Invasive Plants in the Eastern United States*, pp. 117–30. FHTET-2002-04. USDA Forest Service, Morgantown, WV.

Read, D.C. (1962) Notes on the life history of *Aleochara bilineata* (Gyll.) (Coleoptera: Staphylinidae), and on its potential value as a control agent for the cabbage maggot, *Hylemya brassicae* (Bouché) (Diptera: Anthomyiidae). *The Canadian Entomologist* **94**: 417–24.

Rebek, E.J., Sadof, C.S., and Hanks, L.M. (2005) Manipulating the abundance of natural enemies in ornamental landscapes with floral resource plants. *Biological Control* **33**: 203–16.

Reddiex, B., Hickling, G.J., Norbury, G.L., and Frampton, C.M. (2002) Effects of predation and rabbit haemorrhagic disease on population dynamics of rabbits (*Oryctolagus cuniculus*) in North Canterbury, New Zealand. *Wildlife Research* **29**: 627–33.

Reeve, J.D. and Murdoch, W.W. (1985) Aggregation by parasitoids in the successful control of the California red scale: a test of theory. *Journal of Animal Ecology* **54**: 797–816.

Reeve, J.D. and Murdoch, W.W. (1986) Biological control by the parasitoid *Aphytis melinus*, and population stability of the California red scale. *Journal of Animal Ecology* **55**: 1069–82.

Reimold, R.J. and Shealy, Jr., M.H. (1976) Chlorinated hydrocarbon pesticides and mercury in coastal young-of-the-year finfish, South Carolina and Georgia, 1972–74. *Pesticides Monitoring Journal* **9**: 170–5.

Rejmánek, M. and Pitcairn, M.J. (2002) When is eradication of exotic pest plants a realistic goal? In: Veitch, C.R. and Cout, M.N. (eds.), *Turning the Tide: the Eradication of Invasive Species*, pp. 249–53. IUCN SSC Invasive Species Specialist Group. IUCN, Gland and Cambridge. www.hear.org/articles/turningthetide/turningthetide.pdf.

Relini, G., Relini, M., and Torchia, G. (1998) Fish biodiversity in a *Caulerpa taxifolia* meadow in the Ligurian Sea. *Italian Journal of Zoology* **65**: 465–70.

Remaudière, G. and Keller, S. (1980) Revision systematique des genres d'Entomophthoraceae a potentialite entomopathogene. *Mycotaxon* **11**: 323–38.

Renault, S., Stasiak, K., Federici, B., and Bigot, Y. (2005) Commensal and mutualistic relationships of reoviruses with their parasitoid wasp hosts. *Journal of Insect Physiology* **51**: 137–48.

Rhoades, D.F. and Cates, R.G. (1976) Toward a general theory of plant antiherbivore chemistry. *Recent Advances in Phytochemistry* **10**: 168–213.

Ricciardi, A., Whoriskey, F.G., and Rasmussen, J.B. (1996) Impact of the *Dreissena* invasion on native unionid bivalves in the upper St. Lawrence River. *Canadian Journal of Fishereis and Aquatic Science* **53**: 1434–44.

Rice, R.E. and Jones, R.A. (1982) Collections of *Prospaltella perniciosi* Tower (Hymenoptera: Aphelinidae) on San Jose scale (Homoptera: Diaspididae) pheromone traps. *Environmental Entomology* **11**: 876–80.

Rice, M.E. and Wilde, G.E. (1988) Experimental evaluation of predators and parasitoids in suppressing greenbugs (Homoptera: Aphididae) in sorghum and wheat. *Environmental Entomology* **17**: 836–41.

Richards, O.W. and Waloff, N. (1954) Studies on the biology and population dynamics of British grasshoppers. Anti-locust Bulletin no. 17.

Richards, O.W., Waloff, N., and Spradberry, J.P. (1960) The measurement of mortality in an insect population in which recruitment and mortality widely overlap. *Oikos* **11**: 306–10.

Richardson, B.J., Baverstock, P.R., and Adams, M. (1986) *Allozyme Electrophoresis*. Academic Press, Orlando, FL.

Richardson, D.M. (1998) Forestry trees as invasive aliens. *Conservation Biology* **12**: 18–26.

Richardson, D.M., Macdonald, I.A.W., Holmes, P.M., and Cowling, R.M. (1992) Plant and animal invasions. In: Cowling, R. (ed.), *The Ecology of Fynbos: Nutrients, Fire and Diversity*, pp. 271–308. Oxford University Press, Cape Town.

Riechert, S.E. and Bishop, L. (1990) Prey control by an assemblage of generalist predators: spiders in garden test systems. *Ecology* **71**: 1441–50.

Riechert, S.E. and Lockley, T. (1984) Spiders as biological control agents. *Annual Review of Entomology* **29**: 299–320.

Rizke, R.M. and Rizki, T.M. (1990) Parasitoid virus-like particles destroy *Drosophila* cellular immunity. *Proceedings of the National Academy of Sciences USA* **87(21)**: 8388–92.

Rizzo, D.M. and Garbelotto, M. (2003) Sudden oak death: endangering California and Oregon forest ecosystems. *Frontiers in Ecology and the Environment* **1(5)**: 197–204.

Roberts, D.W. and Wraight, S.P. (1986) Current status on the use of insect pathogens as biological agents in agriculture: fungi. In: Samson, R.A., Vlak, J.M., and Peters, D. (eds.), *Fundamental and Applied Aspects of Invertebrate Pathology*, pp. 510–13. Proceedings of the 4th International Colloquium of Invertebrate Pathology, August 18–22, 1986, Veldhoven, The Netherlands. Ponsen and Looijen, Wageningen.

Robin, M.R. and Mitchell, W.C. (1987) Sticky traps for monitoring leafminers *Liriomyza sativae* and *Liriomyza trifolii* (Diptera: Agromyzidae) and their associated hymenopterous parasites in watermelon. *Journal of Economic Entomology* **80**: 1345–7.

Robinson, A.J. and Holland, M.K. (1995) Testing the concept of virally vectored immunosterilization for the control of wild rabbit and fox populations in Australia. *Australian Veterinary Journal* **72**: 65–8.

Robinson, A.J., Jackson, R., Kerr, P., Merchant, J., Parer, I., and Pech, R. (1997) Progress towards using a recombinant myxoma virus as a vector for fertility control in rabbits. *Reproduction, Fertility and Development* **9**: 77–83.

Robinson, G.S. (1975) *Macrolepidoptera of Fiji and Rotuma*, pp. 321–2. Classey, Faringdon.

Robinson, M.T. and Hoffmann, A.A. (2002) The pest status and distribution of three cryptic blue oat mite species (*Penthaleus* spp.) and redlegged earth mites (*Halotydeus destructor*) in southeastern Australia. *Experimental and Applied Acarology* **25**: 699–716.

Robinson, T. and Westbury, H. (1996) The Australian and New Zealand calicivirus disease program. In: *ESVV Symposium on Caliciviruses, Abstracts of Oral and Poster*

Presentations, p. 5. European Society for Veterinary Virology, University of Reading, Reading.

Rodda, G.H., Fritts, T.H., and Chiszar, D. (1997) The disappearance of Guam's wildlife. *BioScience* **47**: 565–74.

Rodda, G.H., Fritts, T.H., McCoid, M.J., and Campbell, E.W. (1999) An overview of the biology of the brown treesnake (*Boiga irregularis*), a costly introduced pest on Pacific Islands. In: Rodda, G.H., Sawai, Y., Chiszar, D., and Tanaka, H. (eds.), *Problem Snake Management: the Habu and the Brown Treesnake*, pp. 44–80. Comstock Publishing, Ithaca, NY.

Rodger, J.C. (1997) Likely targets for immunocontraception in marsupials. *Reproduction, Fertility and Development* **9**: 131–6.

Roduner, M., Cuperus, G., Mulder, P., Stritzke, J., and Payton, M. (2003) Successful biological control of the musk thistle in Oklahoma using the musk thistle head weevil and the rosette weevil. *American Entomologist* **49(2)**: 112–20.

Roehrdanz, R.L., Reed, D.K., and Burton, R.L. (1993) Use of polymerase chain reaction and arbitrary primers to distinguish laboratory-raised colonies of parasitic Hymenoptera. *Biological Control* **3**: 199–206.

Rogers, C.E. (1985) Extrafloral nectar: entomological implications. *Bulletin of the Entomological Society of America* **31**: 15–20.

Rohani, P., Godfray, H.C.J., and Hassell, M.P. (1994) Aggregation and the dynamics of host-parasitoid systems: a discrete-generation model with within-generation redistribution. *American Naturalist* **144**: 491–509.

Rojas, M.G., Morales-Ramos, J.A., and King, E.G. (1999) Response of *Catolaccus grandis* (Hymenoptera: Pteromalidae) to its natural host after ten generations on a factitious host, *Callosobruchus maculatus* (Coleoptera: Bruchidae). *Environmental Entomology* **28**: 137–41.

Roland, J. (1988) Decline in winter moth populations in North America: direct versus indirect effect of introduced parasites. *Journal of Animal Ecology* **57**: 523–31.

Roland, J. (1994) After the decline: what maintains low winter moth density after successful biological control? *Journal of Animal Ecology* **63**: 392–8.

Roland, J. and Embree, D.G. (1995) Biological control of the winter moth. *Annual Review of Entomology* **40**: 475–92.

Roland, J., Evans, W.G., and Myers, J.H. (1989) Manipulation of oviposition patterns of the parasitoid *Cyzenis albicans* (Tachinidae) in the field using plant extracts. *Journal of Insect Behaviour* **2**: 487–503.

Room, P.M. (1980) Biological control of weeds – modest investments can give large returns. In: *Proceedings of the Australian Agronomy Conference: Pathways to Productivity*, p. 291, April, 1980, Gratton, Queensland. Australian Institute of Agricultural Science, Melbourne, Victoria.

Room, P.M. (1990) Ecology of a simple plant-herbivore system: biological control of salvinia. *Trends in Ecology and Evolution* **5(3)**: 74–9.

Room, P.M. and Thomas, P.A. (1985) Nitrogen and establishment of a beetle for biological control of the floating weed salvinia in Papua New Guinea. *Journal of Applied Ecology* **22**: 139–56.

Room, P.M., Harley, K.L.S., Forno, I.W., and Sands, D.P.A. (1981) Successful biological control of the floating weed salvinia. *Nature* **294**: 78–80.

Root, R.B. (1973) Organization of plant-arthropod association in simple and diverse habitats: the fauna of collards (*Brassica oleracea*). *Ecological Monographs* **43**: 95–124.

Roots, C. (1976) *Animal Invaders*. Universe Book, New York.

Rose, K.E., Louda, S.M., and Rees, M. (2005) Demographic and evolutionary impacts of native and invasive herbivores on *Cirsium canescens*. *Ecology* **86**: 453–65.

Rose, M. (1990) Rearing and mass rearing of natural enemies. In: Rosen, D. (ed.), *Armored Scale Insects: their Biology, Natural Enemies, and Control*, vol. 4B, pp. 263–87. Elsevier, Amsterdam.

Rose, M. and DeBach, P. (1992) Biocontrol of *Parabemisia myricae* (Kuwana) (Homoptera: Aleyrodidae) in California. *Israel Journal of Entomology* **25–26**: 73–95.

Rosen, D. and DeBach, P. (1979) *Species of Aphytis of the World (Hymenoptera: Aphelinidae)*. Israel Universities Press, Jerusalem, and Dr. W. Junk, The Hague.

Rosenheim, J.A. (1998) Higher-order predators and the regulation of insect herbivore populations. *Annual Review of Entomology* **43**: 421–47.

Rosenheim, J.A. (2005) Intraguild predation of *Orius tristicolor* by *Geocoris* spp. and the paradox of irruptive spider mite dynamics in California cotton. *Biological Control* **32**: 172–9.

Rosenheim, J.A. and Hoy, M.A. (1986) Intraspecific variation in levels of pesticide resistance in field populations of a parasitoid, *Aphytis melinus* (Hymenoptera: Aphenlinidae): the role of past selection pressures. *Journal of Economic Entomology* **79**: 1161–73.

Rosenheim, J.A. and Rosen, D. (1991) Foraging and oviposition decisions in the parasitoid *Aphytis lingnanenesis*: distinguishing the influences of egg load and experience. *Journal of Animal Ecology* **60**: 873–93.

Rosenheim, J.A., Kaya, H.K., Ehler, L.E., Marois, J.J., and Jaffee, B.A. (1995) Intraguild predation among biological-control agents: theory and evidence. *Biological Control* **5**: 303–35.

Rosenheim, J.A., Limburg, D.D., Colfer, R.G., Letourneau, D.K., and Andow, D.A. (1999) Impact of generalist predators on a biological control agent, *Chrysoperla carnea*: direct observations. *Ecological Applications* **9**: 409–17.

Ross, D.J., Johnson, C.R., and Hewitt, C.L. (2003) Assessing the ecological impacts of an introduced seastar: the importance of multiple methods. *Biological Invasions* **5**: 3–21.

Rothschild, G. (1966) A study of a natural population of *Conomelus anceps* Germar (Homoptera: Delphacidae), including observation on predation using the precipitin test. *Journal of Animal Ecology* **35**: 413–34.

Roush, R.T. (1990a) Genetic variation in natural enemies: critical issues for colonization in biological control. In: Mackauer, M., Ehler, L.E., and Roland, J. (eds.), *Critical Issues in Biological Control*, pp. 263–88. Intercept, Andover.

Roush, R.T. (1990b) Genetic considerations in the propagation of entomophagous species. In: Baker, R.R. and Dunn, P.E. (eds.), *New Directions in Biological Control: Alternatives for Suppressing Agricultural Pests and Disease*, pp. 373–87. Alan R. Liss, New York.

Roush, R.T. and Hoy, M.A. (1981) Genetic improvement of *Metaseiulus occidentalis*: selection with methomyl, dimethoate, and carbaryl and genetic analysis of carbaryl resistance. *Journal of Economic Entomology* **74**: 138–41.

Royama, T. (1981) Evaluation of mortality factors in insect life table analysis. *Ecological Monographs* **5**: 495–505.

Royama, T. (1984) Population dynamics of the spruce budworm *Choristoneura fumiferana*. *Ecological Monographs* **54**: 429–62.

Royama, T. (1992) *Analytic Population Dynamics*. Chapman and Hall, London.

Rudzki, S. (1995) Escaped rabbit calicivirus highlights Australia's chequered history of biological control. *Search* **26**: 287.

Ruesink, W.G. (1975) Estimating time-varying survival of arthropod life stages from population density. *Ecology* **56**: 244–7.

Russell, E.P. (1989) Enemies hypothesis: a review of the effect of vegetational diversity on predatory insects and parasitoids. *Environmental Entomology* **18**: 590–9.

Rutz, D.A. (1986) Parasitoid monitoring and impact evaluation in the development of filth fly biological control programs for poultry farms. *Miscellaneous Publications of the Entomological Society of America* **61**: 45–51.

Rutz, D.A. and Axtell, R.C. (1979) Sustained releases of *Muscidifurax raptor* (Hymenoptera: Pteromalidae) for house fly (*Musca domestica*) control in two types of caged-layer poultry houses. *Environmental Entomology* **8**: 1105–10.

Rutz, D.A. and Axtell, R.C. (1980) House fly (*Musca domestica*) parasites (Hymenoptera: Pteromalidae) associated with poultry manure in North Carolina. *Environmental Entomology* **9**: 175–80.

Rutz, D.A. and Axtell, R.C. (1981) House fly (*Musca domestica*) control in broiler-breeder poultry houses by pupal parasites (Hymenoptera: Pteromalidae): indigenous parasite species and releases of *Muscidifurax raptor*. *Environmental Entomology* **10**: 343–5.

Rutz, D.A. and Patterson, R.S. (eds.) (1990) *Biocontrol of Arthropods Affecting Livestock Poultry*. Westview Press, Boulder, CO.

Ryan, J., Ryan, M.F., and McNaeidhe, F. (1980) The effect of interrow plant cover on populations of the cabbage root fly *Delia brassicae* (Wied.). *Journal of Applied Ecology* **17**: 31–40.

Ryoo, M.I. (1996) Influence of the spatial distribution pattern of prey among patches and spatial coincidence on the functional and numerical response of *Phytoseiulus persimilis* (Acarina, Phytoseiidae). *Journal of Applied Entomology* **120**: 187–92.

Sabelis, M.W. (1992) Predatory arthropods. In: Crawley, M.J. (ed.) *Natural Enemies: the Population Biology of Predators, Parasites and Diseases*, pp. 225–64. Blackwell Science, Oxford.

Sabelis, M.W. and Van de Baan, H.E. (1983) Location of distant spider mite colonies by phytoseiid predators: demonstration of specific kairomones emitted by *Tetranychus urticae* and *Panonychus ulmi*. *Entomologia Experimentalis et Applicata* **33**: 303–14.

Sagarra, L.A., Vincent, C., and Stewart, R.K. (2001) Suitability of nine mealybug species (Homoptera: Pseudococcidae) as hosts for the parasitoid *Anagyrus kamali* (Hymenoptera: Encyrtidae). *Florida Entomologist* **84**: 112–16.

Sailer, R. (1978) Our immigrant insect fauna. *Bulletin of Entomological Society of America* **24**: 3–11.

Sailer, R.I. (1983) History of insect introductions. In: Wilson, G.L. and Graham, C.L. (eds.), *Exotic Plant Pests and North American Agriculture*, pp. 15–38. Academic Press, New York.

Saiyed, H., Dewan, A., Bhatnager, V. et al. (2003) Effect of endosulfan on male reproductive development. *Environmental Health Perspectives* **111**: 1958–62.

Salas, J. and Frank, J.H. (2001) Development of *Metamasius callizona* (Coleoptera: Curculionidae) on pineapple stems. *Florida Entomologist* **84**: 123–6.

Samuels, K.D.Z., Pinnock, D.E., and Bull, R.M. (1990) Scarabaeid larvae control in sugarcane using *Metarhizium anisopliae*. *Journal of Invertebrate Pathology* **55**: 135–7.

Samways, M.J. (1988) Comparative monitoring of red scale, *Aonidiella aurantii* (Mask.) (Hom., Diaspididae) and its *Aphytis* spp. (Hym., Aphelinidae) parasitoids. *Journal of Applied Entomology* **105**: 483–9.

Samways, M.J. (1990) Ant assemblage structure and ecological management in citrus and subtropical fruit orchards in southern Africa. In: van der Meer, R.K., Jaffe, K., and Cedeno, A. (eds.), *Applied Myrmecology, a World Perspective*, pp. 570–87. Westview Press, Boulder, CO.

Samways, M.J., Nel, M., and Prins, A.J. (1982) Ants (Hymenoptera: Formicidae) foraging in citrus trees and attending honeydew producing Homoptera. *Phytophylactica* **14**: 155–7.

Sands, D.P.A. (1997) The safety of biological control agents: assessing their impact on beneficial and other non-target hosts. *Memoires of the Museum of Victoria* **56**: 611–16.

Sands, D.P.A. and Combs, M. (1999) Evaluation of the Argentinian parasitoid, *Trichopoda giacomellii* (Diptera: Tachinidae), for biological control of *Nezara viridula* (Hemiptera: Pentatomidae) in Australia. *Biological Control* **15**: 19–24.

Sato, S., Yasuda, H., and Evans, E.W. (2005) Dropping behavior of larvae of aphidophagous ladybirds and its effects on incidence of intraguild predation: interactions between the intraguild prey, *Adalia bipunctata* (L.) and *Coccinella septempunctata* (L.), and the intraguild predator, *Harmonia axyridis* Pallas. *Ecological Entomology* **30**: 220–4.

Saunders, G.R. and Giles, J.R. (1977) A relationship between plagues of the house mouse *Mus musculus* (Rodentia: Muridae) and prolonged periods of dry weather in south-eastern Australia. *Australian Journal of Wildlife Research* **4**: 241–7.

Saunders, G., Kay, B., Mutze, G., and Choquenot, D. (2002) Observations on the impacts of rabbit haemorrhagic disease on agricultural production values in Australia. *Wildlife Research* **29**: 605–13.

Saunders, G., Berghout, M., Kay, M., Triggs, B., Van de Ven, R., and Winstanley, R. (2004) The diet of foxes (*Vulpes vulpes*) in southeastern Australia and the potential effects of rabbit haemorrhagic disease. *Wildlife Research* **31**: 13–18.

Schaefer, P.W., Dysart, R.J., Flanders, R.V., Burger, T.L., and Ikebe, K. (1983) Mexican bean beetle (Coleoptera: Coccinellidae) larval parasite *Pediobius foveolatus* (Hymenoptera: Eulophidae) from Japan: field release in the United States. *Environmental Entomology* **12**: 852–4.

Schaefer, P.W., Dysart, R.J., and Specht, H.B. (1987) North American distribution of *Coccinella septempunctata* (Coleoptera: Coccinellidae) and its mass appearance in coastal Delaware. *Environmental Entomology* **16**: 368–73.

Schaffelke, B., Murphy, N., and Uthicke, S. (2002) Using genetic techniques to investigate the sources of the invasive alga *Caulerpa taxifolia* in three new locations in Australia. *Marine Pollution Bulletin* **44**: 204–10.

Schaffner, U. and Muller, C. (2001) Exploitation of the fecal shield of the lily leaf beetle, *Lilioceris lilii* (Coleoptera: Chrysomelidae), by the specialist parasitoid *Lemophagus pulcher* (Hymenoptera: Ichneumonidae). *Journal of Insect Behavior* **14**: 739–57.

Schat, M. and Blossey, B. (2005) Influence of natural and simulated leaf beetle herbivory on biomass allocation and plant architecture of purple loosestrife (*Lythrum salicaria* L.) *Environmental Entomology* **34**: 906–14.

Scheffer, S.J. and Grissell, E.E. (2003) Tracing the geographic origin of *Megastigmus transvaalensis* (Hymenoptera: Torymidae): an African wasp feeding on a South American plant in North America. *Molecular Ecology* **12**: 415–21.

Scheffer, S.J., Giblin-Davis, R.M., Taylor, G.S. et al. (2004) Phylogenetic relationships, species limits, and host specificity of gall-forming *Fergusonina* flies (Diptera: Fergusoninidae) feeding on *Melaleuca* (Myrtaceae). *Annals of the Entomological Society of America* **97**: 1216–21.

Scheffer, S.J., Lewis, M.L., and Joshi, R.C. (2006) DNA barcoding applied to invasive leafminers (Diptera: Agromyzidae) in the Philippines. *Annals of the Entomological Society of America* **99**: 204–10.

Schellhorn, N.A., Lane, C.P., and Olson, D.M. (2005) The co-occurrence of an introduced biological control agent (Coleoptera: *Coccinella septempunctata*) and an endangered butterfly (Lepidoptera: *Lycaeides melissa samuelis*). *Journal of Insect Conservation* **9**: 41–7.

Schettler, T., Soloman, G., and Valenti, M. (1999) *Generations at Risk, Reproductive Health and the Environment*. MIT Press, Cambridge, MA.

Schloesser, D.W. (1995) Introduced species, zebra mussels in North America. In: *Encyclopedia of Environmental Biology*, vol. 2, pp. 337–56. Academic Press, San Diego, CA.

Schlotterer, C. (2000) Evolutionary dynamics of microsatellite DNA. *Chromosoma* **109**: 365–71.

Schmidt, J.M. and Smith, J.J.B. (1986) Correlations between body angles and substrate curvature in the parasitoid wasp *Trichogramma minutum*: a possible mechanism of host radius measurement. *Journal of Experimental Biology* **125**: 271–85.

Schmidt, J.M. and Smith, J.J.B. (1987) Measurement of host curvature by the parasitoid wasp *Trichogramma minutum*, and its effect on host examination and progeny allocation. *Journal of Experimental Biology* **129**: 151–64.

Schmidt, J.M., Cardé, R.T., and Vet, L.E.M. (1993) Host recognition by *Pimpla instigator* F. (Hymenoptera: Ichneumonidae): preferences and learned responses. *Journal of Insect Behavior* **6**: 1–11.

Schmidt, S., Naumann, I.D., and De Barro, P.J. (2001) *Encarsia* species (Hymenoptera: Aphelinidae) of Australia and the Pacific Islands attacking *Bemisia tabaci* and *Trialeurodes vaporariorum* (Hemiptera: Aleyrodidae): a pictorial key and descriptions of four new species. *Bulletin of Entomological Research* **91**: 369–87.

Schneider, H., Borgemeister, C., Sétamou, M. et al. (2004) Biological control of the larger grain borer, *Prostephanus truncatus* (Horn) (Coleoptera: Bostrichidae) by its predator *Teretrius nigrescens* (Lewis) (Coleoptera: Histeridae) in Togo and Benin. *Biological Control* **30**: 241–55.

Schnepf, H.E. and Whiteley, H.R. (1981) Cloning and expression of the *Bacillus thuringiensis* crystal protein gene in *Escherichia coli*. *Proceedings of the National Academy of Sciences USA* **78**: 2893–7.

Schnepf, E., Crickmore, N., Van Rie, J., Lereclus, D., Baum, J., and Feitelson, J. (1998) *Bacillus thuringiensis* and its pesticidal proteins. *Microbiology and Molecular Biology Reviews* **62**: 775–806.

Schoen, L. (2000) The use of open rearing units or "banker plants" against *Aphis gossypii* Glover in protected courgette and melon crops in Roussillon (south of France). *Bulletin OILB/SROP* **23(1)**: 181–6.

Scholz, D. and Höller, C. (1992) Competition for hosts between hyperparasitoids of aphids, *Dendrocerus laticeps* and *Dendrocerus carpenteri* (Hymenoptera: Megaspilidae): the benefit of interspecific host discrimination. *Journal of Insect Behavior* **5**: 289–300.

Schonbeck, H. (1988) Biological control of aphids on wild cherry. *Allgemeine Forstzeitschrift* **34**: 944.

Schoonhoven, L.M. (1962) Diapause and the physiology of host-parasite synchronization in *Bupalus pinarius* L. (Geometridae) and *Eucarcelia rutilla* Vill. (Tachinidae). *Archives Neerlandais de Zoologie* **15**: 111–73.

Schroeder, D. (1985) The search for effective biological control agents in Europe. 1. Diffuse and spotted knapweed. In: Delfosse, E.S. (ed.), *Proceedings of the VI International Symposium on Biological Control of Weeds*, pp. 103–9, August 19–24, 1984, Vancouver, Canada. Agriculture Canada, Ottawa.

Schroeder, D. and Goeden, R.D. (1986) The search for arthropod natural enemies of introduced weeds for biological control – in theory and practice. *Biocontrol News and Information* **7**: 147–55.

Schroer, S., Sulistyanto, D., and Ehler, R.-U. (2005) Control of *Plutella xylostella* using polymer-formulated *Steinernema carpocapsae* and *Bacillus thuringiensis* in cabbage fields. *JEN* **129**: 198–204.

Schweizer, H., Morse, J.G., Luck, R.F., and Forster, L.D. (2002) Augmentative releases of a parasitoid (*Metaphycus* sp. nr. *flavus*) against citricola scale (*Coccus pseudomagnoliarum*) on oranges in the San Joaquin Valley of California. *Biological Control* **24**: 153–66.

Schweizer, H., Morse, J.G., and Luck, R.F. (2003a) Evaluation of *Metaphycus* spp. for suppression of black scale (Homoptera: Coccidae) on southern California citrus. *Environmental Entomology* **32**: 377–86.

Schweizer, H., Luck, R.F., and Morse, G. (2003b) Augmentative releases of *Metaphycus* sp. nr. *flavus* against citricola scale on oranges in the San Joaquin Valley of California: are early releases better than late ones? *Journal of Economic Entomology* **96**: 1375–87.

Scott, J.G., Rutz, D.A., and Walcott, J. (1988) Comparative toxicity of seven insecticides to adult *Spalangia cameroni* Perkins. *Journal of Agricultural Entomology* **5**: 139–45.

Scott, J.G., Geden, C.J., Rutz, D.A., and Liu, N. (1991) Comparative toxicity of seven insecticides to immature stages of *Musca domestica* (Diptera: Muscidae) and two of its important biological control agents, *Muscidifurax raptor* and *Spalangia cameroni* (Hymenoptera: Pteromalidae). *Journal of Economic Entomology* **84**: 776–9.

Scott, J.K. (1992) Biology and climatic requirements of *Perapion antiquum* (Coleoptera: Apionidae) in southern Africa: implications for the biological control of *Emex* spp. in Australia. *Bulletin of Entomological Research* **82**: 399–406.

Scott, J.K. and Yeoh, P.B. (1998) Host range of *Brachycaudus rumexicolens* (Patch), an aphid associated with the Polygonaceae. *Biological Control* **13**: 135–42.

Scott, M.E. (1987) Regulation of mouse colony abundance by *Heligmosomoides polygyrus*. *Parasitology* **95**: 111–24.

Scott, M.E. and Dobson, A. (1989) The role of parasites in regulating host abundance. *Parasitology Today* **5**: 176–83.

Sears, M.K., Helmich, R.L., Stanley-Horn, D.E. et al. (2001) Impact of Bt corn pollen on monarch butterfly populations: a risk assessment. *Proceedings of the National Academy of Sciences USA* **98**: 11937–42.

Secord, D. (2003) Biological control of marine invasive species: cautionary tales and land-based lessons. *Biological Invasions* **5**: 117–31.

Seehausen, O., Witte, F., Katunzi, E.F., Smits, J., and Bouton, N. (1997) Patterns of the remnant cichlid fauna in southern Lake Victoria. *Conservation Biology* **11**: 890–904.

Seife, C. (1996) A harebrained scheme. *Scientific American* **274**: 24–6.

Seixas, C.D.S., Barreto, R.W., Freitas, L.G., Maffia, L.A., and Monteiro, F.T. (2004) *Ditylenchus drepanocercus* (Nematoda), a potential biological control agent for *Miconia calvescens* (Melastomataceae): host-specificity and epidemiology. *Biological Control* **31**: 29–37.

Selkoe, K.A. and Toonen, R.J. (2006) Microsatellites for ecologists: a practical guide to using and evaluating microsatellite markers. *Ecology Letters* **9**: 615–29.

Sengonca, C. and Frings, B. (1989) Enhancement of the green lacewing *Chrysoperla carnea* (Stephens), by providing artificial facilities for hibernation. *Turkiye Entomoloji Dergisi* **13(4)**: 245–50.

Sengonca C., Khan, I.A., and Blaeser, P. (2004) The predatory mite *Typhlodromus pyri* (Acari: Phytoseiidae) causes feeding scars on leaves and fruit of apple. *Experimental and Applied Acarology* **33**: 45–53.

Shadduck, J.A., Singer, S., and Lause, S. (1980) Lack of mammalian pathogenicity of entomocidal isolates of *Bacillus sphaericus*. *Environmental Entomology* **9**: 403–7.

Shah, M.A. (1982) The influence of plant surfaces on the searching behavior of coccinellid larvae. *Entomologia Experimentalis et Applicata* **31**: 377–80.

Shamin, M., Baig, M., Datta, R.K., and Gupta, S.K. (1994) Development of a monoclonal antibody-based sandwich ELISA for the detection of nuclear polyhedra of nuclear polyhedorosis virus infection in *Bombyx mori* L. *Journal of Invertebrate Pathology* **63**: 151–6.

Shapiro, M. and Robertson, J.L. (1990) Laboratory evaluation of dyes as ultraviolet screens for the gypsy moth (Lepidoptera: Lymantriidae) nuclear polyhedrosis virus. *Journal of Economic Entomology* **83**: 168–72.

Shapiro, M. and Robertson, J.L. (1992) Enhancement of gypsy moth (Lepidoptera: Lymantriidae) baculovirus activity by optical brightners. *Journal of Economic Entomology* **85**: 1120–1224.

Shapiro-Ilan, D.I., Stuart, R.J., and McCoy, C.W. (2005) Targeted improvement of *Steinernema carpocapsae* for control of the pecan weevil, *Curculio caryae* (Horn) (Coleoptera: Curculionidae), through hybridization and bacterial transfer. *Biological Control* **34**: 215–21.

Shapiro-Ilan, D.I., Gouge, D.H., Piggott, S.J., and Fife, J.P. (2006) Application technology and environmental considerations for use of entomopathogenic nematodes in biological control. *Biological Control* **38**: 124–33.

Sharov, A.A. (1996) Modeling insect dynamics. In: Korpilahti, E., Mukkela, H., and Salonen, T. (eds.) *Caring for the Forest: Research in a Changing World*, pp. 293–303. Congress Report, vol. II, IUFRO XXth World Congress,

August 6–12, 1995, Tampere, Finland. Gummerus Printing, Jyvaskyla.

Sharov, A.A. and Colbert, J.J. (1994) *Gypsy Moth Life System Model. Integration of Knowledge and a User's Guide*. Virginia Polytechnic Institute and State University, Blacksburg, VA.

Shaw, M. and Huddleston, T. (1991) Classification and biology of braconid wasps (Hymenoptera: Braconidae). *Handbooks for the Identification of British Insects* **7(11)**: 1–126.

Shaw, S.R. (1988) Euphorine phylogeny: the evolution of diversity in host-utilization by parasitoid wasps (Hymenoptera: Braconidae). *Ecological Entomology* **13**: 323–35.

Shea, K. and Chesson, P. (2002) Community ecology as a framework for biological invasions. *Trends in Ecology and Evolution* **17**: 170–6.

Shea, K. and Kelly, D. (1998) Estimating biocontrol agent impact with matrix models: *Carduus nutans* in New Zealand. *Ecological Applications* **8**: 824–32.

Shea, K., Possingham, H.P., Murdoch, W.W., and Roush, R. (2002) Active adaptive management in insect pest and weed control: intervention with a plan for learning. *Ecological Applications* **12**: 927–36.

Shea, K., Kelly, D., Sheppard, A.W., and Woodburn, T.L. (2005) Context-dependent biological control of an invasive thistle. *Ecology* **86**: 3174–81.

Shea, K., Sheppard, A., and Woodburn, T. (2006) Seasonal life-history models for the integrated management of the invasive weed nodding thistle *Carduus nutans* in Australia. *Journal of Applied Ecology* **43**: 517–26.

Sheehan, K.A. (1989) Models for the population dynamics of *Lymantria dispar*. In: Wallner, W.E. and McManus, K.A. (eds.), *Proceedings, Lymantriidae: a Comparison of Features of New and Old World Tussock Moths*, pp. 533–47. June 26–July 1, 1986, New Haven, Connecticut. USDA Forest Service General Technical Report NE-123. USDA Forest Service, Morgantown, WV.

Sheehan, W. (1986) Response by specialist and generalist natural enemies to agroecosystem diversification: a selective review. *Environmental Entomology* **15**: 456–61.

Sheehan, W. and Shelton, A.M. (1989) Parasitoid response to concentration of herbivore food plants: finding and leaving plants. *Ecology* **70**: 993–8.

Sheehan. W., Wäckers, F.L., and Lewis, W.J. (1993) Discrimination of previously searched, host-free sites by *Microplitis croceipes* (Hymenoptera: Braconidae). *Journal of Insect Behavior* **6**: 323–31.

Sheldon, S.P. and Creed, R.P. (1995) Use of a native insect as a biological control for an introduced weed. *Ecological Applications* **5**: 1122–32.

Shellam, G.R. (1994) The potential of murine cytomegalovirus as a viral vector for immunocontraception. *Reproduction, Fertility and Reproduction* **6**: 401–9.

Shelton, A.M., Zhao, J.-Z., and Roush, R.T. (2002) Economic, ecological, food safety, and social consequences of the deployment of Bt transgenic plants. *Annual Review of Entomology* **47**: 845–81.

Shenhmar, M. and Brar, K.S. (1996) Evaluation of *Trichogramma chilonis* Ishii (Hymenoptera: Trichogrammatidae) for the control of *Chilo auricilius* Dudgeon on sugarcane. *Indian Journal of Plant Protection* **24(1/2)**: 47–9.

Shenhmar, M., Brar, K.S., Bakhetia, D.R.C., and Singh, J. (1998) Tricho-capsules: a new technique for release of the egg parasitoids – trichogrammatids. *Insect Environment* **4(3)**: 95.

Shenhmar, M., Singh, J., Singh, S.P., Brar, K.S., and Singh, D. (2003) Effectiveness of *Trichogramma chilonis* Ishii for the management of *Chilo auricilius* Dudgeon on sugarcane in different sugar mill areas of the Punjab. In: *Proceedings of the Symposium of Biological Control of Lepidopteran Pests*, pp. 333–5, July 17–18, 2002, Bangalore, India. Society for Biocontrol Advancement, Bangalore.

Shenk, T.M., White, G.C., and Burnham, K.P. (1998) Sampling-variance effects on detecting density dependence from temporal trends in natural populations. *Ecological Monographs* **68**: 445–64.

Shepard, M.H., Rapusas, R., and Estano, D.B. (1989) Using rice straw bundles to conserve beneficial arthropod communities in ricefields. *International Rice Research News* **14(5)**: 30–1.

Sheppard, A.W. (1999) Which test? A mini review of test usage in host specificity testing. In: Withers, T.M., Barton Browne, L., and Stanley, J. (eds.), *Host Specificty Testing in Australiasia: Towards Improved Assays for Biological Control*, pp. 60–9. Department of Natural Resources, Indooroopilly, Queensland.

Sheppard, A.W., Aeschlimann, J.-P., Sagliocco, J.-L., and Vitou, J. (1991) Natural enemies and population stability of the winter-annual *Carduus pycnocephalus* L. in Mediterranean Europe. *Acta Oecologia* **12**: 707–26.

Sheppard, A.W., van Klinken, R.D., and Heard, T.A. (2005) Scientific advances in the analysis of direct risks of weed biological control agents to nontarget plants. *Biological Control* **35**: 215–26.

Shetlar, D.J., Suleman, P.E., and Georgis, R. (1988) Irrigation and use of entomogenous nematodes, *Neoaplectana* spp. and *Heterorhabditis heliothidis* (Rhabditida: Steinernematidae and Heterorhabditidae) for control of Japanese beetle (Coleoptera: Scarabaeidae) grubs in turfgrass. *Journal of Economic Entomology* **81**: 1318–22.

Shililu, J.I., Tewolde, G.M., Brantly, E. et al. (2003) Efficacy of *Bacillus thuringiensis, Bacillus sphaericus* and temephos for managing *Anopheles* larvae in Eritrea. *Journal of the American Mosquito Control Association* **19(3)**: 251–8.

Shimoda, T., Takabayashi, J., Ashihara, W., and Takafuji, A. (1997) Response of a predatory insect, *Scolothrips takahashii* toward herbivore-induced plants volatiles under laboratory and field conditions. *Journal of Chemical Ecology* **23**: 2033–48.

Shimoda, T., Ozawa, R., Sano, K., Yano, E., and Takabayashi, J. (2005) The involvement of volatile infochemicals from spider mites and from food-plants in prey location of the generalist predatory mite *Neoseiulus californicus*. *Journal of Chemical Ecology* **31**: 2019–32.

Shipp, J.L. and Ramakers, P.M.J. (2004) Biological control of thrips on vegetable crops. In: Heinz, K.M., Van Driesche, R.G., and Parrella, M.P. (eds.), *Biocontrol in Protected Culture*, pp. 265–76. Ball Publishing, Batavia, IL.

Shipp, J.L. and Wang, K. (2006) Evaluation of *Dicyphus hesperus* (Heteroptera: Miridae) for biological control of *Frankliniella occidentalis* (Thysanoptera: Thripidae) on greenhouse tomato. *Journal of Economic Entomology* **99**: 414–20.

Shipp, J.L., Ward, K.I., and Gillespie, T.J. (1996) Influence of temperature and vapor pressure deficit on the rate of predation by the predatory mite, *Amblyseius cucumeris*, on *Frankliniella occidentalis*. *Entomologia Experimentalis et Applicata* **78**: 31–8.

Shivik, J.A., Savarie, P.J., and Clark, L. (2002) Aerial delivery of baits to brown treesnakes. *Wildlife Society Bulletin* **30**: 1062–7.

Shonouda, M.L., Bombosch, S., Shalaby, A.M., and Osman, S.I. (1998) Biological and chemical characterization of a kairomone excreted by the bean aphid, *Aphis fabae* Scop. (Hom., Aphididae) and its effect on the predator *Metasyrphus corollae* Fabr. I. Isolation, identification, and bioassay of aphid kairomone. *Journal of Applied Entomology* **122**: 15–23.

Shrewsbury, P.M. and Smith-Fiola, D.C. (2000) Evaluation of green lacewings for suppressing azalea lace bug populations in nurseries. *Journal of Environmental Horticulture* **18**: 207–11.

Shulman, S. (1995) Immunological reactions and infertility. In: Kurpisz, M. and Fernandez, N. (eds.), *Immunology of Human Reproduction*, pp. 53–78. BIOS Scientific Publishers, Oxford.

Siegel, J.P. and Shadduck, J.A. (1990a) Clearance of *Bacillus sphaericus* and *Bacillus thuringiensis* ssp. *israelensis* from mammals. *Journal of Economic Entomology* **83**: 347–55.

Siegel, J.P. and Shadduck, J.A. (1990b) Mammalian safety of *Bacillus sphaericus*. In: de Barjac, H. and Sutherland, D.J. (eds.), *Bacterial Control of Mosquitoes and Black Flies: Biochemistry, Genetics, and Application of Bacillus thuringiensis israelensis and Bacillus sphaericus*, pp. 321–31. Rutgers University Press, New Brunswick, NJ.

Siegel, J.P. and Shadduck, J.A. (1990c) Mammalian safety of *Bacillus thuringiensis israelensis*. In: Barjac, H. and Sutherland, D.J. (eds.), *Bacterial Control of Mosquitoes and Black Flies: Biochemistry, Genetics, and Application of Bacillus thuringiensis israelensis and Bacillus sphaericus*, pp. 202–17. Rutgers University Press, New Brunswick, NJ.

Siegel, J.P. and Shadduck, J.A. (1992) Testing the effects of microbial pest control agents on mammals. In: Levin, M.A., Seidler, R.J., and Rogul, M. (eds.), *Microbial Ecology: Principles, Methods, and Applications*, pp. 745–59. McGraw-Hill, New York.

Siemann, E. and Rogers, W.E. (2001) Genetic differences in growth of an invasive tree species. *Ecology Letters* **4**: 514–18.

Silva, I.M.M.S., Honda, J., van Kan, F. et al. (1999) Molecular differentiation of five *Trichogramma* species occurring in Portugal. *Biological Control* **16**: 177–84.

Simberloff, D. and Stiling, P. (1996) Risks of species introduced for biological control. *Biological Conservation* **78**: 185–92.

Simberloff, D. and Von Holle, B. (1999) Positive interactions of nonindigenous species: invasional meltdown? *Biological Invasions* **1**: 21–32.

Simberloff, D. and Gibbons, L. (2004) Now you see them, now you don't! – Population crashes of established introduced species. *Biological Invasions* **6**: 161–72.

Simberloff, D., Schmitz, D.C., and T.C. Brown (eds.) (1997) *Strangers in Paradise*. Island Press, Washington, D.C.

Sime, K. (2002) Chemical defense of *Battus philenor* larvae against attack by the parasitoid *Trogus pennator*. *Ecological Entomology* **27**: 337–45.

Simmonds, F.J. and Bennett, F.D. (1966) Biological control of *Opuntia* spp. by *Cactoblastis cactorum* in the Leeward Islands (West Indies). *Entomophaga* **11**: 183–9.

Simmons, A.T. and Gurr, G.M. (2004) Trichome-based host plant resistance of *Lycopersicon* species and the biocontrol agent *Mallada signata*: are they compatible? *Entomologia Experimentalis et Applicata* **113**: 95–101.

Simmons, A.T. and Gurr, G.M. (2005) Trichomes of *Lycopersicon* species and their hybrids: effects on pests and natural enemies. *Agricultural and Forest Entomology* **7**: 265–76.

Simmons, E.G. (1998) *Alternaria* themes and variations (224–225). *Mycotaxon* **68**: 417–27.

Simon, C., Frati, F., Beckenbach, A., Crespi, B., Liu, H., and Flook, P. (1994) Evolution, weighting, and phylogenetic utility of mitochondrial gene sequences and a compilation of conserved polymerase chain reaction primers. *Annals of the Entomological Society of America* **87**: 651–701.

Sinclair, A.R.E. (1996) Mammal populations: fluctuation, regulation, life history theory and their implications for conservation. In: Floyd, R.B., Sheppard, A.W., and De Barro, P.J. (eds.), *Frontiers of Population Ecology*, pp. 127–54. CSIRO Publishing, Collingwood, Victoria.

Singer, S. (1987) Current status of the microbial larvicide *Bacillus sphaericus*. In: Maramorosch, K. (ed.), *Biotechnology in Invertebrate Pathology and Cell Culture*, pp. 133–63. Academic Press, New York.

Singer, S. (1990) Introduction to the study of *Bacillus sphaericus* as a mosquito control agent. In: de Barjac, H. and Sutherland, D.J. (eds.), *Bacterial Control of Mosquitoes and Blackflies: Biochemistry, Genetics, and Applications of Bacillus thuringiensis and Bacillus sphaericus*, pp. 221–7. Rutgers University Press, New Brunswick, NJ.

Singhal, R.C., Gupta, M.R., and Dev Narayan (2001) Eco-friendly approach for minimizing populations of sugarcane stalk borer (*Chilo auricilius*) in the Tarai belt of Uttar Pradesh, India. In: *Proceedings of the XXIVth Congress*, vol. 2, pp. 374–7, September 17–21, 2001, Brisbane, Australia. Australian Society of Sugarcane Technologists, Mackay.

Singleton, G.R. (1989) Population dynamics of an outbreak of house mice (*Mus domesticus*) in the mallee wheatlands of Australia – hypothesis of plague formation. *Journal of Zoology, London* **219**: 495–515.

Singleton, G.R. and Spratt, D.M. (1986) The effects of *Capillaria hepatica* (Nematoda) on natality and survival to weaning in BALB/c mice. *Australian Journal of Zoology* **34**: 677–81.

Singleton, G.R. and McCallum, H.I. (1990) The potential of *Capillaria hepatica* to control mouse plagues. *Parasitology Today* **6**: 190–3.

Singleton, G.R. and Chambers, L.K. (1996) A manipulative field experiment to examine the effect of *Capillaria hepatica* (Nematoda) on wild mouse populations in southern Australia. *International Journal for Parasitology* **26**: 383–98.

Singleton, G.R., Spratt, D.M., Barker, S.C., and Hodgson, P.F. (1991) The geographic distribution and host range of *Capillaria hepatica* (Bancroft) (Nematoda) in Australia. *International Journal for Parasitology* **21**: 945–57.

Singleton, G.R., Chambers, L.K., and Spratt, D.M. (1995) An experimental field study to examine whether *Capillaria hepatica* (Nematoda) can limit house mouse populations in eastern Australia. *Wildlife Research* **22**: 31–53.

Singleton, G.R., Brown, P.R., Pech, R.P., Jacob, J., Mutze, G.J., and Krebs, C.J. (2005) One hundred years of eruptions of house mice in Australia – a natural biological curio. *Biological Journal of the Linnean Society* **84**: 617–27.

Sjogren, R.D. and Legner, E.F. (1989) Survival of the mosquito predator *Notonecta unifasciata* (Hemiptera: Notonectidae) embryos at low thermal gradients. *Entomophaga* **34**: 201–8.

Skellum, J.G. (1952) Studies in statistical ecology. I. Spatial pattern. *Biometrika* **39**: 346–62.

Skirvin, D.J. (2004) Virtual plant models of predatory mite movement in complex plant canopies. *Ecological Modeling* **171**: 301–13.

Slotta, T.A.B., Foley, M.E., and Horvath, D. (2005) Development of polymorphic markers for *Cirsium arvense*, Canada thistle, and their amplification in closely related taxa. *Molecular Ecology Notes* **5**: 917–19.

Smart, L.E., Stevenson, J.H., and Walters, J.H.H. (1989) Development of field trial methodology to assess short-term effects of pesticides on beneficial arthropods in arable crops. *Crop Protection* **8**: 169–80.

Smith, D. and Papacek, D.F. (1991) Studies of the predatory mite *Amblyseius victoriensis* (Acarina: Phytoseiidae) in citrus orchards in south-east Queensland: control of *Tegolophus australis* and *Phyllocoptruta oleivora* (Acarina: Eriophyidae), effect of pesticides, alternative host plants and augmentative release. *Experimental and Applied Acarology* **12**: 195–217.

Smith, D., Beattie, G.A.C., and Broadley, R. (eds.) (1997) *Citrus Pests and their Natural Enemies*. Department of Primary Industries, Brisbane, Queensland.

Smith, F.E. (1961) Density-dependence in the Australian thrips. *Ecology* **42**: 403–7.

Smith, G. (1994) Parasite population density is regulated. In: Scott, M.E. and Smith, G. (eds.), *Parasitic and Infectious Diseases, Epidemiology and Ecology*, pp. 47–63. Academic Press, San Diego, CA.

Smith, G. and Dobson, A.P. (1992) Sexually transmitted diseases in animals. *Parasitology Today* **8**: 159–66.

Smith, G., Walmsley, A., and Polkinghorne, I. (1997) Plant-derived immunocontraceptive vaccines. *Reproduction, Fertility and Development* **9**: 85–9.

Smith, H.S. (1935) The role of biotic factors in the determination of population densities. *Journal of Economic Entomology* **28**: 873–98.

Smith, H.S. and Armitage, H.M. (1926) Biological control of mealybugs in California. *California State Department of Agriculture Monthly Bulletin* **9**: 104–64.

Smith, J.M. (1957) Effects of the food plant of California red scale, *Aonidiella aurantii* (Mask.) on reproduction of its hymenopterous parasites. *The Canadian Entomologist* **89**: 219–30.

Smith, L. (2006) Cause and effect, and how to make a better biocontrol agent. *Biological Control News and Information* **27(2)**: 28N–30N.

Smith, L. (2007) Physiological host range of *Ceratapion basicorne*, a prospective biological control agent of *Centaurea solstitialis* (Asteraceae). *Biological Control* **41**: 120–33.

Smith, R.A. and Nordlund, D.A. (2000) Mass rearing technology for biological control agents of *Lygus* spp. *Southwestern Entomologist* (suppl. 23): 121–7.

Smith, S.M. (1996) Biological control with *Trichogramma:* advances, successes, and potential of their use. *Annual Review of Entomology* **41**: 375–406.

Smith, S.M., Carrow, J.R., and Laing, J.E. (eds.) (1990) Inundative release of the egg parasitoid *Trichogramma minutum* (Hymenoptera: Trichogrammatidae) against forest insect pests such as spruce budworm *Choristoneura fumiferana* (Lepidoptera: Tortricidae): the Ontario Project 1982–1986. *Memoirs of the Entomological Society of Canada* **153**: 1–87.

Smith, S.M., van Frankenhuyzen, K., Nealis, V.G., and Bourchier, R.S. (2001) *Choristoneura fumiferana* (Clemens), eastern spruce budworm (Tortricidae). In: Mason, P. and Huber, J. (eds.), *Biological Control Programmes in Canada, 1981–2000*, pp. 58–68. CABI Publishing, Wallingford.

Snyder, A.E. and Wise, D.H. (2001) Antipredator behavior of spotted cucumber beetles (Coleoptera: Chrysomelidae) in response to predators that pose varying risks. *Environmental Entomology* **29**: 35–42.

Snyder, C., Young, J., Smith, D., Lemarie, D., Ross, R., and Bennett, R. (1998) *Influence of Eastern Hemlock Decline on Aquatic Biodiversity of Delaware Water Gap National Recreation Area*. Final Report of the USGS Biological Resources Division, Leetown Science Center. Aquatic Ecology Laboratory. http://ael.er.usgs.gov/groups/gis/hemlock/dewa.html.

Snyder, C., Young, J., Smith, D., Lemarie, D., Ross, R., and Bennet, R. (2004) *Stream Ecology Linked to Eastern Hemlock Decline in Delaware Water Gap National Recreation Area.* www.lsc.usgs.gov/aeb/2048-03/dewa.asp.

Snyder, W.E. and Ives, A.R. (2001) Generalist predators disrupt biological control by a specialist parasitoid. *Ecology* **82**: 705–16.

Snyder, W.E., Ballard, S.N., Yang, S. et al. (2004) Complementary biocontrol of aphids by the ladybird beetle *Harmonia axyridis* and the parasitoid *Aphelinus asychis* on greenhouse roses. *Biological Control* **20**: 229–35.

Soares, A.O., Coderre, C., and Schandrel, H. (2004) Dietary self-selection behavior by adults of the aphidophagous ladybeetle *Harmonia axyridis* (Coleoptera: Coccinellidae). *Journal of Animal Ecology* **73**: 474–86.

Sobhian, R., Ryan, F.J., Khamraev, A., Pitcairn, M.J., and Bell, D.E. (2003) DNA phenotyping to find a natural enemy in Uzbekistan for California biotypes of *Salsola tragus* L. *Biological Control* **28**: 222–8.

Sobhian, R., McClay, A., Hasan, S., Peterschmitt, M., and Hughes, R.B. (2004) Safety assessment and potential of *Cecidophyes rouhollahi* (Acari, Eriophyidae) for biological control of *Galium spurium* (Rubiaceae) in North America. *Journal of Applied Entomology* **128**: 258–66.

Solbrig, O.T. (1981) Studies on the population biology of the genus *Viola*. II. The effect of plant size on fitness in *Viola sororia*. *Evolution* **35**: 1080–93.

Solomon, M.E. (1949) The natural control of animal populations. *Journal of Animal Ecology* **18**: 1–35.

Solomon, M.G., Easterbrook, M.A., and Fitzgerald, J.D. (1993) Mite-management programmes based on organophosphate-resistant *Typholodromus pyri* in U.K. apple orchards. *Crop Protection* **12**: 249–54.

Solow, A.R. and Steele, J.H. (1990) On sample size, statistical power, and the detection of density dependence. *Journal of Animal Ecology* **59**: 1073–6.

Soper, R.S., Shewell, G.E., and Tyrrell, D. (1976) *Colcondamyia auditrix* nov. sp. (Diptera: Sarcophagidae), a parasite which is attracted by the mating song of its host, *Okanagana rimosa* (Homopteraa: Cicadidae). *The Canadian Entomologist* **108**: 61–8.

Sopp, P.I. (1987) Quantification of predation by polyphagous predators on *Sitobion avenae* (Homoptera: Aphididae) in winter wheat using ELISA. Ph.D. Dissertation, University of Southampton, Southampton.

Southwood, T.R.E. (1978) *Ecological Methods with Particular Reference to the Study of Insect Populations*, 2nd edn. Chapman and Hall, London.

Southwood, T.R.E. and Comins, H.N. (1976) A synoptic population model. *Journal of Animal Ecology* **45**: 949–65.

Southwood, T.R.E. and Jepson, W.F. (1962) Studies on the populations of *Oscinella frit* L. (Diptera: Chloropidae) in the oat crop. *Journal of Animal Ecology* **31**: 481–95.

Spacie, A. (1992) Testing the effects of microbial agents on fish and crustaceans. In: Levin, M.A., Seidler, R.J., and Rogul, M. (eds.), *Microbial Ecology: Principles, Methods, and Applications*, pp. 707–28. McGraw-Hill, New York.

Spafford Jacob, H. and D.T. Briese (eds.) (2003) *Improving the Selection, Testing, and Evaluation of Weed Biological Control Agents.* Proceedings of the CRC for Australian Weed Management Biological Control of Weeds Symposium and Workshop, September 13, 2002, University of Western Australia, Perth. CRC for Australian Weed Management Technical Series No. 7.

Speyer, E.R. (1927) An important parasite of the greenhouse whitefly (*Trialeurodes vaporariorum*) Westwood. *Bulletin of Entomological Research* **17**: 301–8.

Spielman, D. and Frankham, R. (1992) Modeling problems in conservation genetics using captive *Drosophila* populations: improvement of reproductive fitness due to immigration of one individual into small partially inbred populations. *Zoo Biology* **11**: 343–51.

Spratt, D.M. and Singleton, G.R. (1986) Studies of the life cycle infectivity and clinical effects of *Capillaria hepatica* (Bancroft) (Nematoda) in mice, *Mus musculus*. *Australian Journal of Zoology* **34**: 663–75.

Stage, D.A. and Petersen, J.J. (1981) Mass release of pupal parasites for control of stable flies and house flies in confined feedlots in Nebraska. In: Patterson, R.S. (ed.), *Status of Biological Control of Filth Flies*, pp. 52–8. Proceedings of a Workshop, February 4–5, 1981, Gainesville, Florida. USDA-ARS, New Orleans, LA.

Stahly, D.P. and Klein, M.G. (1992) Problems with *in vitro* production of spores of *Bacillus popilliae* for use in biological control of the Japanese beetle. *Journal of Invertebrate Pathology* **60**: 283–91.

Stam, P.A. and Elmosa, H. (1990) The role of predators and parasites in controlling populations of *Earias insulana*, *Heliothis armigera*, and *Bemisia tabaci* on cotton in the Syrian Arab Republic. *Entomophaga* **35**: 315–27.

Stamm Katovich, E.J. (1999) Effect of *Galerucella* spp. on survival of purple loosestrife (*Lythrum salicaria*) roots and crowns. *Weed Science* **47**: 360–5.

Stamp, N.E. (1982) Behavioral interactions of parasitoids and the Baltimore checkerspot caterpillars (*Euphydryas phaeton*). *Environmental Entomology* **11**: 100–4.

Stansly, P.A., Sánchez, P.A., Rodríguez, J.M. et al. (2004) Prospects for biological control of *Bemisia tabaci* (Homoptera: Aleyrodidae) in greenhouse tomatoes of southern Spain. *Crop Protection* **23**: 701–12.

Stapel, J.O., Cortesero, A.M., de Moraes, C.M., Tumlinson, J.H., and Lewis, W.J. (1997) Extrafloral nectar, honeydew, and sucrose effects on searching behavior and efficiency of *Microplitis croceipes* (Hymenoptera: Braconidae) in cotton. *Environmental Entomology* **26**: 617–23.

Starý, P. (1970) *Biology of Aphid Parasites (Hymenoptera: Aphidiidae) with Respect to Integrated Control.* Dr. W. Junk, N.V. Publishers, The Hague.

Stastny, M., Schaffner, U., and Belle, E. (2005) Do vigour of introduced populations and escape from specialist herbivores contribute to invasiveness? *Journal of Ecology* **93**: 27–37.

Steinhaus, E.A. (ed.) (1963) *Insect Pathology: An Advanced Treatise*, vol. 2. Academic Press, New York.

Stewart, C.A., Chapman, R.B., Barrington, A.M., and Frampton, C.M.A. (1999) Influence of temperature on adult longevity, oviposition and fertility of *Agasicles hygrophila* Selman & Vogt (Coleoptera: Chrysomelidae). *New Zealand Journal of Zoology* **26**: 191–7.

Stewart, L.M.D., Hirst, M., Ferber, M.L., Merryweather, A.T., Cayley, P.J., and Possee, R.D. (1991) Construction of an improved baculovirus insecticide containing an insect-specific toxin gene. *Nature* **352**: 85–8.

Steyn, J.J. (1958) The effect of ants on citrus scales at Letaba, South Africa. *Proceedings of the 10th International Congress of Entomology* **4**: 589–94.

Stiling, P. (1989) Exotics: biological invasions. *Florida Wildlife* **43(5)**: 13–16.

Stiling, P. and Rossi, A.M. (1997) Experimental manipulations of top-down and bottom-up factors in a tri-trophic system. *Ecology* **78**: 1602–6.

Stiling, P., Moon, D., and Gordon, D. (2004) Endangered cactus restoration: mitigating the non-target effects of a biological control agent (*Cactoblastis cactorum*) in Florida. *Restoration Ecology* **12**: 605–10.

Stireman, III, J.O. (2002) Host location and selection cues in a generalist tachinid parasitoid. *Entomologia Experimentalis et Applicata* **103**: 23–34.

Stireman, III, J.O., O'Hara, J.E., and Wood, D.M. (2006) Tachinidae: evolution, behavior, and ecology. *Annual Review of Entomology* **51**: 525–55.

Stoetzel, M.B. (2002) History of the introduction of *Adelges tsugae* based on voucher specimens in the Smithsonian Institute national collection of insects. In: Onken, B., Reardon, R., and Lashomb, L. (eds.), *Proceedings: Hemlock Woolly Adelgid in the Eastern United States Symposium*, p. 12, February 5–7, 2002, East Brunswick, NJ. USDA Forest Service and New Jersey Agricultural Experiment Station Publication, East Brunswick, NJ.

Stoltz, D.B. (1993) The polydnavirus life cycle. In: Beckage, N.E., Thompson, S.N., and B.A. Federici (eds.), *Parasites and Pathogens of Insects, Volume I. Parasites*, pp. 167–87. Academic Press, New York.

Stoltz, D.B. and Vinson, S.B. (1979) Viruses and parasitism in insects. *Advances in Virus Research* **24**: 125–71.

Storey, G.K., McCoy, C.W., Stenzel, K., and Andersch, W. (1990) Conidiation kinetics of the mycelial granules of *Metarhizium anisopliae* (BIO 1020) and its biological activity against different soil insects. In: Pinnock, D.E. (ed.), Vth *International Colloquium on Invertebrate Pathology and Microbial Control*, pp. 320–4, August 20–24, 1990, Adelaide, Australia. Department of Entomology, University of Adelaide, Glen Osmond, South Australia.

Story, J.M. (1985) First report of the dispersal into Montana of *Urophora quadrifasciata* (Diptera: Tephritidae), a fly released in Canada for biological control of spotted and diffuse knapweed. *The Canadian Entomologist* **117**: 1061–2.

Story, J.M. and Anderson, N.L. (1978) Release and establishment of *Urophora affinis* (Diptera: Tephritidae) on spotted knapweed in western Montana. *Environmental Entomology* **7**: 445–8.

Story, J.M., Coombs, E.M., and Piper, G.L. (2004a) Spotted knapweed, *Centaurea stoebe* ssp. *micranthos* (= *C. maculosa*). In: Coombs, E.M., Clark, J.K., Piper, G.L., and Cofrancesco, A.F. (eds.), *Biological Control of Invasive Plants in the United States*, pp. 204–5. Oregon State University Press, Corvallis, OR.

Story, J.M., Coombs, E.M., and Piper, G.L. (2004b) *Pterolonche inspersa*. In: Coombs, E.M., Clark, J.K., Piper, G.L., and A.F. Cofrancesco (eds.), *Biological Control of Invasive Plants in the United States*, pp. 221–2. Oregon State University Press, Corvallis, OR.

Story, J.M., Callan, N.W., Corn, J.G., and White, L.J. (2006) Decline of spotted knapweed density at two sites in western Montana with large populations of the introduced root weevil, *Cyphocleonus achates* (Fahraeus). *Biological Control* **38**: 227–32.

Stouthamer, R. (1993) The use of sexual versus asexual wasps in biological control. *Entomophaga* **38**: 3–6.

Stouthamer, R., Luck, R.F., and Hamilton, W.D. (1990) Antibiotics cause parthenogenetic *Trichogramma* (Hymenoptera: Trichogrammatidae) to revert to sex. *Proceedings of the National Academy of Sciences USA* **87**: 2424–7.

Stouthamer, R., Hu, J., van Kan, F.J.P.M., Platner, G.R., and Pinto, J.D. (1999) The utility of internally transcribed spacer 2 DNA sequences of the nuclear ribosomal gene for distinguishing sibling species of *Trichogramma*. *BioControl (Dordrecht)* **43**: 421–40.

Stouthamer, R., Jochemsen, P., Platner, G.R., and Pinto, J.D. (2000a) Crossing incompatibility between *Trichogramma minutum* and *T. platneri* (Hymenoptera: Trichogrammatidae): implications for application in biological control. *Environmental Entomology* **29**: 832–7.

Stouthamer, R., Gai, Y., Koopmanschap, A.B., Platner, G.R., and Pinto, J.D. (2000b) ITS-2 sequences do not differ for the closely related species *Trichogramma minutum* and *T. platneri*. *Entomologia Experimentalis et Applicata* **95**: 105–11.

Stowell, L.J. (1991) Submerged fermentation of biological herbicides. In: TeBeest, D.O. (ed.), *Microbial Control of Weeds*, pp. 225–61. Chapman and Hall, New York.

Strand, M.R. and Vinson, S.B. (1982) Behavioral response of the parasitoid *Cardiochiles nigriceps* to a kairomone. *Entomologia Experimentalis et Applicata* **31**: 308–15.

Strand, M.R. and Vinson, S.B. (1983a) Host acceptance behavior of *Telenomus heliothidis* (Hymenoptera: Scelionidae) toward *Heliothis virescens* (Lepidoptera: Noctuidae). *Annals of the Entomological Society of America* **76**: 781–5.

Strand, M.R. and Vinson, S.B. (1983b) Factors affecting host recognition and acceptance in the egg parasitoid

Telenomus heliothidis (Hymenoptera: Scelionidae). *Environmental Entomology* **12**: 1114–19.

Strand, M.R. and Vinson, S.B. (1983c) Analysis of an egg recognition kairomone of *Telenomus heliothidis* (Hymenoptera: Scelionidae). Isolation and host function. *Journal of Chemical Ecology* **9**: 423–32.

Strasser, H., Vey, A., and Butt, T.M. (2000) Are there any risks in using entomopathogenic fungi for pest control, with particular reference to the bioactive metabolites of *Metarhizium, Tolypocladium,* and *Beauveria* species? *Biocontrol Science and Technology* **10**: 717–35.

Strong, D.R. and Pemberton, R.W. (2000) Biological control of invading species: risk and reform. *Science* **288**: 1969–70.

Strong, D.R., Lawton, J.H., and Southwood, R. (1984) *Insects on Plants – Community Patterns and Mechanisms*. Harvard University Press, Cambridge, MA.

Strong, W.B. and Croft, B.A. (1995) Inoculative release of phytoseiid mites (Acarina: Phytoseiidae) into the rapidly expanding canopy of hops for control of *Tetranychus urticae* (Acarina: Tetranychidae). *Environmental Entomology* **24**: 446–53.

Stronge, D.C., Fordham, R.A., and Minot, E.O. (1997) The foraging ecology of feral goats *Capra hircus* in the Mahoenui giant weta reserve, southern King Country, *New Zealand Journal of Ecology* **21**: 81–8.

Stubbs, M. (1980) Another look at prey detection by coccinellids. *Ecological Entomology* **5**: 179–82.

Sturm, M.M., Sterling, W.L., and Hartstack, A.W. (1990) Role of natural mortality in boll weevil (Coleoptera: Curculionidae) management programs. *Journal of Economic Entomology* **83**: 1–7.

Sugimoto, T., Shimono, Y., Hata, Y., Naki, A., and Yahara, M. (1988) Foraging for patchily distributed leaf-miners by the parasitoid *Dapsilarthra rufiventris* (Hymenoptera: Braconidae). III. Visual and acoustic cues to a close range patch-location. *Applied Entomology and Zoology* **23**: 113–21.

Suh, C.P.-C., Orr, D.B., van Duyn, J.W., and Borchert, D.M. (2000a) *Trichogramma exiguum* (Hymenoptera: Trichogrammatidae) releases in North Carolina cotton: evaluation of heliothine pest suppression. *Journal of Economic Entomology* **93**: 1127–36.

Suh, C.P.-C., Orr, D.B., and van Duyn, J.W. (2000b) *Trichogramma* releases in North Carolina cotton: why releases fail to suppress heliothine pests. *Journal of Economic Entomology* **93**: 1137–45.

Summy, K.R., Gilstrap, F.E., Hart, W.G., Caballero, J.M., and Saenz, I. (1983) Biological control of citrus blackfly (Homoptera: Aleyrodidae) in Texas. *Environmental Entomology* **12**: 782–6.

Summy, K.R., Morales-Ramos, J.A., and King, E.G. (1995) Suppression of boll weevil (Coleoptera: Curculionidae) infestations on south Texas cotton by augmentative releases of the parasite *Catolaccus grandis* (Hymenoptera: Pteromalidae). *Biological Control* **5**: 523–9.

Summy, K.R., Greenberg, S.M., Morales-Ramos, J.A., and King, E.G. (1997) Suppression of boll weevil infestations (Coleoptera: Curculionidae) occurring on fallow-season cotton in southern Texas by augmentative releases of *Catolaccus grandis* (Hymenoptera: Pteromalidae). *Biological Control* **9**: 209–15.

Sunderland, K.D. (1988) Quantitative methods for detecting invertebrate predation occurring in the field. *Annals of Applied Biology* **112**: 201–24.

Surles, W.W. and Kok, L.T. (1977) Ovipositional preference and synchronization of *Rhinocyllus conicus* with *Carduus nutans* and *C. acanthoides*. *Environmental Entomology* **6**: 222–4.

Sutherst, R.W. (1991) Predicting the survival of immigrant insect pests in new environments. *Crop Protection* **10**: 331–3.

Sutherst, R.W. (2000) Change and invasive species: a conceptual framework. In: Mooney, H.A. and Hobbs, R.J. (eds.), *Invasive Species in a Changing World*, pp. 211–40. Island Press, Washington, D.C.

Sutherst, R.W. and Maywald, G.F. (1985) A computerised system for matching climates in ecology. *Agriculture, Ecosystems and the Environment* **13**: 281–99.

Sutherst, R.W. and Maywald, G.F. (2005) A climate model of the red imported fire ant, *Solenopsis invicta* Buren (Hymenoptera: Formicidae): implications for invasion of new regions, particularly Oceania. *Environmental Entomology* **34**: 317–35.

Sutherst, R.W., Maywald, G.F., and Russell, B.L. (2000) Estimating vulnerability under global change: modular modeling of pests. *Agriculture, Ecosystems, and the Environment* **82**: 303–19.

Sutherst, R.W., Maywald, G.F., Bottomley, W., and Bourne, A. (2004) *CLIMEX v2 – User's Guide*. Hearne Scientific Software, Melbourne.

Symondson, W.O.C. (2002) Molecular identification of prey in predator diets. *Molecular Ecology* **11**: 627–41.

Symondson, W.O.C., Sunderland, K.D., and Greenstone, M.H. (2002) Can generalist predators be effective biocontrol agents? *Annual Review of Entomology* **47**: 561–94.

Syrett, P., Briese, D.T., and Hoffmann, J.H. (2000) Success in biological control of terrestrial weeds by arthropods. In: Gurr, G. and Wratten, S. (eds.), *Biological Control: Measures of Success*, pp. 189–230. Kluwer Academic Press, San Diego, CA.

Szentkirályi, F. (2001) Ecology and habitat relationships. In: McEwen, P., New, T.R., and Whittington, A.E. (eds.), *Lacewings in the Crop Environment*, pp. 82–115. Cambridge University Press, Cambridge.

Tabashnik, B.E., Cushing, N.L., Finson, N., and Johnson, M.W. (1990) Field development of resistance to *Bacillus thuringiensis* in diamondback moth (Lepidoptera: Plutellidae). *Journal of Economic Entomology* **83**: 1671–6.

Tabor, P. and Susott, A.W. (1941) Zero to thirty mile-a-minute seedlings. *Soil Conservation* **8**: 61–5.

Takahashi, S., Hajika, M., Takabyashi, J., and Fukui, M. (1990) Oviposition stimulants in the coccoid cuticular waxes of *Aphytis yanonensis* DeBach and Rosen. *Journal of Chemical Ecology* **16**: 1657–65.

Talhouk, A.S. (1991) On the management of the date palm and its arthropod enemies in the Arabian Penisula. *Journal of Applied Entomology* **111**: 514–20.

Tallmon, D.A., Luikart, G., and Waples, R.S. (2004) The alluring simplicity and complex reality of genetic rescue. *Trends in Ecology and Evolution* **19**: 489–96.

Tanada, Y. and Kaya, H.K. (1993) *Insect Pathology*. Academic Press, San Diego, CA.

Tanigoshi, L.K., Fargerlund, J., Nishio-Wong, J.Y., and Griffiths, H.J. (1985) Biological control of citrus thrips, *Scirtothrips citri* (Thysanoptera: Thripidae) in southern California citrus groves. *Environmental Entomology* **14**: 733–41.

Tanwar, R.K., Ashok Varma, and Singh, M.R. (2003) Evaluation of different integrated control tactics for management of major insect pests of sugarcane in central Uttar Pradesh. *Indian Journal of Sugarcane Technology* **18(1/2)**: 64–9.

Tauber, C.A., Johnson, J.B., and Tauber, M.J. (1992) Larval and developmental characteristics of the endemic Hawaiian lacewing, *Anomalochrysa frater* (Neuroptera: Chrysopidae). *Annals of the Entomological Society of America* **85**: 200–6.

Tauber, M.J., Tauber, C.A., and Gardescu, S. (1993) Prolonged storage of *Chrysoperla carnea* (Neuroptera: Chrysopidae). *Environmental Entomology* **22**: 843–8.

Tatchell, C.M. and Payne, C.C. (1984) Field evaluation of a granulosis virus for control of *Pieris rapae* (Lep.: Pieridae) in the United Kingdom. *Entomophaga* **29**: 133–44.

Taylor, C.M. and Hastings, A. (2005) Allee effects in biological invasions. *Ecology Letters* **8**: 895–908.

Taylor, G.S. (2004) Revision of *Fergusonina* Malloch gall flies (Diptera: Fergusoninidae) from *Melaleuca* (Myrtaceae). *Invertebrate Systematics* **18**: 251–90.

Taylor, R.H. and Thomas, B.W. (1993) Rats eradicated from rugged Breaksea Island (170 ha), Fiordland, New Zealand. *Biological Conservation* **65**: 191–8.

Taylor, R.H., Kaiser, G.W., and Drever, M.C. (2000) Eradication of Norway rats for recovery of seabird habitat on Langara Island, British Columbia. *Restoration Ecology* **8**: 151–60.

Tedders, W.L. and Schaefer, P.W. (1994) Release and establishment of *Harmonia axyridis* (Coleoptera: Coccinellidae) in the southeastern United States. *Entomological News* **105(4)**: 228–43.

Telenga, N.A. and Schepetilnikova, V.A. (1949) *A Manual for Breeding and Application of Trichogramma in Agricultural Pest Control*. Izdatelstvo Akademie Nauk Ukranian SSR, Kiev.

Telford, S.R. (1999) The possible use of haemogregarine parasites in the biological control of the brown treesnake (*Boiga irregularis*) and the habu (*Trimeresurus flavovirdis*). In: Rodda G.H., Sawai, Y., Chiszar, D., and Tanaka, H. (eds.), *Problem Snake Management: the Habu and the Brown Treesnake*, pp. 384–90. Comstock Publishing Associates, Ithaca, NY.

Templeton, G.E. (1992) Use of *Colletotrichum* strains as mycoherbicides. In: Bailey, J.A. and Jeger, M.J. (eds.), *Colletotrichum: Biology, Pathology, and Control*, pp. 358–80. CABI Publishing, Wallingford.

Tewksbury, L., Gold, M.S., Casagrande, R.A., and Kenis, M. (2005) Establishment in North America of *Tetrastichus setifer* Thomson (Hymenoptera: Eulophidae), a parasitoid of *Lilioceris lilii* (Coleoptera: Chrysomelidae). In: Hoddle, M.S. (compiler), *Proceedings of the Second International Symposium on Biological Control of Arthropods*, pp. 142–3, Davos, Switzerland. USDA-FS Forest Health Technology Team, Morgantown, WV.

Thaman, R.R. (1974) *Lantana camara*: its introduction, dispersal and impact on islands of the tropical Pacific Ocean. *Micronesica* **10**: 17–39.

Thang, M.H., Mochida, O., Morallo-Rejesus, B., and Robles, R.P. (1987) Selectivity of eight insecticides to the brown planthopper, *Nilaparvata lugens* (Stål) (Homoptera: Delphacidae), and its predator, the wolf spider *Lycosa pseudoannulata* Boes. et Str. (Araneae: Lycosidae). *Philippine Entomologist* **7**: 51–6.

Thibaut, T. and Meinesz, A. (2000) Are the Mediterranean ascoglossan mollusks *Oxynoe olivacea* and *Lobiger serradifalci* suitable agents for a biological control against the invading tropical alga *Caulerpa taxifolia*? *Comptes Rendus de L'Académie des Science. Série III, Sciences de la Vie* **323**: 477–88.

Thibaut, T., Meinesz, A., Amade, P. et al. (2001) *Elysia subornata* (Mollusca) a potential control agent of the alga *Caulerpa taxifolia* (Chlorophyta) in the Mediterranean Sea. *Journal of Marine Biology Association* **81**: 497–504.

Thiele, H.U. (1977) *Carabid Beetles in their Environments*. Springer-Verlag, Berlin.

Thiery, I., Hamon, S., Dumanoir, V.C., and de Barjac, H. (1992) Vertebrate safety of *Clostridium bifermentans* serovar *malaysia*, a new larvicidal agent for vector control. *Journal of Economic Entomology* **85**: 1618–23.

Thomas, M. (1990) Diversification of the arable ecosystem to control natural enemies of cereal aphids. *Game Conservancy Review* **21**: 68–9.

Thomas, M.B., Wratten, S.D., and Sotherton, N.W. (1991) Creation of "island" habitats in farmland to manipulate populations of beneficial arthropods: predator densities and emigration. *Journal of Applied Ecology* **28**: 906–17.

Thomas, M.B., Mitchell, H.J., and Wratten, S.D. (1992) Abiotic and biotic factors influencing the winter distribution of predatory insects. *Oecologia* **89**: 78–84.

Thomas, P. (2000) *Trees: Their Natural History*. Cambridge University Press, Cambridge.

Thomas, P.A. and Room, P.M. (1986) Taxonomy and control of *Salvinia molesta*. *Nature* **320**: 581–4.

Thompson, C.R. and Habeck, D.H. (1989) Host specificity and biology of the weevil *Neohydronomus affinis* (Coleoptera: Curculionidae), a biological control agent of *Pistia stratiotes*. *Entomophaga* **34**: 299–306.

Thompson, L.C., Kulman, H.M., and Hellenthal, R.A. (1979) Parasitism of the larch sawfly by *Bessa harveyi* (Diptera: Tachinidae). *Annals of the Entomological Society of America* **72**: 468–71.

Thompson, S.N. (1981) *Brachymeria lasus*: culture *in vitro* of a chalcid insect parasite. *Experimental Parasitology* **52**: 414–18.

Thompson, W.R. (1924) La theorie mathematique de l'action des parasites entomophages et le facteur du hazard. *Annales de la Faculté des Sciences de Marseille* **2**: 69–89.

Thorbeck, P. and Bilde, T. (2004) Reduced numbers of generalist arthropod predators after crop management. *Journal of Applied Ecology* **41**: 526–38.

Thresher, R.E., Werner, M., Hoeg, J.T. et al. (2000) Developing the options for managing marine pests: specificity trials on the parasitic castrator, *Sacculina carcini*, against the European crab, *Carcinus maenas*, and related species. *Journal of Experimental Marine Biology and Ecology* **254**: 37–51.

Thulin, C.-G., Simberloff, D., Barun, A., McCraken, G., Pascal, M., and Isalm, M.A. (2006) Genetic divergence in the small Indian mongoose (*Herpestes auropunctatus*), a widely distributed invasive species. *Molecular Ecology* **15**: 3947–56.

Tilmon, K.J. and Hoffmann, M.P. (2003) Biological control of *Lygus lineolaris* by *Peristenus* spp. in strawberry. *Biological Control* **26**: 287–92.

Tinzaara, W., Gold, C.S., Dicke, M., and van Huis, A. (2005) Olfactory responses of banana weevil predators to volatiles from banana pseudostem tissue and synthetic pheromone. *Journal of Chemical Ecology* **31**: 1537–53.

Tisdell, C. (1990) Economic impact of biological control of weeds and insects. In: Mackauer, M., Ehler, L.E., and Roland, J. (eds.), *Critical Issues in Biological Control*, pp. 301–16. Intercept, Andover.

Tisdell, C.A., Auld, B.A., and Menz, K.M. (1984) On assessing the value of biological control of weeds. *Protection Ecology* **6**: 169–79.

Toepfer, S. and Kuhlmann, U. (2006) Constructing life-tables for the invasive maize pest *Diabrotica virgifera* (Col., Chrysomelidae) in Europe. *Journal of Applied Entomology* **130**: 193–205.

Tomley, A.J. and Evans, H.C. (2004) Establishment of, and preliminary impact studies on, the rust, *Maravalia cryptostegiae*, of the invasive alien weed, *Cryptostegia grandiflora* in Queensland, Australia. *Plant Pathology* **53**: 475–84.

Torchin, M.E., Lafferty, K.D., and Kuris, A.M. (2001) Release from parasites as natural enemies: increased performance of a globally introduced marine crab. *Biological Invasions* **3**: 333–45.

Torgersen, T.R., Thomas, J.W., Mason, R.R., and van Horn, D. (1984) Avian predators of Douglas-fir tussock moth, *Orgyia pseudotsugata* (McDunnough) (Lepidoptera: Lymantriidae) in southwestern Oregon. *Environmental Entomology* **13**: 1018–22.

Torres, J.B., Ruberson, J.R., and Adang, M.J. (2006) Expression of *Bacillus thuringiensis* Cry1Ac protein in cotton plants, acquisition by pests and predators: a tritrophic analysis. *Agricultural and Forest Entomology* **8**: 191–202.

Tostawaryk, W. (1971) Relationship between parasitism and predation of diprionid sawflies. *Annals of the Entomological Society of America* **64**: 1424–7.

Tothill, J.D., Taylor, T.H.C., and Paine, R.W. (1930) *The Coconut Moth in Fiji: a History of its Control by Means of Parasites*. Imperial Bureau of Entomology, London.

Townes, H. (1969) The genera of Ichneumonidae, Parts 1, 2, and 3. *Memoirs of the American Entomological Institute* nos. 11, 12, and 13.

Townes, H. (1988) The more important literature on parasitic Hymenoptera. In: *Advances in Parasitic Hymenoptera Research. Proceedings of the Second Conference on the Taxonomy and Biology of Parasitic Hymenoptera*, pp. 491–518, November 19–21, 1987, Gainesville, Florida. E.J. Brill, New York.

Tracewski, K.T., Johnson, P.C., and Eaton, A.T. (1984) Relative densities of predaceous Diptera (Cecidomyiidae, Chamaemyiidae, Syrphidae) and their apple aphid prey in New Hampshire, USA, apple orchards. *Protection Ecology* **6**: 199–207.

Treacy, M.F., Benedict, J.H., Segers, J.C., Morrison, R.K., and Lopez, J.D. (1986) Role of cotton trichome density in bollworm (Lepidoptera: Noctuidae) egg parasitism. *Environmental Entomology* **15**: 365–8.

Treacy, M.F., Benedict, J.H., Lopez, J.D., and Morrison, R.K. (1987) Functional response of a predator (Neuroptera: Chrysopidae) to bollworm (Lepidoptera: Noctuidae) eggs on smoothleaf, hirsute, and pilose cottons. *Journal of Economic Entomology* **80**: 376–9.

Triplehorn, C.A. and Johnson, N.F. (2005) *Borror and DeLong's Introduction to the Study of Insects*. Thomson Brooks/Cole, Belmont, CA.

Trotter, D.M., Kent, R.A., and Wong, M.P. (1991) Aquatic fate and effect of carbofuran. *Critical Reviews in Environmental Control* **21**: 137–76.

Trujillo, E.E. (1985) Biological control of Hamakua pa-makani with *Cercosporella* sp. in Hawaii. In: Delfosse, E.S. (ed.), *Proceedings of the VIth International Symposium on Biological Control of Weeds*, pp. 661–71, August 19–25, 1984. Vancouver, British Columbia. Agriculture Canada, Ottawa.

Trujillo, E.E., Latterell, F.M., and Rossi, A.E. (1986) *Colletotrichum gloeosporioides*, a possible biological control agent for *Clidemia hirta* in Hawaiian forests. *Plant Disease* **70**: 974–6.

Trumble, J. and Alvarado-Rodriguez, B. (1998) Trichogrammatid egg parasitoids as a component in the management of vegetable-crop insect pests. In: Ridgway, R.L., Hoffmann, M.P., Inscoe, N.N., and Glenister, C. (eds.), *Mass-reared Natural Enemies: Application, Regulation, and Needs*, pp. 158–84. Thomas Say Publications in Entomology: Proceedings. Entomological Society of America, Lanham, MD.

Tsutsui, N.D., Suarez, A.V., Holway, D.A., and Case, T.J. (2001) Relationships among native and introduced populations of the Argentine ant (*Linepithema humile*) and the source of introduced populations. *Molecular Ecology* **10**: 2151–61.

Tumlinson, J.H., Lewis, W.J., and Vet, L.E.M. (1993) How parasitic wasps find their hosts. *Scientific American* **22**: 100–6.

Tuomi, J., Niemelä, P., Haukioja, E., Sirén, S., and Neuvonen, S. (1984) Nutrient stress: an explanation for plant anti-herbivore responses to defoliation. *Oecologia* **61**: 208–10.

Turchin, P. (1990) Rarity of density dependence or population regulation with lags? *Nature* **344**: 660–3.

Turlings, T.C.J., Tumlinson, J.H., Eller, F.J., and Lewis, W.J. (1991) Larval-damaged plants: source of volatile synomones that guide the parasitoid *Cotesia marginiventris* to the micro-habitat of its hosts. *Entomologia Experimentalis et Applicata* **58**: 75–82.

Turnbull, A.L. and Chant, P.A. (1961) The practice and theory of biological control in Canada. *Canadian Journal of Zoology* **39**: 697–753.

Turner, C.E. (1994) Host specificity and oviposition of *Urophora sirunaseva* (Hering) (Diptera: Tephritidae), a natural enemy of yellow starthistle. *Proceedings of the Entomological Society of Washington* **96**: 31–6.

Turner, C.E., Pemberton, R.W., and Rosenthal, S.S. (1987) Host utilization of native *Cirsium* thistles (Asteraceae) by the introduced weevil *Rhinocyllus conicus* (Coleoptera: Curculionidae) in California. *Environmental Entomology* **16**: 111–15.

Turner, C.E., Sobhian, R., and Maddox, D.M. (1990) Host specificity studies of *Chaetorellia australis* (Diptera: Tephritidae), a prospective biological control agent for yellow starthistle, *Centaurea solstitialis* (Asteracea). In: Delfosse, E.S. (ed.), *Proceedings of the VIIth International Symposium on Biological Control of Weeds*, pp. 231–6. Istituto Sperimentale per la Patologia Vegetal, Ministero dell'Agricoltura e delle Foreste, Rome.

Turner, C.E., Center, T.D., Burrows, D.W., and Buckingham, G.R. (1998) Ecology and management of *Melaleuca quinquenervia*, an invader of wetlands in Florida, USA. *Wetlands Ecology and Management* 5: 165–78.

Turnock, W.J., Wise, I.L., and Matheson, F.O. (2003) Abundance of some native coccinellines (Coleoptera: Coccinellidae) before and after the appearance of *Coccinella septempunctata*. *The Canadian Entomologist* **135**: 391–404.

Tyndale-Biscoe, C.H. (1991) Fertility control in wildlife. *Reproduction, Fertility and Development* **3**: 339–43.

Tyndale-Biscoe, C.H. (1994a) The CRC for biological control of vertebrate pest populations: fertility control of wildlife for conservation. *Pacific Conservation Biology* **1**: 160–2.

Tyndale-Biscoe, C.H. (1994b) Virus-vectored immunocontraception of feral mammals. *Reproduction, Fertility and Development* **6**: 281–7.

Tyndale-Biscoe, C.H. (1995) Vermin and viruses: risks and benefits of viral-vectored immunosterilization. *Search* **26**: 239–44.

U.S. Congress, Office of Technology Assessment (1993) *Harmful Non-indigenous Species in the United States*. OTA-F-565. U.S. Government Printing Office, Washington, D.C.

Udayagiri, S., Welter, S.C., and Norton, A.P. (2000a) Biological control of *Lygus hesperus* with inundative releases of *Anaphes iole* in a high cash value crop. *Southwestern Entomologist* (Suppl. 23): 27–38.

Udayagiri, S., Norton, A.P., and Welter, S.C. (2000b) Integrating pesticide effects with inundative biological control: interpretation of pesticide toxicity curves for *Anaphes iole* in strawberries. *Entomologia Experimentalis et Applicata* **95**: 87–95.

Udvardy, M.D.F. (1969) *Dynamic Zoogeography*. Van Nostrand Reinhold Co., New York.

Unruh, T.R., White, W., Gonzalez, D., Gordh, G., and Luck, R.F. (1983) Heterozygosity and effective size in laboratory populations of *Aphidius ervi* (Hymenoptera, Aphidiidae). *Entomophaga* **28**: 245–58.

USDA-FS (2004) HWA distribution map. www.fs.fed.us/na/morgantown/fhp/hwa/maps/ hwa_1_20_04.jpg.

USEPA (2001) Santa Cruz Island Primary Restoration Plan Draft Environmental Impact Statement, Channel Islands National Park, Santa Barbara County, California; Notice of Availability. www.epa.gov/EPA-IMPACT/2001/March/Day-09/i5948.htm.

Utida, S. (1957) Cyclic fluctuations of population density intrinsic to the host-parasite system. *Ecology* **38**: 442–9.

Uygur, S., Smith, L., Uygur, F.N., Cristofaro, M., and Balciunas, J. (2005) Field assessment in land of origin of host specificity, infestation rate and impact of *Ceratapion basicorne* a prospective biological control agent of yellow starthistle. *Biocontrol* **50**: 525–41.

Vaeck, M., Reynaerts, A., Hofte, H. et al. (1987) Transgenic plants protected from insect attack. *Nature* **328**: 33–7.

Valicente F.H. and O'Neil, R.J. (1995) Effects of host plants and prey on selected life history characteristics of *Podisus maculiventris*. *Biological Control* **5**: 449–61.

Van, T.K., Rayachhetry, M.B., and Center, T.D. (2000) Estimating above-ground biomass of *Melaleuca quinquenervia* in Florida, USA. *Journal of Aquatic Plant Management* **38**: 62–7.

van Alphen, J.J.M. (1988) Patch-time allocation by insect parasitoids: superparasitism and aggregation. In: de Jong, G. (ed.), *Population Genetics and Evolution*, pp. 125–221. Springer-Verlag, Berlin.

van Alphen, J.J.M. and van Harsel, H.H. (1982) Host selection by *Asobara tabida* Nees (Bracondiae: Alysiinae), a larval parasitoid of fruit inhabiting *Drosophila* species. III. Host species selection and functional response. In: van Alphen, J.J.M., *Foraging Behaviour of Asobara tabida, a Larval Parasitoid of Drosophilidae*, pp. 61–93. Ph.D. Dissertation, University of Leiden, Leiden.

van Alphen, J.J.M. and Galis, F. (1983) Patch time allocation and parasitization efficiency of *Asobara tabida* Nees, a larval parasitoid of *Drosophila*. *Journal of Animal Ecology* **52**: 937–52.

van Alphen, J.J.M. and Vet, L.E.M. (1986) An evolutionary approach to host finding and selection. In: Waage, J.K. and Greathead, D. (eds.) (1986) *Insect Parasitoids*, pp. 23–61. Academic Press, London.

van Alphen, J.J.M. and Jervis, M.A. (1996) Foraging behavior. In: Jervis, M. and Kidd, N. (eds.), *Insect Natural Enemies: Practical Approaches to their Study and Evaluation*, pp. 1–62. Chapman and Hill, London.

van Belkum, A., Kluytmans, J., van Leeuwen, W. et al. (1995) Multicenter evaluation of arbitrarily primed PCR for typing of *Staphylococcus aureus* strains. *Journal of Clinical Microbiology* **33**: 1537–47.

van Bergeijk, K.E., Bigler, F., Kaashoek, N.K., and Pak, G.A. (1989) Changes in host acceptance and host suitability as an effect of rearing *Trichogramma maidis* on a factitious host. *Entomologia Experimentalis et Applicata* **52**: 229–38.

Vandenberg, J.D. (1990) Safety of four entomopathogens for caged adult honey bees (Hymenoptera: Apidae). *Journal of Economic Entomology* **83**: 755–9.

van den Berg, M.A. (1977) Natural enemies of certain Acacias in Australia. In: *Proceedings of the 2nd National Weeds Conference of South Africa*, pp. 75–82, February 2–4, 1977, Stellenbosch, South Africa. A.A. Balkema, Cape Town.

van den Berg, H. (1993) Natural control of *Helicoverpa armigera* in smallholder crops in East Africa. Ph.D. Dissertation, Department of Entomology, University of Wageningen, Wageningen.

van den Berg, M.A., Höppner, G., and Greenland, J. (2000) An economic study of the biological control of the spiny blackfly, *Aleurocanthus spiniferus* (Hemiptera: Aleyrodidae), in a citrus orchard in Swaziland. *Biocontrol Science and Technology* **10**: 27–32.

van den Bosch, R., Lagace, C.F., and Stern, W.M. (1967) The interrelationship of the aphid *Acyrthosiphon pisum* and its parasite, *Aphidius smithi*, in a stable environment. *Ecology* **48**: 993–1000.

van den Bosch, R., Frazer, R.D., Davis, C.S., Messenger, P.S., and Hom, R. (1970) *Trioxys pallidus*: an effective new walnut aphid parasite from Iran. *California Agriculture* **24(6)**: 8–10.

van den Meiracker, R.A.F., Hammond, W.N.O., and van Alphen, J.J.M. (1990) The role of kairomones in prey finding by *Diomus* sp. and *Exochomus* sp., two coccinellid predators of the cassava mealybug, *Phenococcus manihoti*. *Entomologia Experimentalis et Applicata* **56**: 209–17.

van der Zweerde, W. (1990) Biological control of aquatic weeds by means of phytophagous fish. In: Pieterse, A.H. and Murphy, K.J. (eds.), *Aquatic Weeds: the Ecology and Management of Nuisance Aquatic Vegetation*, pp. 201–21. Oxford University Press, New York.

van de Vrie, M. and Boersma, A. (1970) The influence of the predaceous mite *Typhlodromus* (*A.*) *potentillae* (Garman) on the population development of *Panonychus ulmi* (Koch) on apple grown under various nitrogen conditions. *Entomophaga* **15**: 291–304.

van de Vrie, M., McMurtry, J.A., and Huffaker, C.B. (1972) Ecology of tetranychid mites and their natural enemies: a review. III. Biology, ecology, and pest status, and host-plant relations of tetranychids. *Hilgardia* **41(13)**: 343–432.

van Dijken, M.J. and Waage, J.K. (1987) Self and conspecific superparasitism by the egg parasitoid *Trichogramma evanescens*. *Entomologia Experimentalis et Applicata* **43**: 183–92.

Van Driesche, J. and Van Driesche, R.G. (2000) *Nature Out of Place: Biological Invasions in a Global Age*. Island Press, Covelo, CA.

Van Driesche, J.P. and Van Driesche, R.G. (2001) Guilty until proven innocent: preventing nonnative species invasions. *Conservation Biology in Practice* **2(1)**: 8–17.

Van Driesche, R.G. (1983) The meaning of percent parasitism in studies of insect parasitoids. *Environmental Entomology* **12**: 1611–22.

Van Driesche, R.G. (1988) Field levels of encapsulation and superparasitism for *Cotesia glomerata* (L.) (Hymen.: Braconidae) in *Pieris rapae* (L.) (Lep.: Pieridae). *Journal of the Kansas Entomological Society* **61**: 328–31.

Van Driesche, R.G. (1993) Methods for the field colonization of new biological control agents. In: Van Driesche, R.G. and Bellows, Jr., T.S. (eds.), *Steps in Classical Arthropod Biological Control*, pp. 67–86. Proceedings of the Thomas Say Publications in Entomology. Entomological Society of America, Lanham, MD.

Van Driesche, R.G. (1994) Classical biological control of environmental pests. *Florida Entomologist* **77**: 20–33.

Van Driesche, R.G. and Gyrisco, G.G. (1979) Field studies of *Microctonus aethiopodes*, a parasite of the adult alfalfa weevil, *Hypera postica*, in New York. *Environmental Entomology* **8**: 238–44.

Van Driesche, R.G. and Taub, G. (1983) Impact of parasitoids on *Phyllonorycter* leafminers infesting apple in Massachusetts, USA. *Protection Ecology* **5**: 303–17.

Van Driesche, R.G. and Hulbert, C. (1984) Host acceptance and discrimination by *Comperia merceti* (Compere) (Hymenoptera: Encrytidae) and evidence for an optimal density range for resource utilization. *Journal of Chemical Ecology* **10**: 1399–1409.

Van Driesche, R.G. and Carey, E. (1987) *Opportunities for Increased Use of Biological Control in Massachusetts*. Massachusetts Agricultural Experiment Station Bulletin 718. Massachusetts Agricultural Experiment Station, Amherst, MA.

Van Driesche. R.G. and Bellows, Jr., T.S. (1988) Use of host and parasitoid recruitment in quantifying losses from parasitism in insect populations with reference to *Pieris rapae* and *Cotesia glomerata*. *Ecological Entomology* **13**: 215–22.

Van Driesche, R.G. and Bellows, Jr., T.S. (1993) *Steps in Classical Arthropod Biological Control*. Proceedings of the Thomas Say Publications in Entomology. Entomological Society of America, Lanham, MD.

Van Driesche, R.G. and Bellows, Jr., T.S. (1996) *Biological Control*. Chapman and Hall, New York.

Van Driesche, R.G. and Hoddle, M. (1997) Should arthropod parasitoids and predators be subject to host range testing when used as biological control agents? *Agriculture and Human Values* **14**: 211–26.

Van Driesche, R.G. and Nunn, C. (2002) Establishment of a Chinese strain of *Cotesia rubecula* (Hymenoptera: Braconidae) in the northeastern United States. *Florida Entomologist* **85**: 386–8.

Van Driesche, R.G. and Lyon, S. (2003) Commercial adoption of biological control-based IPM for whiteflies in poinsettia. *Florida Entomologist* **86**: 481–3.

Van Driesche, R.G. and Nunn, C. (2003) Status of euonymus scale in Massachusetts fourteen years after release of *Chilocorus kuwanae* (Coleoptera: Coccinellidae). *Florida Entomologist* **86**: 384–5.

Van Driesche, R.G. and Reardon, R. (eds.) (2004) *Assessing Host Ranges for Parasitoids and Predators Used for Classical Biological Control: a Guide to Best Practice*. FHTET-2004-03. USDA Forest Service, Morgantown, WV.

Van Driesche, R.G., Bellows, Jr., T.S., Ferro, D.N., Hazzard, R., and Maher, A. (1989) Estimating stage survival from recruitment and density data, with reference to egg mortality in the Colorado potato beetle, *Leptinotarsa decemlineata* (Say) (Coleoptera: Chrysomelidae). *The Canadian Entomologist* **121**: 291–300.

Van Driesche, R.G., Ferro, D.N., Carey, E., and Maher, A. (1990) Assessing augmentative releases of parasitoids using the "recruitment method," with reference to *Edovum puttleri*, a parasitoid of the Colorado potato beetle (Coleoptera: Chrysomelidae). *Entomophaga* **36**: 193–204.

Van Driesche, R.G., Elkinton, J.S., and Bellows, Jr., T.S. (1994) Potential use of life tables to evaluate the impact of parasitism on population growth of the apple blotch leafminer (Lepidoptera: Gracillariidae). In: Maier, C. (ed.), *Integrated Management of Tentiform Leafminers, Phyllonorycter (Lepidoptera: Gracillariidae) spp. in North American Apple Orchards*, pp. 37–51. Thomas Say Publications in Entomology. Entomological Society of America, Lanham, MD.

Van Driesche, R.G., Idoine, K., Rose, M., and Bryan, M. (1998a) Release, establishment and spread of Asian natural enemies of euonymus scale (Homoptera: Diaspididae) in New England. *Florida Entomologist* **81**: 1–9.

Van Driesche, R.G., Idoine, K., Rose, M., and Bryan, M. (1998b) Evaluation of the effectiveness of *Chilocorus kuwanae* (Coleoptera: Coccinellidae) in suppressing euonymus scale (Homoptera: Diaspididae). *Biological Control* **12**: 56–65.

Van Driesche, R.G., Lyon, S.M., Hoddle, M.S., Roy, S., and Sanderson, J.P. (1999) Assessment of cost and performance of *Eretmocerus eremicus* (Hymenoptera: Aphelinidae) for whitefly (Homoptera: Aleyrodidae) control in commercial poinsettia crops. *Florida Entomologist* **82**: 570–94.

Van Driesche, R.G., Hoddle, M.S., Lyon, S., and Sanderson, J.P. (2001) Compatibility of insect growth regulators with *Eretmocerus eremicus* (Hymenoptera: Aphelinidae) for whitefly control (Homoptera: Alyerodidae) control on poinsettia: II. Trials in commercial poinsettia crops. *Biological Control* **20**: 132–46.

Van Driesche, R.G., Blossey, B., Hoddle, M., Lyon, S., and Reardon, R. (eds.) (2002a) *Biological Control of Invasive Plants in the Eastern United States*. FHTET-2002-04. USDA Forest Service, Morgantown, WV.

Van Driesche, R.G., Lyon, S., Smith, T., and P. Lopes. (2002b) Use of *Amblyseius cucumeris* in greenhouse bedding plants for thrips control – is mechanical application better? *Preceedings of the Working Groups Meeting*, May 6–9, 2002, Victoria, British Columbia, Canada. *IOBC WPRS Bulletin* **25(1)**: 273–6.

Van Driesche, R.G., Lyon, S., Jacques, K., Smith, T., and Lopes, P. (2002c) Comparative cost of chemical and biological whitefly control in poinsettia: is there a gap? *Florida Entomologist* **85**: 488–93.

Van Driesche, R.G., Lyon, S., and Nunn, C. (2006a) Compatibility of spinosad with predaceous mites (Acari: Phytoseiidae) used to control of western flower thrips (Thysanoptera: Thripidae) in greenhouse crops. *Florida Entomoloist* **89**: 396–401.

Van Driesche, R.G., Lyon, S., Stanek, III, E.J., Bo Xu, and Nunn, C. (2006b) Evaluation of efficacy of *Neoseiulus cucumeris* for control of western flower thrips in spring bedding crops. *Biological Control* **36**: 203–15.

van Essen, F.W. and Hembree, S.C. (1980) Laboratory bioassay of *Bacillus thuringiensis israelensis* against all instars of *Aedes aegpti* and *Aedes taeniorhynchus* larvae. *Mosquito News* **40(3)**: 424–31.

van Klinken, R.D. (2000) Host specificity testing: why do we do it and how we can do it better? In: Van Driesche, R.G., Heard, T., McClay, A., and Reardon, R. (eds.), *Proceedings of Session: Host Specificity Testing of Exotic Arthropod Biological Control Agents – the Biological Basis for Improvement in Safety*, pp. 54–68. FHTET-99-1. USDA Forest Service, Morgantown, WV.

van Klinken, R.D., Fichera, G., and Cordo, H. (2003) Targeting biological control across diverse landscapes: the release, establishment, and early success of two insects on mesquite (*Prosopis* spp.) insects in Australian rangelands. *Biological Control* **26**: 8–20.

van Lenteren, J.C. (1989) Implementation and commercialization of biological control in western Europe. Proceedings and Abstracts. International Symposium of Biological Control Implementation. *North American Plant Protection Bulletin* **6**: 50–70.

van Lenteren, J.C. (1991) Ecounters with parasitized hosts: to leave or not to leave a patch. *Netherlands Journal of Zoology* **41**: 144–57.

van Lenteren, J.C. (1995) Integrated pest management in protected crops. In: Dent, D.R. (ed.), *Integrated Pest*

Management: Principles and Systems Development, pp. 311–43. Chapman and Hall, London.

van Lenteren, J.C. (2000a) A greenhouse without pesticides: fact or fantasy? *Crop Protection* **19**: 375–84.

van Lenteren, J.C. (2000b) Measures of success in biological control of arthropods by augmentation of natural enemies. In: Gurr, G. and Wratten, S. (eds.), *Measures of Success in Biological Control*, pp. 77–103. Kluwer Academic Publishers, Dordrecht.

van Lenteren, J.C. (2003) *Quality Control and Production of Biological Control Agents: Theory and Testing Procedures.* CABI Publishing, Wallingford.

van Lenteren, J.C. and Woets, J. (1988) Biological and integrated control in greenhouses. *Annual Review of Entomology* **33**: 239–69.

van Lenteren, J.C. and Loomans, A.J.M. (2006) Environmental risk assessment: methods for comprehensive evaluation and quick scan. In: Bigler, F., Babendreir, D., and Kuhlmann, U. (eds.), *Environmental Impact of Invertebrates for Biological Control of Arthropods*, pp. 254–72. CABI Publishing, Wallingford.

van Lenteren, J.C., Woets, J., Van Der Poel, N. et al. (1977) Biological control of the greenhouse whitefly *Trialeurodes vaporariorum* (Westwood) (Homoptera: Aleyrodidae) by *Encarsia formosa* Gahan (Hymenoptera: Aphelinidae) in Holland, an example of successful applied ecological research. *Mededlingen Faculteit Landbouwwetenschappen, Rijksuniversiteit Gent* **42**: 1333–42.

van Lenteren, J.C., Babendreier, D., Bigler, F. et al. (2003) Environmental risk assessment of exotic natural enemies used in inundative biological control. *BioControl* **48**: 3–38.

van Lenteren, J.C., Bale, J., Bigler, F., Hokkanen, H.M.T., and Loomans, A.J.M. (2006a) Assessing risks of releasing exotic biological control agents of arthropod pests. *Annual Review of Entomology* **51**: 609–34.

van Lenteren, J.C., Cock, M.J.W., Hoffmeister, T.S., and Sands, D.P.A. (2006b) Host specificity in arthropod biological control, methods for testing and interpretation of the data. In: Bigler, F., Babendreir, D. and Kuhlmann, U. (eds.), *Environmental Impact of Invertebrates for Biological Control of Arthropods*, pp. 38–63. CABI Publishing, Wallingford.

van Rensburg, P.J.J., Skinner, J.D., and van Aarde, R.J. (1987) Effects of feline panleucopaenia on the population characteristics of feral cats on Marion Island. *Journal of Applied Ecology* **24**: 63–73.

van Veen, F.F.J., Morris, R.J., and Godfray, H.C.J. (2006) Apparent competition, quantitative food webs, and the structure of phytophagous insect communities. *Annual Review of Entomology* **51**: 187–208.

van Wilgen, B.W., de Wit, M.P., Anderson, H.J. et al. (2004) Costs and benefits of biological control of invasive alien plants: case studies from South Africa. *South African Journal of Science* **100**: 113–22.

van Winkelhoff, A.J. and McCoy, C.W. (1984) Conidiation of *Hirsutella thompsonii* var. *synnematosa* in submerged culture. *Journal of Invertebrate Pathology* **43**: 59–68.

van Zon, J.C.J. (1977) Status of biotic agents, other than insects or pathogens, as biocontrols. In: Freeman, T.E. (ed.), *Proceedings of the IVth International Symposium on Biological Control of Weeds*, pp. 245–50, August 30–September 2, 1976, Gainesville, Florida. Institute of Food and Agricultural Sciences, University of Florida, Gainesville, FL.

Vargas-Camplis, J., Cortez, E.M., del Bosque, L.A.R., and Coleman, R.J. (2000) Impact of *Catolaccus grandis* Burks (Hymenoptera: Pteromalidae) field release on cotton boll weevil in the Huasteca region of Mexico. In: Dugger, P. and Richter, D. (ed.), *2000 Proceedings Beltwide Cotton Conferences*, vol. 2, pp. 1195–7, January 4–8, 2000, San Antonio, Texas National Cotton Council, Memphis, TN.

Varley, G.C. and Gradwell, C.R. (1960) Key factors in insect population studies. *Journal of Animal Ecology* **29**: 399–401.

Varley, G.C. and Gradwell, G.R. (1968) Population models for the winter moth. In: Southwood, T.R.E. (ed.), *Symposia of the Royal Entomological Society of London No. 4: Insect Abundance*, pp. 132–42. Blackwell Scientific Publications, Oxford.

Varley, G.C. and Gradwell, C.R. (1970) Recent advances in insect population dynamics. *Annual Review of Entomology* **15**: 1–24.

Varley, G.C. and Gradwell, C.R. (1971) The use of models and life tables in assessing the role of natural enemies. In: Huffaker, C.B. (ed.), *Biological Control*, pp. 93–110. Plenum Press, New York.

Varley, G.C., Gradwell, G.R., and Hassell, M.P. (1973) *Insect Population Ecology*. Blackwell Scientific Publications, Oxford.

Vavre, F., de Jong, J.H., and Stouthamer, R. (2004) Cytogenetic mechanism and genetic consequences of thelytoky in the wasp *Trichogramma cacoeciae*. *Heredity* **93**: 592–6.

Vazquez, R.J., Porter, S.D., and Briano, J.A. (2004) Host specificity of a new biotype of the fire ant decapitating fly *Pseudacteon curvatus* from northern Argentina. *Environmental Entomology* **33**: 1436–41.

Veitch, C.R. (1985) Methods of eradicating feral cats from offshore islands in New Zealand. In: Moors, P.J. (ed.), *Conservation of Island Birds: Case Studies for the Management of Threatened Island Species*, pp. 125–41. Proceedings of a symposium held at the XVIII ICBP World Conference in Cambridge, 1982. ICBP Technical Publication no. 3.

Veitch, C.R. and Clout, M.N. (eds.) (2002) *Turning the Tide: the Eradication of Invasive Species.* IUCN SSC Invasive Species Specialist Group. IUCN, Gland and Cambridge; www.hear.org/articles/turningthetide/turningthetide.pdf.

Velu, T.S. and Kumaraswami, T. (1990) Studies on "skip row coverage" against bollworm damage and parasite emergence in cotton. *Entomon* **15**: 69–73.

Venditti, M.E. and Steffey, K.L. (2003) Field effects of Bt corn on the impact of parasitoids and pathogens on European corn borer in Illinois. In: Van Driesche, R.G. (ed.), *Proceedings of the First International Symposium on Biological Control of Arthropods*, pp. 278–83, January 14–18, 2002, Honolulu, Hawaii. USDA Forest Service, Morgantown, WV.

Vennila, S. and Easwaramoorthy, S. (1997) Disc gel electrophoresis in evaluating spiders for their predatory role in sugarcane ecosystem. *Journal of Biological Control* **9**: 123–4.

Vera, M.T., Rodriguez, R., Segura, D.F., Cladera, J.L., and Sutherst, R.W. (2002) Potential geographical distribution of the Mediterranean fruit fly, *Ceratitis capitata* (Diptera: Tephritidae), with emphasis on Argentina and Australia. *Environmental Entomology* **31**: 1009–22.

Vercher, R., Costa-Comelles, J., Marzal, C., and Gracía-Marí, F. (2005) Recruitment of native parasitoid species by the invading leafminer *Phyllocnistis citrella* (Lepidoptera: Gracillariidae) on citrus in Spain. *Environmental Entomology* **34**: 1129–38.

Vere, D.T., Jones, R.E., and Saunders, G. (2004) The economic benefits of rabbit control in Australian temperate pastures by the introduction of rabbit haemorrhagic disease. *Agricultural Economics* **30**: 143–55.

Verhulst, P.F. (1838) Notice sur la loi que la population suit dans son accroissement. *Correspondance Mathematique et Physique* **10**: 113–21.

Versfeld, D.B. and van Wilgen, B.W. (1986) Impact of woody aliens on ecosystem properties. In: MacDonald, I.A.W., Kruger, F.J., and Ferrar, A.A. (eds.), *The Ecology and Management of Biological Invasions in Southern Africa*, pp. 239–46. Oxford University Press, Cape Town.

Vet, L.E.M. (1985) Response to kairomones by some alysiine and eucoilid parasitoid species (Hymenoptera). *Netherlands Journal of Zoology* **35**: 486–96.

Vet, L.E.M. and Bakker, K. (1985) A comparative functional approach to the host detection behaviour of parasitic wasps. 2. A quantitative study on eight eucoilid species. *Oikos* **44**: 487–98.

Vet, L.E.M. and Dicke, M. (1992) Ecology of infochemical use in a tritrophic level context. *Annual Review of Entomology* **37**: 141–72.

Vet, L.E.M., Wäckers, F.L., and Dicke, M. (1991) How to hunt for hiding hosts: the reliability-detectability problem in foraging parasitoids. *Netherlands Journal of Zoology* **41**: 202–13.

Vet, L.E.M., Lewis, W.J., and Cardé, R.T. (1995) Parasitoid foraging and learning. In: Cardé, R.T. and Bell, W.J. (eds.), *Chemical Ecology of Insects*, pp. 65–101. Chapman and Hall, New York.

Vickery, W.L. (1991) An evaluation of bias in *k*-factor analysis. *Oecologia* **85**: 413–18.

Viggiani, G. (1984) Bionomics of the Aphelinidae. *Annual Review of Entomology* **29**: 257–76.

Villablanca, F.X., Roderick, G.K., and Paliumbi, S.R. (1998) Invasion genetics of the Mediterranean fruit fly: variation in multiple nuclear introns. *Molecular Ecology* **7**: 547–60.

Villaneuva, R.T. and Childers, C.C. (2004) Phytoseiidae increase with pollen deposition on citrus leaves. *Florida Entomologist* **87**: 609–11.

Vink, C.J., Philips, C.B., Mitchell, A.D., Winder, L.M., and Cane, R.P. (2003) Genetic variation in *Microctonus aethiopoides* (Hymenoptera: Braconidae). *Biological Control* **28**: 251–64.

Vinson, S.B. (1976) Host selection by insect parasitoids. *Annual Review of Entomology* **21**: 109–33.

Vinson, S.B. (1981) Habitat location. In: Nordlund, D.A., Jones, R.J., and Lewis, W.J. (eds.), *Semiochemicals, Their Role in Pest Control*, pp. 51–77. John Wiley and Sons, New York.

Vinson, S.B. (1984) How parasitoids locate their hosts: a case of insect espionage. In: Lewis, T. (ed.), *Insect Communication*, pp. 325–48, Academic Press, London.

Vinson, S.B. (1990) Potential impact of microbial insecticides on beneficial arthropods in the terrestrial environment. In: Laird, M., Lace, L.A., and Davidson, E.W. (eds.), *Safety of Microbial Insecticides*, pp. 43–64. CRC Press, Boca Raton, FL.

Vinson, S.B. (1991) Chemical signals used by parasitoids. *Redia* **74**: 15–42.

Vinson, S.B. (1999) Parasitoid manipulation as a plant defense strategy. *Annals of the Entomological Society of America* **92**: 812–28.

Vinson, S.B. and Guillot, F.S. (1972) Host marking: source of a substance that results in host discrimination in insect parasitoids. *Entomophaga* **17**: 241–5.

Vinson, S.B. and Iwantsch, G.F. (1980) Host suitability for insect parasitoids. *Annual Review of Entomology* **25**: 397–419.

Vitousek, P.M. (1986) Biological invasions and ecosystem properties: can species make a difference?. In: Mooney, H.A. and Drake, J.A. (eds.), *Ecology of Biological Invasions of North America and Hawaii*, pp. 163–76. Springer-Verlag, New York.

Vitousek, P.M. (1990) Biological invasions and ecosystem process: towards an integration of population biology and ecosystem studies. *Oikos* **57**: 7–13.

Vitousek, P.M., D'Antonio, C.M., Loope, L.L., and Westbrooks, R. (1996) Biological invasions as global environmental change. *American Scientist* **84**: 468–78.

Voegele, J.M. (1989) Biological control of *Brontispa longissima* in Western Samoa: an ecological and economic evaluation. *Agriculture, Ecosystems, and Environment* **27**: 315–29.

Vogler, W. and Lindsay, A. (2002) The impact of the rust fungus *Maravalia cryptostegiae* on three rubber vine (*Cryptosegia grandiflora*) populations in tropical Queensland. In: Jacob, H.S., Dodd, J., and Moore, J.H. (eds.), *Proceedings of the 13th Australian Weeds Conference: Weeds "Threats Now and Forever?"*, pp. 180–2, September 8–13, 2002, Perth, Australia. Plant Protection Society of Western Australia, Perth.

Vogt, H. (1994) Pesticides and beneficial organisms. *Bulletin of IOBC/WPRS* **17(10)**: 1–178.

Völkl, W. (1994) The effect of ant-attendance on the foraging behaviour of the aphid parasitoid *Lysiphlebus cardui*. *Oikos* **70**: 149–55.

Volterra, V. (1926) Fluctuations in the abundance of a species considered mathematically. *Nature* **118**: 558–60.

Vorley, V.T. and Wratten, S.D. (1987) Migration of parasitoids (Hymenoptera: Braconidae) of cereal aphids (Hemiptera: Aphididae) between grassland, early-sown cereals and late-sown cereals in southern England. *Bulletin of Entomological Research* **77**: 555–68.

Vos, P., Hogers, R., Bleeker, M. et al. (1995) AFLP: a new technique for DNA fingerprinting. *Nucleic Acids Research* **23**: 4407–14.

Waage, J.K. (1978) Arrestment responses of the parasitoid *Nemeritis canescens* to a contact chemical produced by its host, *Plodia interpunctella*. *Physiological Entomology* **3**: 135–46.

Waage, J.K. (1979) Foraging for patchily distributed hosts by the parasitoid *Nemeritis canescens*. *Journal of Animal Ecology* **48**: 353–71.

Waage, J.K. (1983) Aggregation in field parasitoid populations: foraging time allocation by a population of *Diadegma* (Hymneoptera: Ichneumonidae). *Ecological Entomology* **8**: 447–53.

Waage, J.K. (1986) Family planning in parasitoids: adaptive patterns of progeny and sex allocation. In: Waage, J.K. and Greathead, D. (eds.), *Insect Parasitoids*, pp. 63–95. Academic Press, London.

Waage, J.K. (1989) The population ecology of pest-pesticide-natural enemy interactions. In: Jepson, P.C. (ed.), *Pesticides and Non-Target Invertebrates*, pp. 81–93. Intercept, Wimborne.

Waage, J.K. (1990) Ecological theory and the selection of biological control agents. In: Mackauer, M. and Ehler, L.E. (eds.), *Critical Issues in Biological Control*, pp. 135–57. Intercept, Andover.

Waage, J.K. and Lane, J.A. (1984) The reproductive strategy of a parasitic wasp. II. Sex allocation and local mate competition in *Trichogramma evaescens*. *Journal of Animal Ecology* **53**: 417–26.

Waage, J.K. and Greathead, D. (eds.) (1986) *Insect Parasitoids*. Academic Press, London.

Wäckers, F.L. and Lewis, W.J. (1994) Olfactory and visual learning and their combined influence on host site location by the parasitoid *Microplitis croceipes* (Cresson). *Biological Control* **4**: 105–12.

Wäckers, F.L., van Rijn, P.C.J., and J. Bruin (eds.) (2005) *Plant-provided Food for Carnivorous Insects: a Protective Mutualism and its Applications*. Cambridge University Press, Cambridge.

Wagner, D.L., Peacock, J.W., Carter, J.L., and Talley, S.E. (1996) Field assessment of *Bacillus thuringiensis* on non-target Lepidoptera. *Environmental Entomology* **25**: 1444–54.

Wahid, M.B., Ismail, S., and Kamarudin, N. (1996) The extent of biological control of rats with barn owls, *Tyto alba javanica*, in Malaysian oil palm plantations. *The Planter* **72**: 5–18.

Wainhouse, D., Wyatt, T., Phillips, A. et al. (1991) Responses of the predator *Rhizophagous grandis* to host plant derived chemicals in *Dendroctonus micans* larval frass in wind tunnel experiments (Coleoptera: Rhizophagidae, Scolytidae). *Chemoecology* **2**: 53–63.

Waite, G.K. (2001) Managing spider mites in field-grown strawberries using *Phytoseiulus persimilus* and the "pest-in-first" technique. In: Halliday, R.B., Walter, D.E., Proctor, H.C., Norton, R.A., and Colloff, M.J. (eds.), *Acarology: Proceedings of the 10th International Congress*, pp. 381–6. CSIRO Publishing, Collingwood, Victoria.

Wallace, M.S. and Hain, F.P. (2000) Field surveys and evaluation of native and established predators of the hemlock woolly adelgid (Homoptera: Adelgidae) in the southeastern United States. *Environmental Entomologist* **29**: 638–44.

Walter, D. and Proctor, H. (1999) *Mites: Ecology, Evolution, and Behavior*. CABI Publishing, New York.

Walter, D.E. and O'Dowd, D.J. (1992) Leaf morphology and predators: effects of leaf domatia on the abundance of predatory mites (Acari: Phytoseiidae). *Environmental Entomology* **21**: 478–84.

Wan, F.H. and Harris, P. (1997) Use of risk analysis for screening weed biocontrol agents: *Altica carduorum* Guer. (Coleoptera: Chrysomelidae) from China as a biocontrol agent of *Cirsium arvense* (L.) Scoop. in North America. *Biocontrol Science and Technology* **7**: 299–308.

Wan, F.H., Ma, J., Guo, J.Y., and You, L.S. (2003) Integrated control effects of *Epiblema stenuana* (Lepidoptera: Tortricidae) and *Ostrinia orientalis* (Lepidoptera: Pyralidae) against ragweed, *Ambrosia artemisiifolia* (Compositae). *Acta Entomologica Sinica* **46**: 473–8.

Wang, B.-D., Ferro, D.N., and Hosmer, D.W. (1999) Effectiveness of *Trichogramma ostrinae* and *T. nubilale* for controlling the European corn borer *Ostrinia nubilalis* in sweet corn. *Entomologia Experimentalis et Applicata* **91**: 297–303.

Wang, S. (2001) Research progress in *Trichogramma* mass rearing by using artificial host eggs. *Plant Protection Technology and Extension* **21**: 40–1.

Wang, X.G. and Keller, M.A. (2005) Patch allocation by the parasitoid *Diadegma semiclausum* (Hymenoptera: Ichneumonidae). II. Effects of host density and distribution. *Journal of Insect Behavior* **18**: 171–86.

Wang, Z.Y., He, K.L., Zhao, J.Z., and Zhou, D.R. (2003) Implementation of integrated pest management in China. In: Maredia, K.M., Dakouo, D., and Mota-Sanchez, D. (eds.), *Integrated Pest Management in the Global Arena*, pp. 197–207. CABI Publishing, Wallingford.

Wang, Z., He, K., and Yan, S. (2005) Large-scale augmentative biological control of Asian corn borer using *Trichogramma* in China: a success story. In: Hoddle, M.S. (ed.), *2nd International Symposium on Biological Control of Arthropods*, pp. 487–94, September 12–16, 2005, Davos, Switzerland. FHTET-2005-08. USDA Forest Service, Morgantown, WV.

Wapshere, A.J. (1970) The assessment of the biological control potential of organisms for controlling weeds: introduction to the subject. In: Simmonds, F.J. (ed.), *Proceedings of the First International Symposium on Biological Control of Weeds*, pp. 79–80, 6–8 March 1969, Delemont, Switzerland. Commonwealth Institute of Biological Control Miscellaneous Publication no. 1. Commonwealth Agricultural Bureaux, Farnham Royal, Slough.

Wapshere, A.J. (1974a) A strategy for evaluating the safety of organisms for biological control of weeds. *Annals of Applied Biology* **77**: 201–11.

Wapshere, A.J. (1974b) Host specificity of phytophagous organisms and the evolutionary centres of plant genera or sub-genera. *Entomophaga* **19**: 301–9.

Wapshere, A.J. (1985) Effectiveness of biological control agents for weeds: present quandries. *Agriculture, Ecosystems and Environment* **13**: 261–80.

Wapshere, A.J. (1989) A testing sequence for reducing the rejection of potential biological control agents for weeds. *Annals of Applied Biology* **114**: 515–26.

Wapshere, A.J., Delfosse, E.S., and Cullen, J.M. (1989) Recent developments in biological control of weeds. *Crop Protection* **8**: 227–50.

Wardle, A.R. and Borden, J.H. (1989) Learning of an olfactory stimulus associated with a host microhabitat by *Exeristes roborator. Entomologia Experimentalis et Applicata* **52**: 271–9.

Wardle, A.R. and Borden, J.H. (1990) Learning of host microhabitat by *Exeristes roborator* (F.) (Hymenoptera: Ichneumonidae). *Journal of Insect Behavior* **3**: 251–63.

Waterhouse, D.F. (1998) *Biological Control of Insect Pests: Southeast Asian Prospects*. ACIAR, Canberra.

Waterhouse, D.F. and Norris, K.R. (1987) *Biological Control, Pacific Prospects.* Australian Centre for International Agricultural Research, Inkata Press, Melbourne.

Waterhouse, D.F. and Sands, D.P.A. (2001) *Classical Biological Control of Arthropods in Australia.* CSIRO Entomology, Australian Centre for International Agricultural Research, Canberra.

Waterhouse, G.M. (1973) Entomophthorales. In: Ainsworth, G.C., Sparrow, F.K., and Sussman, A.S. (eds.), *The Fungi: An Advanced Treatise*, vol. 4B, pp. 219–29. Academic Press, New York.

Waters, W.E., Brooz, A.T., and Pschorn-Walcher, H. (1976) Biological control of pests of broad-leaved forests and woodlands. In: Huffaker, C.B. and Messenger, P.S. (eds.), *Theory and Practice of Biological Control*, pp. 313–36. Academic Press, New York.

Watson, A.K. (1991) The classical approach with plant pathogens. In: TeBeest, D.O. (ed.), *Microbial Control of Weeds*, pp. 3–23. Chapman and Hall, New York.

Watson, A.K. and Renney, A.J. (1974) The biology of Canadian weeds. 6. *Centaurea diffusa* and *C. maculosa*. *Canadian Journal of Plant Science* **54**: 687–701.

Watson, A.K. and Sackston, W.E. (1985) Plant pathogen containment (quarantine) facility at McDonald College. *Canadian Journal of Plant Pathology* **7**: 177–80.

Watt, K.E.F. (1964) Density dependence in population fluctuations. *The Canadian Entomologist* **96**: 1147–8.

Way, M.J., Cammell, M.E., Bolton, B., and Kanagaratnam, P. (1989) Ants (Hymenoptera: Formicidae) as egg predators of coconut pests, especially in relation to biological control of the coconut caterpillar, *Opisina arenosella* Walker (Lepidoptera: Xyloryctidae) in Sri Lanka. *Bulletin of Entomological Research* **79**: 219–33.

Webb, R.E., White, G.B., Thorpe, K.W., and Talley, S.E. (1999) Quantitative analysis of a pathogen-induced premature collapse of a "leading edge" gypsy moth (Lepidoptera: Lymantriidae) population in Virginia. *Journal of Entomological Science* **34**: 84–100.

Weeks, A.R., Velten, R., and Stouthamer, R. (2003) Incidence of a new sex-ratio-distorting endosymbiotic bacterium among arthropods. *Proceedings of the Royal Society of London Series B Biological Sciences* **270**: 1857–65.

Wehling, W.F. and Piper, G.L. (1988) Efficacy diminution of the rush skeletonweed gall midge, *Cystiphora schmidti* (Diptera: Cecidomyiidae), by an indigenous parasitoid. *Pan-Pacific Entomologist* **64**: 83–5.

Welton, J.S. and Ladle, M. (1993) The experimental treatment of the blackfly *Simulium posticatum* in the Dorset Stour using the biologically produced insecticide *Bacillus thuringiensis* var. *israelensis*. *Journal of Applied Ecology* **30**: 772–82.

Wenziker, K.J., Calver, M.C., and Woodburn, T.L. (2003) Laboratory trials of the efficacy of the crown weevil *Mortadelo horridus* (Coleoptera: Curculionidae) for the biological control of slender thistles *Carduus pycnocephalus* and *C. tenuiflorus* in southwestern Australia. *Biocontrol Science and Technology* **13**: 655–70.

Weppler, R.A., Luck, R.F., and Morse, J.G. (2003) Studies on rearing *Metaphycus helvolus* (Hymenoptera: Encyrtidae) for augmentative release against black scale (Homoptera: Coccidae) on citrus in California. *Biological Control* **28**: 118–28.

Wermelinger, B., Oertli, J.J., and Delucchi, V. (1985) Effect of host plant nitrogen fertilization on the biology of the two-spotted spider mite, *Tetranychus urticae. Entomologia Experimentalis et Applicata.* **38**: 23–8.

Weseloh, R.M. (1972) Influence of gypsy moth egg mass dimensions and microhabitat distribution on parasitization by *Ooencyrtus kuwanai. Annuals of the Entomological Society of America* **65**: 64–9.

Weseloh, R.M. (1974) Host recognition by the gypsy moth larval parasitoid *Apanteles melanoscelus. Annals of the Entomological Society of America* **67**: 583–7.

Weseloh, R.M. (1990) Simulation of litter residence times of young gypsy moth larvae and implications for predation by ants. *Entomologia Experimentalis et Applicata* **57**: 215–21.

West, R.J. and Kenis, M. (1997) Screening four exotic parasitoids as potential controls for the eastern hemlock looper, *Lambdina fisclaria fiscellaria* (Guenée) (Lepidoptera: Geometridae). *The Canadian Entomologist* **129**: 831–41.

Westigard, H. and Moffitt, H.R. (1984) Natural control of the pear psylla (Homoptera: Psyllidae): impact of mating disruption with the sex pheromone for control of the codling moth (Lepidoptera: Tortricidae). *Journal of Economic Entomology* **77**: 1520–3.

Whalon, M.E. and Wingerd, B.A. (2003) Bt: mode of action and use. *Archives of Insect Biochemistry and Physiology* **54**: 200–11.

Whalon, M.E., Croft, B.A., and Mowry, T.M. (1982) Introduction and survival of susceptible and pyrethroid-resistant strains of *Amblyseius fallacis* (Acari: Phytoseiidae) in a Michigan apple orchard. *Environmental Entomology* **11**: 1096–9.

Wharton, R.A. (1993) Bionomics of the Braconidae. *Annual Review of Entomology* **38**: 121–43.

Wharton, R.A., Marsh, P.M., and Sharkey, M. (eds.) (1997) *Manual of the New World Genera of the Family Braconidae (Hymenoptera)*. Special Publication Number 1 of the International Society of Hymenopterists. The International Society of Hymenopterists, Washington, D.C.

Wheeler, G.S. (2005) Maintenance of a narrow host range by *Oxyops vitiosa*, a biological control agent of *Melaleuca quinquenervia*. *Biochemical Systematics and Ecology* **33**: 365–83.

Wheeler, G.S. and Center, T.D. (2001) Impact of the biological control agent *Hydrellia pakistanae* (Diptera: Ephydridae) on the submersed aquatic weed *Hydrilla verticillata* (Hydrocharitaceae). *Biological Control* **21**: 168–81.

Wheeler, G.S., Massey, L.M., and Southwell, I.A. (2002) Antipredator defense of biological control agent *Oxyops vitiosa* is mediated by plant volatiles sequestered from the host plant *Melaleuca quinquenervia*. *Journal of Chemical Ecology* **28**: 297–315.

Wheeler, G.S., Massey, L.M., and Southwell, I.A. (2003) Dietary influences on terpenoids sequestered by the biological control agent *Oxyops vitiosa*: effect of plant volatiles from different *Melaleuca quinquenervia* chemotypes and laboratory host species. *Journal of Chemical Ecology* **29**: 189–208.

Whistlecraft, J.W. and Lepard, I.J.M. (1989) Effect of flooding on survival of the onion fly *Delia antiqua* (Diptera: Anthomyiidae) and two parasitoids, *Aphaereta pallipes* (Hymenoptera: Braconidae) and *Aleochara bilineata* (Coleoptera: Staphylinidae). *Proceedings of the Entomological Society of Ontario* **120**: 43–7.

Whitcomb, W.H. (1981) The use of predators in insect control. In: Pimentel, D. (ed.), *CRC Handbook of Pest Management in Agriculture*, vol. II, pp. 105–23. CRC Press, Boca Raton, FL.

Whitcomb, W.H. and Bell, K. (1964) Predaceous insects, spiders, and mites of Arkansas cotton fields. *Bulletin of the Arkansas Experiment Station* no. 690.

White, E.B., DeBach, P., and Garber, M.J. (1970) Artificial selection for genetic adaptation to temperature extremes in *Aphytis lignanenesis* Compere. *Hilgardia* **40**: 161–92.

White, I.M. and Korneyev, V.A. (1989) A revision of the Western Palaearctic species of *Urophora* Robineau-Desvoidy (Diptera: Tephritidae). *Systematic Entomology* **14**: 327–74.

White, T.C.R. (1993) *The Inadequate Environment: Nitrogen and the Abundance of Animals*. Springer-Verlag, New York.

Whitfield, J.B. (1990) Parasitoids, polydnaviruses, and endosymbiosis. *Parasitology Today* **6**: 381–4.

Wickremasinghe, M.G.V. and van Emden, H.F. (1992) Reactions of adult female parasitoids, particularly *Aphidius rhopalosiphi*, to volatile chemical cues from the host plants of their aphid prey. *Physiological Entomology* **17**: 297–304.

Wiedenmann, R.N. and O'Neil, R.J. (1990) Effects of low rates of predation on selected life-history characteristics of *Podisus maculiventris* (Say) (Heteroptera: Pentatomidae). *The Canadian Entomologist* **122**: 271–83.

Wiedenmann, R.N. and O'Neil, R.J. (1992) Searching strategy of the arthropod generalist predator, *Podisus maculiventris*. *Environmental Entomology* **21**: 1–10.

Wiedenmann, R.N., Legaspi, J.C., and O'Neil, R.J. (1996) Impact of prey density and facultative plant feeding on the life history of the predator, *Podisus maculiventris* (Heteroptera: Pentatomidae), pp. 94–118. In: Alomar, O. and Wiedenmann, R.N. (eds.), *Zoophytophagous Heteropterans*. Proceedings of Thomas Say Publications, Entomological Society of America, Lanham, MD.

Wilder, J.W., Voorhis, N., Colbert, J.J., and Sharov, A. (1994) A three variable differential equation model for gypsy moth population dynamics. *Ecological Modelling* **72**: 229–50.

Wilkes, A. (1947) The effects of selective breeding on the laboratory propagation of insect parasites. *Proceedings of the Royal Society of London Series B Biological Sciences* **134**: 227–45.

Will, K.W. and Rubinoff, D. (2004) Myth of the molecule: DNA barcodes for species cannot replace morphology for identification and classification. *Cladistics* **20**: 47–55.

William, C.B. (1918) The food habits of the mongoose in Trinidad. *Bulletin of the Agriculture Department of Trinidad and Tobago* **17**: 167–86.

Williams, C.L., Goldson, S.L., Baird, D.B., and Bullock, D.W. (1994) Geographical origin of an introduced insect pest, *Listronotus bonariensis* (Kuschel), determined by RAPD analysis. *Heredity* **72**: 412–19.

Williams, D.A., Overholt, W.A., Cuda, J.P., and Hughes, C.R. (2005) Chloroplast and microsatellite DNA diversities reveal the introduction history of Brazilian peppertree (*Schinus terebinthifolius*) in Florida. *Molecular Ecology* **14**: 3643–56.

Williams, D.F. and Banks, W.A. (1987) *Psuedacteon obtusus* (Diptera: Phoridae) attacking *Solenopsis invicta* (Hymenoptera: Formicidae) in Brazil. *Psyche* **94**: 9–13.

Williams, D.F., Oi, D.H., Porter, S.D., Pereira, R.M., and Briano, J.A. (2003) Biological control of imported fire ants (Hymenoptera: Formicidae). *American Entomologist* **49(3)**: 150–63.

Williams, D.W. and Liebhold, A.M. (1995) Detection of delayed density dependence: effects of autocorrelation in an exogenous factor. *Ecology* **76**: 1005–8.

Williams, D.W., Fuester, R.W., Balaam, W.W., Chianese, R.J., and Reardon, R.C. (1992) Incidence and ecological relationships of parasitism in larval populations of *Lymantria dispar*. *Biological Control* **2**: 35–43.

Williams, K.S. and Myers, J.H. (1984) Previous herbivore attack of red alder may improve food quality for fall webworm larvae. *Oecologia* **63**: 166–70.

Williams, M.R. (1999) *Cotton Crop Losses*. www.msstate.edu/Entomology/CTNLOSS/1998loss.html.

Williams, S.L. and Schroeder, S.L. (2003) Eradication of the invasive seaweed *Caulerpa taxifolia* by chlorine bleach. *Marine Ecology, Progress Series* **272**: 69–76.

Williamson, M. (1991) Biocontrol risks. *Nature* **353**: 394.

Williamson, M. (1993) Invaders, weeds, and the risk from genetically modified organisms. *Experientia* **49**: 219–24.

Williamson, M. (1996) *Biological Invasions*. Chapman and Hall, London.

Willis, A.J. and Memmott, J. (2005) The potential for indirect effects between a weed, one of its biocontrol agents and natuve herbivores: a food web approach. *Biological Control* **35**: 299–306.

Wilson, L.T., Pickett, C.H., Flaherty, D.L., and Bates, T.A. (1989) French prune trees: refuge for grape leafhopper parasite. *California Agriculture* **43(2)**: 7–8.

Winder, J.A. and Harley, K.L.S. (1983) The phytophagous insects on lantana in Brazil and their potential for biological control in Australia. *Tropical Pest Management* **29**: 346–62.

Winder, L. (1990) Predation of the cereal aphid *Sitobion avenae* by polyphagous predators on the ground. *Ecological Entomology* **15**: 105–10.

Withers, T.M. and Barton Browne, L. (2004) Behavioral and physiological processes affecting outcomes of host range testing. In: Van Driesche, R.G. and Reardon, R. (eds), *Assessing Host Ranges for Parasitoids and Predators Used for Classical Biological Control: a Guide to Best Practice*, pp. 40–55. FHTET-2004-03. USDA Forest Service, Morgantown, WV.

Withgott, J. (2002) California tries to rub out the monster of the lagoon. *Science* **295**: 2201–2.

Woets, J. and van Lenteren, J.C. (1976) The parasite-host relationship between *Encarsia formosa* (Hym., Aphelinidae) and *Trialeurodes vaporariorum* (Hom., Aleyrodidae). VI. Influence of the host plant on the greenhouse whitefly and its parasite *Encarsia formosa*. *IOBC/WPRS Bulletin* **4**: 151–64.

Wolf, F.T. (1988) Entomophthorales and their parasitism of insects. *Nova Hedwigia* **46**: 121–42.

Wood, H.A. and Granados, R.R. (1991) Genetically engineered baculoviruses as agents for pest control. *Annual Review of Microbiology* **45**: 69–87.

Wood, H.A., Hughes, P.R., and Shelton, A. (1994) Field studies of the co-occlusion strategy with a genetically altered isolated of the *Autographica californica* nuclear polyhedrosis virus. *Environmental Entomology* **23**: 211–19.

Woodburn, T.L. (1993) Host specificity testing, release and establishment of *Urophora solstitialis* (L.) (Diptera: Tephritidae), a potential biological control agent for *Carduus nutans* L., in Australia. *Biocontrol Science and Technology* **3**: 419–26.

Woodring, J.L. and Kaya, H.K. (1988) *Steinernematid and Heterorhabditid Nematodes: a Handbook of Techniques*. Southern Cooperative Series Bulletin 331, Arkansas Agricultural Experimental Station, Fayetteville, AR.

Woods, S. and Elkinton, J.S. (1987) Bimodal patterns of mortality from nuclear polyhedrosis virus in gypsy moth (*Lymantria dispar*) populations. *Journal of Invertebrate Pathology* **50**: 151–7.

Wootton, J.T. (1994) The nature and consequences of indirect effects in ecological communities. *Annual Review of Ecology and Systematics* **25**: 443–66.

Work, T.T., McCullough, D.G., Cavey, J.F., and Komosa, R. (2005) Arrival rate of nonindigenous insect species into the United States through foreign trade. *Biological Invasions* **7**: 323–32.

Wraight, S.P., Molloy, D., and Jamback, H. (1981) Efficacy of *Bacillus sphaericus* strain 1593 against the four instars of laboratory reared and field collected *Culex pipiens pipiens* and laboratory reared *Culex salinarius*. *The Canadian Entomologist* **113**: 379–86.

Wratten, S.D. (1987) The effectiveness of native natural enemies. In: Burn, A.J., Croaker, T.H., and Jepson, P.C. (eds), *Integrated Pest Management*, pp. 89–112. Academic Press, London.

Wratten, S., Berndt, L., Gurr, G., Tylianakis, J., Fernando, P., and Didham, R. (2002) Adding floral diversity to enhance parasitoid fitness and efficacy. In: Van Driesche, R.G. (ed.), *Proceedings of the First International Symposium on Biological Control of Arthropods*, pp. 211–14, January 14–18, 2002, Honolulu, Hawaii. FHTET-03-05. USDA Forest Service, Morgantown, WV.

Wright, M.G., Kuhar, T.P., Hoffmann, M.P., and Chenus, S.A. (2002) Effect of inoculative releases of *Trichogramma ostriniae* on populations of *Ostrinia nubilalis* and damage to sweet corn and field corn. *Biological Control* **23**: 149–55.

Wright, M.G., Hoffmann, M.P., Kuhar, T.P., Gardner, J., and Pitcher, S.A. (2005) Evaluating risks of biological control introductions: a probabilistic risk-assessment approach. *Biological Control* **35**: 338–47.

Wright, R.J., Villani, M.G., and Agudelo-Silva, F. (1988) Steinernematid and heterorhabditid nematodes for control of larval European chafers and Japanese beetles (Coleoptera: Scarabaeidae) in potted yew. *Journal of Economic Entomology* **81**: 152–7.

Wright, W.H. (1973) Geographical distribution of schistosomes and their intermediate hosts. In: Ansari, N. (ed.), *Epidemiology and Control of Schistosomiasis (Bilharziasis)*, pp. 32–249. University Park Press, Baltimore, MD.

Xia, J.Y., Cui, J.J., Ma, L.H., Dong, S.X., and Cui, X.F. (1999) The role of transgenic *Bt* cotton in integrated pest management. *Acta Gossypii Sinica* **11**: 57–64.

Xu, F.Y. and Wu, D.X. (1987) Control of bamboo scale insects by intercropping rape in the bamboo forest to attract coccinellid beetles. *Chinese Journal of Biological Control* **5**: 117–19.

Yaninek, J.S. and Bellotti, A.C. (1987) Exploration for natural enemies of cassava green mites based on agrometeorogical criteria. In: Rijks, D. and G. Mathys (eds.), *Proceedings of the Seminar on Agrometeorology and Crop Protection in the Lowland Humid and Subhumid Tropics*, pp. 69–75, July 7–11, 1986, Cotonou, Benin. World Meteorological Organization, Geneva.

Yaninek, S. and Hanna, R. (2003) Cassava green mite in Africa – a unique example of successful classical biological control of a mite pest on a continental scale. In: Neuenschwander, P., Borgemeister, C., and Langewald, J. (eds.), *Biological Control in IPM Systems in Africa*, pp. 61–75. CABI Publications, Wallingford.

Yara, K. (2005) Identification of *Torymus sinensis* and *T. beneficus* (Hymenoptera: Torymidae), introduced and indigenous parasitoids of the chestnut gall wasp, *Dryocosmus kuriphilus* (Hymenoptera: Cynipidae), using the ribosomal ITS2 region. *Biological Control* **36**: 15–21.

Yelenik, S.G., Stock, W.D., and Richardson, D.M. (2004) Ecosystem level impacts of invasive *Acacia saligna* in the South African fynbos. *Restoration Ecology* **12**: 44–51.

Ylönen, H. (2001) Rodent plagues, immunocontraception and the mousepox virus. *Trends in Ecology and Evolution* **16**: 418–20.

Yong, T.H. (2003) Nectar-feeding by a predatory ambush bug (Heteroptera: Phymatidae) that hunts on flowers. *Annals of the Entomological Society of America* **96**: 643–51.

York, G.T. (1958) Field tests with the fungus *Beauveria* sp. for control of the European corn borer. *Iowa State College Journal of Science* **33**: 123–9.

Young, J., Van Manen, F., and Ross, R. (1998) Modeling stand vulnerability and biological impacts of the hemlock woolly adelgid. Study Plan Number 2055. USGS, Leetown Science Center, Kearneysville, WV.

Young, S.Y. and Yearian, W.C. (1986) Formulation and application of baculoviruses. In: Granados, R.R. and B.A. Federici (eds.), *The Biology of Baculoviruses: Volume II. Practical Application for Insect Control*, pp. 157–79. CRC Press, Boca Raton, FL.

Yu, D. and Horstmann, K. (1997) A catalogue of world Ichneumonidae (Hymenoptera) Part 1. *Memoirs of the American Entomological Institute* **58(1)**.

Yu, G. (2001) The coccinellids (Coleoptera) predaceous on adelgids, with notes on the biocontrol of the hemlock woolly adelgid (Homoptera: Adelgidae). *Special Publication of the Japanese Coleoptera Society, Osaka* **1**: 297–304.

Yu, G., Montgomery, M.E., and Yao, D. (2000) Lady beetles (Coleoptera: Coccinellidae) from Chinese hemlocks infested with the hemlock woolly adelgid, *Adelges tsugae* Annand (Homoptera: Adelgidae). *Coleopterists Bulletin* **54**: 154–99.

Zane, L., Bargelloni, L., and Patarnello, T. (2002) Strategies for microsatellite isolation: a review. *Molecular Ecology* **11**: 1–16.

Zangerl, A.R. and Berenbaumer, M.R. (2005) Increase in toxicity in an invasive weed after reassociation with its coevolved herbivore. *Proceedings of the National Academy of Sciences USA* **102(43)**: 15529–32.

Zangger, A., Lys, J.A., and Nentwig, W. (1994) Increasing the availability of food and the reproduction of *Poecilus cupreus* in a cereal field by strip-management. *Entomologia Experimentalis et Applicata* **71**: 111–20.

Zchori-Fein, E. and Perlman, S.J. (2004) Distribution of the bacterial symbiont *Cardinium* in arthropods. *Molecular Ecology* **13**: 2009–16.

Zchori-Fein, E., Gottlieb, Y., Kelly, S.E. et al. (2001) A newly discovered bacterium associated with parthenogenesis and a change in host selection behavior in parasitoid wasps. *Proceedings of the National Academy of Sciences USA* **98**: 12555–60.

Zeddies, J., Schaab, R.P., Neuenschwander, P., and Herren, H.R. (2001) Economics of biological control of cassava mealybug in Africa. *Agricultural Economics* **24**: 209–11.

Zelazny, B., Lolong, A., and Crawford, A.M. (1990) Introduction and field comparison of baculovirus strains against *Oryctes rhinoceros* (Coleoptera: Scarabaeidae) in Maldives. *Environmental Entomology* **19**: 1115–21.

Zelger, R. (1996) The population dynamics of the cockchafer in South Tyrol since 1980 and measures applied for control. *IOBC/WPRS Bulletin* **19(2)**: 109–13.

Zhang, A. and Olkowski, W. (1989) Ageratum cover crop aids citrus biocontrol in China. *The IPM Practitioner* **11(9)**: 8–10.

Zhang, N.X. and Li, Y.X. (1989) An improved method of rearing *Amblyseius fallacis* (Acari: Phytoseiidae) with plant pollen. *Chinese Journal of Biological Control* **5**: 149–52.

Zhang, Y. and Shipp, J.L. (1998) Effect of temperature and vapor pressure deficit on the flight activity of *Orius insidiosus* (Hemiptera: Anthocoridae). *Environmental Entomology* **27**: 736–42.

Zhang, Z.H., Gao, S., Zhang, G.Y. et al. (2000) Using *Metarhizium flavoviridae* oil spray to control grasshoppers in inner Mongolia grassland. *Chinese Journal of Biological Control* **16**: 49–52.

Zhang, Z., Ye, G.-Y., and Hu, C. (2004) Effects of venom from two pteromalid wasps, *Pteromalus puparum* and *Nasonia vitripennis* (Hymenoptera: Pteromalidae), on the spreading, viability and encapsulation capacity of *Pieris rapae* hemocytes. *Acta Entomologica Sinica* **47**: 551–61.

Zheng, L., Zhou, Y., and Song, K. (2005) Augmentative biological control in greenhouses: experiences from China. In: Hoddle, M.S. (compiler), *Proceedings of the Second International Symposium on Biological Control of Arthropods, Davos Switzerland*, pp. 538–45. USDA-FS Forest Health Technology Team, Morgantown, WV.

Zhi-Qiang Zhang (1992) The use of beneficial birds for biological pest control in China. *Biocontrol News and Information* **13(1)**: 11N–16N.

Zhou, L., Bailey, K.L. Chen, C.Y., and Keri, M. (2005) Molecular and genetic analyses of geographic variation in isolates of *Phoma macrostoma* used for biological weed control. *Mycologia* **97**: 612–20.

Zhu, Y.-C., Burd, J.D., Elliott, N.C., and Greenstone, M.H. (2000) Specific ribosomal DNA marker for early polymerase chain reaction detection of *Aphelinus hordei* (Hymenoptera: Aphelinidae) and *Aphidius colemani* (Hymenoptera: Aphidiidae) from *Diuraphis noxia* (Homoptera: Aphididae). *Annals of the Entomological Society of America* **93**: 486–91.

Zilahi-Balogh, G.M.G., Kok, L.T., and Salom, S.M. (2002) Host specificity of *Laricobius nigrinus* Fender (Coleoptera: Derodontidae), a potential biological control agent of the hemlock woolly adelgid, *Adelges tsugae* Annand (Homoptera: Adelgidae). *Biological Control* **24**: 192–8.

Zilahi-Balogh, G.M.G., Humble, L.M., Lamb, A.B., Salom, S.M., and Kok, L.T. (2003a) Seasonal abundance and synchrony between *Laricobius nigrinus* (Coleoptera: Derodontidae) and its prey, the hemlock woolly adelgid (Hemiptera: Adelgidae). *The Canadian Entomologist* **135**: 103–15.

Zilahi-Balogh, G.M.G., Salom, S.M., and Kok, L.T. (2003b) Development and reproductive biology of *Laricobius nigrinus*, a potential biological control agent of *Adelges tsugae*. *Biocontrol* **48**: 293–306.

Zilahi-Balogh, G.M.G., Shipp, J.L., Cloutier, C., and Brodeur, J. (2006) Influence of light intensity, photoperiod, and temperature on the efficacy of two aphelinid parasitoids of the greenhouse whitefly. *Environmental Entomology* **35**: 581–9.

Zimmerman, E.C. (1994) *Australian Weevils, vol. I. Anthribidae to Attelabidae*. CSIRO Publishing, Melbourne.

Zimmermann, G. (1986) Insect pathogenic fungi as pest control agents. In: Franz, J.M. (ed.), *Biological Plant and Health Protection: Biological Control of Plant Pests and of Vectors of Human and Animal Diseases*, pp. 217–31. International Symposium of the Akademie der Wissenschaften und der Literatur, November 15–17, 1984, Mainz, Germany. *Fortschritte der Zoologie* vol. 32. Gustav Fischer Verlag, Stuttgart.

Zimmermann, G.G., Moran, V.C., and Hoffmann, J.H. (2001) The renowned cactus moth, *Cactoblastis cactorum* (Lepidoptera: Pyralidae): its natural history and threat to native *Opuntia* floras in Mexico and the United States of America. *Florida Entomologist* **84**: 543–51.

Zimmermann, O. (2004) Use of *Trichogramma* wasps in Germany: present status of research and commercial application of egg parasitoids against lepidopterous pests for crop and storage protection. *Gesunde Pflanzen* **56(6)**: 157–66.

Zwölfer, H. and Harris, P. (1971) Host specificity determination of insects for biological control of weeds. *Annual Review of Entomology* **16**: 159–78.

Zwölfer, H. and Brandl, R. (1989) Niches and size relationships in Coleoptera associated with Cardueae host plants; adaptations to resource gradients. *Oecologia* **78**: 60–8.

Zwölfer, H. and Harris, P. (1984) Biology and host specificity of *Rhinocyllus conicus* (Froel.) (Col., Curculionidae), a successful agent for biocontrol of the thistle *Carduus nutans* L. *Zeitschrift für Angewandte Entomologie* **97**: 36–62.

INDEX

Note: Page numbers in *italics* refer to Figures; those in **bold** to Tables.